Neuromethods

Series Editor
Wolfgang Walz
University of Saskatchewan
Saskatoon, Canada

For further volumes:
http://www.springer.com/series/7657

Biochemical Approaches for Glutamatergic Neurotransmission

Editors

Sandrine Parrot

Lyon Neuroscience Research Center, University Lyon 1, Lyon, France

Luc Denoroy

Lyon Neuroscience Research Center, University Lyon 1, Lyon, France

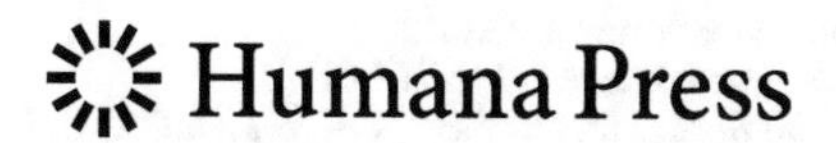

Editors
Sandrine Parrot
Lyon Neuroscience Research Center
University Lyon 1
Lyon, France

Luc Denoroy
Lyon Neuroscience Research Center
University Lyon 1
Lyon, France

ISSN 0893-2336 ISSN 1940-6045 (electronic)
Neuromethods
ISBN 978-1-4939-8423-7 ISBN 978-1-4939-7228-9 (eBook)
DOI 10.1007/978-1-4939-7228-9

Cover Illustration: An organotypic slice of the hippocampus where neurons were biolistically labeled with LiGluN and tdTomato.

Printed on acid-free paper

This Humana Press imprint is published by Springer Nature
The registered company is Springer Science+Business Media LLC
The registered company address is: 233 Spring Street, New York, NY 10013, U.S.A.

Preface to the Series

Experimental life sciences have two basic foundations: concepts and tools. The *Neuromethods* series focuses on the tools and techniques unique to the investigation of the nervous system and excitable cells. It will not, however, shortchange the concept side of things as care has been taken to integrate these tools within the context of the concepts and questions under investigation. In this way, the series is unique in that it not only collects protocols but also includes theoretical background information and critiques which led to the methods and their development. Thus it gives the reader a better understanding of the origin of the techniques and their potential future development. The *Neuromethods* publishing program strikes a balance between recent and exciting developments like those concerning new animal models of disease, imaging, in vivo methods, and more established techniques, including, for example, immunocytochemistry and electrophysiological technologies. New trainees in neurosciences still need a sound footing in these older methods in order to apply a critical approach to their results.

Under the guidance of its founders, Alan Boulton and Glen Baker, the *Neuromethods* series has been a success since its first volume published through Humana Press in 1985. The series continues to flourish through many changes over the years. It is now published under the umbrella of Springer Protocols. While methods involving brain research have changed a lot since the series started, the publishing environment and technology have changed even more radically. Neuromethods has the distinct layout and style of the Springer Protocols program, designed specifically for readability and ease of reference in a laboratory setting.

The careful application of methods is potentially the most important step in the process of scientific inquiry. In the past, new methodologies led the way in developing new disciplines in the biological and medical sciences. For example, Physiology emerged out of Anatomy in the nineteenth century by harnessing new methods based on the newly discovered phenomenon of electricity. Nowadays, the relationships between disciplines and methods are more complex. Methods are now widely shared between disciplines and research areas. New developments in electronic publishing make it possible for scientists that encounter new methods to quickly find sources of information electronically. The design of individual volumes and chapters in this series takes this new access technology into account. Springer Protocols makes it possible to download single protocols separately. In addition, Springer makes its print-on-demand technology available globally. A print copy can therefore be acquired quickly and for a competitive price anywhere in the world.

Wolfgang Walz

Preface

Glutamate is the main excitatory neurotransmitter in the brain, but its biological role is not limited to neurotransmission since it has many other functions in both the central nervous system and the peripheral organs, such as its paramount involvement in protein and energy metabolisms, its roles in brain maturation and plasticity as well as in immune response. Alteration of glutamatergic transmission in humans can be tremendously involved in neurological disorders, such as stroke, epilepsy, amyotrophic lateral sclerosis, Parkinson's and Alzheimer's diseases, and in psychiatric diseases, such as schizophrenia, depression, and addiction, to name but a few ailments.

As compared to other neurotransmitters, such as monoamines for instance, glutamate fulfills atypically the classical 1980s criteria known for labeling it as a neurotransmitter. As examples of atypical features, the precursors and synthesis enzymes of glutamate are not always found in glutamatergic neurons, and its target receptors or transporters are not exclusively located in the glutamatergic neurons, reflecting that glutamate is also a gliotransmitter in addition to be a neurotransmitter. The fact that glutamate biochemistry is complicated explains that its study is not obvious and requires the utilization of methods able to discriminate the neuronal part from the metabolic pool of that ubiquitous compound, to distinguish the contribution of each glutamatergic protein, or to analyze the molecular dynamics of the ultra-rapid glutamatergic transmission. Compartmentation, identification of neurons, quantification of enzymes, mechanisms of transporters, and dynamics of receptors, as well as knowledge about pharmacology are prerequisites to succeed to decipher the role of glutamate in the nervous system transmission, in normal conditions, but also in physiopathological states.

This book presents established mastered techniques, as well as up-to-date recent developments in representative biochemical approaches to study glutamatergic neurotransmission, providing key insights to understand glutamate biochemistry and methodological details to use well tried and tested protocols. All the major approaches, still used nowadays or emerging, are covered including tracing of neuronal pathways, functional or spectroscopic imaging, fluorescent sensing, optogenetic or pharmacological tools, extracellular neurochemistry in experimental or clinical models, as well as the use of invertebrates as models of glutamatergic synapse.

Briefly, Chap. 1 is devoted to the identification of glutamatergic neurons in a discrete brain area and Chap. 2 presents the finest tools to study dynamics of single ionotropic receptors. Chaps. 3 and 4 deal with in vivo imaging of metabotropic glutamate receptors with positron emission tomography and with the study of glutamate metabolism by proton/carbon nuclear magnetic resonance, respectively. The use of glutamate fluorescent sensors for studying neurotransmission (Chap. 5) or monitoring glio-releases (Chap. 6) is reviewed and detailed. The drugs altering glutamate transport or metabotropic receptor functions are reviewed in a comprehensive way in Chaps. 7 and 8, respectively. The use of the invertebrate neuromuscular junction as a model of glutamatergic synapse in neuroscience is emphasized in Chap. 9. Useful information about optogenetic approaches with photo-switchable or light-sensitive sensors can be found in Chaps. 10 and 11. Four chapters are devoted to the in vivo extracellular glutamate monitoring in preclinical research using biosensors (Chap. 11, again) or microdialysis sampling with the subsequent analytical techniques such as microchip electrophoresis (Chap. 12), capillary electrophoresis and liquid chromatography (Chap. 13), or isotopic determination when determining glutamate uptake (Chap. 14). Methods for assessing brain glutamatergic neurotransmission in clinical research are not forgotten in the three last chapters of this book: quantification of the glutamatergic enzymes in cellular fractions from human brain (Chap. 15), potential brain glutamatergic biomarkers in biofluids (Chap. 16), and extracellular glutamate monitoring in human brain and spinal cord (Chap. 17).

To the reader, more than two thirds of the chapters detail step-by-step their procedures for an easier use in other labs or, simply, to appreciate the deciphering methods, their advantages, and their limits. Addressed to any young or expert scientist interested in this transmitter, both excitatory and quite exciting, all the contributions include a detailed review of the literature on methodological and neurobiological aspects, which provides a critical up-to-date overview, as well as a solid strengthening of what is glutamate biochemistry in 2017.

Biochemical Approaches for Glutamatergic Neurotransmission is an extensive and various resource of 17 chapters from molecular and cellular biology or biochemistry, as well as analytical biochemistry or neuroimaging. This book provides suitable information for the current researches and methods on glutamate neurochemistry.

Lyon, France

Luc Denoroy
Sandrine Parrot

Contents

Contributors

YI AI • *Department of Anatomy and Neurobiology, University of Kentucky Medical Center, Lexington, KY, USA*
SHAI BERLIN • *Helen Wills Neuroscience Institute, University of California, Berkeley, CA, USA; The Ruth and Bruce Rappaport Faculty of Medicine, Technion—Israel Institute of Technology, Haifa, Israel*
IRINA S. BOKSHA • *Gamaleya Center of Epidemiology and Microbiology of the Ministry of Health of the Russian Federation, Moscow, Russia; Federal State Budgetary Scientific Institution, Mental Health Research Centre, Moscow, Russia*
GERARD J. BROUSSARD • *Department of Biochemistry and Molecular Medicine, University of California Davis, Davis, CA, USA; Neuroscience Graduate Group, University of California Davis, Davis, CA, USA*
JASON J. BURMEISTER • *Department of Anatomy and Neurobiology, University of Kentucky Medical Center, Lexington, KY, USA*
NAURA CHOULNAMOUNTRI • *INSERM U1217, CNRS UMR5310, Institut NeuroMyoGène, Université Claude Bernard Lyon 1, Lyon, France*
DELPHINE COLLIN-CHAVAGNAC • *Laboratoire de Biochimie et Biologie Moléculaire, Groupement Hospitalier Sud, Centre de Biologie et de Pathologie Sud, Hospices Civils de Lyon, Pierre-Bénite, France*
ROBIN L. COOPER • *Department of Biology and Center for Muscle Biology, University of Kentucky, Lexington, KY, USA*
LUC DENOROY • *CNRS UMR5292, INSERM U1028, University Lyon 1, Lyon Neuroscience Research Center, NeuroDialyTics Unit, Lyon, France*
JULIEN DUPUIS • *Université de Bordeaux, CNRS, Interdisciplinary Institute for Neuroscience, UMR 5297, Bordeaux, France*
ANDRÉIA CRISTINA KARKLIN FONTANA • *Department of Pharmacology and Physiology, Drexel University College of Medicine, Philadelphia, PA, USA*
GREG A. GERHARDT • *Department of Anatomy and Neurobiology, University of Kentucky Medical Center, Lexington, KY, USA*
KAREN J. GREGORY • *Drug Discovery Biology, Monash Institute of Pharmaceutical Sciences and Department of Pharmacology, Monash University, Parkville, VIC, Australia*
LAURENT GROC • *Université de Bordeaux, CNRS, Interdisciplinary Institute for Neuroscience, UMR 5297, Bordeaux, France*
KENJI HASHIMOTO • *Division of Clinical Neuroscience, Chiba University Center for Forensic Mental Health, Chiba, Japan*
ANDREAS HEUER • *Lund University, Lund, Sweden*
MICHAEL HOGARD • *Department of Chemistry, University of Kansas, Lawrence, KS, USA; Ralph N. Adams Institute for Bioanalytical Chemistry, University of Kansas, Lawrence, KS, USA*
PETER HUETTL • *Department of Anatomy and Neurobiology, University of Kentucky Medical Center, Lexington, KY, USA*

EHUD Y. ISACOFF • *Helen Wills Neuroscience Institute, University of California, Berkeley, CA, USA; Department of Molecular and Cell Biology, University of California, Berkeley, CA, USA; Physical Bioscience Division, Lawrence Berkeley National Laboratory, Berkeley, CA, USA*
JOHAN JAKOBSSON • *Lund University, Lund, Sweden*
RUQIANG LIANG • *Department of Biochemistry and Molecular Medicine, University of California Davis, Davis, CA, USA*
MARTIN LUNDBLAD • *Lund University, Lund, Sweden*
SUSAN M. LUNTE • *Department of Bioengineering, University of Kansas, Lawrence, KS, USA; Department of Chemistry, University of Kansas, Lawrence, KS, USA; Department of Pharmaceutical Chemistry, University of Kansas, Lawrence, KS, USA; Ralph N. Adams Institute for Bioanalytical Chemistry, University of Kansas, Lawrence, KS, USA*
FRANÇOIS MAINGRET • *Université de Bordeaux, CNRS, Interdisciplinary Institute for Neuroscience, UMR 5297, Bordeaux, France*
BRIAN P. MCGREW • *Department of Biochemistry and Molecular Medicine, University of California Davis, Davis, CA, USA*
PATRICK MERTENS • *Department of Neurosurgery, Department of Anatomy - INSERM 1028, Hospices Civils de Lyon, Lyon 1 University, Lyon, France*
MARISELA MORALES • *National Institute on Drug Abuse, Intramural Research Program, Neuronal Networks Section, Baltimore, MD, USA*
ELIZABETA B. MUKAETOVA-LADINSKA • *Institute of Neuroscience, Newcastle University and Northumberland, Tyne and Wear NHS Foundation Trust, Newcastle upon Tyne, UK*
NATHAN OBORNY • *Department of Bioengineering, University of Kansas, Lawrence, KS, USA; Ralph N. Adams Institute for Bioanalytical Chemistry, University of Kansas, Lawrence, KS, USA*
SANDRINE PARROT • *CNRS UMR5292, INSERM U1028, University Lyon 1, Lyon Neuroscience Research Center, NeuroDialyTics Unit, Lyon, France*
OLIVIER PASCUAL • *INSERM U1217, CNRS, UMR5310, Institut NeuroMyoGène, Université Claude Bernard Lyon 1, Lyon, France*
CHRISTOPHE PLACE • *ENS de Lyon, Université Claude Bernard Lyon 1, CNRS, Laboratoire de Physique, Lyon, France*
FRANCOIS POMERLEAU • *Department of Anatomy and Neurobiology, University of Kentucky Medical Center, Lexington, KY, USA*
TATYANA A. PROKHOROVA • *Federal State Budgetary Scientific Institution, Mental Health Research Centre, Moscow, Russia*
JORGE E. QUINTERO • *Department of Anatomy and Neurobiology, University of Kentucky Medical Center, Lexington, KY, USA*
BERNARD RENAUD • *INSERM U1028, CNRS UMR5292, Université Lyon 1, Lyon Neuroscience Research Center, Lyon, France*
OLGA K. SAVUSHKINA • *Federal State Budgetary Scientific Institution, Mental Health Research Centre, Moscow, Russia*
KATHY SENGMANY • *Drug Discovery Biology, Monash Institute of Pharmaceutical Sciences and Department of Pharmacology, Monash University, Parkville, VIC, Australia*
SELENA MILICEVIC SEPHTON • *Department of Clinical Neurosciences, Wolfson Brain Imaging Centre, University of Cambridge, Cambridge, UK*
JUN SHEN • *Section on Magnetic Resonance Spectroscopy, NIMH/NIH, Bethesda, MD, USA*
JOHN T. SLEVIN • *Department of Anatomy and Neurobiology, University of Kentucky Medical Center, Lexington, KY, USA*

FUYUKO TAKATA-TSUJI • *INSERM U1217, CNRS UMR5310, Institut NeuroMyoGène, Université Claude Bernard Lyon 1, Lyon, France; Department of Pharmaceutical Care and Health Sciences, Faculty of Pharmaceutical Sciences, Fukuoka University, Fukuoka, Japan*

ELENA B. TERESHKINA • *Federal State Budgetary Scientific Institution, Mental Health Research Centre, Moscow, Russia*

LIN TIAN • *Department of Biochemistry and Molecular Medicine, University of California Davis, Davis, CA, USA; Neuroscience Graduate Group, University of California Davis, Davis, CA, USA*

JOSHUA S. TITLOW • *Department of Biochemistry, University of Oxford, Oxford, UK*

MONIQUE TOURET • *Lyon Neuroscience Research Center, INSERM U1028, CNRS UMR5292, Université Lyon 1, Lyon, France*

ELIZABETH K. UNGER • *Department of Biochemistry and Molecular Medicine, University of California, Davis, CA, USA*

SHILIANG ZHANG • *National Institute on Drug Abuse, Intramural Research Program, Neuronal Networks Section, Baltimore, MD, USA*

Abbreviations

[^{18}F]FDG	[^{18}F]-labeled glucose
^{1}H-MRS	Proton magnetic resonance spectroscopy
1P	One-photon
2-ME	2-mercaptoethanol
2P	Two photons
2PA	Two photons absorption
6-OHDA	6-hydroxydopamine
7TMD	Seven transmembrane domains
AAD	α-amino adipic acid
AAT	Aspartate aminotransferase
AATc	Cytoplasmic aspartate aminotransferase
AATm	Mitochondrial aspartate aminotransferase
AAV	Adeno-associated virus
Acc	Nucleus accumbens
Ach	Acetylcholine
ACN	Acetonitrile
aCSF	Artificial cerebrospinal fluid
AD	Alzheimer's disease
AIND	Acute ischemic neurological deficit
ALS	Amyotrophic lateral sclerosis
AMPA	α-amino-3-hydroxy-5-methyl-4-isoxazolepropionic acid
AMPA-R	α-amino-3-hydroxyl-5-methyl-4-isoxazole-propionate receptor
AP	Antero-posterior
ApoE	Apolipoprotein E
AR	Autoradiography
Ara-C	Cytosine arabinosine
ASD	Autism spectrum disorders
Asp	Aspartate
AT	Axon terminal
ATCM	Allosteric ternary complex model
ATP	Adenosine triphosphate
AU	Arbitrary unit
Aβ	Amyloid beta peptide
BA	Brodmann area
BBB	Blood brain barrier
BBS	Borate buffered saline
BD	Bipolar disorder
BES	*N*,*N*-Bis(2-hydroxyethyl)-2-aminoethanesulfonic acid
BG	Benzylguanine
BGE	Background electrolyte
B_{max}	Concentration of binding sites
BP_{ND}	Binding potential
BSA	Bovine serum albumin

bvFTD	Behavioral variant of fronto temporal dementia
C	Control
CaM	Calmodulin
CBI	Cyanobenzo[f]isoindol
CE	Capillary electrophoresis
CEF	Ceftriaxone
CFSE	5-carboxyfluorescein *N*-succinimidyl ester
cGMP	Cyclic guanosine monophosphate
ChR2	Channelrhodopsin
CMV	Cytomegalovirus
CNS	Central nervous system
COC	Cyclic olefin copolymer
CPEC	Circular polymerase extension cloning
cpGFP	Circularly permutated green fluorescent protein
CPP	Cerebral perfusion pressure
CPu	Caudate putamen
CRD	Cysteine rich domain
CREB	cAMP response element-binding protein
CSF	Cerebrospinal fluid
CT	Carboxy terminus
CT	Computerized tomography
DAB	3,3-diaminobenzidine-4 HCl
DAPI	4',6-diamidine-2'-phenylindole
D-Asp	D-aspartate
DBS	Deep brain stimulation
ddH_2O	Double distilled water
DEPC	Diethylpyrocarbonate
DG	Dentate gyrus
DH	Dorsal horn
DHCR24	3β-hydroxysteroid-Δ24 reductase
DHK	Dihydrokainate
DI	Deionized
diH_2O	Deionized water
DIND	Delayed ischemic neurological deficit
DLPFC	Dorsolateral prefrontal cortex
DM	Dextromethorphan
DMEM	Dulbecco's modified eagle medium
DOPAC	3,4-dihydroxyphenylacetic acid
DREZ	Dorsal root entry zone
DREZO	DREZotomy or microcoagulation of the dorsal root entry zone
D-Ser	D-serine
DV	Dorso ventral
EAAC1	Excitatory amino acid carrier 1
EAAT1-5	Human excitatory amino acid transporter subtypes 1-5
EAAT2	Human glutamate transporter 2
EAATs	Excitatory amino acid transporters
ECF	Extracellular fluid
ECL	Enhanced chemiluminescence

ECoG	Electrocorticography
EEG	Electroencephalography
$eEPSC_{NMDA}$	NMDA-dependent evoked excitatory post synaptic currents.
EGF	Epidermal growth factor
eGFP	Enhanced green fluorescent protein
EJP	Excitatory junction potential
ELISA	Enzyme-linked immunosorbent assay
EM-CCD	Electron-multiplying charge coupled device
EOF	Electroosmotic flow
EOS	Glutamate optical sensor
EPP	Error-prone PCR
EPSC	Excitatory postsynaptic current
EPSP	Excitatory postsynaptic potential
ER	Endoplasmic reticulum
ERK1/2	Extracellular signal-regulated kinase 1/2
EYFP	Enhanced yellow fluorescent protein
FBS	Fetal bovine serum
FG	Fluoro-Gold
FiO_2	Fractional inspired oxygen
FISH	Fluorescence in situ hybridization
FITC	Fluorescein isothiocyanate
FLIPE	Fluorescent indicator protein for glutamate
FMRP	Fragile X mental retardation protein
FP	Fluorescent protein
f_P	Plasma function
FRET	Förster resonance energy transfer
FSCE	Free solution capillary electrophoresis
FTP	Frontotemporal dementia
FXS	Fragile X syndrome
GABA	γ-aminobutyric acid
GAD	Glutamic acid decarboxylase
GAERS	Genetic absence epilepsy rats from Strasbourg
GDH	Glutamate dehydrogenase
GECI	Genetically encoded calcium indicator
GFAP	Glial fibrillary acidic protein
GFP	Green fluorescent protein
GINA	Genetically encoded indicator of neural activity
GLAST	Glutamate and aspartate transporter
Glc	Glucose
Gln	Glutamine
GLS1	Kidney-type glutaminase
GLS2	Liver-type glutaminase
GLT-1	Rat glutamate transporter 1
GltI	Bacterial glutamate/aspartate-binding protein
GltPh	Glutamate transporter homolog from *Pyrococcus horikoshii*
Glu	Glutamate
GluDH	Glutamate dehydrogenase
GluN1-SEP	GluN1-super-ecliptic pHluorin

GluNRs	Glutamate receptors of the NMDA kind
GluOx	Glutamate oxidase
GluSnFR	Glutamate-sensing fluorescent reporter
GLUT	Glutamate transporter
Glx	Combination of glutamate and glutamine
Gly	Glycine
GPCR	G protein-coupled receptors
GPi	Internal globus pallidus
GS	Glutamine synthetase
GSH	Glutathione
GSK3β	Glycogen synthase kinase 3 beta
GSLP	Glutamine synthetase-like protein
HAND	HIV-associated neurocognitive disorder
HBBS	Hank's balanced salt solution
HD	Huntington's disease
HEK	Human embryonic kidney
hGDH2	Homologous GDH isoenzyme
HILIC	Hydrophilic interaction liquid chromatography
HP	Hairpin loop
HPLC	High performance liquid chromatography
HP-β-CD	Hydroxypropy-β-cyclodextrine
HRP	Horseradish peroxidase
HSB	Hybrid structure-based
hSyn	Human synapsin
HTS	High throughput screening
HV	High voltage
ICH	Intracerebral hemorrhage
ICP	Intracerebral pressure
ICU	Intensive care unit
ICV	Intracerebroventricular
iGluR	Ionotropic glutamate receptor
iGluSnFR	Intensity-based glutamate-sensing fluorescent reporter
IL-1β	Interleukin-1β
IPA	Isopropyl alcohol
IPTG	Isopropyl β-D-1-thiogalactopyranoside
IR	Immunoreactivity
IRES	Internal ribosome entry site
ISH	In situ hybridization
ITR	Inverted terminal repeats
JAK	Janus kinase
JAK/STAT	Janus kinase/signal transducer and activator of transcription
JAK2	Janus-activated kinase-2
KA	Kainate
KA-R	Kainic acid receptor
K_D	Dissociation constant
K_i	Inhibition constant
Lac	Lactate
L-Asp	L-aspartate

LBD	Ligand-binding domain
LC	Locus coeruleus
L-DOPA	L-3,4-dihydroxyphenylalanine
LED	Light-emitting diode
LID	L-DOPA induced dyskinesia
LIF	Laser-induced fluorescence
LiGluNRs	Light-gated glutamate receptors of the NMDA kind
LIVBP	Leucine/isoleucine/valine binding protein
LME	L-leucine methyl ester
LOD	Limit of detection
LOQ	Limit of quantification
LPS	Lipopolysaccharide
LR	LR white embedding medium
L-Ser	L-serine
LSOS	(*RS*)-2-amino-3-(1- hydroxy-1,2,3-triazol-5-yl) propionate and L-serine-*O*-sulfate
LTD	Long-term depression
LTF	Long-term facilitation
LTP	Long-term potentiation
MAG	Maleimide-azobenzene-glutamate
MAN	Mannitol
MAPK	Mitogen-activated protein kinases
MBP	Maltose-binding protein
MCA	Middle cerebral artery
MCAO	Middle cerebral artery occlusion
MCS	Multiple cloning site
MD	Microdialysis
MDD	Major depression disorder
MD-ME	Microdialysis coupled to microchip electrophoresis
MDPV	3,4-methylenedioxypyrovalerone
ME	Microchip electrophoresis
MEA	Microelectrode arrays
MEKC	Micellar electrokinetic chromatography
MEM	Minimum essential medium
mGlu	Metabotropic glutamate receptors
mGluR	Metabotropic glutamate receptors
ML	Mediolateral
mPD	Meta-phenylenediamine
MPTP	1-methyl-4-phenyl-1,2,3,6-tetrahydropyridine
MRI	Magnetic resonance imaging
MRM	Multiple reaction monitoring
MS	Mass spectrometry
MS-153	[R]-[-]-5-methyl-1-nicotinoyl-2-pyrazoline
MSD	Mean square displacement
MSG	Monosodium glutamate
mTLE	Medial temporal lobe epilepsy
MWCO	Molecular weight cut off
NAA	*N*-acetyl-L-aspartate

NAAG	*N*-acetylaspartylglutamate
NAC	*N*-acetylcysteine
nAcc	Nucleus accumbens
NADPH	Nicotinamide adenine dinucleotide phosphate
NAL	Neutral allosteric ligand
NAM	Negative allosteric modulator
NBD-F	4-fluoro-7-nitro-2,1,3-benzoxadiazole
NCS	Newborn calf serum
NDA	Naphthalene-2,3-dicarboxaldehyde
NEB	New England Biolabs
NEC	Non-epileptic control
NF-κB	Nuclear factor kappa-light-chain-enhancer of activated B cells
NGS	Normal goat serum
NMDA	*N*-methyl-D-aspartate
NMDAR	*N*-methyl-D-aspartate receptor
NMDG	*N*-methyl-D-glucamine
NMJ	Neuromuscular junction
NMO	Neuromyelitis optica
NMR	Nuclear magnetic resonance
NOE	Nuclear Overhauser Enhancement
NT	Amino terminus
OAA	Oxalacetate
OD	Optical density
OGD	Oxygen and glucose deprivation
OPA	*o*-phthalaldehyde
OPA	Optical parametric amplifier
PACAP	Pituitary adenylate cyclase-activating peptide
PAG	Phosphate-activated glutaminase
PAGE	Polyacrylamide gel electrophoresis
PAM	Positive allosteric modulators
PaO_2	Arterial oxygen partial pressure
PB	Phosphate buffer
PBP	Periplasmic binding protein
PBR	Peripheral benzodiazepine receptor
PBS	Phosphate buffered saline
PBSm	Dulbecco's phosphate buffered saline modified
PBST	Phosphate buffered saline with Tween
$PbtO_2$	Brain tissue partial pressure of oxygen
PCB	Printed circuit board
PCL	Photochromic ligands
PCP	Phencyclidine
PCR	Polymerase chain reaction
PD	Human idiopathic Parkinson's disease
PDC	L-*trans*-pyrrolidine-2,4-dicarboxylate
PDH	Pyruvate dehydrogenase
PDMS	Polydimethylsiloxane
PEA	Phosphoethanolamine
PEG	Polyethylene glycol

PES	Polyethersulfone
PET	Positron emission tomography
PFA	Paraformaldehyde
PHA-L	Phytohemagglutinin-L
PI	Phosphoinositide
PI3K	Phosphatidylinositol 3-kinase
PKC	Protein kinase C
PLA_2	Phospholipase A_2
PLC	Phospholipase C
PLL	Poly-L-lysine
PLP	Pyridoxal 5'-phosphate
PM	Plasma membrane
PMA	Phorbol 12-myristate 13-acetate
PMI	Post mortem interval
PMMA	Poly(methyl methacrylate)
PMSF	Phenylmethylsulfonyl fluoride
PMT	Photomultiplier tube
POCE	Proton-observed, carbon-edited
PRESS	Point-resolved spectroscopy
PSD	Postsynaptic density
Pt	Platinum
Pt/Ir	Platinum/iridium
PTL	Photoswitchable tethered ligand
PTP	Post-tetanic potentiation
Pyr	Pyruvate
QD	Quantum dots
qPCR	Quantitative polymerase chain reaction
rAAV	Recombinant adeno associated virus
rCBF	Regional cerebral blood flow
red GECI	Red genetically-encoded Ca^{2+}-indicator
ROI	Region of interest
RT	Room temperature
RT-PCR	Reverse transcription-poly chain reaction
RU	Relative unit
SAH	Subarachnoid hemorrhage
SAR	Structure–activity relationship
SD	Standard deviation
SDS	Sodium dodecyl sulfate
SDS-PAGE	Sodium dodecyl sulfate polyacrylamide gel electrophoresis
Ser	Serine
SHMT	Serine hydroxymethyltransferase
Snifit	SNAP-tag-based indicator proteins with a fluorescent intramolecular tether
SNR	Signal-to-noise ratio
SOD1	Superoxide dismutase 1
SPT	Single particle tracking
SSEP	Somatosensory evoked potential
STAT3	Signal transducers and activators of transcription 3
STEAM	Stimulated echo acquisition mode

STF	Short-term facilitation
STN	Subthalamic nucleus
SUV	Standard uptake value
Sz	Schizophrenia
TAC	Time activity curve
TBI	Traumatic brain injury
TBOA	Threo-beta-benzyloxyaspartate
TBS	TRIS buffered saline
TC	Transit compartment
TCA	Tricarboxylic acid
TE	Tris EDTA
TEA	Tetraethylammonium
TEVC	Two-electrode voltage clamp
TFB-TBOA	(3S)-3-[[3-[[4-(Trifluoromethyl)benzoyl]amino]phenyl]methoxy]-L-aspartic acid
TH	Tyrosine hydroxylase
THA	Threo-β-hydroxyaspartate
TIRF	Total internal reflection fluorescence
TLC	Thin layer chromatography
TM	Transmembrane
TQ	Triple quadrupole
TSC	Tuberous sclerosis complex
TSCI	Traumatic spinal cord injury
TTAB	Tetradecyltrimethylammonium bromide
TTX	Tetrodotoxin
UAA	Unnatural amino acids
UHPLC	Ultra-high performance liquid chromatography
UTR	Untranslated region
V_D	Volume of distribution
VFT	Venus flytrap
VGAT	Vesicular GABA transporter
VGluT	Vesicular glutamate transporter
VMAT	Vesicular monoamine transporter
V_{ND}	Non-displaceable volume of distribution
VOI	Volume of interest
V_T	Total volume of distribution
VTA	Ventral tegmental area
WAY 213613	(*N*-[4-(2-Bromo-4,5-difluorophenoxy)phenyl]-L-asparagine)
WT	Wild type
α-KG	α-ketoglutarate
γ-GTP	γ-glutamyltranspeptidase
$\Delta F/Fmax$	Maximal fluorescence changes over basal fluorescence

Chapter 1

Identification of Glutamatergic Neurons

Shiliang Zhang and Marisela Morales

Abstract

Glutamate is a major excitatory neurotransmitter in the mammalian central nervous system that gets accumulated into synaptic vesicles by vesicular glutamate transporters (vGluTs). Three isoforms of vGluTs have been cloned (vGluT1, vGluT2, and vGluT3), and shown by biochemical studies to selectively transport glutamate. The cloning of vGluTs has facilitated the anatomical and functional analysis of glutamatergic neurons, and while there are commercially available specific antibodies against vGluTs that label axonal terminals of glutamatergic neurons, the vGluT1 and vGluT2 proteins are undetectable in most cell bodies of glutamatergic neurons. Thus cellular detection of transcripts encoding vGluT1 or vGluT2 is so far the only available and reliable method to label cell bodies of glutamatergic neurons in wild-type animals. However, advances in viral cell-specific tagging of vGluTs neurons in transgenic mice has greatly facilitated the identification and manipulation of glutamatergic neurons. We will describe a multidisciplinary approach including neuronal track tracing, in situ hybridization, immuno-fluorescence, and immuno-electron microscopy that has allowed us the identification of glutamatergic neurons within the ventral tegmental area.

Key words Glutamate, Vesicular glutamate transporters, vGluT2, Dopamine, In situ hybridization, Fluorescence microscopy, Electron microscopy, Ventral tegmental area, Nucleus accumbens

1 Introduction

Glutamate (Glu), a major excitatory neurotransmitter in the mammalian central nervous system (CNS), is transported into synaptic vesicles by a distinct family of vesicular Glu transporters (vGluTs) membrane proteins. Three isoforms of vGluTs (vGluT1, vGluT2, and vGluT3) are present in the CNS. While vGluT1 and vGluT2 are expressed in glutamatergic neurons [1–10], vGluT3 is found in GABAergic neurons, cholinergic interneurons, serotonergic neurons, and glia [11–13]. However, a recent study has shown that the vGluT3 neurons provide excitatory glutamatergic inputs to the ventral tegmental area (VTA) [14]. Because vGluT1 and vGluT2 are Glu transporters restricted to known glutamatergic neurons, their presence has become a reliable molecular phenotypic marker to identify glutamatergic neurons. vGluTs proteins are highly concentrated in synaptic vesicles in axonal terminals of glutamatergic

Sandrine Parrot and Luc Denoroy (eds.), *Biochemical Approaches for Glutamatergic Neurotransmission*, Neuromethods, vol. 130, DOI 10.1007/978-1-4939-7228-9_1,

neurons [6, 7, 9], but undetectable in most of cell bodies of vGluT1 and vGluT2 neurons. Thus, cellular detection of transcripts encoding vGluT1 or vGluT2 is so far the only available and reliable method to label cell bodies of these glutamatergic neurons.

For over 50 years the VTA has been considered a dopaminergic structure. However, we had provided anatomical and electrophysiological evidence showing that glutamatergic neurons are major neuronal subpopulations within the VTA. By a combination of in situ hybridization and immunohistochemistry, we had demonstrated that the VTA contains cells expressing transcripts encoding vGluT2, but not vGluT1 or vGluT3 [15, 16]. We found that some of these VTA vGluT2 neurons co-express tyrosine hydroxylase (TH) [16], which is the limiting enzyme for the production of catecholamines and is present in VTA dopamine neurons [17]. We had also found that a subset of VTA vGluT2 neurons co-express GABAergic markers and these neurons co-release Glu and γ-aminobutyric acid (GABA) in the lateral habenula [18]. To determine the neurotransmitter properties of VTA dual vGluT2-TH neurons, we had combined Cre recombinase-dependent viral tracing methodology which allowed the tagging of VTA-vGluT2 neurons and their terminals within the nucleus accumbens (nAcc). By confocal and electron microscopy, we found that fibers from VTA dual vGluT2-TH neurons have distinct pools of vesicles that contain dopamine or Glu, each pool of vesicles are segregated into different micro-domains within a shared axon [19]. By a combination of electrophysiology and voltammetry, we had demonstrated that these vGluT2-TH axons release both dopamine and Glu [19]. Here we provide the detailed guidelines on how to apply neuronal track tracing, in situ hybridization, immuno-fluorescence and immuno-electron microscopy to study neuronal sites of Glu release.

1.1 In Situ Hybridization

To determine whether glutamatergic neurons are present within the VTA we used radioactive riboprobes to identify cellular expression of transcripts encoding vGluT1, vGluT2, or vGluT3. We found that, within the VTA, the detected cells expressed vGluT2 mRNA, but not vGluT1 or vGluT3 mRNA. By a combination of in situ hybridization (to detect vGluT2 mRNA) and immunocytochemistry (to detect TH), we determined that a subset of vGluT2 neurons co-express TH-immunoreactivity (Fig. 1). By a combination of retrograde tract tracing, immunocytochemistry, and in situ hybridization, we detected that two subpopulations of neurons expressing vGluT2 mRNA innervate the nAcc (Fig. 2).

1.2 Fluorescence Microscopy

A fluorescence microscope works via filters and dichroic mirrors, which separate the excitation light (shorter wavelength) from the emission light (longer wavelength). Advances in fluorescent markers, multiple filters, and spectrally resolved detectors have allowed

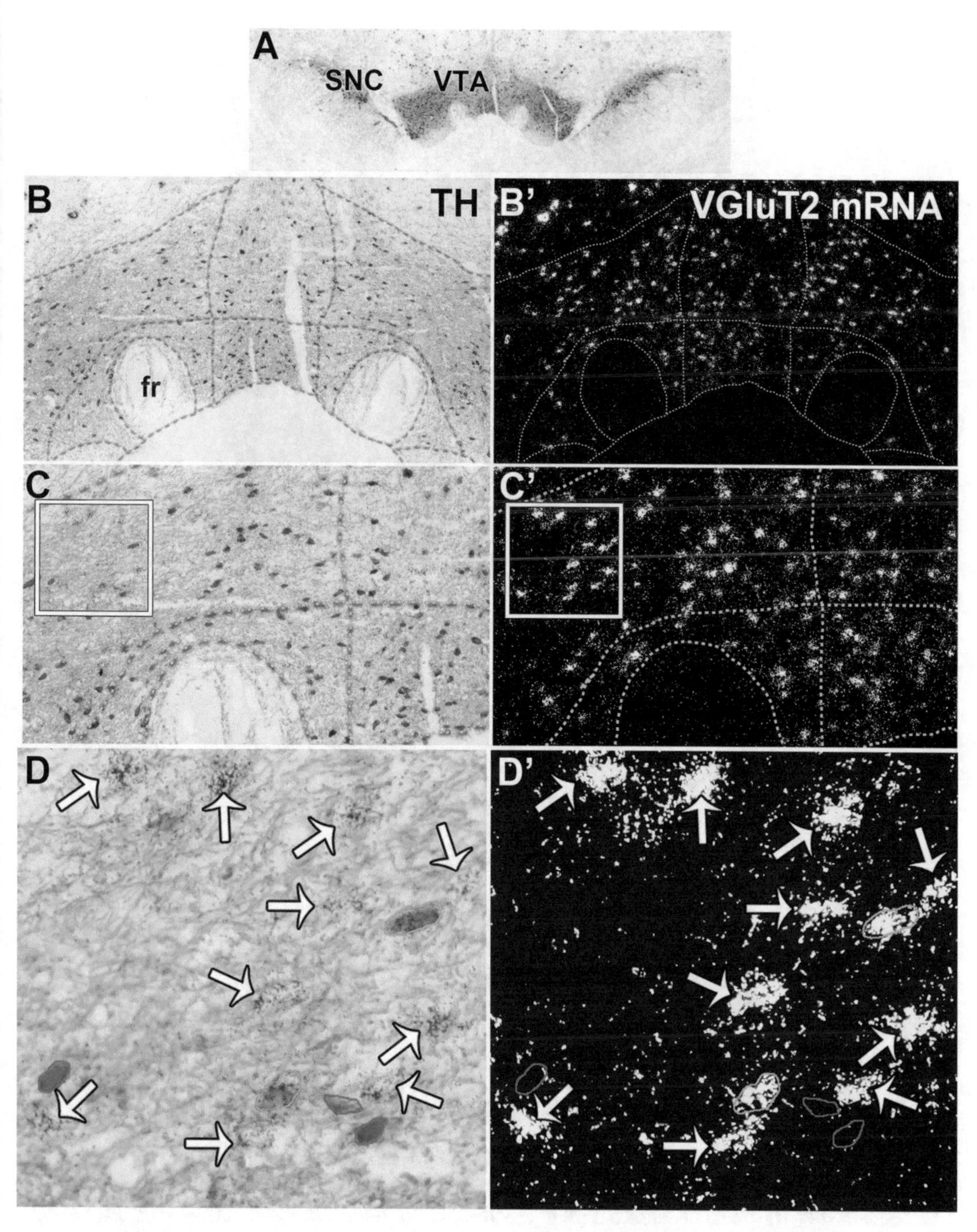

Fig. 1 Two subpopulations of neurons expressing vGluT2 mRNA in the VTA. (**a**) Low magnification of a 5 μm thick coronal section with TH immunoreactivity in the SNC and VTA. (**b**–**d**′) Correspond to the same section at different magnifications showing TH immunoreactivity under bright-field microscopy (*brown* in **b**, **c**, **d**) or vGluT2 mRNA under epiluminescence microscopy (*silver grains* in **b**′, **c**′, **d**′). In (**c**) and (**c**′) the *boxes* delimit areas at higher magnification in (**d**) and (**d**′) showing two neurons coexpressing TH and vGluT2 mRNA (*blue outlines*), ten neurons expressingvGluT2 mRNA and lacking TH immunoreactivity (*arrows*), and three neurons with TH immunoreactivity without vGluT2 mRNA (*purple outlines*). *SNC* substantia nigra compacta, *fr* fasciculus retroflexus. (Modified from [16])

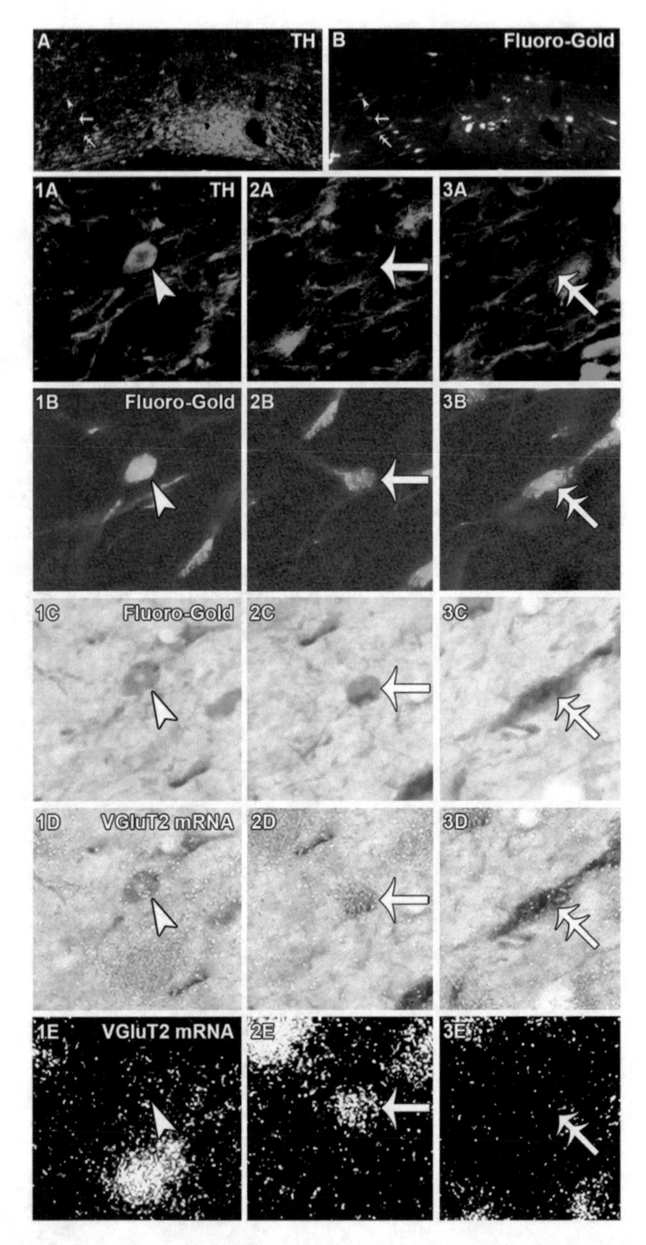

Fig. 2 Two subpopulations of neurons expressing vGluT2 mRNA innervate the nAcc. (**a**) and (**b**) showing VTA coronal section (18 μm thickness) at low magnification from a rat that was injected with Fluoro-Gold into the nAcc. Detection of TH immunofluorescent (**a**) or Fluoro-Gold (**b**) in the VTA. Cells (*arrowheads*, *long arrows*, *double-headed arrows*) in (**a**) and (**b**) are shown in (**1a**)–(**3e**) at higher magnification: **1a**–**3a** (TH immunofluorescent), **1b**–**3b** (FG fluorescent), **1c**–**3c** (FG detection with an anti-FG antibody), **1d**–**3d** (vGluT2 mRNA seen as aggregates of *green grains*), and **1e**–**3e** (vGluT2 mRNA seen as aggregates of *white grains*). **1a**–**e**, FG-labeled cell containing TH immunoreactivity and lacking vGluT2 mRNA (*arrowheads*). **2a**–**e**, FG-labeled cell lacking TH immunoreactivity but containing vGluT2 mRNA (long arrows). **3a**–**e**, FG-labeled cell lacking both TH and vGluT2 mRNA (*double-headed arrows*). (Modified from [16])

observations of distinct neuronal markers within specific neuronal compartments. By using fluorescence microscopy together with three-dimensional reconstruction, we found that within the nAcc the VTA dual vGluT2-TH neurons segregate vGluT2 and TH to distinct subcellular compartments located within the same axon (Fig. 3). However, the resolution provided by the fluorescence microscope (typical resolution of 200 nm) does not allow the identification of the type of synapse established by these dual vGluT2-TH axons. Thus immuno-electron microscopy studies are necessary to determine the ultrastructural properties of the synaptic composition of dual vGluT2-TH axons.

1.3 Immuno-Electron Microscopy

Electron microscopy utilizes an electron beam with a far smaller wavelength than light. As a result, the resolution of a standard transmission electron microscope is about 0.2 nm. To date, electron microscopic analyses have provided important information regarding sites of Glu release at the ultrastructural level. For example, by using pre-embedding and post-embedding immunolabeling techniques, we detected within the nAcc vesicular monoamine transporter (VMAT2) immunoreactivity in the axon segment proximal to a vGluT2-terminal that is making an asymmetric synapse on the head of a dendritic spine (Fig. 3).

Our ultrastructural findings provide evidence that within the nAcc the storage of each dopamine and Glu takes place in distinct vesicular pools enriched in different axonal micro-domains within the same axon. To further confirmed the results obtained by ultrastructural techniques, we determined the possible coexistence of vGluT2 and VMAT2 at the vesicular level by immunolabeling and by co-immunoprecipitation of vGluT2 and VMAT2 from vesicles obtained from rat nAcc synaptosomes, We conclude that accumulation of dopamine by VMAT2 occurs in vesicles different from those that accumulate Glu by vGluT2 (Fig. 4).

2 Materials

2.1 In Situ Hybridization

2.1.1 Cryosection Collection

1. Adult Sprague–Dawley male rats (300–350 g body weight).
2. 0.1 mol/L phosphate buffer (PB), pH 7.3: dissolve 33.67 g NaH_2PO_4 and 7.7 g NaOH to 1 L ddH_2O (see **Note 1**).
3. Chloral hydrate (80 mg/mL); 12%, 14%, and 18% sucrose solutions in 0.1 mol/L PB; and 4% paraformaldehyde (PFA) in 0.1 mol/L PB (see **Note 2**).
4. CM3050 S cryostat (Leica Microsystems Inc., Baffalo Grove, IL).

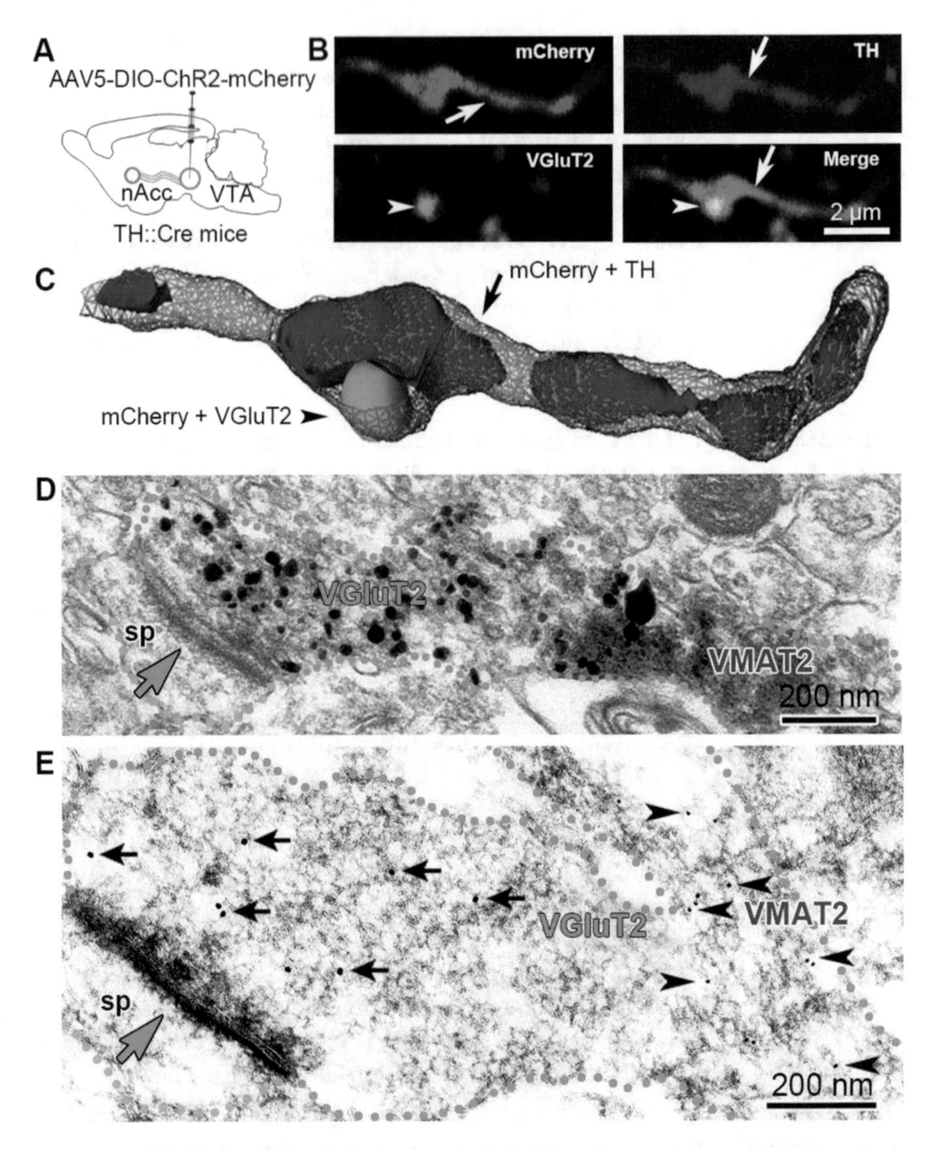

Fig. 3 vGluT2-TH neurons segregate vGluT2 and TH to distinct subcellular compartments in the same axon. (**a**) Schematic representation of nAcc inputs from VTA neurons expressing mCherry under the control of the TH promoter. (**b**, **c**) Fluorescence detection of mCherry-IR (*red*), TH-IR (*blue*) and vGluT2-IR (*green*) in the nAcc. (**b**) In the nAcc, mCherry (under the control of the TH promoter) was detected throughout the axon (*red*). In contrast, vGluT2-IR was restricted to a terminal-like structure (*arrowheads*). TH-IR was present in segments in the mCherry$^+$ axon (*arrows*). (**c**) Segregation among one vGluT2 terminal-like structure and TH axon segments in the same nAcc axon from a vGluT2-TH neuron is better seen in this three-dimensional reconstruction from Z-stack confocal microscopy images of triple labeled nAcc. (**d**) Pre-embedding detection of vGluT2-IR in axon terminal (AT) lacking VMAT2, which establish asymmetric synapses (*green arrow*) on dendritic spines (*orange outline*). vGluT2-IR (*gold particles*) was confined to the AT. The contiguous axon segments to these vGluT2 terminals contained VMAT2-IR (*scattered dark material*). (**e**) Post-embedding detection of vGluT2-IR (18 nm *gold particles*, *black arrows*) in an AT lacking VMAT2-IR and establishing an asymmetric synapse (*green arrow*) on a dendritic spine (*orange outline*). The contiguous axon segment to this vGluT2$^+$ terminal contained VMAT2-IR (12 nm *gold particles*, *black arrowheads*). (Modified from [19])

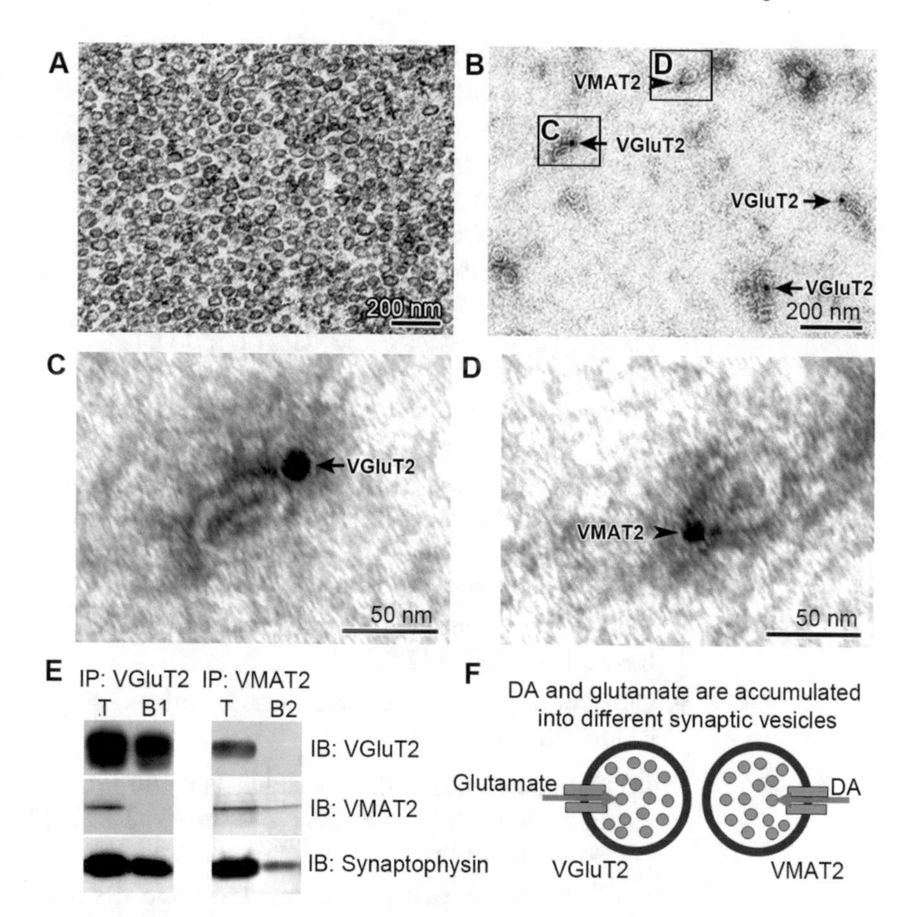

Fig. 4 vGluT2 and VMAT2 localize to distinct subpopulations of synaptic vesicles (wild-type rats). (**a**) Electron micrograph showing the purity and integrity of nAcc-isolated synaptic vesicles used for either dual detection of vGluT2-IR and VMAT2-IR (**b–d**) or co-immunoprecipitation of vGluT2 and VMAT2 (**e**). (**b–d**) Detection of vGluT2-IR (*arrows*, 18 nm *gold particles*) or VMAT2-IR (*arrowheads*, 12 nm *gold particles*) associated with purified synaptic vesicles. (**e**) Western blots of proteins from isolated vesicles before immunoprecipitation (IP, T) and after IP with antibodies against vGluT2 (IP: vGluT2, B1) or VMAT2 (IP: VMAT2, B2). Western blots were immunolabeled (IB) with antibodies to vGluT2, VMAT2 or the vesicular marker synaptophysin. The vesicular nature of each fraction was confirmed by the detection of synaptophysin. vGluT2 and VMAT2 were present in the total pool of vesicles (T). In contrast, vGluT2 was detected only in the sample IP with antibody to vGluT2 (IP: vGluT2), and VMAT2 was detected only in the sample IP with antibody to VMAT2. F. A model showing accumulation of dopamine by VMAT2 in a vesicle different from the one that accumulates Glu by vGluT2. (Modified from [19])

2.1.2 In Situ Hybridization

1. Diethylpyrocarbonate (DEPC)-treated ddH_2O and DEPC-treated 0.1 mol/L PB buffer.
2. Section incubation solution: 0.1 mol/L PB containing 0.5% Triton X-100.
3. 0.2 mol/L HCl; 0.25% acetic anhydride in 0.1 mol/L triethanolamine; 4% PFA in 0.1 mol/L PB; and 4 μg/mL RNase A.
4. Hybridization buffer (50% formamide; 10% dextran sulphate; 5× Denhardt's solution; 0.62 mol/L NaCl; 50 mmol/L DTT;

10 mmol/L EDTA; 20 mmol/L PIPES, pH 6.8; 0.2% SDS; 250 μg/mL salmon sperm DNA; 250 μg/mL tRNA).

5. [^{35}S]- and [^{33}P]-labeled single-stranded antisense or sense of rat vGluT1 (nucleotides 53–2077; GenBank accession number NM-053859.1), vGluT2 (nucleotides 317–2357; GenBank accession number NM-053427), or vGluT3 (nucleotides 1–1729; GenBank accession number BC117229.1) probes at 10^7 cpm/mL.
6. Plasmids that contained the vGluT1 and vGluT2 were generously provided by Dr. Robert H. Edwards (University of California, San Francisco).
7. 1× saline sodium citrate (SSC) and 0.1× SSC. To make 20× SSC, dissolve 175.3 g of NaCl and 88.2 g of sodium citrate in 800 mL of ddH_2O. Adjust the pH to 7.0 with a few drops of 1 mol/L HCl. Adjust the volume of 1 L with additional ddH_2O. Sterilize by autoclaving.
8. Tris buffer saline (TBS) buffer (20 mmol/L Tris–HCl, 0.5 mol/L NaCl, pH 8.2).
9. Chromium aluminum-coated slides. To subbed slides, the procedures are as follows:
   (a) Fill metal racks with new slides: Put 400 mL each of the following solutions in clear glass boxes: 10% HCl in 70% ethanol; distilled H_2O and 95% ethanol.
   (b) Dip racks in each solution for 1–5 s and dry in oven over night.
   (c) Prepare subbing solution: heat 1 L distilled H_2O to 60 °C; add 4 g gelatin and mix with magnetic stir bar; once gelatin is dissolved, add 0.4 g chromium potassium sulfate.
   (d) Filter 500 mL of the solution and pour into clear glass slide box.
   (e) Dip each rack for 10 s, gently rake slides to disperse bubbles (see **Note 3**).
   (f) Blot edge of rack with a paper towel and replenish chromium aluminum solution with filtered stock as necessary.
   (g) Dry in oven for 2 days and store slides in original boxes.
10. Mounting solution (ethanol-gelatin solution): 0.3% gelatin in 40% Ethanol. To make 2× ethanol-gelatin solution, heat 1 L ddH_2O to 50 °C, add 6 g gelatin mix, cool to room temperature, add 1 L of 80% ethanol to gelatin solution, add 250 μL HCl. Before use, make 1× working solution, add 100 mL 2× ethanol-gelatin solution, 100 mL ddH_2O, and one drop HCl.
11. Coverslip slides. The following procedure is applied to preserve tissue sections for a long period of time:

(a) Dehydration: serial ethanol 30%, 60%, 90%, 95%, 100%, 100% 2 min for each, then 100% CitriSolv or histoclear 2 × 10 min.

(b) Coverslipping: leave slides in CitriSolv prior to coverslipping (do not let the slides dry). Remove the slides from CitriSolv one at a time. Blot excess of CitriSolv (it is okay if the slides are still wet). Place 50–100 μL Permount along one edge of the slide. Place coverslip on the slide. Wipe back of slide and look for bubbles. Squeeze out any bubbles with forceps and let the slide dry under a fume hood (see **Note 4**).

(c) Clean up: after the slides have dried (1–2 days), clean off excess Permount using CitriSolv. Allow the cleaned slides to dry before putting them into storage boxes.

12. LS6500 Multipurpose Scintillation Counter (Beckman Coulter Inc., Brea, CA) for measuring the radioactive samples.
13. Typhoon FLA 7000 (GE Healthcare Life Sciences, Pittsburgh, PA) for scanning radioisotopic labels.
14. Bx53 Turboscan Microscope (Olympus Life Science, Center Valley, PA) for collecting in situ hybridization images and VS120 Virtual Slide Microscope (Olympus Life Science, Center Valley, PA) for collecting in situ hybridization images (see **Note 5**).

2.1.3 Combination of In Situ Hybridization and Immunocytochemistry

1. All materials in the above Sect. 2.1.2 for single in situ hybridization.
2. Blocking buffer: 4% bovine serum albumin (BSA) and 0.3% Triton X-100 in 0.1 mol/L PB buffer.
3. Anti-TH mouse monoclonal antibody (1:500, EMD Millipore, Billerica, MA); Biotinylated goat-anti-mouse IgG antibody (1:200, Vector Laboratories, Burlingame, CA); Avidin-biotinylated horseradish peroxidase (ABC kit, Vector Laboratories, Burlingame, CA).
4. 0.05% 3,3-diaminobenzidine-4 HCl (DAB); 0.003% hydrogen peroxide (H_2O_2).
5. Ilford K.5 nuclear tract emulsion (1:1 dilution in double distilled water; Polysciences Inc., Warrington, PA).

2.1.4 Retrograde Tract Tracing and In Situ Hybridization

1. All materials in the above Sect. 2.1.2 for single in situ hybridization.
2. 300–350 g Sprague Dawley male rats.
3. Chloral hydrate (3 mL/kg, i.p.); Saline solution; Retrograde tracer Fluoro-Gold (FG; 1% in cacodylate buffer, pH 7.5).
4. Rabbit anti-FG antibody (AB153; 1:500; EMD Millipore, Billerica, MA); Anti-TH mouse monoclonal antibody (MAB318; 1:500; EMD Millipore, Billerica, MA).

5. RNasin ribonuclease inhibitor (40 U/μL stock; 5 μL/mL of buffer; Promega, Madison, WI).
6. [^{35}S]- and [^{33}P]-labeled single-stranded antisense or sense of rat vGluT2.
7. ABC kit (Vector Laboratories, Burlingame, CA); 0.05% DAB; 0.003% H_2O_2.
8. Ilford K.5 nuclear tract emulsion (1:1 dilution in double distilled water; Polysciences Inc., Warrington, PA).
9. DM LB 100T microscope (Leica Microsystems Inc., Buffalo Grove, IL).

2.2 Fluorescence Microscopy

2.2.1 Animals and Surgical Procedures

1. Adult Sprague–Dawley male rats (300–350 g body weight); Transgenic mice or rats (e.g., TH-Cre mice, vGluT2-Cre mice).
2. Adeno-associated virus vector (AAV) encoding ChR2 tethered to mCherry under the Camk2a promoter (AAV5-CaMKII-ChR2-mCherry, 0.2 μL).
3. Picospritzer III (Parker-Hannifin Filtration, Halethorpe, MD); Stereotaxic apparatus (David Kopf Instruments, Tujunga, CA).
4. The AAV5-DIO-ChR2-mCherry viral vector (AAV-EF1a-DIO-hChR2(H134R)-mCherry-WPRE-pA, 3×10^{12} genomes/mL) from the UNC Vector Core Facility.

2.2.2 Animal Perfusion and Vibratome Sectioning

1. Wild-type adult Sprague–Dawley male rats, *Phaseolus vulgaris* leukoagglutinin (PHA-L) or virus injected rats or transgenic mice.
2. Chloral hydrate (80 mg/mL); 0.1 mol/L PB (pH 7.3); 0.9% saline solution; 2% PFA in 0.1 mol/L PB.
3. Fixative solution: 4% PFA + 0.15% glutaraldehyde + 15% picric acid in 0.1 mol/L PB (see **Note 6**). Prepare 1 L of 8% PFA in 2× PB (see **Note 2**), filter the solution and bring it to final volume of 2 L (to obtain 4% PFA in 1× PB) by adding 300 mL of picric acid, 12 mL of 25% glutaraldehyde (16216; Electron Microscopy Sciences, Hatfield, PA), and 688 mL of ddH_2O.
4. 1000 U/mL of heparin in saline solution: add 550 mg of heparin per 100 mL of 0.9% saline solution.
5. Storage solution for vibratome sections: 25% sucrose and 10% glycerol in ddH_2O with 2 mmol/L NaN_3.
6. VT1000 S vibratome (Leica Microsystems Inc., Buffalo Grove, IL).

2.2.3 Fluorescence Microscopy

1. 0.1 mol/L PB (pH 7.3); Blocking solution: 4% BSA and 0.3% Triton X-100 in 0.1 mol/L PB buffer.
2. Primary antibodies: Mouse-anti-mCherry (632543; 1:500; Clontech Laboratories, Inc., Mountain View, CA); Mouse-

anti-GFP (632380; 1:500; Clontech Laboratories, Inc., Mountain View, CA); Mouse-anti-TH (MAB318; 1:1000; EMD Millipore, Billerica, MA); Rabbit-anti-TH (AB152; 1:1000; EMD Millipore, Billerica, MA); Sheep-anti-TH (AB1542; 1:1000; EMD Millipore, Billerica, MA); Guinea pig-anti-vGluT2 (vGluT2-GP-Af240-1; 1:500; Frontier Institute Co., Ltd., Japan); Rabbit-anti-VGAT (VGAT-Rb-Af500; 1:500; Frontier Institute Co., Ltd., Japan).

3. Donkey fluorescent secondary antibodies (Jackson Immuno-research Laboratories Inc., West Grove, PA).
4. Vectashield mounting medium (H1000; Vector Laboratories, Burlingame, CA).
5. Olympus FV1000 Confocal System (Olympus, Center Valley, PA).
6. Imaris microscopy software (Bitplane Inc., South Windsor, CT); Amira 3D software for life sciences (FEI, Hillsboro, OR).

2.3 Immuno-Electron Microscopy

1. 0.1 mol/L PB (pH 7.3); 1% sodium borohydride in 0.1 mol/L PB; 1.5% glutaraldehyde in 0.1 mol/L PB.
2. Blocking solution: 1% normal goat serum (NGS) and 4% BSA in 0.1 mol/L PB supplemented with 0.02% saponin.
3. Primary antibodies: Mouse-anti-mCherry (632543; 1:500; Clontech Laboratories, Inc., Mountain View, CA); Mouse-anti-GFP (632380; 1:500; Clontech Laboratories, Inc., Mountain View, CA); Mouse-anti-TH (MAB318; 1:1000; EMD Millipore, Billerica, MA); Rabbit-anti-TH (AB152; 1:1000; EMD Millipore, Billerica, MA); Sheep-anti-TH (AB1542; 1:1000; EMD Millipore, Billerica, MA); Guinea pig-anti-vGluT2 (vGluT2-GP-Af240-1; 1:500; Frontier Institute Co., Ltd., Japan).
4. Biotinylated goat-anti-mouse, or anti-rabbit, or anti-guinea pig IgG antibody (1:200, Vector Laboratories, Burlingame, CA).
5. Anti-mouse-IgG coupled to 1.4 nm gold (2001; 1:100; Nanoprobes, Yaphank, NY), anti-guinea pig-IgG Fab' fragment coupled to 1.4 nm gold (2055; 1:100; Nanoprobes, Yaphank, NY), or anti-rabbit-IgG coupled to 1.4 nm gold (2003; 1:100; Nanoprobes, Yaphank, NY).
6. ABC kit (Vector Laboratories, Burlingame, CA); 0.05% DAB; 0.003% H_2O_2; HQ silver enhancement kit (2012; Nanoprobes, Yaphank, NY); 0.5% osmium tetroxide in 0.1 mol/L PB (see **Note 7**); 1% uranyl acetate (19481; Ted Pella, Inc., Redding, CA) in 70% ethanol (see **Note 8**).
7. A series of graded ethanol: 30%, 50%, 70%, 90%, 100%, 100%, 100%.

2.3.1 Animals and Surgical Procedures (See Sect. 2.2.1)

2.3.2 Animal Perfusion and Vibratome Sectioning (See Sect. 2.2.2)

2.3.3 Pre-Embedding Immuno-Electron Microscopy

8. Propylene oxide (18601; Ted Pella, Inc., Redding, CA); Durcupan ACM, Epoxy Resin (14040; Electron Microscopy Sciences, Hatfield, PA).
9. Blue M oven (Thermo Fisher Scientific, Waltham, MA); MVX10 microscope (Olympus Life Science, Center Valley, PA).
10. TEM razor blades (Tousimis, Rockville, MD); UC-7 ultramicrotome (Leica Microsystems Inc., Buffalo Grove, IL); Glass knife maker (Leica Microsystems Inc., Buffalo Grove, IL); Diamond knife (Diatome, Hatfield, PA).
11. 5% uranyl acetate (19481; Ted Pella, Inc., Redding, CA) in ddH_2O (see **Note 8**).
12. Sato's lead [20]: Weigh out 0.1 g of lead nitrate, 0.1 g of lead citrate, 0.1 g of lead acetate, and 0.2 g of sodium citrate. Add 8.2 mL of degassed distilled water to the above mixture of chemicals in a 15 mL Falcon tube and shake vigorously for 1 min. The solution looks very milky. Add 1.8 mL of freshly made 4% NaOH. The solution becomes clear except for some large white grains at the bottom of the tube. Filter the solution with a 0.22 μm syringe driven filter unit. It is ready for use (see **Note 9**).
13. Tecnai G^2 12 transmission electron microscope (FEI, Hillsboro, OR) equipped with a digital micrograph 3.4 camera (Gatan, Pleasanton, CA).

2.3.4 Post-Embedding Immuno-Electron Microscopy

1. 0.1 mol/L PB (pH 7.3); 0.25% tannic acid in ddH_2O (see **Note 10**); 2% uranyl acetate in ddH_2O (see **Note 8**); 1× PB saline-Tween 20 (PBST) buffer; 0.05 mol/L glycine in 1× PBS buffer.
2. A series of graded ethanol: 30%, 50%, 70%, 90%, 100%, 100%, 100%.
3. LR white embedding medium (14380; Electron Microscopy Sciences, Hatfield, PA).
4. Lindberg Blue M oven (Thermo Fisher Scientific, Waltham, MA); MVX10 microscope (Olympus Life Science, Center Valley, PA).
5. TEM razor blades (Tousimis, Rockville, MD); UC-7 ultramicrotome (Leica Microsystems Inc., Buffalo Grove, IL).
6. Blocking solution: 2% NGS and 2% BSA in 1× PBS-Tween 20 buffer.
7. Primary antibodies in the blocking buffer: Guinea pig-anti-vGluT2 (vGluT2-GP-Af240-1; 1:20; Frontier Institute Co., Ltd., Japan); Goat-anti-VMAT2 (EB06558; 1:25; Everest Biotech Ltd., United Kingdom).

8. Secondary antibodies in the blocking buffer: Donkey-anti-goat 12 nm colloidal gold (705-205-147; 1:20; Jackson ImmunoResearch Laboratories, West Grove, PA); Donkey-anti-guinea pig 18 nm colloidal gold (706-215-148; 1:10; Jackson ImmunoResearch Laboratories, West Grove, PA).
9. 2% glutaraldehyde in 1× PBS-Tween 20 buffer; 5% uranyl acetate (19,481; Ted Pella, Inc., Redding, CA) in ddH_2O (see **Note 8**).
10. Sato's lead (see Sect. 2.3.3).
11. Tecnai G² 12 transmission electron microscope (FEI, Hillsboro, OR) equipped with a digital micrograph 3.4 camera (Gatan, Pleasanton, CA).

2.3.5 Immuno-Gold Detection within the Isolated Synaptic Vesicles

1. Male Sprague-Dawley rats (6 weeks of age).
2. Sigmacote (SL2-100 mL; Sigma-Aldrich, St. Louis, MO) (see **Note 11**); 1 mol/L KOH solution; 1 mol/L tartaric acid; 5 mmol/L HEPES buffer (pH 7.4). The pH is adjusted with KOH or tartaric acid.
3. 0.2 mol/L Phenylmethylsulfonyl fluoride (PMSF) in 100% ethanol or 2 propanol (500× stock solution). Final working solution is 0.1 mmol/L. Keep in −20 °C for 2 months.
4. SB 320 buffer (0.32 mol/L sucrose in 5 mmol/L HEPES buffer, pH 7.4); SBI: SB 320 buffer containing 1× protein inhibitor (11697498001; Roche Diagnostics Corporation, Indianapolis, IN) and PMSF. Every 50 mL SB 320 buffer, dissolve one tablet protein inhibitor and add 0.2 mol/L PMSF 100 μL before use.
5. Osmotic shock solution: ddH_2O containing 1× protein inhibitor (11697498001; Roche Diagnostics Corporation, Indianapolis, IN) and PMSF.
6. 0.25 mol/L potassium HEPES (pH 6.5); 1 mol/L neutral L-(+)-tartaric acid dipotassium salt buffer (pH 7.5); 1 mol/L $MgSO_4$ (1000× stock solution).
7. Freshly made 4% PFA in ddH_2O; 2% osmium tetroxide in 0.1 mol/L PB (see **Note 7**); 4% or 5% uranyl acetate (19481; Ted Pella, Inc., Redding, CA) in ddH_2O (see **Note 8**); Propylene oxide (18601; Ted Pella, Inc., Redding, CA); Durcupan ACM, Epoxy Resin (14040; Electron Microscopy Sciences, Hatfield, PA); A series of graded ethanol: 30%, 50%, 70%, 90%, 100%, 100%, 100%.
8. Blue M oven (Thermo Fisher Scientific, Waltham, MA); MVX10 microscope (Olympus Life Science, Center Valley, PA); UC-7 ultramicrotome (Leica Microsystems Inc., Buffalo Grove, IL).

9. Sato's lead (see Sect. 2.3.3).
10. 1 mol/L K^+-tartrate/EDTA. To make 20 mL of 1 mol/L K^+-tartrate/EDTA buffer, to dissolve 1.19 g HEPES in 16 mL, adjust pH with KOH or tartaric acid to 7.4. Add 4.7 g K^+-tartrate and 0.03722 g EDTA.
11. Freshly made 4% PFA in 0.1 mol/L PB; 1 mol/L ethanolamine-HCl; TBS buffer: 0.1 mol/L Tris/HCl (pH 7.4) containing 0.3 mol/L NaCl; Blocking buffer: 10% newborn calf serum (NCS) in TBS buffer.
12. Primary antibodies in 1% NCS/TBS buffer: Guinea-pig anti-vGluT2 (vGluT2-GP-Af240-1; 1:20; Frontier Institute Co., Ltd., Japan); Goat-anti-VMAT2 (EB06558; 1:25; Everest Biotech Ltd., United Kingdom).
13. Secondary antibodies in 1% NCS/TBS buffer containing 5 mg/10 mL polyethylene glycol (PEG): Donkey-anti-goat 12 nm colloidal gold (705-205-147; 1:20; Jackson Immuno-Research Laboratories, West Grove, PA); Donkey-anti-guinea pig 18 nm colloidal gold (706-215-148; 1:10; Jackson ImmunoResearch Laboratories, West Grove, PA).
14. 0.1 mol/L PB buffer; 2.5% glutaraldehyde in 0.1 mol/L PB; 1% osmium tetroxide in 0.1 mol/L PB (see **Note 7**).
15. 1% and 5% uranyl acetate (19,481; Ted Pella, Inc., Redding, CA) in ddH_2O (see **Note 8**).
16. Avanti J-25 centrifuge (Beckman Coulter Inc., Brea, CA); Optima MAX-XP ultracentrifuge (Beckman Coulter Inc., Brea, CA); Tecnai G^2 12 transmission electron microscope (FEI, Hillsboro, OR) equipped with a digital micrograph 3.4 camera (Gatan, Pleasanton, CA).

2.3.6 Co-Immuno-precipitation and Immunoblotting

1. Dynabeads protein A immunoprecipitation kit (10006D, Thermo Fisher Scientific, Waltham, MA); Dynabeads protein G immunoprecipitation kit (10007D, Thermo Fisher Scientific, Waltham, MA).
2. Primary antibodies: Guinea pig-anti-vGluT2 (AB2251; EMD Millipore, Billerica, MA); Mouse-anti-vGluT2 (MAB5504; EMD Millipore, Billerica, MA); Goat-anti-VMAT2 (EB06558; 1:25; Everest Biotech Ltd., United Kingdom); Mouse-anti-synaptophysin p38 (MAB638; 1:4000;EMD Millipore, Billerica, MA).
3. NuPAGE Novex 4–12% Bis-Tris Protein Gels, 1.0 mm, 12-well (NP0322BOX; Thermo Fisher Scientific, Waltham, MA); NuPAGE MOPS SDS Running Buffer (20×) (NP0001; Thermo Fisher Scientific, Waltham, MA); NuPAGE LDS Sample Buffer (4×) (NP0007; Thermo Fisher Scientific, Waltham, MA); NuPAGE Transfer Buffer (20×) (NP0006; Thermo Fisher Scientific, Waltham, MA).

4. Immun-Blot PVDF/Filter Paper Sandwiches (1620239; Bio-Rad, Hercules, CA); Precision Plus Protein Dual Color Standards (1610374; Bio-Rad, Hercules, CA).
5. 1× PBS and 1× PBST; Odyssey blocking buffer (Ready-to-use formulation in PBS, Li-cor Biosciences, Lincoln, NE).
6. Secondary antibodies: IRDye 800CW or IRDye 680RD (Li-cor Biosciences, Lincoln, NE).
7. Quick Western Kit (926-68100; Li-cor Biosciences, Lincoln, NE).
8. Odyssey imaging system (Li-cor Biosciences, Lincoln, NE).

3 Methods

3.1 In Situ Hybridization

3.1.1 Cryosection Collection

1. All animal procedures were approved by the NIDA Animal Care and Use Committee.
2. Adult Sprague–Dawley male rats were anaesthetized with chloral hydrate and perfused transcardially with 4% PFA in 0.1 mol/L PB, pH 7.3. Leave brains in 4% PFA for 2 h at 4 °C. Rinse brains with 0.1 mol/L PB and transfer sequentially to 12%, 14% and 18% sucrose solutions in 0.1 mol/L PB.
3. Cut coronal serial sections of 5, 12, or 16 μm thickness using a CM3050 S cryostat.

3.1.2 Single In Situ Hybridization

1. Rinse with 0.1 mol/L DEPC-treated PB for 2 × 5 min, treat with 0.2 mol/L HCl for 10 min, rinse with PB for 2 × 5 min and then acetylate in 0.25% acetic anhydride in 0.1 mol/L triethanolamine, pH 8.0 for 10 min. Rinse sections with 0.1 mol/L DEPC-treated PB for 2 × 5 min, postfix with 4% PFA for 10 min, then rinse with 0.1 mol/L DEPC-treated PB.
2. Hybridize in the hybridization buffer containing [^{35}S]- and [^{33}P]-labeled single-stranded antisense or sense of rat vGluT1, vGluT2 or vGluT3 probes at 10^7 cpm/mL for at 55 °C 16 h.
3. Treat sections with 4 μg/mL RNase A at 37 °C for 1 h, wash with 1× SSC, 50% formamide at 55 °C for 1 h, and with 0.1× SSC at 68 °C for 1 h. Rinse sections with TBS buffer.
4. Mount sections on coated slides, air dry, dip in Ilford K.5 nuclear tract emulsion and expose in the dark at 4 °C for several weeks prior to development.

3.1.3 Combination of In Situ Hybridization and Immunocytochemistry

1. Use Coronal free-floating sections (12 μm thick); Process radioactive single in situ hybridization (for detection of vGluT2 mRNA). Rinse with 0.1 mol/L DEPC-treated PB and incubate sections in the blocking buffer for 1 h.

2. Incubate sections with anti-TH mouse monoclonal antibody overnight at 4 °C. Rinse sections with 0.1 mol/L DEPC-treated PB for 3 × 10 min.
3. Process sections with an ABC kit. Incubate sections at room temperature (RT) in a 1:200 dilution of the biotinylated secondary antibody for 1 h. Rinse sections with 0.1 mol/L DEPC-treated PB for 3 × 10 min.
4. Incubate with avidin-biotinylated horseradish peroxidase at RT for 1 h. Rinse sections with 0.1 mol/L DEPC-treated PB for 3 × 10 min. Develop peroxidase reaction with 0.05% DAB and 0.003% H_2O_2.
5. Mount sections on coated slides, air dry. Dip in Ilford K.5 nuclear tract emulsion and expose in the dark at 4 °C for several weeks prior to development.

3.1.4 Retrograde Tract Tracing and In Situ Hybridization

1. Anesthetize male Sprague Dawley rats with chloral hydrate in a physiological saline solution.
2. Fix deeply anesthetized rats in a stereotaxic apparatus. Deliver the retrograde tracer Fluoro-Gold (FG) bilaterally into the nAcc with pipettes lowered at a 10° angle in the coronal plane. The FG is delivered iontophoretically through a stereotaxically positioned glass micropipette (inner tip diameter between 40 μm) by applying 5 μA current in 5 s pulses at 10 s intervals for 20 min. After each injection, the micropipette is left in place for an additional 10 min to prevent backflow of tracer up the injection track (see **Note 12**).
3. One week after FG injections, perfuse the rats as indicated above Sect. 3.1.1.
4. Cut coronal serial sections of 18 μm thickness using a CM3050 S cryostat.
5. Incubate sections with a mixture of rabbit-anti-FG antibody and the mouse monoclonal anti-TH antibody in the antibody buffer supplemented with RNasin ribonuclease inhibitor. Rinse sections with 0.1 mol/L DEPC-treated PB for 3 × 5 min.
6. Incubate sections in biotinylated goat-anti-rabbit antibody and fluorescein-conjugated donkey-anti-mouse antibody in DEPC-treated PB supplemented with RNasin ribonuclease inhibitor for 1 h at 30 °C. Rinse sections with 0.1 mol/L DEPC-treated PB for 3 × 5 min.
7. Transfer sections to 4% PFA in 0.1 mol/L DEPC-treated PB. Visualize by epifluorescence with a Leica DM LB microscope to identify FG- or TH-labeled neurons. Rinse sections with 0.1 mol/L DEPC-treated PB for 3 × 5 min. Rinse sections with the blocking buffer. Rinse sections with 0.1 mol/L DEPC-treated PB for 2 × 5 min. Treat with 0.2 mol/L HCl

for 10 min. Rinse sections with 0.1 mol/L DEPC-treated PB for 2 × 5 min.

8. Acetylate in 0.25% acetic anhydride in 0.1 mol/L triethanolamine, pH 8.0, for 10 min. Rinse sections with 0.1 mol/L DEPC-treated PB for 2 × 5 min.
9. Postfix with 4% PFA for 10 min. Incubate sections in hybridization buffer at 55 °C for 2 h.
10. Hybridize in the hybridization buffer containing [^{35}S]- and [^{33}P]-labeled single-stranded antisense or sense of rat vGluT2 at 55 °C for 16 h.
11. Treat sections with 4 μg/mL RNase A at 37 °C for 1 h, wash with 1× SSC, 50% formamide at 55 °C for 1 h, and with 0.1× SSC at 68 °C for 1 h. Incubate in avidin-biotinylated horseradish peroxidase at room temperature for 1 h. Rinse sections with 0.1 mol/L DEPC-treated PB for 3 × 10 min.
12. Develop peroxidase reaction with 0.05% DAB and 0.003% H_2O_2.
13. Mount sections on coated slides, air dry. Dip in Ilford K.5 nuclear tract emulsion and expose in the dark at 4 °C for several weeks prior to development.

3.2 Fluorescence Microscopy

3.2.1 Animals and Surgical Procedures

1. Iontophoretically inject the anterograde tracer PHA-L in 0.01 mol/L sodium phosphate buffer, pH 7.8 into the VTA (bregma AP −5.2, ML ±0.8, DV −8.4) of male Sprague-Dawley rats (350–420 g body weight) through a stereotaxically positioned glass micropipette (inner tip diameter of 20 μm, 5 μA current, 5 s on/off for 15 min).
2. Deliver viral vector AAV encoding ChR2 tethered to mCherry under the Camk2a promoter (AAV5-CaMKII-ChR2-mCherry, 0.2 μL) into the VTA by pressure through a glass micropipette attached to a Picospritzer III.
3. Inject male TH-Cre mice (Jackson Laboratory) or vGluT2-Cre mice (from O. Kiehn) (background: C57BL/6J mouse, 25–30 g body weight) into the VTA with Cre-inducible recombinant AAV encoding ChR2 tethered to mCherry (TH-ChR2-mCherry mice or vGluT2-ChR2-mCherry mice).
4. Anesthetize rats or mice with isoflurane and fix animals in a stereotaxic apparatus for viral injections.
5. Inject the AAV5-DIO-ChR2-mCherry viral vector (AAV-EF1a-DIO-hChR2(H134R)-mCherry-WPRE-pA, 3 × 1012 genomes/mL) into the VTA of TH-Cre mice (0.4 μL) or VTA of vGluT2-Cre mice (0.2 μL).
6. Each injection was bilaterally with a NanoFil syringe (with 35 gauge needle, WPI) into the VTA (bregma AP −3.4, ML ±0.2, DV −4.3 for mice; bregma AP −5.2, ML ±0.8, DV −8.4 for rats).

7. Leave the micropipette in place for an additional 10 min after each injection (see **Note 12**).
8. Rats are double housed, and all mice are housed in groups of up to four animals per cage in the animal rooms at 22 °C under a 12-h light/dark cycle (light on at 7 a.m.), with ad libitum access to food and water.

3.2.2 Animal Perfusion and Vibratome Sectioning

1. For anatomical studies, animals are deeply anesthetized with chloral hydrate (Sprague-Dawley rats without any manipulation, PHA-L rats, TH-ChR2-mCherry mice, and vGluT2-ChR2-mCherry mice).
2. Perfuse animals injected with the tracer (PHA-L rats, 2 weeks after injection) or viral vector (TH-ChR2-mCherry mice, vGluT2-ChR2-mCherry mice; 6–8 weeks after injections) for fluorescence microscopic or immuno-electron microscopic studies. Perfuse first with heparin solution follow by fixative solution (see **Note 13**).
3. Leave brains in the fixative solution at 4 °C for 2 h. Replace the fixative solution with 2% PFA and postfix the brains at 4 °C overnight. Rinse brains with 0.1 mol/L PB.
4. Cut brains into coronal serial sections (40 μm thick for mice, 50 μm thick for rats) with a VT1000 S vibratome. Collect six series of sequential sections in six wells plates.
5. Transfer coronal serial sections to the storage solution and leave on the shaker overnight at 4 °C. Transfer coronal serial sections to the Corning external thread cryogenic vials containing the new storage solution. Leave the transferred sections in the cryogenic vials on the shaker at room temperature for 1 h. Transfer the cryogenic vials containing sections to liquid nitrogen for 1 min for fast freezing. Store the sections at −80 °C.

3.2.3 Fluorescence Microscopy

1. Rinse coronal brain sections with 0.1 mol/L PB for 3 × 10 min. Incubate sections from TH-ChR2-mCherry mice or vGluT2-ChR2-mCherry mice with the blocking solution for 1 h.
2. Incubate sections with cocktails of primary antibodies overnight at 4 °C. Rinse sections with 0.1 mol/L PB for 3 × 10 min.
3. Incubate the corresponding cocktails of donkey fluorescent secondary antibodies at room temperature for 2 h. Rinse sections with 0.1 mol/L PB for 3 × 10 min.
4. Mount sections with Vectashield mounting solution on the slides.
5. Collect fluorescent images with Olympus FV1000 Confocal System. Sequentially take images with different lasers with 100× oil immersion objectives and set at 0.2 μm each of Z-stack.

3.3 Immuno-Electron Microscopy

3.3.1 Animals and Surgical Procedures (See Sect. 3.2.1)

3.3.2 Animal Perfusion and Vibratome Sectioning (See Sect. 3.2.2)

3.3.3 Pre-Embedding Immuno-Electron Microscopy

1. Rinse vibratome brain sections with 0.1 mol/L PB for 4 × 10 min. Incubate sections with 1% sodium borohydride in PB for 30 min to inactivate free aldehyde groups (see **Note 14**). Rinse sections with 0.1 mol/L PB for 4 × 10 min.
2. Incubate sections with the blocking solution for 30 min. Incubate sections with cocktails of primary antibodies overnight at 4 °C. Rinse sections with 0.1 mol/L PB for 4 × 10 min.
3. Incubate the corresponding cocktails of biotinylated and 1.4 nm nanogold conjugates secondary antibodies overnight at 4 °C. Rinse sections with 0.1 mol/L PB for 4 × 10 min.
4. Process sections with an ABC kit for 1–2 h. Add two drops of A in 5 mL of 0.1 mol/L PB and mix, then add two drops of B and keep rotating for 30 min before use. Rinse sections with 0.1 mol/L PB for 4 × 10 min.
5. Postfix sections with 1.5% glutaraldehyde for 10 min to keep gold particles intact. Rinse sections with 0.1 mol/L PB for 2 × 10 min. Transfer sections to 24-well plate. Keep one section per well. Rinse the sections with ddH_2O for 5 × 1 min.
6. Process sections by silver enhancement of the gold particles with the Nanoprobe Silver Kit for 7 min at room temperature (see **Note 15**). Rinse sections with ddH_2O for 5 min. Rinse sections with 0.1 mol/L PB for 4 × 10 min.
7. Postfix sections with 0.5% osmium tetroxide in the fume hood at room temperature for 25 min. Rinse sections with ddH_2O for 2 × 10 min.
8. Dehydrate sections with a series of graded ethanol (30%, 50%, 70%, 90%, 100%, 100%, and 100%) for 10 min each. It is okay to leave the sections in 70% ethanol overnight. Usually we leave the sections in 70% ethanol containing 1% uranyl acetate for 1 h. Use a new unopened bottle of 100% ethanol for each experiment (see **Note 16**). Rinse sections with propylene oxide for 2 × 10 min to remove the residual ethanol.
9. Infiltrate sections with Durcupan ACM epoxy resin for overnight (see **Note 17**). Transfer the sections onto a plastic sheet and drop the fresh Durcupan ACM epoxy resin on the sections and cover with another plastic sheet (we called plastic sandwich). Put the plastic sandwich between two glass slides and hold to make it flat using two binder clips. Bake the plastic sandwich units in a 60 °C oven for 48 h.
10. Remove the top plastic sheet and use the scalpel blades to cut the small pieces of target brain areas. Transfer each of them to the cap of flat embedding capsule. Stick each piece to the prepared resin holder by using the freshly made resin. Bake the resin holder together with samples in a 60 °C oven for 24–48 h.

11. Use a glass knife maker to make glass knives following the manual. Make a "boat" for each glass knife with a GKB (Glass Knife Boats) plastic and nail polish.
12. Perform coarse sectioning with the glass knife. Sections will float on the water within the boat. Once you can see the section containing samples, change to a diamond knife and start to continue the serial sections. Collect 100–150 serial sections onto slots with carbon-formvar supporting film. Store the slot in the grid storage box.
13. Line the bottom of glass Petri dish with parafilm. Add drops of 5% filtered uranyl acetate onto the parafilm. Stain the grid for 10–15 min with 5% filtered uranyl acetate. Gently rinse the grid with degassed ddH_2O and place it back to the grid storage box until dry.
14. Stain with Sato's lead as described by Sato [20] for 3 min. This is done in a covered glass Petri dish in the presence of NaOH pellets. Do not stain more than five grids at one time. Carefully rinse the grid with degassed ddH_2O and place it back into the grid storage box.
15. Screen sections under a transmission electron microscope (see **Note 18**).

3.3.4 Post-Embedding Immuno-Electron Microscopy

1. Cut the vibratome brain sections to small pieces in the target brain areas and rinse them with 0.1 mol/L PB for 3 × 10 min. Rinse sections with ddH_2O for 3 × 5 min.
2. Incubate sections with 0.25% tannic acid in ddH_2O for 5 min. Rinse sections with ddH_2O for 3 × 5 min. Incubate sections with 2% uranyl acetate in ddH_2O for 30 min.
3. Dehydrate sections with a series of graded ethanol (30%, 50%, 70%, 90%, 100%, 100%, and 100%) for 10 min each.
4. Infiltrate sections in LR white embedding medium (LR white)/100% ethanol (1:1) for 1–2 h. Aliquot LR white and leave at room temperature to warm up at least 30 min before use. Infiltrate sections in LR white/100% ethanol (2:1) for 1–2 h. Infiltrate sections in LR white for overnight. Change sections into flat capsule containing new LR white. Close the capsule cap and avoid air bubbles. Bake the sections in a 60 °C vacuum oven for 48 h.
5. Perform coarse and fine ultra-sectioning as described above (see Sect. 3.3.3). Collect serial sections onto Ni-mesh 200 with carbon-formvar supporting film. Leave overnight for air dry. Rinse sections with PBST for 2 × 2 min.
6. Incubate sections with 0.05 mol/L glycine in PBS buffer for 15 min to inactivate residual aldehyde groups present after aldehyde fixation. Incubate sections in blocking buffer for 1 h.

7. Incubate sections with goat anti-VMAT2 primary antibodies overnight at 4 °C. Rinse sections with PBST for 5 × 5 min.
8. Incubate sections in donkey-anti-goat 12 nm colloidal gold secondary antibody at RT for 1 h. Rinse sections with PBST for 3 × 5 min.
9. Postfix sections with 2% glutaraldehyde in PBST for 5 min. Rinse sections with PBST for 3 × 5 min.
10. Incubate sections in blocking buffer for 1 h. Incubate sections with guinea pig-anti-vGluT2 primary antibodies overnight at 4 °C. Rinse sections with PBST for 5 × 5 min.
11. Incubate sections in donkey-anti-guinea pig 18 nm colloidal gold secondary antibody at RT for 1 h. Rinse sections with PBST for 3 × 5 min.
12. Postfix sections with 2% glutaraldehyde in PBST for 5 min. Rinse sections with PBST for 3 × 5 min. Rinse sections with ddH_2O for 5 × 2 min.
13. Counterstain sections with 5% uranyl acetate for 3 min and Sato's lead as described by Sato [20] for 1 min.
14. Screen sections under a transmission electron microscope.

3.3.5 Immuno-Gold Detection Within the Isolated Synaptic Vesicles

The protocol for immuno-gold detection within the isolated synaptic vesicle is modified from Erickson et al. [21], Teng et al. [22], Boulland et al. [23] and Kadota and Kadota [24].

1. Decapitate rats and dissect out brains and quickly place them in ice-cold SB320 buffer. Drain off the SB320 buffer and dissect the slices containing the nAcc. Bring the slices to the ice-cold SB320I.
2. Dissect the nAcc out from the slices and transfer the nAcc materials in 8 mL ice-cold SB320I to a homogenizer and homogenize with 12 strokes.
3. Centrifuge the homogenate at 2000 × *g* for 10 min. Collect the supernatant and centrifuge at 10,000 × *g* for 30 min. The pellet contains the enriched synaptosomal fraction.
4. Osmotic shock for 5 min after adding 7 mL of ddH_2O containing 1× protease inhibitor. Homogenize with five strokes on ice. Readjust the osmolarity by adding 0.25 mol/L potassium HEPES (pH 6.5) and neutral 1.0 mol/L potassium tartrate (pH 7.5) in 1/10 volume (if add 7 mL of ddH_2O containing 1× protease inhibitor to 2 mL of synaptosome, add 900 μL of 0.25 mol/L potassium HEPES (pH 6.5) and 900 μL of neutral 1.0 mol/L potassium tartrate (pH 7.5)).
5. Centrifuge the preparation at 20,000 × *g* for 20 min to remove the mitochondria.

6. For immuno-gold labeling and electron microscopy analysis, collect the supernatant and dilute with 1 M HEPES/K^+ tartrate to a final concentration of 0.1 mol/L and process for immuno-gold EM. For co-immunoprecipitation and western blot experiments, collect the supernatant and centrifuge at 55,000 × *g* for 60 min.
7. Collect the supernatant and add $MgSO_4$ buffer (final concentration 1 mmol/L), centrifuge at 100,000 × *g* for 45 min. Resuspend the pellets containing the isolated synaptic vesicles with vesicle assay buffer containing 5 mmol/L HEPES, 0.32 mol/L sucrose, and protease inhibitors. Freeze the synaptic vesicles in liquid nitrogen and store at −80 °C until use.
8. To determine the purity of the synaptic vesicle preparation, fix the pellets harvested in the last step with 2% osmium tetroxide for 1 h. Rinse pellets with ddH_2O and postfix with 12% glutaraldehyde in ddH_2O overnight at 4 °C.
9. Rinse pellets with ddH_2O and stain with 4% uranyl acetate for 30 min. Rinse pellets with ddH_2O and dehydrate in a series of graded ethanol for 10 min each. Embed the pellets with Durcupan ACM epoxy resin. Bake the resin-embedded sections in a 60 °C oven for 48 h. Cut sections at 60–70 nm thickness with a UC-7 ultramicrotome using a diamond knife. Collect the serial sections on the formvar-coated slot grids and counterstain with uranyl acetate and lead (see Sect. 3.3.3).
10. Screen sections under a transmission electron microscope.
11. Thaw the vesicle aliquot from step 9 in this section. Dilute vesicles with 1× K^+-Tartrate/EDTA (1/2, 1/10, 1/20). Put the diluted vesicles onto formvar-coated grids (10 or 20 μL). Wait until most of the fluid has evaporated, but do not allow grids to dry. Usually need 1 h.
12. Fix vesicles with 4% PFA in a set of glass Petri dish overnight at 4 °C. Rinse 3 times with 0.1 mol/L PB.
13. Incubate in ethanolamine-HCl for 10 min. Rinse three times with 0.1 mol/L PB. Incubate in the blocking buffer 10% NCS/TBS for 1 h.
14. Incubate with the primary antibody goat-anti-VMAT2 in 1% NCS/TBS overnight at 4 °C. Rinse with 1% NCS/TBS.
15. Incubate with the secondary antibody donkey-anti-goat 12 nm colloidal gold in 1% NCS/TBS containing 5 mg/10 mL PEG at RT for 2 h. Rinse three times with TBS.
16. Postfix with 2.5% glutaraldehyde in 0.1 PB for 10 min. Rinse three times with 0.1 mol/L PB.
17. Incubate in the blocking buffer 10% NCS/TBS for 1 h. Incubate with the primary antibody guinea pig-anti-vGluT2 in 1% NCS/TBS overnight at 4 °C. Rinse with 1% NCS/TBS.

18. Incubate with the secondary antibody donkey-anti-guinea pig 18 nm colloidal gold in 1% NCS/TBS containing 5 mg/10 mL PEG at RT for 2 h. Rinse three times with TBS.
19. Postfix with 2.5% glutaraldehyde in 0.1 mol/L PB for 10 min. Rinse three times with ddH_2O.
20. Postfix with 1% osmium tetroxide in 0.1 mol/L PB for 30 min. Rinse three times with ddH_2O. Rinse with three drops of 1% uranyl acetate and immediately dry on filter paper.
21. Screen the samples under a transmission electron microscope.

3.3.6 Co-Immunoprecipitation and Immunoblotting

1. Perform co-immunoprecipitation following the manual of Dynabeads protein A/G immunoprecipitation kit.
2. Incubate primary antibodies (guinea pig-anti-vGluT2 or goat-anti-VMAT2) with Dynabeads with rotation for 10 min. Wash with binding and washing buffer and resuspend with synaptic vesicle solution from the Sect. 3.3.4.
3. Incubate the mixture with rotation at 20–25 °C for 100 min. Transfer the supernatant to a clean tube for further analysis. Wash the Dynabeads-Antibody-Vesicles complex three times by using washing buffer and then gently resuspend with the elution buffer.
4. Load aliquots of the original synaptic vesicle solution, Dynabeads-Antibody-Vesicles elution, and the above supernatant in a NuPAGE Novex 4–12% Bis-Tris Protein Gel.
5. Apply the primary antibodies (mouse-anti-vGluT2, goat-anti-VMAT2, and mouse anti-synaptophysin) in the western blots of these samples.
6. Apply the secondary antibodies (IRDye 800CW or IRDye 680RD).
7. Visualize the images using Odyssey imaging system.

3.4 Data Analysis

3.4.1 In Situ Hybridization

1. Trace the VTA subdivisions according to Paxinos and Watson [25]. Single- and double-labeled neurons were observed within each traced region at high power (20× objective lens) and marked electronically.
2. The cells expressing vGluT2, TH, or vGluT2-TH transcripts were counted separately.
3. A neuron was considered double-labeled when its soma was purple and contained an aggregation of silver particles over the purple cell but in the same focal field. Three observers counted labeled cells.
4. The background was evaluated from slides hybridized with sense probes. Images were opened and processed with an Adobe Photoshop 7 program (Adobe Systems Incorporated, Seattle, WA).

3.4.2 Fluorescence Microscopy

1. Analyze Z-stacks of confocal images using Imaris microscopy software.
2. Analyze the same confocal images to obtain 3-D reconstruction of putative synapses with the Amira 3D software.
3. Perform all statistical analysis with GraphPad prism 5.

3.4.3 Immuno-Electron Microscopy

1. Analyze serial ultrathin sections of the VTA (bregma −4.92 mm to −6.48 mm) and nAcc (bregma 2.76 mm to 0.96 mm) from male Sprague-Dawley rats and serial ultrathin sections of the VTA (bregma −2.92 mm to −3.88 mm) and nAcc (bregma 1.94 mm to 0.86 mm) from male TH-ChR2-mCherry mice and male vGluT2-ChR2-mCherry mice.
2. Synaptic contacts were classified according to their morphology and immunolabel, and photographed at a magnification of 6800–13,000×. The morphological criteria used for identification and classification of cellular components observed in these thin sections were as previously described. Type I synapses, here referred as asymmetric synapses, were defined by the presence of contiguous synaptic vesicles within the presynaptic axon terminal and a thick postsynaptic density (PSD) greater than 40 nm. Type II synapses, here referred as symmetric synapses, were defined by the presence of contiguous synaptic vesicles within the presynaptic axon terminal and a thin PSD [25]. Serial sections were obtained to determine the type of synapse.
3. In the serial sections, a terminal containing greater than five immunogold particles were considered as immunopositive terminal.
4. Adjust pictures to match contrast and brightness by using Adobe Photoshop (Adobe Systems Incorporated, Seattle, WA).

4 Notes

1. To make 1 L of 0.1 mol/L 1× PB, dissolve 33.67 g NaH_2PO_4 and 7.7 g NaOH to 1 L ddH_2O (pH is 7.3–7.4. It is not necessary to use pH meter to adjust pH). You can also make 2× PB as a stock solution, dilute to 1× PB with ddH_2O before use. The recipe is different from the classic sodium phosphate buffer, but it is convenient to make and the buffer works well for in situ hybridization and immuno-electron microscopic studies.
2. Usually prepare of 2× fixative (8% PFA in 2× PB buffer). For 1 L of fixative, heat 800 mL of ddH_2O and add 8.0 g of NaOH, stir until dissolved. Add 80 g of PFA, stir until dissolved. Add 33.66 g of NaH_2PO_4, stir until dissolved. Cool to room

temperature. Filter the solution. Bring to a final volume to 1 L. For working solution, dilute 1 L of 2× solution (8% PFA in 2× PB buffer) with 1 L of ddH_2O to obtain 4% PFA in 0.1 mol/L PB. To obtain an optimal tissue preservation, 4% PFA must be freshly prepared within a few hours prior to the perfusion.

3. To make subbed slides, the most critical point is to avoid bubbles when dipping rack for 10 s. Gently rake slides to disperse bubbles and blot edge of rack with a paper towel. The bubbles on the slides will affect the image quality.
4. If the bubbles cannot be removed, put the slide back into CitriSolv solution for clearing and re-cover the slides.
5. Bx53 Turboscan Research Microscope is designed for optimal fluorescence imaging, using UIS2 components that set new standards in precision and clarity. Automation enables the imaging of large areas in high magnification and also facilitates multicolor fluorescence. We use this microscope to collect high magnification images for in situ hybridization studies. To quickly scan huge numbers of images, we use Virtual Slide Microscope, which allows for manually loading one and six standard slides, respectively, along with any associated metadata. Designed for high throughput research and pathology, the VS120-L100-W system features a highly dependable, robustly designed slide loader for up to 100 slides.
6. The role of the picric acid in this solution is to slowly penetrate into the tissue and cause coagulation of proteins by forming salts with basic proteins.
7. Osmium tetroxide is highly toxic and is a rapid oxidizer. Exposure to the vapor can cause severe chemical burns to the eyes, skin, and respiratory tract. Wear nitrile gloves (osmium can penetrate latex gloves) and eye protection. Do not open any vials of osmium tetroxide outside of the fume hood.
8. Uranyl acetate is both radioactive and toxic. For better staining, filter the uranyl acetate with a 0.22 μm filter before use. Uranyl acetate helps to increase membrane contrast.
9. Usually, the staining solution is ineffective after 3 days. For good staining, always use freshly prepared lead staining solution in each experiment.
10. Tannic acid fixation can improve the resolution of the ultrastructure.
11. Glass or plastic ware should be siliconized with Sigmacote. In a fume hood, evenly cover the plastic or glass surface and recover to solution to apply to the next piece of glassware or plasticware. Let the treated pieces stand overnight in the fume hood. Microcentrifuge tubes and glass pipets can be autoclaved before use.

12. After each injection, do not remove the micropipette. The additional at least 10 min is to prevent backflow of tracer up the injection track. A lot of new researchers skip this step and cause some contamination for the injection to other brain areas.
13. Try to start perfusion within 30 s after cutting the diaphragm. Otherwise the postsynaptic density and mitochondria won't be well preserved. Usually apply 50 mL of heparin, 200 mL of fixative solution per mouse and 150 mL of heparin, 500 mL of fixative solution per rat.
14. Sodium borohydride solution needs to be prepared just before use. Gas bubbles forms when sodium borohydride reacts with water. Don't add too much sodium borohydride solution to avoid the sections to stick with bubbles on the top wall of well. Keep shaking slowly.
15. Make aliquots for A, B, and C in the silver enhancement kit to centrifuge tubes. Cover tubes with foil and keep in −20 °C. Take the aliquots out of the freezer and put them at room temperature before use. Aliquots should be frozen and thawed once.
16. If some water remains after dehydration, the resin will not polymerize properly and sectioning of the embedded samples won't be possible.
17. For preparation of Durcupan ACM epoxy resin, use pipettes to measure 10 mL of A, 10 mL of B. Mix on the nutator for 10 min. Then add 0.3 mL of C and mix on the nutator for 10 min. Add 0.2 mL of D and mix on the nutator for 30 min it will be ready to use.
18. Get properly trained before using the transmission electron microscope. Take care not to burn the supporting film by abruptly increasing the voltage of the filament or switching from high to low magnifications without decondensing the electron beam.

5 Discussion and Conclusion

We used radioactive riboprobes to identify cellular expression of transcripts encoding vGluT1, vGluT2, or vGluT3 within the VTA. We detected cells expressing vGluT2 mRNA (but not vGluT1 or vGluT3) within the VTA. By a combination of in situ hybridization (to detect vGluT2 mRNA) and immunocytochemistry (to detect TH), we determined that a subset of vGluT2 neurons co-express TH-immunoreactivity. By a combination of retrograde tract tracing, immunocytochemistry and in situ hybridization, we detected that two subpopulations of neurons expressing vGluT2 mRNA innervate the nAcc. The use of radioactive riboprobes for mRNA detection offers: (1) high sensitivity and specificity for detection of mRNA; (2) the detected signal is stable and permanent; (3) and allows the processing of large number of samples simultaneously. However, compared to the advanced Cell Diagnostics (ACD) RNAscope method, this method does not require the precautions associated with the use of radioactive material, and the procedure is fast. However, this method is not practical for the processing of large number of samples simultaneously.

A fluorescence microscope has allowed observations of distinct neuronal markers within specific neuronal compartments. By using fluorescence microscopy together with three-dimensional reconstruction, we found that within the nAcc the VTA dual vGluT2-TH neurons segregate vGluT2 and TH to distinct subcellular compartments located within the same axon. However, the resolution provided by the fluorescence microscope does not allow the identification of the type of synapse established by these dual vGluT2-TH axons. Thus immuno-electron microscopy studies are necessary to determine the ultrastructural properties of the synaptic composition of dual vGluT2-TH axons. Researchers in all different research fields commonly use the combination of fluorescence microscopy and electron microscopy. However, the new technology in correlative light and electron microscopy (CLEM) is a trend in the future. In this technique, low resolution 'overview' images from the light microscope are used to identify specific areas of interest that can be studied at higher resolution in the electron microscope using the same sample.

Acknowledgements

We thank Bing Liu, Hui-ling Wang and Tsuyoshi Yamaguchi for their help in processing brain tissue in in situ hybridization studies, and Rong Ye for her help in the immuno-electron microscopic studies. This work was supported by the Intramural Research Program of the National Institute on Drug Abuse, US National Institutes of Health (IRP/NIDA/NIH).

References

1. Bellocchio EE, Hu HL, Pohorille A, Chan J, Pickel VM, Edwards RH (1998) The localization of the brain-specific inorganic phosphate transporter suggests a specific presynaptic role in glutamatergic transmission. J Neurosci 18(21):8648–8659
2. Bellocchio EE, Reimer RJ, Fremeau RT Jr, Edwards RH (2000) Uptake of glutamate into synaptic vesicles by an inorganic phosphate transporter. Science 289(5481):957–960
3. Takamori S, Rhee JS, Rosenmund C, Jahn R (2000) Identification of a vesicular glutamate transporter that defines a glutamatergic phenotype in neurons. Nature 407(6801): 189–194
4. Takamori S, Rhee JS, Rosenmund C, Jahn R (2001) Identification of differentiation-associated brain-specific phosphate transporter as a second vesicular glutamate transporter (VGLUT2). J Neurosci 21(22):RC182

5. Bai LQ, Xu H, Collins JF, Ghishan FK (2001) Molecular and functional analysis of a novel neuronal vesicular glutamate transporter. J Biol Chem 276(39):36764–36769. doi:10.1074/jbc.M104578200
6. Fremeau RT, Troyer MD, Pahner I, Nygaard GO, Tran CH, Reimer RJ, Bellocchio EE, Fortin D, Storm-Mathisen J, Edwards RH (2001) The expression of vesicular glutamate transporters defines two classes of excitatory synapse. Neuron 31(2):247–260. doi:10.1016/S0896-6273(01)00344-0
7. Fujiyama F, Furuta T, Kaneko T (2001) Immunocytochemical localization of candidates for vesicular glutamate transporters in the rat cerebral cortex. J Comp Neurol 435(3):379–387. doi:10.1002/Cne.1037
8. Hayashi M, Otsuka M, Morimoto R, Hirota S, Yatsushiro S, Takeda J, Yamamoto A, Moriyama Y (2001) Differentiation-associated Na+-dependent inorganic phosphate cotransporter (DNIPI) is a vesicular glutamate transporter in endocrine glutamatergic systems. J Biol Chem 276(46):43400–43406. doi:10.1074/jbc.M106244200
9. Herzog E, Bellenchi GC, Gras C, Bernard V, Ravassard P, Bedet C, Gasnier B, Giros B, El Mestikawy S (2001) The existence of a second vesicular glutamate transporter specifies subpopulations of glutamatergic neurons. J Neurosci 21(22):RC181
10. Varoqui H, Schafer MKH, Zhu HM, Weihe E, Erickson JD (2002) Identification of the differentiation-associated Na+/P-I transporter as a novel vesicular glutamate transporter expressed in a distinct set of glutamatergic synapses. J Neurosci 22(1):142–155
11. Gras C, Herzog E, Bellenchi GC, Bernard V, Ravassard P, Pohl M, Gasnier B, Giros B, El Mestikawy S (2002) A third vesicular glutamate transporter expressed by cholinergic and serotoninergic neurons. J Neurosci 22(13):5442–5451
12. Fremeau RT, Burman J, Qureshi T, Tran CH, Proctor J, Johnson J, Zhang H, Sulzer D, Copenhagen DR, Storm-Mathisen J, Reimer RJ, Chaudhry FA, Edwards RH (2002) The identification of vesicular glutamate transporter 3 suggests novel modes of signaling by glutamate. Proc Natl Acad Sci U S A 99(22):14488–14493. doi:10.1073/pnas.222546799
13. Schafer MK, Varoqui H, Defamie N, Weihe E, Erickson JD (2002) Molecular cloning and functional identification of mouse vesicular glutamate transporter 3 and its expression in subsets of novel excitatory neurons. J Biol Chem 277(52):50734–50748. doi:10.1074/jbc.M206738200
14. Qi J, Zhang S, Wang HL, Wang H, de Jesus Aceves Buendia J, Hoffman AF, Lupica CR, Seal RP, Morales M (2014) A glutamatergic reward input from the dorsal raphe to ventral tegmental area dopamine neurons. Nat Commun 5:5390. doi:10.1038/ncomms6390
15. Yamaguchi T, Sheen W, Morales M (2007) Glutamatergic neurons are present in the rat ventral tegmental area. Eur J Neurosci 25(1):106–118. doi:10.1111/j.1460-9568.2006.05263.x
16. Yamaguchi T, Wang HL, Li XP, Ng TH, Morales M (2011) Mesocorticolimbic glutamatergic pathway. J Neurosci 31(23):8476–8490. doi:10.1523/Jneurosci.1598-11.2011
17. Li XP, Qi J, Yamaguchi T, Wang HL, Morales M (2013) Heterogeneous composition of dopamine neurons of the rat A10 region: molecular evidence for diverse signaling properties. Brain Struct Funct 218(5):1159–1176. doi:10.1007/s00429-012-0452-z
18. Root DH, Mejias-Aponte CA, Zhang SL, Wang HL, Hoffman AF, Lupica CR, Morales M (2014) Single rodent mesohabenular axons release glutamate and GABA. Nat Neurosci 17(11):1543–1551. doi:10.1038/nn.3823
19. Zhang SL, Qi J, Li XP, Wang HL, Britt JP, Hoffman AF, Bonci A, Lupica CR, Morales M (2015) Dopaminergic and glutamatergic microdomains in a subset of rodent mesoaccumbens axons. Nat Neurosci 18(3):386–392. doi:10.1038/nn.3945
20. Sato T (1968) A modified method for lead staining of thin sections. J Electron Microsc 17(2):158–159
21. Erickson JD, Masserano JM, Barnes EM, Ruth JA, Weiner N (1990) Chloride ion increases [3H]dopamine accumulation by synaptic vesicles purified from rat striatum: inhibition by thiocyanate ion. Brain Res 516(1):155–160
22. Teng L, Crooks PA, Dwoskin LP (1998) Lobeline displaces [3H]dihydrotetrabenazine binding and releases [3H]dopamine from rat striatal synaptic vesicles: comparison with d-amphetamine. J Neurochem 71(1):258–265
23. Boulland JL, Jenstad M, Boekel AJ, Wouterlood FG, Edwards RH, Storm-Mathisen J, Chaudhry FA (2009) Vesicular glutamate and GABA transporters sort to distinct sets of vesicles in a population of presynaptic terminals. Cereb Cortex 19(1):241–248. doi:10.1093/cercor/bhn077
24. Kadota K, Kadota T (1973) Isolation of coated vesicles, plain synaptic vesicles, and flocculent material from a crude synaptosome fraction of guinea pig whole brain. J Cell Biol 58(1):135–151
25. Paxinos G, Watson C (2007) The rat brain in stereotaxic coordinates. Elsevier, Amsterdam

Chapter 2

Single Nanoparticle Tracking: A Method for Investigating the Surface Dynamics of Glutamate Receptors

Julien Dupuis, François Maingret, and Laurent Groc

Abstract

The spatiotemporal organization of neurotransmitter receptors within synapses is a critical determinant of synaptic transmission and adaptation, and thus of information processing and storage in the brain. Long considered as immobile at the plasma membrane, glutamate ionotropic receptors—the principal mediators of fast excitatory neurotransmission—are instead highly dynamic. Indeed, the recent development of single molecule imaging techniques has revealed that their number, composition, and distribution at the neuronal surface are constantly regulated through a combination of exo-/endocytosis processes and lateral diffusion in and out of synaptic sites within the membrane plane, allowing a fine control of the intensity of synaptic transmissions. Among these techniques, quantum dot-based single nanoparticle tracking methods provide unique means to explore the surface behavior of individual receptors either in vitro or ex vivo. Here, we describe the experimental procedures to perform single nanoparticle tracking in primary dissociated neuronal cultures, organotypic hippocampal preparations, and acute brain slices. We also provide insights on how these methods can be used to investigate the dynamics of glutamate receptors at the plasma membrane, and to explore their interactions with surface partners.

Key words Glutamate, Receptor, Quantum dot, Single particle tracking, Lateral diffusion, Surface trafficking, Synaptic plasticity, Dissociated neurons, Organotypic cultures, Acute brain slices

1 Introduction

Glutamate (Glu), the main excitatory neurotransmitter in the brain, is released within the synaptic cleft by the presynaptic neuron and targets two receptor families: metabotropic Glu receptors and ionotropic Glu receptors. Members of this last family can be divided into three different classes based both on their pharmacology and Glu binding sites: alpha-amino-3-hydroxy-5-methyl-4-isoxazolepropionic acid (AMPA) receptors, kainate receptors, and *N*-methyl-D-aspartate (NMDA) receptors [1]. Most of the ionotropic Glu receptors are concentrated in the postsynaptic density which is located just in front of the presynaptic Glu release sites and their activation leads to the depolarization of the postsynaptic neuron. Plasticity of this synaptic transmission was

Sandrine Parrot and Luc Denoroy (eds.), *Biochemical Approaches for Glutamatergic Neurotransmission*, Neuromethods, vol. 130, DOI 10.1007/978-1-4939-7228-9_2, © Springer Science+Business Media LLC 2018

quickly identified as the molecular substrate underlying learning and memory [2]. In that sense, synaptic changes could occur both at presynaptic and postsynaptic sites through modulation in Glu release and gating properties of the ionotropic Glu receptors respectively [3–5]. At first, the postsynaptic density containing neurotransmitter receptors, associated proteins and scaffold cytoskeleton elements was believed to be stable and immobile over time. This set-in-stone picture of the postsynaptic element held for decades until one established that synaptic Glu receptors are subject to a rapid and constant turnover through exo- and endocytosis processes [6–9]. This was thought to be the only way to regulate the number and composition of synaptic Glu receptors and once inserted at the plasma membrane the synaptic receptors were still considered as immobile. On the contrary, it has been demonstrated that synaptic receptors are highly dynamic and that lateral diffusion within the plasma membrane at the cell surface plays a key role in modifying receptor numbers at synapses [10–12]. This profoundly changed our view of neurotransmitter signaling which is now believed to be the result from a dynamic equilibrium between synaptic, extrasynaptic, and intracellular compartments [13], and there is now accumulating evidence that lateral diffusion of Glu receptors accounts for synaptic plasticity [14–16].

This major breakthrough in our understanding of synaptic organization and function was achieved thanks to the development of single molecule detection and super-resolution microscopy. These powerful imaging methods, saluted by the 2014 Nobel Prize in chemistry, combine the ability to switch fluorescent dyes on and off and to localize them with very high temporal and spatial precision in order to bypass the diffraction limit and reach the nanoscale resolution [17–23]. Altogether, they have led to a better understanding of glutamatergic synapse morphology [24, 25] and protein nanoscale clustering both at presynaptic [26] and postsynaptic sites [27, 28].

While most of these super resolution imaging techniques rely on the “photoswitching” or “photobleaching” of individual fluorophores [29], single particle tracking takes advantage of “photostable” nanoparticles to directly measure individual surface receptor movements [30]. The most commonly used nanoparticles are Quantum Dots (QD). These are relatively small (10–20 nm) semiconductor nanocrystals composed of various elements such as a core of heavy metal (e.g. CdSe) beneath a shell of ZnS and functionalized for biological applications [31]. In addition to a robust photostability, QD exhibit distinctive properties such as high brightness, blinking behavior, large Stokes shift, and narrow emission spectrum [31]. For biological applications in neurosciences, QD are usually complexed with an antibody and bound to an epitope located on the extracellular domain of the designated neurotransmitter receptor of interest. Due to their high signal to noise ratio, QD can be detected with a pointing accuracy of approximately 30 nm at the surface of neurons. Single nanoparticle tracking is therefore a method of choice to investigate the surface dynamics of neurotransmitter receptors within synaptic compartments.

In this book chapter, we provide the reader with guidelines on how to perform single nanoparticle tracking of Glu receptors in dissociated cultured neurons. We also particularly focus our attention on the recently described experimental steps which have allowed to extend the tracking of neurotransmitter receptors in various intact brain network preparations [32–34].

2 Materials

2.1 Media and Buffers

Depending on the preparation of interest, adapted media should be prepared as specified below. The most commonly used media for single-particle tracking applications on primary dissociated neuronal cultures are complete Neurobasal medium and Tyrode's solution.

(a) Complete Neurobasal medium. Complete Neurobasal medium is obtained by supplementing 500 mL of Neurobasal medium (Gibco, Life Technologies, Thermo Fisher Scientific Inc., USA) with 10 mL of SM1 serum-free supplement (stemcell technologies, Vancouver, CA) and 2 mmol/L glutamine (Sigma-Aldrich, St Louis, USA). Complete Neurobasal medium can be stored at 4 °C for several weeks under sterile conditions.

(b) Tyrode's solution. Tyrode's solution is a physiological saline which contains in mmol/L: 105 NaCl, 5 KCl, 2 $MgCl_2$, 2 $CaCl_2$, 12 D-glucose, 25 HEPES (all chemicals from Sigma-Aldrich, St Louis, USA). pH is adjusted to 7.4 with NaOH. Tyrode's solution can be filtered and stored at 4 °C for several weeks.

Organotypic hippocampal preparations and acute brain slices are dissected from fresh brain tissue using the following dissection media.

(c) Organotypic slices dissection medium. The dissection medium for organotypic hippocampal preparations is a hyperosmotic saline prepared from ultrapure water which contains in mmol/L: 175 sucrose, 25 D-glucose, 50 NaCl, 2.5 KCl, 0.5 $CaCl_2$, 0.66 KH_2PO_4, 2 $MgCl_2$, 0.28 $MgSO_4{\cdot}7H_2O$, 0.85 $Na_2HPO_4{\cdot}12H_2O$, 2.7 $NaHCO_3$, 0.4 HEPES, 2×10^{-5}% phenol red (all chemicals from Sigma-Aldrich, St Louis, USA). pH is adjusted to 7.3 with NaOH. Organotypic slices dissection medium can be filtered and stored at 4 °C for several weeks.

(d) Acute slices dissection medium. The dissection medium for acute brain slices is a hyperosmotic saline prepared from ultrapure water which contains in mmol/L: 250 sucrose, 2 KCl, 7 $MgCl_2$, 0.5 $CaCl_2$, 1.15 NaH_2PO_4, 11 D-glucose, 26 $NaHCO_3$ (all chemicals from Sigma-Aldrich, St Louis, USA). The dissection medium is prepared on the day of experiment and constantly gassed with 95% O_2/5% CO_2 to ensure proper oxygenation and pH of the samples (pH 7.35).

Organotypic hippocampal preparations and acute brain slices are usually imaged in HEPES-based and classical artificial cerebrospinal fluid (aCSF), respectively.

(e) HEPES-based aCSF. HEPES-based aCSF is a physiological saline which closely matches the electrolyte concentrations of the cerebrospinal fluid. It is prepared from ultrapure water and contains in mmol/L: 130 NaCl, 2.5 KCl, 2.2 $CaCl_2$, 1.5 $MgCl_2$, 10 HEPES, 10 D-glucose (all chemicals from Sigma-Aldrich, St Louis, USA). pH is adjusted to 7.3 with NaOH. HEPES-based aCSF is filtered and stored at 4 °C for several weeks.

(f) Classical aCSF. aCSF is a physiological saline which closely matches the electrolyte concentrations of the cerebrospinal fluid. It is prepared from ultrapure water and contains in mmol/L: 126 NaCl, 3.5 KCl, 2 $CaCl_2$, 1.3 $MgCl_2$, 1.2 NaH_2PO_4, 25 $NaHCO_3$, 12.1 D-glucose (all chemicals from Sigma-Aldrich, St Louis, USA). aCSF is prepared on the day of experiment and constantly gassed with 95% O_2/5% CO_2 to ensure proper oxygenation and pH of the samples (pH 7.35).

When required, phosphate buffer saline (PBS, Euromedex, France) is commonly used to dilute antibodies or quantum dots. PBS is filtered before use and stored at 4 °C.

2.2 Dissociated Neuronal Cultures

The procedures described here apply to any neuronal preparation. We will deliberately take primary hippocampal cultures as an example since they are the most commonly used in vitro model within the neuroscience community. Briefly, cultures of hippocampal neurons are prepared from embryonic (day 18) Sprague-Dawley rats in sterile conditions under a hood as follows:

(a) Hippocampi are dissected and transferred to a 15 mL Falcon tube filled with pre-heated Hank's Balanced Salt Solution medium (HBBS, Gibco, Life Technologies, Thermo Fisher Scientific Inc., USA).

(b) HBSS medium is then replaced by 5 mL of pre-heated Trypsin/EDTA solution (Gibco, Life Technologies, Thermo Fisher Scientific Inc., USA) for 15 min incubation at 37 °C in order to initiate dissociation. Trypsin/EDTA is then removed and washed twice with 10 mL of pre-heated HBSS medium.

Hippocampi are transferred in 2 mL of pre-heated HBSS and the tissue is dissociated by gently pipetting several times.

(c) Cells are plated on polylysine-precoated glass coverslips in Ø 60 mm petri dishes (Falcon, Corning Inc., USA) at a density of 350 × 10^3 cells per dish. Coverslips are maintained at 37 °C/5% CO_2 for 72 h in 5 mL of a 3% horse serum-supplemented Neurobasal medium. This medium is then replaced by pre-heated and equilibrated serum-free Neurobasal medium. Cultures can be maintained at 37 °C/5% CO_2 for up to 20 days.

(d) For imaging experiments requiring the tracking of recombinant membrane proteins, neurons can be transfected with plasmids of interest at 7–10 days in vitro using the Effecten transfection kit (Qiagen, Germany) according to the manufacturer's instructions (Fig. 1a).

2.3 Organotypic Hippocampal Cultures

Organotypic hippocampal cultures are prepared from postnatal day 5 to 7 Sprague-Dawley rats in sterile conditions under a hood as follows:

(a) Hippocampi are dissected and transferred to organotypic slice dissection medium.

(b) 350-μm-thick sagittal hippocampal slices are obtained using a McIlwain tissue chopper and are placed in preheated (37 °C) dissection medium.

(c) After 25 min of incubation, slices are transferred on white FHLC membranes (0.45 μm) set on Millicell cell culture inserts (Millipore, 0.4 mm; Ø 30 mm) and cultured for up to 14 days on multi-well plates at 35 °C/5% CO_2 in a culture medium composed of 50% basal medium Eagle (Gibco, Life Technologies, Thermo Fisher Scientific Inc., USA), 25% HBSS medium (with $MgCl_2/CaCl_2$), 25% heat-inactivated horse serum, 0.45% D-glucose, 1 mmol/L L-glutamine. The medium is changed every 2–3 days.

(d) For imaging experiments requiring the tracking of recombinant membrane proteins, single neurons can be electroporated with plasmids of interest at 4–6 days in vitro using single-cell electroporation, as previously described [35]. Briefly, individual CA1 pyramidal neurons are electroporated to transfer cDNA encoding for the construct of interest (e.g. GluN1-super-ecliptic pHluorin, GluN1-SEP, and cytosolic enhanced green fluorescent protein, eGFP, to identify transfected cells). Plasmids (5 μg total) are dissolved in a filtered cesium-based solution containing in mmol/L: 135 cesium-methanesulfonate, 8 NaCl, 10 HEPES, 0.2 ethylene glycol-bis(β-aminoethyl ether)-*N,N,N′,N′*-tetraacetic acid, 4 Na_2ATP, 0.33 Na_3GTP, 5 tetraethylammonium chloride (all chemicals from Sigma-Aldrich,

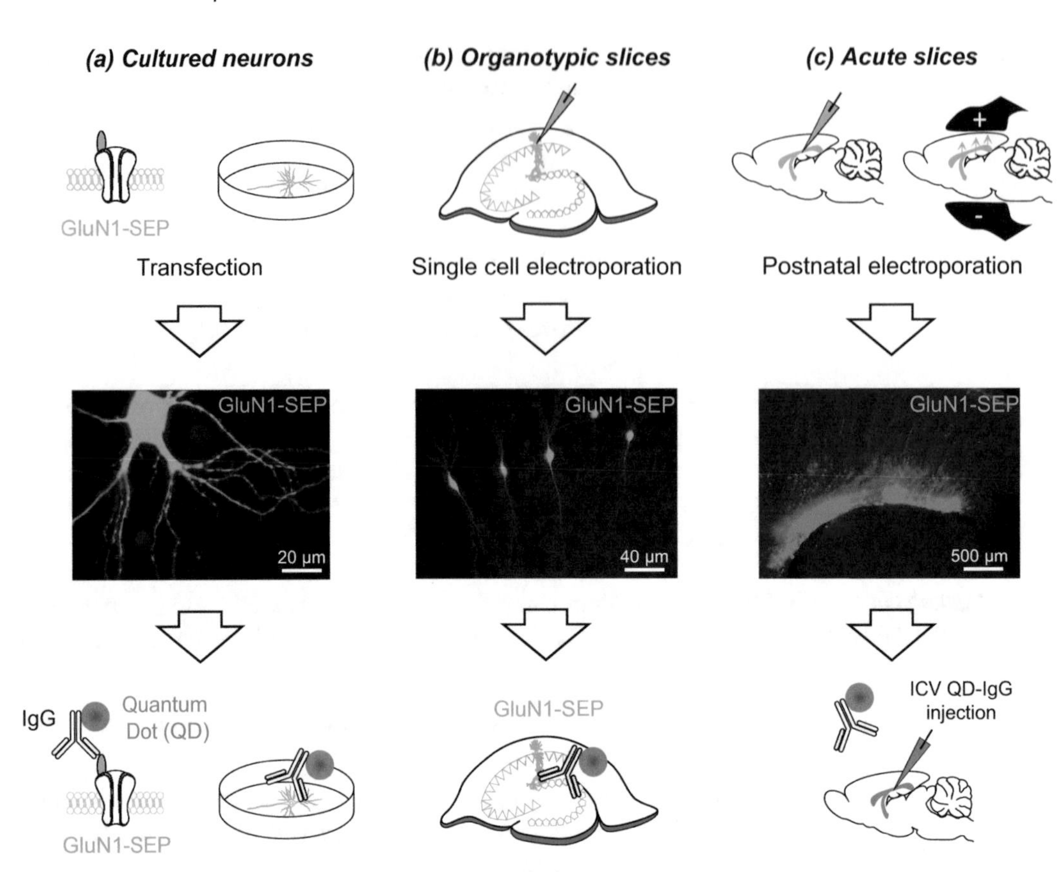

Fig. 1 Single nanoparticle tracking of individual exogenous genetically modified NMDA glutamate receptors in dissociated neuronal cultures, organotypic and acute brain slices. (**a**) Cultured hippocampal neurons were transfected at 10 days in vitro to express a recombinant version of the NMDAR GluN1 obligatory subunit tagged with Super-Ecliptic pHluorin (GluN1-SEP). Two to three days after transfection, cells were incubated with quantum dots functionalized with mouse polyclonal anti-GFP antibodies. (**b**) Individual CA1 pyramidal neurons from 4–6 days in vitro hippocampal slices were electroporated to express recombinant GluN1-SEP. Two to three days after electroporation, organotypic slices were carefully detached from their inserts and incubated with quantum dots functionalized with anti-GFP antibodies. (**c**) Postnatal day 0–1 rats were injected in the cerebral ventricles with cDNA coding for recombinant GluN1-SEP, then electroporated using forceps-type circular electrodes to allow gene transfer to cortical neurons. Three days after electroporation, pups received an intracerebroventricular (ICV) injection of quantum dots (in *red*) functionalized with anti-GFP antibodies before sacrifice and brain slice preparation

St Louis, USA). pH is adjusted to 7.3 with cesium hydroxide (CsOH). This solution is supplemented with 10 μL of filtered endotoxin-free buffer TE (Tris-EDTA, Qiagen), then centrifuged twice to pull down potential debris (10,000 rpm, 15 min, 4 °C) and used to fill 5- to 6-MΩ borosilicate patch pipettes. Electroporation is performed in 2 mL of pre-warmed (37 °C/5% CO_2) HEPES-based aCSF. Plasmid transfer is allowed by the delivery of 50 μs-width square-pulses at 100 Hz (1 s duration; −14 V current amplitude; Fig. 1b).

2.4 Acute Brain Slices

Acute brain slices are prepared from postnatal day 4 to 5 Sprague-Dawley rats as follows:

(a) 350-μm-thick sagittal hippocampal slices are obtained using a vibratome (VT1200S, Leica Microsystems) in ice-cold acute slices dissection medium.

(b) Slices are then transferred to classical aCSF and left for equilibration for 30 min at 33 °C, then 30 min at room temperature.

(c) For imaging experiments requiring the tracking of recombinant membrane proteins, post-natal electroporation with plasmids of interest is performed in post-natal day 0 or day 1 pups, as previously described [36] and following ethical and animal care guidelines (Fig. 1c). Pups are anesthetized by hypothermia and the injection coordinates to reach the cerebral ventricles are set by drawing an imaginary line between the eye and lambda, identifying the midpoint, and retroceding 2 mm from that midpoint. Injection of a mixture containing 2 μg of the DNA constructs of interest (e.g. GluN1-SEP, eGFP) diluted in 8 μL of PBS and 1 μL of Fast Green (Sigma-Aldrich, St Louis, USA) is performed in both lateral ventricles at a depth of 2.6 mm under cold illumination. The success of the injection is evaluated by the ventricular dispersion of the Fast Green dye. Successfully injected pups are electroporated using forceps-type circular electrodes (7-mm platinum Tweezertrode BTX, Harvard apparatus) covered with conductive gel. Electrodes can be positioned either with the positive pole under the throat and the negative on top of the head to electroporate hippocampal neurons, or the opposite configuration to electroporate cortical neurons. Five electrical pulses (150 V, 50 ms duration, 1 s interval between pulses) are delivered with a pulse generator (BTX Harvard apparatus ECM830) and pups are immediately reanimated on a thermal blanket at 37 °C before being placed back with the mother.

2.5 Antibodies and Quantum Dots

Single-particle tracking approaches necessitate highly specific monoclonal or polyclonal antibodies targeting extracellular epitopes of the protein of interest. As an example, NMDA Glu receptors can be tracked using rabbit polyclonal antibodies directed against an N-terminal extracellular epitope corresponding to residues 385–399 of the endogenous GluN1 obligatory subunit (Alomone Labs, #AGC-001) [14]. When no reliable antibody against an endogenous target is available, another option consists in expressing and tracking recombinant proteins incorporating specific tags (e.g. GFP, flag, myc). Typically, SEP- or GFP-tagged NMDA receptors can be efficiently tracked using mouse anti-GFP IgG1κ (clones 7.1 and 13.1, Roche, #11814460001) or rabbit anti-GFP IgG2a (clone 3E6, Molecular Probes, # A-11120) monoclonal antibodies [34].

Single Particle Tracking (SPT) approaches also require bright and photostable nanoparticles functionalized with secondary antibodies of which QD are the most commonly used. QD are now commercially available (see Thermo Fisher Scientific catalogue) with different species emitting fluorescence at specific wavelengths spanning the entire visible spectrum (e.g. 525, 565, 585, 605, 655, 705, 800 nm) allowing simultaneous multicolor imaging if necessary. Primary antibodies and secondary antibody-coupled QD should always be handled with care and kept on ice.

QD labeling on dissociated neurons and organotypic slices requires the following materials:

(a) Cell culture incubator set at 37 °C/5% CO_2.

(b) Twelve-well culture plate.

(c) 50 mL of Bovine Serum Albumin (BSA)-supplemented (1%) Neurobasal medium, Tyrode's solution or HEPES-based aCSF.

(d) 1 mL of PBS.

(e) Appropriate primary antibody and QD-coupled secondary antibody (see examples listed above for glutamate NMDA receptors).

(f) 1000, 200, and 2.5 μL micropipettes.

(g) Tweezers.

Warning: Neurobasal medium, Tyrode's solution, and HEPES-based aCSF should be pre-heated and equilibrated in an incubator (37 °C/5% CO_2) before use.

For specificity concerns, QD tracking in acute slices is so far restricted to recombinant receptors. Indeed, expression of exogenous fluorescent labels (e.g. GFP) and recombinant receptors only allows to evaluate if the trajectories recorded in densely packed tissue match electroporated neurons and to estimate the specificity of this labeling through the calculation of enrichment factors (see Ref. 47). QD labeling on acute brain slices requires the following materials:

(a) Sterile 0.3 mL insulin syringes (Becton Dickinson, USA).

(b) Cold-light illumination.

(c) Mix of 2 μL of appropriate primary antibody and 2 μL of QD-coupled secondary antibody, qs 30 μL with sterile PBS. As an example, labeling of SEP-tagged NMDA receptors can be performed with 2 μL of mouse anti-GFP IgG1κ (clones 7.1 and 13.1, Roche, #11814460001) or rabbit anti-GFP IgG2a (clone 3E6, Molecular Probes, # A-11120) monoclonal antibodies mixed with 2 μL of QD655 goat F(ab′)2 anti-rabbit IgG (Invitrogen, #Q-11421MP) or QD655 goat F(ab′)2 anti-mouse IgG (Invitrogen, #Q-11021MP), respectively, qs 30 μL with PBS.

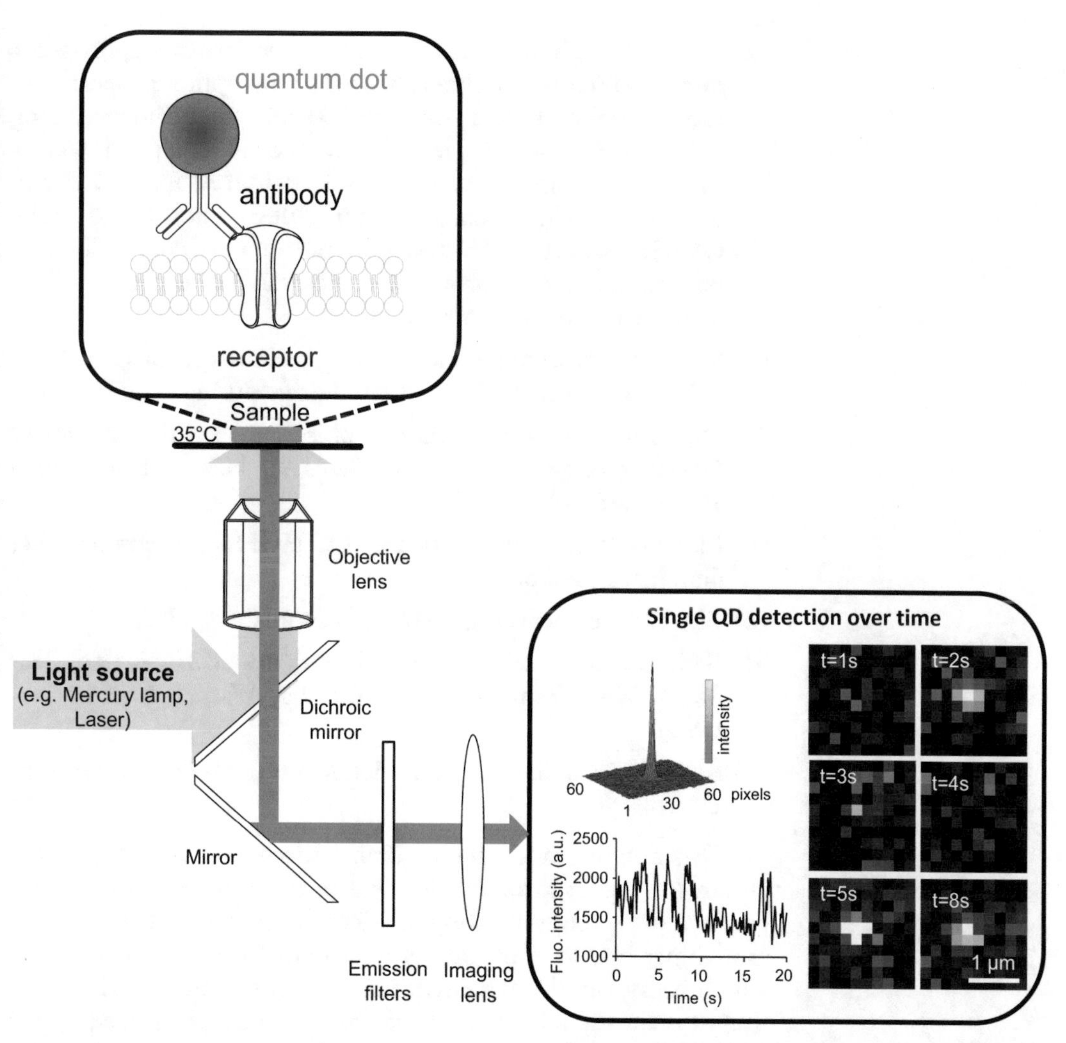

Fig. 2 Schematic microscopy setting to perform single nanoparticle tracking in live brain cells. Single nanoparticles (quantum dots) are associated with antibodies directed against an extracellular epitope of the receptor of interest. The sample is illuminated by means of a mercury lamp or laser light source. In return, the fluorescence signals emitted by quantum dot are separated from the excitation light source through a dichroic mirror and collected through an EMCCD camera. The fluorescence intensity of the quantum dots can be plotted as a function of time in order to visualize their characteristic blinking behavior

(d) Ice.

(e) Thermal blanket set at 37 °C.

2.6 Imaging Setups

Depending on the preparation of interest, adequate microscopy setups have to be used to perform single particle tracking. Below are listed the main features required to perform SPT experiments in dissociated neuronal cultures and organotypic or acute brain slices.

Dissociated neuronal cultures—Dissociated cultures present the major advantage of a two-dimensional geometry, which allows single particle tracking in classical epifluorescence configurations with the following elements (see Fig. 2):

(a) Inverted epifluorescence microscope with appropriate excitation/emission filters adapted to the optical properties of the fluorophores. As an example, QD655 can be imaged using a Nikon Eclipse Ti inverted microscope equipped with a BrightLine® single-band filter set optimized for QDot 655 nanocrystals (Semrock, #QD655-C-000), allowing to excite QD655 between 415 and 450 nm and to collect light at 655 nm using an appropriate combination of excitation filter, dichroic mirror, and emission filter.

(b) Vibration isolation table supporting the microscope (e.g. TMC Vibration Control, USA) to avoid drifts during acquisitions.

(c) Oil-immersion magnification objectives in the range of 60–100× (e.g. Plan Apo λ 60×/1.40 NA or Plan Apo λ 100×/1.45 NA, Nikon).

(d) Mercury lamp (e.g. IntensiLight C-HGFIE precentered fiber illuminator, Nikon).

(e) EM-CCD camera (e.g. Evolve, Photometrics, USA).

(f) Bath heating system to maintain samples at 37 °C (e.g. TC-324B/344B Temperature Controller, Warner Instruments, USA).

(g) Acquisition software (e.g. MetaMorph, Molecular Devices, USA).

Organotypic and acute brain slices—Unlike dissociated cultures, imaging three-dimensional thick tissue samples requires confocal microscopy allowing to select specific focal planes, as well as high performance imaging systems in order to collect signals in a high background-noise environment. Thus, accurate QD detection in brain slices is usually performed on microscopes equipped with spinning disk devices, as described below:

(a) Inverted fluorescence microscope (e.g. Leica DMI6000, Leica Microsystems) equipped with a Nipkow spinning disk unit (e.g. CSU-X1 confocal scanning unit, Yokogawa), a live cell chamber keeping the temperature at 37 °C, a 561 nm diode laser line for excitation purposes and appropriate emission filters to collect light from QD655 (e.g. 650–800 nm bandpass filter).

(b) Vibration isolation table supporting the microscope (e.g. TMC Vibration Control, USA) to avoid drifts during acquisitions.

(c) Oil-immersion magnification objectives in the range of 60–100× (e.g., HCX PL APO 63×/1.40-060 NA, Leica Microsystems).

(d) EM-CCD camera (e.g. Evolve, Photometrics, USA).

(e) Acquisition software (e.g. MetaMorph, Molecular Devices, USA).

3 Methods

3.1 Quantum Dots Labeling

Dissociated neuronal cultures—Labeling of cultured neurons is generally performed in complete Neurobasal medium or Tyrode's solution. Both media should be supplemented with BSA 1% to decrease unspecific antibodies binding and transferred to a 50 mL flask pre-incubated at 37 °C/5% CO_2 at least 1 h before the experiment, with the cap of the flask slightly loose in order to allow gas exchange and medium equilibration. The successive steps of the labeling procedure are the following:

(a) Fill the wells of a 12-well plate with 2 mL of BSA-supplemented medium. The first four wells should be used to wash the coverslip after incubation with antibodies, the four following to wash them after incubation with the QD, and the four last serve as stocks of equilibrated medium. Always keep the 12-well plate in the incubator to ensure proper equilibration and temperature of the medium.

(b) Prepare the primary antibody solution by diluting the appropriate amount of antibody in BSA-supplemented medium. Depending on the antibody used, the dilution factor may vary. As an example, rabbit anti-GluN1 polyclonal antibodies (Alomone Labs, #AGC-001) used to track endogenous NMDA Glu receptors are generally applied at 1:200 dilution, while mouse anti-GFP IgG1κ (clones 7.1 and 13.1, Roche, #11814460001) or rabbit anti-GFP IgG2a (clone 3E6, Molecular Probes, #A-11120) monoclonal antibodies used to track SEP- or GFP-tagged recombinant NMDA receptors are rather used at 1:5000 and 1:10,000 dilutions, respectively. The antibody solution is gently homogenized.

(c) Stretch a piece of Parafilm over the lid of the 12-well plate and put a 100 μL droplet of primary antibody solution over the Parafilm. Carefully grab a coverslip of cultured neurons with a thin forceps and flip it over the droplet, with the cells facing the solution. Incubate for 10 min at 37 °C/5% CO_2 to allow antibody labeling.

(d) During the primary antibody incubation, prepare the secondary antibody-coupled QD solution by diluting commercial QD stocks 1:10,000 in BSA-supplemented medium. A convenient way to do so is to prepare a 1:10 pre-dilution in PBS, then to dilute this intermediate solution 1:1000 in the BSA-supplemented medium. The QD solution must be vortexed before use to avoid aggregation of the nanoparticles.

(e) After 10 min, wash the coverslip four times by successively transferring it to the dedicated wells containing BSA-supplemented medium (10 s in each well).

(f) Put a 100 μL droplet of the 1:10,000 secondary antibody-coupled QD on the Parafilm and flip the coverslip over the droplet, with the cells facing the solution. Incubate for 10 min at 37 °C/5% CO_2 to allow QD labeling of the primary antibodies. If mitochondrial labeling is used as a synaptic marker, a 30 s step incubation with Mitotracker (Green, Red or Orange depending on the wavelength of interest; Thermo Fisher Scientific) at 1:10,000 in BSA-supplemented medium can be performed at the end of the 10 min.

(g) Wash the coverslip four times again by successively transferring it to the dedicated wells containing BSA-supplemented medium (10 s in each well).

(h) Place the coverslip in an appropriate imaging chamber (e.g. Ludin chamber or Quick Exchange Chamber, Warner Instruments) and fill with 500 μL of BSA-free medium.

(i) Place the chamber under the microscope in the environmental conditions required for the desired experiment (typically at 37 °C/5% CO_2).

Organotypic slices—Unlike cultured neurons, labeling of organotypic slices is performed in HEPES-based aCSF. As for Tyrode and Neurobasal media, HEPES-based aCSF is supplemented with BSA 1% to decrease unspecific antibody binding and transferred to a 50 mL flask pre-incubated at 35 °C/5% CO_2 at least 1 h before the experiment, with the cap of the flask slightly loose in order to allow gas exchange and medium equilibration. When tracking recombinant receptors, the labeling steps described below should be performed 2–3 days after the single-cell electroporation procedure to allow sufficient surface expression.

(a) Fill the wells of a 12-well plate with 2 mL of medium supplemented with BSA 1%. The first two wells should remain empty and be used to wash the coverslip after incubation with antibodies and with QD. The others should serve as stocks of equilibrated medium. Always keep the 12-well plate in the incubator to ensure proper equilibration and temperature of the medium.

(b) Prepare the primary antibody solution by diluting the appropriate amount of antibody in BSA-supplemented medium. Depending on the antibody used, the dilution factor may vary. As an example, rabbit anti-GluN1 polyclonal antibodies (Alomone Labs, #AGC-001) used to track endogenous NMDA Glu receptors are generally applied at 1:200 dilution, while mouse anti-GFP IgG1κ (clones 7.1 and 13.1, Roche, #11814460001) or rabbit anti-GFP IgG2a (clone 3E6, Molecular Probes, #A-11120) monoclonal antibodies used to track SEP- or GFP-tagged recombinant NMDA receptors are rather used at 1:5000 and 1:10,000 dilutions, respectively. The antibody solution is gently homogenized.

(c) Carefully detach organotypic slices from their culture membranes in a 200 μL droplet of medium using a scalpel. Transfer the slice to a well containing 100 μL of primary antibody solution. Incubate for 10 min at 35 °C/5% CO_2 to allow antibody labeling.

(d) During the primary antibody incubation, prepare the secondary antibody-coupled QD solution by diluting commercial QD stocks 1:10,000 in BSA-supplemented medium. The QD solution must be vortexed before use to avoid aggregation of the nanoparticles.

(e) After 10 min, remove the primary antibody solution and carefully wash the slice by adding 500 μL of fresh BSA-supplemented medium and gently pipetting up and down (repeat five times).

(f) Fill the well with a 500 μL droplet of the 1:10,000 secondary antibody-coupled QD and incubate for 10 min at 35 °C/5% CO_2 to allow QD labeling of the primary antibodies.

(g) Wash the slice by adding 500 μL of fresh BSA-supplemented medium and gently pipetting up and down (repeat five times).

(h) Place the slice in an appropriate imaging chamber (e.g. Ludin chamber or Quick Exchange Chamber, Warner Instruments), fill with 500 μL of BSA-free medium, and maintain the slice in place using a slice anchor (Warner Instruments).

(i) Place the chamber under the microscope in the environmental conditions required for the desired experiment (typically at 37 °C/5% CO_2).

Acute slices—Because of tissue density issues preventing antibody and QD to reach their targets, receptor labeling in acute slices cannot be performed through classical incubations as for cultured neurons or organotypic slices. Instead, QD and antibodies are premixed and delivered in vivo to the lateral ventricles in order to allow them to freely enter brain tissue. For specificity concerns, QD tracking in acute slices is so far restricted to recombinant receptors expressed through post-natal electroporation (see Sect. 2.4, step c and Fig. 1b), and the labeling steps described below should be performed 3–4 days after the single-cell electroporation.

(a) Prepare the QD/antibody premix by diluting the appropriate amount of QD and antibody in sterile PBS. Depending on the QD and antibody used, the dilution factor may vary. As an example, labeling of SEP- or GFP-tagged recombinant NMDA receptors can be performed by mixing 2 μL QD655 goat F(ab′)2 antirabbit IgG (Invitrogen, #Q-11421MP) or antimouse IgG (Invitrogen, #Q11022MP) with 2 μL of rabbit anti-GFP IgG2a (clone 3E6, Molecular Probes, #A-11120) or mouse anti-GFP IgG1κ (clones 7.1 and 13.1, Roche, #11814460001) antibodies and 26 μL of PBS (final volume of 30 μL). The premix solution is gently homogenized and incubated at 37 °C.

(b) After 30 min, electroporated post-natal day 4 pups are anesthetized by hypothermia and injected with QD/antibody premix in the lateral ventricles (~15 μL per pup). Coordinates for the injection site are obtained as for post-natal electroporation (see Sect. 2.4, step c), and the depth of injection is adjusted to 3.5 mm. Reanimate the pups on a heating blanket and place them back with their mother.

(c) Two hours after injection, sacrifice the pups and prepare 350-μm-thick brain slices as described in Sects. 2.4, steps a and b. Store the slices at room temperature in aCSF (gassed with 95% O_2/5% CO_2 to ensure proper oxygenation and pH).

(d) Place the slice in a Ludin chamber constantly perfused with aCSF, and maintain the slice in place using a slice anchor (Warner Instruments).

Place the chamber under the microscope in the environmental conditions required for the desired experiment (typically at 37 °C/5% CO_2).

3.2 Imaging

Dissociated neuronal cultures—Single molecule tracking on dissociated neurons can be performed using a standard epifluorescence microscope equipped with a camera that is sensitive and fast enough to detect individual QD at a fast sampling rate (typically 20 Hz). The proper sets of filters adapted to your labels should be used (e.g. BrightLine® single-band filter set optimized for QDot 655 nanocrystals; Semrock, #QD655-C-000).

(a) Choose a neuronal field of interest. This choice is usually made based on the quality of the bright field image, on the labeling or on the expression level of fluorescent recombinant receptors or proteins.

(b) Ensure that the QD density per neuronal field is high enough to allow a decent statistical analysis, but low enough to avoid crossing of particles which would impede trajectory reconnection during the analysis step. If necessary, adjust the QD density by changing antibody and/or QD concentrations during the incubation process.

(c) Record images of the fluorescent dyes or recombinant proteins used (e.g. Mitotracker Green or PSD-95-GFP used as synaptic markers), as well as a bright field image.

(d) Record an image stream of QD signals (500–1000 consecutive frames is usually enough) at a high sampling rate (20–30 images/s). Avoid imaging for more than 20–25 min, as photodamage increases with time and QD might get internalized through endocytosis.

Since most of these procedures apply to the three types of preparations described in this chapter, we will now briefly focus on the specificities of organotypic and acute brain slices.

Organotypic and acute slices—Single molecule imaging on brain slices can be performed in similar conditions as those described for cultured neurons. However, unlike dissociated cultures, imaging three-dimensional tissue samples requires confocal microscopy to select specific focal planes, and is usually performed on microscopes equipped with spinning disk devices. To note, single-particle tracking in slices can also be performed using an upright microscope equipped with water-immersion objectives [32], which allows parallel electrophysiological recordings if required.

(a) Choose a field of interest by selecting regions with electroporated neurons expressing the fluorescent recombinant receptors of interest (e.g. GluN1-SEP-NMDAR).

(b) Record images of the fluorescent dyes or recombinant proteins used (e.g. Mitotracker, PSD-95-GFP, GluN1-SEP), as well as a bright field image.

(c) Record an image stream of QD signals (500–1000 consecutive frames is usually enough) at a high sampling rate (20–30 images/s). Avoid imaging for more than 20–25 min, as photodamage increases with time.

Because of tissue density, efficiently washing away unspecifically bound QD/antibody complexes is often difficult. Thus, single-molecule imaging in organotypic and acute brain slices is associated with lower levels of specificity and fewer on-target trajectories as compared to what is observed in dissociated neurons, and special care should be taken at cleaning steps during the analysis process.

3.3 Image Analysis

Analysis steps are almost identical for dissociated neurons, organotypic and acute brain slices, and the procedures described below apply for all three preparations. Image processing for single particle tracking is a delicate process which should be performed with the assistance of imaging experts. Several commercial softwares allow automated particle detection including Imaris (Bitplane) or Volocity (PerkinElmer). While some laboratories develop their own codes and in-house softwares, a solution is to use available or personal scripts and plug-ins running in Metamorph (Molecular Devices), Matlab (MathWorks), or ImageJ (NIH).

(a) The first step of single particle tracking analysis is the automated detection of QD positions in each frame of the stream. The most accurate way to perform this step consists in detecting intensity maxima and fitting the point-spread function of each signal with two-dimensional Gaussians to obtain the position of the QD. The pointing accuracy of this procedure is mainly given by the number of photons collected per signal during exposure time. Due to the brightness and high signal-to-noise ratio of QD, the typical accuracy of single-particle tracking experiments can reach up to 30 nm [30]. If markers of interest are labeled or electroporated beforehand (e.g. synaptic markers such as Mitotracker or PSD-95-GFP), QD can be precisely located within specific membrane compartments—i.e. synaptic, perisynaptic, and extrasynaptic compartments—on each frame of a stream. Identified QD positions can then be projected on a single image to provide a high-resolution distribution of the successive locations of receptor/QD complexes over the duration of the acquisition.

(b) Once the detection completed, 2D trajectories of single receptor/QD complexes in the plane of focus can be reconstructed by correlation analysis between consecutive images using a Vogel algorithm which allows reconnecting individual QD positions from one frame to another (Fig. 3a). To note, a typical feature of QD is their ability to blink and remain in a dark state for a certain amount of time, which may affect this reconnection step. Up to a certain point, blinking can be taken into account in the reconnection by setting conditions to the linkage of trajectories that are adequate to the experimental configuration. In particular, automated blinking correction should define a maximum number of frames to reconnect two time-separated points of a trajectory, and a maximum physical distance allowed for the displacement during the blink.

(c) Diffusion parameters, such as instantaneous diffusion coefficients, synaptic dwell time, or surface explored are obtained from the mean square displacement (MSD) versus time plot (see Fig. 3). For a trajectory of N positions $(x(t),y(t))$ with times from $t = 0$ to $t = N * \Delta t$ the MSD corresponding to the time interval $\tau = n * \Delta t$ can be calculated as follows:

$$\sum_{N-n}^{i=1} \frac{\left[x\left((i+n)^{*}\Delta t\right)-x\left(i^{*}\Delta t\right)\right]^{2}+\left[y\left((i+n)^{*}\Delta t\right)-y\left(i^{*}\Delta t\right)\right]^{2}}{N-n}$$

The instantaneous diffusion coefficient D is calculated for each trajectory from linear fits of the first four points of the mean square displacement versus time function $\mathrm{MSD}(\tau)$ using the following equation (see Fig. 3b and c):

$$\mathrm{MSD}(\tau) = r^{2}(\tau) = 4D\tau.$$

The negative curvature of the MSD versus time plot informs on the level of confinement of receptors at the cell surface (Fig. 3c). Using appropriate labels (e.g. Mitotracker Green or PSD-95-GFP used as synaptic markers), one may plot MSD versus time as a function of the successive compartments explored by surface receptors (i.e. extrasynaptic, perisynaptic, synaptic; Fig. 3d), which allows to decipher if specific regulations occur in one or the other.

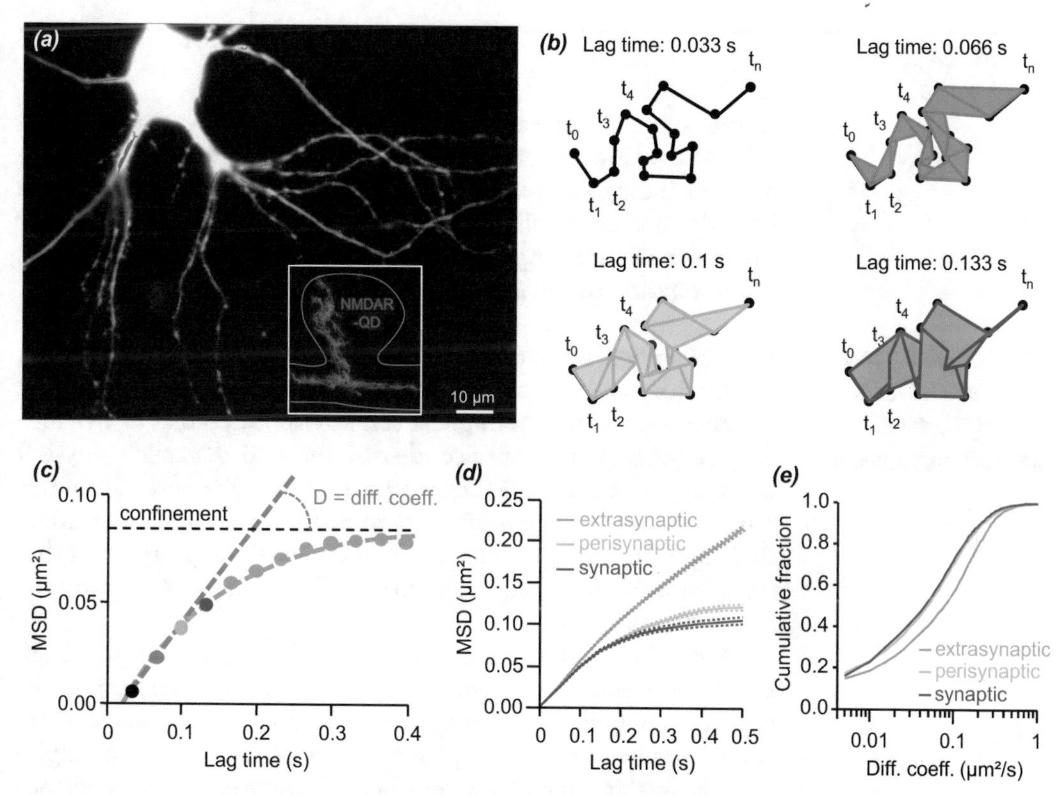

Fig. 3 NMDA receptors surface dynamics properties in live brain cells. (**a**) Trajectories of quantum dot-labeled surface NMDA receptors on the dendrites of a hippocampal neuron (20 Hz acquisition, 30 s duration). The *inset* shows an example of a typical trajectory exploring a dendritic spine. (**b**) Representative examples of mean square displacements for lag times of 33 ms (*black*; surface explored between two acquisition steps), 66 ms (*red*; surface explored between three acquisition steps), 100 ms (*green*; surface explored between four acquisition steps), and 133 ms (*blue*; surface explored between five acquisition steps) for an artificial trajectory. The corresponding points are reported on the mean square displacement (MSD) versus plot described in (**c**). The curvature of the plot reflects the level of confinement of receptor movements, and instantaneous diffusion coefficients (diff. coeff., *D*) are calculated from a linear fit to the first four points of the plot. (**d**) Typical mean square displacement (MSD) versus time plot of NMDA receptor trajectories in the different surface compartments (extrasynaptic, perisynaptic, synaptic). The perisynaptic compartment is arbitrarily defined as a 320 nm annulus surrounding the synaptic area. (**e**) Cumulative distributions of NMDA receptor instantaneous surface diffusion coefficients in the different surface compartments. The initial point of the distributions represents the immobile fraction of the population of receptors ($D < 0.005$ µm²/s)

Importantly, depending on experimental configurations (e.g. 20 Hz acquisition rate), all receptor/QD complexes showing instantaneous diffusion coefficients below limit values defined by their system (e.g. 0.005 µm²/s) should be arbitrarily considered as immobile. These fractions of immobile receptors appear as the ordinate to the first point of the cumulative distributions of instantaneous diffusion coefficients, which allow visualizing the global behavior of receptor populations compartment by compartment (Fig. 3e).

4 Discussion

Importantly, single particle tracking has proved a very powerful tool to investigate surface interactions between membrane proteins, including Glu receptors and their partners. We briefly discuss how the use of specific tools and the development of multicolor single-particle tracking strategies have allowed us to improve our understanding of cell surface interaction dynamics.

4.1 Combining SPT and Chemical Tools to Study Surface Interaction Dynamics

Understanding the mechanisms through which neurons integrate signals from multiple neurotransmitter systems is a major challenge in neuroscience. Interestingly, it recently appeared that by dynamically regulating their surface distribution and dynamics, physical interactions between NMDA receptors and other neurotransmitter receptors provide a powerful way to reciprocally and rapidly affect their signaling, synaptic content, and composition, suggesting that surface interactions may represent the first level of neurotransmission integration [37].

Over the past few years, a number of chemical tools have been developed either to reinforce (antibody-based crosslinking) or disrupt (biomimetic competing peptides, pharmacological agents) these interactions. The implementation of such tools in single-particle tracking paradigms has helped uncover where receptor–receptor interactions occur at the surface of neurons, and how modulations of these interactions affect their respective distributions and functions. As an example, combining high-resolution imaging and interaction modulations with specific agonists and/or disrupting peptides mimicking interaction sites revealed that dopamine D1 and NMDA receptors form perisynaptic clusters and that D1R/NMDAR complexes disruption leads to a rapid surface redistribution of both partners which eventually increases NMDAR synaptic content and favors long-term synaptic potentiation. Moreover, preventing this redistribution process through antibody-based cross-linking resulted in an impaired ability for synapses to express synaptic plasticity [14, 38, 39]. Importantly, this rapid and bidirectional surface crosstalk between receptors sets the plasma membrane as the primary stage of the dopamine–glutamate interplay.

Thus, combined with single-particle tracking approaches, the recent mapping of receptor–receptor interaction sites and extensions of the toolbox of biomimetic modulators to other partners of NMDAR such as EphBRs, alpha7-nAChR or mGluR5 [40–43] should allow us to improve our understanding of how surface dialogue between neurotransmitter receptors affect their trafficking and signaling properties and eventually contribute to neurotransmission integration.

4.2 Multicolor SPT: A New Approach to Study Receptor Interactions

Another application of high-resolution imaging to study surface interactions recently emerged from the development of multicolor single-particle tracking [44–47]. As an example, two-color single-particle tracking in A431 cells was used to visualize the homodimerization of the human epidermal growth factor receptor erbB1 and quantify the kinetics of transitions between free, co-confined, and dimerized states of the receptor [45]. Importantly, it revealed that ligand bound dimers are more stable than their unliganded counterparts, that dimerization decreases the surface dynamics of receptors dramatically, and that manipulation of ErbB1 affects their mobility at the cell surface, demonstrating that multicolor imaging is a powerful tool to investigate receptor interaction dynamics and how these affect receptor signaling. Moreover, the development of high-speed hyperspectral microscopy has extended this approach to simultaneous acquisitions of up to eight different species of QD at 27 Hz with a precision of ~10 nm [48], allowing to perform high resolution monitoring of multiple oligomerization processes of membrane proteins at the same time.

5 Conclusions

Single nanoparticle tracking has provided the means to investigate the surface dynamics of glutamate receptors in dissociated neurons. This has shed new lights on Glu receptor organization and trafficking at the plasma membrane between synaptic and extrasynaptic compartments and drawn attention on unsuspected mechanisms regulating synaptic transmission and plasticity. Over the past few years, single nanoparticle tracking in neurons has been extended to organotypic and acute brain slices. Although still in its infancy, single nanoparticle tracking in brain tissue preparations opens new perspectives to decipher the dynamic organization and regulation of Glu receptors in neuronal networks. Improving this approach will require the development of smaller and brighter fluorescent nanoprobes in order to allow accurate detection in high background noise environments due to light scattering, absorption, and tissue autofluorescence. Tackling this ambitious challenge should help us break the current frontiers of our knowledge in neurosciences and understand how neurotransmitter receptor surface organization and dynamics contribute to brain functions and diseases.

Acknowledgements

This work was supported by the Centre National de la Recherche Scientifique, the Human Frontier Science Program, the Agence Nationale de la Recherche, the Fondation pour la Recherche

Médicale and the Conseil Régional d'Aquitaine. We thank the cell culture facility of the institute, the animal facility of the University of Bordeaux, the Bordeaux Imaging Center for technical support and lab members for constructive discussions.

References

1. Traynelis SF, Wollmuth LP, McBain CJ, Menniti FS, Vance KM, Ogden KK, Hansen KB, Yuan H, Myers SJ, Dingledine R (2010) Glutamate receptor ion channels: structure, regulation, and function. Pharmacol Rev 62(3):405–496. doi:10.1124/pr.109.002451
2. Bliss TV, Lomo T (1973) Long-lasting potentiation of synaptic transmission in the dentate area of the anaesthetized rabbit following stimulation of the perforant path. J Physiol 232(2):331–356
3. Bear MF, Malenka RC (1994) Synaptic plasticity: LTP and LTD. Curr Opin Neurobiol 4(3):389–399
4. Bliss TV, Collingridge GL (1993) A synaptic model of memory: long-term potentiation in the hippocampus. Nature 361(6407):31–39. doi:10.1038/361031a0
5. Scannevin RH, Huganir RL (2000) Postsynaptic organization and regulation of excitatory synapses. Nat Rev Neurosci 1(2):133–141. doi:10.1038/35039075
6. Bredt DS, Nicoll RA (2003) AMPA receptor trafficking at excitatory synapses. Neuron 40(2):361–379
7. Carroll RC, Beattie EC, von Zastrow M, Malenka RC (2001) Role of AMPA receptor endocytosis in synaptic plasticity. Nat Rev Neurosci 2(5):315–324. doi:10.1038/35072500
8. Collingridge GL, Isaac JT, Wang YT (2004) Receptor trafficking and synaptic plasticity. Nat Rev Neurosci 5(12):952–962. doi:10.1038/nrn1556
9. Song I, Huganir RL (2002) Regulation of AMPA receptors during synaptic plasticity. Trends Neurosci 25(11):578–588
10. Borgdorff AJ, Choquet D (2002) Regulation of AMPA receptor lateral movements. Nature 417(6889):649–653. doi:10.1038/nature00780
11. Dahan M, Levi S, Luccardini C, Rostaing P, Riveau B, Triller A (2003) Diffusion dynamics of glycine receptors revealed by single-quantum dot tracking. Science 302(5644):442–445. doi:10.1126/science.1088525
12. Tardin C, Cognet L, Bats C, Lounis B, Choquet D (2003) Direct imaging of lateral movements of AMPA receptors inside synapses. EMBO J 22(18):4656–4665. doi:10.1093/emboj/cdg463
13. Choquet D, Triller A (2013) The dynamic synapse. Neuron 80(3):691–703. doi:10.1016/j.neuron.2013.10.013
14. Dupuis JP, Ladepeche L, Seth H, Bard L, Varela J, Mikasova L, Bouchet D, Rogemond V, Honnorat J, Hanse E, Groc L (2014) Surface dynamics of GluN2B-NMDA receptors controls plasticity of maturing glutamate synapses. EMBO J 33(8):842–861. doi:10.1002/embj.201386356
15. Ehlers MD, Heine M, Groc L, Lee MC, Choquet D (2007) Diffusional trapping of GluR1 AMPA receptors by input-specific synaptic activity. Neuron 54(3):447–460. doi:10.1016/j.neuron.2007.04.010
16. Heine M, Groc L, Frischknecht R, Beique JC, Lounis B, Rumbaugh G, Huganir RL, Cognet L, Choquet D (2008) Surface mobility of postsynaptic AMPARs tunes synaptic transmission. Science 320(5873):201–205. doi:10.1126/science.1152089
17. Betzig E, Patterson GH, Sougrat R, Lindwasser OW, Olenych S, Bonifacino JS, Davidson MW, Lippincott-Schwartz J, Hess HF (2006) Imaging intracellular fluorescent proteins at nanometer resolution. Science 313(5793):1642–1645. doi:10.1126/science.1127344
18. Hell SW (2007) Far-field optical nanoscopy. Science 316(5828):1153–1158. doi:10.1126/science.1137395
19. Hess ST, Girirajan TP, Mason MD (2006) Ultra-high resolution imaging by fluorescence photoactivation localization microscopy. Biophys J 91(11):4258–4272. doi:10.1529/biophysj.106.091116
20. Huang B, Babcock H, Zhuang X (2010) Breaking the diffraction barrier: super-resolution imaging of cells. Cell 143(7):1047–1058. doi:10.1016/j.cell.2010.12.002
21. Maglione M, Sigrist SJ (2013) Seeing the forest tree by tree: super-resolution light microscopy meets the neurosciences. Nat Neurosci 16(7):790–797. doi:10.1038/nn.3403
22. Rust MJ, Bates M, Zhuang X (2006) Sub-diffraction-limit imaging by stochastic optical reconstruction microscopy (STORM). Nat Methods 3(10):793–795. doi:10.1038/nmeth929

23. Sahl SJ, Moerner WE (2013) Super-resolution fluorescence imaging with single molecules. Curr Opin Struct Biol 23(5):778–787. doi:10.1016/j.sbi.2013.07.010
24. Nagerl UV, Willig KI, Hein B, Hell SW, Bonhoeffer T (2008) Live-cell imaging of dendritic spines by STED microscopy. Proc Natl Acad Sci U S A 105(48):18982–18987. doi:10.1073/pnas.0810028105
25. Tonnesen J, Katona G, Rozsa B, Nagerl UV (2014) Spine neck plasticity regulates compartmentalization of synapses. Nat Neurosci 17(5):678–685. doi:10.1038/nn.3682
26. Kittel RJ, Wichmann C, Rasse TM, Fouquet W, Schmidt M, Schmid A, Wagh DA, Pawlu C, Kellner RR, Willig KI, Hell SW, Buchner E, Heckmann M, Sigrist SJ (2006) Bruchpilot promotes active zone assembly, Ca^{2+} channel clustering, and vesicle release. Science 312(5776): 1051–1054. doi:10.1126/science.1126308
27. MacGillavry HD, Song Y, Raghavachari S, Blanpied TA (2013) Nanoscale scaffolding domains within the postsynaptic density concentrate synaptic AMPA receptors. Neuron 78(4):615–622. doi:10.1016/j.neuron.2013.03.009
28. Nair D, Hosy E, Petersen JD, Constals A, Giannone G, Choquet D, Sibarita JB (2013) Super-resolution imaging reveals that AMPA receptors inside synapses are dynamically organized in nanodomains regulated by PSD95. J Neurosci 33(32):13204–13224. doi: 10.1523/JNEUROSCI.2381-12.2013
29. Godin AG, Lounis B, Cognet L (2014) Super-resolution microscopy approaches for live cell imaging. Biophys J 107(8):1777–1784. doi:10.1016/j.bpj.2014.08.028
30. Groc L, Lafourcade M, Heine M, Renner M, Racine V, Sibarita JB, Lounis B, Choquet D, Cognet L (2007) Surface trafficking of neurotransmitter receptor: comparison between single-molecule/quantum dot strategies. J Neurosci 27(46):12433–12437. doi:10.1523/JNEUROSCI.3349-07.2007
31. Michalet X, Pinaud FF, Bentolila LA, Tsay JM, Doose S, Li JJ, Sundaresan G, Wu AM, Gambhir SS, Weiss S (2005) Quantum dots for live cells, in vivo imaging, and diagnostics. Science 307(5709):538–544. doi:10.1126/science.1104274
32. Biermann B, Sokoll S, Klueva J, Missler M, Wiegert JS, Sibarita JB, Heine M (2014) Imaging of molecular surface dynamics in brain slices using single-particle tracking. Nat Commun 5:3024. doi:10.1038/ncomms4024
33. Varela JA, Dupuis JP, Etchepare L, Espana A, Cognet L, Groc L (2016) Targeting neurotransmitter receptors with nanoparticles in vivo allows single-molecule tracking in acute brain slices. Nat Commun 7:10947. doi:10.1038/ncomms10947
34. Varela JA, Ferreira JS, Dupuis JP, Durand P, Bouchet D, Groc L (2016) Single nanoparticle tracking of [Formula: see text]-methyl-d-aspartate receptors in cultured and intact brain tissue. Neurophotonics 3(4):041808. doi:10.1117/1.NPh.3.4.041808
35. Haas K, Sin WC, Javaherian A, Li Z, Cline HT (2001) Single-cell electroporation for gene transfer in vivo. Neuron 29(3):583–591
36. Ito H, Morishita R, Iwamoto I, Nagata K (2014) Establishment of an in vivo electroporation method into postnatal newborn neurons in the dentate gyrus. Hippocampus 24(12):1449–1457. doi:10.1002/hipo.22325
37. Ladepeche L, Dupuis JP, Groc L (2014) Surface trafficking of NMDA receptors: gathering from a partner to another. Semin Cell Dev Biol 27:3–13. doi:10.1016/j.semcdb.2013.10.005
38. Ladepeche L, Dupuis JP, Bouchet D, Doudnikoff E, Yang L, Campagne Y, Bezard E, Hosy E, Groc L (2013) Single-molecule imaging of the functional crosstalk between surface NMDA and dopamine D1 receptors. Proc Natl Acad Sci U S A 110(44):18005–18010. doi:10.1073/pnas.1310145110
39. Lee FJ, Xue S, Pei L, Vukusic B, Chery N, Wang Y, Wang YT, Niznik HB, Yu XM, Liu F (2002) Dual regulation of NMDA receptor functions by direct protein-protein interactions with the dopamine D1 receptor. Cell 111(2):219–230
40. Dalva MB, Takasu MA, Lin MZ, Shamah SM, Hu L, Gale NW, Greenberg ME (2000) EphB receptors interact with NMDA receptors and regulate excitatory synapse formation. Cell 103(6):945–956
41. Li S, Li Z, Pei L, Le AD, Liu F (2012) The alpha7nACh-NMDA receptor complex is involved in cue-induced reinstatement of nicotine seeking. J Exp Med 209(12):2141–2147. doi:10.1084/jem.20121270
42. Nolt MJ, Lin Y, Hruska M, Murphy J, Sheffler-Colins SI, Kayser MS, Passer J, Bennett MV, Zukin RS, Dalva MB (2011) EphB controls NMDA receptor function and synaptic targeting in a subunit-specific manner. J Neurosci 31(14):5353–5364. doi:10.1523/JNEUROSCI.0282-11.2011
43. Perroy J, Raynaud F, Homburger V, Rousset MC, Telley L, Bockaert J, Fagni L (2008) Direct interaction enables cross-talk between ionotropic and group I metabotropic glutamate receptors. J Biol Chem 283(11):6799–6805. doi:10.1074/jbc.M705661200

44. Kawashima N, Nakayama K, Itoh K, Itoh T, Ishikawa M, Biju V (2010) Reversible dimerization of EGFR revealed by single-molecule fluorescence imaging using quantum dots. Chemistry 16(4):1186–1192. doi:10.1002/chem.200902963
45. Low-Nam ST, Lidke KA, Cutler PJ, Roovers RC, van Bergen en Henegouwen PM, Wilson BS, Lidke DS (2011) ErbB1 dimerization is promoted by domain co-confinement and stabilized by ligand binding. Nat Struct Mol Biol 18(11):1244–1249. doi:10.1038/nsmb.2135
46. Steinkamp MP, Low-Nam ST, Yang S, Lidke KA, Lidke DS, Wilson BS (2014) erbB3 is an active tyrosine kinase capable of homo- and heterointeractions. Mol Cell Biol 34(6):965–977. doi:10.1128/MCB.01605-13
47. Vu TQ, Lam WY, Hatch EW, Lidke DS (2015) Quantum dots for quantitative imaging: from single molecules to tissue. Cell Tissue Res 360(1):71–86. doi:10.1007/s00441-014-2087-2
48. Cutler PJ, Malik MD, Liu S, Byars JM, Lidke DS, Lidke KA (2013) Multi-color quantum dot tracking using a high-speed hyperspectral line-scanning microscope. PLoS One 8(5):e64320. doi:10.1371/journal.pone.0064320

Chapter 3

Positron Emission Tomography of Metabotropic Glutamate Receptors

Selena Milicevic Sephton

Abstract

Positron emission tomography (PET) is a noninvasive molecular imaging technique which shows continually growing applications in both preclinical and clinical research besides its well established use in routine clinical practice. PET is a highly multidisciplinary technique requiring specialized knowledge and skills from teams of Chemists, Physicists, and Biologists/Medical staff. In this chapter, we discuss PET methodology through explanation of the general principles governing PET, the requirements for successful PET imaging and finally examples of application of PET in studying metabotropic glutamate receptors (mGlu). Unlike, groups II and III, group I of mGlu receptors has been extensively studied for which reason herein, $mGlu_5$ and $mGlu_1$ will be the focus of this chapter.

Key words Positron emission tomography, Metabotropic glutamate receptors, Molecular imaging, Radiotracer, Imaging probe, MicroPET

1 Introduction

Positron emission tomography (PET) is a noninvasive molecular imaging technique based upon injection of a radiolabeled compound (i.e., radiotracer, imaging probe or radioligand) into the blood stream, whereby the radiotracer is transported to its biological target. An interaction between radiotracer and the target tissue or organ results in accumulation of radioactivity which is recorded and visualized as an image [1, 2].

The principle of PET lies in the process of annihilation between the positron emitted from the radiotracer and an electron from the living tissue resulting in the formation of two gamma rays (γ-rays, with 511 keV energy each) which are detected and measured (Fig. 1). Positrons are emitted with the high kinetic energy from the nucleus of the radiotracer, and as the positron travels through the tissue it subsequently loses its energy. Positron with significantly reduced kinetic energy interacts with a neighboring electron to form a specie called a positronium which decays by annihilation

Sandrine Parrot and Luc Denoroy (eds.), *Biochemical Approaches for Glutamatergic Neurotransmission*, Neuromethods, vol. 130, DOI 10.1007/978-1-4939-7228-9_3,

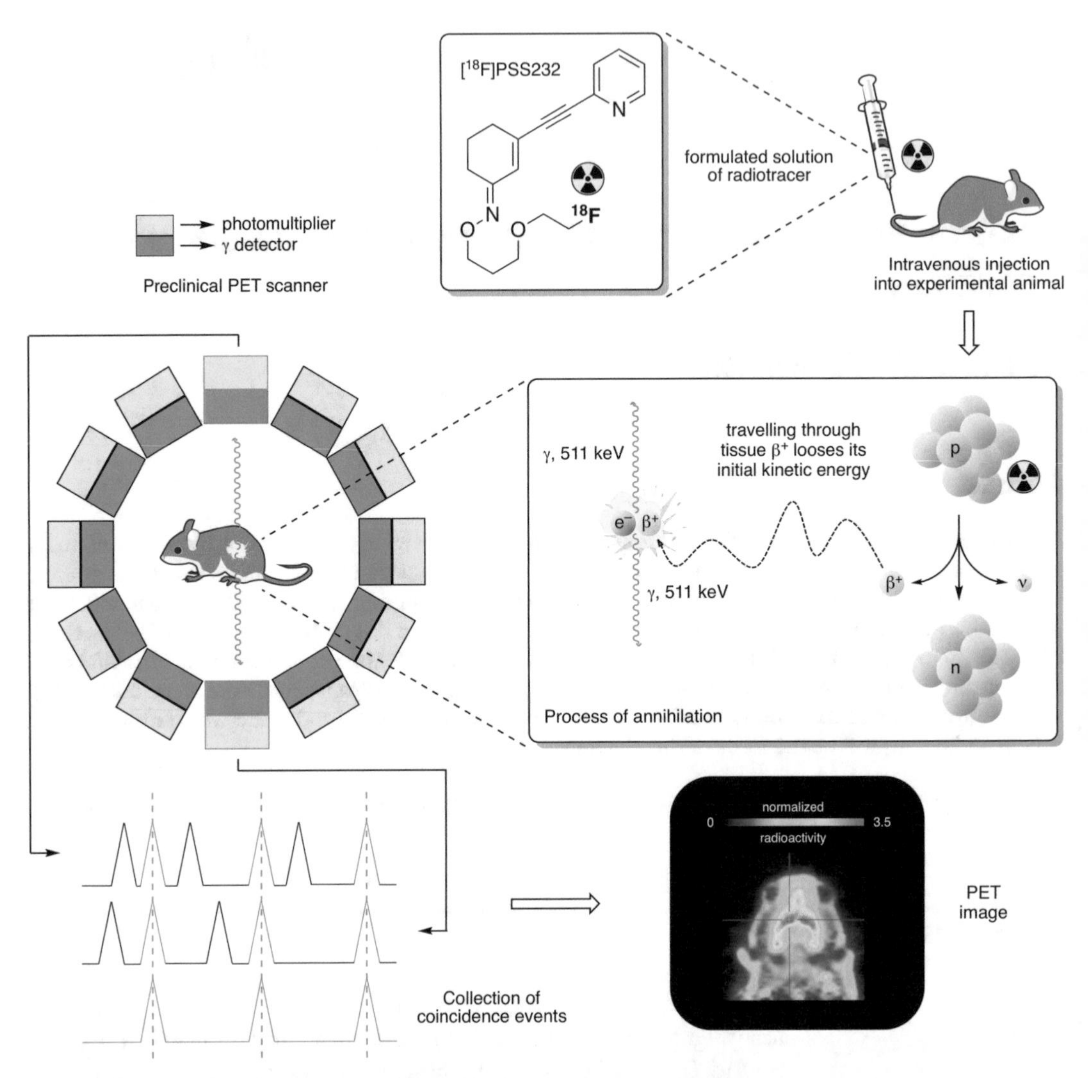

Fig. 1 Principle of PET: Formulated solution of radiotracer is injected into the experimental animal (e.g., rat). Radionuclide (e.g., ^{18}F) emits positrons (β^+) which travel through the tissue and lose their initially very high kinetic energy. Slowed down positrons interact with electrons (e^-) from the tissue to initially form positronium which decays by annihilation resulting in the formation of two γ-rays of the same energy. Only those γ-rays which travel at the opposite angles (180°) and reach the detectors at the same time are measured and then converted to optical signals to afford the PET image after reconstruction

to emit two γ-rays [1]. These γ-rays, traveling in equal but opposite directions (180°) are counted as they reach each of the detectors as a pair in coincidence. Additionally, the pairs of detectors are organized in a ring-like pattern (Fig. 1) thus allowing for measurement of radioactivity from a number of angles and radial distances [3]. This data enables tomographic images to be reconstructed to provide a visualization of the distribution of radioactivity in specific regions.

While the principles of PET remain the same, there is a plethora of possibilities for imaging various biological targets or metabolic processes in living systems. In addition to that, there lies the possibility for development of chemically varied imaging probes in order to achieve higher selectivity and specificity towards these biological targets [1–3]. This in turn makes PET a unique and powerful molecular imaging technique despite its relatively low spatial resolution (3–8 mm for clinical PET scanners and 1–2 mm for microPET preclinical scanners). Although [^{18}F]-labeled glucose ([^{18}F]FDG) is by far the most clinically applied PET radiotracer, tremendous efforts have been made for the development of improved PET radiotracers which will target very specific biomarkers allowing for earlier diagnosis and personalized medicine. Consequently, PET has seen many successful applications in Oncology, Neurology, and Cardiology [2].

As a functional imaging technique, PET is commonly employed to visualize and quantify metabolic processes (e.g., glucose metabolism with [^{18}F]FDG). Although [^{18}F]FDG can be used to for instance, monitor brain tumor therapy, the limitation of [^{18}F]FDG is due to an already rapid glucose metabolism in the healthy brain. For these reasons, in Neurology, targeting various receptors by PET has been widely utilized [4]. This is enabled by the significantly large difference in the receptor expression between healthy and diseased state as measured by PET [5]. The interaction between the ligand and the receptor is highly selective and specific and allows for measurement of the receptor's distribution, density, and activity by PET [2].

The group of receptors in focus herein are metabotropic glutamate receptors (mGlu receptors) which together with the ionotropic receptor (e.g., *N*-methyl-D-aspartate receptors) regulate glutamate (Glu) which is a major neurotransmitter in the mammalian system [6–10]. Metabotropic Glu receptors are responsible for slow neurotransmission of Glu and eight subtypes of mGlus (1–8) are divided into three groups I ($mGlu_1$ and $mGlu_5$), II ($mGlu_2$ and $mGlu_3$), and III ($mGlu_4$, $mGlu_6$, $mGlu_7$ and $mGlu_8$). The subtypes are determined based on pharmacological profile, sequence homology, and signal transduction pathways of each mGlu [6]. In these terms, group I of mGlu receptors is distinctly unique in that $mGlu_1$ and $mGlu_5$ regulate excitatory neurotransmission by modulating excitability postsynaptically. Unlike the other two groups, group I mGlus also shows distinct expression patterns in the brain [11]. As indicated on the schematic in Fig. 2, mGlu receptors are characterized as having seven transmembrane domains (7TMD). In the case of $mGlu_1$ and $mGlu_5$ the extracellular hydrophilic N-terminal domain is extraordinarily large and necessary for Glu binding. Glutamate or the ligand are trapped within

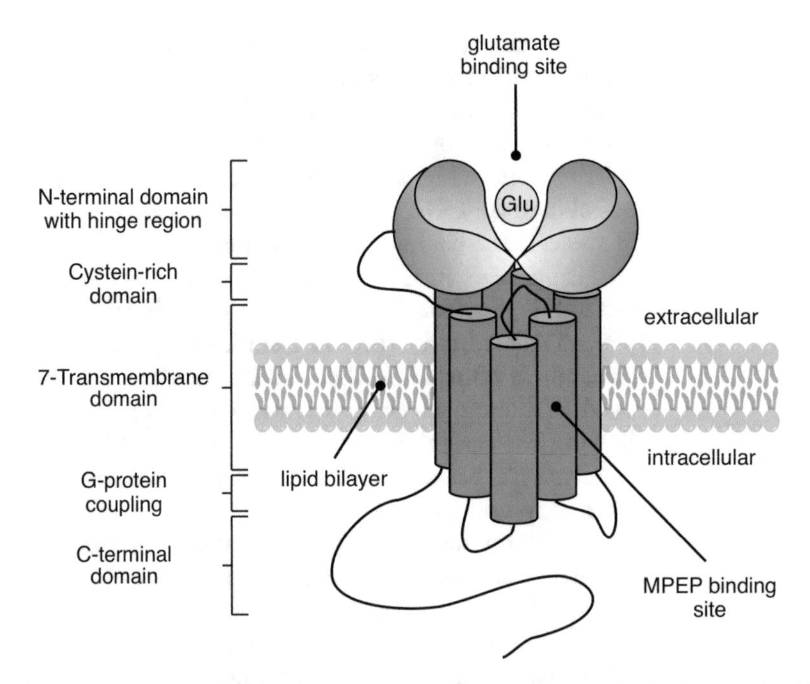

Fig. 2 Schematic representation of $mGlu_5$ receptor in the cell membrane: $mGlu_5$ is characterized by the 7TMD (*red*) and extraordinarily large extracellular hydrophilic *N*-terminal domain (*blue*), which is the binding site for Glu (*yellow*) and other ligands

two globular domains with a hinge region enabled by a change in their conformation (Fig. 2) [9].

Due to their role in the regulation of Glu transmission they are involved with several disorders of the central nervous system (CNS) such as drug addiction, depression, schizophrenia, and neuropathic pain. They are also implicated in Parkinson's, Alzheimer's diseases, and Fragile X syndrome [12–18]. Because of their involvement in the mechanisms of CNS disorders, mGlu receptors have been biological targets of interest for the development of respective PET imaging probes [11, 12, 19]. To date, owing to its distinct properties, the most studied subtype is group I, with the particular emphasis on $mGlu_5$ [19, 20]. For this reason, most examples will concern $mGlu_5$ receptor, and thereafter $mGlu_1$ [21]. Other mGlu subtypes will be mentioned to the extent they have been studied by PET.

2 Materials

Three key components for successful PET imaging can be identified as follows:

1. *Radiotracer*—compound (e.g., small synthetic organic molecule, large biomolecule, etc.) radiolabeled with a positron emitting radionuclide (e.g., carbon-11, fluorine-18, copper-64,

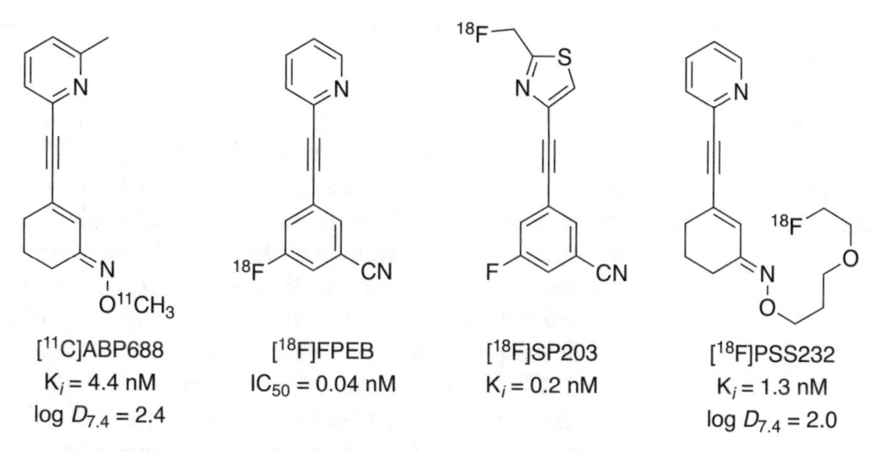

Fig. 3 Structures of the most commonly employed $mGlu_5$ radiotracers

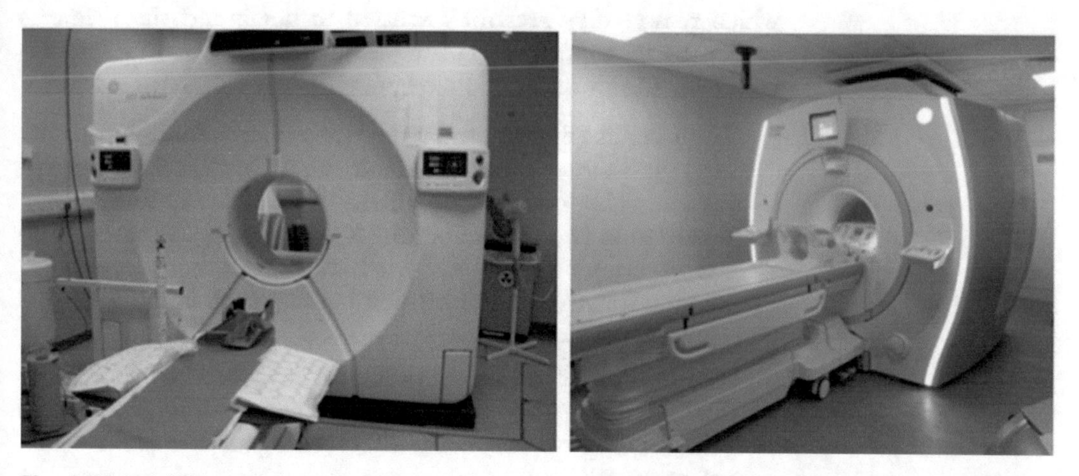

Fig. 4 Clinical PET scanners at the Wolfson Brain Imaging Centre at the University of Cambridge: Old GE Healthcare PET Advance scanner (*left*) and new state-of-the-art GE Healthcare Signa 3T PET/MR clinical scanner (*right*). Pictures are courtesy of WBIC team and superintendent research radiographer Ms. Victoria Lupson

gallium-68, Fig. 3) prepared as an injectable formulation solution (e.g., saline or polyethylene glycol (PEG200)) [2].

2. *Scanner*—large clinical PET scanners are often combined with computed tomography (CT) or Magnetic Resonance Imaging (MRI); small animal preclinical microPET scanners commonly merged with CT or MRI.
3. *Subject*—human subject (clinical PET, please refer to Fig. 4) or animal: small animal (e.g., rodent) or primate (e.g., marmoset) for preclinical PET.

3 Methods

3.1 Development and Selection of Appropriate Radiotracer

The first step in PET imaging is the selection of an appropriate PET imaging probe for the desired biological target [2]. The selection of the appropriate radiotracer adheres to several criteria which ensure the radiotracer is suitable for the desired study. The current pool of radiotracers is continually expanding however the need for development of new imaging probes is justified by the ever growing demand for highly selective and specific imaging probes.

The development of new PET radiotracer combines tools from Medicinal Chemistry, Organic Chemistry and Radiochemistry. The process can be summarized as indicated in Fig. 5.

There are several criteria which make a radiotracer applicable to PET imaging. Methods used in assessing applicability of a radiotracer and its characterization exceed the scope of this chapter for which reason they are not discussed in detail and the reader is guided to appropriate literature (see below). A good PET radiotracer will ideally have the following properties [1, 20]:

- High target affinity. Ideally the radiotracer would bind to the biological target with high affinity to allow clear image visualization. The measure of affinity is expressed through the dissociation (K_D) or inhibition (K_i) constants [22] based upon the experimental protocol used to assess affinity (saturation binding for K_D and competition binding experiment for K_i).
- High selectivity for the target. When a radiotracer is selective, visualization of off-target processes is reduced to minimum. The selectivity for the target is particularly important when pharmacological properties of two targets are similar (i.e., $mGlu_5$ vs. $mGlu_1$).
- Low nonspecific binding. The background activity decreases the overall image quality and significant nonspecific binding increases background.
- Permeability to the blood brain barrier (BBB). Radiotracer must be able to access the target by having the appropriate properties to permeate the BBB, including appropriate lipophilicity. The polarity of a radiotracer is one factor that determines its passage across the BBB. It is expressed as log D or log P value and the value obtained is heavily dependent on the experimental method used for its determination (e.g., shake-flask vs. high performance liquid chromatography (HPLC) method) [23–25].
- Know metabolism. Possible radiometabolites can significantly affect a PET image (e.g., they may have different affinity and selectivity to the parent compound), for which reason metabolic products should be known and studied.

PRECLINICAL DEVELOPMENT AND EVALUATION

IDENTIFICATION OF ANALOGUES

Biological target of interest is identified. Design and selection of possible chemical structure is made based on results from throughput studies or computational modelling.

↓

ANALOGUES SYNTHESIS

Using the knowledge and skills from Medicinal and Organic Chemistry, libraries of derivatives around the lead compound are prepared.

↓

***IN VITRO* EVALUATION**

Series of *in vitro* assays are conducted and the data analyzed to select a potential candidate based on various criteria (e.g., high affinity, good lipophilicity).

↓

RADIOSYNTHESIS

Appropriate radiolabelling precursor is prepared and used to develop an efficient and robust radiosynthetic method to access PET radionuclide-containing analogue.

↓

***IN VIVO* EVALUATION (SMALL-ANIMAL PET)**

Series of *in vivo* experiments are designed and executed, including blocking experiments to establish selectivity/specificity, biodistribution, metabolite analysis, kinetic PET data analysis. Proven good *in vivo* properties secure future clinical studies.

↓

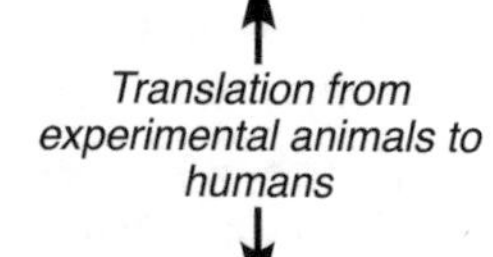

SETTING THE STAGE FOR CLINICAL STUDIES

Radiosynthesis is optimized for automation on synthetic modules under GMP conditions. Validation of radiolabelling method is conducted to enable robust and reproducible synthesis. Toxicology studies are perfomed and ethical approval sought.

↕

CLINICAL EVALUATION AND APPLICATION

HUMAN STUDIES

Once the radiotracer has been established under preclinical setting, first in-man PET studies are conducted to determine viability of investigated radiotracer for PET imaging in human subjects. Test-retest studies provide level of variability observed amongst subjects and are important criteria in establishing new PET imaging probe for application in humans.

↓

ROUTINE APPLICATION

Established radiotracer is produced and used routinely. Application of imaging probe goes beyond diagnosis and therapy monitoring and extends onto clinical research studies.

Fig. 5 Process of the development of novel PET imaging probe from bench to bedside

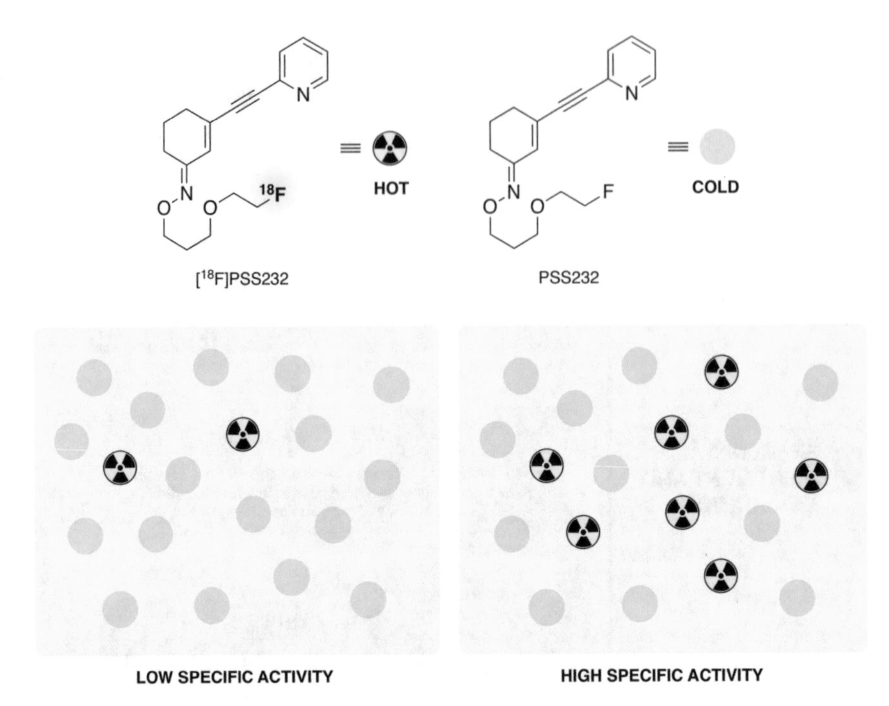

Fig. 6 Schematic representation of specific activity: ratio of cold (stable isotope) and hot (radioactive isotope) molecules. Better image quality is achieved with the larger number of hot molecules

- Choice of radionuclide. Most PET radiotracers are radiolabeled with either carbon-11 or fluorine-18, which have a physical half-life of 20 and 110 min, respectively. A shorter physical half-life limits the application of the radiotracer.
- High specific activity. Specific activity is the measure of radio-labeled compound relative to cold compound present in the radiotracer sample (Fig. 6). It is expressed in Bq/mol and calculated by dividing the activity of a sample with the mass summed for all radioactive (hot) and stable (cold) isotopes [26]. The higher the specific activity the better quality of PET image can be obtained. Low specific activity presents a risk of pharmacological drug effect due to the larger amount required for dosing. Higher specific activity is also desired when binding potential is low for similar reasons.

Further consideration in the development of the imaging probe include:

- Concentration of binding sites (B_{max}). An excess of target sites is desired whereby B_{max} exceeds K_D of the radiotracer. The ratio of B_{max} over K_D is calculated as binding potential [20, 27]. The higher the binding potential the better the target to nontarget ratio and therefore the higher the image quality.

- Reproducible radiolabeling methods. Radiolabeling method used for the preparation of a radiotracer should be robust, fast, and well-established to allow for a reliable synthesis. It should also conform to automation, which is particularly important for translation to clinical setting.

Considering the selection criteria discussed above several radiotracers, which adhere to those requirements are now well-established PET agents for imaging mGlu receptors [19, 20]. There is a plethora of evidence of the exploratory work leading to the development of these imaging probes which presents an invaluable knowledge base. However, herein priority will be given to those imaging agents routinely applied in preclinical and clinical practice.

3.1.1 PET Imaging Agents for Group I of mGlus

$mGlu_1$

Based on their pharmacological profile, two receptors $mGlu_1$ and $mGlu_5$ belong to the same group 1 of mGlu receptors and show similar properties for which reason design and development of a selective PET radiotracer for imaging either of the two is particularly challenging [28], however, there are differences. Expression of $mGlu_1$ in the brain is highest in the cerebellum, and moderate or low in the thalamus, striatum, and cerebral cortex [29]. They are involved with spatial and associative learning, synaptic plasticity in the hippocampus and cerebellum and for this reason play a role in several diseases such as stroke, epilepsy, pain, cerebellar ataxia, anxiety, mood disorders, and neurodegenerative disorders [15, 21, 28]. Despite the well-founded rationale as to the need to develop a $mGlu_1$ PET imaging probe, only three PET candidates have been applied in a clinical setting to date: [^{11}C]LY2428703, [^{11}C]ITMM, and [^{18}F]FIMX (Table 1).

The list of $mGlu_1$ PET radiotracer candidates in Table 1 is not exhaustive; however none of the imaging probes have been widely employed. Based on their in vitro evaluation the two most promising carbon-11 radiotracers, [^{11}C]-LY2428703 [30] and [^{11}C]ITMM [31, 32], failed as PET imaging agents for human application because of low brain uptake. Studies with [^{11}C]-LY2428703 in monkeys showed low uptake in all brain regions albeit higher in cerebellum than other regions but no uptake reduction was established upon blocking. In human subjects, low uptake was found in all brain regions. This was explained by high protein plasma binding due to high lipophilicity of the compounds. Higher brain uptake was observed for [^{11}C]ITMM, however, the kinetics of the uptake were slow reaching maximum only >40 min post-injection (p.i.). Furthermore, very slow washout from the brain was found and considering the short physical half-life of carbon-11 (i.e., 20 min), [^{11}C]ITMM proved to be unsuitable as a PET radiotracer for imaging $mGlu_1$. Only recently first-in-man PET studies were performed with [^{18}F]FMIX, which showed good radioactivity uptake in the $mGlu_1$-rich brain regions (i.e., the cerebellum) with

Table 1
Clinical and preclinical PET radiotracer candidates for imaging $mGlu_1$ receptors: a summary of their properties log $D_{7.4}$ expresses lipophilicity; *AR* autoradiography, *NR* not reported, *NAM* negative allosteric modulator

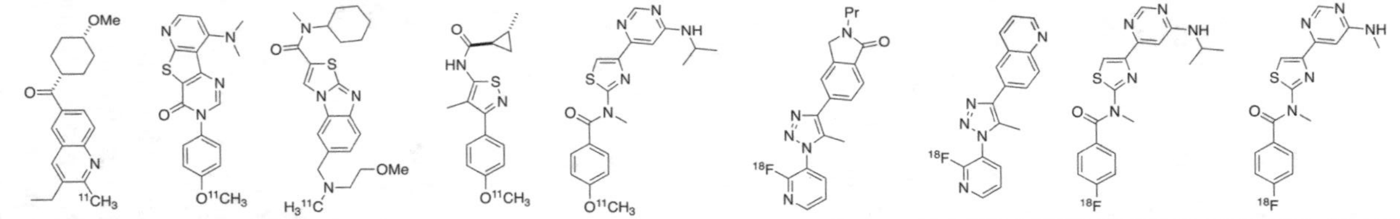

Radiotracer	[11C]JNJ16567083	[11C]MMTP	[11C]YM202074	[11C]LY2428703	[11C]ITMM	[18F]MK1312	[18F]FPTQ	[18F]FITM	[18F]FIMX
	Antagonist	NAM	Antagonist	Antagonist	NR	NR	NR	NAM	Antagonist
K_i/K_d (nM)	0.87	–	4.80	–	12.6	0.4 (monkey) 0.84 (human)	–	–	–
IC_{50}	–	–	–	8.9 nM	–	–	1.4 nM (mouse) 3.6 nM (human)	5.1 nM	1.8 nM (human)
log $D_{7.4}$	–	–	2.7	4.02	–	2.3	–	1.46	2.5
AR (*in vitro*)	–	–	Heterogeneous	Heterogeneous	Heterogeneous	Hetereogenous	Heterogeneous	Heterogeneous	–
AR (*ex vivo*)	–	–	Distribution not in agreement with the *in vitro* data	–	–	–	–	86-91% reduction in uptake under blocking conditions	–
In vivo data (preclinical)	Selective and specific to $mGlu_1$; 81% specific binding in cerebellum	In baboons, highest uptake in cerebellum	Less than 10% of intact radiotracer 30 min p.i. in rats	Displacable signal in rats; one radiometabolite	Relatively slow uptake kinetics; for a 90-min scan, uptake in cerebellum irrevirsible	Heterogeneous distribution, concentration dependent blockade in monkeys; 17% intact radiotracer, 90 min p.i.	Good uptake in cerebellum (SUV of 2.3); it has radiometabolite	High *in vivo* stability in monkey brain; slow uptake kinetics	Excellent uptake in monkey brain; specific and selective to $mGlu_1$
In vivo data (clinical)	–	–	–	Low brain uptake; high binding to plasma proteins speculated	Low brain uptake and slow washout	–	–	–	High brain uptake in cerebellum; faster washout; requires 120-170 min scan

Table 2
Summary of properties of the four most potent $mGlu_5$ PET radiotracers
log $D_{7.4}$ expresses lipophilicity; *AR* autoradiography, *NR* not reported

Radiotracer	[11C]ABP688	[18F]SP203	[18F]FPEB	[18F]PSS232
	Antagonist	Antagonist	Antagonist	NR
K_i/K_d (nM)	1.7	0.2	0.15	–
IC_{50}	–	36 pM	–	1.3 nM (rat) 1.1 nM (human)
log $D_{7.4}$	2.4	–	–	2.0
AR (*in vitro*)	–	–	Heterogeneous	Heterogeneous
AR (*ex vivo*)	In agreement with $mGlu_5$ expression; no accumulation in $mGlu_5$ k/o mice	–	–	–
In vivo data (preclinical)	80% selective binding under blocking conditions; biodistribution data in agreement with *in vitro* data	In rats and monkeys showed defluorination	77% specific binding under blocking conditions; highest uptake 20-30 min p.i.	84% selective binding under blocking conditions; highest uptake 5 min p.i.; 91% intact radiotracer 20 min p.i. in the brain
In vivo data (clinical)	Expected $mGlu_5$ expression profile; 25% intact radiotracer present 60 min p.i.	Defluorination found to a small extent, it can be accounted for in PET kinetic calculations	Despite poor radiochemical yields successfully used in clinical practice; test-retest variability <10%	–

the signal significantly reduced under blocking conditions [33]. The uptake kinetics of [^{18}F]FMIX were shown to be fast with the peak at 10 min p.i., followed by the rapid washout from the brain. Stable distribution volumes in the cerebellum were only reached in a 170 min-long scan, which presents a disadvantage of [^{18}F]FMIX in that it limits multiple scans. While [^{18}F]FMIX shows promise for application in human clinical practice, at present, further studies are needed to establish a PET radiotracer for imaging $mGlu_1$.

$mGlu_5$

In sharp contrast to the struggle in finding a $mGlu_1$ PET radiotracer, [^{11}C]ABP688 [34, 35], a PET probe for imaging $mGlu_5$ has reached the status of state-of-the-art imaging agent (Fig. 3) [19, 20, 36].

Table 2 summarizes the main conclusions of in vitro and in vivo studies for the four most prominent $mGlu_5$ PET radiotracers, three of which have been evaluated in human subjects.

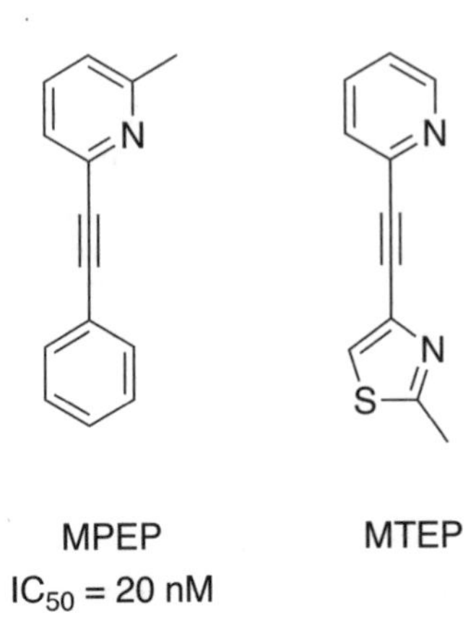

Fig. 7 Structures of MPEP and MTEP—lead compounds in the development of [^{11}C]ABP688

The development of $mGlu_5$ PET probes was facilitated by the discovery of a synthetic molecule MPEP (Fig. 7) as an antagonist of $mGlu_5$ in a throughput screen [37]. MPEP does not bind at the conserved Glu binding site, instead it acts via a noncompetitive mechanism. Carbon-11 radiolabeled MPEP was prepared and evaluated preclinically only to show nonspecific binding in the hippocampus and striatum. An analogue of MPEP $mGlu_5$ antagonist was explored, MTEP (Fig. 7) and performed similarly poorly in vivo [20].

Subsequent more elaborate structure–activity relationship (SAR) studies finally enabled the identification of the ABP688 scaffold in 2006 by the Ametamey group [34, 35]. Preclinical evaluation of [^{11}C]ABP688 paved the way to clinical application which showed the expected radioactivity distribution and good test-retest variability amongst subjects. The disadvantage of [^{11}C]ABP688 lies in the short-lived carbon-11 radiolabel which limits its application to centers with on-site cyclotron facilities. A search for a fluorinated analogue of [^{11}C]ABP688 lead to three other PET radiotracers. Developed by the Pike group [^{18}F]SP203, a derivative of MTEP, demonstrated good in vitro properties however a high level of defluorination was observed in monkey studies [38, 39]. High initial uptake in under 30 min p.i. was compromised by the accumulation in the bone. However, in human subjects, only negligible defluorination was found suggesting that [^{18}F]SP203 can be used under clinical conditions. This was particularly facilitated by applying the equilibrium method by performing bolus plus constant infusion instead of the standard bolus injection followed by the kinetic modeling [40]. A MPEP structural analogue [^{18}F]FPEB showed excellent properties, both preclinically and clinically, including $<10\%$ test-retest variability [41, 42]. In contrast to [^{11}C]ABP688 (peak uptake at ca. 5 min) [35], [^{18}F]FPEB has slower uptake kinetics (peak uptake at ca. 20–30 min) and slower washout from the brain as established from the time activity curves (TACs) obtained for various brain regions [41].

Radiosynthesis of [^{18}F]FPEB is low yielding which was considered to be a disadvantage; however automation of the radiosynthesis allowed for higher amounts of starting activity to be used, thus leading to higher amounts of formulated radiotracer. A direct analogue of [^{11}C]ABP688, [^{18}F]PSS232 was radiolabeled with fluorine-18 in a high yielding one step reaction, and showed excellent in vitro and in vivo properties in rodents [43–45]. Highest uptake was observed early at 5 min p.i., and [^{18}F]PSS232 demonstrated fast washout from the brain (TACs in Fig. 8b). The radioactivity was successfully displaced with $mGlu_5$ antagonist MMPEP when injected 40 min after the injection of [^{18}F]PSS232 (Fig. 8c). Figure 8a shows PET/CT image of [^{18}F]PSS232 in male Wistar rat in transversal, sagittal, and coronal view with the striatum in the crosshairs. The highest accumulation of activity is indicated by strong red color in the $mGlu_5$-rich regions. Blue color in cerebellum shows no radioactivity accumulation in agreement with regional distribution of $mGlu_5$.

Application of [^{18}F]PSS232 in human subjects is ongoing in the Ametamey group.

3.1.2 PET Imaging Agents for Group II of mGlus

Discovered at the beginning of 1991, mGlu receptors are a relatively young family of receptors, which is a likely cause for the limited research examples focused on addressing group I, but not groups II and III. Only recently, reports on the development of a PET radiotracer for imaging $mGlu_2$ have emerged [19, 20]. Expression of $mGlu_2$ is primarily in the presynaptic nerve terminals. They negatively modulate Glu and GABA release and clinical studies have shown that stimulation of $mGlu_2$ has a therapeutic effect by reducing levels of Glu in the synaptic cleft [46].

$mGlu_2$

First investigated $mGlu_2$ PET radiotracer, was [^{11}C]CMG (Fig. 9) which failed to penetrate the BBB [47]. To overcome this, a prodrug approach was employed in which dimethyl ester [^{11}C]CMGDE was used to successfully enter the brain. However, [^{11}C]CMGDE had an uptake peak at 2 min p.i., and showed poor selectivity between $mGlu_2$ and $mGlu_3$ in addition to only 20–30% decrease under blocking conditions, for which reason it was not further investigated. Another potential candidate, LY341495 was radiolabeled with tritium and despite excellent binding potency, it demonstrated low uptake which peaked at 2 min p.i., suggesting that it cannot be used as a PET radiotracer for $mGlu_2$ [48]. In 2012, a conference report appeared describing a new $mGlu_2$ PET radiotracer [^{11}C]JNJ42491293 which has been employed in human subjects [49, 50]. Due to scarcity of reported data, the full potential of [^{11}C]JNJ42491293 as a PET imaging agent cannot be anticipated. It has been suggested that [^{11}C]JNJ42491293 shows

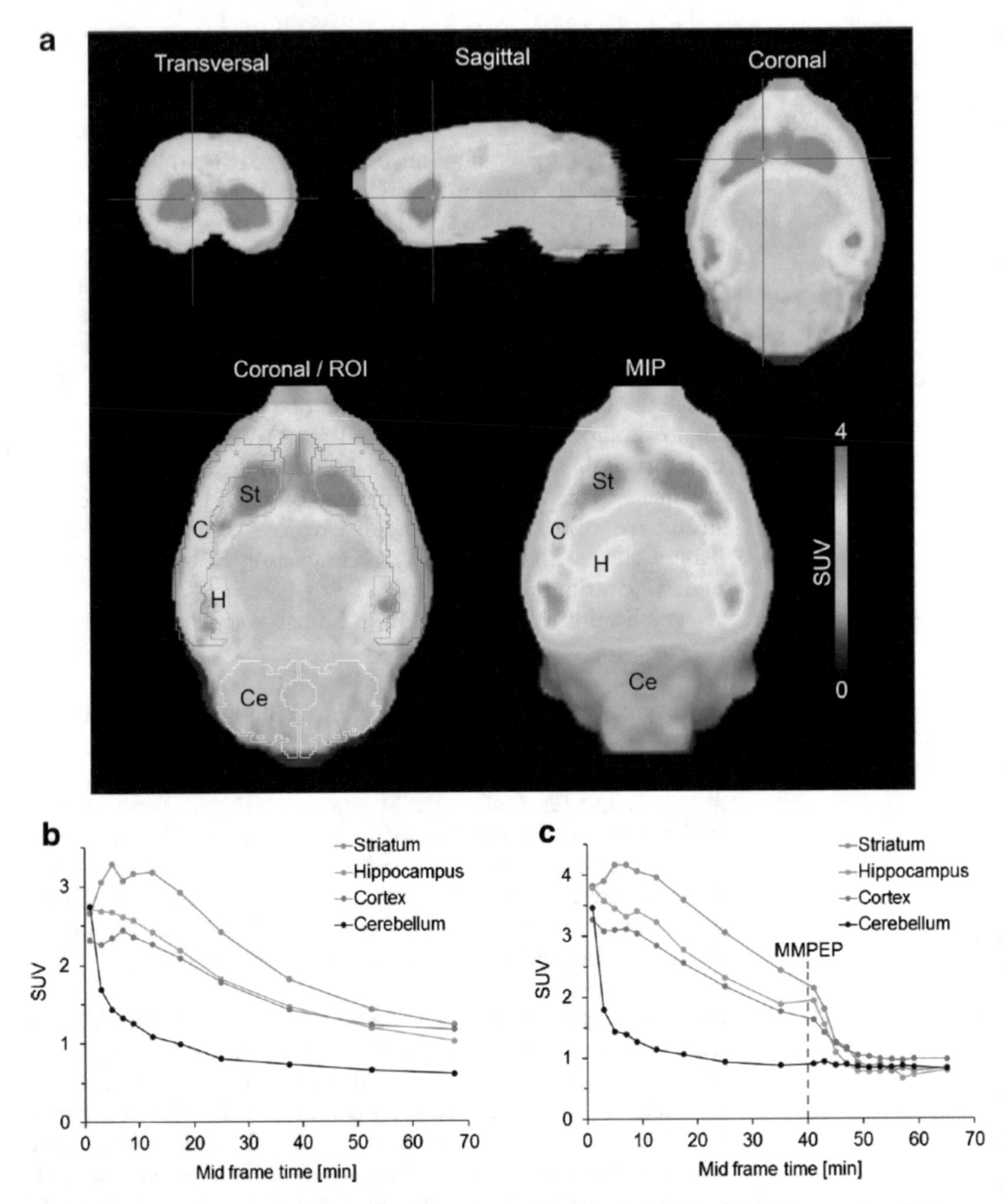

Fig. 8 (**a**) PET images of [^{18}F]PSS232 in a Wistar rat after injection of 35 MBq. *Colored images* from PET scan are superimposed on an MRI T2 template. The images are averaged for 2–30 min p.i. Striatum is in crosshairs, in different view (transversal, sagittal, and coronal) in *top images*. *Bottom row* shows ROIs for quantitative analysis. *MIP* maximal intensity projection. (**b**, **c**) TACs of [^{18}F]PSS232 in different brain regions in either control (**b**) or after displacement with 1 mg/kg of MMPEP 40 min p.i. (**c**). Taken with permission from Springer [44]

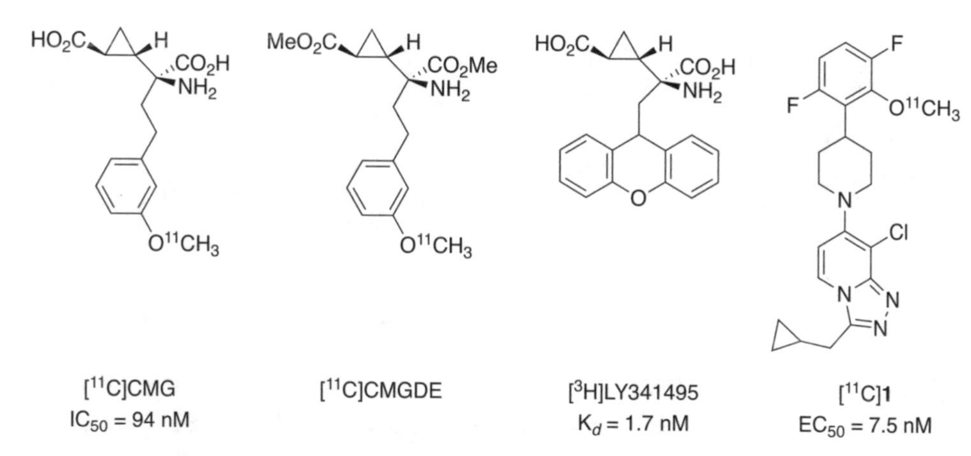

Fig. 9 Structures of some $mGlu_2$ PET radiotracer candidates

moderate uptake with peak activity at 30 min p.i., and slow washout kinetics. In the same year, a positive allosteric modulator (PAM) for $mGlu_2$, carbon-11 labeled triazolopyridine derivative ([^{11}C]**1**, Fig. 9) was described and evaluated in a preclinical study to show highest uptake at 20–30 min p.i., and it was successfully blocked [51]. The difficulty in analysis of in vivo data was due to a lack of reference region required for any kinetic calculations. No further studies have been published on the properties of [^{11}C]**1**.

mGlu$_3$

To date, studies on the development of PET radiotracers for imaging $mGlu_3$ have not been reported.

3.1.3 PET Imaging Agents for Group III of mGlus

Similar to group II of mGlu receptors, group III has been little investigated. Group III of mGlu receptors is the largest of the three, with four members all located in the presynaptic nerve terminals at the active zone of neurotransmitter release. Three, $mGlu_4$, $mGlu_7$ and $mGlu_8$ are negatively coupled to adenylyl cyclase, whereas, $mGlu_6$ is positively coupled to cGMP phosphodiesterase [19].

mGlu$_4$

The most studied receptor subtype from the group III, $mGlu_4$ is highly expressed in the cerebellum. It has been demonstrated that $mGlu_4$ plays a role in regulating neurotransmission in both direct and indirect Parkinson disease pathways for which reason it has received some interest for the development of therapeutic agents [52].

Figure 10 depicts several preclinically investigated potential PET radiotracers for imaging $mGlu_4$. Poor selectivity over other mGlu receptors was observed when [^{11}C]Me-PHCCC was applied [53, 54]. Established PAM of $mGlu_4$, [^{11}C]APX88178 showed extremely low specific binding in vitro [55]. While [^{11}C]ML128 penetrated BBB, activity uptake was low with poor 22–28% activity reduction under blocking [56]. A very fast washout of <20 min

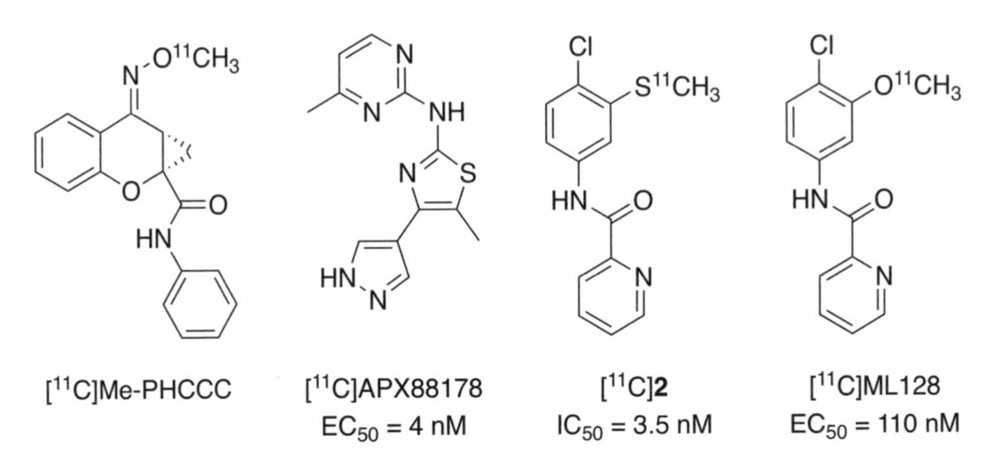

Fig. 10 Explored PET radiotracer candidates for imaging $mGlu_4$

p.i., was found limiting for the quality of obtained images. Chemical modification of [^{11}C]ML128 lead to [^{11}C]**2** (Fig. 10), for which activity uptake was higher with the peak at 2 min p.i., and slower washout warrants potential for possible further application [57].

$mGlu_6$

To date, a PET radiotracer for imaging $mGlu_6$ has not been reported.

$mGlu_7$

Facing the same difficulties of selectivity and activity uptake development of PET imaging probe for $mGlu_7$ receptors has seen only modest success. Furthermore, the limiting step is seen in the lack of efforts in identifying new lead structures.

In analogy to groups II and III, $mGlu_7$ is localized presynaptically, however it is the most abundant. The activation of $mGlu_7$ is achieved by Glu released from the presynaptic terminal [58]. The role of $mGlu_7$ is both in regulating inhibitory GABAergic transmission as well as glutamatergic synaptic responses.

One investigated PET radiotracer candidate for imaging $mGlu_7$ was [^{11}C]MMPIP (Fig. 11) [59] which showed extremely low brain uptake, with immediate washout from the brain in a preclinical study in rats.

$mGlu_8$

A PET radiotracer has not been evaluated for imaging $mGlu_8$ to date.

3.2 PET Imaging as a Means of Quantification of mGlu Receptors

3.2.1 PET Scanner

Whether PET imaging is performed with the aim of a research study or as a routine diagnostic tool, the primary difference in methods employed for PET imaging is in terms of its application (Fig. 4) [60]. Preclinical PET scanners or microPET scanners are employed for imaging of small animals, e.g., rodents. Clinical scanners are used for PET imaging of human subjects [3]. MicroPET scanners are smaller in size and have higher resolution.

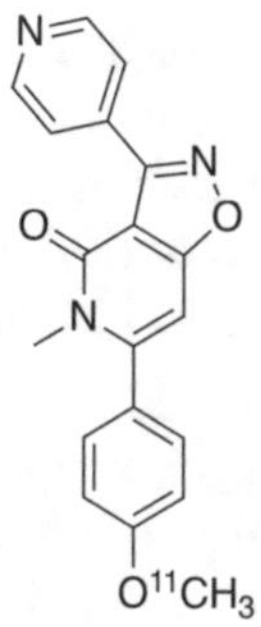

$O^{11}CH_3$

$[^{11}C]$MMPIP
$K_B = 30$ nM

Fig. 11 Structure of $[^{11}C]$MMPIP

In recent years, both preclinical and clinical PET scanners have seen significant improvements in design and construction. One of the critical parameters in advancing quality of images in the modern PET scanners is application of certain detectors. Detection of photons, which are the result of annihilation is challenging because not all photons in coincidence will be detected. Some will be absorbed by the neighboring tissue, and some will be scattered, resulting in only one of the photons reaching the detector or a difference in arrival time and occasionally two photons from two different annihilation events will reach the detectors at almost exact time, all affecting the image quality [61]. A more detailed discussion about different types of detectors and contributions they have made in obtaining better PET images is beyond the scope of the present chapter.

3.2.2 PET Measurement

Preclinical

In a typical preclinical PET measurement using a rodent, the following steps are employed as exemplified in $[^{18}F]$PSS232 study [44, 45]:

- From where it is housed, the animal is selected, removed from its cage, and weighted. Short transfer of the animal is conducted in the red box (i.e., to simulate nocturnal environment).
- The animal is anesthetized employing one of the commonly used methods (e.g., with gaseous isoflurane) [62, 63]. A need for an anesthesia in animals steams from the requirement of keeping the subject stationary whilst in the PET scanner.
- The tail of the anesthetized animal is then warmed to facilitate intravenous injection (e.g., by placing the tail of the animal in the warm water or by using a blanket) [64]. Correct and precise injection of the radiotracer ascertains to the accuracy of PET data.

- Typically a cannula is fitted and secured into the tail vein to accommodate injection of the radiotracer solution. In a PET experiment, it is saline which is injected first, to establish a smooth flow of the liquid through the vein, which is then followed by the injection of the radiotracer solution.
- Based on the data of specific activity of the radiotracer and body weight of the animal, the amount of the radiotracer to be injected is calculated and measured in the syringe (i.e., total radioactivity of the syringe is recorded).
- The animal is then moved onto the PET scanner bed and secured. Anesthesia inlet is secured, and animal body parameters are closely monitored (e.g., body temperature via rectal probe, respiration rate using pneumatic pillow, blood pressure using tail cuff sensor, heart rate using electrocardiogram, oxygen saturation in blood via pulse oxymetry). It is essential to maintain steady basic bodily functions of the animal whilst under anesthesia. For instance, a small drop in body temperature can have fatal results.
- In the subsequent step, the animal is moved inside the scanner, it is injected with saline first, followed by the radiotracer solution. At the same time point when the radiotracer is to enter the tail vein, collection of PET data starts.
- Once the collection of data is complete, the animal is removed from the scanner, lines from the body parameter instruments are removed, and the animal is transferred to its cage.
- In case of metabolite studies, blood sampling is performed. For a short-term blood sampling (e.g., 1 day experiment), temporary catheters are used. The amount of blood which can be removed will depend on the species and the size of the animal [64]. In an experiment, several time points will be planned at which small sample will be taken and then further analyzed (e.g., HPLC or radio-thin layer chromatography).
- In the case of biodistribution studies, at the end of the scan, or depending on the protocol, after the injection, the animal is sacrificed; target organs are removed, and measured in the gamma counter. For the brain, further dissection is performed to separate different brain regions.
- If the PET scanner is connected with a CT, CT scan (for anatomical orientation) is conducted at the end of the PET scan.

Clinical

A typical human PET scan requires the following steps:

- No anesthesia is necessary.
- The subject is explained the procedure and allowed time to prepare.

- Subject is seated and a tracer solution is injected into the vein commonly using an intravenous catheter inserted in the arm. The tracer is then given the time to travel through the body (ca. 60 min).
- Subject is then asked to lie on the scanner bed, and all necessary adjustments are made to allow subject as much comfort as possible.
- Scanner bed is moved into the scanner and the data collection is then started.
- At the end of the scan, the subject may leave immediately.

3.2.3 PET Data Analysis

While detailed PET data analysis requires an independent chapter and many excellent reviews have covered the topic [65–67], herein focus will be only on several basic parameters used for the PET data analysis. The data analysis follows analogous principles for both preclinical and clinical studies. Advances in the analysis of PET data have significantly contributed in overall improved field of PET particularly through various mathematical models, which further provide more in depth information from a single experiment. To assist with the analysis of PET data, many user-friendly computer software programs have been developed (e.g., PMOD software), which require specialized knowledge to a lesser degree.

Most commonly employed basic terms in PET kinetic analysis are:

- Compartmental models (e.g., with one or two transit compartments (TC)—1TC, 2TC) [67]. Due to complexity of living organisms, representation of various biochemical processes and pharmacokinetics as well as pharmacodynamics of the injected radiotracer is challenging in mathematical terminology. For these reasons, various models have been developed, which simplify biochemical events from the radiotracer injection to its excretion. The most commonly employed compartmental models, assume each individual event (e.g., delivery, uptake, binding) and all of the radiotracer molecules at any given time exist in one of the compartments (Fig. 12). Compartments refer not only to physical location of the molecule of radiotracer (e.g., synapse) but also its chemical state (e.g., whether it is bound, metabolized). The compartmental model is not static in that it accounts for changes and movement of the molecules of radiotracer from one compartment to the other. This way a rate constant (k) is defined and represents rate of change of concentration of radiotracer in one given compartment.
- As indicated in Fig. 12, radiotracer (R), after being intravenously injected through blood, travels to target tissue where it crosses the membrane and can either be bound to its biological

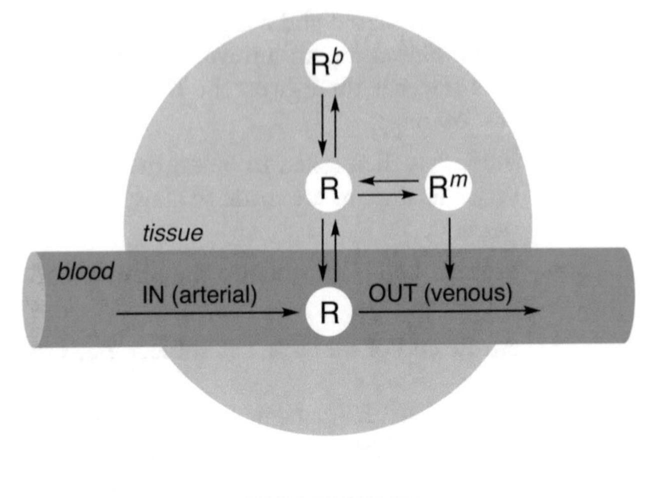

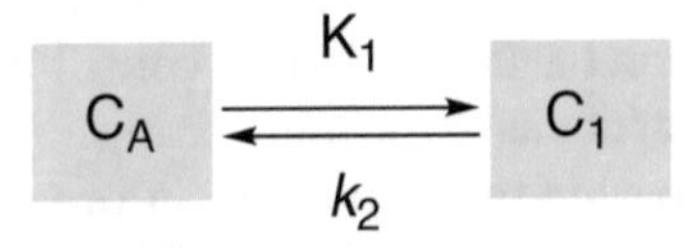

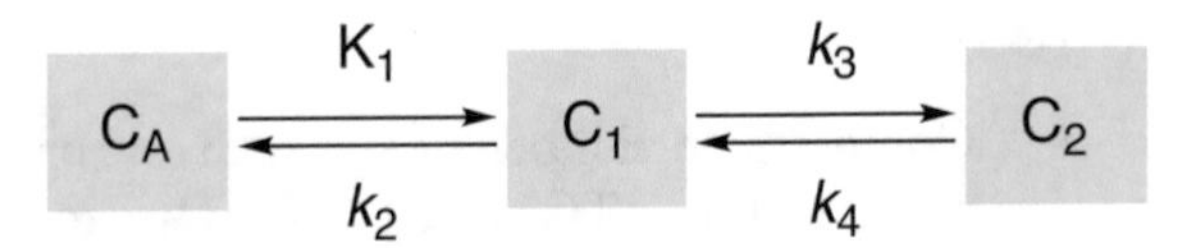

Fig. 12 Compartmental models are mathematical expression of pharmacokinetics and pharmacodynamics of radiotracer. *Legend*: *R* radiotracer, R^b radiotracer bound, R^m radiotracer metabolized, K_1, k_2–k_4 rate constants, C_P concentration of radiotracer in arterial blood, C_1–C_3 concentration of radiotracer in different compartments. *Note*: If the input values for concentration of radiotracer (C_P) are known, blood is not considered to be a compartment. However, when these values are predicted by the model, blood is accounted as a compartment

target (e.g., $mGlu_5$) or it gets metabolized and then excreted from the tissue through outflow of venous blood [67]. One-tissue compartmental model is applied for a radiotracer with a reversible tissue uptake. On the other hand, a two-tissue compartmental model is applied for a radiotracer which enters tissue from the blood but it is then metabolized (k_3) or excreted through blood (k_4). The two compartments therefore represent unmetabolized and metabolized radiotracer.

- Typically, the arterial plasma TAC is used as the input function to the compartmental model used for the estimation of parameters.
- Logan plot is a graphical analysis technique which was developed for reversible receptor systems thus allowing estimation of V_T.

- Volume of distribution (V_D) is a ratio of tissue concentration to the blood concentration at the equilibrium [65, 67]. The equilibrium is reached when the concentration of tracer which can diffuse in the tissue compartment is constant. This can be achieved under the assumption that concentration of the tracer in the blood stays constant. The total volume of distribution (V_T) is employed for receptor-binding radiotracers and is defined as ratio of total radiotracer in tissue over radiotracer in plasma at the equilibrium [67]. Nondisplaceable volume of distribution (V_{ND}) refers to volume of distribution estimated from the reference regions.
- Plasma function (f_P) is plasma free fraction (plasma protein binding) from blood sampling, freely dissolved in tissue water.
- Binding potential (BP_{ND}) is the ratio of B_{max} and K_d. In vitro, BP is equal to ratio of specifically bound radiotracer to its free concentration. In vivo, due to endogenous radiotracer binding not all receptors are available for which reason $B_{available}$ is used.
- Standard uptake value (SUV). SUV is the ratio of concentration of radiotracer at given time and injected activity at given time divided by the body weight.
- Region of interest (ROI). ROI is a region selected during the analysis of PET image, (e.g., hippocampus). The disadvantage of using ROI lies in its high variability from an image to an image which originates from selecting/drawing regions manually.
- Volume of interest (VOI). Analogous to ROI, VOI is selected during the PET analysis and is three-dimensional ROI.

The mathematical interpretation of PET data allows for many complex biochemical processes to be simplified and provides means of understanding their mechanisms. The parameters defined above are used as quantitative assessment of the PET image. As such, the application of the methods (e.g., 1-TC vs 2-TC) and parameters (e.g., SUV, ROI) provides an insight into distribution, binding, and excretion of the radiotracer in the living tissue as well as functional information about the biological target radiotracer is binding to (e.g., location, occupancy). These parameters also allow for a direct comparison between different regions within the same organism (e.g., radiotracer expression in hippocampus vs. cerebellum in the brain) but also amongst different subjects in both preclinical and clinical imaging (e.g., how does treated animal compares to control).

3.3 Application of PET in Preclinical and Clinical Practice Using Examples of $mGlu_5$ Imaging Agents

As indicated above, there are only few clinically applied mGlu radiotracers. The clinical application in the first instance is primarily for the purposes of in vivo characterization of those radiotracers in human subjects. At the time this chapter was written, it was primarily [^{11}C]ABP688, PET radiotracer for imaging $mGlu_5$ which has been employed in clinical research studies. Several examples of its application both preclinically and clinically will be next discussed. Some of the recent examples of preclinical studies using [^{18}F]FPEB will also be depicted.

3.3.1 PET Imaging of $mGlu_5$: Preclinical Applications

In 2012, Elmenhorst et al. performed test retest in rats using PET with [^{11}C]ABP688 to show no significant difference in test and retest measurements and provided evidence of reproducible and reliable quantification of $mGlu_5$ using [^{11}C]ABP688 [68].

It has been shown in animal models that psychiatric disorders can be induced in offspring by the parental microglial induced neuroinflammation. Arsenault et al. used PET imaging of $mGlu_5$ with [^{18}F]FPEB to investigate the relationship between $mGlu_5$ and peripheral benzodiazepine receptor (PBR) with respect to regulation of glial function. An animal model where mice are prenatally treated with LPS has been established as an inflammatory model of psychiatric disorders. Using PET imaging the authors showed that activation in $mGlu_5$ reduced expression of PBR, in healthy controls but not in LPS-treated mice [17].

In a different study it was shown by Brownell et al. that progression of amyotrophic lateral sclerosis (ALS), a progressive neurodegenerative motor neuron disorder by PET using [^{18}F]FPEB [18]. They compared the data with the data obtained from PET imaging of PBR to show increased BP with disease progression. The animal model used was SOD1-G93A gene expressing mice since mutation of this gene has been implicated with the ALS pathology. The study confirmed the role of Glu and neuroinflammation in ALS suggesting that excessive Glu is likely to contribute inflammation in neurodegenerative processes in ALS.

Preclinical PET study of $mGlu_5$ using [^{11}C]ABP688 was employed to monitor changes in $mGlu_5$ expression during epileptogenesis in a pilocarpine-induced epilepsy rat model. In the model, for more than 3 weeks, recurrent spontaneous seizures were observed. Simplified reference tissue model was used to estimate BP_{ND} and healthy controls were compared to epilepsy model (Fig. 13).

Choi et al. demonstrated reduction in BP_{ND} in hippocampus and amygdala in the chronic epilepsy rat model (Fig. 13). In the acute period BP_{ND} was reduced in the whole brain, whereas in the chronic period it was normalized except hippocampus and amygdala [69].

PET imaging with [^{11}C]ABP688 has also been used to determine changes in $mGlu_5$ expression between control and ceftriaxone (CEF)-treated rats. CEF is known as the activator of Glu transporter (EAAT2) which decreases extracellular levels of Glu. Zimmer et al. applied simplified reference tissue method to estimate BP_{ND}. In CEF-treated animals increase of BP_{ND} in thalamic ventral anterior was observed suggesting that availability of $mGlu_5$ allosteric binding sites is sensitive to extracellular concentrations of Glu [70].

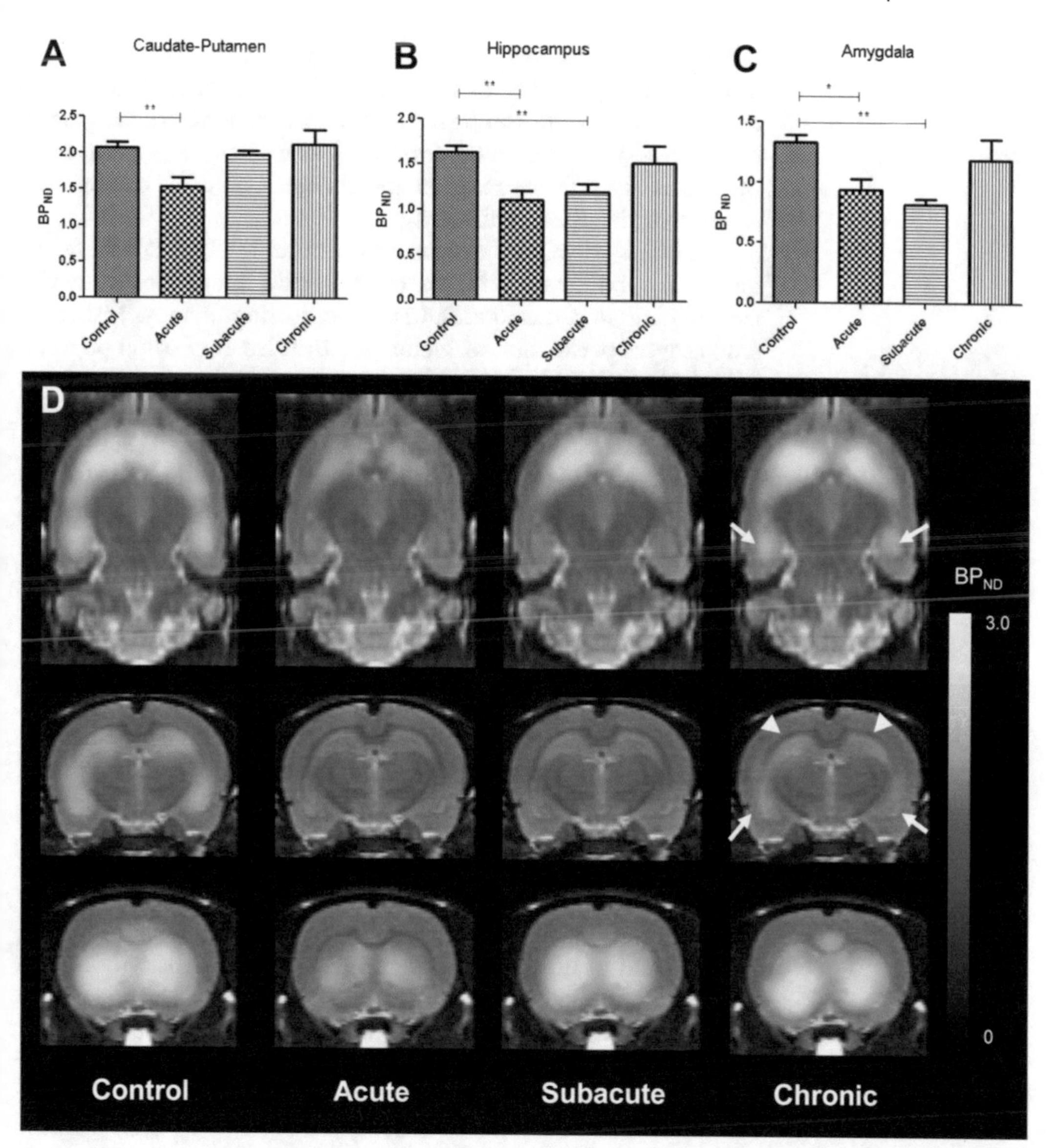

Fig. 13 (**a–c**) BP_{ND} obtained from PET images using VOI-based analysis of chronic epilepsy rats and controls show decreased dependence on time periods after status epilepticus. In acute period it was decreased globally, whereas in subacute period BP_{ND} decreased only in hippocampus and amygdala, but not caudate-putamen. In chronic period BP_{ND} did not show difference between epilepsy model rats and control. (**d**) BP_{ND} parametric maps which are in agreement with VOI-based analysis. Taken with permission through Open Access of *PLOSONE* [69]

Mathews et al. conducted a study in which they employed [^{11}C]ABP688 PET of mGlu$_5$ in baboons to establish if [^{11}C] ABP688 competes with fenobam for the same binding site [71]. Fenobam is a negative allosteric modulator of mGlu$_5$ and has shown promise as a potential drug for treatment of psychiatric disorder such as Fragile X. If fenobam binds similarly to [^{11}C]

ABP688 in a dose-dependent and saturable manner, this in turn leads to the possibility of using PET for occupancy-drug plasma concentration relationships in the brain. Plasma reference graphical method was used to determine V_T as well as BP_{ND} using cerebellum as the reference region and showed that PET can be used to characterize therapeutic drug.

In 2011, Miyake et al. reported another [^{11}C]ABP688 investigation in baboons with the aim of using PET imaging to study Glu transmission in disease state and for monitoring therapy. Test retest studies were performed with controls and also employing pharmacological challenge with *N*-acetylcysteine (NAC) which increases extrasynaptic Glu release. The results of analysis showed NAC induced decrease in BP_{ND} and no change from test to retest [72]. To extend on these finding Sandiego et al. reported [^{11}C]ABP688 PET study in rhesus monkeys using both 50 and 100 mg/kg on NAC to induce Glu release extrasynaptically [73]. Sandiego et al. used various methods to determine V_T as well as BP_{ND} to show increasing trend from test to retest BP_{ND}. Furthermore, NAC-treated animals had increased V_T and decreased BP_{ND} showing disagreement between the two studies.

3.3.2 PET Imaging of mGlu$_5$: Clinical Applications

In 2011, DeLorenzo et al. reported test-retest studies in healthy human subjects using [^{11}C]ABP688. They successfully applied unconstrained two-compartmental model and showed increased binding in retest study [74]. This was attributed to varying Glu level presumably due to the stress of the PET scan. The same group reported the application of [^{11}C]ABP688 to establish no significant difference in elderly subjects with major depression disorder (MDD) and healthy volunteers in 2015. Lower density of mGlu$_5$ was previously reported with the MDD patients with average age of 40.8 suggesting that further investigation was needed to answer whether it was the pathophysiology of MDD or expression of mGlu$_5$ in elderly which was different [75]. Despite the requirement for subsequent study, PET imaging of mGlu$_5$ suggested possibly different mechanism of MDD dependent on the age group.

Recent study by DuBois et al. employed 31 subjects to establish effects of gender and age on regional distribution of mGlu$_5$ in the brain [76]. The study employed advanced partial volume correction method and calculated BP_{ND} through a simplified reference tissue model with cerebellum as the reference region. In the development of a new tracer and its establishment in the human clinical practice, there are often data for healthy male subjects only. Preclinical studies have previously shown gender differences and study detailing the relationship between [^{11}C]ABP688 binding and smoking, indicated lower radioactivity uptake in female nonsmokers compared to male subjects. However, present study [76] clearly showed no age or gender difference in [^{11}C]ABP688 uptake between healthy volunteers.

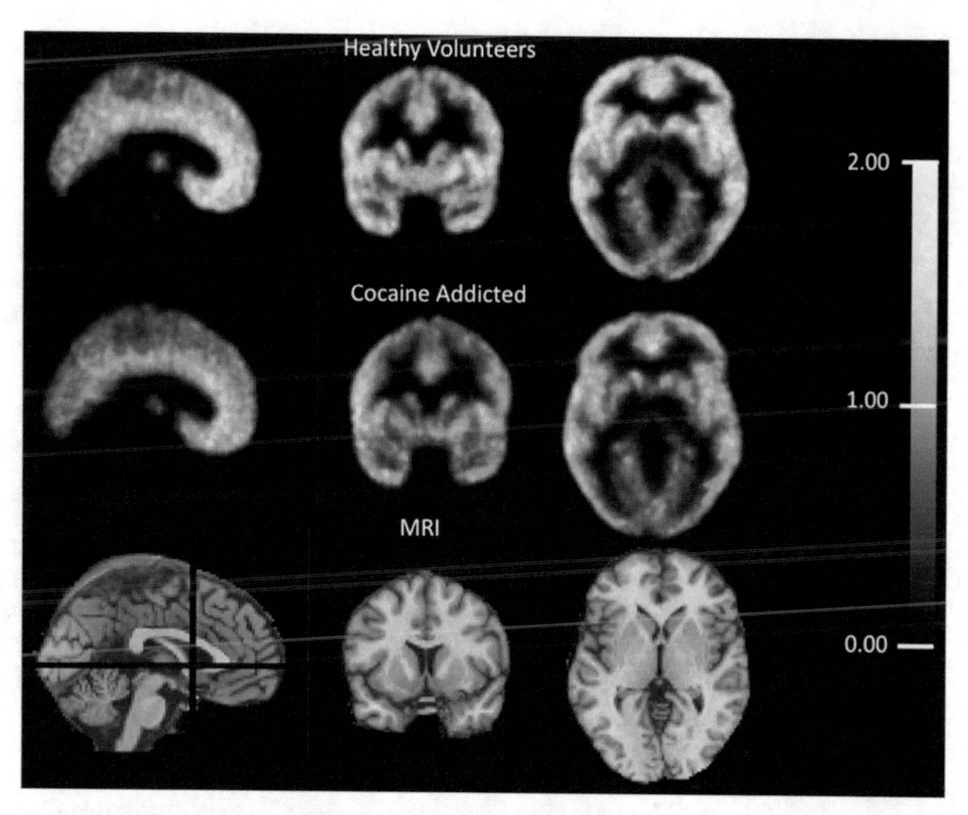

Fig. 14 Average BP_{ND} images of [^{11}C]ABP688 in healthy controls and cocaine-addicted subjects. Taken with permission from Society of Biological Psychiatry [78]

N-Methyl-D-aspartate receptor antagonist, ketamine, can produce a rapid antidepressant response in patients with treatment-resistant depression [77]. Administration of ketamine in humans results in Glu release and in 2015 DeLorenzo et al. hypothesized that this would be reflected in reduction in radioactivity uptake of [^{11}C]ABP688 due to a mechanism different to direct competition. In ten investigated subjects two scans were performed one before and one during the ketamine administration. Arterial blood sampling allowed for input functions to be derived. The study showed significant reduction in radioactivity uptake in the scan performed during the ketamine administration in agreement with the hypothesis [77]. Also observed was significant individual variation in ketamine response.

In 2014, Martinez et al. investigated effects of cocaine exposure on Glu signaling by PET [78]. Preclinical studies have shown that cocaine exposure leads to the disruption of Glu homeostasis, reduction in $mGlu_5$ expression and reduction in Glu turnover. Total of 30 participants (15 healthy controls and 15 cocaine-addicted volunteers) was subjected to PET imaging with [^{11}C]ABP688 and the results clearly showed reduction in radioactivity uptake in striatum (Fig. 14). Similarly cortical and subcortical regions showed reduced binding of [^{11}C]ABP688 of cocaine-addicted subjects in comparison to healthy controls [78]. This was demonstrated by the clear reduction of yellow coloration in the $mGlu_5$-rich regions of the images, where yellow represented the highest radioactivity accumulation. In the data analysis authors determined ROIs and used cerebellum as the reference region.

In the same year Milella et al. reported a PET study with [^{11}C]ABP688 in cocaine-addicted subjects [79]. The goal of the study was to quantify regional $mGlu_5$ density in recently abstinent cocaine-addicted participants. From the PET scans, BP_{ND} was determined in 18 subjects (nine healthy controls and nine cocaine-dependent participants) to show significantly lower BP_{ND} in the striatum. It was also demonstrated that $mGlu_5$ availability is related to abstinence and lower BP_{ND} was observed with longer abstinence [79].

Kagedal et al. employed PET to estimate $mGlu_5$ occupancy of a novel drug AZD2066 which is a negative allosteric modulator (NAM) of $mGlu_5$. In vitro studies have demonstrated AZD2066 to inhibit $mGlu_5$ and in vivo in rats, AZD2066 was capable of crossing the BBB and displacing [^{3}H]methoxymethyl-MTEP and [^{11}C]ABP688. In this study, Kagedal et al. aimed to quantify by PET the relationship between concentration of AZD2066 in plasma and $mGlu_5$ occupancy in the brain using [^{11}C]ABP688. For this they employed and compared two analysis methods; one developed by Lassen and the other one based on nonlinear mixed effects model [80]. The results of the data showed decrease in V_T with increasing concentration of AZD2066 (Fig. 15). The reduction in radioactivity uptake (i.e., binding of [^{11}C]ABP688) in the images is demonstrated by reduction in red coloration of $mGlu_5$-rich regions. Most of the variability in V_T was explained by variability between individuals and occasions in V_{ND}.

Leuzy et al. used PET imaging of $mGlu_5$ with [^{11}C]ABP688 to demonstrate changes in $mGlu_5$ expression between healthy subjects and subjects affected with behavioral variant of frontotemporal dementia (bvFTD). In the analysis they employed cerebellum as the reference region for estimation of BP_{ND} and confirmed reduction in [^{11}C]ABP688 binding and therefore decreased $mGlu_5$ density in bvFTD [81].

3.4 Conclusions

In the present chapter, basic principles of PET methodology were discussed and its application in the study of mGlu receptors has been described. From its humble beginnings, facing the challenges of the requirements for specialized knowledge and skill set as well as financial burdens, PET has developed into a practical research and diagnostic tool. User-friendly software packages have enabled easier data analysis and interpretation, making PET a method of choice particularly due to its sensitivity and possibility of functional imaging in vivo. Advancements within Organic and Medicinal Chemistry have enabled access to many different chemical

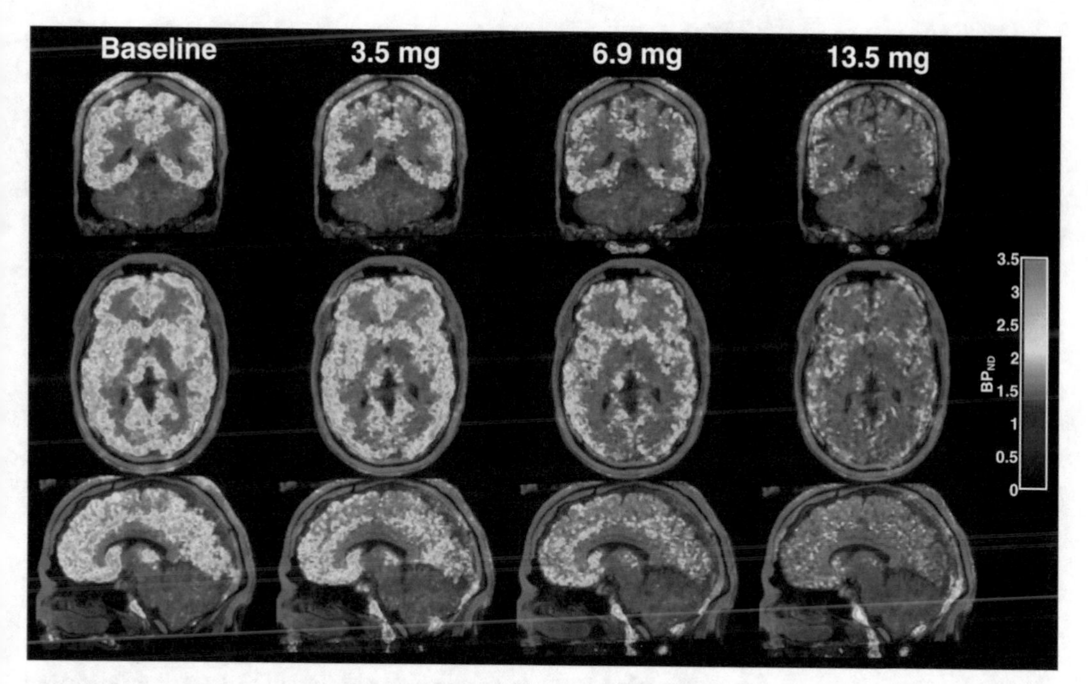

Fig. 15 BP_{ND} parametric images of dose-dependent binding of [^{11}C]ABP688. Radioactivity uptake in $mGlu_5$-rich regions was significantly decreased with increased amount of AZD2066 administered. BP_{ND} images are overlaid on anatomical MR images. Taken with permission from Elsevier [80]

scaffolds leading to very specific and selective radiotracers. One such example is PET imaging of $mGlu_5$ in which case both carbon-11 ([^{11}C]ABP688) and fluorine-18 ([^{18}F]FPEB) radiotracers have been established and successfully used in human clinical practice. To date, more commonly employed [^{11}C]ABP688 has seen applications in various neuropsychiatric disorders or for the occupancy studies. In turn, this has contributed to better understanding of the $mGlu_5$ family of receptors. Other seven members of the mGlu subclass are awaiting their radiotracer development. At present, $mGlu_1$ has witnessed most progress.

In conclusion, PET methodology has contributed greatly to the advancement of knowledge base on mGlu receptors enabling better understanding of involvement of mGlu receptors in various psychiatric and neurological conditions. PET of $mGlu_5$ receptors has led to their visualization and quantification in human subjects and paved the way to allow receptor density mapping of other members of the mGlu family.

4 Notes

4.1 Development and Selection of Appropriate Radiotracer

- Development of radiotracers is often heavily varied due to various techniques and methods employed in their evaluation dependent upon research group. For this reason, comparing two radiotracers solely on the basis of literature data often is not sufficient and commonly evaluation of two radiotracers is conducted in a direct comparison studies.

- The physical half-life of 110 min, makes fluorine-18 a preferred radionuclide particularly with respect to practical aspects of PET technique. Longer half-life allows more flexibility with the length of the radiosynthesis, and it enables radiotracer to be shipped off-site. PET imaging with carbon-11 labeled probes is limited by the requirement of the on-site cyclotron. On the other hand, carbon-11 radiolabeled PET imaging probes, are chemically and structurally closer to their nonlabeled analogues. Introduction of fluorine atom into the molecule can have significant effect on physicochemical properties of investigated derivatives. With respect to chemistry of radiolabeling, introduction of carbon-11 label is typically performed using carbon-11 labeled CO_2 directly from the cyclotron or alternatively via methylation where $[^{11}C]CO_2$ is first converted to carbon-11 methyliodide or methyltriflate. Electrophilic or nucleophilic substitutions are two primary pathways for assembly of fluorine-18 labeled PET molecules. By far, nucleophilic substitution is more commonly employed due to demonstrated low specific activity obtained from the production of electrophilic fluoride. There is a continuous growth in new chemical methods which enable more varied and more efficient introduction of radionuclides, especially amongst Organic Chemistry community.
- While there exist several methods for lipophilicity determination of drugs, the determination of lipophilicity of the given radiotracer is most commonly accomplished using one of the two methods. A traditional shake-flask method [23, 24] in which the radiotracer is partitioned between *n*-octanol and phosphate buffer (pH 7.4) phases. The two phases are well shaken and separated and activity of each measured. The log $D_{7.4}$ is therefore determined as logarithmic value of the ratio of the amount of the radiotracer present in octanol over buffer phase. Another method employed is HPLC based, whereby the HPLC retention time is plotted against the log D values for several known compounds with established log D values to form a calibration curve. Under same HPLC conditions, retention time of the radiotracer cold reference is then established, and log D determined from the calibration curve.
- From the overview of PET radiotracers for imaging various mGlu receptors, it is also evident that the purpose of research study for which the imaging probe is required, will determine selection of the PET probe. For instance, fast washout from the brain leads to reduction in the image quality. On the other hand, long retention of the radiotracer in the brain, is not applicable to a carbon-11 radiolabeled probe. Furthermore, if

the occupancy of the receptor in the brain is low, radiotracers with high binding affinity to target and those prepared with high specific activity will afford better quality images.

4.2 PET Measurement

- Depending on the type of the experiment, at the beginning of the scan it may be required to additionally inject the blocker or displacer solution. For some studies blocker/displacer solution is injected before the solution of radiotracer, whereas occasionally, blocking/displacing is desired at a certain time point post-injection.
- Maintaining steady body temperature of the animal is a critical parameter in ensuring wellbeing of the animal and subsequently accuracy of the data obtained. Regulation of body temperature is practically most commonly accomplished by application of a heating blanket. The animal is placed on the blanket which regulates the temperature and the temperature of the animal is monitored with the rectal probe. Fueger et al. reported on an eight degree temperature drop (from 32 to 24 °C) if the animal was kept under anesthesia without the heating blanket (i.e., at ambient temperature) for period of 30 min [82]. The fatal temperature at which animal undergoes hypothermia and dies, is dependent upon specie and also time of exposure [83].

4.3 PET Imaging of mGlu$_5$: Preclinical Applications

- One critical aspect of preclinical studies is a selection of a good animal model. Discussion of various animal models exceeds the scope of this chapter and reader is directed to relevant literature as cited in examples provided.

References

1. Ametamey SM, Honer M, Schubiger PA (2008) Molecular Imaging with PET. Chem Rev 108:1501–1516
2. Sephton SM, Ametamey SM (2013) Positron emission tomography agents. Future Sci. doi:10.4155/EBO.12.504
3. Ziegler SI (2005) Positron emission tomography: principles, technology, and recent developments. Nucl Phys A 752:679–687
4. Zhang Y, Fox GB (2012) PET imaging for receptor occupancy: meditations on calculation and simplification. J Biomed Res 26: 69–76
5. Heiss W-D, Herholz K (2006) Brain receptor imaging. J Nucl Med 47:302–312
6. Conn PJ, Pin J-P (1997) Pharmacology and functions of metabotropic glutamate receptors. Annu Rev Pharmacol Toxicol 37:205–237
7. Masu M, Tanabe Y, Tsuchida K et al (1991) Sequence and expression of a metabotropic glutamate receptor. Nature 349: 760–765
8. Niswender CM, Conn PJ (2010) Metabotropic glutamate receptors: physiology, pharmacology, and disease. Annu Rev Pharmacol Toxicol 50:295–322
9. Pin J-P, Duvoisin R (1995) The metabotropic glutamate receptors: structure and functions. Neuropharmacology 34:1–26
10. Tanabe Y, Masu M, Ishii T et al (1992) A family of metabotropic glutamate receptors. Neuron 8:169–179
11. Nicoletti F, Bockaert J, Collingridge GL et al (2011) Metabotropic glutamate receptors: from the workbench to the bedside. Neuropharmacology 60:1017–1041

12. Pillai RLI, Tipre DN (2016) Metabotropic glutamate receptor 5 – a promising target in drug development and neuroimaging. Eur J Nucl Med Mol Imaging 43:1151–1170
13. Poels EM, Kegeles LS, Kantrowitz JT et al (2014) Imaging glutamate in schizophrenia: review of findings and implications for drug discovery. Mol Psychiatry 19:20–29
14. Sanchez-Pernaute R, Wang J-Q, Kuruppu D et al (2008) Enhanced binding of metabotropic glutamate receptor type 5 (mGluR5) PET tracers in the brain of parkinsonian primates. Neuroimage 42:248–251
15. Swanson CJ, Bures M, Johnson MP et al (2005) Metabotropic glutamate receptors as novel targets for anxiety and stress disorders. Nat Rev Drug Discov 4:131–144
16. Tokunaga M, Seneca N, Shin R-M et al (2009) Neuroimaging and physiological evidence for involvement of glutamatergic transmission in regulation of the striatal dopaminergic system. J Neurosci 29:1887–1896
17. Arsenault D, Coulombe K, Zhu A et al (2015) Loss of metabotropic glutamate receptor 5 function on peripheral benzodiazepine receptor in mice prenatally exposed to LPS. PLoS One 10:e0142093
18. Brownell A-L, Kuruppu D, Kil K-E et al (2015) PET imaging studies show enhanced expression of mGluR5 and inflammatory response during progressive degeneration in ALS mouse model expressing SOD1-G93A gene. J Neuroinflammation 12:217
19. Majo VJ, Prabhakaran J, Mann JJ et al (2013) PET and SPECT tracers for glutamate receptors. Drug Discov Today 18:173–184
20. Mu L, Ametamey SM (2014) Current radioligands for the PET imaging of metabotropic glutamate receptors. PET and SPECT of neurobiological systems. Springer, Berlin, Heidelberg, pp 409–443
21. Ribeiro MF, Paquet M, Cregan SP et al (2010) Group I metabotropic glutamate receptor signalling and its implication in neurological disease. CNS Neurol Disord Drug Targets 9:574–595
22. Eckelman WC, Kilbourn MR, Mathis CA (2006) Discussion of targeting proteins in vivo: in vitro guidelines. Nucl Med Biol 33:449–451
23. Kerns EH, Di L (2008) Drug-like properties: concepts, structure design and methods: ADME to toxicity optimization. Academic Press, London
24. Wenlock MC, Potter T, Barton P et al (2011) A method for measuring the lipophilicity of compounds in mixtures of 10. J Biomol Screen 16:348–355
25. Rutkowska E, Pajak K, Jóźwiak K (2013) Lipophilicity-methods of determination and its role in medicinal chemistry. Acta Pol Pharm 70:3–18
26. De Goeij JJM, Bonardi ML (2005) How do we define the concepts specific activity, radioactive concentration, carrier, carrier-free and no-carrier-added? J Radioanal Nucl Chem 263:13–18
27. Eckelman WC, Mathis CA (2006) Targeting proteins in vivo: in vitro guidelines. Nucl Med Biol 33:161–164
28. Ferraguti F, Crepaldi L, Nicoletti F (2008) Metabotropic glutamate 1 receptor: current concepts and perspectives. Pharmacol Rev 60:536–581
29. Steckler T, Oliveira AFM, Van Dyck C et al (2005) Metabotropic glutamate receptor 1 blockade impairs acquisition and retention in a spatial water maze task. Behav Brain Res 164:52–60
30. Zanotti-Fregonara P, Barth VN, Zoghbi SS et al (2013) 11C-LY2428703, a positron emission tomographic radioligand for the metabotropic glutamate receptor 1, is unsuitable for imaging in monkey and human brains. EJNMMI Res 3:47
31. Toyohara J, Sakata M, Oda K et al (2013) Initial human PET studies of metabotropic glutamate receptor type 1 ligand 11C-ITMM. J Nucl Med 54:1302–1307
32. Toyohara J, Sakata M, Fujinaga M et al (2013) Preclinical and the first clinical studies on [11C]ITMM for mapping metabotropic glutamate receptor subtype 1 by positron emission tomography. Nucl Med Biol 40:214–220
33. Zanotti-Fregonara P, Xu R, Zoghbi SS et al (2016) The PET radioligand 18F-FIMX images and quantifies metabotropic glutamate receptor 1 in proportion to the regional density of its gene transcript in human brain. J Nucl Med 57:242–247
34. Ametamey SM, Kessler LJ, Honer M et al (2006) Radiosynthesis and preclinical evaluation of 11C-ABP688 as a probe for imaging the metabotropic glutamate receptor subtype 5. J Nucl Med 47:698–705
35. Ametamey SM, Treyer V, Streffer J et al (2007) Human PET studies of metabotropic glutamate receptor subtype 5 with 11C-ABP688. J Nucl Med 48:247–252
36. Mu L, Schubiger AP, Ametamey SM (2010) Radioligands for the PET imaging of metabotropic glutamate receptor subtype 5 (mGluR5). Curr Top Med Chem 10:1558–1568
37. Gasparini F, Lingenhöhl K, Stoehr N et al (1999) 2-Methyl-6-(phenylethynyl)-pyridine (MPEP), a potent, selective and systemically

active mGlu5 receptor antagonist. Neuropharmacology 38:1493–1503

38. Shetty HU, Zoghbi SS, Sime FG et al (2008) Radiodefluorination of 3-fluoro-5-(2-(2-[18F](fluoromethyl)-thiazol-4-yl)ethynyl) benzonitrile ([18F]SP203), a radioligand for imaging brain metabotropic glutamate subtype-5 receptors with positron emission tomography, occurs by glutathionylation in rat brain. J Pharmacol Exp Ther 327:727–735
39. Siméon FG, Brown AK, Zoghbi SS et al (2007) Synthesis and simple ^{18}F-labeling of 3-fluoro-5-(2-(2-(fluoromethyl)thiazol-4-Yl) ethynyl)benzonitrile as a high affinity radioligand for imaging monkey brain metabotropic glutamate subtype-5 receptors with positron emission tomography. J Med Chem 50: 3256–3266
40. Kimura Y, Simeon FG, Zoghbi SS et al (2012) Quantification of metabotropic glutamate subtype 5 receptors in the brain by an equilibrium method using 18F-SP203. NeuroImage 59: 2124–2130
41. Wong DF, Waterhouse R, Kuwabara H et al (2013) 18F-FPEB, a PET radiopharmaceutical for quantifying metabotropic glutamate 5 receptors: a first-in-human study of radiochemical safety, biokinetics, and radiation dosimetry. J Nucl Med 54:388–396
42. Wang J-Q, Tueckmantel W, Zhu A et al (2007) Synthesis and preliminary biological evaluation of 3-[(18)F]fluoro-5-(2-pyridinylethynyl)benzonitrile as a PET radiotracer for imaging metabotropic glutamate receptor subtype 5. Synapse 61:951–961
43. Sephton SM, Mu L, Dragic M et al (2013) Synthesis and in vitro evaluation of E- and Z-geometrical isomers of PSS232 as potential metabotropic glutamate receptors subtype 5 (mGlu 5) binders. Synth 45:1877–1885
44. Sephton SM, Herde AM, Mu L et al (2014) Preclinical evaluation and test–retest studies of [18F]PSS232, a novel radioligand for targeting metabotropic glutamate receptor 5 (mGlu5). Eur J Nucl Med Mol Imaging 42:128–137
45. Müller Herde A, Keller C, Milicevic Sephton S et al (2015) Quantitative positron emission tomography of mGluR5 in rat brain with [^{18} F]PSS232 at minimal invasiveness and reduced model complexity. J Neurochem 133: 330–342
46. Patil ST, Zhang L, Martenyi F et al (2007) Activation of mGlu2/3 receptors as a new approach to treat schizophrenia: a randomized phase 2 clinical trial. Nat Med 13:1102–1107
47. Wang J-Q, Zhang Z, Kuruppu D et al (2012) Radiosynthesis of PET radiotracer as a prodrug for imaging group II metabotropic glutamate receptors in vivo. Bioorg Med Chem Lett 22:1958–1962
48. Waterhouse RN, Schmidt ME, Sultana A et al (2003) Evaluation of [3H]LY341495 for labeling group II metabotropic glutamate receptors in vivo. Nucl Med Biol 30:187–190
49. Celen S, Koole M, Alcazar J et al (2012) Preliminary biological evaluation of [11C] JNJ42491293 as a radioligand for PET imaging of mGluR2 in brain. J Nucl Med 53:286–286
50. Laere KV, Koole M, Hoon J d et al (2012) Biodistribution, dosimetry and kinetic modeling of [11C]JNJ-42491293, a PET tracer for the mGluR2 receptor in the human brain. J Nucl Med 53:355–355
51. Andrés J-I, Alcázar J, Cid JM et al (2012) Synthesis, evaluation, and radiolabeling of new potent positive allosteric modulators of the metabotropic glutamate receptor 2 as potential tracers for positron emission tomography imaging. J Med Chem 55:8685–8699
52. Fuchigami T, Nakayama M, Yoshida S (2015) Development of PET and SPECT probes for glutamate receptors. Sci World J 2015:716514
53. Wang J-Q, Kuruppu D, Brownell A-L (2008) Radiosynthesis of (±)-N-(P-[11C]tolyl)-7-(hydroxyimino)cyclopropa[b]chromen-1a-carboxamide ([11C]methyl-PHCCC) as a specific mGluR4 PET ligand. J Nucl Med 49:288
54. Engers DW, Niswender CM, Weaver CD et al (2009) Synthesis and evaluation of a series of heterobiarylamides that are centrally penetrant metabotropic glutamate receptor 4 (mGluR4) positive allosteric modulators (PAMs). J Med Chem 52:4115–4118
55. Fujinaga M, Yamasaki T, Nengaki N et al (2016) Radiosynthesis and evaluation of 5-methyl-N-(4-[11C]methylpyrimidin-2-Yl)-4-(1H-pyrazol-4-Yl)thiazol-2-amine ([11C] ADX88178) as a novel radioligand for imaging of metabotropic glutamate receptor subtype 4 (mGluR4). Bioorg Med Chem Lett 26: 370–374
56. Kil K-E, Zhang Z, Jokivarsi K et al (2013) Radiosynthesis of N-(4-chloro-3-[11C]methoxyphenyl)-2-picolinamide ([11C] ML128) as a PET radiotracer for metabotropic glutamate receptor subtype 4 (mGlu4). Bioorg Med Chem 21:5955–5962
57. Kil K-E, Poutiainen P, Zhang Z et al (2016) Synthesis and evaluation of N-(methylthiophenyl)picolinamide derivatives as PET radioligands for metabotropic glutamate receptor subtype 4. Bioorg Med Chem Lett 26:133–139

58. Nakamura M, Kurihara H, Suzuki G et al (2010) Isoxazolopyridone derivatives as allosteric metabotropic glutamate receptor 7 antagonists. Bioorg Med Chem Lett 20:726–729
59. Yamasaki T, Kumata K, Yui J et al (2013) Synthesis and evaluation of [11C]MMPIP as a potential radioligand for imaging of metabotropic glutamate 7 receptor in the brain. EJNMMI Res 3:54
60. Hutchins GD, Miller MA, Soon VC et al (2008) Small animal PET imaging. ILAR J 49: 54–65
61. Blokland JA, Trindev P, Stokkel MP et al (2002) Positron emission tomography: a technical introduction for clinicians. Eur J Radiol 44:70–75
62. Alstrup AKO, Smith DF (2013) Anaesthesia for positron emission tomography scanning of animal brains. Lab Anim 47:12–18
63. Hildebrandt IJ, Su H, Weber WA (2008) Anesthesia and other considerations for in vivo imaging of small animals. ILAR J 49:17–26
64. Kuntner C, Stout D (2014) Quantitative preclinical PET imaging: opportunities and challenges. Front Phys 2:12
65. Morris ED, Enders CJ, Schmidt KC et al (2004) Kinetic modeling in positron emission tomography. In: Wernick MN, Aarsvold JN (eds) Emission tomography, 1st edn. Elsevier, Amsterdam, pp 499–540
66. Schmidt KC, Turkheimer FE (2004) Kinetic modeling in positron emission tomography. Q J Nucl Med 46:70–85
67. Valk PE, Bailey DL, Townsend DW, Maisey MN (2003) Positron emission tomography: basic science and clinical practice. Springer, London, pp 147–149
68. Elmenhorst D, Aliaga A, Bauer A et al (2012) Test-retest stability of cerebral mGluR5 quantification using [11C]ABP688 and positron emission tomography in rats. Synapse 66:552–560
69. Choi H, Kim YK, Oh SW et al (2014) In vivo imaging of mGluR5 changes during epileptogenesis using [11C]ABP688 PET in pilocarpine-induced epilepsy rat model. PLoS One 9:e92765
70. Zimmer ER, Parent MJ, Leuzy A et al (2015) Imaging in vivo glutamate fluctuations with [(11)C]ABP688: a GLT-1 challenge with ceftriaxone. J Cereb Blood Flow Metab 35: 1169–1174
71. Mathews WB, Kuwabara H, Stansfield K et al (2014) Dose-dependent, saturable occupancy of the metabotropic glutamate subtype 5 receptor by fenobam as measured with [^{11}C] ABP688 PET imaging. Synapse 68:565–573
72. Miyake N, Skinbjerg M, Easwaramoorthy B et al (2011) Imaging changes in glutamate transmission in vivo with the metabotropic glutamate receptor 5 tracer [11C] ABP688 and N-acetylcysteine challenge. Biol Psychiatry 69:822–824
73. Sandiego CM, Nabulsi N, Lin S-F et al (2013) Studies of the metabotropic glutamate receptor 5 radioligand [^{11} C]ABP688 with N-acetylcysteine challenge in rhesus monkeys. Synapse 67:489–501
74. DeLorenzo C, Kumar JSD, Mann JJ et al (2011) In vivo variation in metabotropic glutamate receptor subtype 5 binding using positron emission tomography and [11C]ABP688. J Cereb Blood Flow Metab 31:2169–2180
75. DeLorenzo C, Sovago J, Gardus J et al (2015) Characterization of brain mGluR5 binding in a pilot study of late-life major depressive disorder using positron emission tomography and [11C]ABP688. Transl Psychiatry 5:e693
76. Dubois JM, Rousset OG, Rowley J et al (2016) Characterization of age/sex and the regional distribution of mGluR5 availability in the healthy human brain measured by high-resolution [11C]ABP688 PET. Eur J Nucl Med Mol Imaging 43:152–162
77. DeLorenzo C, DellaGioia N, Bloch M et al (2015) In vivo ketamine-induced changes in [11C]ABP688 binding to metabotropic glutamate receptor subtype 5. Biol Psychiatry 77: 266–275
78. Martinez D, Slifstein M, Nabulsi N et al (2014) Imaging glutamate homeostasis in cocaine addiction with the metabotropic glutamate receptor 5 positron emission tomography radiotracer [11C]ABP688 and magnetic resonance spectroscopy. Biol Psychiatry 75:165–171
79. Milella MS, Marengo L, Larcher K et al (2014) Limbic system mGluR5 availability in cocaine dependent subjects: a high-resolution PET [11C]ABP688 study. Neuroimage 98: 195–202
80. Kågedal M, Cselényi Z, Nyberg S et al (2013) A positron emission tomography study in healthy volunteers to estimate mGluR5 receptor occupancy of AZD2066 — estimating occupancy in the absence of a reference region. NeuroImage 82:160–169
81. Leuzy A, Zimmer ER, Dubois J et al (2016) In vivo characterization of metabotropic glutamate receptor type 5 abnormalities in behavioral variant FTD. Brain Struct Funct 221: 1387–1402
82. Fueger BJ, Czernin J, Hildebrandt I et al (2006) Impact of animal handling on the results of 18F-FDG PET studies in mice. J Nucl Med 47:999–1006
83. Gordon CJ (1993) Temperature regulation in laboratory rodents. Cambridge University Press, Cambridge. doi:10.1017/CBO9780511565595

Chapter 4

NMR Spectroscopy of Brain Glutamate Function

Jun Shen

Abstract

Glutamate has fundamentally important functions in the CNS. It is the principal excitatory neurotransmitter as well as a key metabolite linking carbon and nitrogen metabolism. The dual roles of glutamate as a neurotransmitter and metabolite are intricately connected. Its unusually high concentration and rapid turnover in brain make it accessible to both proton and ^{13}C NMR spectroscopy methods. This has enabled researchers to study the various functions of glutamate in basic neuroscience and brain disorders noninvasively and in vivo. Here, we provide an overview of proton and heteronuclear nuclear magnetic resonance (NMR) spectroscopy techniques and their applications to studying glutamate metabolism and glutamatergic neurotransmission in brain with emphasis placed on recent progress.

Key words Magnetic resonance, Radiofrequency, Isotope labeling, Neuroenergetics, Glutamate–glutamine neurotransmitter cycle, Metabolic modeling

1 Introduction

1.1 Glutamate Functions as a Neurotransmitter and a Metabolite in Brain

A prominent feature of amino acid neurotransmitters such as glutamate (Glu) is their high concentration in neutral tissues. Glutamate has been known to be one of the most abundant intracellular amino acids in the brain of mammals. The free Glu pool size in majority of brain regions is very high (5–10 mmol/L) [1], making it readily detectable using various in vivo NMR spectroscopy techniques [2–5]. Unlike other neurotransmitters, amino acid neurotransmitters such as Glu are ubiquitous and participate in numerous biochemical processes including intermediary metabolism and protein turnover. In particular, Glu serves as a source of energy and ammonia, linking the metabolism of carbon and nitrogen [6]. Because of the crucial roles of Glu in glutamatergic processes and energy metabolism NMR spectroscopy has been widely used to study the functions of Glu in basic neuroscience as well as in various brain disorders [7–12].

The neurotransmission and metabolism of Glu are intimately connected. Although Glu is rapidly synthesized from glucose in

Sandrine Parrot and Luc Denoroy (eds.), *Biochemical Approaches for Glutamatergic Neurotransmission*, Neuromethods, vol. 130, DOI 10.1007/978-1-4939-7228-9_4, © Springer Science+Business Media LLC 2018

neural tissues the biochemical process for replenishing the neurotransmitter Glu after Glu release predominantly relies on the glutamate–glutamine cycle between neurons and astroglial cells (see Fig. 1a and b; [1, 13–17]). As a zwitterionic molecule, Glu cannot cross cell membranes via self-diffusion. It is well understood that glial uptake of neurotransmitter Glu plays crucial roles in regulating the extracellular concentration of Glu in the brain [18]. A major role for Glu transporters is to limit the concentration of free Glu in the extracellular space, preventing excessive stimulation of various Glu receptors [19, 20]. Excessive activation of Glu receptors by Glu can lead to a variety of severe pathological conditions due to excitotoxicity. Inactivation of the released neurotransmitter Glu is primarily by uptake via high-affinity Na^+-dependent Glu transport systems. Glutamate taken up into astroglial cells is either converted into glutamine (Gln) by glutamine synthetase, which is exclusively located in glial cells [1, 15, 21] or oxidized by its assimilation into the tricarboxylic acid (TCA) cycle operating in the mitochondria of astroglial cells. Glutamine, once formed, is readily discharged from glial cells. The released Gln then enters nerve terminals by Gln low affinity transport systems or by simple diffusion. There, Gln is reconverted into Glu by phosphate-activated glutaminase. The replenished Glu pool can be again utilized for Glu vesicular release or assimilated into neuronal TCA cycle. Although this glutamate–glutamine neurotransmitter cycle was proposed several decades ago regarding it as a significant metabolic flux has been an important recent development. Over the past two decades, because of the rapid technological advances of in vivo ^{13}C and ^{15}N NMR spectroscopy technology, the glutamate–glutamine cycling flux was quantified in vivo first in anesthetized rat brain [22–24] and then in human brain [10, 16, 25, 26]. The recent in vivo NMR spectroscopy studies of Glu metabolism and neurotransmission have established that the glutamate–glutamine neurotransmitter cycle between glutamatergic neurons and astroglia is a major metabolic flux, reflecting synaptic Glu release [27, 28]. A large amount of recent in vivo NMR spectroscopy evidence strongly supports the notion that the glutamate–glutamine cycling flux is directly coupled to neuroenergetics [10, 29].

The various enzymes involved in Glu synthesis, degradation, and exchange mainly with α-ketoglutarate control the overall concentration of Glu measured by proton NMR spectroscopy [2] of brain. The activities of these enzymes are constantly adjusted by modulation of the concentration of various ligands including allosteric effectors. Coordination among different compartments, especially the neuronal and astroglial compartments via the glutamate–glutamine neurotransmitter cycle, also regulates the total Glu (as well as Gln) concentrations [30]. Important metabolic couplings exist between various cells and subcellular compartments through the use of shared substrates and the exchange of several

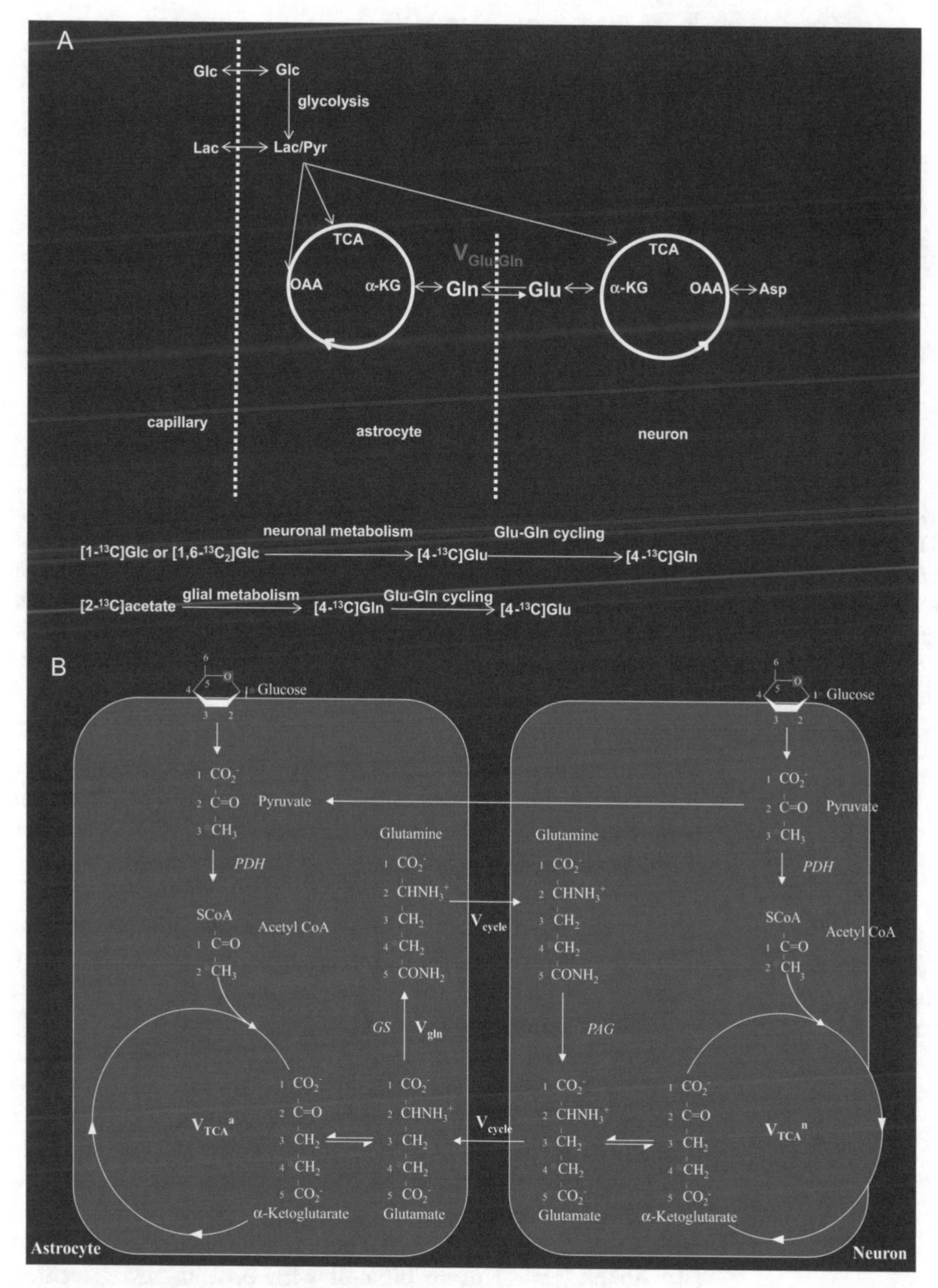

Fig. 1 (**a**) Schematic illustration of the glutamate–glutamine cycle between neurons and astroglia and glucose metabolism. Released neurotransmitter glutamate is transported from the synaptic cleft by surrounding astroglial end processes. In astroglia, glutamate is converted into glutamine by glutamine synthetase. Glutamine is then released by the astroglia, transported into the neurons, and converted back into glutamate by glutaminase, which completes the cycle. *Glc* glucose. (**b**) The glutamate–glutamine cycle with positions of molecules explicitly labeled. *Asp* aspartate, *Pyr/Lac* pyruvate/lactate, *OAA* oxaloacetate, *α-KG* α-ketoglutarate, *Glu* glutamate, *Gln* glutamine, *PDH* pyruvate dehydrogenase, *GS* glutamine synthase, *PAG* phosphate-activated glutaminase, *TCA* tricarboxylic acid cycle, V_{gln} glutamine synthesis flux, V_{TCA}^{a} astroglial tricarboxylic acid cycle flux, V_{cycle} glutamate–glutamine cycling flux, V_{TCA}^{n} neuronal tricarboxylic acid cycle flux

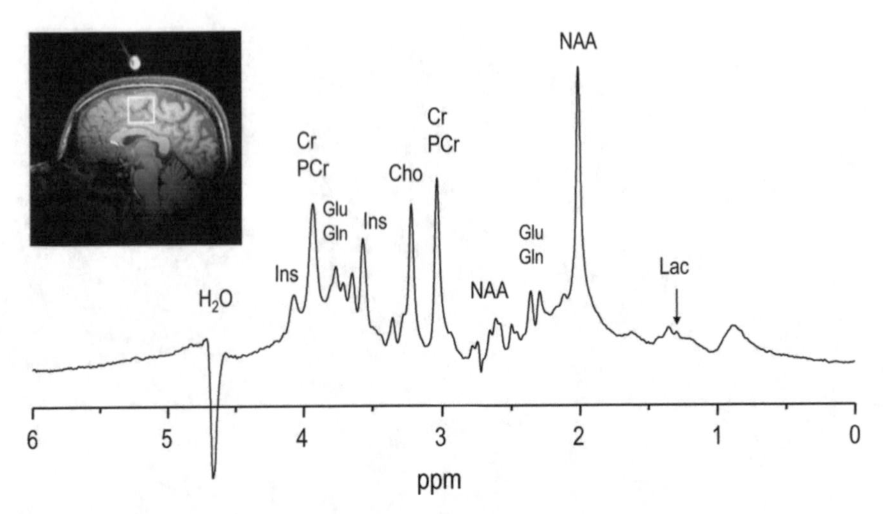

Fig. 2 In vivo short echo time proton NMR spectrum from the frontoparietal region in the human brain acquired at 3 T. The spectrum was acquired with a PRESS sequence (echo time = 26 ms, relaxation delay = 2 s, number of averages = 64). The location of the spectroscopy voxel (3 cm × 3 cm × 3 cm) is marked with a square on a scout image shown in the inset. *Ins* myo-inositol, *Cr* creatine, *PCr* phosphocreatine, *Glu* glutamate, *Gln* glutamine, *Cho* choline-containing compounds, *NAA* N-acetylaspartate, *Lac* lactate. The overlapping glutamate and glutamine C2 protons at ~3.75 ppm and C4 protons at ~2.40 ppm are labeled. Reproduced from Choi et al., 2004 [32]

metabolic intermediates such as Glu, Gln, and γ-amino butyric acid (GABA) [1, 10]. Because Glu also acts as the major excitatory neurotransmitter in the central nervous system (CNS), the neurotransmission of Glu is intimately connected to its metabolism.

1.2 Proton NMR Spectroscopy of Brain Glutamate Concentration

The proton NMR spectrum of Glu is entirely composed of scalarly coupled resonances [2–5, 31]. Glutamate and glutamine are often referred to collectively as Glx because, at magnetic field strengths accessible to most clinical studies (e.g., 1.5 and 3 T), NMR spectroscopy signals of Glu and Gln strongly overlap with each other at all of their resonant positions in the commonly used short echo time spectra [2, 31]. A short echo time (26 ms; [32]) single voxel proton NMR spectrum [4, 5] of human brain is displayed in Fig. 2. Moreover, due to strong scalar coupling effects, their proton NMR spectra appear different at different echo times and different magnetic field strengths [31, 33]. These factors make their observation and quantification more difficult with proton NMR techniques than many other resonance signals such as those of N-acetylaspartate, creatine, and choline-containing compounds. Precisely because of the difficulties and importance of accurately measuring brain Glu concentration much effort has been devoted to developing novel NMR spectroscopy techniques to better measure Glu (and Gln, as the two are intimately connected via the glutamate–glutamine neurotransmitter cycle) by reducing or eliminating spectral interferences from overlapping metabolite and macromolecule signals and by improving quantification using internal and external standards. Accompanying the rapid methodological advances proton NMR spectroscopy of Glu has evolved into a major tool for studying glutamatergic function in the CNS (e.g., [34–38]).

1.3 Heteronuclear NMR Spectroscopy of Brain Glutamate Metabolism and Neurotransmission

The “static” Glu concentration measured by proton NMR spectroscopy is a result of complex interplay between Glu anabolism and its catabolism. After administering exogenous ^{13}C-labeled glucose (e.g., 1-[^{13}C]-glucose, 2-[^{13}C]-glucose or uniformly labeled glucose) ^{13}C labels are readily incorporated into Glu, and subsequently into Gln via the glutamate–glutamine neurotransmitter cycle (Fig. 1; [10]). The rapid exchange between Glu and the TCA cycle intermediate α-ketoglutarate links Glu to oxidative metabolism in both neurons and astrocytes. In vivo ^{13}C NMR spectroscopy measures the rate of oxidative metabolism based on rapid labeling of Glu after administering ^{13}C-labeled substrates such as [1-^{13}C]-glucose or [1,6-$^{13}C_2$]-glucose or [U-$^{13}C_6$]-glucose [39]. Because of the large size of the brain Glu pool, which is predominantly located in glutamatergic neurons, Glu effectively acts as a ^{13}C label trapping pool for the neuronal TCA cycle through rapid exchange with mitochondrial α-ketoglutarate. The kinetics of ^{13}C label incorporation into neuronal Glu contains the quantitative information of the neuronal TCA cycle flux rate. Therefore, in vivo ^{13}C NMR spectroscopy [31] of Glu allows noninvasive assessment of the predominantly neuronal oxidative metabolism in brain. An example of aliphatic ^{13}C NMR spectrum acquired from nonhuman primate brain [40] is shown in Fig. 3.

Investigations of glucose metabolism in brain by in vivo ^{13}C NMR spectroscopy found rapid and highly significant labeling of Gln [41], which predominantly resides in the metabolically less active astroglial cells. This surprising result indicates rapid transfer of ^{13}C labels from the predominantly neuronal Glu compartment to the predominantly astroglial Gln compartment. Contrary to the concept of a small, metabolically inactive neurotransmitter Glu pool a large number of in vivo NMR spectroscopy studies of human and animal brains using various ^{13}C-labeling strategies have consistently shown that the glutamate–glutamine neurotransmitter cycle is a highly significant metabolic flux between neurons and astroglia. Rapid and significant labeling of Gln C2–C4 has been consistently reproduced by all studies employing either direct ^{13}C detection or the more sensitive indirect proton detection after infusing [1-^{13}C]-glucose or [1,6-$^{13}C_2$]-glucose, which predominantly labels neuronal Glu due to the high energy demand of neuronal activities. Similarly, rapid labeling of Gln C5 was observed using [2-^{13}C]-glucose or [2,5-$^{13}C_2$]-glucose infusion and direct ^{13}C detection of carboxylic/amide carbons [42, 43]. ^{15}N NMR spectroscopy [2] studies of $^{15}NH_3$ labeling of Glu and Gln in a hyperammonemia model are also consistent with a significant

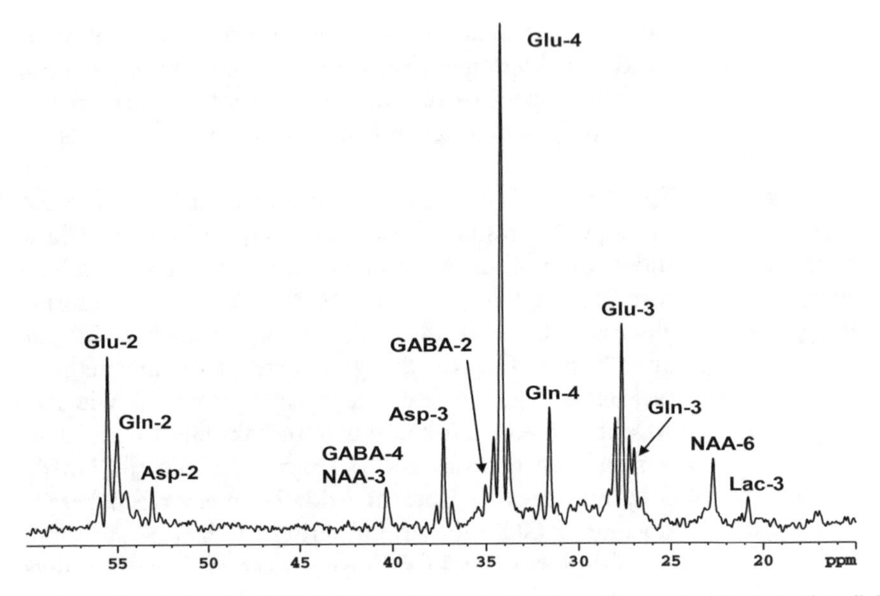

Fig. 3 Three-dimensionally localized ^{13}C NMR spectrum acquired from the monkey brain in the aliphatic ^{13}C region of 15–60 ppm with infusion of [1-^{13}C]-glucose. The spectrum was accumulated from 80 to 140 min after the start of [1-^{13}C]-glucose infusion. Homonuclear ^{13}C–^{13}C couplings from glutamate C3–C4, glutamine C3–C4 and aspartate C2–C3 were observed. ^{13}C labels progressively replaces glutamate and glutamine C4, C3 and C2 as well as aspartate C3 and C2 as the tricarboxylic acid cycle turns. Reproduced from Li et al. (2005) [40]. *Glu* glutamate, *Gln* glutamine, *Asp* aspartate, *Lac* lactate, *NAA* *N*-acetylaspartate, *GABA* γ-aminobutyrate

trafficking between neurons and astroglial cells via the glutamate–glutamine neurotransmitter cycle [9, 23]. While infusion of [2-^{13}C]-glucose or [2,5-$^{13}C_2$]-glucose NMR spectroscopy data acquisition follows the ^{13}C labels from Glu to Gln, the Gln to Glu half of the glutamate–glutamine neurotransmitter cycle can be directly probed by utilizing either the anaplerotic pathway or the glia-specific substrate acetate. Rapid transfer of ^{13}C labels from Gln C3 and C2 to Glu C3 and C2 was detected in vivo after administering [2-^{13}C]-glucose or [2,5-$^{13}C_2$]-glucose [44]. When the glia-specific substrate acetate is administered there is a rapid labeling of predominantly glial Gln followed by rapid transfer of ^{13}C labels from glial Gln to neuronal Glu [26, 45, 46]. In vivo ^{13}C- and ^{15}N NMR spectroscopy studies have proved that glutamate–glutamine cycling is a major metabolic pathway with a flux rate substantially greater than those suggested by early in vitro studies of cell cultures and brain slices.

In addition to static concentration and metabolic fluxes, NMR spectroscopy can also measure certain chemical reactions catalyzed by specific enzymes in vivo [47]. The aspartate aminotransferase reaction catalyzes the interconversion of Glu and oxaloacetate with α-ketoglutarate and aspartate at near equilibrium conditions: glutamate + oxaloacetate ↔ α-ketoglutarate + aspartate. Both cytoplasmic and mitochondrial aspartate aminotransferase isozymes are

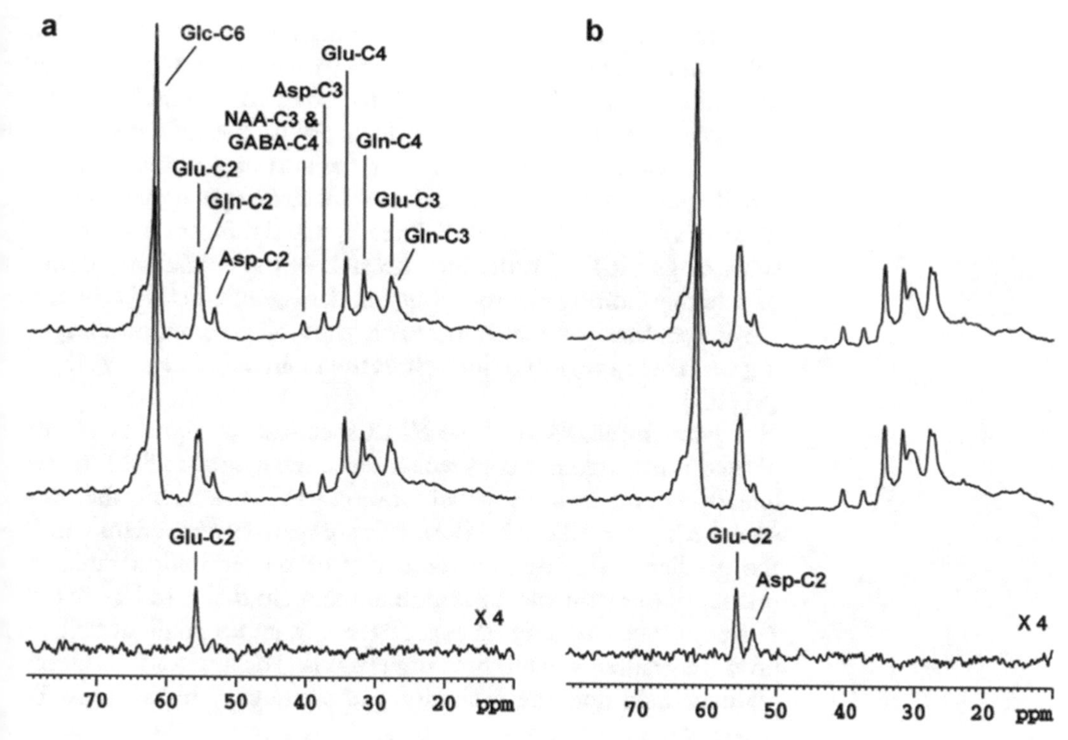

Fig. 4 Magnetization transfer effect catalyzed by aspartate aminotransferase reaction glutamate + oxaloacetate ↔ α-ketoglutarate + aspartate in brain. (**a**) α-Ketoglutarate ↔ glutamate half reaction with continuous ^{13}C radiofrequency saturation pulse centered on α-ketoglutarate C2 at 206.0 ppm (**b**) oxaloacetate ↔ aspartate half reaction with continuous ^{13}C radiofrequency saturation pulse centered on oxaloacetate C2 at 201.3 ppm. *Upper traces*: control spectra; *middle traces*: magnetization transfer spectra; *lower traces*: difference spectra with intensity scale: ×4. All spectra were acquired from anesthetized adult rats at 11.7 T with relaxation delay = 10 s and number of scans = 512. *Asp* aspartate, *GABA* γ-aminobutyric acid, *Glc* glucose, *Glu* glutamate, *NAA N*-acetylaspartate. Reproduced from Xu et al. (2009) [50] with permission from Elsevier

components of the malate-aspartate shuttle, which maintains cytosolic and mitochondrial redox states in the brain [48]. The magnetization transfer effect [2] catalyzed by aspartate aminotransferase reaction was discovered and quantified in vivo using ^{13}C NMR spectroscopy ([49, 50]; see Fig. 4).

2 Materials and Instrumentation

2.1 Hardware for In Vivo NMR Spectroscopy

In vivo NMR spectroscopy is performed using magnetic resonance imaging (MRI) spectrometers or scanners [5, 31]. A clinical MRI system comprises a magnet, shim coils, gradient coils, a radiofrequency transmitter/receiver system and a computer to perform the function of equipment control, acquisition, processing, and storage of data. Essentially all modern in vivo NMR spectroscopy work relies on super-conducting magnets which produce static magnetic fields that are highly uniform in space and stable in time. Correcting field inhomogeneities using additional weaker static magnetic fields generated by specialized shim coils to "shim" or "tweak" the static field generated by the super conductor as well as field distortions caused by other structures including the body per se is essential for NMR spectroscopy. This is because spectroscopy signals such as those of Glu are easily overwhelmed by any significant background static magnetic field variations. Because of the crucial importance of a highly homogeneous background magnetic field for in vivo NMR spectroscopy many automatic shimming techniques targeting a restricted region of interest in the brain have been developed [51, 52].

Spatial localization of the NMR spectroscopy signal is accomplished using gradient coils which cause the magnetic field to vary linearly as a function of spatial location [31]. As a result, the resonant frequencies of the NMR signals are spatially dependent within the gradient. Varying the frequency of the excitation radiofrequency pulses controls the region in the brain that is to be excited. A spectrometer or scanner must have x, y, and z gradient coils to produce gradients in all three dimensions. The application of each gradient field and the excitation radiofrequency pulses must be properly sequenced, or timed. By applying a gradient in the z direction, for example, one can change the resonant frequency required to excite a slab perpendicular to the z direction. Therefore, by sequentially exciting x, y, and z slabs the pulse sequence can be designed to only permit signals originated at the intersection of the three perpendicular slabs to emerge in the acquired spectrum, leading to the so-called single voxel spectroscopy [53]. Alternatively, spatial location of a slab can be imaged as in MRI except that each pixel contains a NMR spectrum [54, 55].

While most proton NMR spectroscopy can use standard MRI scanners and radiofrequency coils made for MRI applications, in order to directly detect heteronuclear NMR signal the hardware system must be capable of handling two frequencies. Ideally, spectrometers with two independent and programmable channels are preferred for heteronuclear NMR spectroscopy experiments. Many commercial MRI scanners, however, are only equipped with a single broadband channel. In this case, a stand-alone proton decoupler must be used to provide radiofrequency pulses for Nuclear Overhauser Enhancement (NOE) and proton decoupling [43]. A radiofrequency coil assembly that can operate at both proton NMR frequency and the frequency of a heteronucleus such as ^{13}C is also needed. Often, the heteronucleus NMR coil is a simple, circular surface coil that can be placed adjacent to the area of interest for optimal sensitivity although development of more sophisticated phased-array ^{13}C coils is also being pursued. For direct heteronuclear NMR spectroscopy work, the proton coil generally consists of one or two larger surface coils whose sensitive volume is always

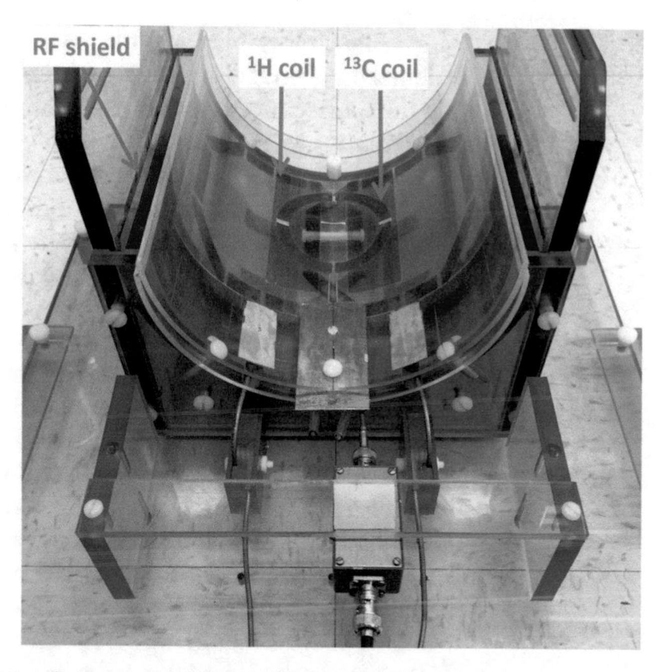

Fig. 5 A radiofrequency coil assembly for direct ^{13}C NMR spectroscopy at 7 T. The ^{13}C coil, the proton coil (two overlapping octagonal loops), and the radiofrequency shield are mounted on the bottom surface of the upper, middle, and lower tubes, respectively. Each proton loop has a single-tuned proton cable trap constructed using the RG-316 cable. A ^{13}C/^{1}H dual-tuned cable trap, built inside a RF shielded box, is connected to the ^{13}C coil. Adapted from Li et al. (2016) [57]

made similar or somewhat larger than that of the heteronuclear radiofrequency coil [56, 57]. For a typical radiofrequency coil assembly for ^{13}C NMR spectroscopy of human brain, see Fig. 5. For indirect ^{13}C NMR spectroscopy the protons attached to ^{13}C are detected instead for better spatial localization and sensitivity. Then, the configuration of the proton and ^{13}C coils is reversed [58].

2.2 Human Subject Preparation

The preparation of human subjects for most proton NMR spectroscopy is similar to that of structural or functional MRI. Nevertheless, preparation of a patient for proton NMR spectroscopy is generally more demanding than for routine structural or functional MRI because most of the spectroscopy scans are performed after locating the desired voxel(s) on MRI images. Any patient motion occurring during NMR spectroscopy scans not only degrades spectral editing, water suppression, and spectral resolution, but may change the position of the patient being scanned such that the acquired NMR spectra are obtained from a location different from the intended one. Good cooperation (sometimes even sedation) of subjects may be necessary to guarantee success for certain disease studies.

Since the natural abundance of ^{13}C is only 1.1% exogenous ^{13}C labeled substrates need to be administered to make in vivo ^{13}C NMR spectroscopy feasible. Because glucose is the primary substrate for cerebral energy metabolism under normal physiological conditions, it is the most commonly used substrate in ^{13}C NMR spectroscopy studies. In addition to ^{13}C-labeled glucose, other substrates, such as acetate, β-hydroxybutyrate, ethanol, and lactate, can also be used to take advantage of their different metabolic pathways [59, 60].

^{13}C-labeled glucose can be administered either intravenously or orally although intravenous (i.v.) administration is generally preferred for quantitative metabolic flux measurement because i.v. infusion protocol can be designed to quickly raise the ^{13}C fractional enrichment of glucose to a high and stable value, providing a close to ideal input function for subsequent quantitative metabolic analysis [10, 61]. Administering glucose also raises plasma glucose to hyperglycemic levels. In response, endogenous insulin is secreted from the pancreas to stimulate glucose uptake in the liver, an undesirable process that increases the cost of in vivo ^{13}C NMR spectroscopy of brain. In order to suppress this natural but unwanted response, somatostatin, a hormone that inhibits insulin release, is often co-infused. For i.v. infusion studies, two i.v. lines are typically placed in the antecubital vein of each arm, one for infusion of ^{13}C labeled substrates and the other for periodic blood sampling to monitor total plasma glucose level and measure ^{13}C fractional enrichment of plasma glucose. In comparison, the input plasma glucose ^{13}C fractional enrichment rises more slowly following oral administration, making extraction of metabolic flux rates more complicated and therefore less reliable [62]. For oral glucose administration, one i.v. line is still required for periodic blood withdrawing to measure plasma glucose concentration and ^{13}C isotopic enrichment of blood glucose.

2.3 Animal Preparation

The description here uses rats as an example although the preparation of other types of animals is often similar and can be found in the literature [63]. Either commercially available or in-house developed animal holders, often with integrated radiofrequency coil assembly [64], are used for positioning, radiofrequency probe tuning, physiological monitoring, and intervention. For ^{13}C NMR spectroscopy, one femoral vein is cannulated for the continuous infusion of ^{13}C-labeled substrate. A second femoral vein could be cannulated for administration of other chemicals such as i.v. anesthetics. One artery is also cannulated for periodic sampling of arterial blood to measure plasma gases and glucose concentrations using a blood analyzer. ^{13}C fractional enrichment of arterial blood can also be measured. Rectal temperature is monitored and maintained at normal body temperature using an external heat source. End-tidal CO_2, tidal pressure of ventilation, and heart rate are usually monitored throughout the study. After surgery, anesthesia is maintained either by i.v. infusion of α-chloralose (initial dose: 80 mg/kg supplemented with a constant infusion of 26.7 mg/kg/h throughout the experiment), or oral ventilation with a mixture of 70% N_2O, 30% O_2 and, for example, 1.5% isoflurane [22, 46, 65] or other anesthetics.

2.4 ^{13}C-Labeled Substrates

One of the most important considerations in ^{13}C NMR spectroscopy is the choice of ^{13}C-labeled substrate as this largely determines the metabolic pathways to be measured. Given that glucose is the primary fuel source of cerebral energy metabolism under most situations, glucose with ^{13}C labels at different positions has been the most commonly used substrate in in vivo ^{13}C NMR spectroscopy studies [66]. A variety of other substrates including the glia-specific acetate, the ketone body β-hydroxybutyrate, ethanol, Gln, and lactate can also be used for ^{13}C NMR spectroscopy experiments [26, 59, 60, 67–71].

2.4.1 Glucose

The majority of ^{13}C NMR spectroscopy studies in vivo have used [1-^{13}C]-glucose as the exogenous substrate. During the first turn of the TCA cycle, ^{13}C label from [1-^{13}C]-glucose is incorporated to [4-^{13}C]-α-ketoglutarate and subsequently to [4-^{13}C]-Glu (see Fig. 6), which is readily detectable in the aliphatic spectral region of ^{13}C NMR spectrum if scalp lipid signals are effectively suppressed. Transfer of carbon from glucose to Glu and eventually to carbon dioxide is a major part of the most active energy-producing metabolic pathways in brain under normal physiological conditions, namely glycolysis and the subsequent neuronal TCA cycle. The labeling pattern of metabolites generated by [1,6-$^{13}C_2$]-glucose is identical to that by [1-^{13}C]-glucose. Since both carbons from the 1- and the 6-position of glucose are incorporated into the 4-position of Glu, detection sensitivity using [1,6-$^{13}C_2$]-glucose is twice as that of [1-^{13}C]-glucose although [1,6-$^{13}C_2$]-glucose is commercially far more expensive than [1-^{13}C]-glucose.

Although [2-^{13}C]-glucose can be used to follow the cerebral anaplerotic pathway by measuring ^{13}C label incorporation into Glu C2 and C3 resonating in the spectral regions of aliphatic CH and CH_2 carbons [44], incorporation of ^{13}C labels from exogenous [2-^{13}C]-glucose into the carboxylic carbon of Glu and the amide carbon of Gln via the pyruvate dehydrogenase reaction is the primary route of [2-^{13}C]-glucose metabolism in brain. As shown in Fig. 6, when exogenous [1-^{13}C]-glucose is administered as the exogenous ^{13}C-labeled substrate, the ^{13}C labels at glucose C1 are incorporated into pyruvate C3 through the glycolysis pathway. Via the pyruvate dehydrogenase reaction, [3-^{13}C]-pyruvate is converted into [2-^{13}C]-acetylCoA, and via the TCA cycle it is incorporated into α-ketoglutarate at C4. By aspartate aminotransferase and other enzymes, the ^{13}C labels originally at the glucose C1 position

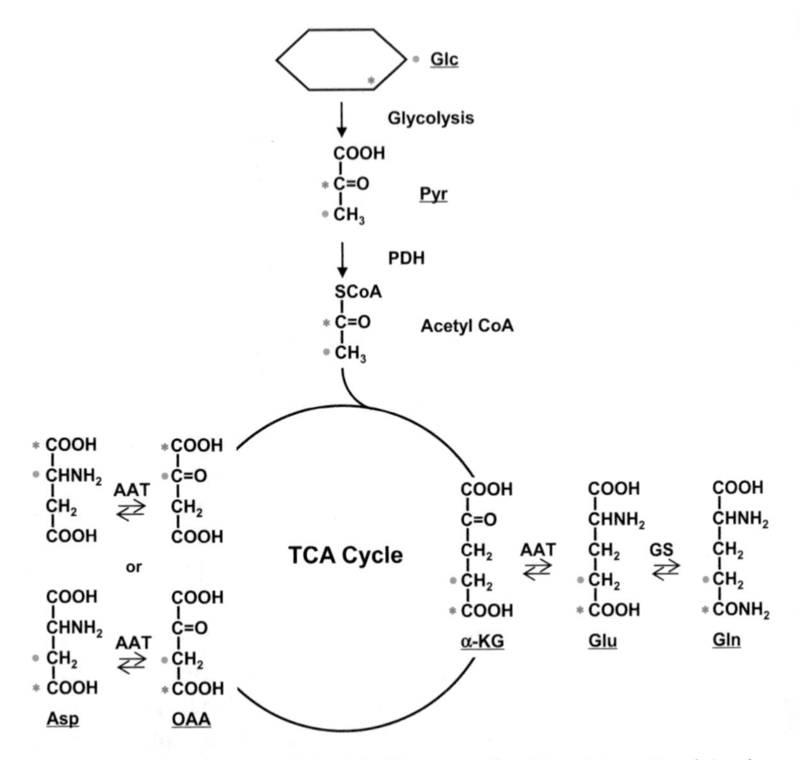

Fig. 6 Schematic diagram of the label incorporation into glutamate, glutamine, aspartate, and GABA from ^{13}C-labeled glucose. For clarity and because the turnovers of glutamate C4 from [1-^{13}C]-glucose, and glutamate C5 from [2-^{13}C]-glucose reflect the same de novo incorporation of exogenous ^{13}C labels into the TCA cycle, only the first turn of the TCA cycle is drawn. ^{13}C labels that originate from glucose C2 carbon are denoted by an *asterisk* (*, in *red color*), and ^{13}C labels that originate from glucose C1 carbon are denoted by a *filled circle* (•, in *blue color*). *AAT* aspartate aminotransferase, *Asp* aspartate, *Glc* glucose, *Gln* glutamine, *Glu* glutamate, *GS* glutamine synthetase, *α-KG* α-ketoglutarate, *OAA* oxaloacetate, *PDH* pyruvate dehydrogenase, *Pyr* pyruvate. Adapted from Li et al. (2007) [42]

are incorporated into Glu C4, and subsequently into Gln C4. Figure 6 shows that the C1 and C2 carbons of glucose are incorporated into the C4 and C5 positions of Glu as a single entity with the original covalent bond between them intact. Therefore, the turnover of C4 and C5 on Glu reflects the same rate of de novo incorporation of ^{13}C labels into Glu regardless of the number of turns of the TCA cycle because every turn involves replacing the current C4 and C5 by acetylCoA. Therefore, the kinetics of ^{13}C incorporation into Glu C5 and Gln C5 from exogenous [2-^{13}C]-glucose is identical to that of ^{13}C incorporation into Glu C4 and Gln C4 from exogenous [1-^{13}C]-glucose [42].

The use of [U-$^{13}C_6$]-glucose can be advantageous in many scenarios. For indirect ^{13}C NMR spectroscopy the detection sensitivity from administering [U-$^{13}C_6$]-glucose is approximately twice that of [1-^{13}C]-glucose. For direct ^{13}C NMR detection [U-$^{13}C_6$]-glucose generates well-defined doublets in the carboxylic/amide region due to homonuclear ^{13}C–^{13}C scalar couplings [43, 59]. A judiciously chosen singlet-generating second substrate such as [1-^{13}C]-acetate, therefore, can be used to measure two metabolic pathways simultaneously based on different isotopomers [59].

2.4.2 Acetate

Acetate is a glia-specific substrate [72, 73]. Prolonged incubation of neuron culture with ^{13}C-labeled acetate showed no enrichment of ^{13}C labels in Glu, Gln, or GABA [4]. Exogenous [2-^{13}C]-acetate is metabolized via glial acetate-CoA ligase and TCA cycle, and subsequently labels Gln at its C4 position. Glutamine is predominantly localized in glia. ^{13}C label carried by Gln enters neuronal compartments via the glutamate–glutamine cycling flux to produce ^{13}C-labeled Glu, which is predominantly located in glutamatergic neurons. Because of the high specificity of acetate metabolism [2-^{13}C]-acetate has been used to quantitatively characterize glial metabolism in the human brain [26, 67, 74, 75], in animal brain [45, 46, 68, 76, 77] and to study Glu metabolism in diseases of glial origin [78, 79]. Detecting carboxylic/amide carbons after coadministering [U-$^{13}C_6$]-glucose and [1-^{13}C]-acetate allows probing neuronal and glial metabolism simultaneously based on different isotopomer patterns [59].

3 Methods

3.1 Proton NMR Spectroscopy

3.1.1 Low Field Techniques

Proton NMR spectroscopy has been widely used to measure metabolite concentrations in vivo including Glu. Early studies relied on short echo time single voxel methods such as PRESS (point-resolved spectroscopy) and STEAM (stimulated echo acquisition mode) spectroscopy to acquire signals of all major metabolites [53, 80]. Because of severe spectral overlap in the crowded short echo time proton spectra the entire spectrum is fitted together using a spectral model consisting of major metabolites [81, 82]. The detection of Glu, however, is often hampered by the severe spectral overlap at clinical magnetic field strengths (see Fig. 2). The proton NMR spectrum of Glu is entirely composed of J-coupled resonances [31]. Glutamate is very similar to Gln in molecular structure, and both are characterized by the coupled spins of C2–C4 hydrogen nuclei. At the prevalent field strengths of 1.5 or 3.0 T for clinical studies, magnetic resonance signals of Glu and Gln almost completely overlap with each other at all of their chemical shifts. Therefore, it is very challenging to reliably separate Glu and Gln by spectral fitting of very short echo time NMR spectra. Instead, the resonances are usually assigned to a combination of Glu and Gln, often referred to as Glx [36]. Due to the obvious connection between Glu and Gln via the glutamate–glutamine

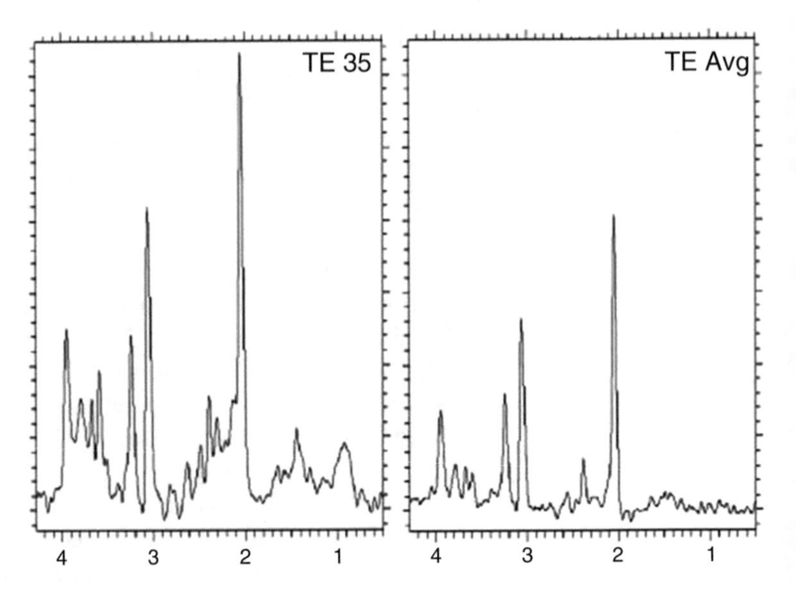

Fig. 7 Comparison between short echo time (35 ms) PRESS (*left*) and echo time-averaged PRESS spectrum (*right*). The echo time-averaged PRESS spectrum was sampled with 16 increments of 10 ms starting at TE = 35 ms with four averages per echo time; 128 total averages were collected for both short echo time and echo time-averaged acquisitions. The glutamate C4 proton at 2.35 ppm in the echo time-averaged spectrum is relatively free of spectral interferences. Reproduced from Hurd et al. (2004) [35] with permission from John Wiley & Sons, Inc.

neurotransmitter cycle and the related enzyme reactions, Glu and Gln levels can change in opposite directions (e.g., in hyperammonemia [83]). The combined Glu and Gln signal Glx therefore has reduced sensitivity as a disease marker.

As a result, many proton NMR spectroscopy methods have been proposed to improve the detection of Glu [84–91]. One of the methods that has found numerous clinical applications for NMR spectroscopy of brain Glu is echo time-averaged spectroscopy at the magnetic field strength of 3 T [35, 37, 92]. In this method, a number of one-dimensional PRESS spectra are acquired at variable but equally spaced echo times, and then averaged to generate a single one-dimensional, echo time-averaged spectrum. In this one-dimensional spectrum spectral peaks for the C4 protons of Glu at 2.35 ppm and the C2 protons of Glx at ~3.75 ppm are relatively free of spectral interferences that plague short echo time spectra (see Fig. 7; [35]). While echo time-averaging is a simple and reliable method for measuring Glu it undesirably cancels out the C4 signal of Gln, leaving few resonance signals to determine Gln concentrations. Note that simultaneous determination of Glu and Gln is very useful in most clinical applications because of their interconnection via the glutamate–glutamine neurotransmitter cycle. Instead of additive averaging, different strategies, with varied degree of success, have also been suggested for recovering the resonance signal of the Gln C4 protons from echo time-averaged proton NMR spectra acquired in vivo (e.g., [93]).

An echo time-averaged spectrum is essentially the one-dimensional cross section of a two-dimensional *J*-resolved spectrum at $J = 0$. Recently, a new technique (*J* modulated spectroscopy; [34]) was introduced to spectrally separate Glu and Gln at 3 T from the one-dimensional cross section of the two-dimensional J-resolved spectrum at $J = 7.5$ Hz. The transverse relaxation times (T_2s) of metabolites are first determined using standard echo time-averaged spectroscopy with different starting echo times and then incorporated into the spectral model for determining both Glu and Gln. In vivo data in Fig. 8 shows that the resonance signals of Glu and Gln were clearly spectrally separated around 2.35 ppm, allowing quantitative measurement of Glu and Gln simultaneously at 3 T using the same echo time averaging dataset.

3.1.2 High Field Techniques

While the C4 proton resonances of Glu (2.35 ppm), Gln (2.45 ppm), and the glutamyl moiety of glutathione (GSH) (2.54 ppm) overlap each other at low- and mid-magnetic fields (1.5–4.7 T), at high fields such as 7 T the C4 proton resonances of Glu, Gln, and GSH are spectrally resolved as chemical shift dispersion is proportional to magnetic field strength [31]. The weakening of strong scalar couplings at high magnetic field strengths also contributes to spectral separation of these otherwise overlapped signals. However, the Gln C4 multiplet at 2.45 ppm and the GSH glutamyl C4 multiplet at 2.54 ppm are still overlapped by the much larger signals from the *N*-acetylaspartate (NAA) aspartyl C3 multiplet centered at 2.49 ppm. Choi et al. [94] have proposed using an echo time-optimized PRESS sequence at 7 T to measure Glu, Gln, and GSH, relying on chemical shift offsets [95–97] to suppress overlapping signals from the aspartyl moiety of NAA. Because of inhomogeneous distribution of metabolites in vivo which may compromise NAA suppression a more effective technique [98] for suppressing spectral interference from the aspartyl moiety of NAA was proposed that uses a selective radiofrequency pulse placed at the resonance frequency of the NAA aspartyl CH proton at 4.38 ppm. This pulse selectively alters the *J*-evolution of the NAA aspartyl CH_2 multiplet at 2.49 ppm. The flip angle of this suppression pulse along with the echo time of the PRESS sequence were optimized to suppress the NAA aspartyl multiplet at 2.49 ppm in a single shot without conventional spectral editing. Figure 9 shows proton NMR spectra acquired at 7 T from the prefrontal cortex

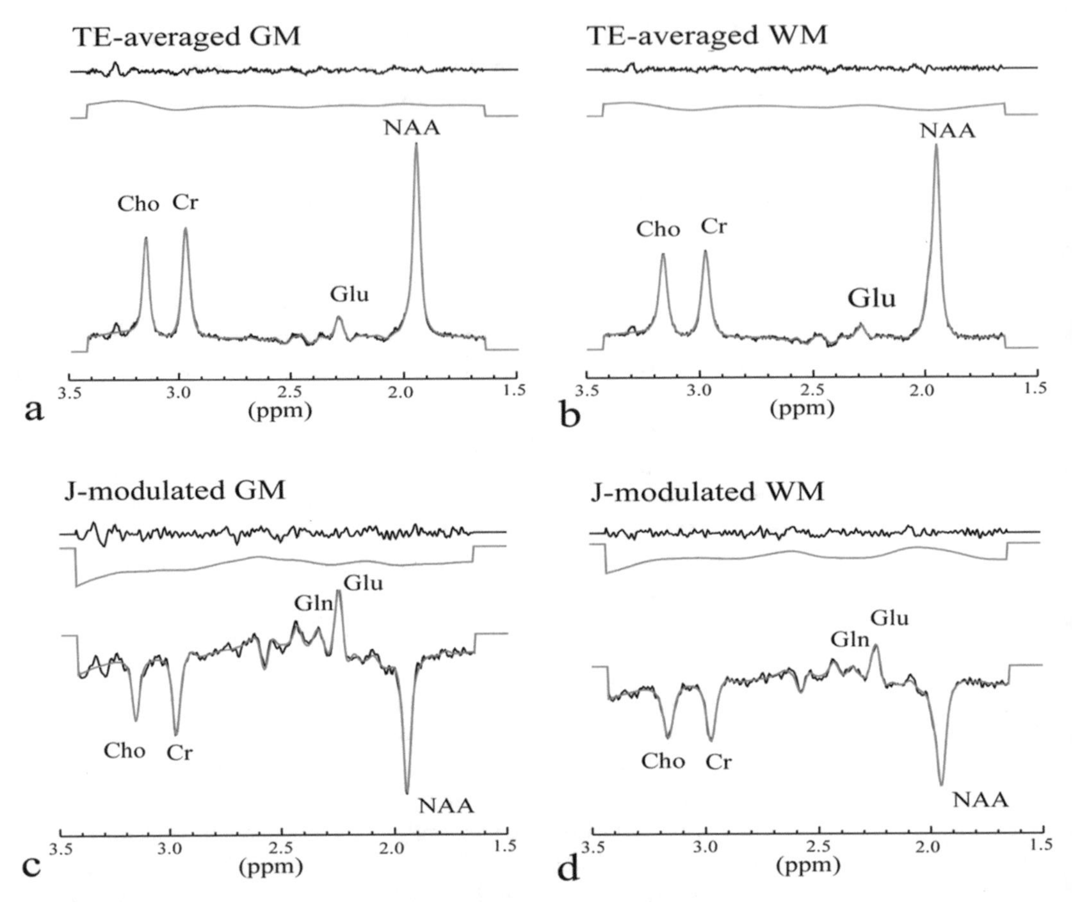

Fig. 8 Examples of (**a**, **b**) echo time-averaged spectra and (**c**, **d**) J-modulated spectra acquired from gray matter-dominant voxel (**a**, **c**) and white matter-dominant voxel (**b**, **d**) in the anterior cingulate cortex of a human subject. The echo time-averaged spectra and the J-modulated spectra both show higher glutamate (Glu) levels in the gray matter-dominant voxel. Higher glutamine levels in the gray matter-dominant voxel were revealed by the two J-modulated spectra (**c**, **d**). The J-modulated spectra (**c**, **d**) were scaled up by ×2. The fitted spectra and the baselines are displayed in *red*, along with the fit residuals. *TE* echo time, *GM* gray matter, *WM* white matter. Reproduced from Zhang et al. (2016) [38]

and right frontal cortex voxels (2 cm × 2 cm × 2 cm) of a human subject using a *J*-suppression pulse and PRESS echo time optimization. Glutamate, Gln, and GSH are spectrally resolved in a single scan as compared to echo-time averaging which requires 32–128 scans to resolve Glu from overlapping resonances [35].

3.1.3 Interpretation of Brain Glutamate Levels Measured by Proton NMR Spectroscopy

Glutamate plays multiple roles in the brain as such measuring Glu noninvasively using NMR spectroscopy provides important insights into its functions in normal brain and in a variety of brain disorders. Because Glu is the primarily excitatory neurotransmitter in the CNS alteration in the overall Glu level may affect the crucial balance between excitation and inhibition. Caution, however, is

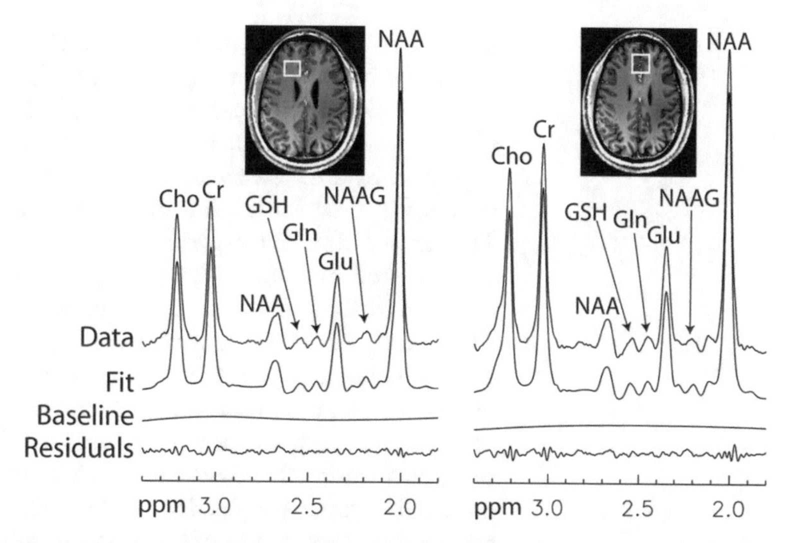

Fig. 9 Proton NMR spectra acquired at 7 T from the frontal cortex gray matter- and white matter-dominant voxels (2 cm × 2 cm × 2 cm) of a human subject using a J-suppression pulse and echo time optimization. The echo time of the first spin echo of the PRESS sequence is 69 ms; the echo time of the second spin echo is 37 ms. J suppression pulse flip angle = 90°. Number of averages = 128. Linear combination fitting basis spectra of glutamate, glutamine, and glutathione are displayed on the bottom rows. Spectral data between 1.8 and 3.3 ppm were used in data fitting. *Cho* total choline, *Cr* total creatine, *NAA N*-acetylaspartate, *GSH* glutathione, *Gln* glutamine, *Glu* glutamate, *NAAG N*-acetylaspartylglutamate

needed when attributing the chronic effects of brain disorders on the total concentrations of Glu to changes in the glutamate–glutamine neurotransmitter cycling flux because many factors could contribute to or explain an altered total Glu level measured by proton NMR spectroscopy. In particular, changes in aspartate aminotransferase and glutamate dehydrogenase activities are known to modulate the total concentration of Glu detected by proton NMR spectroscopy. A more informative characterization of Glu function in brain can be accomplished using ^{13}C NMR spectroscopy methods to quantify the glutamate–glutamine neurotransmitter cycling rate [10] and/or to measure the magnetization transfer effect of aspartate aminotransferase activity [49] that plays a major role in Glu metabolism. In the following, two examples of altered total Glu concentration are discussed.

Under many pathophysiological conditions (e.g., hypoglycemia) total Glu concentration decreases as a result of reduced overall glucose consumption. This phenomenon can be attributed to a decreased delivery of glucose and subsequently an increase in oxidation of endogenous substrates including Glu. With its high concentration in brain tissue Glu, like Gln, aspartate (Asp) and GABA, represents an alternative fuel source [99]. The decrease in Glu concentration during hypoglycemia is associated with an increase in Asp concentration [100, 101]. During acute hypoglycemia the brain can lose up to 10 μmol/g of Glu and elevate Asp concentration by the same amount [102]. The concerted change in Glu and Asp concentration during hypoglycemia can be explained by a massive shift in the aspartate aminotransferase reaction in the direction of Glu consumption and Asp formation [1, 6]. This large shift in the aspartate aminotransferase reaction results in an increased production of α-ketoglutarate, generating oxidative energy with conversion of α-ketoglutarate into oxaloacetate in the TCA cycle. Oxidative deamination by glutamate dehydrogenase also contributes to the reduction of total Glu concentration during hypoglycemia, accompanied by an increase in ammonia level. This scenario agrees with the known shift of the redox state towards oxidation during hypoglycemia which shifts the reversible glutamate dehydrogenase reaction to the formation of α-ketoglutarate, resulting in a reduction of Glu concentration.

Because of its role as a powerful excitatory neurotransmitter, changes in Glu levels can cause cell dysfunction during severe hypoxic stress. During hypoxic hypoxia an increase in Glu and GABA as well as a decrease in Asp are detected [103]. The increase in Glu concentration is due to a mismatch between the rate of glycolysis and the rate of oxidative metabolism, leading to an accumulation of pyruvate and a reduction of both mitochondrial and cytoplasmic NAD^+, and subsequently causing a redox-dependent rise in the malate/oxaloacetate ratio. The increase in total Glu concentration can be explained by the rise in the malate/oxaloacetate ratio which triggers a shift of the aspartate aminotransferase reaction in the direction of Glu formation and Asp consumption [1].

3.2 ^{13}C NMR Spectroscopy

^{13}C NMR spectroscopy is the only noninvasive method that can measure both neuroenergetics and Glu neurotransmitter cycling in brain [10, 104]. ^{13}C NMR spectroscopy, in combination with the administration of ^{13}C-labeled substrates, allows researchers to monitor the kinetics of ^{13}C labels incorporation from ^{13}C-labeled substrates into various carbon positions of metabolites especially Glu and Gln [31]. The versatile ^{13}C NMR spectroscopy experiments can be performed in several different ways, providing multiple means by which Glu metabolism and neurotransmission can be studied. These include: direct (aliphatic or carboxylic/amide) or indirect data acquisition, choice of substrate (glucose, acetate, lactate, β-hydroxybutyrate, ethanol, and label positions as discussed in Sect. 2.4), and method of analysis (labeling kinetics and isotopic steady state). Below is an overview of each of the different methods with emphasis on key technical aspects.

3.2.1 Direct ^{13}C NMR Spectroscopy

The early effort of ^{13}C NMR spectroscopy of brain Glu focused on direct excitation and detection of aliphatic carbon signals of Glu (C2–C4) after administering [1-^{13}C] or [1,6-$^{13}C_2$]-glucose

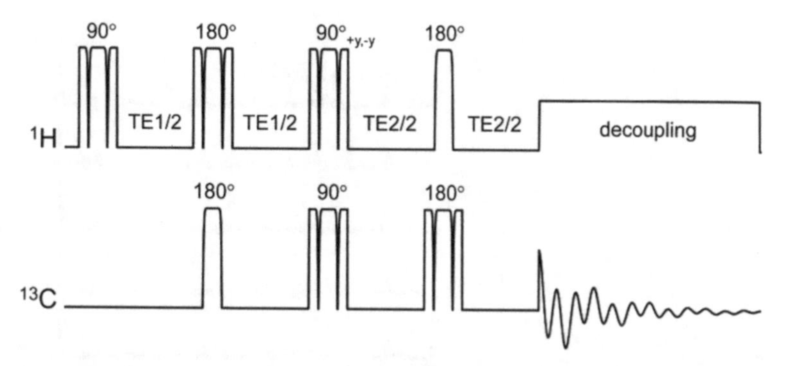

Fig. 10 Polarization sequence based on adiabatic full passage and BIR-4 pulses (B1 independent rotation using four segments). The acquisition phase is phase-cycled in concert with the second proton BIR-4 90° pulse to acquire the transferred polarization from proton to ^{13}C and cancel the thermal equilibrium ^{13}C magnetization. To mitigate radiofrequency field inhomogeneity, all radiofrequency pulses are made adiabatic using either adiabatic full passage or composite adiabatic pulses (BIR-4). For broadband decoupling of aliphatic carbons during signal acquisition WALTZ-4 or adiabatic decoupling sequences are commonly used for low field strength human studies or animal studies. Reproduced from de Graaf et al. (2011) [39] with permission from John Wiley & Sons, Inc.

[105, 106]. Although the major spectral interference in direct ^{13}C NMR spectroscopy of aliphatic carbons comes from scalp lipids there is very little background signal of indigenous ^{13}C Glu because the natural abundance of ^{13}C is approximately 1.1%. Signal contamination from scalp lipids can be suppressed or eliminated using single voxel localization combined with additional outer-volume suppression [41]. Significant sensitivity enhancement was achieved using heteronuclear polarization transfer techniques such as INEPT [107] or DEPT [108] to excite protons attached to ^{13}C [16, 40]. As described in Sect. 2.1, the radiofrequency coils used in common ^{13}C NMR spectroscopy work render significant radiofrequency field inhomogeneity. For multi-pulse sequences such as INEPT and DEPT, radiofrequency field inhomogeneity, if uncorrected, quickly leads to a major loss of signal. To mitigate this problem, adiabatic pulses that are invariant under radiofrequency field inhomogeneity [109–111] and/or half-volume proton coils that provide a less inhomogeneous sensitive volume have been developed and successfully implemented for ^{13}C NMR spectroscopy of brain [16, 56]. A typical adiabatic INEPT pulse sequence for direct ^{13}C NMR spectroscopy is shown in Fig. 10. As seen in Fig. 11 following i.v. infusion of [1,6-$^{13}C_2$]-glucose ^{13}C labeling of Glu and Gln progresses over the course of infusion. The kinetics of ^{13}C label incorporation into Glu and Gln can be explained using a two-compartment metabolic model of the glutamate–glutamine neurotransmitter cycle (see Sect. 3.2.3).

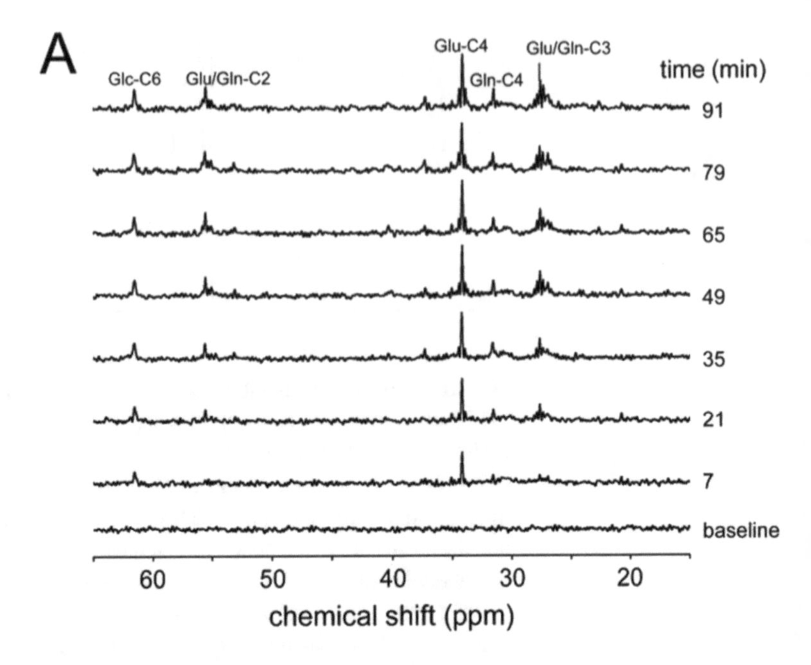

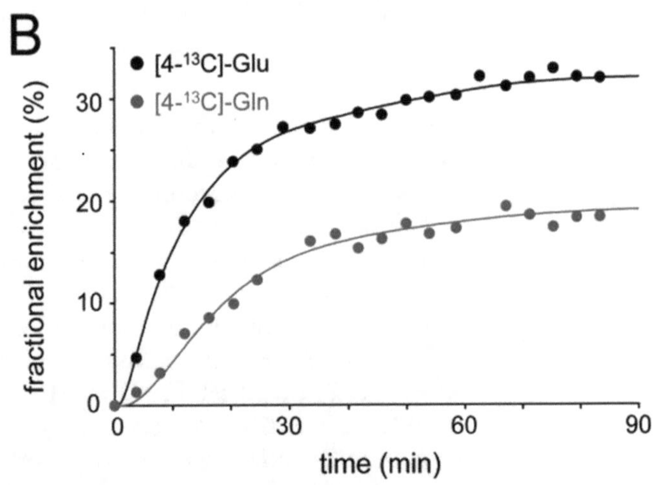

Fig. 11 (**a**) Time resolved proton-decoupled, ^{13}C NMR spectra acquired from rat brain in vivo following the intravenous infusion of [1,6-$^{13}C_2$]-glucose. Spectra are acquired using an adiabatic INEPT sequence from a voxel of 200 μL at 7.05 T. (**b**) Turnover curves of [4-^{13}C]-glutamate and [4-^{13}C]-glutamine. *Dots* represent measured fractional enrichments, whereas the *solid line* represents the best fit to metabolic model. *Glu* glutamate, *Gln* glutamine, *Glc* glucose. Reproduced from de Graaf et al. (2011) [39] with permission from John Wiley & Sons, Inc.

For ^{13}C NMR spectroscopy, it is necessary to decouple the large ^{1}H-^{13}C scalar couplings ($^{1}J_{CH}$: 125–145 Hz) to obtain a useful signal-to-noise ratio (SNR) and spectral resolution in vivo at magnetic field strengths accessible to most clinical studies (1.5–4 T). The most common proton decoupling method for in

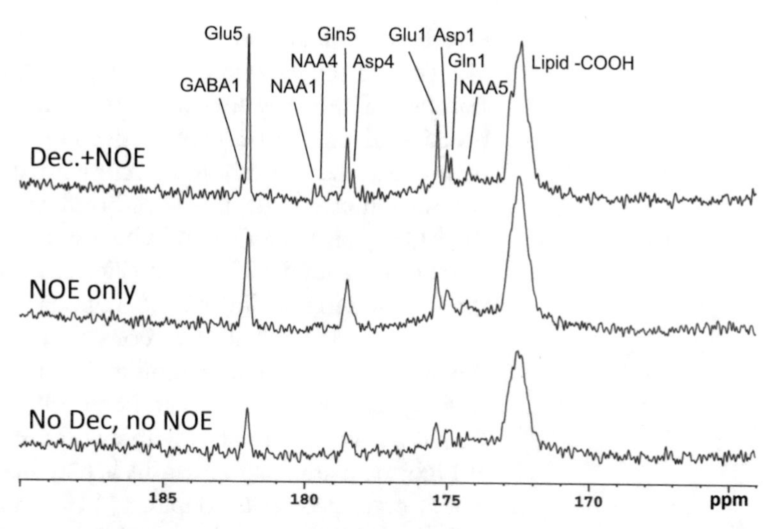

Fig. 12 Carboxylic/amide ^{13}C Spectra obtained from human brain under the following conditions: (*lower trace*) with neither nuclear Overhauser effect (NOE) nor decoupling, (*middle trace*) with NOE only, and (*upper trace*) with both NOE and decoupling. Each spectrum was averaged over 104 scans. The upper trace spectrum used decoupling power and average transmit power of less than 35 watts (W) and 3.6 W, respectively. The spectra were collected over the time period from 60 to 100 min after [2-^{13}C]-glucose infusion began. Using the peak amplitude of glutamate C5 in lower trace as a reference, the peak amplitude of glutamate C5 was increased on average, by a factor of 2.3 when NOE was turned on, and by a factor of 4.3 when both NOE and decoupling were turned on. *Asp4* aspartate C4, *Asp1* aspartate C1, *GABA1* gamma-aminobutyric acid C1, *Gln1* glutamine C1, *Gln5* glutamine C5, *Glu5* glutamine C5, *Glu1* glutamate C1, *NAA* *N*-acetylaspartate. Adapted from Li et al. (2016) [57]

vivo ^{13}C NMR spectroscopy almost exclusively employs the WALTZ-4 sequence [112] because of its superb performance under the condition of severe radiofrequency field inhomogeneity generated by surface transceiver coils [113, 114]. Since chemical shift dispersion is proportional to static magnetic field strength, the proton decoupling bandwidth and therefore the decoupling radiofrequency field strength required for broadband decoupling increase linearly with static magnetic field strength. That is, the radiofrequency power required for broadband proton decoupling increases as a function of the square of static field strength. With the advent of high-field clinical magnets (e.g., 7 T), proton decoupling is no longer deemed a viable option for in vivo ^{13}C NMR spectroscopy because of the prohibitively large radiofrequency power deposition required for effective heteronuclear decoupling. Therefore, detection of the carboxylic/amide carbons with broadband stochastic proton decoupling at very low radiofrequency power [42, 57] provides an attractive alternative to conventional direct ^{13}C NMR spectroscopy techniques especially at high magnetic fields. Figure 12 [57] compares the effects of nuclear

Overhauser enhancement and broadband stochastic proton decoupling on carboxylic/amide ^{13}C spectra acquired from human brain. Because of the very low radiofrequency power requirements the broadband stochastic proton decoupling only used 3.6 W, far below the safety threshold recommended by FDA.

Also interestingly, in the carboxylic/amide ^{13}C spectral region, the ^{13}C signals from Glu and Gln are not overlapped by scalp lipids resonating around 172.5 ppm (Fig. 12), a dramatic departure from the situation in the aliphatic ^{13}C regions where Glu and Gln C4 carbons as well as methylene carbons of scalp lipids co-resonate. With the lack of lipid contamination and the fact that the unprotonated carboxylic/amide carbons can be decoupled using stochastic radiofrequency waveforms at very low radiofrequency power levels even at high magnetic field strengths it may become feasible to perform whole brain proton decoupling [115] and using phased array ^{13}C coils to increase sensitivity and spatially localize the carboxylic/amide ^{13}C signals to cortical gray matter according to irregularly shaped natural anatomical boundaries [116, 117].

3.2.2 Indirect ^{13}C NMR Spectroscopy

By excitation and detection of protons attached to ^{13}C, the much larger gyromagnetic ratio of protons leads to a significantly higher sensitivity for indirect ^{13}C NMR spectroscopy techniques. In addition, because protons attached to both ^{12}C and ^{13}C can be detected, measurement of the all-important ^{13}C fractional enrichments, which are necessary for metabolic modeling, is straightforward. The most commonly used indirect ^{13}C NMR spectroscopy technique for studying brain Glu is the so-called POCE (proton-observed, carbon-edited [118–121]) method. It is based on different spin evolution between protons attached to ^{12}C and ^{13}C, only the latter is affected by a ^{13}C refocusing pulse. At the echo time of $1/{}^{1}J_{CH}$ spin evolution due to ^{1}H–^{13}C scalar coupling is completely refocused without the 180° radiofrequency refocusing pulse on the ^{13}C channel such that heteronuclear coupled spins are in-phase with respect to proton spins attached to ^{12}C. With the 180° refocusing radiofrequency pulse on the ^{13}C channel, however, protons attached to ^{13}C are antiphase with respect to protons attached to ^{12}C. This differentiation allows one to separate protons based on the isotopic identity they are attached to, therefore, leading to indirect detection of ^{13}C (as well as ^{12}C). Figure 13 [122] shows the effects of electrical forepaw stimulation on the rate of ^{13}C incorporation from [1,6-$^{13}C_2$]-glucose into Glu and Gln in the somatosensory cortex of α-chloralose anesthetized rats. Time course of POCE spectra following i.v. infusion of [1,6-$^{13}C_2$]-glucose shows faster Glu turnover, therefore, higher neuronal TCA cycle rate, during functional stimulation of the brain.

At high magnetic field strength (e.g., 7 T) it is also possible to omit ^{13}C-editing for indirect ^{13}C NMR spectroscopy [123] due to the much improved spectral resolution associated with high

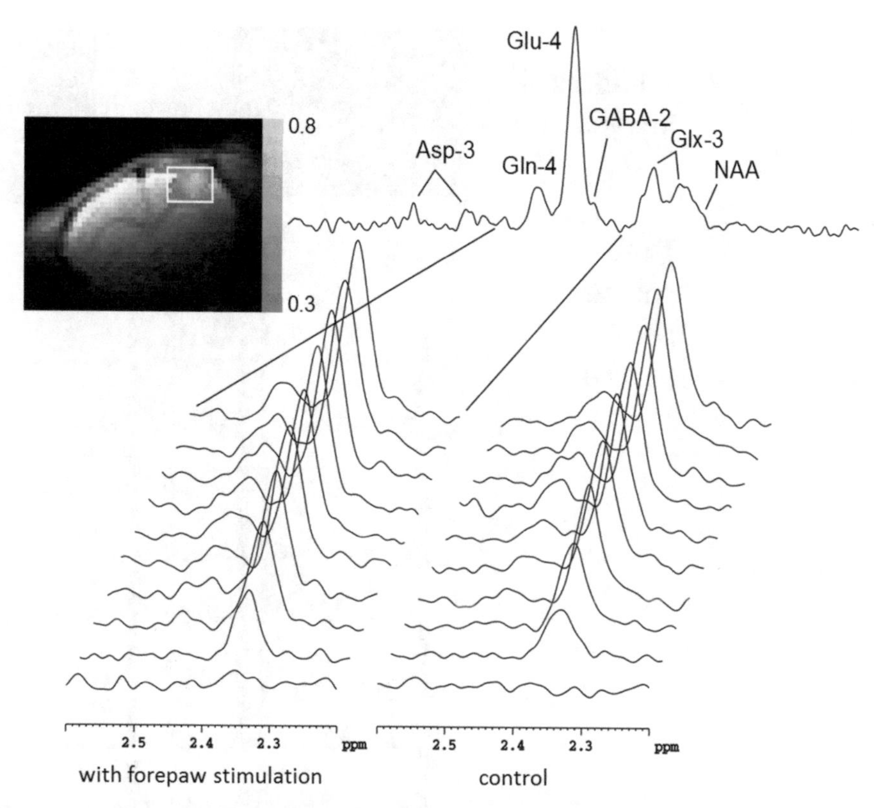

Fig. 13 Time course of proton-detected, ^{13}C-edited (POCE) spectra following intravenous infusion of [1,6-$^{13}C_2$] glucose with 128 averages per spectrum. *Left stack*: with electrical forepaw stimulation; *right stack*: control without forepaw stimulation. An expanded view of the endpoint spectrum in the left stack was also shown. The [4-^{13}C]glutamine signal at 2.46 ppm and the [2-^{13}C]GABA signal at 2.30 ppm are spectrally resolved in vivo from the target [4-^{13}C]-glutamate signal at 2.35 ppm. *Asp-3* [3-^{13}C]-aspartate, *GABA-2* [2-^{13}C]-γ-aminobutyrate, *Gln-4* [4-^{13}C]-glutamine, *Glu-4* [4-^{13}C]-glutamate, *Glx-3* [3-^{13}C]-glutamate + [3-^{13}C]-glutamine, *NAA-6*: [6-^{13}C]-*N*-acetylaspartate. *Inset*: blood oxygenation level-dependent activation map overlaid on raw single-shot coronal echo planar image. A color intensity scale was used for cross-correlation coefficient. Slice thickness = 2 mm. FOV = 20 mm × 20 mm with a 64 × 64 matrix corresponding to an in-plane resolution of 312 μm × 312 μm. The spectroscopic voxel (3.5 mm × 2.5 mm × 4 mm, 35 μl) was marked with a *rectangular box*. The turnover of glutamate C4 was found to be significantly faster during forepaw stimulation. Reproduced from Yang et al. (2005) [122] with permission from Elsevier

magnetic field strength. Figure 14 shows time-course proton spectra acquired from a 2 cm × 2 cm × 2 cm voxel placed in the medial prefrontal cortex of human brain at 7 T during [U-$^{13}C_6$]-glucose infusion at an echo time of 106 ms with additional radiofrequency suppression to suppress the aspartyl moiety of NAA [98]. As infusion progressed Glu peak at 2.35 ppm ([4-^{12}C]-glutamate) became smaller as the replacement of ^{12}C by ^{13}C splits the original proton signal by ^{1}H–^{13}C scalar coupling. At 76–82 min after infusion started the [4-^{12}C]-glutamate peak was less than half of the original peak in the baseline spectrum. In the absence of ^{13}C decoupling, the major peak of [4-^{13}C]-glutamate is located at

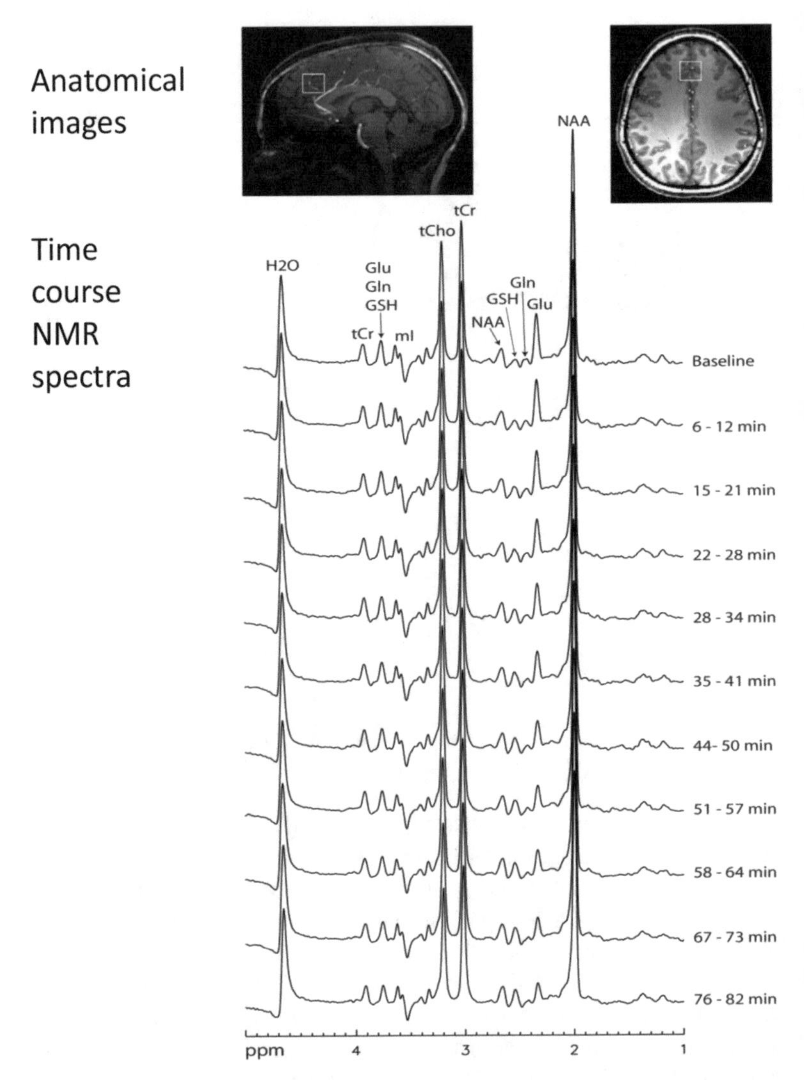

Fig. 14 Time-course proton spectra acquired from a 2 cm × 2 cm × 2 cm voxel placed in the medial prefrontal cortex of human brain at 7 T during [U-$^{13}C_6$]-glucose infusion at an echo time of 106 ms with additional radiofrequency suppression to suppress the aspartyl moiety of NAA (An et al. 2015). The spectrum acquired before the start of infusion is plotted on top and labeled as "Baseline." As infusion progressed glutamate peak at 2.35 ppm ([4-^{12}C]-glutamate) became smaller as replacement of ^{12}C by ^{13}C splits the original proton signal. In the absence of ^{13}C decoupling, the major peak of [4-^{13}C]-glutamate is located at 2.56 ppm, which overlaps the GSH peak. It grew larger during the course of [U-$^{13}C_6$]-glucose infusion. *tCr* total creatine, *Glu* glutamate, *Gln* glutamine, *GSH* glutathione, *tCho* total choline, *NAA* *N*-acetylaspartate. Adapted from An et al. (2015) [98]

2.56 ppm, which overlaps the GSH peak. The 2.56 ppm peak grew larger during the course of i.v. [U-$^{13}C_6$]-glucose infusion as more and more ^{13}C labels are incorporated into Glu C4. Instead of ^{13}C editing, here both the decrease of the original Glu C4 signal as well as the increase of the 2.56 ppm signal can be used to quantify ^{13}C incorporation into Glu from [U-$^{13}C_6$]-glucose.

3.2.3 Metabolic Modeling of Glutamate Metabolism and Neurotransmission

The rapid labeling of Glu C4 carbon upon administering [^{13}-C]-glucose is primarily due to oxidative metabolism in the metabolically highly active neurons. The biochemical processes for replenishing the neurotransmitter Glu after Glu release involve the glutamate–glutamine neurotransmitter cycle (Fig. 1a and b). The existence of this cycle was supported by multiple findings including: (1) biochemical and autoradiographic studies clearly demonstrated that Glu is selectively accumulated by glial cells and rapidly converted into Gln [20, 123–125]; (2) glutaminase, which catalyzes the conversion of Gln into Glu, is predominantly localized in nerve terminals which contains no glutamine synthetase, the latter catalyzes the conversion of Glu into Gln and is found to be exclusively localized to glial cells [1, 11]; (3) Gln preferentially enters neurons where it is converted into Glu [126, 127]; (4) in vivo ^{13}C NMR spectroscopy studies of Glu metabolism found rapid and highly significant labeling of Gln following administering ^{13}C labeled glucose [16, 25, 41]. Because Gln is predominantly located in the metabolically less active astroglial cells, the in vivo ^{13}C NMR spectroscopy findings indicate rapid transfer of ^{13}C labels from the predominantly neuronal Glu compartment to the predominantly astroglial Gln compartment. The dual roles of Glu as the principal neurotransmitter in the CNS and as a key metabolite linking carbon and nitrogen metabolism thus make it possible to probe Glu neurotransmitter cycling using ^{13}C NMR spectroscopy by measuring the labeling kinetics of Glu and Gln.

In order to obtain absolute metabolic fluxes (e.g., neuronal TCA cycle and glutamate–glutamine neurotransmitter cycling rates) a metabolic model [128–132] is used to fit the kinetics of ^{13}C label incorporation from exogenous ^{13}C-labeled substrates into Glu and Gln (and Asp if its labeling can be measured accurately). The most popular model that explains the labeling of Glu and Gln determined by in vivo ^{13}C NMR spectroscopy is the neuron-astroglia two-compartment glutamate–glutamine neurotransmitter cycling model [1, 22, 39]. The time dependence of each ^{13}C-labeled metabolic pool is given by a differential equation describing the metabolic fluxes in and out of that pool. The modeling specifics depend on the chosen exogenous ^{13}C-labeled substrate, its primary metabolic pathways, and ^{13}C label positions on the exogenous substrate. For example, for [1-^{13}C]-glucose or [1,6-$^{13}C_2$]-glucose infusion studies it is important to explicitly incorporate isotopic dilution of Gln by competing pathways into the two-compartment model to avoid artificially reducing the sensitivity of the glutamate–glutamine neurotransmitter cycling rate to the kinetics of ^{13}C label incorporation into Glu and Gln [133]. For [2-^{13}C]-acetate infusion studies ^{13}C from acetate labels glial Gln first [26]. The subsequent labeling of neuronal Glu by the glutamate–glutamine neurotransmitter cycle competes with isotopic dilution of Glu by the dominant neuronal TCA cycle. The glutamate–glutamine neurotransmitter cycling rate to neuronal TCA cycle rate can be readily calculated using the fractional enrichments of Glu and Gln at isotopic steady state [26].

Like other statistical models, metabolic modeling of Glu metabolism and neurotransmission can lead to error due to under- or over-parameterization of the model [133, 134]. A familiar example is spectral fitting in quantification of the crowded short echo time proton NMR spectra of brain (Fig. 2). While under parameterization of the spectral model leads to obvious fitting residuals because an underparametrized model cannot adequately describe the data, a less appreciated fact is that over-parameterization can also cause significant errors including strong artificial correlation between fitting parameters [135]. In the case of [2-^{13}C]-acetate infusion experiments described above addition of flux terms to which the data are insensitive can lead to increased uncertainly in the glutamate–glutamine neurotransmitter cycling flux rate [134].

3.2.4 Interpretation of Altered Glutamate–Glutamine Neurotransmitter Cycling

As outlined in Sect. 3.1.3, many possible factors such as changes in redox states and the activities of enzymes involved in Glu metabolism could lead to altered total concentration of Glu measured by proton NMR spectroscopy. Therefore, a much better characterization of glutamatergic function could be achieved using administration of exogenous ^{13}C-labeled substrates and dynamic in vivo ^{13}C NMR spectroscopy methods to measure the turnover of Glu and Gln. Dysfunction of Glu and Glu hyperactivity are thought to be involved in many brain disorders such as major depression, schizophrenia, Alzheimer's diseases, Parkinson's disease, amyotrophic lateral sclerosis, and dementia (e.g., [135]). Although tissue-specific defects in glial transporter genes leading to impaired Glu uptake and abnormal activation of glutamatergic pathways have been identified in certain cases it is not yet understood what role of the total concentration of Glu, as measured by proton NMR spectroscopy, plays in these disorders. For example, abnormality in glutamatergic regulation is considered to be a major factor in the initiation, spread, and maintenance of seizure activities in many types of epilepsy. Topiramate, an effective antiepileptic drug, exerts its anticonvulsant action by selectively antagonizing AMPA and kainate receptors [136, 137]. Preclinical studies have convincingly shown that Glu release is significantly augmented during seizure activities. Many patients with complex partial epilepsy or temporal lobe epilepsy have neuronal loss and sclerosis, especially in the mesial hippocampus. A recent study using in vitro ^{13}C NMR spectroscopy analysis of epileptogenic hippocampal tissue excised from patients infused with [2-^{13}C]-glucose found that the glutamate–glutamine neurotransmitter cycling flux rate was significantly reduced in epilepsy patients with hippocampal sclerosis compared with those with minimal neuronal loss [138]. It was suggested that the measured lower rate of glutamate–glutamine neurotransmitter cycling resulting from a failure in glial Glu uptake could account for slow Glu clearance from synaptic cleft, which leads to excitotoxicity. Damage caused by excitotoxicity therefore may explain why the concentration of total Glu (and that of NAA, a putative neuronal marker) measured from human epileptic cortex excised for neurosurgical therapy of focal epilepsy is reduced, rather than elevated ([139, 140]).

4 Conclusion

There have been ever-increasing interests in the functions of Glu in human neurophysiology (e.g., neurotransmission, aging, and pain), mental disorders and disease processes (e.g., brain tumors, Alzheimer's disease, chronic hepatic encephalopathy, schizophrenia, major depression). Noninvasive NMR spectroscopic methods, both proton and ^{13}C, continue to see major technological advances, especially with the increasing popularity of commercial high magnetic field 7 T scanners. These NMR spectroscopy techniques will continue to be indispensable in our understanding of the role of Glu in neuroenergetics and neurotransmission. ^{13}C NMR spectroscopy, in particular, offers a unique window into the metabolism, neurotransmission, and associated specific enzyme activities of brain Glu that goes far beyond the static metabolite levels accessible to proton NMR spectroscopy. The observation and characterization of the TCA cycle and the glutamate–glutamine neurotransmitter cycles as well as major enzymes in Glu metabolism provides highly detailed insights into the function of Glu in brain. These insights are fundamental to our understanding of various brain diseases and may potentially yield specific targets for therapeutic intervention.

Acknowledgements

This work is supported by the intramural research program of the NIH, NIMH.

References

1. Erecinska M, Silver IA (1990) Metabolism and role of glutamate in mammalian brain. Prog Neurobiol 35:245–296
2. Gunther H (1992) NMR spectroscopy, basic principles, concepts, and applications in chemistry, 2nd edn. John Wiley & Sons, New York
3. Braun S, Kalinowski HO, Berger S (1998) 150 and more basic NMR experiments, a practical guide, 2nd edn. Wiley-VCH, Weinheim
4. Slichter CP (1996) Principles of magnetic resonance, 3rd edn. Springer, Berlin
5. Derome AE (1987) Modern NMR techniques for chemistry research. Pergamon Press, Oxford
6. Siesjö BK (1978) Brain energy metabolism. John Wiley & Sons, Chichester, UK
7. Ende G (2015) Proton magnetic resonance spectroscopy: relevance of glutamate and GABA to neuropsychology. Neuropsychol Rev 25(3):315–325
8. Gigante AD, Bond DJ, Lafer B et al (2012) Brain glutamate levels measured by magnetic resonance spectroscopy in patients with bipolar disorder: a meta-analysis. Bipolar Disord 14(5):478–487
9. Kanamori K (2016) In vivo N-15 MRS study of glutamate metabolism in the rat brain. Anal Biochem 16:30263–30269
10. Rothman DL, De Feyter HM, de Graaf RA et al (2011) 13C MRS studies of neuroenergetics and neurotransmitter cycling in humans. NMR Biomed 24:943–957
11. Spencer AE, Uchida M, Kenworthy T et al (2014) Glutamatergic dysregulation in pediatric psychiatric disorders: a systematic review of the magnetic resonance spectroscopy literature. J Clin Psychiatry 75(11):1226–1241
12. Treen D, Batlle S, Mollà L et al (2016) Are there glutamate abnormalities in subjects at high risk mental state for psychosis? A review of the evidence. Schizophr Res 171(1–3): 166–175
13. Van den Berg CJ, Krzalic L, Mela P et al (1969) Compartmentation of glutamate metabolism in brain. Evidence for the existence of two different tricarboxylic acid cycles in brain. Biochem J 113:281–290
14. Berl S, Nicklas WJ, Clarke DD (1970) Compartmentation of citric acid cycle metabolism in brain: labelling of glutamate, glutamine, aspartate and GABA by several radioactive tracer metabolites. J Neurochem 17:1009–1015
15. Hertz L (1979) Functional interactions between neurons and astrocytes: I. Turnover and metabolism of putative amino acid transmitters. Prog Neurobiol 13:277–323
16. Shen J, Petersen KF, Behar KL et al (1999) Determination of the rate of the glutamate/glutamine cycle in the human brain by in vivo 13C NMR. Proc Natl Acad Sci U S A 96:8235–8240
17. Shen J (2006) 13C magnetic resonance spectroscopy studies of alterations in glutamate neurotransmission. Biol Psychiatry 59: 883–887
18. Niciu MJ, Kelmendi B, Sanacora G (2012) Overview of glutamatergic neurotransmission in the nervous system. Pharmacol Biochem Behav 100(4):656–664
19. Danbolt NC, Storm-Mathisen J, Kanner BI (1992) An [Na+-K+]coupled l-glutamate transporter purified from rat brain is located in glial cell processes. Neuroscience 51: 295–310
20. Rothstein JD, Dykes-Hoberg M, Pardo CA et al (1996) Knockout of glutamate transporters reveals a major role for astroglial transport in excitotoxicity and clearance of glutamate. Neuron 16(3):675–686
21. Ottersen OP, Zhang N, Walberg F (1992) Metabolic compartmentation of glutamate and glutamine: morphological evidence obtained by quantitative immunocytochemistry in rat cerebellum. Neuroscience 46: 519–534
22. Sibson NR, Dhankhar A, Mason GF et al (1997) In vivo 13C NMR measurements of cerebral glutamine synthesis as evidence for glutamate-glutamine cycling. Proc Natl Acad Sci U S A 94:2699–2704
23. Shen J, Sibson NR, Cline G, Behar KL et al (1998) 15N-NMR spectroscopy studies of ammonia transport and glutamine synthesis in the hyperammonemic rat brain. Dev Neurosci 20:434–443
24. Patel AB, de Graaf RA, Mason GF et al (2004) Glutamatergic neurotransmission and neuronal glucose oxidation are coupled during intense neuronal activation. J Cereb Blood Flow Metab 24(9):972–985
25. Gruetter R, Seaquist ER, Kim S et al (1998) Localized In Vivo 13C-NMR Of Glutamate Metabolism In The Human Brain: Initial Results At 4 Tesla. Dev Neurosci 20 380–388
26. Lebon V, Petersen KF, Cline GW et al (2002) Astroglial contribution to brain energy

metabolism in humans revealed by 13C nuclear magnetic resonance spectroscopy elucidation of the dominant pathway for neurotransmitter glutamate repletion and measurement of astrocytic oxidative metabolism. J Neurosci 22:1523–1531

27. Shen J, Rothman DL (2002) Magnetic resonance spectroscopic approaches to studying neuronal: glial interactions. Biol Psychiatry 52:694–700
28. Gruetter R (2002) In vivo 13C NMR studies of compartmentalized cerebral carbohydrate metabolism. Neurochem Int 41(2–3):143–154
29. Sibson NR, Dhankhar A, Mason GF et al (1998) Stoichiometric coupling of brain glucose metabolism and glutamatergic neuronal activity. Proc Natl Acad Sci U S A 95: 316–321
30. Stelmashook EV, Isaev NK, Lozier ER et al (2011) Role of glutamine in neuronal survival and death during brain ischemia and hypoglycemia. Int J Neurosci 121:415–422
31. de Graaf RA (1998) In vivo NMR spectroscopy. John Wiley & Sons, Chichester, UK
32. Choi IY, Lee SP, Merkle H et al (2004) Single-shot two-echo technique for simultaneous measurement of GABA and creatine in the human brain in vivo. Magn Reson Med 51(6):1115–1121
33. Govindaraju V, Young K, Maudsley AA (2000) Proton NMR chemical shifts and coupling constants for brain metabolites. NMR Biomed 13(3):129–153
34. Cox IJ (1996) Development and applications of in vivo clinical magnetic resonance spectroscopy. Prog Biophys Mol Biol 65(1–2): 45–81
35. Hurd R, Sailasuta N, Srinivasan R et al (2004) Measurement of brain glutamate using TE-averaged PRESS at 3T. Magn Reson Med 51:435–440
36. Ramadan S, Lin A, Stanwell P (2013) Glutamate and glutamine: a review of in vivo MRS in the human brain. NMR Biomed 26(12):1630–1646
37. Srinivasan R, Sailasuta N, Hurd R et al (2005) Evidence of elevated glutamate in multiple sclerosis using magnetic resonance spectroscopy at 3 T. Brain 128(Pt 5):1016–1025
38. Zhang Y, Shen J (2016) Simultaneous quantification of glutamate and glutamine by J-modulated spectroscopy at 3 Tesla. Magn Reson Med 76(3):725–732
39. de Graaf RA, Rothman DL, Behar KL (2011) State of the art direct 13C and indirect 1H-[13C] NMR spectroscopy in vivo. A practical guide. NMR Biomed 24:958–972
40. Li S, Chen Z, Zhang Y et al (2005) In vivo single-shot, proton-localized ^{13}C MRS of rhesus monkey brain. NMR Biomed 18: 560–569
41. Gruetter R, Novotny EJ, Boulware SD et al (1994) Localized 13C NMR spectroscopy in the human brain of amino acid labeling from D-[1-13C]glucose. J Neurochem 63(4): 1377–1385
42. Li S, Yang J, Shen J (2007) Novel strategy for cerebral 13C MRS using very low RF power for proton decoupling. Magn Reson Med 57(2):265–271
43. Li S, Zhang Y, Ferraris Araneta M et al (2012) In vivo detection of ^{13}C isotopomer turnover in the human brain by sequential infusion of ^{13}C labeled substrates. J Magn Reson 218: 16–21
44. Sibson NR, Mason GF, Shen J et al (2001) In vivo 13C NMR measurement of neurotransmitter glutamate cycling, anaplerosis and TCA cycle flux in rat brain during. J Neurochem 76:975–989
45. Deelchand DK, Nelson C, Shestov AA et al (2009) Simultaneous measurement of neuronal and glial metabolism in rat brain in vivo using co-infusion of [1,6-13C2]glucose and [1,2-13C2]acetate. J Magn Reson 196(2): 157–163
46. Yang J, Shen J (2009) Elevated endogenous GABA concentration attenuates glutamate-glutamine cycling between neurons and astroglia. J Neural Transm (Vienna) 116(3):291–300
47. Leibfritz D, Dreher W (2001) Magnetization transfer MRS. NMR Biomed 14(2):65–76
48. McKenna MC, Waagepetersen HS, Schousboe A et al (2006) Neuronal and astrocytic shuttle mechanisms for cytosolic-mitochondrial transfer of reducing equivalents: current evidence and pharmacological tools. Biochem Pharmacol 71:399–407
49. Shen J (2005) In vivo carbon-13 magnetization transfer effect. Detection of aspartate aminotransferase reaction. Magn Reson Med 54:1321–1326
50. Xu S, Shen J (2009) Studying enzymes by in vivo C magnetic resonance spectroscopy. Prog Nucl Magn Reson Spectrosc 55(3): 266–283
51. Gruetter R (1993) Automatic, localized in vivo adjustment of all first- and second-order shim coils. Magn Reson Med 29(6):804–811
52. Shen J, Rothman DL, Hetherington HP et al (1999) Linear projection method for automatic slice shimming. Magn Reson Med 42(6):1082–1088

53. Bottomley PA (1987) Spatial localization in NMR spectroscopy in vivo. Ann N Y Acad Sci 508:333–348
54. Brateman L (1986) Chemical shift imaging: a review. AJR Am J Roentgenol 146(5): 971–980
55. Juchem C, Logothetis NK, Pfeuffer J (2005) High-resolution 1H chemical shift imaging in the monkey visual cortex. Magn Reson Med 54:1541–1546
56. Adriany G, Gruetter R (1997) A half-volume coil for efficient proton decoupling in humans at 4 tesla. J Magn Reson 125(1):178–184
57. Li S, An L, Yu S et al (2016) (13)C MRS of human brain at 7 Tesla using [2-(13)C]glucose infusion and low power broadband stochastic proton decoupling. Magn Reson Med 75(3):954–961
58. van Eijsden P, Behar KL, Mason GF et al (2010) In vivo neurochemical profiling of rat brain by 1H-[13C] NMR spectroscopy: cerebral energetics and glutamatergic/GABAergic neurotransmission. J Neurochem 112(1): 24–33
59. Xiang Y, Shen J (2011) Simultaneous detection of cerebral metabolism of different substrates by in vivo ^{13}C isotopomer MRS. J Neurosci Methods 198(1):8–15
60. Hertz L, Rothman DL (2016) Glucose, lactate, β-hydroxybutyrate, acetate, GABA, and succinate as substrates for synthesis of glutamate and GABA in the glutamine-glutamate/GABA cycle. Adv Neurobiol 13:9–42
61. Ross BD, Lin A, Harris K et al (2003) Clinical experience with 13C MRS in vivo. NMR Biomed 16:358–369
62. Mason GF, Falk Petersen K, de Graaf RA et al (2003) A comparison of (13)C NMR measurements of the rates of glutamine synthesis and the tricarboxylic acid cycle during oral and intravenous administration of [1-(13)C] glucose. Brain Res Brain Res Protoc 10(3): 181–190
63. Lee D, Marcinek D (2009) Noninvasive in vivo small animal MRI and MRS: basic experimental procedures. J Vis Exp 20(32). pii: 1592. doi:10.3791/1592
64. Li S, Shen J (2005) Integrated RF probe for in vivo multinuclear spectroscopy and functional imaging of rat brain using an 11.7 Tesla 89 mm bore vertical microimager. MAGMA 18(3):119–127
65. Ennis K, Deelchand DK, Tkac I et al (2011) Determination of oxidative glucose metabolism in vivo in the young rat brain using localized direct-detected ^{13}C NMR spectroscopy. Neurochem Res 36(11):1962–1968
66. Beckmann N, Turkalj I, Seelig J et al (1991) ^{13}C NMR for the assessment of human brain glucose metabolism in vivo. Biochemistry 30:6362–6366
67. Bluml S, Moreno-Torres A, Shic F et al (2002) Tricarboxylic acid cycle of glia in the in vivo human brain. NMR Biomed 15:1–5
68. Cerdan S, Kunnecke B, Seelig J (1990) Cerebral metabolism of [1,2-13C2]acetate as detected by in vivo and in vitro 13C NMR. J Biol Chem 265:12916–12926
69. Yang J, Li SS, Bacher J et al (2007) Quantification of cortical GABA-glutamine cycling rate using in vivo magnetic resonance signal of [2-13C]GABA derived from glia-specific substrate [2-13C]acetate. Neurochem Int 50(2):371–378
70. Boumezbeur F, Petersen KF, Cline GW et al (2010) The contribution of blood lactate to brain energy metabolism in humans measured by dynamic 13C nuclear magnetic resonance spectroscopy. J Neurosci 30(42): 13983–13991
71. Bagga P, Behar KL, Mason GF et al (2014) Characterization of cerebral glutamine uptake from blood in the mouse brain: implications for metabolic modeling of 13C NMR data. J Cereb Blood Flow Metab 34(10): 1666–1672
72. Muir D, Berl S, Clarke DD (1986) Acetate and fluoroacetate as possible markers for glial metabolism in vivo. Brain Res 380:336–340
73. Waniewski RA, Martin DL (1998) Preferential utilization of acetate by astrocytes is attributable to transport. J Neurosci 18:5225–5233
74. Sonnewald U, Westergaard N, Schousboe A et al (1993) Direct demonstration by [13C] NMR spectroscopy that glutamine from glia is a precursor for GABA synthesis in neurons. Neurochem Int 22:19–29
75. Gulanski BI, De Feyter HM, Page KA et al (2013) Increased brain transport and metabolism of acetate in hypoglycemia unawareness. J Clin Endocrinol Metab 98(9): 3811–3820
76. Chateil J, Biran M, Thiaudiere E et al (2001) Metabolism of [1-13C]glucose and [2-13C] acetate in the hypoxic rat brain. Neurochem Int 38:399–407
77. Hassel B, Sonnewald U, Fonnum F (1995) Glial-neuronal interactions as studied by cerebral metabolism of [2-13C]acetate and [1-13C]glucose: an ex vivo 13C NMR spectroscopic study. J Neurochem 64:2773–2782
78. Garcia-Espinosa MA, Garcia-Martin ML, Cerdan S (2003) Role of glial metabolism in diabetic encephalopathy as detected by high

resolution 13C NMR. NMR Biomed 16: 440–449

79. Sonnewald U, Kondziella D (2003) Neuronal glial interaction in different neurological diseases studied by ex vivo 13C NMR spectroscopy. NMR Biomed 16:424–429
80. Frahm J, Bruhn H, Gyngell ML et al (1989) Localized high-resolution proton NMR spectroscopy using stimulated echoes: initial applications to human brain in vivo. Magn Reson Med 9(1):79–93
81. Provencher SW (1993) Estimation of metabolite concentrations from localized in vivo proton NMR spectra. Magn Reson Med 30:672–679
82. Michaelis T, Merboldt KD, Bruhn H et al (1993) Absolute concentrations of metabolites in the adult human brain in vivo: quantification of localized proton MR spectra. Radiology 187:219–227
83. de Graaf AA, Deutz NE, Bosman DK et al (1991) The use of in vivo proton NMR to study the effects of hyperammonemia in the rat cerebral cortex. NMR Biomed 4:31–37
84. Lee HK, Yaman A, Nalcioglu O (1995) Homonuclear J-refocused spectral editing technique for quantification of glutamine and glutamate by 1H NMR spectroscopy. Magn Reson Med 34:253–259
85. Pan JW, Mason GF, Pohost GM et al (1996) Spectroscopic imaging of human brain glutamate by water-suppressed J-refocused coherence transfer at 4.1 T. Magn Reson Med 36:7–12
86. Thompson RB, Allen PS (1998) A new multiple quantum filter design procedure for use on strongly coupled spin systems found in vivo: its application to glutamate. Magn Reson Med 39:762–771
87. Thomas MA, Yue K, Binesh N et al (2001) Localized two-dimensional shift correlated MR spectroscopy of human brain. Magn Reson Med 46:58–67
88. Schulte RF, Trabesinger AH, Boesiger P (2005) Chemical-shift-selective filter for the in vivo detection of J-coupled metabolites at 3T. Magn Reson Med 53:275–281
89. Yahya A, Madler B, Fallone BG (2008) Exploiting the chemical shift displacement effect in the detection of glutamate and glutamine (Glx) with PRESS. J Magn Reson 191:120–127
90. Choi C, Coupland NJ, Bhardwaj PP et al (2006) Measurement of brain glutamate and glutamine by spectrally selective refocusing at 3 Tesla. Magn Reson Med 55:997–1005
91. Soher BJ, Pattany PM, Matson GB et al (2005) Observation of coupled 1H metabolite resonances at long TE. Magn Reson Med 53:1283–1287
92. Ernst T, Jiang CS, Nakama H et al (2010) Lower brain glutamate is associated with cognitive deficits in HIV patients: a new mechanism for HIV-associated neurocognitive disorder. J Magn Reson Imaging 32:1045–1105
93. Prescot AP, Richards T, Dager SR et al (2012) Phase-adjusted echo time (PATE)-averaging 1H MRS: application for improved glutamine quantification at 2.89 T. NMR Biomed 25: 1245–1252
94. Choi CH, Dimitrov IE, Douglas D et al (2010) Improvement of resolution for brain coupled metabolites by optimized H-1 MRS at 7 T. NMR Biomed 23:1044–1052
95. Slotboom J, Mehlkopf AF, Bovee WMMJ (1994) The effects of frequency-selective Rf pulses on J-coupled spin-1/2 systems. J Magn Reson A 108:38–50
96. Yablonskiy DA, Neil JJ, Raichle ME et al (1998) Homonuclear J coupling effects in volume localized NMR spectroscopy: pitfalls and solutions. Magn Reson Med 39:169–178
97. Maudsley AA, Govindaraju V, Young K et al (2005) Numerical simulation of PRESS localized MR spectroscopy. J Magn Reson 173: 54–63
98. An L, Li S, Murdoch JB et al (2015) Detection of glutamate, glutamine, and glutathione by radiofrequency suppression and echo time optimization at 7 tesla. Magn Reson Med 73(2):451–458
99. Pascual JM, Carceller F, Roda JM et al (1998) Glutamate, glutamine, and GABA as substrates for the neuronal and glial compartments after focal cerebral ischemia in rats. Stroke 29:1048–1056
100. Engelsen B, Fonnum F (1983) Effects of hypoglycemia on the transmitter pool and the metabolic pool of glutamate in rat brain. Neurosci Lett 42:317–322
101. Wong KL, Tyce GM (1983) Glucose and amino acid metabolism in rat brain during sustained hypoglycemia. Neurochem Res 8: 401–415
102. Lewis LD, Ljunggren B, Norberg K et al (1974) Changes in carbohydrate substrates, amino acids and ammonia in the brain during insulin-induced hypoglycemia. J Neurochem 23(4):659–671
103. Siesjö BK, Borgström L, Jóhannsson H et al (1976) Cerebral oxygenation in arterial hypoxia. Adv Exp Med Biol 75:335–342

104. Kanamatsu T, Tsukada Y (1994) Measurement of amino acid metabolism derived from [1-13C]glucose in the rat brain using 13C magnetic resonance spectroscopy. Neurochem Res 19:603–612
105. Lapidot A, Gopher A (1994) Cerebral metabolic compartmentation. Estimation of glucose flux via pyruvate carboxylase/pyruvate dehydrogenase by 13C NMR isotopomer analysis of d-[U-13C]glucose metabolites. J Biol Chem 269:27198–27208
106. Bachelard H (1998) Landmarks in the application of 13C-magnetic resonance spectroscopy to studies of neuronal/glial relationships. Dev Neurosci 20(4–5):277–288
107. Morris GA, Freeman R (1979) Enhancement of nuclear magnetic resonance signals by polarization transfer. J Am Chem Soc 101: 760–762
108. Doddrell DM, Pegg DT, Bendall MR (1982) Distortionless enhancement of NMR signals by polarization transfer. J Magn Reson 48:323–327
109. Baum J, Tycko R, Pines A (1985) Broadband and adiabatic inversion of a two-level system by phase-modulated pulses. Phys Rev A 32: 3435–3447
110. Silver MS, Joseph RI, Hoult DI (1984) Highly selective $\pi/2$ and π pulse generation. J Magn Reson 59:347–351
111. Garwood M, Ke Y (1991) Symmetric pulses to induce arbitrary flip angles with compensation for RF inhomogeneity and resonance offsets. J Magn Reson 94:511–525
112. Shaka AJ, Keeler J (1984) Broadband spin decoupling in isotropic liquids. Prog NMR Spectrosc 19:47–64
113. Shaka AJ, Barker PB, Freeman R (1986) Cycling sidebands in broadband decoupling. J Magn Reson 67:396–401
114. de Graaf RA (2005) Theoretical and experimental evaluation of broadband decoupling techniques for in vivo nuclear magnetic resonance spectroscopy. Magn Reson Med 53(6):1297–1306
115. Li S, Zhang Y, Wang S et al (2010) ^{13}C MRS of occipital and frontal lobes at 3 T using a volume coil for stochastic proton decoupling. NMR Biomed 23:977–985
116. An L, Warach S, Shen J (2011) Spectral localization by imaging using multielement receiver coils. Magn Reson Med 66(1):1–10
117. An L, Shen J (2015) Image-guided spatial localization of heterogeneous compartments for magnetic resonance. Med Phys 42(9): 5278–5286
118. Rothman DL, Behar KL, Hetherington HP et al (1985) ^{1}H-Observe/^{13}C-decouple spectroscopic measurements of lactate and glutamate in the rat brain in vivo. Proc Natl Acad Sci U S A 82:1633–1637
119. Pfeuffer J, Tkac I, Choi IY et al (1999) Localized in vivo ^{1}H NMR detection of neurotransmitter labeling in rat brain during infusion of [1-^{13}C] D-glucose. Magn Reson Med 41:1077–1083
120. Pan JW, Stein DT, Telang F et al (2000) Spectroscopic imaging of glutamate C4 turnover in human brain. Magn Reson Med 44(5):673–679
121. de Graaf RA, Brown PB, Mason GF et al (2003) Detection of [1,6-$^{13}C_2$]-glucose metabolism in rat brain by in vivo ^{1}H-[^{13}C]-NMR spectroscopy. Magn Reson Med 49: 37–46
122. Yang J, Shen J (2006) Increased oxygen consumption in the somatosensory cortex of alpha-chloralose anesthetized rats during forepaw stimulation determined using MRS at 11.7 Tesla. Neuroimage 32(3):1317–1325
123. Boumezbeur F, Besret L, Valette J et al (2004) NMR measurement of brain oxidative metabolism in monkeys using ^{13}C-labeled glucose without a ^{13}C radiofrequency channel. Magn Reson Med 52:33–40
124. Schousboe A (2003) Role of astrocytes in the maintenance and modulation of glutamatergic and GABAergic neurotransmission. Neurochem Res 28:347–352
125. Gether U, Andersen PH, Larsson OM et al (2006) Neurotransmitter transporters: molecular function of important drug targets. Trends Pharmacol Sci 27:375–383
126. Duce IR, Keen P (1983) Selective uptake of [3H]glutamine and [3H]glutamate into neurons and satellite cells of dorsal root ganglia in vitro. Neuroscience 8:861–866
127. Rae C, Hare N, Bubb WA et al (2003) Inhibition of glutamine transport depletes glutamate and GABA neurotransmitter pools: further evidence for metabolic compartmentation. J Neurochem 85:503–514
128. Dzubow LM, Garfinkel D (1970) A simulation study of brain compartments. II. Atom-by-atom simulation of the metabolism of specifically labeled glucose and acetate. Brain Res 23:407–417
129. van den Berg CJ, Garfinkel D (1971) A simulation study of brain compartments. Metabolism of glutamate and related substances in mouse brain. Biochem J 123: 211–218

130. Mason GF, Rothman DL, Behar KL et al (1992) NMR determination of the TCA cycle rate and alpha-ketoglutarate/glutamate exchange rate in rat brain. J Cereb Blood Flow Metab 12:434–447
131. Mason GF, Gruetter R, Rothman DL et al (1995) Simultaneous determination of the rates of the TCA cycle, glucose utilization, alpha-ketoglutarate/glutamate exchange, and glutamine synthesis in human brain by NMR. J Cereb Blood Flow Metab 15:12–25
132. Shen J, Rothman DL, Behar KL et al (2009) Determination of the glutamate-glutamine cycling flux using two-compartment dynamic metabolic modeling is sensitive to astroglial dilution. J Cereb Blood Flow Metab 29(1): 108–118
133. Shen J (2013) Modeling the glutamate-glutamine neurotransmitter cycle. Front Neuroenergetics 5:1
134. Zhang Y, Shen J (2014) Smoothness of in vivo spectral baseline determined by mean-square error. Magn Reson Med 72(4): 913–922
135. Lin AP, Shic F, Enriquez C et al (2003) Reduced glutamate neurotransmission in patients with Alzheimer's disease – an in vivo 13C magnetic resonance spectroscopy study. MAGMA 16:29–42
136. Costa J, Fareleira F, Ascenção R et al (2011) Clinical comparability of the new antiepileptic drugs in refractory partial epilepsy: a systematic review and meta-analysis. Epilepsia 52: 1280–1291
137. Petroff OA, Errante LD, Rothman DL et al (2002) Glutamate-glutamine cycling in the epileptic human hippocampus. Epilepsia 43: 703–710
138. Sherwin AL (1999) Neuroactive amino acids in focally epileptic human brain: a review. Neurochem Res 24:1387–1395
139. Pan JW, Spencer DD, Kuzniecky R et al (2012) Metabolic networks in epilepsy by MR spectroscopic imaging. Acta Neurol Scand 126(6):411–420
140. Pan JW, Duckrow RB, Gerrard J et al (2013) 7T MR spectroscopic imaging in the localization of surgical epilepsy. Epilepsia 54(9): 1668–1678

Chapter 5

Imaging Glutamate with Genetically Encoded Fluorescent Sensors

Gerard J. Broussard, Elizabeth K. Unger, Ruqiang Liang, Brian P. McGrew, and Lin Tian

Abstract

Superimposed on the vast and complex synaptic network is a largely invisible set of chemical inputs, such as neurotransmitters and neuromodulators, that exert a profound influence on brain function across many structures and temporal scales. Thus, the determination of the spatiotemporal relationships between these chemical signals with synaptic resolution in the intact brain is essential to decipher the codes for transferring information across circuitry and systems. Recent advances in imaging technology have been employed to determine the extent of spatial and temporal neurotransmitter dynamics in the brain, especially glutamate, the most abundant excitatory neurotransmitter. Here, we discuss recent imaging approaches, particularly with a focus on the design and application of genetically encoded indicator iGluSnFR, in analyzing glutamate transients in vitro, ex vivo, and in vivo.

Key words Glutamate, iGluSnFR, Fluorescent sensor, Genetically encoded indicators of neural activity, Protein engineering, Fluorescent functional imaging

1 Introduction

1.1 Glutamate Signaling in the Central Nervous System

Glutamate (Glu) is an amino acid used in a variety of contexts across all known kingdoms of life. In central nervous systems (CNS), Glu acts as the primary excitatory neurotransmitter [1]. As such, glutamatergic signaling forms the basis for information transfer in the brain. In addition to its role as key mediator of synaptic information relay, Glu signaling can induce de novo growth of functional synapses [2] and is also crucial in modifying synaptic strength during phenomena such as long-term potentiation and long-term depression [3]. Glutamate also serves as an essential component of the complex signaling interactions between neurons and astrocytes, the most prominent glial cell type in the CNS [4].

Gerard J. Broussard and Elizabeth K. Unger contributed equally to this work.

Sandrine Parrot and Luc Denoroy (eds.), *Biochemical Approaches for Glutamatergic Neurotransmission*, Neuromethods, vol. 130,
DOI 10.1007/978-1-4939-7228-9_5, © Springer Science+Business Media LLC 2018

Finally, dysregulation of Glu has been implicated in the pathogenesis of acute neurological insults, such as traumatic brain injury and stroke [5], and degenerative neurological disorders, including multiple sclerosis, Alzheimer disease, Huntington disease, Parkinson disease, amyotrophic lateral sclerosis, and others [6].

Glutamate signaling has a correspondingly wide range of spatial and temporal dynamics. Following action potential-evoked vesicular release, synaptic Glu concentrations rapidly increase to the low millimolar range [7]. Typically, clearance of synaptic Glu occurs with a time constant of decay of around 1 ms and is handled by a combination of diffusion and active reuptake by a family of excitatory amino acid transporters, which are primarily localized to nearby astrocytes [7, 8]. While the distance between synapses in many brain regions is typically less than 1 μm [9], spillover of Glu during synaptic release can potentially affect neighboring cells over 10 μm away [10]. Meanwhile, Glu tone within the extra-synaptic space is primarily regulated by astrocytes and sits at a concentration within the high nanomolar to low micromolar range, dependent on local patterning of Glu release and reuptake sites [11].

Glutamate receptors are broken into two primary subtypes based on whether they are mediated by metabotropic or ionotropic Glu receptors (mGluRs or iGluRs, respectively). The human mGluR family is made up of eight G-protein-coupled receptors, subdivided into three groups based on mechanism and pharmacology. Group I mGluRs exert their action by activating phospholipase C, while mGluRs in groups II and III inhibit production of cyclic AMP [12]. Alternatively, the rapid excitatory signaling at synapses is mediated by iGluRs, which are ligand-gated cation channels classically categorized into three pharmacologically identified subtypes: *N*-methyl-D-aspartate (NMDA) receptors, α-amino-3-hydroxy-5-methyl-4-isoaxazole propionate (AMPA) receptors, and kainate receptors [13].

Physiological consequences of activation of Glu receptors are dictated by receptor affinity, kinetics, and localization. Affinities of these various receptor types range over about three orders of magnitude, with the highest affinities (~1–10 μmol/L) found among the NMDA receptors and most mGluRs, while the lowest affinity receptors are non-NMDA ionotropic receptors and mGluR7, with affinities of 500 μmol/L and 1000 μmol/L, respectively [14, 15]. Activation kinetics of Glu receptors range from ones of milliseconds for non-NMDA receptors to tens of milliseconds for all other receptor subtypes [14, 16, 17]. In the CNS, ionotropic Glu receptors localize primarily to synaptic and perisynaptic sites, while mGluRs can be found in a range of different neuronal cellular compartments [18], as well as studding the surface of glial cells, such as astrocytes [19].

Given the importance of glutamatergic signaling in the brain, there is a pressing need for technology to enable direct and precise measurements of Glu signaling at the synaptic and microcircuit levels across various temporal scales in behaving animals.

1.2 Current Approaches for Direct Measurement of Glutamate in the Brain

Analytic chemistry has provided useful insights about the concentrations of Glu in the brain via microdialysis [20], glutamate oxidase-enabled cyclic voltammetry [21], and semisynthetic sensors [22–24]. However, microdialysis measurement surface areas are typically in the range of 1000 μm², while concentration measurements can be taken on intervals ranging from ones to tens of minutes [20]. Whereas cyclic voltammetry does allow for rapid temporal precision, typical probe size is in the range of 30–250 μm² [25, 26]. This approach also lacks cellular resolution, can potentially lose sensitivity in biological tissues, and can be confounded by other potential sources of naturally occurring electroactive molecules [27]. Both methods thus fall far short of the spatial resolution required to track Glu changes in the subcellular domains in vivo.

Alternatively, microscopy-based measurements can provide the temporal and spatial resolution required to image Glu transients in intact tissue. Advances in the speed, depth penetration, and spatial resolution of fluorescence microscopy techniques have extended our ability to measure glutamatergic signaling noninvasively. Optical recordings from hundreds to thousands of neuronal elements across large tissue areas and volumes are now possible with single-cell or single-synapse resolution, both in vitro and in vivo. One-photon imaging with ultrafast cameras permits very rapid (>1 kHz) subcellular-resolution optical recordings from millimeter-scale surface regions. In contrast, multiphoton microscopy allows noninvasive functional imaging from up to ~1 mm tissue depths across near-millimeter fields of view, with subsecond temporal and submicron spatial resolution. In addition, both one- and multiphoton imaging can be performed in behaving animals and through endoscopes or fiber optics, allowing measurements from deep brain structures, such as the striatum or brainstem. As imaging technologies become more developed and available for routine laboratory applications, high-quality optical probes have also been developed in parallel (for details, see Sect. 1.5).

1.3 Imaging Probes for Glutamate

To directly measure Glu with molecular specificity, an array of Glu probes including small molecule-protein hybrids and genetically encoded sensors, have been developed to be compatible with fluorescent microscopy. For example, Glu optical sensor (EOS) is a hybrid sensor constructed from the S1S2 Glu-binding domain of the AMPA receptor GluR2 subunit and a small-molecule dye [22]. One of the EOS variants exhibited an increase of 48% in $\Delta F/F_{max}$ (maximal fluorescence changes over basal fluorescence) upon Glu binding (K_d = 1.57 μmol/L) when measured at the surface of HeLa cells and about a 1–2% fluorescence increase in mouse somatosensory cortical neurons following limb movement. However, the use of such sensors generally requires application of an exogenous cofactor, such as a small molecule dye [22], limiting their utility for in vivo deployment and barring their use in experiments requiring chronic measurements.

More recently, Snifit (SNAP-tag-based indicator proteins with a fluorescent intramolecular tether)-based sensors for optical measurement of Glu have been developed. A Glu-Snifit is a fusion protein with three components, an ionotropic Glu receptor 5 (iGluR5), a CLIP-tag with a synthetic donor fluorophore, and a SNAP-tag bearing another synthetic acceptor fluorophore conjugated with O6-benzylguanine (BG)-polyethylene glycol (PEG) 11-Cy5-glutamate as an antagonist [23]. In the absence of Glu, the Snifit adopts a closed conformation and Förster resonance energy transfer (FRET) efficiency increases. In the presence of Glu, the intramolecular antagonist is displaced to shift Snifit toward an open state with a resultant decrease in FRET efficiency. The Snifit-iGluR5 showed a ΔR_{max} (maximal fluorescence change in donor-acceptor ratio) of 1.93 and an apparent K_d of 12 μmol/L for Glu in HEK 293T cells. Further optimization and characterization are likely needed to allow in situ detection of physiological levels of Glu release and uptake by living neurons and intact circuitry.

1.4 Genetically Encoded Fluorescent Indicators for Neural Activity

Genetically encoded indicators of neural activity (GINAs) are a set of tools that permit noninvasive, direct, specific, and long-term measurements of changes in calcium, voltage, Glu, and vesicular release over various temporal scales within genetically specified cells or synapses, and within or across neural circuits of live animals (reviewed in [28, 29]).

GINAs are proteins comprising a sensor and a fluorescent-reporter domain that can act as optical readouts of signals propagated within the nervous system. Conformational changes, upon binding of an analyte by the sensor domain, drive a fractional modulation of the reporter-domain fluorescence output (often referred to as $\Delta F/F$). Applications of these sensors have facilitated large-scale recording of neural activity in genetically identified neurons or glial cells. In addition, genetically encoded sensors without overlapped spectra have been engineered to allow multiplex imaging or to combine with optogenetic actuators, which has opened many possibilities for sophisticated analysis of neural activity. Genetically encoded sensors can be selectively targeted to genetically defined cell types or subcellular locations, such as axons or dendritic spines, or to cells with specific anatomical connectivity. Finally, these sensors can be stably expressed over a long time (from days to several months), allowing the study of how neural-activity patterns change with learning, development, or disease progression.

The first GINAs to gain widespread use for in vivo detection of neural activity were genetically encoded calcium indicators (GECIs) designed to detect action potential-related calcium fluxes. This subset of GINAs includes single-fluorescent protein (FP)-based GCaMP family and FRET-based Cameleon family [30–34]. For example, in the scaffold of GCaMPs, calmodulin (CaM) binds the RS20 peptide from smooth muscle myosin light chain kinase in the presence of calcium; this coupling reverses when calcium is absent. The sensor domain transduces conformational changes of calcium binding to a change in the fluorescence intensity through its coupling with the circularly permutated green fluorescent protein (GFP) (cpGFP; see [35]). The crystal structure of GCaMP2 has revealed that the rearrangement of the CaM/RS20 domain upon calcium binding brings an arginine at position 377 into proximity to coordinate the chromophore of cpGFP into a deprotonated bright state [36]. Iterative improvements have since been made to this original design through rational design combined with directed evolution, which has produced a series of widely applicable sensors, GCaMP3/5/6, for calcium imaging in awake, behaving animals [34, 37–41]. Studies have also been conducted examining changes to calcium dynamics with temporal scales ranging from single milliseconds [42, 43] to learning-associated [44] and plasticity [45] changes unfolding over months. While optimization of calcium indication continues with the focus on faster kinetics, superior brightness, and sensitivity for detecting subthreshold activity, the genetically encoded voltage sensors have also been intensively optimized to facilitate voltage imaging both at mesoscopic and cellular resolutions in living behaving mammals [46].

1.5 Design of Genetically Encoded Fluorescent Indicators for Glutamate

Despite many advantages presented by GECIs as a tool for neuroscience, in certain contexts, a GINA designed to directly detect Glu is superior or complementary. For example, coupling between calcium influx and Glu release from presynaptic compartments can be highly nonlinear, such as during the process of synaptic facilitation or depression [47]. A Glu sensor expressed in postsynaptic cells could therefore more accurately reflect the influence of presynaptic activation on receiving-cell physiology. Additionally, in experiments involving bulk imaging of neural tissue, GECIs report primarily spiking activity in cells within the imaged volume. Whereas a Glu sensor grants experimental access to subthreshold synaptic activation [48]. Finally, not all Glu-initiated neural signaling events involve obligate calcium influx (see [49]). The ability to perform multiplex measurement of Glu signaling combined with measurement of calcium, voltage, and circuitry manipulation tools is even more important.

To date, protein-engineering efforts have led to the development of a series of single-FP or FRET-based Glu sensors consisting of the bacterial glutamate/aspartate-binding protein, GltI (also called YbeJ). GltI is the periplasmic component of the ATP-binding–cassette transporter complex for aspartate and Glu in *Escherichia coli* (*E. coli*; [50]). GltI is a member of the periplasmic binding protein (PBP) superfamily, which is typically composed of two domains that undergo a Venus flytrap-like hinge-twist motion upon ligand binding [51]. This conformational change serves as the basis for Glu sensitivity in FRET-based Glu sensors, including fluorescent indicator protein for Glu FLIPE [52] and SuperGluSnFR [53], or the single-FP-based iGluSnFR [54].

To date, iGluSnFR is the only Glu indicator that has been widely used for in vivo imaging. The single-FP-based iGluSnFR was developed by inserting cpGFP, the same cpGFP developed for GCaMP2 [31], into Gltl. Upon Glu binding, conformational adjustments result in changes in the chromophore environment, thus transforming the ligand-binding event into fluorescent intensity changes. A critical challenge in the construction of iGluSnFR (and indeed, any sensor with similar design properties) is to determine the optimal insertion site in Gltl for cpGFP. One approach to rational design of the insertion position is analysis of change in the dihedral-bond angle of alpha carbons (i.e., bond angle formed by four sequential amino acids) upon ligand binding (for details, please see Sect. 2). To maximize the $\Delta F/F$ and kinetics in response to Glu, linker regions between Gltl and cpGFP were optimized via site-saturated mutagenesis. Because of the close proximity of the linkers to the chromophore, those linkers are well situated to modify chromophore-solvent access and the stability of apo- and bound conformations [30, 33, 55]. A sizable portion of the possible 160,000 variants was screened from each cpGFP insertion site.

The final design of iGluSnFR produced a sensor with greater $\Delta F/F$ (~4.5) and affinity (K_d ~ 4 μmol/L) to Glu than previously available genetically encoded Glu sensors. Additionally, the "on rate" for binding was too fast to be determined by stopped-flow cytometry (~6 ms resolution), suggesting that the kinetics of iGluSnFR may be sufficiently rapid to faithfully track even the fast component of Glu transients at the synaptic cleft [7].

The superior intrinsic properties of iGluSnFR permit in vivo applications. iGluSnFR has proven to be useful in tracking phenomena at spatial scales ranging from subcellular to mesoscopic imaging of entire brain regions [48–64]. As with all genetic tools, iGluSnFR can be targeted to genetically defined cell populations, and has thus also proven ideal for unraveling the contributions to dynamics of the local brain circuitry made by specific neural

[56, 57] and even glial cells [65–69]. Finally, the superior kinetics of iGluSnFR, coupled with the rapidity of Glu transients themselves, have allowed wide-field imaging with greater temporal precision than is possible with current calcium indicators [48].

1.6 Systematization of Sensor Characterization and Validation In Vitro, Ex Vivo, and In Vivo

While GINAs have made massive strides in the past decade, further development is an ongoing task. The ultimate goal of GINA development is to obtain an appropriate signal-to-noise ratio (SNR) that matches the system studied for in vivo preparations. Because of imaging depth and motion artifacts, in vivo imaging has very demanding signal-to-noise requirements. To maximize the SNR and to achieve the best imaging outcomes, the intrinsic properties of sensors, including photophysical properties, specificity, kinetics, and affinity, must be matched to the physiological properties of measured signaling events, including the amplitude and transient size, time course, and frequency, as well as to the imaging system properties, such as speed, resolution, sensitivity, and depth penetration. In addition, it is essential to optimize the expression level for minimizing long-term cellular expression effects, such as clumping and off-target fluorescence species and interference with endogenous signaling pathways. End users must choose the most appropriate sensors and imaging systems; therefore, calibrating and comparing each sensor's performance under the same experimental conditions is critical.

Recently, a highly optimized pipeline for novel sensor development has been established in our lab and others, from concept to protein purification and validation in cultured cells, brain slices, and in vivo (Fig. 1) [28, 70, 71]. For probe development, in silico-aided design along with high-throughput genetic cloning strategies are becoming a standard set of tools for the modern protein engineer. To balance between throughput and in vivo predictability, with the goal of finally demonstrating the capability of these sensors in living animals, we carry out a systematic characterization

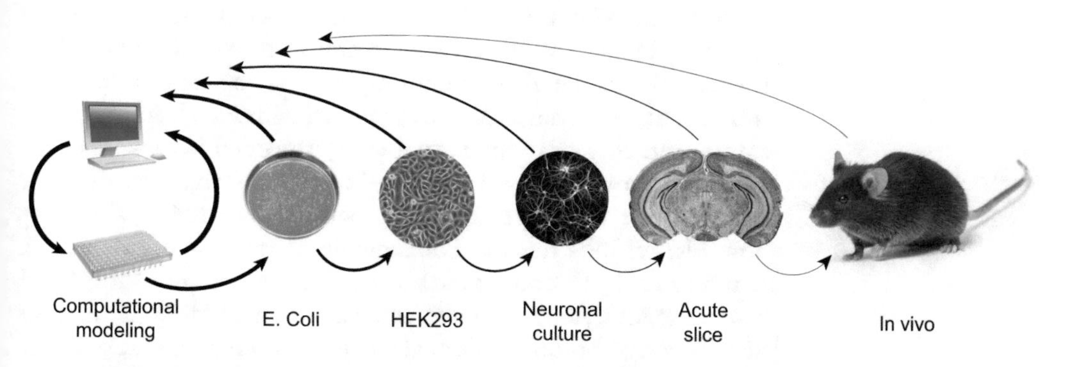

Fig. 1 Iterative workflow for testing variants in increasingly complex biological systems from computational design, *E. coli*, HEK293 cells, dissociated neuronal culture, acute brain slice, and in vivo

in *E. coli*-derived protein, in mammalian cells, in dissociated neuronal culture, and in acute brain slices and in vivo. This multiple-layer characterization also provides feedback for improving overall design and generation of libraries to iteratively optimize the intrinsic properties of the sensors. It is critical that the entire design process is best thought of as a closed loop where each step in the process informs earlier steps. Through iterative design and experimental validation, we expect to obtain lead sensors with the required design specifications and SNR for in vivo imaging. Here, we discuss each of these steps for developing Glu sensors, which can be broadly utilized for other sensor development.

2 Materials and Methods

2.1 Modeling Optimal Insertion Site of cpGFP with Rosetta

Single-wavelength, genetically encoded, fluorescent sensors undergo specific conformational changes induced by ligand-protein binding, resulting in a fluorescence readout. Molecular modeling is an indispensable tool to guide the engineering of this process. For this purpose, we use Rosetta, a suite of software libraries for macromolecular modeling [72].

For ligands having a natural bacterial PBP, a cpGFP insertion site can be predicted by maximizing the difference of C-alpha torsion between the backbones of the apo- and ligand-bound PBP scaffolds with the UCSF Chimera program [73]. The resultant structure of the cpGFP insertion can be simulated with rosetta_cm.xml script using the structures of the PBP (for example, Gltl PDB ID: 2VHA) and cpGFP (PDB ID: 3EVP) as templates [74]. As the Glu-unbound state of Gltl is not available, homologous analysis with maltose-binding protein (MBP) was used to predict the insertion site of cpGFP in Gltl (Fig. 2). Similar to GltI, MBP is a PBP family member, and the two proteins share a high degree of homology.

The psi/phi torsion changes between apo- and ligand-bound conformations of PBP scaffold protein can be calculated by extracting these torsions using Chimera. These values can be accessed from the Render by Attribute submenu in the Structure Analysis submenu under the Tools menu. After exporting these values into a text file, the difference can be calculated with Excel. Importantly, attention to correct residue numbering is needed because the two sequences between apo- and ligand-bound conformations may not necessarily be the same. cpGFP should be inserted at sites that show a large alpha torsion. For example, in the case of iMaltSnFR an amino acid position was identified near the hinge of the MBP that undergoes large torsional changes upon binding of maltose [55]. Through homologous analysis, it was predicted that position 249 in Gltl would be a potential insertion site for cpGFP (Fig. 2c).

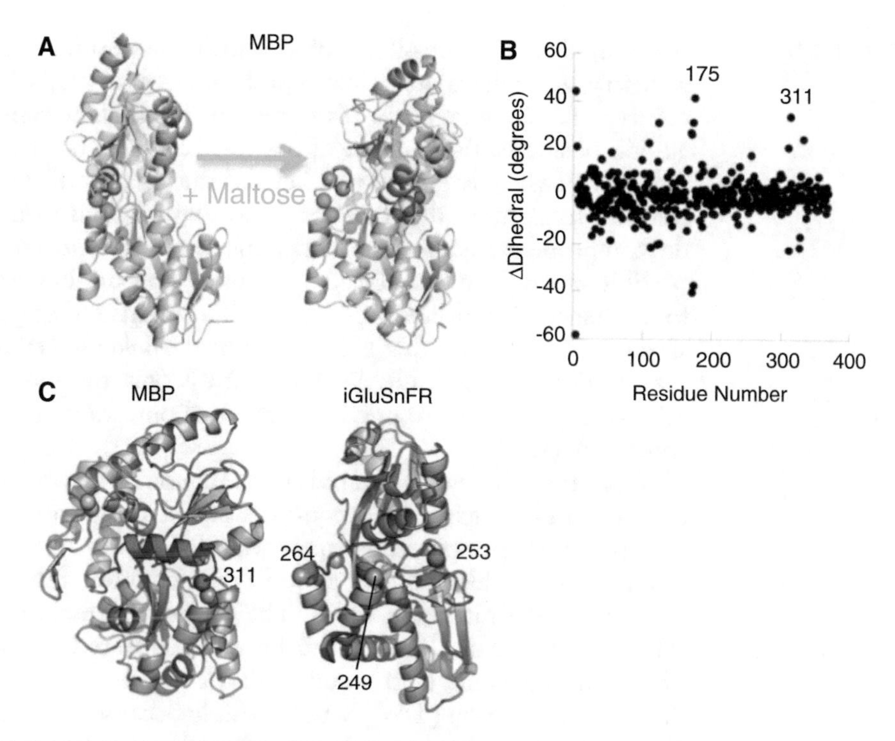

Fig. 2 (**a**) Crystal structure of iMaltSnFR with and without maltose bound, and (**b**) C-α-torsion at each residue (reproduced with permission from Marvin et al. (2011) [55]). (**c**) Comparison between iMaltSnFR and iGluSnFR (reproduced with permission from Marvin et al. (2013) [54])

Based on this initial site combined with optimization, insertion of cpGFP into position 253 of GltI yielded the basis for the final iGluSnFR design.

2.2 Directed Evolution

Computer modeling is an excellent way to begin the process of redesigning a protein; however, not all physical properties of the protein are, or even can be, explored. Even with sophisticated computer modeling algorithms such as the ones available in Rosetta, fully predicting the most-beneficial mutations for a specific protein function is still difficult. The idea of directed evolution is to introduce mutations in an iterative manner, while applying selection pressure to gradually enhance specific properties of the protein, be it thermal stability, brightness, binding affinity, and so forth. Mutations may be completely random, as in error-prone polymerase chain reaction (PCR), or they may be targeted to certain positions deemed most likely to yield the most promising results, as in site-saturated mutagenesis. Selection pressure is then applied by subjecting the variants to a test and promoting only the best performers to the next round of mutagenesis (Fig. 3).

2.2.1 Site-Directed Mutagenesis

Full sampling of all possible amino acids at all positions within a protein is generally an infeasible approach. For example, iGluSnFR consists of 562 amino acids. If we were to sample all 20 residues at all 562 positions, the number of possible combinations would be incomprehensibly large and well beyond the capacity of any lab to test, not to mention that most of those combinations would produce improperly folded or nonfunctional proteins. The number of possible variants can be reduced by focusing only on sites expected to influence ligand-binding-induced fluorescence change. The strength of this approach is exemplified by the linker screen used in the creation of iGluSnFR. By focusing on only the four amino acids connecting GltI to cpGFP, libraries of only 1.6×10^5 candidates were created.

Traditionally, site-saturated mutagenesis is performed by creating primers with degenerate codons at the position to be mutated. For example, the primer will include 10–15 matching bases, then the degenerative codon NNK or NNS, followed by another 10–15 matching bases. The purpose of including a K (G or T) or S (G or C) in the third position is to reduce the possibility of producing a stop codon (TAA or TGA). Eliminating the third stop codon (TAG) is not possible because both methionine and tryptophan only have one codon: ATG and TGG, respectively; thus, G must be a possibility in the third position to cover all 20 amino acids. The advantage of such a strategy is that a single oligo can be created that contains codons for all 20 amino acids. One major criticism of this strategy is that there is not equal probability of randomly selecting each amino acid, and some residues will be over-represented, while others will be under-represented. For instance, using either an NNK or NNS primer, there are 32 possible codons, with three possible codons for leucine, serine, and arginine; two possible codons for valine, glycine, alanine, threonine, and proline; and one possible codon for each of the rest. An alternative strategy is to generate one primer for each amino acid [75], which ensures equal probability of each one but is more costly and time-consuming. However, duplicate variants are expected within a screen, even one with equal probability distribution. Thus, to ensure all 20 amino acids are represented, many more samples are tested than the number of possibilities. For a typical library at one position with 20 possible amino acids, 96 colonies of variants are picked and tested. Even with an unequal probability distribution, this setup is usually sufficient to obtain at least one instance of each amino acid. The results of the screen should not depend on the number of times the amino acid is tested; the only consideration should be how many colonies need to be tested to ensure each amino acid is represented in the dataset.

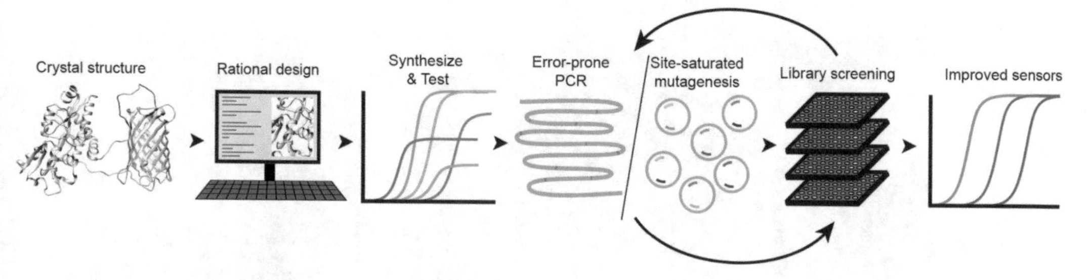

Fig. 3 Workflow for performing directed evolution

There are several methods for introducing mutations, insertions, or deletions into the template plasmid (Fig. 4). If only one or two positions are being interrogated, simple methods can be employed, such as QuikChange® mutagenesis. With QuikChange®, complementary primers containing the mutated codon are used in a PCR, which then linearly amplifies the plasmid while incorporating the mutation. The template is then digested away using DpnI, and the nicked plasmids are transformed to bacteria. The efficiency of QuikChange® is heavily impacted by the formation of primer dimers. One solution is to perform separate parallel reactions with a single primer each [76]. Another solution is to offset the primers such that they have a shorter region of homology with a melting temperature lower than that of the annealing temperature of the PCR [77]. A similar method is site-directed mutagenesis with blunt-end ligation, where primers are designed to amplify the entire plasmid, while introducing the mutation at the end of one or both of the primers. Because there is no homology between the primers, primer dimer is not an issue and the plasmid is linearized and exponentially amplified in a PCR. This method makes it easy to introduce insertions, because additional basepairs at the end of the primer do not affect the binding affinity. One concern is that longer primers occasionally will be missing the last basepair or two, and thus the sequence must be confirmed afterward. This method requires phosphorylated primers and a high-performance polymerase that does not leave an A-tail, such as Phusion (NEB) or Hotstart Taq (NEB). Alternatively, the added A overhang from a traditional Taq-based PCR may be digested away using Klenow (NEB). The template DNA is destroyed by DpnI digestion, and the ends of the linearized plasmid are ligated together and transformed to bacteria. While blunt-end ligation can be more efficient than QuikChange®, it requires the added steps and cost of phosphorylating the primers.

While QuikChange® and blunt-end ligation are traditionally used to introduce one mutation at time, for libraries at two distant positions, we typically use circular polymerase extension cloning (CPEC) [78]. Two sets of complementary primers are designed, each spanning one of the positions to be mutated. In the first PCR

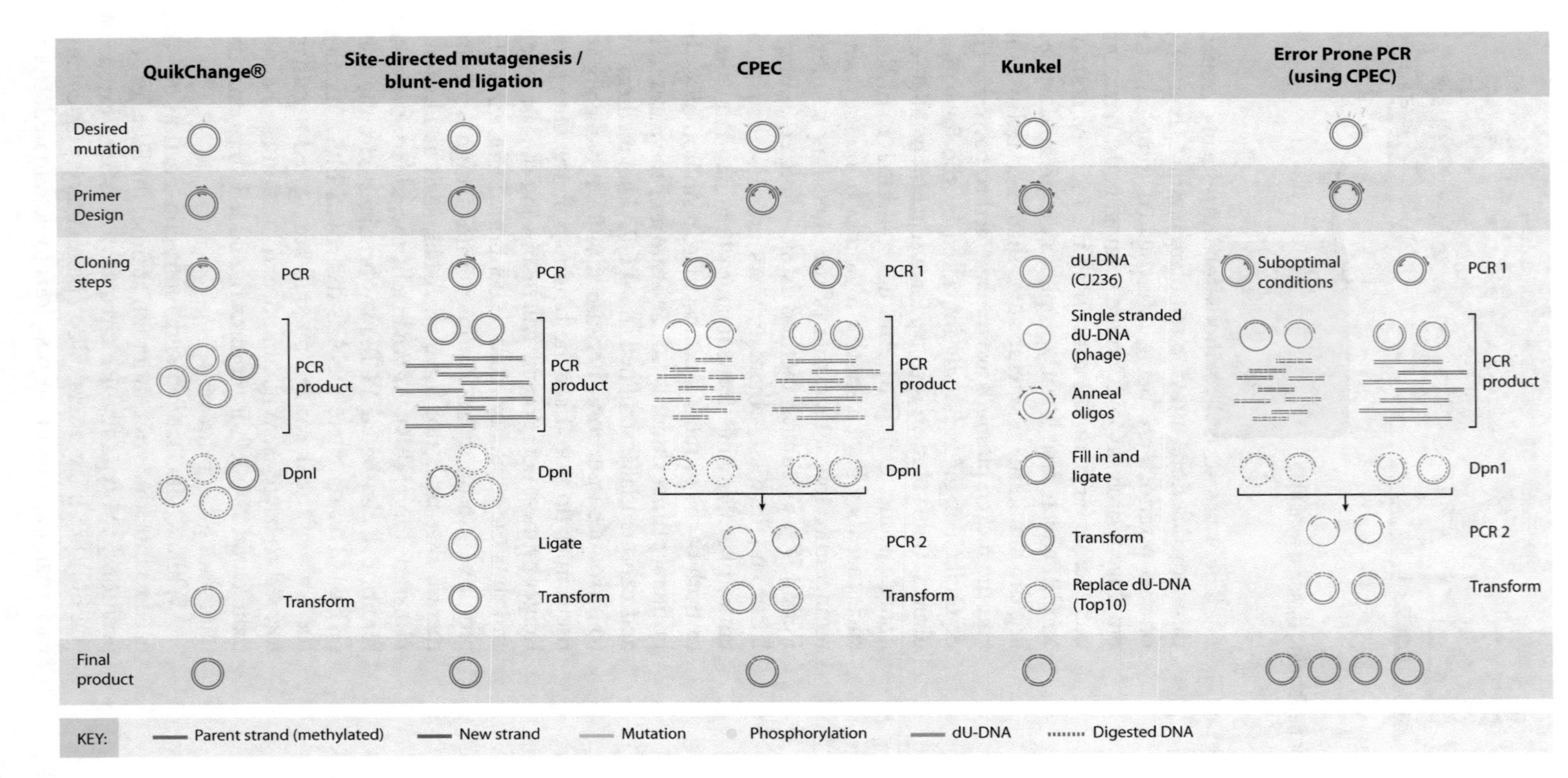

Fig. 4 Comparison of different cloning methods used to introduce random or targeted mutations into a genetically encoded sensor

step, noncomplementary primers are added to two separate reactions, such that each reaction will exponentially amplify a section of the plasmid between the mutated positions. Template DNA is digested away, and then the products are combined in the second PCR step, wherein the two reactions are mixed. Because the initial primers were designed to be complementary, each amplified region can now serve as a primer for the remainder of the plasmid. There is no amplification at this step, so fewer cycles need to be run. Nicked plasmids are then transformed into bacteria. Both QuikChange® and CPEC are robust and easy to use, however, it must be noted that the probability of two matching primers coming together in the same plasmid is quite low. Thus, we rely on the mismatch repair machinery within the bacteria to correct these errors and to ligate the nicked plasmids.

While there are several methods for introducing mutations at more than two positions, one of the most efficient methods is Kunkel mutagenesis [79]. In this method, unidirectional primers are designed for each mutation (one primer per mutation). The plasmid is transformed into a strain of bacteria lacking a uracil-degrading enzyme, CJ236 (−UNG). The bacteria are then infected with a bacteriophage (M13K07), which creates single-stranded, uracil-containing DNA. Single-stranded dU-DNA is harvested, and phosphorylated oligos containing each mutation are annealed to the template. The gaps are then filled in and ligated together using a T7 polymerase and a T4 ligase. Note that oligos must be annealed on the same strand as the f1 origin for the process to work. The plasmid is then transformed to a normal bacterial strain capable of degrading uracil (+UNG), for example Top10 (Thermo Fisher), DH5α (NEB), or HB101 (Thermo Fisher), which will degrade the uracil-containing parent strand and replace it using the mutated strand to create a double-stranded, mutated plasmid. We have successfully introduced as many as 18 separate mutations in a single round of Kunkel mutagenesis with a surprisingly high success rate. Generating the single-stranded dU-DNA can be time-consuming, but is much faster than repeated rounds of QuikChange®. Including an additional oligo that destroys a restriction site and then digesting with that restriction enzyme allows selection of only those plasmids that have incorporated the new primers [80]. In many cases, most or all of the primers will bind in a single plasmid.

An alternative method of Kunkel mutagenesis skips the generation of single-stranded dU-DNA. In this method, phosphorylated oligos are annealed directly to the parent plasmid. Then the gaps are filled in by a polymerase and ligated together as above. The parent plasmid is digested away using DpnI, and a single primer is added to the remaining strand in a single-step PCR to produce the complementary strand. The nicked plasmid is then transformed to bacteria.

2.2.2 Error-Prone PCR

Rather than introducing mutations at specific locations, introducing mutations at random is often advantageous, especially when computational models for rational design are not possible. One method for doing so is error-prone PCR (EPP), where the PCR conditions are altered to be suboptimal, which produces higher error rates [81]. Any combination of adding excess $MgCl_2$ or $MnCl_2$, adding unequal concentrations of each nucleotide, and using a low-fidelity polymerase can achieve this result. Often, these mutations will be single-point mutations but may also be deletions or insertions, which may lead to frameshifts. To introduce mutations only to the gene being interrogated rather than the entire plasmid, the gene may be amplified by EPP and then inserted as a library into the vector via ligation cloning, CPEC (step 2), or Gateway® cloning [82]. Although the efficiency of producing improvements in the engineered protein is lower than with site-directed mutagenesis, this method allows for the exploration of many more possibilities and the potential for unpredicted benefits.

2.3 High-Throughput Bacterial Screening

2.3.1 Determining In Vitro ΔF/F

In silico design methods result in a large number of potentially suitable target protein structures. However, their true utility must be tested in vitro. We performed our initial screening on a bacterial platform, which is fast and fairly easy. We began with a bacterial expression vector, pRSET-A (Thermo Fisher), which contains a conditional promoter, and a 6×-his tag for protein purification. Plasmid libraries are transformed to a bacterial strain lacking several protein degradation enzymes, BL21(DE3)(Thermo Fisher). This strain contains a T7 RNA polymerase, which can be induced by addition of Isopropyl β-D-1-thiogalactopyranoside (IPTG) once the culture reaches log-phase growth. The competency of BL21 cells is somewhat less than that of other bacterial strains; thus, to obtain a sufficient number of colonies, first amplifying libraries in a high competency strain may be necessary before transforming to BL21. BL21 also shows leaky expression of the protein, which may be a disadvantage if a toxic protein is being expressed. If leaky expression occurs, modified strains such as BL21(DE3)pLysS (Thermo Fisher) are available. The pLysS strain has the added advantage of being able to lyse the cells by repeated freeze-thaw cycles because of the lysogenic enzymes released during each cycle. However, for our purposes, leaky expression is an advantage because IPTG is also somewhat toxic to cells, and with leaky expression, IPTG is not needed at all. Once the bacteria reaches log-phase growth, cultures are simply grown for a long time (72–96 h) at a lower temperature (18 °C), which provides sufficient protein for screening or purification purposes. In addition, leaky expression is useful for colony selection subjected to screening.

Candidate colonies can be derived from libraries of variants grown on one or several agar plates, depending on the size of the library. If selection pressure can be applied at this step, the throughput of the screening will greatly increase. For example, iGluSnFR is a Glu sensor with increased brightness in the presence of Glu. Given that Glu is in the growth medium, colonies could be selected under fluorescent illumination based on their brightness. In this way, thousands of colonies can be screened on a single agar plate in a single step. Not all libraries are amenable to such screening, and in that case, colonies can be picked at random. Selected colonies can then be grown in 1 mL of culture medium in 96-well format deep-well plates. Cultures can be grown as above, first at 37 °C and then at 18 °C, until the desired protein production level is reached (72 h).

When screening, obtaining a quick, reliable, quantifiable readout of the performance of each variant is necessary. Given that the only relevant output of our Glu sensor is its fluorescence upon Glu binding, we test our variants in a simple fluorescence assay with and without Glu. $\Delta F/F$ can then be calculated from these values. Bacterial colonies can be pelleted, washed, and lysed using B-PER Complete (Thermo Fisher). The cell lysate is transferred to a 96-well plate with an optically clear bottom and read on a fluorescent plate reader (Tecan). Reading the same wells both before and after ligand addition is important to control for pipetting-error differences between wells. Testing control wells with no ligand added is also important to control for differences between reads. Top-scoring variants from each plate are regrown and retested as purified protein with multiple concentrations of ligand to confirm the improved performance. Only variants with the best fold-change in fluorescence are then promoted to the next round of evolution.

2.3.2 Photophysical Characterization

After initial screening by expressing the library variants in *E. coli* and measuring changes in fluorescence in the presence and absence of ligands in the lysate, the lead variants will be further purified and subjected to systematic photophysical characterization, under both 1- and 2-photon illumination, as well as ligand-binding specificity measurement. Sensors that perform well under one-photon (1P) excitation do not always translate to good probes for in vivo experiments. One issue that arises is that the 1P molar extinction coefficient of a fluorophore frequently does not serve as a good predictor of 2-photon (2P) cross-section [83]. Testing the brightness of any candidate sensors under 2P excitation becomes necessary to benchmark candidate sensors. We follow methods modified from Makarov et al. and Drobizhev et al. [83, 84] for determining fluorescence lifetime, quantum yield, and 2P cross section.

Fluorescence lifetime measurements can be taken on a synchroscan streak camera (Hamamatsu). Excitation can be performed using the vertically-polarized second harmonic of a Ti:sapphire oscillator (Coherent). Resultant fluorescence can then be collected at 90° relative to the excitation light path and focused onto the streak camera slit using an appropriate objective. Scattered excitation

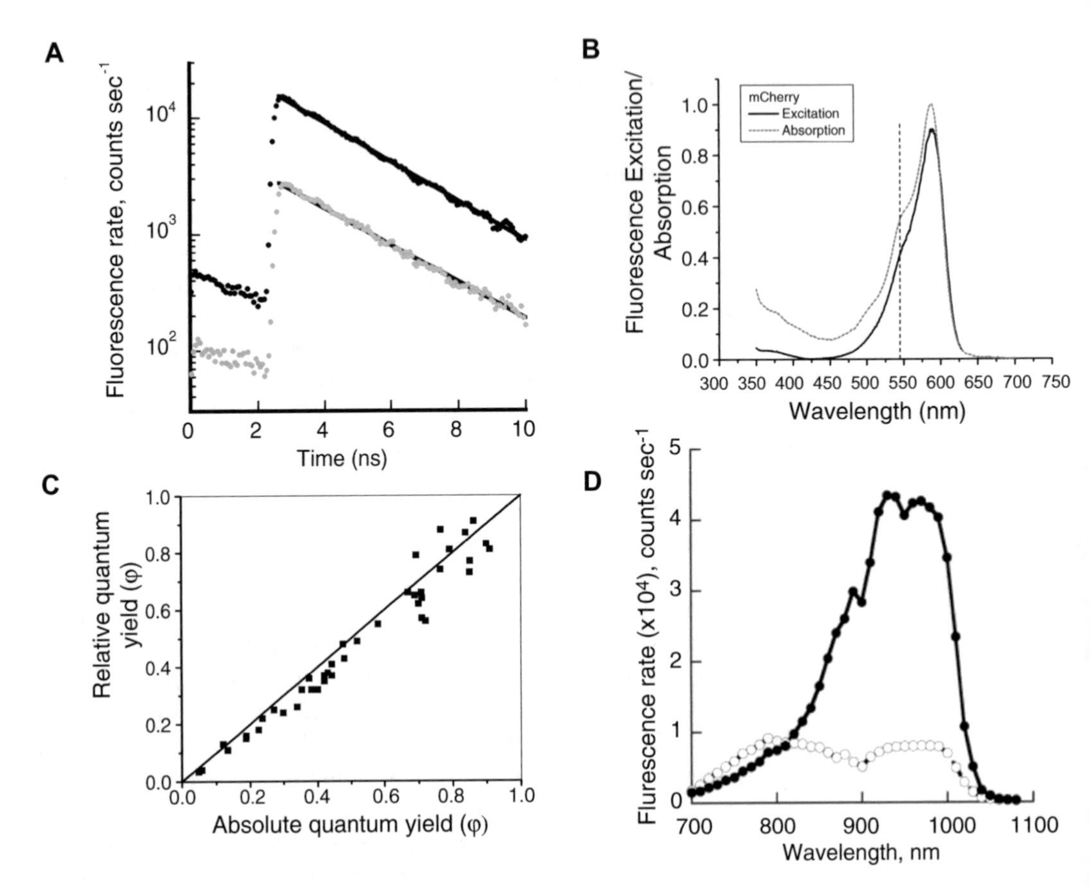

Fig. 5 Determination of the 2-photon cross section of fluorescent protein-based sensors. (**a**) Calculation of the fluorescence lifetime of iGluSnFR in both the apo- (*lower trace*) and bound (*upper trace*) states. (**b**) Graphical description of the normalization procedure used to calculate optical density. In this case, the excitation spectrum of the red fluorescent protein mCherry is normalized to the absorption spectrum of the same sample. (**c**) Scatterplot demonstrating that use of the relative method for determination of quantum yield of proteins leads to systematically lower values than use of the absolute method. Each point indicates values obtained for a single protein. (**d**) Two photon excitation spectra of iGluSnFR in the apo- (*white*) and bound (*black*) states. Reproduced with permission from Marvin et al. (2013) [54] (**a**, **d**), Drobizhev et al. (2009) [100] (**b**), and Drobizhev et al. (2011) [83] (**c**)

photons should be excluded from the detector by utilizing a short-pass filter inserted into the detection light path before the streak camera slit. For Glu sensors, fluorescence decay values should be determined for both apo and bound conformations. For iGluSnFR itself, these values are mono-exponential with a decay constant of around 2.5 ns in each conformation (Fig. 5a) [54]. For sensor variants, a log-linear plot of the photon counts over time will be fit by a line if decay is mono-exponential. The fluorescence time constant can then be fit as a simple exponential decay (Fig. 5a). If the resultant line is not satisfactorily fit by a line, it is likely that more than a single time constant controls the fluorescence lifetime.

In such an instance, the superposition of two (or potentially more) exponential decay processes may serve to better describe the kinetics of fluorescence decay.

Fluorescence quantum yield can be measured either relative to a standard sample or in absolute terms. For determining relative quantum yield, standards with spectral properties that roughly match the properties of the tested sensor variant should be used. For example, sensors with spectral properties similar to iGluSnFR can be compared to Fluorescein in an alkaline (pH 11) solution. Red-shifted sensor variants can be compared with Rhodamine B in methanol, or tested against Rhodamine 6G in ethanol. Fluorescence emission and excitation spectra of purified proteins and reference fluorophores are measured on a luminescence spectrometer (PerkinElmer). Fluorescence spectra are corrected for changes to illumination intensity and detection sensitivity at different wavelengths. Corrected spectra of the sample, $F(\lambda)$, and the reference, $F_R(\lambda)$, can then be integrated over the wavelength range and normalized by optical density of the sample at the excitation wavelength, $OD(\lambda_{exc})$, and for the reference, $OD_R(\lambda_{exc})$. In each case, keeping the maximum OD low (<0.07) is important to avoid concentration-dependent filtering effects. Note that for fluorescent proteins, calculating OD from the measured absorption spectrum introduces artifact as scattering, and absorption by immature fluorophores contribute strongly at lower wavelengths. Therefore, to approximate the OD for our protein-based sensors, we use the emission spectrum values normalized such that the long wavelength edge overlaps the absorption spectrum (Fig. 5b). By so doing, we isolate that region of the spectra where contributions from scattering and immature fluorophore absorption are minimized. One can then use the value of the normalized excitation spectrum at λ_{exc} as the $OD(\lambda_{exc})$. Finally, the fluorescence quantum yield of the sensor, ϕ, can be calculated from the reference quantum yield, ϕ_R, as:

$$\phi = \frac{OD_R(\lambda_{exc})n^2\int_0^\infty F(\lambda)d\lambda}{OD(\lambda_{exc})n_R^2\int_0^\infty F_R(\lambda)d\lambda}\phi_R$$

where n and n_R are the refractive indices of the protein and reference solutions, respectively.

Absolute quantum yield values can be calculated by making photon measurements using an integrating sphere device (Hamamatsu). Using this method, all scattered photons that are not absorbed by the sample can be counted, allowing direct experimental access to the number of photons actually absorbed by the sample. Quantum yield is then simply the number of photons emitted divided by the number absorbed by the sample. This method is preferable because it has been shown that the relative method systematically underestimates the true fluorescent quantum yield, presumably due to unaccounted-for scattering contribution to the measurement of OD(λ_{exc}) [83] (Fig. 5c).

For methods developed by Makarov et al., a necessary step in calculating the 2P cross section is determination of the extinction coefficient, ε_m. This calculation can be made using an all-optical approach, which obviates the need for making exact measurements of mature fluorophore concentration. Specifically, the fluorescence lifetime, τ, and quantum yield, φ, which do not depend on knowledge of mature fluorophore concentration, can be related directly to radiative lifetime as $\tau_R = \tau / \phi$. This relationship, along with data taken to determine the fluorescent quantum yield, allows recovery of ε_m by a rearrangement of the Strickler-Berg equation, such that:

$$\varepsilon_m = 3.47 \times 10^8 \frac{\tilde{\nu}_f^{-3}}{n^2 \int \frac{u(\tilde{\nu}) d\tilde{\nu}}{\tilde{\nu}}} \frac{\phi}{\tau}$$

where n is the refractive index of the medium and $u\left(\bar{\nu}\right)$ is the absorption line-shape function,

$u\left(\bar{\nu}\right) = \varepsilon\left(\bar{\nu}\right) / \varepsilon_m$, and $\bar{\nu}_f^{-3}$ is the temporal average of the inverse cubic frequency of fluorescence found from the corrected fluorescence spectrum $F\left(\bar{\nu}\right)$ by:

$$\tilde{\nu}_f^{-3^{-1}} = \int F(\tilde{\nu}) d\tilde{\nu} \Big/ \int \frac{F(\tilde{\nu}) d\tilde{\nu}}{\tilde{\nu}^3}$$

Finally, the 2P absorption cross section (2PA) can be derived using the values of fluorescence lifetime τ, and the extinction coeffiecient ε_m. To determine this quantity, we excite samples with a Ti:sapphire femtosecond oscillator (Coherent), pumped by a constant-wavelength, frequency-doubled Nd:YAG laser (Coherent); a 1 kHz repetition rate Ti:Sapphire femtosecond regenerative amplifier (Coherent); and an optical parametric amplifier, OPA (Quantronix), whose output signal or idler is continuously tunable from 1100 to 1600 nm or 1600 to 2200 nm respectively. We use the second harmonic of the signal for 2P excitation in the λ_{exc} = 550–790 region, the second harmonic of the idler in the 790–1100 nm region, and the fundamental signal in the 1100–1400 nm region. The OPA output signal pulse energy is 100–250 μJ (10–40 μJ after frequency doubling), and the pulse duration is 70–120 fs. All 2PA are then determined relative to a set of well-characterized standards [84] to correct spectral variations in the excitation source. Additionally, all fluorescence measurements should be checked to ensure a quadratic relationship to input excitation intensity to exclude artifacts such as linear absorption of photons within the 1P absorption spectrum. The 2PA cross-section of the sample, σ_2, can then be calculated from the reference value, $\sigma_{2,R}$, as:

$$\sigma_2 = \frac{F_2(\lambda_{reg})C_R\eta_R(\lambda_{reg})}{F_{2,R}(\lambda_{reg})C\eta(\lambda_{reg})}\sigma_{2,R}$$

where λ_{reg} is the registration wavelength at which sample and reference values of 2P excited fluorescence, F_2, are taken; C gives the concentration; and η is the differential quantum yield. For all variables, the R subscript denotes values pertaining to the reference standard. Concentration for sample and reference is then calculated by Beer's Law as $C = {OD_2}/{\varepsilon_m l}$. In this equation, l denotes the cuvette path length, while OD_2 denotes optical density in its spectral maximum. Differential quantum yield of the sample can then be calculated as $\eta(\lambda_{reg}) = {F_2(\lambda_{reg})}/{OD_{abs}}$, where OD_{abs} is calculated for the determination of fluorescence quantum yield of the sample. σ_2 values can be obtained at several points within the 2PA spectrum of the sample. The 2PA spectrum can then be calibrated along its length based on the obtained values of σ_2 (Fig. 5d).

2.4 Microscopes and Imaging Stages

For imaging of cell cultures (heterologous and primary), we use an inverted Zeiss 710 laser scanning confocal microscope equipped with laser lines at 405, 458, 488, 514, 561, and 633 nm. A variety of stage adapters are required to accommodate the different culture systems we use. One stage-adapter system we have found to be particularly useful is an SA-20KZ-AL stage adapter (Warner Instruments, 64-0297) fitted with a QE-1 platform (Warner Instruments, 64-0375). This system accommodates inserts of different sizes, allowing mechanically stable recordings from a variety of culture glassware. In instances where spatial resolution is not paramount, we use a Plan-Apochromat 20×/0.8 NA objective lens (Zeiss, 440640-9903) with the confocal pinhole open to its greatest extent. For higher spatial resolution imaging, for example, when signals near the cell membrane are of interest, we use a Plan-Apochromat 63×/1.4 NA objective lens (Zeiss, 440762-9904) with the pinhole restricted so that the point-spread function of the excitation light through the selected objective passes approximately 1 Airy Disk (i.e., 1 Airy Unit). As a brief aside, Airy unit calculation is made from excitation wavelength as a matter of convenience. This value can be more precisely calculated based on the peak of the emitted fluorescence. This difference can become significant when the imaged fluorophores exhibit a large Stoke's shift. Temporal precision is especially important when imaging iGluSnFR and variants with comparably fast kinetics. For example, fluorescence transients due to spontaneous activity in dissociated neurons are only well resolved at frame rates above 60 Hz, well above the typical 512 × 512 frame rates for our system. With our scanning laser system, tradeoffs can be made between the amount of spatial information collected and temporal precision. For instance, recordings taken at the lowest magnification (0.6×) at the Nyquist limit for resolution for the 63× objective (2316 × 2316) require ~7 s for acquisition of a single image but provide exquisite spatial detail over an extended area (~230 μm). At the other extreme, single-pixel scans can be completed in as little as a few microseconds but provide essentially no spatial information.

Imaging at depth in intact, living animals, particularly in highly scattering mammalian brain tissue, requires 2P microscopy. To test sensors in this imaging modality, we use a SliceScope (Scientifica) outfitted with a resonant (8 kHz) galvo scan head, one gallinium arsenide phosphide (GaAsP) photo-multiplier tube (PMT; Scientifica, S-MDU-PMT-50-65), one conventional PMT (Scientifica, S-MDU-PMT-45), and a Chameleon Ultra II mode-locked Ti:sapphire tunable laser (680–1080 nm; Coherent). Our system is therefore capable of scanning a 512 × 512-pixel image at a rate of ~30 Hz, allowing a dwell time of about 90 ns at each pixel. The image size can thus be reduced along the galvo-scanned axis for a linear decrease in image acquisition time. Indeed, given the fast kinetics of iGluSnFR, an acquisition rate exceeding 100 Hz may be desirable, depending on the speed of the imaged Glu transients themselves. In terms of spatial precision, with our 40× objective, resolution exceeding fine, subcellular neuronal features, such as axons and dendritic spine heads, is easily achieved. Finally, a two-PMT system allows for multiplex imaging of green and red fluorophores. Note, however, that care must be taken to ensure that a single excitation wavelength can effectively drive fluorophores of each color if real-time multiplexing is desirable; otherwise, images can be taken at different excitation wavelengths (a process requiring at least several seconds) and combined offline.

2.5 Imaging-Sensor Variants Expressed in Immortalized Cell Lines

Bacteria-based testing allows rapid screening of libraries of sizes ranging from 10^3–10^5 variants depending on the specifics of the method used. However, protein folding, mature protein localization, maturation temperature, and a host of other differences between bacterial and vertebrate physiology mean that performance in *E. coli* does not necessarily predict performance in mammalian systems. As a middle point between in vitro and in vivo preparations, we used mammalian HEK293 cells and rat hippocampal dissociated neuronal culture to characterize the sensor's intrinsic properties, such as expression level, affinity, and sensitivity to perfused Glu or action potential-triggered Glu release (Fig. 6).

2.5.1 Preparing HEK293T Cells for Imaging Experiments

When testing candidate probes in heterologous cells, we use HEK293T cells (Thermo Fisher, R70007). Generally, these cells can be maintained with a standard culture medium composed of Dulbecco's modified Eagle's medium (Thermo Fisher, 11995-065) supplemented with 10% v/v fetal bovine serum (Thermo Fisher, 10437028) and PenStrep at 50 U/mL (Thermo Fisher, 15070063). Because these cells are quite hardy, we flash freeze them as stocks of one million cells per 1 mL aliquot in the standard culture medium, supplemented with 10% v/v dimethyl sulfoxide (Sigma, D8418). To prepare cells for imaging, we plate onto 12 mm round glass coverslips (Thermo Fisher, 12-545-81). Critically, any culture dish used must have #1.5 glass to be compatible with confocal objectives. When the cells are ~60% confluent, we transfect using Effectene (Qiagen, 301425) and miniprep plasmid DNA (Qiagen, 27106), per the manufacturer's recommended protocols. Depending on the specifics of the promoter and construct, imaging can usually be performed 24–48 h after transfection. Typically, we use a CMV or CAG promoter for expression in HEK293T, since both rapidly (within 12 h post transfection depending on the construct) drive high levels of sensor expression.

2.5.2 Deriving Sensor-Apparent K_d on HEK293T Cells

While screening in *E. coli* serves as a high-throughput front-end screen in sensor development, differences in probe performance are possible when expressed in a mammalian system. For example, as the characterization of iGluSnFR moved from purified *E. coli* lysates to surface expression on mammalian cells, the apparent K_d was notably reduced by greater than an order of magnitude [54]. Such discrepancies demand testing in systems that more closely reflect the eventual system of study.

Characterization on HEK293T cells offers a medium throughput step in determining suitability of a particular candidate sensor for its final application (Fig. 6a and b). It is possible to use this system to easily determine the apparent K_d of a Glu sensor when expressed at cell surface. Within the context of this process, determining the $\Delta F/F$ of the candidate sensor is also possible.

To determine apparent K_d of the sensor, we image sensors transfected into HEK293T cells as described in Sect. 2.5.1. The day of the experiment, we make a $\log_2$ series of Glu dissolved in HEPES-buffered (20 mmol/L) Hanks' balanced salt solution (HHBSS), with pH adjusted to 7.4. The concentrations are chosen to match the expected affinity of the sensor, so iGluSnFR concentrations will range from high nanomolar to several hundred micromolar. We then gently wash the cells three times with HHBSS and mount the coverslip to the stage adapter, which is then affixed to the Zeiss 710 stage. To minimize motion artifact (axial or lateral

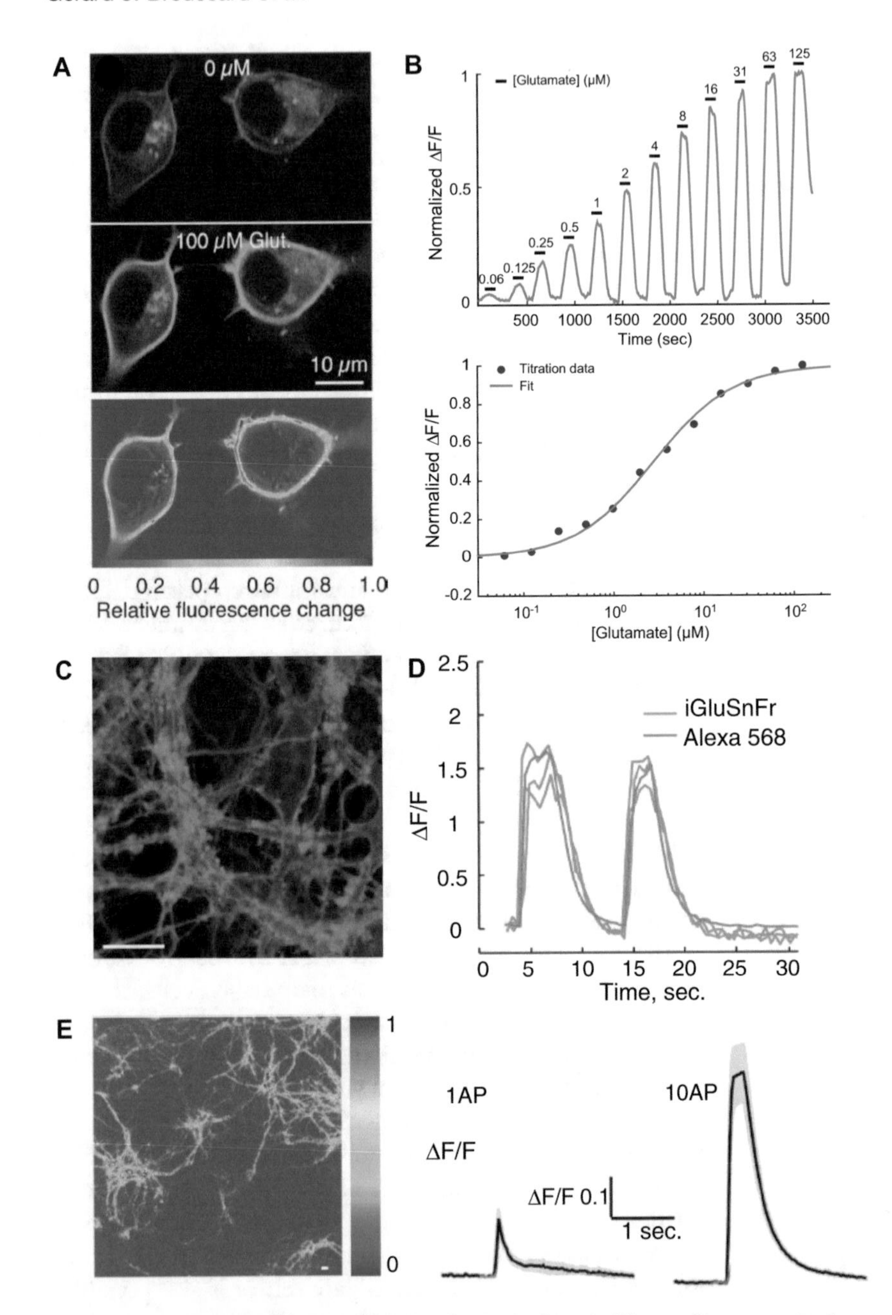

Fig. 6 (**a**) Expression of iGluSnFR on HEK293 cells. (**b**) Perfusion of different Glu concentrations on HEK293 cells to determine the K_d (4 μmol/L). Measurements are from a single ROI. (**c**) Dissociated hippocampal neurons infected with AAV.hSynapsin.iGluSnFR. (**d**) Response of neurons to "puffs" of Glu/AlexaFluor 568 solution. (**e**) Response of neurons to electrical field stimulation. Reproduced with permission from Marvin et al. (2013) [54]. Scale bar = 10 μm. Reproduced with permission from Marvin et al. (2013) [54] (**a**,**c**-**e**)

from cell movement), we image using the 20× objective with the confocal pinhole opened as wide as possible. HHBSS is then perfused onto the chamber using a peristaltic pump set to a perfusion rate of about 1.5 mL/min. We then alternate between washing with HHBSS and applying the Glu titration from the lowest to highest concentration values. For each concentration and wash, we allow 3 min for full equilibration of the stage concentration. During this procedure, we capture images at a rate of one per 10 s with a 512 × 512 pixel size and pixel dwell of around 6 μs.

Data is processed by hand-drawing regions of interest (ROIs) over cells within the image that are assessed as having moved very little by their average intensity projection along the time axis. Averages within these ROIs are taken point-wise following the equation $\frac{\Delta F}{F} = \frac{F - F_0}{F_0}$. Here, F_0 is the background-subtracted average fluorescence during the period prior to the first Glu application, and F is a vector of background-subtracted mean intensities for each time point. To calculate the apparent K_d, we fit the peaks in the resultant traces to a single binding-site isotherm using custom Matlab (Mathworks) routines. $\Delta F/F$ can also be extracted as the maximum point in the curve.

2.6 Imaging-Sensor Variants Expressed in Primary Cell Lines

While immortalized cell lines offer a convenient way to test sensor variants, they lack aspects of neuronal signaling, including exocytotic release of Glu. As a final test before packaging candidate sensors for expression in intact mammalian preparations, we characterize performance in dissociated neuronal culture (Fig. 6c–e).

2.6.1 Preparing Primary Neuronal Cultures

For imaging experiments that require primary dissociated cells, we use hippocampal cells derived from E18 Sprague-Dawley pups (Charles River). One day prior to the dissection, we coat culture dishes (35 mm Petri dishes) with 14 mm #1.5 glass window (MatTek, P35G-0.170-14-C) with a sufficient volume of 10 μg/mL laminin (Sigma, L2020) and 10 μg/mL poly-DL ornithine (Sigma, P0421) to cover the area to which the cells will be seeded. The coating is kept on the culture dishes overnight and then rinsed four times with sterile deionized water.

On the day of the dissection, the pregnant female is deeply anesthetized by isoflurane (Henry Schein, 50562-1) in an induction chamber. We then cervically dislocate the animal and perform a Cesarean section to remove the uterus. Individual pups are removed from the uterus and rapidly decapitated. The heads are placed in sterile HBSS (Thermo Fisher, 14175103) in a 10 cm Petri dish. Heads are then stabilized by poking the tips of Dumont #7 forceps (Fine Science Tools, 11274-20) through the ocular sockets, and the brains are removed by cutting caudal to rostral along the sagittal suture using 3 mm angled Vannas scissors (Fine Science Tools, 15000-00). The brains are removed with a microspatula (Thermo Fisher, 21-401-10) and placed into a 35 mm Petri dish containing HBSS where they are stabilized for dissection by impaling the cerebellum with Dumont #5 forceps (Fine Science Tools, 11251-10).

We perform the hippocampal dissection using a scalpel (Fine Science Tools, 10035-15) to delicately peel back one hemisphere of cortex, gently pulling with the forceps to remove the meninges. We stabilize the hippocampus by placing the forceps onto the surrounding cortex, and then cut away the cortical tissue. Isolated hippocampi are transferred to a 15 mL conical tube containing 2 mL of 0.05% trypsin (Thermo Fisher, 25300120) and placed into a cell culture incubator at 35 °C for 15 min. Following the incubation period, the trypsin is inactivated by adding 10 mL of NeuroBasal (NB; Thermo Fisher, 21103049) supplemented by 2% B27 (Thermo Fisher, 17504044) and 5% fetal bovine serum (FBS). Hippocampi are then transferred to a 15 mL conical tube containing 5 mL HBSS where the tissue is triturated by suctioning into and out of a constricted (~500 μm diameter) Pasteur pipette about 15 times, or until there are no large pieces of tissue remaining. We then allow the debris to settle for a few minutes and transfer the cell-containing supernatant to an empty 15 mL conical tube. Finally, the supernatant is centrifuged at 500 rcf for 3–5 min at room temperature, and the pellet is resuspended in sufficient NB + B27 + FBS for seeding onto the prepared culture dishes.

Cells are allowed to grow undisturbed for ~4 days or until astrocyte proliferation is sufficient to produce a nearly confluent monolayer. At this point, we perform a half change of the medium with fresh media further supplemented by 10 μmol/L 5-fluoro-2′-deoxyuridine (FUDR, Sigma, F0503) final concentration. Neurons so-derived can be transfected using the calcium phosphate method (Thermo Fisher, K278001), per the manufacturer's instructions, between 5 and 10 days in culture. Because primary cells are more sensitive to culture conditions, we generally use an endotoxin-free DNA preparation method such as the ZR plasmid miniprep kit (Zymo Research, D4015). In addition, CMV and CAG promoters drive expression of the sensor in nonneuronal cells, so for neuronal transfections, we instead use the hSynapsin promoter, which results in neuron-specific sensor expression.

2.6.2 Determining Sensor Action-Potential Sensitivity

For these experiments, we first wash the cells three times by HHBSS. The MatTek culture dish is then placed into the stage adapter and affixed to the Zeiss 710 stage, after which we attach electrophysiology insert RC-37FS (Warner Instruments) above the cells. While placing the RC-37FS above the cells, care must be taken to position the insert slowly and at an angle to avoid applying excessive positive pressure down onto the coverslip. We then drive an isolated electrical current controlled by Ephus software (Vidrio Technologies) across the stage to stimulate action potentials in the cultured neurons. Stimuli are delivered as 1 V square-wave pulses with 1 ms duration at 30 Hz. These stimuli drive simultaneous action-potential activation across the stage, resulting in extensive extra-synaptic spillover of Glu [53]. Because the extra-synaptic

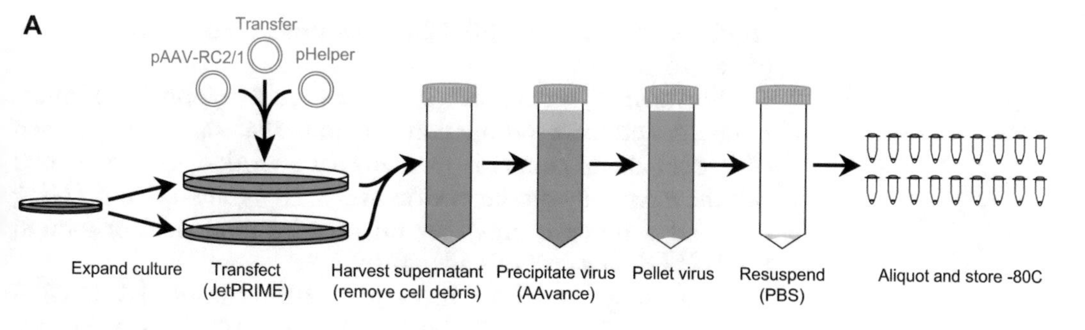

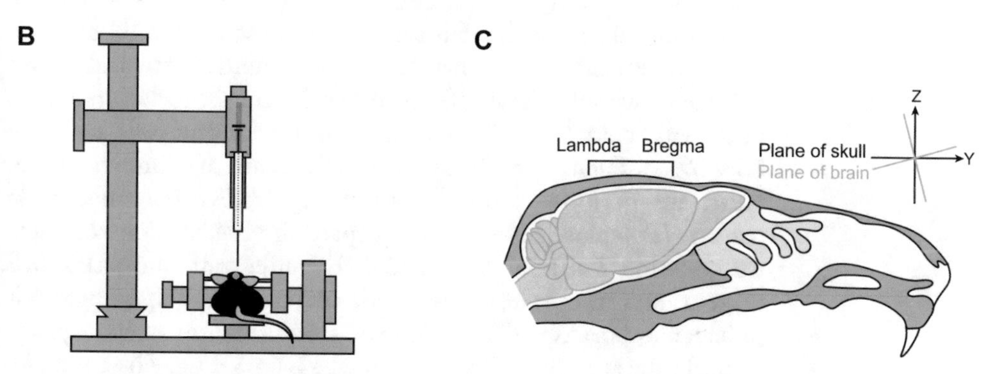

Fig. 7 (**a**) Protocol for making adeno-associated virus (AAV). (**b**) Stereotaxic setup for injecting virus into a mouse brain. (**c**) Position of mouse brain inside the skull, and relative to suture landmarks

spillover is more spatially and temporally extensive than synaptic Glu signaling, we are able to image these events at <20 Hz and capture the shape of the resultant fluorescence waveform. However, it is also possible to capture spontaneous synaptic and extra-synaptic activation by increasing the frame rate to ~60–100 Hz (Fig. 6c).

2.7 In Vivo Characterization of Sensor Variants

Sensor characterization is not complete without expression within the animal system of interest. Here, we discuss procedures for transduction of cells within the intact mouse brain and in vivo imaging.

2.7.1 Virus Preparation

For in vivo transduction, particularly in mammalian experimental systems, recombinant adeno-associated virus (rAAV) has become the vector of choice. In large part, this preference has come about because rAAV does not incorporate into the host-cell genome (preventing possible insertional mutagenesis); drives high, relatively constant levels of the transgene payload; and can be produced at very high titers (10^{12} virion/mL is not an uncommon benchmark) relatively inexpensively (Fig. 7a). As such, it has the potential to serve sensor workflows in the final in vivo phases of testing. For the interested researcher, the only potential caveat is that common academic viral cores, such as those housed at the Universities at Pennsylvania and North Carolina Chapel Hill, require a roughly $2000 USD investment per custom vector packaged in rAAV.

Fortunately, recent years have seen the development of much more cost-effective production methods that can be performed with equipment common in many laboratories. Our particular workflow typically produces virus at 5×10^{12} genomic count (GC; i.e., viral genome counts) per mL, with a total yield of around 5×10^{11} GC at a cost of under $200 USD.

For a single rAAV batch, we plate 10^6 reconstituted AAV293 cells (Agilent Technologies, 240073) onto a 10 cm tissue culture dish using 10 mL of the culture medium described in Sect. 2.5.1. Once these cells have grown to ~60% confluency (typically after 1–2 days), we split these cells onto two 15 cm tissue culture dishes in 20 mL culture medium. After waiting for the cells to reach 50–80% confluency, we change to a fresh 20 mL of culture medium. Next, we perform a transfection using jetPRIME transfection reagent (Polyplus, 114-15). To prepare the transfection complex, we combine 2 mL of the jetPRIME buffer with the jetPRIME reagent and the transfer plasmid, along with helper plasmids pHelper and pAAV-RC2/1 (Agilent, 240071), into an appropriate conical tube at a ratio of 80 μL:17 μg:14.5 μg:8.4 μg. The complete transfection mixture is then applied, 1 mL per tissue culture dish, per the manufacturer's recommendations. After a 72-h incubation period, we collect the supernatant from both plates into a 50 mL conical tube, pellet out any cell debris, add 10 mL AAVanced Concentration Reagent (Systems Bioscience, AAV100A-1), and concentrate the virus per the manufacturer's recommendations.

Critically, it should be noted that different rAAV serotypes have been shown to release from host production cells with different efficiencies [85]. Our production method ensures relatively pure virus by harvesting only from the culture supernatant rather than lysed cells. Therefore, serotypes that do not localize to the supernatant are not suited to our process. The propensity to remain in production cells rather than be released to the supernatant is thought to be driven primarily by the relative affinity of the capsid protein for heparin. Thus, RC helper plasmids containing, for example, the serotype 2 *cap* gene should be avoided because the resultant protein has a high affinity for heparin.

2.7.2 Stereotactic Injection of rAAV

To target the virus to the brain ROI, delivery directly to the brain is necessary because viruses are able to cross the blood–brain barrier only under specific circumstances (Fig. 7b and c). Before surgery, the animal must be anesthetized, which should be confirmed by a toe-pinch test. Common ketamine-based cocktails, such as ketamine/xylazine/acepromazine, do not induce surgical-plane anesthesia, so underdosing and supplementing with an inhaled anesthetic such as isoflurane is preferable. Inhaled anesthetics may also be used without an injectable anesthetic, but must be supplemented with some sort of analgesia during the surgery, such as meloxicam or carprofen, because isoflurane provides no analgesia itself. Topical analgesics, such as lidocaine, are also recommended at the incision site which can be injected subdermally or applied as a jelly to the surface after sterilization of the incision site. Opiate-based analgesics are not recommended before or during the surgery since opiates are respiratory depressants and may cause accidental death during anesthesia. Inhaled anesthetics have the added advantages that the dosing is much easier to adjust during surgery, and they have very fast recovery times (5–10 min). Injectable anesthetics may require several hours before the animal is ambulatory again.

Selection of injection coordinates can be made using a mouse brain atlas such as the one produced by Franklin and Paxinos [86]. Mouse brain atlas coordinates are surprisingly accurate, with only small animal-to-animal variability, particularly for the most common inbred strains, such as C57Bl/6 (Black 6) or 129/SvEv. However, with several inbred mouse strains, the skull is angled relative to the brain; thus, coordinates need to be adjusted to accommodate this tilt. For structures near the surface of the brain, such as the cortex, making any adjustments is not necessary, and surface landmarks such as blood vessels and skull sutures may be used for positioning the needle. However, deeper structures must account for this angle and must also account for any rotation of the skull relative to the manipulating arm of the stereotax. To ensure accurate positioning of the injection needle, it is possible to take measurements from the surface of the skull and use a rotation matrix to recalculate coordinates for drilling and needle insertion (for details, see Note 1).

Once coordinates have been chosen and located relative to the skull surface, a virus is introduced through a small craniotomy, just large enough for the needle to fit through. The needle should be lowered slowly to prevent excess tissue damage and to prevent the tissue from compressing, which would change the location of the target structure. Pulled glass pipettes are thinner and do less damage than steel needles, but cannot reach deeper structures and are at risk of breaking off inside the brain. For shallower structures (0–3 mm), we use glass pipettes, while for deeper structures (3–6 mm), we use 33G steel needles. The virus is then injected slowly (50–100 nL/min) using a high-precision micropump (picospritzer, nanoject, WPI micropump). Waiting a few minutes after the injection is complete allows the virus to diffuse away from the needle tip, lowering the chance of it getting drawn back up the needle track during removal. Remove the needle slowly to minimize tissue damage. After the virus has been delivered, the wound can be closed using either sutures or skin glue (vetbond, dermabond, liquid band-aid). Wound clips are not recommended for the scalp because mice have a tendency to scratch and rip them out.

Stereotaxic surgery can be a time-consuming process and is difficult to master. There are several alternatives, including in utero electroporation, perinatal injection, or tail vein injection. In utero electroporation is a method whereby plasmid is injected directly into the brain of embryos and then the brain is electrically stimulated, either through the uterine wall or using microelectrodes, so that the cells of interest take up the plasmid. This process is not as precise and has a lower survival rate than adult injection, but it is faster and easier to do many animals. Perinatal injection is similar to adult injection, except that the skull is sufficiently soft that the needle can puncture it, and no incision or drilling is necessary. Considerable precision is lost, but speed and facility make it possible to do many more animals. A third option is to package the virus such that it can cross the blood–brain barrier and inject it via the tail vein [87]. This method is quite easy and can theoretically be done at any age; however, the virus will infect most of the brain, so some level of genetic control is recommended. In addition, different packaging plasmids provide slightly different infectivity for different brain regions.

2.7.3 Window and Head Post-Implantation

To view the brain, a craniotomy needs to be performed and a window implanted. This is typically done at the same time as virus injection. The skull should be removed by drilling a doughnut shape larger than the area to be imaged. Drilling causes damage to the brain so any additional drilling should be avoided. The circular piece of skull above the imaging area should be lifted away carefully. The dura can be left intact, and serves as a depth marker during imaging because it displays a stereotyped autofluorescence. Again, avoid damaging the brain in the imaging area. At this point, the remaining surface of the skull should be scuffed up using sharp forceps or another abrasive tool. This creates additional surface area for adhering the cement to the skull later. A circular glass coverslip is then lowered and held in place just above the brain and barely touching the highest point. The angle of the coverslip should be match the most optimal viewing angle for the brain region to be imaged. It is also possible to install a small prism in order to change the imaging angle for access to non-surface structures [88, 89]. Low melting point agarose is then used to fill in the remaining space between the brain and the coverslip. Avoid air bubbles when filling in the gaps. Then the skull should be covered with a thin layer of crazy glue in order to improve the bond of the cement. Only a small amount of glue should be on the agarose and coverslip. Then a custom headpost can be fitted around the coverslip. The headpost should be large enough to accommodate the desired objective, and have protrusions that can be clamped into the headstage to reduce motion artifacts during imaging. The coverslip and headpost are then secured to the skull with black pigmented dental cement. The black pigment helps to reduce light pollution during imaging, particularly for experiments involving visual stimuli. The cement will harden within 30 min, and will fully cure within 24 h. The skin should be closed around the cement to cover any exposed areas of the skull using skin glue, to prevent infection.

2.7.4 Head Mounting and In Vivo Imaging

To collect usable data from the sensor, it is important to obtain a stable image, which be achieved by clamping the head under the microscope using the headpost installed during surgery. Anesthesia is not necessary at this step, but may be used if preferred. The headpost should fit reliably into the clamping mechanism such that individual cells can be identified and imaged multiple times during repeated imaging sessions. The clamp should also be solid enough to overcome any forces created by the animal. One option is to anesthetize the animal during imaging; however, anesthesia may not be compatible with the behavioral paradigm, and different forms of anesthesia may alter the activity of different brain regions. For awake recordings, the animal should be mounted on a moveable surface to enable walking and running, such as a treadmill, a rotating disc, a suspended ball [90], or a mobile homecage [91]. Techniques involving restraint are not recommended because this is highly stressful for the animal, which may alter the results of the experiment, and causes unnecessary duress. If a restraint method is desired, the animal must be well habituated to the restraint, a process that requires 8–10 days of daily restraint, and may be supplemented with regular rewards [92, 93]. Obviously, the more natural the motion (suspended ball or mobile homecage), the less stressful the experiment will be for the animal. We use a suspended ball because of its lower cost of installation. Several habituation sessions may be required before performing the actual experiment. In addition, adding a shield may be necessary to protect the objective from the motion of the animal's tail, as well as stray light sources, such as those introduced by application of visual stimuli.

Even if all forces are eliminated from the animal's walking and running, the animal's heartbeat, breathing, and whisking will still create motion artifacts. Motion in the x–y plane can be easily eliminated by reregistering the images before analysis [94]. Artifacts in the z-plane are more problematic in that they change which objects are within the focal plane. Some solutions for mitigating this issue involve real-time adjustments of the focal plane or broadening the focus so that fluctuations along the axial direction do not affect the image [95].

A properly prepared animal will allow assessment of sensor function in intact tissue. We have successfully imaged spontaneous Glu transients in apical dendrites of cortical pyramidal cells in anesthetized, as well as awake, behaving rodents. In our hands, this spontaneous Glu release was modified during certain behavioral paradigms, as well as under pharmacological manipulation [54] (Fig. 8). Following these and similar procedures, it will be possible

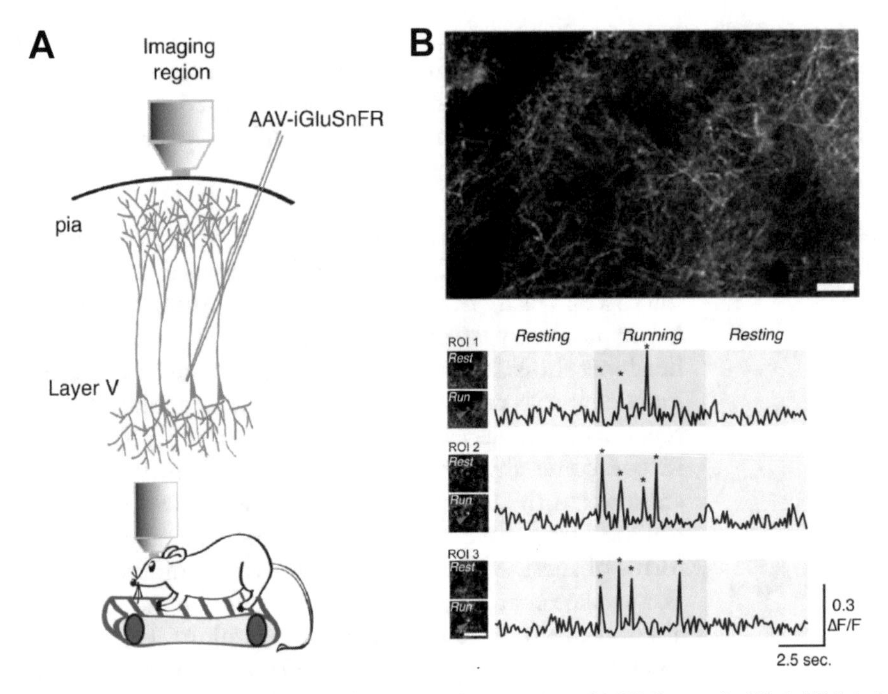

Fig. 8 (a) Schematic illustrating experimental approach for injection of AAV.hSynapsin.iGluSnFR into layer V of primary motor cortex for in vivo transcranial two-photon microscopy. (**b**). Two-photon image of low-density infection of primary motor cortex (forelimb region) and representative ROIs showing that iGluSnFR signals correlate with onset and offset of locomotion. *, glutamate events. Reproduced with permission from Marvin et al. (2013). Scale bar = 10 μm

to determine whether candidate sensors are able to operate in relevant physiological ranges, in terms of both sensitivity and kinetics, to detect the required Glu transients.

3 Perspective

Despite iGluSnFR's broad success in detecting Glu release in in situ and in vivo preparations, further adjustments to its current attributes may yield additional useful probes. For example, tweaks to iGluSnFR's affinity profile will result in probes that can be used in contexts where iGluSnFR currently functions poorly. Particularly, the apparent cell-surface K_d of 4 μmol/L would suggest that the majority of iGluSnFR localized to the synaptic cleft should be saturated as action potential-evoked Glu transients rise briefly (<10 ms) into the millimolar range [7]. Therefore, a variant with decreased affinity may be able to track synaptic Glu more effectively while excluding signal derived from extra-synaptic Glu. Synaptic signal may be further enhanced by targeting the sensor to the synaptic cleft, reducing nonsynaptic fluorescence sources (see, e.g., [96]). In addition, the kinetics of iGluSnFR can be further optimized for "slow" laser-scanning microscopy. The decay dissociation rate constant allows for an elongation of the fluorescence reporter of the Glu transient, increasing photon budget and resultant SNR. Such elongation can be particularly helpful for SNR when utilizing imaging modalities where frame rates are relatively slow, as in the case of laser-scanning microscopy. For example, 2P laser scanning microscopy provides full field frame rates of tens of hertz, a sampling rate too low to effectively capture the upstroke of a synaptic Glu transient. Therefore, it is essential to develop Glu sensors with slower kinetics that enable large-scale monitoring of transient responses at individual synapses in the living animal.

Additionally, as iGluSnFR is fundamentally a GFP-based sensor, it should be possible to shift its spectral attributes. In particular, a red-shifted variant would be desirable because red light is less scattered in intact tissue [97]. Spectral variants can and should be utilized in conjunction with other tool classes. For example, iGluSnFR could be combined experimentally with the red GECIs from the jRCaMP series [98] or optogenetic actuators with red-shifted action spectra, such as ReaChR [99]. Experimental conditions requiring optogenetic multiplexing with iGluSnFR would find particular benefit in a red-shifted iGluSnFR variant.

To date, how the complex patterns of glutamatergic signaling at multiple synapses interact to drive changes in circuit connectivity remains poorly defined. New types of Glu sensors need to be developed that will permit visualizing the history of neural activity at individual synapses in large fields of view during short behavioral epochs. Broad application of these proposed Glu integrators/sensors would transform the activity of transient synaptic inputs into permanently labeled active synapses, thus enabling access to information mapping neural activity to the structure of the neural circuitry underlying a specific physiological process or behavioral task.

4 Note

1. In many stereotaxic frames, precise control of positioning of landmark coordinates is a difficult task. As an alternative, we have developed a strategy that allows recovery of coordinates rotated from the frame of the mouse head into the reference frame of the stereotax. In this derivation, we assume that the mouse frame does not deviate by more than $\pi/2$ radians in any of the three directions from the stereotax frame.

   Prior to inputting coordinates for determining proper rotation, bregma should be set to the coordinates (0, 0, 0). Once that is accomplished, required inputs include the coordinates of lambda, and two points that lie to 2 mm to the right and left, respectively, of the midpoint of the projection of an imaginary line connecting bregma to lambda onto the horizontal plain. The left and right points can then be sampled two millimeters to the left and right of the line along a line perpendicular to the bregma-lambda projection. We then define variables:

$\vec{\lambda} = [\lambda_1, \lambda_2, \lambda_3]$ is the vector from the bregma to the lambda.

$\vec{R}$ is the vector from bregma to the right sampled point (facing forward along $\vec{\lambda}$ toward bregma).

$\vec{L}$ is the vector from bregma to the left sampled point (facing forward along $\vec{\lambda}$ toward bregma).

$\vec{J} = [J_1, J_2, J_3]$ is the vector $\vec{R} - \vec{L}$.

$\vec{N} = \vec{\lambda} \times \vec{J}$ is a normal vector to the horizontal plane in the mouse head frame.

$\vec{K} = [-\lambda_1\lambda_3, -\lambda_2\lambda_3, \lambda_1^2 + \lambda_2^2]$ is a vector orthogonal to $\vec{\lambda}$. It is in the direction that $\vec{N}$ would be if there were no roll.

$\vec{\upsilon}$ will denote the norm of a vector.

θ will denote the angle of yaw; ϕ will denote the angle of pitch; ψ will denote the angle of roll.

Yaw Calculation

$$\theta = \arcsin\left(\frac{\lambda_1}{\sqrt{\lambda_1^2 + \lambda_2^2}}\right)$$

Pitch Calculation

$$\phi = \arcsin\left(\frac{-\lambda_3}{\vec{\lambda}}\right)$$

Roll Calculation

$$\psi = -\operatorname{sgn}(J_3) \cdot \arccos\left(\frac{\vec{K} \quad \vec{N}}{\vec{K} \cdot \vec{N}}\right)$$

where sgn(J_3) is the sign of J_3.

The coordinate transformation from the frame of the mouse's head, $[x, y, z]$ to the frame of the stereotax, $\left[\tilde{x}, \tilde{y}, \tilde{z}\right]$, is as follows:

$$\begin{bmatrix} \cos(\theta) & -\sin(\theta) & 0 \\ \sin(\theta) & \cos(\theta) & 0 \\ 0 & 0 & 1 \end{bmatrix} \cdot \begin{bmatrix} 1 & 0 & 0 \\ 0 & \cos(\phi) & -\sin(\phi) \\ 0 & \sin(\phi) & \cos(\phi) \end{bmatrix} \cdot \begin{bmatrix} \cos(\psi) & 0 & \sin(\psi) \\ 0 & 1 & 0 \\ -\sin(\psi) & 0 & \cos(\psi) \end{bmatrix} \cdot \begin{bmatrix} x \\ y \\ z \end{bmatrix} = \begin{bmatrix} \tilde{x} \\ \tilde{y} \\ \tilde{z} \end{bmatrix}$$

As linear operators it is:

$$\Upsilon(\theta)P(\phi)B(\psi)\vec{x} = \vec{x}$$

The inverse operator taking coordinates in the stereotax frame to the mouse frame is:

$$B(-\psi)P(-\phi)\Upsilon(-\theta)\vec{x} = \vec{x}$$

As an example, we can use these equations to recalculate coordinates for the dorsal lateral geniculate nucleus (dLGN), a brain structure found at ±2.25, −2.30, −2.80 (*x*, *y*, *z* corresponding to mediolateral (ML), anteroposterior (AP), dorsal ventral (DV) directions, respectively). Note that these coordinates have been corrected for the tilt of the skull relative to the brain for our mouse strain (Black 6) as determined by pilot injections. If bregma is set to 0, 0, 0 (*x*, *y*, *z* or ML, AP, ML), and lambda is measured at −0.16, −4.40, −0.64 (*x*, *y*, *z* or ML, AP, ML), then the pitch is calculated to be 8.27° and the yaw is calculated to be −2.08°. We then measure the height of the skull at the points −2.08, −2.13 (left, *x*, *y* or ML, AP) and 1.92, −2.27 (right, *x*, *y* or ML, AP), which are points 2 mm from the center of the line connecting bregma and lambda, and determine that they are −0.45 (*z* or DV) and −0.10 (*z* or DV) respectively. We can now calculate the roll to be −4.95°. Based on this rotation, our corrected drill sites should be at −2.06, −1.77 (left, *x*, *y* or ML, AP) and 2.41, −1.99 (right, *x*, *y* or ML, AP), and we should lower the needle to a depth of −3.28 (*z* or DV) and −2.90 (*z* or DV) respectively.

Acknowledgements

This work is supported by NIH DP2 MH107059 (L.T.), Brain Initiative U01NS090604 (L.T., E.K.U., G.J.B.) and U01NS09058 (R.L.), Rita Allen Foundation (R.L.), Human Frontier Research Program (G.J.B.), and NIH R21NS095325 (B.P.M.). We are grateful for the contributions of Douglas Unger in generating the rotation matrix. We are grateful to Loren Looger, Jonathan Marvin and Philip Borden for their pioneering work in engineering iGluSnFR and critical comments. We also thank Lisa Makhoul for careful reading and discussion of this book chapter.

References

1. Attwell D, Laughlin SB (2001) An energy budget for signaling in the grey matter of the brain. J Cereb Blood Flow Metab 21:1133–1145. doi:10.1097/00004647-200110000-00001
2. Kwon H-B, Sabatini BL (2011) Glutamate induces de novo growth of functional spines in developing cortex. Nature 474:100–104. doi:10.1038/nature09986

3. Lüscher C, Malenka RC (2012) NMDA receptor-dependent long-term potentiation and long-term depression (LTP/LTD). Cold Spring Harb Perspect Biol. doi:10.1101/cshperspect.a005710
4. Volterra A, Liaudet N, Savtchouk I (2014) Astrocyte Ca2+ signalling: an unexpected complexity. Nat Rev Neurosci 15:327–335. doi:10.1038/nrn3725
5. Arundine M, Tymianski M (2004) Molecular mechanisms of glutamate-dependent neurodegeneration in ischemia and traumatic brain injury. Cell Mol Life Sci 61:657–668. doi:10.1007/s00018-003-3319-x
6. Woodroofe N, Amor S (2014) Neuroinflammation and CNS disorders. John Wiley & Sons, West Sussex, UK
7. Clements JD (1996) Transmitter timecourse in the synaptic cleft: its role in central synaptic function. Trends Neurosci 19:163–171. doi:10.1016/S0166-2236(96)10024-2
8. Marcaggi P, Attwell D (2004) Role of glial amino acid transporters in synaptic transmission and brain energetics. Glia 47:217–225. doi:10.1002/glia.20027
9. Ventura R, Harris KM (1999) Three-dimensional relationships between hippocampal synapses and astrocytes. J Neurosci 19:6897–6906
10. Veruki ML, Mørkve SH, Hartveit E (2006) Activation of a presynaptic glutamate transporter regulates synaptic transmission through electrical signaling. Nat Neurosci 9:1388–1396. doi:10.1038/nn1793
11. Moussawi K, Riegel A, Nair S, Kalivas PW (2011) Extracellular glutamate: functional compartments operate in different concentration ranges. Front Syst Neurosci. doi:10.3389/fnsys.2011.00094
12. Ferraguti F, Shigemoto R (2006) Metabotropic glutamate receptors. Cell Tissue Res 326:483–504. doi:10.1007/s00441-006-0266-5
13. Hollmann M, Heinemann S (1994) Cloned glutamate receptors. Annu Rev Neurosci 17:31–108. doi:10.1146/annurev.ne.17.030194.000335
14. Zito K, Scheuss V (2009) NMDA receptor function and physiological modulation. Encyclopedia Neurosci. doi:10.1016/b978-008045046-9.01225-0
15. Conn PJ, Pin JP (1997) Pharmacology and functions of metabotropic glutamate receptors. Annu Rev Pharmacol Toxicol 37:205–237. doi:10.1146/annurev.pharmtox.37.1.205
16. Marcaggi P, Mutoh H, Dimitrov D et al (2009) Optical measurement of mGluR1 conformational changes reveals fast activation, slow deactivation, and sensitization. Proc Natl Acad Sci U S A 106:11388–11393. doi:10.1073/pnas.0901290106
17. Vafabakhsh R, Levitz J, Isacoff EY (2015) Conformational dynamics of a class C G-protein-coupled receptor. Nature 524:497–501. doi:10.1038/nature14679
18. Masugi-Tokita M, Shigemoto R (2007) High-resolution quantitative visualization of glutamate and GABA receptors at central synapses. Curr Opin Neurobiol 17:387–393. doi:10.1016/j.conb.2007.04.012
19. Araque A, Carmignoto G, Haydon PG (2001) Dynamic signaling between astrocytes and neurons. Annu Rev Physiol 63:795–813. doi:10.1146/annurev.physiol.63.1.795
20. Chefer VI, Thompson AC, Zapata A, Shippenberg TS (2009) Overview of brain microdialysis. Curr Protoc Neurosci. Chapter 7:Unit7.1. doi: 10.1002/0471142301.ns0701s47
21. McLamore ES, Mohanty S, Shi J et al (2010) A self-referencing glutamate biosensor for measuring real time neuronal glutamate flux. J Neurosci Methods 189:14–22. doi:10.1016/j.jneumeth.2010.03.001
22. Namiki S, Sakamoto H, Iinuma S et al (2007) Optical glutamate sensor for spatiotemporal analysis of synaptic transmission. Eur J Neurosci 25:2249–2259. doi:10.1111/j.1460-9568.2007.05511.x
23. Brun MA, Tan K-T, Griss R et al (2012) A semisynthetic fluorescent sensor protein for glutamate. J Am Chem Soc 134:7676–7678. doi:10.1021/ja3002277
24. Takikawa K, Asanuma D, Namiki S et al (2014) High-throughput development of a hybrid-type fluorescent glutamate sensor for analysis of synaptic transmission. Angew Chem Int Ed 53:13439–13443. doi:10.1002/anie.201407181
25. Oldenziel WH, Beukema W, Westerink BHC (2004) Improving the reproducibility of hydrogel-coated glutamate microsensors by using an automated dipcoater. J Neurosci Methods 140:117–126. doi:10.1016/j.jneumeth.2004.04.038
26. Rahman MA, Kwon N-H, Won M-S et al (2005) Functionalized conducting polymer as an enzyme-immobilizing substrate: an amperometric glutamate microbiosensor for in vivo measurements. Anal Chem 77:4854–4860. doi:10.1021/ac050558v
27. Hu Y, Mitchell KM, Albahadily FN et al (1994) Direct measurement of glutamate release in the brain using a dual enzyme-based electrochemical sensor. Brain Res 659:117–125
28. Broussard GJ, Liang R, Tian L (2014) Monitoring activity in neural circuits with genetically encoded indicators. Front Mol Neurosci. doi:10.3389/fnmol.2014.00097

29. Lin MZ, Schnitzer MJ (2016) Genetically encoded indicators of neuronal activity. Nat Neurosci 19:1142–1153. doi:10.1038/nn.4359
30. Nakai J, Ohkura M, Imoto K (2001) A high signal-to-noise Ca2+ probe composed of a single green fluorescent protein. Nat Biotechnol 19:137–141. doi:10.1038/84397
31. Tallini YN, Ohkura M, Choi B-R et al (2006) Imaging cellular signals in the heart in vivo: cardiac expression of the high-signal Ca2+ indicator GCaMP2. Proc Natl Acad Sci U S A 103:4753–4758. doi:10.1073/pnas.0509378103
32. Tian L, Hires SA, Mao T et al (2009) Imaging neural activity in worms, flies and mice with improved GCaMP calcium indicators. Nat Methods 6:875–881. doi:10.1038/nmeth.1398
33. Akerboom J, Chen T-W, Wardill TJ et al (2012) Optimization of a GCaMP calcium indicator for neural activity imaging. J Neurosci 32:13819–13840. doi:10.1523/JNEUROSCI.2601-12.2012
34. Chen T-W, Wardill TJ, Sun Y et al (2013) Ultrasensitive fluorescent proteins for imaging neuronal activity. Nature 499:295–300. doi:10.1038/nature12354
35. Baird GS, Zacharias DA, Tsien RY (1999) Circular permutation and receptor insertion within green fluorescent proteins. Proc Natl Acad Sci U S A 96:11241–11246. doi:10.1073/pnas.96.20.11241
36. Wang Q, Shui B, Kotlikoff MI, Sondermann H (2008) Structural basis for calcium sensing by GCaMP2. Structure 16:1817–1827. doi:10.1016/j.str.2008.10.008
37. Petreanu L, Gutnisky DA, Huber D et al (2012) Activity in motor-sensory projections reveals distributed coding in somatosensation. Nature 489:299–303. doi:10.1038/nature11321
38. Shigetomi E, Bushong EA, Haustein MD et al (2013) Imaging calcium microdomains within entire astrocyte territories and endfeet with GCaMPs expressed using adeno-associated viruses. J Gen Physiol 141:633–647. doi:10.1085/jgp.201210949
39. Issa JB, Haeffele BD, Agarwal A et al (2014) Multiscale optical Ca2+ imaging of tonal organization in mouse auditory cortex. Neuron 83:944–959. doi:10.1016/j.neuron.2014.07.009
40. Vanni MP, Murphy TH (2014) Mesoscale transcranial spontaneous activity mapping in GCaMP3 transgenic mice reveals extensive reciprocal connections between areas of somatomotor cortex. J Neurosci 34:15931–15946. doi:10.1523/JNEUROSCI.1818-14.2014
41. Murakami T, Yoshida T, Matsui T, Ohki K (2015) Wide-field Ca2+ imaging reveals visually evoked activity in the retrosplenial area. Front Mol Neurosci. doi:10.3389/fnmol.2015.00020
42. Sun XR, Badura A, Pacheco DA et al (2013) Fast GCaMPs for improved tracking of neuronal activity. Nat Commun 4:2170. doi:10.1038/ncomms3170
43. Helassa N, Zhang X, Conte I et al (2015) Fast-response calmodulin-based fluorescent indicators reveal rapid intracellular calcium dynamics. Sci Rep 5:15978. doi:10.1038/srep15978
44. Huber D, Gutnisky DA, Peron S et al (2012) Multiple dynamic representations in the motor cortex during sensorimotor learning. Nature 484:473–478. doi:10.1038/nature11039
45. Ziv Y, Burns LD, Cocker ED et al (2013) Long-term dynamics of CA1 hippocampal place codes. Nat Neurosci 16:264–266. doi:10.1038/nn.3329
46. Vogt N (2015) Voltage sensors: challenging, but with potential. Nat Methods 12:921–924. doi:10.1038/nmeth.3591
47. Fioravante D, Regehr WG (2011) Short-term forms of presynaptic plasticity. Curr Opin Neurobiol 21:269–274. doi:10.1016/j.conb.2011.02.003
48. Xie Y, Chan AW, McGirr A et al (2016) Resolution of high-frequency mesoscale intracortical maps using the genetically encoded glutamate sensor iGluSnFR. J Neurosci 36:1261–1272. doi:10.1523/JNEUROSCI.2744-15.2016
49. Vyleta NP, Smith SM (2011) Spontaneous glutamate release is independent of calcium influx and tonically activated by the calcium-sensing receptor. J Neurosci 31:4593–4606. doi:10.1523/JNEUROSCI.6398-10.2011
50. Schellenberg GD, Furlong CE (1977) Resolution of the multiplicity of the glutamate and aspartate transport systems of Escherichia coli. J Biol Chem 252:9055–9064
51. Dwyer MA, Hellinga HW (2004) Periplasmic binding proteins: a versatile superfamily for protein engineering. Curr Opin Struct Biol 14:495–504. doi:10.1016/j.sbi.2004.07.004
52. Okumoto S, Looger LL, Micheva KD et al (2005) Detection of glutamate release from neurons by genetically encoded surface-displayed FRET nanosensors. Proc Natl Acad Sci U S A 102:8740–8745. doi:10.1073/pnas.0503274102
53. Hires SA, Zhu Y, Tsien RY (2008) Optical measurement of synaptic glutamate spillover and reuptake by linker optimized glutamate-sensitive fluorescent reporters. Proc Natl Acad Sci U S A 105:4411–4416. doi:10.1073/pnas.0712008105

54. Marvin JS, Borghuis BG, Tian L et al (2013) An optimized fluorescent probe for visualizing glutamate neurotransmission. Nat Methods 10:162–170. doi:10.1038/nmeth.2333
55. Marvin JS, Schreiter ER, Echevarría IM, Looger LL (2011) A genetically encoded, high-signal-to-noise maltose sensor. Proteins 79:3025–3036. doi:10.1002/prot.23118
56. Brunert D, Tsuno Y, Rothermel M et al (2016) Cell-type-specific modulation of sensory responses in olfactory bulb circuits by serotonergic projections from the raphe nuclei. J Neurosci 36:6820–6835. doi:10.1523/JNEUROSCI.3667-15.2016
57. Borghuis BG, Marvin JS, Looger LL, Demb JB (2013) Two-photon imaging of nonlinear glutamate release dynamics at bipolar cell synapses in the mouse retina. J Neurosci 33:10972–10985. doi:10.1523/JNEUROSCI.1241-13.2013
58. Borghuis BG, Looger LL, Tomita S, Demb JB (2014) Kainate receptors mediate signaling in both transient and sustained OFF bipolar cell pathways in mouse retina. J Neurosci 34:6128–6139. doi:10.1523/JNEUROSCI.4941-13.2014
59. Yonehara K, Farrow K, Ghanem A et al (2013) The first stage of cardinal direction selectivity is localized to the dendrites of retinal ganglion cells. Neuron 79:1078–1085. doi:10.1016/j.neuron.2013.08.005
60. Baxter PS, Bell KFS, Hasel P et al (2015) Synaptic NMDA receptor activity is coupled to the transcriptional control of the glutathione system. Nat Commun 6:6761. doi:10.1038/ncomms7761
61. O'Herron P, Chhatbar PY, Levy M et al (2016) Neural correlates of single-vessel haemodynamic responses in vivo. Nature. doi:10.1038/nature17965
62. Bao H, Goldschen-Ohm M, Jeggle P et al (2016) Exocytotic fusion pores are composed of both lipids and proteins. Nat Struct Mol Biol 23:67–73. doi:10.1038/nsmb.3141
63. Poleg-Polsky A, Diamond JS (2016) Retinal circuitry balances contrast tuning of excitation and inhibition to enable reliable computation of direction selectivity. J Neurosci 36:5861–5876. doi:10.1523/JNEUROSCI.4013-15.2016
64. Zhang R, Li X, Kawakami K, Du J (2016) Stereotyped initiation of retinal waves by bipolar cells via presynaptic NMDA autoreceptors. Nat Commun 7:12650. doi:10.1038/ncomms12650
65. Haustein MD, Kracun S, X-H L et al (2014) Conditions and constraints for astrocyte calcium signaling in the hippocampal mossy fiber pathway. Neuron 82:413–429. doi:10.1016/j.neuron.2014.02.041
66. Rosa JM, Bos R, Sack GS et al (2015) Neuron-glia signaling in developing retina mediated by neurotransmitter spillover. eLife 4:e09590. doi:10.7554/eLife.09590
67. Stork T, Sheehan A, Tasdemir-Yilmaz OE, Freeman MR (2014) Neuron-glia interactions through the heartless FGF receptor signaling pathway mediate morphogenesis of drosophila astrocytes. Neuron 83:388–403. doi:10.1016/j.neuron.2014.06.026
68. Parsons MP, Vanni MP, Woodard CL et al (2016) Real-time imaging of glutamate clearance reveals normal striatal uptake in Huntington disease mouse models. Nat Commun 7:11251. doi:10.1038/ncomms11251
69. Poskanzer KE, Yuste R (2016) Astrocytes regulate cortical state switching in vivo. Proc Natl Acad Sci U S A 113:E2675–E2684. doi:10.1073/pnas.1520759113
70. Looger LL, Griesbeck O (2012) Genetically encoded neural activity indicators. Curr Opin Neurobiol 22:18–23. doi:10.1016/j.conb.2011.10.024
71. Akerboom J, Tian L, Marvin J, Looger L (2012) Engineering and application of genetically encoded calcium indicators. In: Martin J-R (ed) Genetically encoded functional indicators. Humana Press, New York, pp 125–147
72. Moretti R, Bender BJ, Allison B, Meiler J (2016) Rosetta and the design of ligand binding sites. Methods Mol Biol 1414:47–62. doi:10.1007/978-1-4939-3569-7_4
73. Pettersen EF, Goddard TD, Huang CC et al (2004) UCSF Chimera—a visualization system for exploratory research and analysis. J Comput Chem 25:1605–1612. doi:10.1002/jcc.20084
74. Bender BJ, Cisneros A, Duran AM et al (2016) Protocols for molecular modeling with Rosetta3 and RosettaScripts. Biochemistry 55:4748–4763. doi:10.1021/acs.biochem.6b00444
75. Hughes MD, Nagel DA, Santos AF et al (2003) Removing the redundancy from randomised gene libraries. J Mol Biol 331:973–979
76. Edelheit O, Hanukoglu A, Hanukoglu I (2009) Simple and efficient site-directed mutagenesis using two single-primer reactions in parallel to generate mutants for protein structure-function studies. BMC Biotechnol 9:61. doi:10.1186/1472-6750-9-61
77. Liu H, Naismith JH (2008) An efficient one-step site-directed deletion, insertion, single and multiple-site plasmid mutagenesis protocol. BMC Biotechnol 8:91. doi:10.1186/1472-6750-8-91
78. Quan J, Tian J (2011) Circular polymerase extension cloning for high-throughput cloning

of complex and combinatorial DNA libraries. Nat Protoc 6:242–251. doi:10.1038/nprot.2010.181
79. Kunkel TA (1985) Rapid and efficient site-specific mutagenesis without phenotypic selection. Proc Natl Acad Sci U S A 82:488–492
80. Huang R, Fang P, Kay BK (2012) Improvements to the Kunkel mutagenesis protocol for constructing primary and secondary phage-display libraries. Methods 58:10–17. doi:10.1016/j.ymeth.2012.08.008
81. Cadwell RC, Joyce GF (1992) Randomization of genes by PCR mutagenesis. Genome Res 2:28–33. doi:10.1101/gr.2.1.28
82. Gruet A, Longhi S, Bignon C (2012) One-step generation of error-prone PCR libraries using Gateway® technology. Microb Cell Factories 11:14. doi:10.1186/1475-2859-11-14
83. Drobizhev M, Makarov NS, Tillo SE et al (2011) Two-photon absorption properties of fluorescent proteins. Nat Methods 8:393–399. doi:10.1038/nmeth.1596
84. Makarov NS, Drobizhev M, Rebane A (2008) Two-photon absorption standards in the 550–1600 nm excitation wavelength range. Opt Express 16:4029–4047. doi:10.1364/OE.16.004029
85. Vandenberghe LH, Xiao R, Lock M et al (2010) Efficient serotype-dependent release of functional vector into the culture medium during adeno-associated virus manufacturing. Hum Gene Ther 21:1251–1257. doi:10.1089/hum.2010.107
86. Paxinos G, Franklin KBJ (2012) Paxinos and Franklin's the mouse brain in stereotaxic coordinates, 4th edn. Academic Press, Amsterdam
87. Deverman BE, Pravdo PL, Simpson BP et al (2016) Cre-dependent selection yields AAV variants for widespread gene transfer to the adult brain. Nat Biotechnol 34:204–209. doi:10.1038/nbt.3440
88. Chia TH, Levene MJ (2009) Microprisms for in vivo multilayer cortical imaging. J Neurophysiol 102:1310–1314. doi:10.1152/jn.91208.2008
89. Low RJ, Gu Y, Tank DW (2014) Cellular resolution optical access to brain regions in fissures: imaging medial prefrontal cortex and grid cells in entorhinal cortex. Proc Natl Acad Sci U S A 111:18739–18744. doi:10.1073/pnas.1421753111
90. Guo ZV, Hires SA, Li N, O'Connor DH, Komiyama T, Ophir E, Huber D, Bonardi C, Morandell K, Gutnisky D, Peron S, Xu N-l, Cox J, Svoboda K (2014) Procedures for behavioral experiments in head-fixed mice. PLoS One 9(2). doi:10.1371/journal.pone.0088678
91. Kislin M, Mugantseva E, Molotkov D, Kulesskaya N, Khirug S, Kirilkin I, Pryazhnikov E, Kolikova J, Toptunov D, Yuryev M, Giniatullin R, Voikar V, Rivera C, Rauvala H, Khiroug L (2014) Flat-floored air-lifted platform: a new method for combining behavior with microscopy or electrophysiology on awake freely moving rodents. J Vis Exp 88:–e51869. doi:10.3791/51869
92. Urbain N, Gervasoni D, Soulière F et al (2000) Unrelated course of subthalamic nucleus and globus pallidus neuronal activities across vigilance states in the rat. Eur J Neurosci 12:3361–3374
93. Urbain N, Salin PA, Libourel P-A et al (2015) Whisking-related changes in neuronal firing and membrane potential dynamics in the somatosensory thalamus of awake mice. Cell Rep 13:647–656. doi:10.1016/j.celrep.2015.09.029
94. Guizar-Sicairos M, Thurman ST, Fienup JR (2008) Efficient subpixel image registration algorithms. Opt Lett 33(2):156–158
95. Wu M, Chen R, Soh J, Shen Y, Jiao L, Wu J, Chen X, Ji R, Hong M (2016) Super-focusing of center-covered engineered microsphere. Sci Rep 6:31637. doi:10.1038/srep31637
96. Granseth B, Odermatt B, Royle SJ, Lagnado L (2006) Clathrin-mediated endocytosis is the dominant mechanism of vesicle retrieval at hippocampal synapses. Neuron 51:773–786. doi:10.1016/j.neuron.2006.08.029
97. Shen Y, Lai T, Campbell RE (2015) Red fluorescent proteins (RFPs) and RFP-based biosensors for neuronal imaging applications. Neurophotonics 2:031203. doi:10.1117/1.NPh.2.3.031203
98. Dana H, Mohar B, Sun Y et al (2016) Sensitive red protein calcium indicators for imaging neural activity. Elife. doi:10.7554/eLife.12727
99. Lin JY, Knutsen PM, Muller A et al (2013) ReaChR: a red-shifted variant of channelrhodopsin enables deep transcranial optogenetic excitation. Nat Neurosci 16:1499–1508. doi:10.1038/nn.3502
100. Drobizhev M, Tillo S, Makarov NS, Hughes TE, Rebane A (2009) Absolute two-photon absorption spectra and two-photon brightness of orange and red fluorescent proteins. J Phys Chem B 113(4):855–859. doi:10.1021/jp8087379

Chapter 6

Measurement of Astrocytic Glutamate Release Using Genetically Encoded Probe Combined with TIRF Microscopy

Fuyuko Takata-Tsuji, Naura Chounlamountri, Christophe Place, and Olivier Pascual

Abstract

Astrocytes modulate brain function such as memory and cognition by processing neuronal activity through the release of transmitters such as glutamate, ATP and D-serine. Various release mechanisms have been proposed depending on pathophysiological conditions. Among them, inverted transport, ion channels mediated release, and vesicular release. To better understand how cellular interactions alter glutamate release, we used live imaging techniques to monitor the dynamics of glutamate release in various cell culture conditions. For that purpose we combined total internal reflection fluorescence (TIRF) microscopy with a genetically encoded fluorescent glutamate sensor (iGluSnFR) which is an intensity-based glutamate-sensing fluorescent reporter that has a high signal-to-noise ratio. This system allows a direct measurement of glutamate in real time. Our goal is to provide guidelines on how to use this approach and to highlight its pros and cons.

Key words Glutamate, Gliotransmitter, Astrocyte, Total internal reflection fluorescence, iGluSnFR, Genetically encoded fluorescent glutamate sensor, ATP, Microglia

1 Introduction

Astrocytes are a subtype of glial cell found in the central nervous system (CNS). Their processes are in direct contact with pre and postsynaptic terminals [1, 2]. This proximity allows them to regulate transmitter and ion concentration of the extracellular space during neuronal activity but also to modulate neuronal activity through the release of transmitters such as glutamate (Glu), ATP (adenosine triphosphate) and D-serine [3, 4]. This bidirectional communication stands on metabotropic purinergic (P2Y1R) and metabotropic Glu receptors (mGluRs) expressed at the membrane of astrocytes [5–9] that are activated by sustained neuronal activity. The activation of these receptors on astrocytes induce a

Sandrine Parrot and Luc Denoroy (eds.), *Biochemical Approaches for Glutamatergic Neurotransmission*, Neuromethods, vol. 130, DOI 10.1007/978-1-4939-7228-9_6,

calcium-dependent release of Glu [5, 9–14] thought to be responsible for the modulation of neuronal activity.

In the hippocampus, these cellular mechanisms have been proposed to be important contributors of synaptic plasticity and to participate to the amplification of pertinent signal in a noisy environment [15–18]. The release of gliotransmitter by astrocytes is thus a central mechanism sustaining important cognitive functions. Gliotransmitter release has been widely studied and many mechanisms have been described. Among them inverted transport, ion channels, hemichannels and vesicular release [19, 20]. The exact modality of release in various physiological conditions is widely debated and should thus be studied thoroughly. Modulation of Glu release is poorly studied due to the lack of appropriate and easy to use tools. Various approaches are available to monitor Glu release with for each their drawbacks. Glu microelectrode biosensors are limited by their sensitivity to Glu which is in the μmolar range [21]. While useful in vivo, biosensors do not allow a detection in simplified models such as slices or cells cultures, likely due to poor access to the cells and the dilution of Glu in the surrounding medium. Enzymatic detection based on autofluorescence of the reduced form of Nicotinamide adenine dinucleotide produced by the Glu dehydrogenase in presence of Glu has been successfully used to monitor Glu release in the culture medium. This latter approach however lacks the cellular resolution [22, 23]. In vivo, microdialysis combined with high-performance liquid chromatography is a conventional method to measure Glu concentration, but it is invasive and provide a low temporal resolution [24]. Capillary electrophoresis allows very good sensitivity and can be used to measure Glu concentration in fluids (e.g., culture medium [25, 26]) but lacks the temporal resolution to monitor fast changes in Glu concentration due to the sampling. Indirect methods also exist, one of them consists to use neurons as detector of Glu release. This technique has high temporal resolutions to estimate Glu dynamics in the CNS, but astrocytic Glu measurement is indirect, hence the experimental findings are difficult to interpret accurately [27]. We propose here to use a fluorescent Glu sensor composed of Glu binding proteins and a fluorophore on astrocytes culture [28]. Genetically encoded Glu indicators, such as fluorescent indicator protein for Glu (FLIPE) [29] and Glu-sensing fluorescent reporter (GluSnFR) [28, 30] can be pretty easily targeted to specific cellular populations like astrocytes and to the subcellular compartment. Furthermore, they allow continuous measurement of Glu concentration fluctuations at the single cell level in a noninvasive manner and to repeat imaging. Appropriate imaging techniques combined to fluorescent sensors will allow improving detection and resolution of the signal. For example, genetically encoded Glu indicator combined with total internal reflection fluorescence (TIRF) microscopy is a powerful tool to determine events at the vicinity of the plasma membrane [6, 31, 32]. Another advantage of TIRF microscopy is the possibility to combine monitoring of Glu concentration with other physiological events such as calcium fluctuations or vesicular release. In this chapter, we described the methods we used to combine TIRF microscopy with iGluSnFR to measure directly astrocytic Glu on mixed astrocyte microglia primary culture.

2 Materials

2.1 Cell Preparation and Recording Conditions

Astrocyte primary cultures were prepared from the cortices of newborn C57BL6 mice (Janvier Labs, Le Genest-St Isle, France) aged 1–2 days [33]. C57BL6 mice pups were anesthetized on ice. Whole brains were removed and placed in Dulbecco's phosphate buffered saline modified (PBSm), PBS containing 0.6% D-glucose (Sigma, St. Louis, MO, USA) and 100 U/mL of penicillin–streptomycin at 4 °C. The cortices of both hemispheres free meninges were isolated, cut into small pieces with a scalpel. The culture tissues were gently triturated using a fire-polished Pasteur pipet in 5 mL of cold PBSm, Dissociated cells were harvested by low-speed centrifugation at 4 °C (800 rpm, for 7 min). Cells were resuspended in growth medium, seeded in a Petri dish poly-L-ornithine coated (Sigma). The cells density plating was approximately 5×10^5 cells/35-mm Petri dish with glass bottom (Ibidi Planegg/Martinsried, Germany) and 35×10^5 cells/10 cm Petri dish. The growth medium consisted of Dulbecco's modified Eagle's medium containing 10% L-glutamine, 100 U/mL of penicillin–streptomycin (DMEMc) and 10% of fetal bovine serum (FBS). FBS purchased from Sigma (F7524) and Thermo Fisher Scientific (Hyclone®, SV30160, Waltham, MA, USA) were supplemented in DMEMc for astrocyte cultures and astrocyte "pure culture" respectively. Culture were incubated in a humid atmosphere of 95% air mixed with 5% CO_2 at 37 ° C. Medium was changed at day 1, 3, and 7 after plating, cells were used between day 14 and 18. All products used for astrocyte cultures came from (Gibco®). Generally at day 6 or 7 astrocytes were confluent, the thin-layer of cells composed of heterogeneous cell population, and microglia were grown from this step. To obtain primary culture of "pure astrocytes", as it is called, it is important to treat the culture by adding cytosine arabinoside (Ara-C, 5 μmol/L) (Sigma) in the medium during 7 days to prevent mitosis. Then, at day 14 the culture were treated with 75 mmol/L of L-leucine methyl ester (LME) (Sigma) for 90 min, to kill residual microglial cells, followed by two rinses with DMEMc [34]. The cultures of "pure astrocytes" were transfected with pCMV.iGluSnFR as detailed below.

To obtain culture of transfected astrocytes with nontransfected microglia (controlled mixed culture), microglia, that are not very adherent cells were removed from basic condition astrocytes cultures at day 16 by shaking (300 rpm for 30 min) the dish. Controlled mixed cultures were obtained by adding freshly collected microglia onto the "pure astrocytes" layer (microglia-astrocyte co-cultures). The co-cultures at day 17 were carried out with TIRF microscopy. Human embryonic kidney (HEK) 293T cells used as control were plated on 35-mm Petri dish (3×10^5/dish) with glass bottom, coated with poly-L-lysine (Sigma) and were cultured in DMEM supplemented 5% FBS. pCMV.iGluSnFR were transfected 48–72 h before imaging.

2.2 iGluSnFR Characteristics

pCMV(MinDis).iGluSnFR was obtained from Addgene (plasmid # 41732) (Cambridge, MA, USA). This construct encodes an N-terminal mouse immunoglobulin κ-chain leader sequence, which directs the protein to the secretory pathway; the GltI253. L1LV/L2NP protein; a myc epitope tag; and at the C-terminus, a platelet derived growth factor receptor transmembrane helix, which anchors the protein to the plasma membrane, displaying it on the extracellular side [26].

2.3 Transfection with pCMV.iGluSnFR

pCMV.iGluSnFR (4 μg) and PLUS™ reagent (4 μL) (Thermo Fisher Scientific) were diluted in Opti-MEM® (150 μL) (Thermo Fisher Scientific). Lipofectamine®LTX (5 μL) (Thermo Fisher Scientific) was diluted in 150 μL of Opti-MEM® (Gibco®). Each solution was incubated for 5 min at room temperature (R.T.). Diluted plasmid was combined with diluted Lipofectamine®LTX (150 μL lipofectamine solution:150 μL plasmid solution) and then incubated for 20 min at R.T. (plasmid-LTX complex). For astrocyte transfections, astrocyte culture medium was changed to culture medium without antibiotics 1 day before transfection. 300 μL of plasmid-LTX complex mixed to 700 μL of culture medium (5% FBS and 10% glutamine in DMEM without antibiotics) was added to the dish containing astrocytes. To transfect HEK293T cells, the 300 μL of plasmid-LTX complex mixed with 700 μL of Opti-MEM® was added to the dish containing HEK293T cells. After cells were incubated for 5 h, 2.0 mL of fresh astrocyte culture medium was added to the dish containing astrocytes. For HEK293T cells plasmid-LTX complex was replaced by 2.0 mL of fresh culture medium. Cells were used 48–72 h after transfection.

2.4 iGluSnFR Imaging with TIRF Microscopy

Culture medium in the dish containing astrocytes or HEK293T cells was removed and replaced by 1.0 mL artificial cerebrospinal fluid (aCSF) (pH = 7.4). The aCSF consisted of (in mmol/L): 135 NaCl, 2.5 KCl, 2.8 $MgCl_2$, 2.6 $CaCl_2$, 10 HEPES, 10 glucose. To image iGluSnFR response, an inverted microscope (DMI6000, Leica Microsystems, Wetzlar, Germany) was used for epifluorescence imaging and TIRF imaging using a high numerical aperture objective (100× oil-immersion, NA = 1.46, Leica). The motorized stage (Märzhäuser, Wetzlar, Germany) was equipped with an insert designed to accommodate a 35-mm Petri dishes and was preheated to 37 °C using a temperature controller (TC02, Ibidi, Martinsried, Germany). Fluorophore excitation was obtained with a laser (Diode laser box, Leica Microsystems) tuned to 488 nm. Fluorescence emission was passed through a 505 nm dichroic mirror and 525BP36 emission filter (T-Quad, Leica Microsystems). The microscope was controlled by the software Leica Application Suite (LAS). The penetration depth was estimated from the LAS software to be of 250 nm for TIRF imaging. The images were acquired at a resolution of 512 × 512 pixels (pixel size 160 nm at 100×) at 1 frame/s for 150 s. The Electron-Multiplying charge coupled device (EM-CCD) camera (C9100, Hamamatsu Photonics, Shizuoka, Japan) was controlled by Wasabi software (Hamamatsu Photonics), exposure time was set at 300 ms and the EM gain at 245. To image iGluSnFR responses to exogenous Glu, 2 μL of Glu was applied for approximately 10 s (femtotips, Eppendorf, Hamburg, Germany) using a motorized platform (InjectMan NI2, Eppendorf) with manual injection (Cell Tram Vario, Eppendorf). In practice, capillary (femtotip) was filled with Glu solutions of various concentrations (50 μL, at 10, 50, or 100 μmol/L) (Sigma) allowing to deliver series of 2 μL puffs. In the same way, 2 μL of ATP (10 μmol/L) (Cell Signaling Technology, Danvers, MA, USA) was puffed onto single astrocyte to image iGluSnFR responses to endogenous Glu.

2.5 iGluSnFR Analysis

The following analyzes were carried out with a public domain software Image J (NIH Image, Bethesda, MD, USA). First, to visualize the fluorescence intensity (F) change of iGluSnFR elicited by stimulation (as shown in Figs. 2, 3, and 4); images 1 s before application and at time t when response was maximal were processed as follows: Intensity of considered images at each time point were subtracted by the reference image taken at t_0 (F_0). To normalize intensity, the subtracted images were divided by the t_0 image. Resulting images consist thus as $(F_t - F_0)/F_0$ ratio images and a pseudo color lookup table was applied to better visualize gray levels. Note that if a decrease in Glu signal occurs at time t, this ratio should be negative. As we are monitoring an increase of Glu this is not a matter for us. If we had to perform similar experiments to monitor a decrease of Glu concentration, one could add an offset value as the measurements are not absolute values.

Second, to quantify and show graphically the change in fluorescence intensity according to time, the sequential change in fluorescence intensity was measured at a position where the response to Glu was high as region of interest, and was expressed as arbitrary units (AU). Fluorescence intensity at each time (F_t) was normalized to the mean intensity of fluorescence during the control period ($F_{baseline}$) (before injection of ATP or Glu) and the Glu response was obtained by calculating $(F_t - F_{baseline})/F_{baseline}$ and referred to as $\Delta F/F$ ratio.

3 Methods

3.1 Visualization of Membrane Proteins Using TIRF Microscopy

The TIRF fluorescence microscopy is a technique that allows to selectively image compartments close to the cell surface at the interface with the coverslip [35]. It has a nanometric resolution due to the evanescent wave which is generated when the incident light is totally internally reflected by the coverslip at the critical angle. Evanescent wave range decays exponentially from the interface, thus fluorophores on the cell membrane and at the interface with the coverslip are selectively excited. This physical property of the evanescent wavelength decreases the noise and unwanted light coming from nonfocal planes. The combination of TIRF microscopy with fluorescent sensors has potential benefits in application requiring imaging of molecular events at the cellular surfaces such as release or uptake of soluble factor (i.e., gliotransmitter and neurotransmitter), molecules binding to receptor.

To measure Glu by fluorescence microscopy techniques, several fluorescent sensors have been developed such as EOS [36, 37], FLIPE [29], SuperGluSnFR [30], and iGluSnFR [28] (Table 1). As these fluorescent sensors are composed of Glu-binding proteins conjugated with a fluorescence readout, these signals arise from changes in Glu concentration. The most established iGluSnFR is genetically encoded and put into various vectors such as plasmid

Table 1
Fluorescent sensor for measurement of Glu

Name	Type of sensor	Glutamate binding protein	Fluorescent probe	Brain slice/ culture	In vivo	Kd Glu (μM)	Δ*F*/*F*
iGluSnFR [26]	Genetically encoded	GltI	cpGFP	Yes/Yes	Yes	4.5[a]–100[b]	4.5
superGluSnFR [28]	Genetically encoded (FRET)	GltI	ECFP/ Venus	–/Yes	–	2.5[b]	0.44
FLIPE [27]	Genetically encoded (FRET)	GltI	ECFP/ Citrine	–/Yes	–	0.6–1000[b,c]	0.15–0.47[c]
EOS [34, 36]	Surface biotinylation	GluR (S1S2)	OGB488	Yes/Yes	Yes	0.15–1.6[c]	0.21–0.48[c]

ECFP enhanced cyan fluorescent protein, *Venus* yellow fluorescent protein called venus, *Citrine* yellow fluorescent protein called citrine, *GluR (S1S2)* S1S2 domain of α-amino-3-hydroxy-5-methyl-4-isoxazolepropionic acid receptor (also known as AMPA receptor GluR2 subunit), *OGB488* Oregon Green 488

[a]In situ affinity of the sensor on the neuron surface

[b]Titration with purified protein in solution

[c]Depends on mutations.

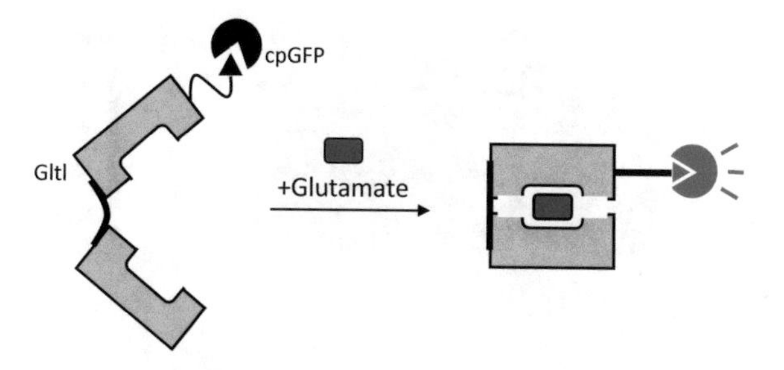

Fig. 1 Schematic representation of the iGluSnFR sensor. iGluSnFR is composed of Gltl (*orange*) a Glu-binding domain which is the scaffold of iGluSnFR bound to the fluorophore, a circularly permutated (cp) GFP (*black*). Glu (*blue*) binding to Gltl induces a conformation change, leading to an increase of the fluorescent intensity of the cpGFP (*green*)

and viruses. It consists in a Glu-binding domain (GltI) and a single fluorescent protein circularly permutated green fluorescent protein (cpGFP). The conformational change in GltI by Glu binding results in changes in the chromophore environment of cpGFP, leading to an increase in fluorescence intensity (Fig. 1) [28]. This alteration of fluorescence intensity allows to estimate the relative change in concentration of Glu next to the sensor. iGluSnFR with cpGFP, which is single wavelength fluorescent protein and have larger signal-to-noise ratio than superGluSnFR [28]. Consequently, iGluSnFR appear to be better suited to measure Glu than FLIPE and superGluSnFR that contain two fluorescent proteins and compared to EOS that has a small-molecule fluorescent dye which have low signal change. Recently published data showed that iGluSnFR robustly and specifically responded to evoked Glu release events in hippocampal cultures, in pyramidal neurons in acute brain slice and do not react to pH changes [28]. Here we provide methods and tools that allow TIRF microscopy combined with iGluSnFR to realize measurements of relative change of Glu concentration with high temporal resolution at the cellular level. In particular, we will focus primarily on the description of the methods to detect Glu release from individual astrocyte.

3.2 Calibration of the iGluSnFR by Measuring Exogenous Glutamate

To evaluate the added value of TIRF microscopy versus epifluorescence to image iGluSnFR signal at the cell membrane, we transfected HEK293T cells, which is a widely used heterologous cell model with round shape. We performed imaging of iGluSnFR using TIRF microscopy and epifluorescence microscopy and measured the fluorescence of HEK293T cells transfected with pCMV. iGluSnFR (Fig. 2a and c). Because of scattered and background noise in epifluorescence the signal-to-noise ratio was lower when compared to TIRF microscopy. As a consequence the fluorescence at the membrane was larger in TIRF microscopy (Fig. 2b and d).

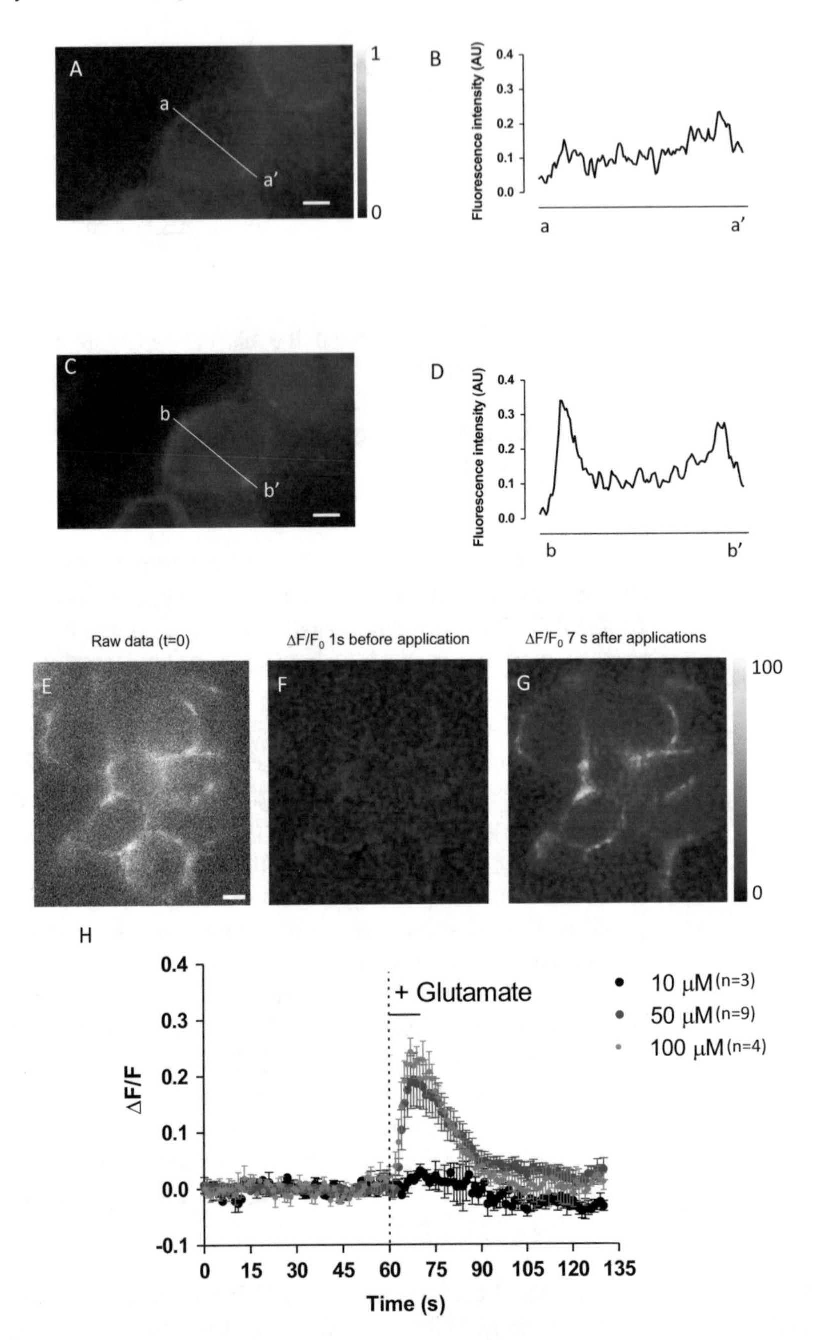

Fig. 2 Expression of iGluSnFR in HEK293T cell. Epifluorescence (**a**) and TIRF (**c**) images of HEK293T cells transfected with iGluSnFR taken in the same *X*, *Y* and *Z* position. These images were processed the same manner, we increased the contrast and carried out a background processing and noise rejection with image J. Scale bar, 5 μm. The profile in (**b**) and (**d**) reflect fluorescent intensity in the regions shown with white lines inserted both in (**a**) and (**c**), respectively. (**e**) Original TIRF image at 0 s of a live HEK293T cell transfected with iGluSnFR. (**f**, **g**) Representative image of iGluSnFR transfected HEK293T cells response just before (**f**) and 7 s after application of 50 μmol/L Glu solution (**g**). Scale bar, 5 μm. (**h**) Intensity profiles in response to Glu (10, 50, 100 μmol/L) (Mean ± SEM). Each response amplitude was shown as normalized $\Delta F/F$. Bar on the graph showed the application period of Glu

As expected, these observations suggested that TIRF microscopy allow a better resolution at the membrane by removing unwanted photons coming from nonfocal planes. Taking those results into account and because iGluSnFR is addressed to the cell membrane, we decided to use TIRF microscopy to monitor Glu release.

To test the response of iGluSnFR to Glu, we directly applied Glu on HEK293T cell transfected with iGluSnFR. Following 60 s of fluorescence baseline recording, 2 μL of Glu (10, 50, or 100 μmol/L) were puffed onto HEK293T cell. We found a very small response with 10 μmol/L and larger responses with 50 μmol/L and 100 μmol/L. However the amplitude of the response with 100 μmol/L Glu was slightly higher than the 50 μmol/L Glu response, likely because the response is not linear for large concentrations and because sensors were saturated. The response to Glu was immediate and reached a maximum at 6 s (peak time) (Fig. 2e–g). This transient and rapid fluorescence increase was observed coincidentally with the application of Glu (Fig. 2h). We found that the response to Glu at the different concentrations had a similar kinetic but the intensity of fluorescence change increased with the concentration. These experiments suggest that measured intensity of the sensors reflects the Glu amount to which the cells are exposed to.

We also performed similar experiments on astrocytes, our cells of interest. Astrocytes were transfected with pCMV.iGluSnFR and we tested the sensitivity and kinetics of iGluSnFR in response to Glu (Fig. 3a–c). When compared to HEK293T cells we found a broader response with a first peak at 6 s and a maximum at 12 s. This biphasic response is likely due to the direct response to the Glu puffed by the pipette followed by a second response that could be due to the Glu released by astrocytes (Fig. 3d).

3.3 Measurement of Glu Release by Astrocytes Using iGluSnFR in TIRF Microscopy

ATP is a well-known transmitter that induces a calcium-dependent Glu release from astrocytes [38]. To quantify the amount of Glu and the kinetic of Glu release by astrocytes, we applied 2 μL ATP (10 μmol/L) to astrocytes transfected with pCMV.iGluSnFR. iGluSnFR expressed on astrocytes was sensitive enough to detect autocrine Glu release (Fig. 4a–c). ATP induced a delayed and slow fluorescence increase that lasted for minutes (Fig. 4d). This result indicates that the kinetics of Glu release following ATP application was very different from the one observed upon Glu application. This could be explained by a different signaling pathway, different release mechanisms, or through the recruitment of neighboring astrocytes; iGluSnFR would then detect auto and paracrine Glu.

In conclusion, TIRF microscopy combined with iGluSnFR allows the monitoring in real time of Glu release and to study the kinetics and the relative Glu concentration changes. Notably, this approach will allow us to better understand the mechanism, roles and dynamics of astrocytic gliotransmission involving Glu.

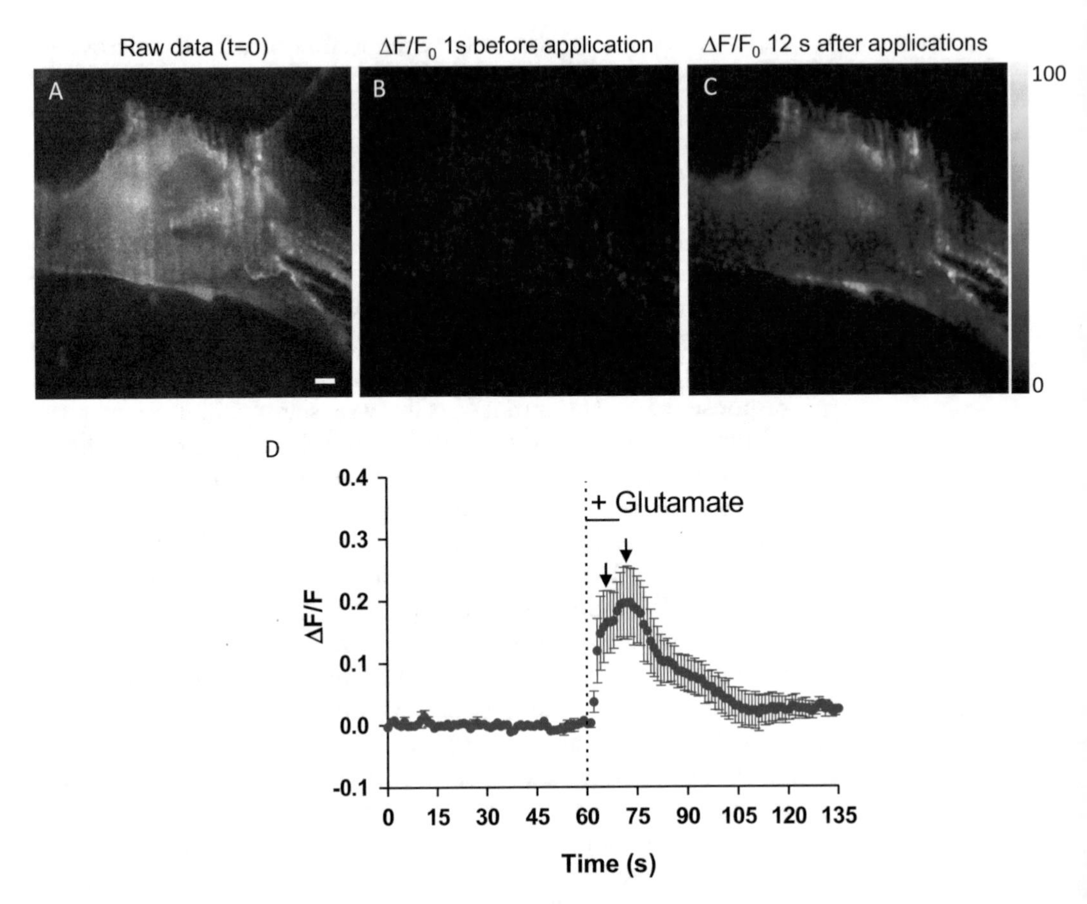

Fig. 3 iGluSnFR response to Glu on astrocyte. Pure astrocytes transfected with iGluSnFR was stimulated with 50 μmol/L Glu. (**a**) Original TIRF images at 0 s of live astrocytes transfected with iGluSnFR. Scale bar, 5 μm. (**b**, **c**) Images were displayed as $\Delta F/F_0$ just before application (**b**) and 12 s after (**c**). (**d**) The fluorescent intensity profile of iGluSnFR was represented as normalized $\Delta F/F$ ($n = 3$, mean ± SEM). Bar in the graph showed the application period of Glu, *arrows* indicate the two peaks observed following Glu application

4 Conclusions

TIRF microscopy has been widely used to observe fluorescent molecule behaviors near or at the plasma membrane. TIRF microscopy combined with the fluorescent sensor for Glu enables the visualization and the relative quantification of Glu release at the cellular level. This method is an attractive tool to better understand the communication between cells and the mechanisms of Glu released by astrocytes.

Using this approach, we were able to reliably measure physiological change in Glu concentration in astrocytes. As shown in our test on HEK293T cells, it was difficult to monitor change in Glu concentration, lower than 10 μmol/L. This relatively low sensitivity could be explained by a dilution effect of Glu when puffed in

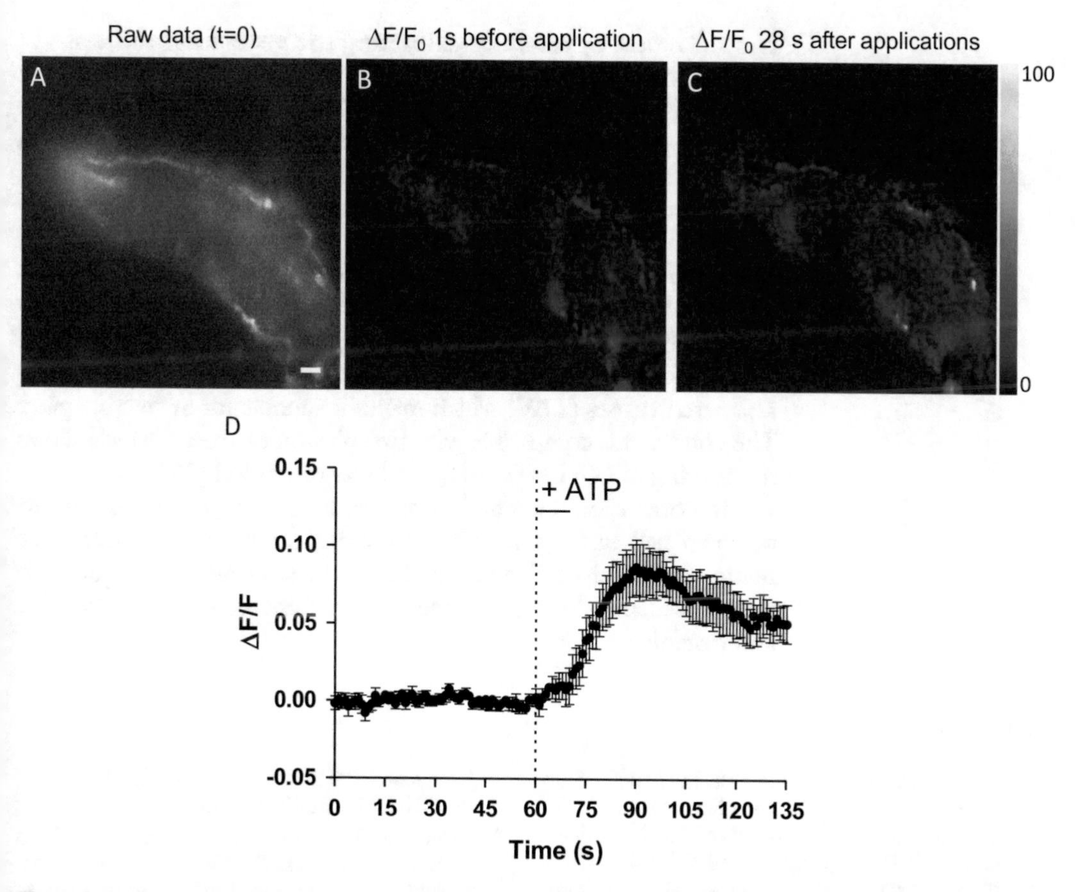

Fig. 4 iGluSnFR response to ATP-induced Glu release from astrocyte. Astrocytes transfected with iGluSnFR. Two microliters ATP was applied onto single astrocyte at 60 s using capillary system. (**a**) Original TIRF image at 0 s of a live astrocyte transfected with iGluSnFR. Scale bar, 5 μm. (**b**, **c**) Images were displayed as $\Delta F/F_0$ just before application (**b**) and 28 s after (**c**). (**d**) Intensity profiles of iGluSnFR response to ATP (10 μmol/L). Each response amplitude was shown as normalized $\Delta F/F$ ($n = 7$, mean ± SEM). Bar in the graph showed the period of application of ATP

the medium, indeed real concentration of Glu at the membrane is likely lower especially because we are monitoring the membrane which is in contact with the coverslip. In our hands, it also seems that we saturated the receptors when using 100 μmol/L Glu concentration. In any case, to monitor endogenous Glu, because the sensor is supposedly right next to the sites of release of Glu, the local quantity of Glu being released by the cell was sufficient to be monitored. So iGluSnFR usage is relevant to monitor broadly, change in Glu concentration at the level of the membrane. It is the main advantage of this technique which is obviously less sensitive than other quantitative techniques such as capillary electrophoresis. It is also important to note that the signal recorded is largely dependent on the expression level of the Glu sensor which makes even more difficult an absolute measurement of Glu concentrations. One may obtain broad determination of absolute Glu concentration by carefully calibrating the system but this was not the purpose of our experiments. It is to note that over the course of our experiments, we found that ATP concentration to stimulate astrocytes should be kept as low as possible as high concentration of ATP (100 μmol/L) was found to quench the GFP signal.

Interestingly this method could be coupled with other techniques such as calcium measurement or other FRET signals and photoactivation techniques. It can also be used with nonfluorescent techniques such as electrophysiology.

Finally, we used plasmids and transfection to obtain expression of iGluSnFR in our cell cultures but the sensor exist also packed in adeno associated viruses (AAV) which makes it suitable for in vivo imaging. The construct is compatible with two photon excitation which allows monitoring of Glu in vivo with a cellular resolution [28].

In conclusion, encoded fluorescent sensors are helpful tools to perform cell biology in complex systems with the advantage of being noninvasive. Thanks to the CRISPR/Cas9 technique, this type of sensors should become easily accessible to many cell type and organisms [39].

References

1. Halassa MM, Fellin T, Haydon PG (2007) The tripartite synapse: roles for gliotransmission in health and disease. Trends Mol Med 13(2):54–63. doi:10.1016/j.molmed.2006.12.005
2. Witcher MR, Park YD, Lee MR, Sharma S, Harris KM, Kirov SA (2010) Three-dimensional relationships between perisynaptic astroglia and human hippocampal synapses. Glia 58(5):572–587. doi:10.1002/glia.20946
3. Bazargani N, Attwell D (2016) Astrocyte calcium signaling: the third wave. Nat Neurosci 19(2):182–189. doi:10.1038/nn.4201
4. Perea G, Navarrete M, Araque A (2009) Tripartite synapses: astrocytes process and control synaptic information. Trends Neurosci 32(8):421–431. doi:10.1016/j.tins.2009.05.001
5. Bezzi P, Carmignoto G, Pasti L, Vesce S, Rossi D, Rizzini BL, Pozzan T, Volterra A (1998) Prostaglandins stimulate calcium-dependent glutamate release in astrocytes. Nature 391(6664):281–285. doi:10.1038/34651
6. Bezzi P, Gundersen V, Galbete JL, Seifert G, Steinhauser C, Pilati E, Volterra A (2004) Astrocytes contain a vesicular compartment that is competent for regulated exocytosis of glutamate. Nat Neurosci 7(6):613–620. doi:10.1038/nn1246
7. Bowser DN, Khakh BS (2004) ATP excites interneurons and astrocytes to increase synaptic inhibition in neuronal networks. J Neurosci 24(39):8606–8620. doi:10.1523/JNEUROSCI.2660-04.2004
8. Domercq M, Brambilla L, Pilati E, Marchaland J, Volterra A, Bezzi P (2006) P2Y1 receptor-evoked glutamate exocytosis from astrocytes: control by tumor necrosis factor-alpha and prostaglandins. J Biol Chem 281(41):30684–30696. doi:10.1074/jbc.M606429200
9. Pasti L, Volterra A, Pozzan T, Carmignoto G (1997) Intracellular calcium oscillations in astrocytes: a highly plastic, bidirectional form of communication between neurons and astrocytes in situ. J Neurosci 17(20):7817–7830
10. Parpura V, Basarsky TA, Liu F, Jeftinija K, Jeftinija S, Haydon PG (1994) Glutamate-mediated astrocyte-neuron signalling. Nature 369(6483):744–747. doi:10.1038/369744a0
11. Parri HR, Gould TM, Crunelli V (2001) Spontaneous astrocytic Ca2+ oscillations in situ drive NMDAR-mediated neuronal excitation. Nat Neurosci 4(8):803–812. doi:10.1038/90507
12. Fellin T, Pascual O, Gobbo S, Pozzan T, Haydon PG, Carmignoto G (2004) Neuronal synchrony mediated by astrocytic glutamate through activation of extrasynaptic NMDA receptors. Neuron 43(5):729–743. doi:10.1016/j.neuron.2004.08.011
13. Angulo MC, Kozlov AS, Charpak S, Audinat E (2004) Glutamate released from glial cells synchronizes neuronal activity in the hippocampus.

J Neurosci 24(31):6920–6927. doi:10.1523/JNEUROSCI.0473-04.2004
14. Kang J, Jiang L, Goldman SA, Nedergaard M (1998) Astrocyte-mediated potentiation of inhibitory synaptic transmission. Nat Neurosci 1(8):683–692. doi:10.1038/3684
15. Letellier M, Park YK, Chater TE, Chipman PH, Gautam SG, Oshima-Takago T, Goda Y (2016) Astrocytes regulate heterogeneity of presynaptic strengths in hippocampal networks. Proc Natl Acad Sci U S A 113(19):E2685–E2694. doi:10.1073/pnas.1523717113
16. Pascual O, Casper KB, Kubera C, Zhang J, Revilla-Sanchez R, Sul JY, Takano H, Moss SJ, McCarthy K, Haydon PG (2005) Astrocytic purinergic signaling coordinates synaptic networks. Science 310(5745):113–116. doi:10.1126/science.1116916
17. Zhang JM, Wang HK, Ye CQ, Ge W, Chen Y, Jiang ZL, Wu CP, Poo MM, Duan S (2003) ATP released by astrocytes mediates glutamatergic activity-dependent heterosynaptic suppression. Neuron 40(5):971–982
18. Serrano A, Haddjeri N, Lacaille JC, Robitaille R (2006) GABAergic network activation of glial cells underlies hippocampal heterosynaptic depression. J Neurosci 26(20):5370–5382. doi:10.1523/JNEUROSCI.5255-05.2006
19. Ben Achour S, Pont-Lezica L, Bechade C, Pascual O (2010) Is astrocyte calcium signaling relevant for synaptic plasticity? Neuron Glia Biol 6(3):147–155. doi:10.1017/S1740925X10000207
20. Zorec R, Araque A, Carmignoto G, Haydon PG, Verkhratsky A, Parpura V (2012) Astroglial excitability and gliotransmission: an appraisal of Ca2+ as a signalling route. ASN Neuro 4(2). doi:10.1042/AN20110061
21. Vasylieva N, Maucler C, Meiller A, Viscogliosi H, Lieutaud T, Barbier D, Marinesco S (2013) Immobilization method to preserve enzyme specificity in biosensors: consequences for brain glutamate detection. Anal Chem 85(4):2507–2515. doi:10.1021/ac3035794
22. Innocenti B, Parpura V, Haydon PG (2000) Imaging extracellular waves of glutamate during calcium signaling in cultured astrocytes. J Neurosci 20(5):1800–1808
23. Zhang Q, Fukuda M, Van Bockstaele E, Pascual O, Haydon PG (2004) Synaptotagmin IV regulates glial glutamate release. Proc Natl Acad Sci U S A 101(25):9441–9446. doi:10.1073/pnas.0401960101
24. Benveniste H, Drejer J, Schousboe A, Diemer NH (1984) Elevation of the extracellular concentrations of glutamate and aspartate in rat hippocampus during transient cerebral ischemia monitored by intracerebral microdialysis. J Neurochem 43(5):1369–1374
25. Masson J, Darmon M, Conjard A, Chuhma N, Ropert N, Thoby-Brisson M, Foutz AS, Parrot S, Miller GM, Jorisch R, Polan J, Hamon M, Hen R, Rayport S (2006) Mice lacking brain/kidney phosphate-activated glutaminase have impaired glutamatergic synaptic transmission, altered breathing, disorganized goal-directed behavior and die shortly after birth. J Neurosci 26(17):4660–4671. doi:10.1523/JNEUROSCI.4241-05.2006
26. Marignier R, Nicolle A, Watrin C, Touret M, Cavagna S, Varrin-Doyer M, Cavillon G, Rogemond V, Confavreux C, Honnorat J, Giraudon P (2010) Oligodendrocytes are damaged by neuromyelitis optica immunoglobulin G via astrocyte injury. Brain 133(9):2578–2591. doi:10.1093/brain/awq177
27. Nedergaard M, Verkhratsky A (2012) Artifact versus reality—how astrocytes contribute to synaptic events. Glia 60(7):1013–1023. doi:10.1002/glia.22288
28. Marvin JS, Borghuis BG, Tian L, Cichon J, Harnett MT, Akerboom J, Gordus A, Renninger SL, Chen TW, Bargmann CI, Orger MB, Schreiter ER, Demb JB, Gan WB, Hires SA, Looger LL (2013) An optimized fluorescent probe for visualizing glutamate neurotransmission. Nat Methods 10(2):162–170. doi:10.1038/nmeth.2333
29. Okumoto S, Looger LL, Micheva KD, Reimer RJ, Smith SJ, Frommer WB (2005) Detection of glutamate release from neurons by genetically encoded surface-displayed FRET nanosensors. Proc Natl Acad Sci U S A 102(24):8740–8745. doi:10.1073/pnas.0503274102
30. Hires SA, Zhu Y, Tsien RY (2008) Optical measurement of synaptic glutamate spillover and reuptake by linker optimized glutamate-sensitive fluorescent reporters. Proc Natl Acad Sci U S A 105(11):4411–4416. doi:10.1073/pnas.0712008105
31. Liu T, Sun L, Xiong Y, Shang S, Guo N, Teng S, Wang Y, Liu B, Wang C, Wang L, Zheng L, Zhang CX, Han W, Zhou Z (2011) Calcium triggers exocytosis from two types of organelles in a single astrocyte. J Neurosci 31(29):10593–10601. doi:10.1523/JNEUROSCI.6401-10.2011
32. Shigetomi E, Khakh BS (2009) Measuring near plasma membrane and global intracellular calcium dynamics in astrocytes. J Vis Exp 26. doi:10.3791/1142
33. Roumier A, Pascual O, Bechade C, Wakselman S, Poncer JC, Real E, Triller A, Bessis A (2008) Prenatal activation of microglia induces delayed impairment of glutamatergic synaptic function.

PLoS One 3(7):e2595. doi:10.1371/journal.pone.0002595

34. Pascual O, Ben Achour S, Rostaing P, Triller A, Bessis A (2012) Microglia activation triggers astrocyte-mediated modulation of excitatory neurotransmission. Proc Natl Acad Sci U S A 109(4):E197–E205. doi:10.1073/pnas.1111098109
35. Poulter NS, Pitkeathly WT, Smith PJ, Rappoport JZ (2015) The physical basis of total internal reflection fluorescence (TIRF) microscopy and its cellular applications. Methods Mol Biol 1251:1–23. doi:10.1007/978-1-4939-2080-8_1
36. Namiki S, Sakamoto H, Iinuma S, Iino M, Hirose K (2007) Optical glutamate sensor for spatiotemporal analysis of synaptic transmission. Eur J Neurosci 25(8):2249–2259. doi:10.1111/j.1460-9568.2007.05511.x
37. Okubo Y, Sekiya H, Namiki S, Sakamoto H, Iinuma S, Yamasaki M, Watanabe M, Hirose K, Iino M (2010) Imaging extrasynaptic glutamate dynamics in the brain. Proc Natl Acad Sci U S A 107(14):6526–6531. doi:10.1073/pnas.0913154107
38. Jeremic A, Jeftinija K, Stevanovic J, Glavaski A, Jeftinija S (2001) ATP stimulates calcium-dependent glutamate release from cultured astrocytes. J Neurochem 77(2):664–675
39. Doudna JA, Charpentier E (2014) Genome editing. The new frontier of genome engineering with CRISPR-Cas9. Science 346(6213):1258096. doi:10.1126/science.1258096

Chapter 7

Drugs to Alter Extracellular Concentration of Glutamate: Modulators of Glutamate Uptake Systems

Andréia Cristina Karklin Fontana

Abstract

Excitatory amino acid transporters maintain low extracellular concentrations of glutamate in the central nervous system (CNS), therefore shaping excitatory signaling. This activity is mediated by glutamate transporters EAAT2/GLT-1, EAAT1/GLAST, EAAC1/EAAT3, EAAT4 and EAAT5. Pharmacological inhibitors of glutamate transport were pivotal to our understanding of intrinsic properties and the function of the transporters. On the other hand, endogenous or exogenous compounds that positively modulate the function of EAAT2/GLT-1 may prevent excitotoxicity and therefore could be developed into therapies for several CNS disorders. Herein, we describe selected compounds that alter extracellular concentration of glutamate by modulating the function of glutamate transporters. Their relevance for further understanding glutamate transport mechanisms and for drug discovery and therapy development is also discussed.

Key words Allosteric modulation, Ceftriaxone, EAAT2, Excitotoxicity, Glutamate uptake, Glutamate transporter, High-throughput screening, Riluzole, Neurodegeneration, Parawixin1

1 Introduction

Glutamate (Glu) is the predominant excitatory amino acid neurotransmitter in the mammalian CNS and is essential for many aspects of normal brain function including cognition, memory, learning, developmental plasticity, and long-term potentiation [1–4].

The termination of Glu neurotransmission is achieved by rapid uptake of the released Glu by presynaptic and astrocytic high affinity sodium-dependent transporters [5, 6]. By regulating the levels of extracellular Glu that have access to Glu receptors, Glu uptake systems hold the potential to effect both normal synaptic signaling and abnormal over-activation of Glu receptors that can trigger excitotoxic pathology.

Five structurally distinct subtypes of plasma glutamate/aspartate transporters have been cloned from animal and human tissue, namely EAAT1 or GLAST, EAAT2 or GLT-1, EAAT3 or EAAC1

Sandrine Parrot and Luc Denoroy (eds.), *Biochemical Approaches for Glutamatergic Neurotransmission*, Neuromethods, vol. 130, DOI 10.1007/978-1-4939-7228-9_7,

(human/rat homologs, respectively), EAAT4 and EAAT5. For reviews discussing the cloning and localization of these transporters, see [7–11].

The Glu transporter subtype EAAT2/GLT-1 is expressed throughout the brain, in the spinal cord, primarily in astrocytes, and also in neurons and oligodendrocytes, contributing to 95% of the total Glu transport activity and 1% of total brain protein in the CNS [12–20]. Therefore, EAAT2/GLT-1 plays a central role in the maintenance of extracellular Glu homeostasis [21–24]. Drugs that modulate EAAT2/GLT-1 expression/activity hence are potentially important for the regulation of extracellular concentration of Glu.

In this regard, also of importance for Glu homeostasis is transporter subtype EAAT1/GLAST, which is expressed in astrocytes at the highest levels in the cerebellum [25], and has been implicated in disease states (further developed and discussed in Sect. 1.2.3). Furthermore, EAAC1/EAAT3, that is expressed in neurons, however at a more modest level (~1% of EAAT2/GLT-1) [26], is a transporter that can also rapidly transport cysteine and modulate the activity of proximal Glu receptors [27].

EAAT4 is enriched in Purkinje cells of the cerebellum and is also found at lower levels in other cells in the brain [28–30] and EAAT5 is expressed in vertebrate retina [31]. Both function as very slow transporters with relatively large associated anion currents, and therefore likely function as inhibitory receptors rather than as a mechanism to clear Glu, and they will not be further discussed in this work.

The emphasis of this chapter will be on exogenous modulators of uptake systems, with focus on pharmacological inhibitors and activators of EAAT1/GLAST, EAAT2/GLT-1 and EAAT3/EAAC1 that affect the extracellular Glu concentration. However other strategies for modulating extracellular Glu concentrations exist but will not be discussed in details here. For discussions on endogenous molecules that can induce activation at the transcriptional level of EAAT2/GLT-1 expression, please see [7, 32–35]. Drugs that inhibit or increase glutamate release at nerve terminals could also alter extracellular concentrations of Glu, however, they are usually nonselective, acting at multiple targets including ion channels [36–38]. Because the emphasis of this chapter is on pharmacological interventions directed at Glu transporters, a discussion on such compounds is beyond our scope. For reviews on glutamatergic medications that rely on targeting release mechanisms or receptor interaction for the treatment of drug and behavioral addictions (including acamprosate, *N*-acetylcysteine, D-cycloserine, lamotrigine, memantine, modafinil, and topiramate), please see [39, 40]. Lastly, other strategies for modulation of extracellular Glu concentration include blood glutamate grabbers, like oxaloacetate and analogs, recombinant glutamate oxaloacetate transaminase

1 and pyruvate. This approach relies on activation of the blood-resident enzyme glutamate-pyruvate transaminase, an enzyme that catalyzes the reversible transformation of pyruvate and Glu to alanine and β-ketoglutarate; therefore, artificially increasing pyruvate and shifting the equilibrium of the reaction to the right side and decreasing Glu levels in the blood, resulting in neuroprotection after stroke [41, 42].

1.1 Structure and Molecular Functioning of Glutamate Transporters

The structure of EAATs has been probed biochemically, and a topology of 8–10 transmembrane (TM) helices with intracellular N- and C-termini was predicted [43–48]. Subsequently, the identification of several crystal structures of a bacterial homolog of Glu transporters Glt_{Ph} has greatly facilitated our understanding of mechanistic aspects of the transport cycle [49–52], setting the stage for structure-based drug discovery. Glt_{Ph} consists of three identical protomers that associate within the membrane to form a bowl-like structure [53]. Each protomer has eight TMs and two helix-turn-helix motifs termed hairpin 1 (HP1) and HP2, that serves as the extracellular gate of the transporter [54] (Fig. 1a). The individual protomers in the trimeric complex are held together by a central core; this core is made of TMs 1, 2, 4, and 5 from each protomer that will associate together, while TMs 3, 6, 7, and 8, along with HP1 and HP2, form a transport domain [10] (shown in green in Fig. 1a and b). Fluorescence spectroscopy studies have demonstrated that substrate binding to the transporter is coupled to the binding of all three Na^+ in a highly cooperative manner at both outward-facing and inward-facing conformations [55] (Fig. 1b and c), and the movement of cations (H^+, and K^+ counter-transport) [56–58], that are necessary for Glu uptake and transporter cycling, as well as anions that are uncoupled from the flux of Glu [59, 60] (not shown). The current model proposes that transport domains translocate substrates by moving across the membrane within a central trimerization scaffold in an 'elevator-like' transport domain motion [61–63].

1.2 Glutamate Excitotoxicity and Pathologies: Involvement of EAATs

Glutamate excitotoxicity is the pathological process of the failure of proper removal of Glu by astrocytes from the synapses. This results in sustained elevation of extracellular Glu levels and excessive activation of postsynaptic Glu receptors resulting in Ca^{2+} influx [64] and activation of a cascade of phospholipases, endonucleases, and proteases such as calpain that can lead to apoptotic or necrotic cell death [65]. In fact, ischemic events in humans and in animals lead to an acute and sustained increase in extracellular Glu concentrations [66–69], indicating a lack of timely clearance by Glu transporters. In excitotoxic states, the extracellular concentrations of Glu may reach a millimolar range, causing degeneration of neurons through excessive stimulation of Glu receptors [8, 70–72].

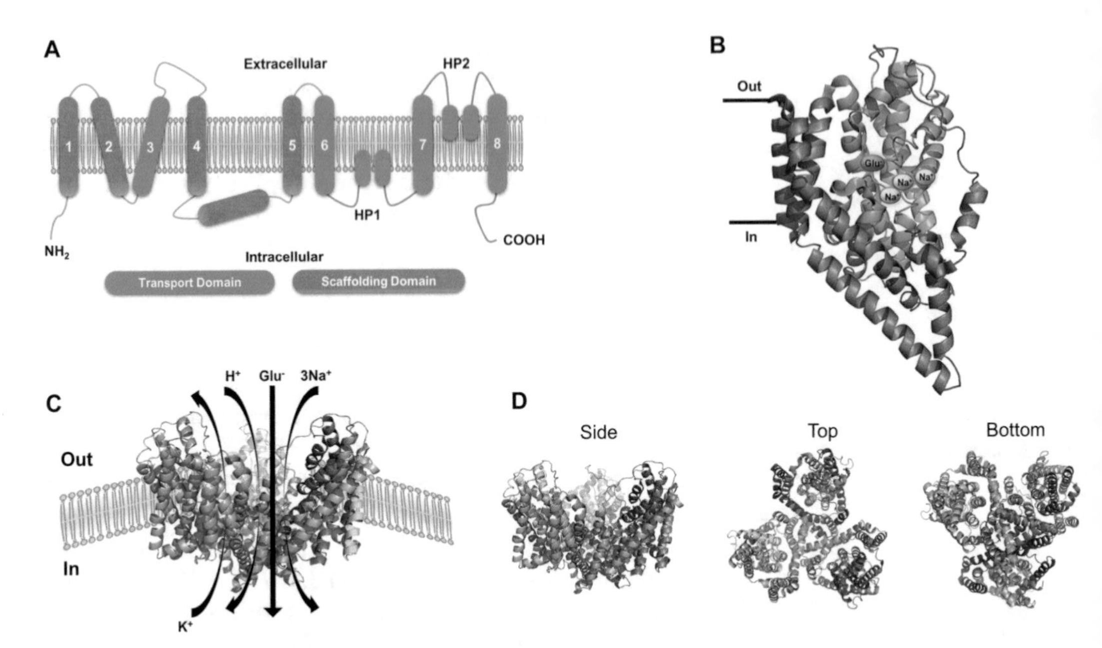

Fig. 1 (**a**) Two-dimensional topology diagram of a single protomer of EAATs, depicting the eight transmembrane domains and hairpin loops. The trimerization (or scaffold) and transport domains are given in *red* and *green*, respectively. (**b**) A single protomer shown in the plane of the membrane, depicting substrate and sodium binding sites in Glt_{Ph}. (**c**) Stoichiometry of EAATs. Picture illustrates the ion coupling stoichiometry of an EAAT, in which the uptake of one glutamate molecule is coupled with the influx of three Na^+, one H^+ and the efflux of one K^+. In addition, EAATs have a glutamate-activated anion conductance, which results in the influx of chloride under physiological conditions (not shown). (**d**) Trimeric structure of Glt_{Ph} viewed in the plane of the membrane with each protomer colored in shades of *red*, *blue*, and *green*. The trimer is shown as *side view*, *top view*, and *bottom view*

EAAT2/GLT-1 appears to be concentrated near synapses, where it acts acutely in a very dynamic manner to shape synaptic transmission by binding released Glu through a process that involves movements of the transporters in the cell membrane [73, 74]. Therefore, tight regulation of Glu concentration by EAAT2/GLT-1 is of great importance for normal Glu neurotransmission and activators of Glu transport can prevent aberrant Glu signaling when EAAT2/GLT-1 function/expression is diminished in injured states. See Fig. 2 for a simplified depiction of a glutamatergic synapse showing localization of EAAT1–3.

Mechanisms involved in Glu-mediated excitotoxicity include downregulation of Glu transporters or Glu efflux via transporter reversal.

1.2.1 Reduced Expression and Function of EAATs

Numerous neurological disorders have reported decreased expression of EAAT2/GLT-1, such as cerebral ischemia [75], ALS [76–78] Alzheimer's disease [79], brain trauma in animals [80], and human patients of traumatic brain injury (TBI) and stroke [81–83]. Consistently, studies with knockout of Glu transporters

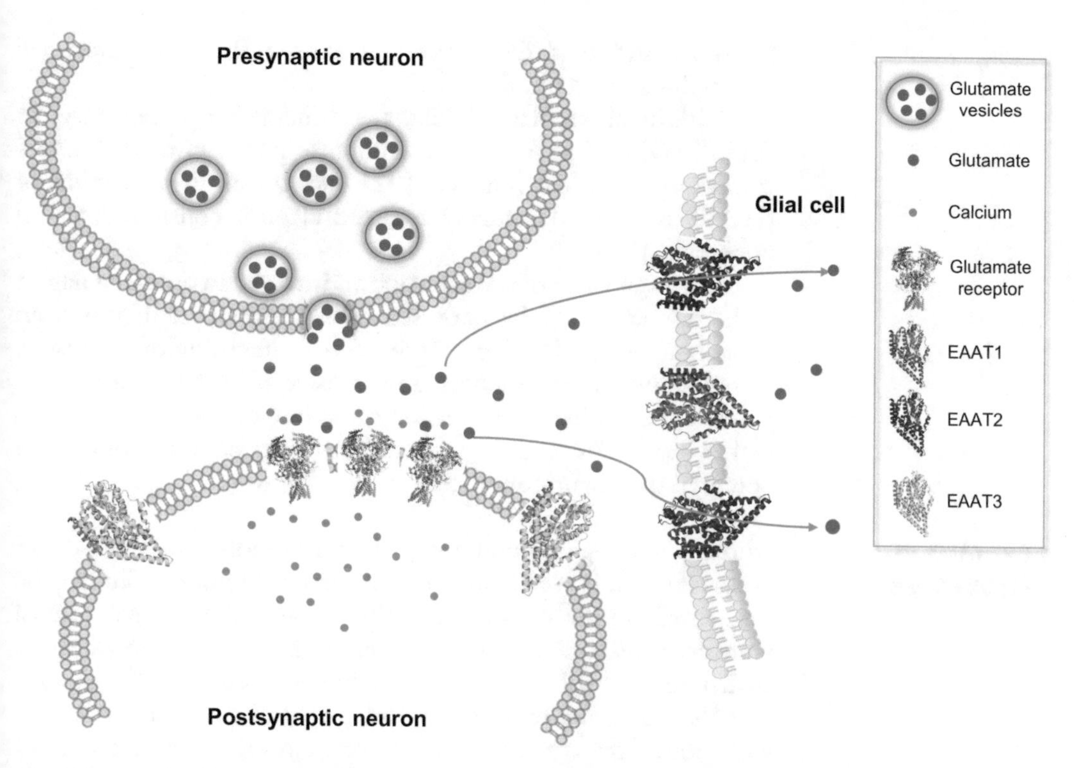

Fig. 2 Schematic diagram of a glutamatergic synapse showing the predominant pre- and postsynaptic locations and glial cell locations of the EAAT subtypes. L-Glutamate is stored in synaptic vesicles at presynaptic terminals. Glutamate (*red dots*) is released into the synaptic cleft, activating postsynaptic glutamate receptors and inducing Ca^{2+} influx (*green dots*). EAATs take up released glutamate, removing it from the synaptic cleft to maintain the extracellular glutamate concentration below neurotoxic levels

reveal a major role for transporter in excitotoxicity and clearance of Glu, especially the astroglial transporters. A significant study has shown that knock-out of the GLT-1 gene resulted in exacerbated damage, with lethal spontaneous seizures and increased susceptibility to acute cortical injury, compared to their wild-type counterparts, following cerebral injury in mice [13]. Later studies confirmed severe disturbances in mice lacking GLT-1 or GLAST [84–86], such as elevated extracellular Glu levels, neurodegeneration and a progressive paralysis [87], exacerbated hippocampal neuronal damage after brain injury [88], and impairment of several essential aspects of neuronal development [89, 90].

On the other hand, deletion of EAAC1/EAAT3 results in depletion of glutathione and neuronal cell loss [91], suggesting that EAAC1 is mostly responsible for providing cysteine for the synthesis of glutathione. Additionally, EAAT3 expression has been shown to be reduced in several disease states, such as bipolar disorder, ischemia, Huntington, and others (reviewed in [27]). However, these processes seem to be related more to cysteine transport and oxidative stress problems, rather than increased extracellular concentration of Glu.

Additional studies highlight a remarkable importance of EAAT1/GLAST in the brain. Antisense knockdown of GLAST compromises retinal function [92] and constitutive deletion of GLAST results in markedly reduced alcohol consumption and preference [93].

Although some of the mechanisms involved in downregulation of EAATs are not fully understood, it is clear that they play an important role in the etiology of neurological diseases. Consequently, there has been an extensive effort to identify compounds that enhance expression/function of these transporters with a goal of decreasing extracellular Glu concentration, as a potential therapeutic approach.

1.2.2 Glutamate Transporter Reversal

Under physiological conditions, the conventional transport direction is inward, however, under excitotoxic conditions, Glu may be transported in the outward direction because the optimal ratio of potassium and sodium concentrations in the extra- and intracellular environments might be altered, resulting in disruption of the gradients that drive the transporters [94–96]. Indeed, raising extracellular potassium concentration evokes a reversed uptake current [97], and massive increases in extracellular potassium and indiscriminate release of Glu have been reported following concussive brain injury [98], lateral fluid percussion [99] and ischemia [100, 101]. Furthermore, the Na^+/K^+-ATPase that maintains these ion gradients are compromised in the injured brain [102, 103].

In this context, it remains to be clarified whether EAAT2/GLT-1 activators will facilitate Glu clearance under excitotoxic conditions, or will rather intensify reverse transport. A study showed that in vitro ischemia led to neuronal damage, however, the ischemia-induced reversal of GLT-1 contributed to the survival of astrocytes themselves, demonstrating that reversal of Glu transporter activity, while damaging to neurons, is also important for the survival of astrocytes under conditions of ischemia and might contribute to later survival of neurons [104]. Also, several strategies involving the application of pharmacological agents such as ceftriaxone, riluzole, and others (as reviewed in [105]) have shown significant protection of nervous tissues against ischemia.

Collectively, these above observations are indications that modulation of Glu transporters as a therapeutic concept is relevant. However, this concept faces the same inherent obstacle as other drugs acting through Glu-mediated mechanisms, such as a significant risk of inducing adverse effects, as glutamatergic signaling is involved in several functions including brain development, cell survival, and synaptic differentiation. For extensive reviews discussing Glu excitotoxicity and acute and chronic pathologies and the implication of EAAT2/GLT-1, see [7, 24, 106–110].

1.2.3 Diseases/Conditions in Which Extracellular Glutamate Concentration Is Altered

In this section, we will briefly discuss some diseases/conditions in which an abnormal extracellular concentration of Glu has been observed.

Stroke: the third leading cause of death in industrialized countries and the most frequent cause of permanent disability in adults worldwide [111]. Several studies have shown that Glu released during ischemia is responsible for cell death [66, 112–116], suggesting that abnormal function of Glu transport plays an essential role in the excitotoxic neurodegeneration. Additionally, decreased Glu transporter expression was observed in ischemic neuronal death, suggesting that a disrupted clearance of Glu contributed to the death [88]. Moreover, studies using ceftriaxone, an upregulator of EAAT2/GLT-1 expression and activity [117–119], and studies using adeno-associated viral (AAV) vector expressing GLT-1 cDNA [120] were shown to reduce the damage caused by brain ischemia, including decrease in lesion size and improvement of behavioral recovery. Treatments for stroke are, to date, restricted to the administration of plasminogen activator to break up blood clots in the arteries of the brain. No treatments are available to halt the progression of secondary damages. Therefore, there is an expectation for future research to explore the potential role of EAAT2/GLT-1 as target for neuroprotection after ischemic stroke.

Epilepsy: a group of disorders characterized by recurrent spontaneous seizures that seemingly result from complex processes involving several neurotransmitter systems including Glu [110, 121, 122], resulting in an imbalance between neuronal excitatory–inhibitory activities that result in epileptogenesis [123, 124]. In patients with medial temporal lobe epilepsy (mTLE) increased extracellular Glu levels were found [125–127] as well as decreased levels of EAAT1 and EAAT2 [128, 129]. Also, decreased levels of GLT-1 were observed in animal models of epilepsy such as pilocarpine- [130], albumin- [131], tuberous sclerosis- [132], and $FeCl_3$-induced models [133] and chest compression-induced audiogenic model [134]. Nevertheless, some studies claim no significant changes in the amount of EAAT2 in the human epileptogenic hippocampus [135, 136], as well in animal models of epilepsy, including kindling [137–139], anticonvulsant ketogenic diet [140], and in spontaneously epileptic rats [141]. On the other hand, mice lacking GLT-1 are prone to exhibit seizures [13], whereas another study reported increased extracellular Glu, but not seizures [87]. Interestingly, increasing GLT-1 expression protected mice against status epilepticus-induced death, neuropathological changes, and chronic seizure development [142, 143]. Recently, it was demonstrated that conditional deletion of GLT-1 protected against fatal epilepsy [144].

Together, these studies indicate that astrocytic Glu uptake plays a crucial role in protecting neurons from hyperexcitability, however, discrepancy in findings suggests that the exact mechanism is still unknown. Despite of this, the notion that modulating EAAT2/GLT-1 as a potential therapeutic approach to treat epilepsy cannot be ruled out. For further information on astrocytic targets (including EAAT2/GLT-1) and new avenues for therapeutic treatment of epilepsy, please refer to [145–147].

Traumatic brain injury (TBI): is a complex pathology of many etiologies [148, 149]. Studies on TBI in humans and in animals have demonstrated an acute increase in tissue Glu concentrations that remain high for up to 5 days in humans, suggesting a lack of timely clearance by Glu transporters [67, 69, 150–153]. Clinical and experimental studies have determined that the resulting Glu-mediated excitotoxicity is a significant contributor to acute post-injury neurodegenerative events [65, 68, 83, 154]. Additionally, decreased GLT-1 expression was reported in animal models of TBI [80], and in TBI in humans [81, 82]. Furthermore, antisense knockdown of GLT-1 in rat exacerbates neuronal damage following brain trauma [88]. Collectively, these studies suggest that decreasing Glu excitotoxicity through targeting GLT-1 could be a strategy for treating TBI.

Amyotrophic lateral sclerosis (ALS): is a debilitating disease, its pathology involves inflammation, oxidant stress, apoptosis, mitochondrial dysfunction, SOD1 (superoxide dismutase 1) protein aggregation and astroglial dysfunction, including dramatic loss of EAAT2/GLT-1 both in the motor cortex and in the spinal cord [76, 77, 155, 156]. Additionally, a marked loss or inactivation of Glu transporters has been shown in several mouse models of ALS [157–159], and in sporadic and familial cases of ALS [160]. Extensive studies using EAAT2/GLT-1 expression activators such as ceftriaxone and others have been performed with hopes to find a therapy for ALS (further explored in Sect. 2.3).

Autism: the etiology of austism spectrum disorder is complex and may involve genetic predisposition, environmental or unknown factors. It is thought that Glu excitotoxicity, mitochondrial dysfunction, and degeneration are key components in autism [161–163]. Recent evidence revealed astroglial dysfunction in the autistic brain [164], and activators of EAAT2/GLT-1 expression improved some symptoms of autism and decreased epilepsy seizures [165], suggesting that EAAT2/GLT-1 may likely be a novel therapeutic strategy for autism.

Huntington's disease (HD): is a devastating progressive autosomal dominant neurodegenerative disorder, with accumulation of the huntingtin protein [166, 167], resulting in oxidative stress and mitochondrial dysfunction [168] and excitotoxicity [169–171]. Decreased EAAT2 mRNAs has been reported in HD patients [172, 173], and in transgenic mouse models of HD [174]. Besides, increase in the functional expression of GLT-1 can improve the behavioral phenotype of a mouse model of HD [175–177]. These studies suggests that changes in EAAT2/GLT-1 expression/function can potentiate or ameliorate the progression of HD, however, effective therapies are yet to be discovered and developed.

Neuropathic pain: neurons responsible for sensing noxious stimuli and conducting pain signals from periphery to the spinal cord are predominantly glutamatergic [178]. Decreases in expression of spinal Glu transporters and Glu uptake have been reported in animal models of pathological pain, with pharmacological inhibition or antisense downregulation of spinal transporters being reported to induce/aggravate pain behaviors [179]. On the other hand, overexpression of GLT-1 in astrocytes in the spinal cord attenuated the induction of inflammatory and neuropathic pain, suggesting that upregulation of spinal EAAT2/GLT-1 could be a novel strategy for the prevention of pathological pain [180–184].

Alzheimer's disease (AD): is a progressive age-related neurodegenerative disorder, characterized by abnormal deposition of fibrillar amyloid β (Aβ) protein, intracellular neurofibrillary tangles, oxidative damage, and neuronal death [185]. Aberrant Glu stimulation and resulting synaptic dysfunction has been proposed as one of several mechanisms by which synapses are damaged in AD [186–188], which has been supported by reports that found reduced EAAT2/GLT-1 expression and function in AD [79, 189–194]. Nonetheless, it is not yet entirely clear whether EAAT2/GLT-1 dysfunction plays a pathogenic role in AD [195, 196] and whether EAAT2/GLT-1 may be used as target for neuroprotective functions in AD pathogenesis remains to be further explored.

Human idiopathic Parkinson's disease (PD): is a progressive neurodegenerative movement disorder that is primarily characterized by degeneration of the dopaminergic neurons that is believed to increase the firing rate of the glutamatergic excitatory projections to the substantia nigra [197–201]. Therefore, Glu-mediated excitotoxicity may be involved in a lethal vicious cycle, which critically contributes to the exacerbation of nigrostriatal degeneration in PD. Studies demonstrate a link between disturbed glutamatergic neurotransmission and Glu transporter functioning in the striatum of an unilateral 6-hydroxydopamine (6-OHDA) animal model for PD [202, 203]. Moreover, it was reported that increased expression of GLT-1 ameliorated locomotor impairments in this animal model of PD [204, 205]. Alternative treatments for dyskinesias, a side effect produced by the major current treatment for PD (L-3,4-dihydroxyphenylalanine or L-DOPA) are needed, and EAAT2/GLT-1 may be a potential target for therapy development.

HIV-associated neurocognitive disorder (HAND): excess Glu release associated with neurotoxicity mediated by gp120, tat (transactivator of transcription), and other HIV proteins is common in the brains of HIV infected individuals [206]. Preliminary evidence with regulation of Glu transporters for potential therapeutic utility for the treatment of HAND is at early stage [207–209].

Additionally, EAAT2/GLT-1 has been implicated in the pathogenesis of major depressive disorders [210, 211], mood disorders [212, 213], substance abuse [214], nicotine dependence [215] and alcohol use disorder [216–220].

Hence, compounds that modulate the function of EAATs, especially EAAT2/GLT-1, could have widespread utility for many CNS disorders.

1.3 EAATs as Target for Modulating Extracellular Glutamate Concentration

Several classes of compounds that target Glu-mediated excitotoxicity, such as NMDAR (*N*-methyl-D-aspartate receptor) antagonists [151, 221–225] and compounds that target calcium influx [226, 227] have been shown to alleviate cellular damage and neurologic deficits. However, these compounds are limited and inefficient, and severe side effects make them less attractive for translation to the clinic. As a result, there are no safe and effective drugs for the prevention or treatment of Glu-mediated excitotoxicity diseases/conditions.

Historically, modulation of EAAT2/GLT-1 expression and/or activity has received surprisingly little attention as putative drug targets, considering the immense therapeutic potential in pharmacological intervention into glutamatergic neurotransmission and the fact that other neurotransmitter transporters are targeted by clinically administered drugs against epilepsy, depression, and attention deficit hyperactivity disorder [228].

In this regard, the concept of pursuing Glu transporters, especially EAAT2/GLT-1, to prevent excitotoxicity has made great progress in recent years. For instance, transgenic approaches for EAAT2/GLT-1 overexpression have confirmed it to be a neuroprotective approach. A study with double transgenic mice created from crossing the ALS mouse model to a mouse model overexpressing GLT-1 resulted in an animal that displays delayed grip strength decline and motor neuron loss and increased life expectancy [229]. Nevertheless, the true potential of EAAT2/GLT-1 as target for therapeutics still remains to be understood, appreciated, and explored [230].

Throughout this chapter we will focus on pharmacological agents that modulate activity and expression of the EAATs, and discuss some perspectives for future identification of novel modulators and therapy possibilities for CNS disorders.

1.3.1 Methods for Identification, Examination, and Validation of Compounds that Modulate EAATs

Regarding approaches for identification of compounds, conventional methods, like medicinal chemistry, enabled the identification of inhibitors such as TBOA and UCPH-101 (Sect. 2.1).

On the other hand, more modern approaches such as high-throughput screening (HTS), have gained popularity over the last decade, together with the availability of libraries for screening for millions of compounds, such as libraries of bioactive small molecules and approved drugs from all major drug classes, or libraries such as the National Institute of Neurological Disorders and Stroke (NINDS) Custom Collection (MicroSource Discovery Systems, CT) comprising 1040 compounds, mostly FDA-approved and marketed drugs. For instance, in a groundbreaking work published in 2005 [231] a blinded screen of 1040 FDA-approved drugs and nutritionals was used and many beta-lactam antibiotics that were potent stimulators of EAAT2/GLT-1 expression were discovered. Later, HTS screenings were useful to identify compounds that increase EAAT2/GLT-1 translation [232], and coupled to structure–activity relationship (SAR) study, pyridazine-based series were further investigated, leading to the identification of a compound, LDN/OSU-0212320, that showed efficacy in several models of CNS disorders [233, 234] (compounds further discussed in Sect. 3.1).

On the other hand, the use of computational homology models that were made possible by the discovery of the Glt_{Ph} structures enabled studies of the transport mechanism. Progress on these fronts have been slower with only a few papers published so far, nevertheless, there are very encouraging results that should aid to resolve outstanding questions on EAATs and other families of transporters/exchangers, through a combined approach involving computational prediction and experimental verification [235].

In addition, critical progress has been made in the understanding of how allosteric modulation of EAAT2/GLT-1 function can be achieved [236], enabling future search for positive allosteric modulators using in in silico screening approaches. These approaches have been used successfully in the past for the identification of novel modulators for ionotropic Glu receptor iGluA2 [237] and for context-dependent GluN2B-selective inhibitors of NMDA receptor function [238].

Regarding approaches to examine and validate compounds that modulate Glu uptake systems, a common one involves the use of radiolabeled substrates or inhibitors. Even though this approach has limitations (such as cost and radiation hazards), it is nonetheless a very attractive method for testing compounds and for low and medium-throughput screenings. Another approach takes advantage of the Na^+-dependence of Glu transporters, as the Glu transport system, when activated by Glu, causes rapid intracellular Na^+ concentration changes, enabling real time monitoring of transporter activity using methods such as patch clamping [239] or fluorescence microscopy using the sodium-binding sensitive probes [240]. Moreover, existing tools for quantitative measurement of rapid Glu transients in intact preparations typically exhibit poor signal-to-noise ratio, kinetics, and targetability. Enzymes such as glutamate dehydrogenase or glutamate oxidase can be coupled to secondary readouts such as NADH fluorescence [241] or current through a microelectrode [242].

In this regard, another emerging promising tool for studying Glu transport in vitro and in vivo is fluorescence-based high-throughput assay. Recently, an intensity-based Glu-sensing fluorescent reporter ("iGluSnFR") with signal-to-noise ratio and kinetics appropriate for in vivo imaging was designed and validated for visualizing Glu release by neurons and astrocytes in increasingly intact neurological systems [243]. Detailed information about this technology is found on Chap. 5.

2 Inhibitors of EAATs

The study of the intrinsic properties and function of the EAATs and the subsequent discovery of pharmacological agents opened the doors for drug development using transporters as targets [244]. The success of identifying inhibitors and substrates have been based upon the concept that restricting the spatial positions that can be occupied by required functional groups can serve to enhance both the potency and selectivity of the analogues. Useful pharmacological probes have emerged through the introduction of additional functional groups (e.g., methyl, hydroxyl, benzyloxy) onto the acyclic backbone of Glu and aspartate (Asp), as well as through the exploitation of novel ring systems (e.g., pyrrolidine-, cyclopropyl-, azole-, oxazole-, and oxazoline-based analogues) to conformationally lock the position of the amino and carboxyl groups [245].

In this regard, a very important and pivotal series of medicinal chemistry work is illustrated by the development of the potent EAAT inhibitor TBOA (threo-ß-benzyloxyaspartate) and analogs [246–249]. TBOA was the first nontransportable blocker for all subtypes of EAATs ever identified; hence, this work revolutionized the field. This discovery facilitated the elucidation of several functions of the EAATs, as further discussed below.

While there are selective inhibitors for EAAT1/GLAST, such as UCPH-101 [250] and for EAAT2/GLT-1, such as DHK [251], there are none for EAAT3/EAAC1. Though ß-benzo-Asp analogs show a slight preference towards EAAT3 [252], this preference is not true selectivity. We will herein divide this section into competitive (2.1) and noncompetitive inhibitors (2.2) and inhibitors of expression of EAATs (2.3), rather than inhibitors for each transporter subtype. See Fig. 3 for a depiction of substrates, competitive and noncompetitive inhibitors of EAATs.

2.1 Competitive Inhibitors

Competitive inhibitors bind to the same binding sites as Glu thereby competing with Glu for binding to the transporter protein. Some competitive inhibitors are substrates of the Glu transporter and they inhibit Glu uptake by being transported instead of Glu, and, importantly, they are also exchanged with internal Glu and thereby induce Glu release [253, 254].

Substrates

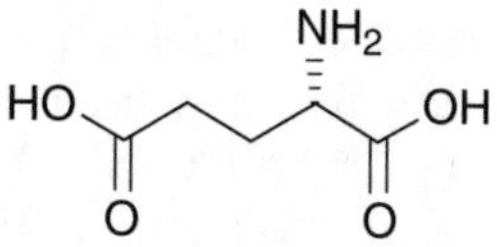

L-glutamic acid

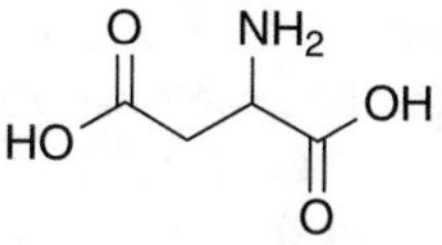

Aspartic acid

Competitive inhibitors

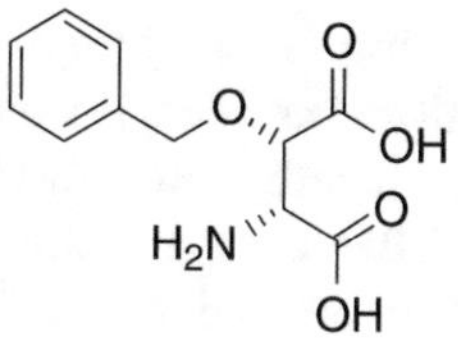

TBOA

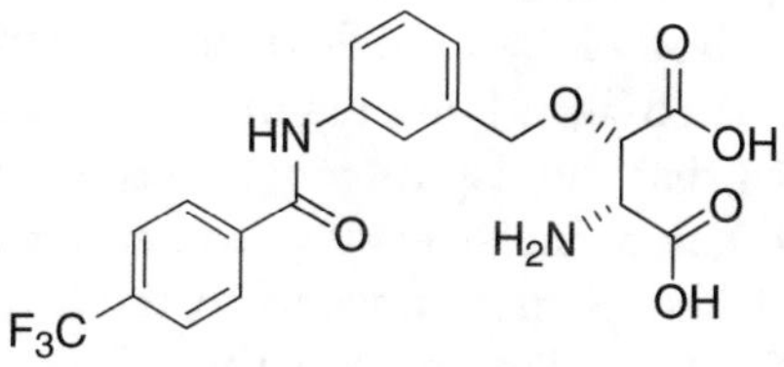

TF-TBOA

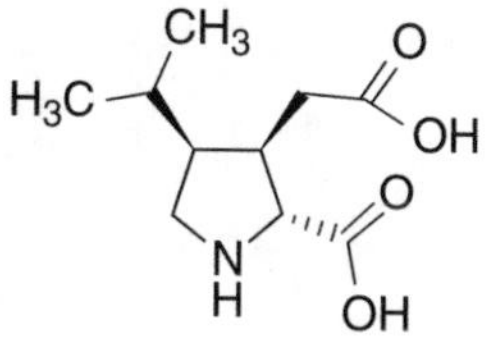

DHK

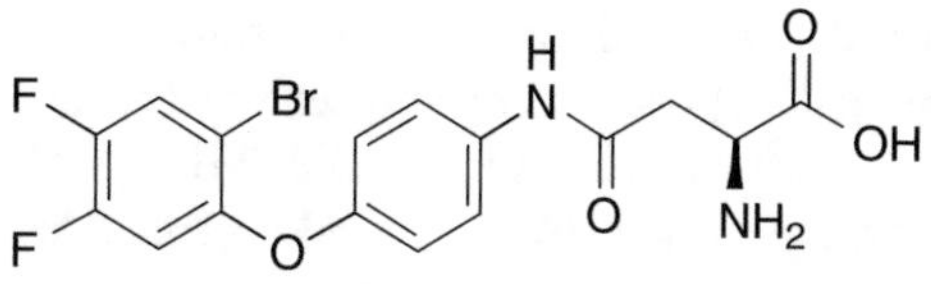

WAY-213613

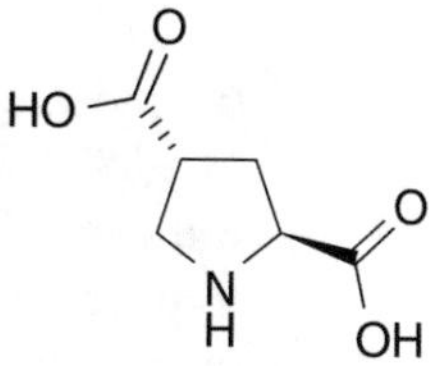

PDC

Non-competitive inhibitors

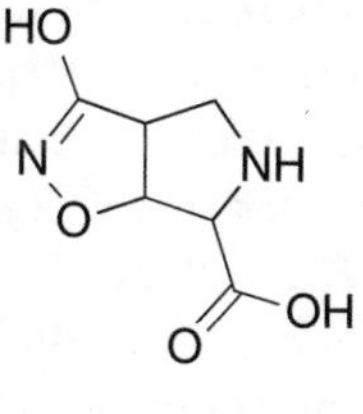

HIP-B

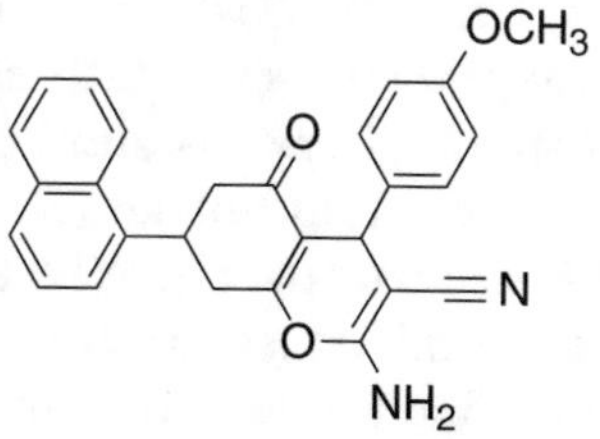

UCPH-101

Fig. 3 Structures of substrates: L-glutamate and aspartate; competitive inhibitors: TBOA (DL-threo-β-Benzyloxyaspartic acid), TFB-TBOA ((3S)-3-[[3-[[4-(Trifluoromethyl)benzoyl]amino]phenyl]methoxy]-L-aspartic acid), DHK (Dihydrokainic acid), WAY 213613 (*N*-[4-(2-Bromo-4,5-difluorophenoxy)phenyl]-L-asparagine), PDC (L-trans-Pyrrolidine-2,4-dicarboxylic acid) and noncompetitive inhibitors: HIP-B (3-hydroxy-4,5,6,6a-tetrahydro-3aH-pyrrolo[3,4-d]isoxazole-6-carboxylic acid) and UCPH-101 (2-Amino-5,6,7,8-tetrahydro-4-(4-methoxyphenyl)-7-(naphthalen-1-yl)-5-oxo-4H-chromene-3-carbonitrile). Not shown: Noncompetitive inhibitor HIP-A (3-hydroxy-4,5,6,6a–tetrahydro-3aH-pyrrolo[3,4-d]isoxazole-4-carboxylic acid)

TBOA binds to the transporter like Glu, however, it is too large to be moved through the transporter to the intracellular space. Nontransportable blockers are desirable because, unlike competitive substrates, they do not cause ion flux and heteroexchange. TBOA blocks EAAT1, EAAT2 and EAAT3 with IC_{50} values of 70, 6, and 6 μmole/L, respectively, displaying a high selectivity for EAATs over ionotropic and metabotropic Glu receptors [249]. Studies have shown that blockade of EAAT2/GLT-1 by TBOA resulted in marked neurotoxicity through the activation of NMDARs due to an increased extracellular Glu level [253, 255–257]. Later, TBOA was also critical in determining conformational changes during transport and the mechanism of Glu transport. Using homology models of the bacterial homolog Glt_{Ph}, fluorescent-based ligand binding assays and computational studies, it was shown that the Asp moiety of TBOA binds to the substrate binding site in a similar manner to Asp. However, TBOA has an additional benzyl ring attached to the β-carbon that props hairpin 2 (HP2) open. This exposes the substrate-binding site to the extracellular solution and disrupts the binding of sodium. Therefore, these studies with TBOA suggested that HP2 may serve as an extracellular gate of the transporter [10].

TFB-TBOA [(3S)-3-[[3-[[4-(Trifluoromethyl)benzoyl]amino] phenyl]methoxy]-L-aspartic acid] is a potent and selective inhibitor of glial transporters EAAT1/GLAST and EAAT2/GLT-1 (IC_{50} values are 22, 17 and 300 nmol/L for EAAT1, EAAT2 and EAAT3 respectively), with no effect on EAAT4 and EAAT5, or a wide range of neuronal receptors and transporters [259]. This compound has been shown to have a higher affinity for the transporter as compared to TBOA, and studies demonstrated it attenuates Glu-stimulated intracellular Na^+ elevation in cultured astrocytes (IC_{50} = 43 nmol/L) [240] and induces severe convulsions in vivo [260]. Hence, TFB-TBOA is to be considered as an extremely valuable tool to study Glu transport and neuron–glia interactions [240].

Dihydrokainate (DHK) is another competitive inhibitor that inhibits L-Glu and L-Asp uptake (K_i = 23 μmol/L for EAAT2), with 130-fold selectivity over EAAT1 and EAAT3 (K_i > 3 mmol/L) [251]. Because of this selectivity over EAAT2, DHK has been used as a tool to examine the relative contributions of the transporters to uptake in various preparations [251, 258]. Recordings of transporter evoked currents from astrocytes and brain slices show that ~30–50% of the current is blocked by concentrations of DHK that should selectively reduce EAAT2/GLT-1-mediated activity by 80–90% [17, 19]. This suggests that EAAT1/GLAST makes a fairly substantial contribution to Glu clearance even in the forebrain. Together, these studies strongly suggest that the bulk of Glu clearance in brain is mediated by the astroglial pools of EAAT1/GLAST and EAAT2/GLT-1.

WAY213613(N-[4-(2-Bromo-4,5-difluorophenoxy)phenyl]-L-asparagine), a potent, nonsubstrate EAAT2/GLT-1 inhibitor that displays >44-fold selectivity over EAAT1/GLAST and EAAT3/EAAC1 and exhibits no activity towards ionotropic and metabotropic Glu receptors [5, 245, 251, 260–262]. Importantly, this compound has facilitated the understanding of protein kinase C (PKC) involvement in the regulation of surface trafficking of GLT-1b (an isoform of GLT-1) and EAAC1 in glutamatergic cerebellar granule cells [263].

In contrast, other potent inhibitors of subtypes EAAT1–4, such as threo-3-hydroxyaspartate and L-transpyrrolidine-2, 4-dicarboxylate (L-trans-2,4-PDC, or PDC) are themselves transported [244, 264, 265]. In this regard, it has been shown that acutely induced dysfunction of EAATs in the rat by single unilateral injection of the PDC triggers a neurodegenerative process mimicking several PD features [266].

2.2 Noncompetitive Inhibitors

Noncompetitive (or allosteric) inhibitors act by binding EAATs at a site different from the orthosteric substrate binding site, often resulting in a conformational change that affects transporter dynamics. Allosteric modulation represents a novel approach of targeting EAATs, as the development of pharmacological tools for the EAATs has mainly focused on the synthesis of competitive inhibitors (usually substrate analogs).

HIP-A (3-hydroxy-4,5,6,6a-tetrahydro-3aH-pyrrolo[3,4-d] isoxazole-4-carboxylic acid) and *HIP-B* (3-hydroxy-4,5,6, 6a-tetrahydro-3aH-pyrrolo[3,4-d]isoxazole-6-carboxylic acid), are two conformationally constrained Asp and Glu analogs, nontransportable, potent, and noncompetitive inhibitors of Glu uptake, with IC_{50} values of ~17 μmol/L, similar to TBOA [267]. In organotypic rat hippocampal slices and mixed mouse brain cortical cultures, HIP-A was neuroprotective under ischemic conditions at 10–30 μmol/L, but not TBOA at the same concentration or 100 μmol/L HIP-A. This suggests that, under ischemia, EAATs mediate both release (reverse transport) and uptake of Glu, and the neuroprotection with the lower HIP-A concentrations may indicate a selective inhibition of the reverse transport, proposing a new strategy for neuroprotective action [268]. Intriguingly, HIP-B has been shown to be more potent in inhibiting forward transport compared to reverse transport, suggesting a new paradigm of Glu transporter inhibition that is based on targeting of a regulatory site [269].

On the other hand, selective allosteric inhibitors of EAAT1/GLAST were recently discovered, whereas selective noncompetitive inhibitors of EAAT2/GLT-1 and EAAT3/EAAC1 have not been identified to date. The discovery of the first class of subtype-selective inhibitors of EAAT1/GLAST came from a structure–activity relationship (SAR) study that addressed the influence of substitutions at the 4- and 7-positions of the parental skeleton

2-amino-5-oxo-5,6,7,8-tetrahydro-4H-chromene-3-carbonitrile. This study identified a potent and selective inhibitor of EAAT1/GLAST (that displays >400-fold selectivity over EAAT2/GLT-1 and EAAT3/EAAC1, making it a highly valuable pharmacological tool [270]), as previous compounds of EAAT1/GLAST inhibitors, such as (*RS*)-2-amino-3-(1-hydroxy-1,2,3-triazol-5-yl) propionate and L-serine-*O*-sulfate (LSOS), was much less potent and selective over the transporter subtypes [271].

The lead compound, *UCPH-101* (2-Amino-5,6,7, 8-tetrahydro-4-(4-methoxyphenyl)-7-(naphthalen-1-yl)-5-oxo-4H-chromene-3-carbonitrile) induces a long-lasting inactive state of EAAT1/GLAST. Its binding site in EAAT1/GLAST has been delineated to target a predominantly hydrophobic crevice in the trimerization domain of the EAAT1/GLAST monomer, and the inhibitor is demonstrated to inhibit the uptake through the monomer that it binds to exclusively and not to affect substrate translocation through the other monomers in the EAAT1/GLAST trimer. The allosteric mode of UCPH-101 inhibition underlines the functional importance of the trimerization domain of the EAAT and demonstrates the feasibility of modulating transporter function through ligand binding to regions distant from its transport domain [250].

In addition to the selective inhibitors of EAATs discussed herein, several other less selective inhibitors have been identified. However, research on these compounds has been limited and the most recent studies have been focused on the more selective and potent compounds mentioned above. Examples of these less selective inhibitors are: cis-ACBD (cis-1-Aminocyclobutane-1,3-dicarboxylic acid), L-CCG-lll ((2S,1′S,2′R)-2-(Carboxycyclopropyl)glycine, Chicago Sky Blue 6B, 7-Chlorokynurenic acid sodium salt, 7-Chlorokynurenic acid, Evans Blue tetrasodium salt, (±)-threo-3-Methylglutamic acid, MPDC, and SYM 2081.

2.3 Inhibitors of the Expression of EAATs

It has been known for many years that the activity of transporters can be regulated by many means, such as phosphorylation, sulfhydryl oxidation, arachidonic acid, pharmacological agents, among others [272, 273]. Additionally, the expression of Glu transporters may be downregulated in certain disease states, and this alteration is likely to regulate neuronal excitation (as discussed in Sect. 1.2.1). Since the mechanisms of regulation of the EAATs expression are not well understood, synthetic modulators of expression could be useful tools to comprehend the functions of the EAATs and how they are regulated in physiological and disease states. However, we should be careful on the interpretation of these studies as the inhibitors of EAATs expression usually act by indirect mechanisms that may involve alteration of several cellular pathways. See Fig. 4 for a representation of the inhibitors of EAAT expression discussed herein.

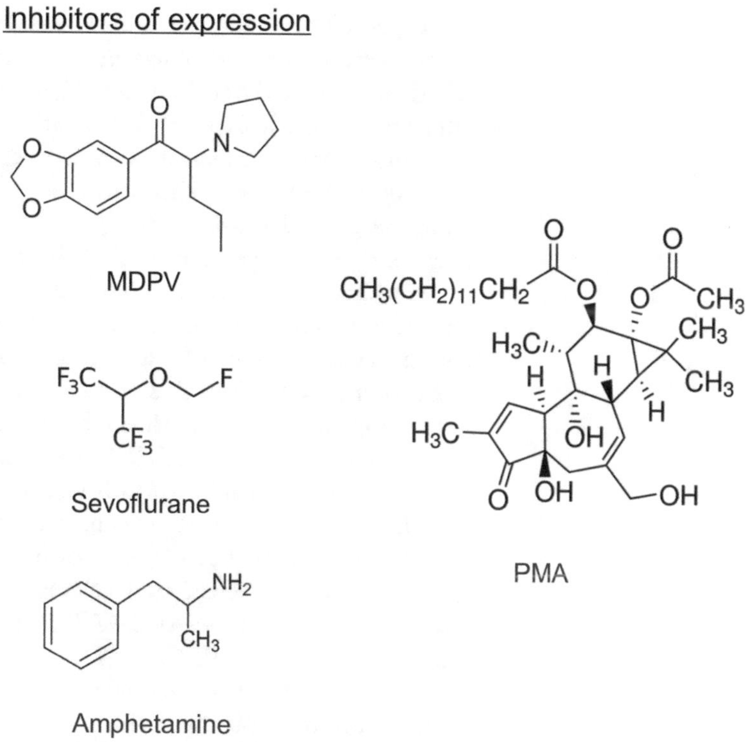

Inhibitors of expression

MDPV

Sevoflurane

Amphetamine

PMA

Fig. 4 Inhibitors of EAATs expression, including synthetic cathinone MDPV (3,4-methylenedioxypyrovalerone), sevoflurane [(1,1,1,3,3,3-Hexafluoro-2-(fluoromethoxy)propane], amphetamine [(RS)-1-phenylpropan-2-amine] and PMA (phorbol 12-myristate 13-acetate). Note that, depending on the specific type of EAAT, PMA may act as activator (Sect. 2.3). Not shown in this figure: thrombin and Oncostatin M (OSM), proteins that inhibit EAAT expression

Synthetic cathinone MDPV (3,4-methylenedioxypyrovalerone): MDPV is a stimulant of the cathinone class which acts as a norepinephrine-dopamine reuptake inhibitor and it was also observed to downregulate GLT-1 expression and produce rewarding and locomotor-activating effects. GLT-1 protein expression in the nucleus accumbens, but not prefrontal cortex, was decreased following withdrawal from repeated MDPV treatment. Ceftriaxone, an activator of EAAT2/GLT-1 expression (further discussed in Sect. 3.1), when administered prophylactically, attenuated the development of MDPV-induced sensitization of repetitive movements in rats. This demonstrates dysregulation of corticolimbic Glu transport systems during withdrawal from chronic MDPV exposure, and suggest that activating expression of GLT-1 disrupts behavioral effects of MDPV [214].

Sevoflurane: is a general anesthetic that was shown to inhibit GLAST and Glial Fibrillary Acidic (GFAP) protein expression in hippocampal astrocytes. General anesthetics may induce neurotoxicity, and especially with children and infants. In this study, postnatal day 7 rats were treated with sevoflurane for 6 h and expressions of GFAP and GLAST were robustly decreased on days 1, 3, 7, and 14 after sevoflurane inhalation. Inactivation of the Janus Kinase/signal transducer and activator of transcription (JAK/STAT) pathway possibly contributes to this effect of sevoflurane. Moreover, this study suggests that astrocytic dysfunction induced by sevoflurane may contribute to its neurotoxicity in the developing brain [274].

Amphetamine, a potent CNS stimulant through mechanisms generally attributed to their ability to regulate extracellular dopamine concentrations, was also shown to be associated with adaptations in glutamatergic signaling. A study demonstrated that amphetamine enters dopamine neurons through the dopamine transporter, and stimulates endocytosis of EAAT3/EAAC1 in dopamine neurons, in a process that is dynamin- and Rho-mediated [275].

Phosphorylation and modulation of brain glutamate transporters by PKC has been subject of investigation for a long time, and it has become clear that activation of PKC has differential effects in different types of EAATs. The use of activators of the signal transduction of PKC, such as diesters of phorbol, and inhibitors, such as Bisindolylmaleimide II, have helped elucidate mechanisms involved in this regulation.

PMA (phorbol 12-myristate 13-acetate), an activator of PKC, was shown to decrease the activity and cell surface expression of GLT-1, a process that was dependent upon clathrin-mediated endocytosis. These studies identified mechanism by which the levels of GLT-1 could be rapidly downregulated via lysosomal degradation. It is possible that this mechanism will contribute to the loss of EAAT2/GLT-1 observed after acute insults to the CNS [276, 277].

In the context of the involvement of the glutamatergic system in the effects of antidepressant therapies, a study showed PMA to decrease EAAT3 expression. Doxepin and imipramine, but not fluoxetine, inhibited EAAT3 activity in *Xenopus* oocytes, at clinically relevant concentrations. PKC may be involved in the effects of doxepin on EAAT3, but is not involved in the effects of imipramine at the concentrations studied [278].

Intriguingly, in other systems, EAAT3/EAAC1 activity and trafficking to the cell surface is increased by activation of protein kinase C. In studies with C6 glioma cells, a model system that endogenously expresses EAAT3, preincubation with PMA was shown to increase transport activity two- to threefold. An inactive phorbol ester did not stimulate Glu transport activity, and the PKC inhibitors chelerythrine and bisindolylmaleimide II blocked the stimulation caused by PMA. The putative PI3K inhibitor, wortmannin, decreased Glu uptake and cell surface expression of EAAC1. Although wortmannin did not block the effects of PMA on activity, it prevented the PMA-induced increase in cell surface expression. These studies revealed a clustering of EAAC1 at cell surface after treatment with PMA, suggesting that the trafficking of EAAC1 is regulated by two independent signaling pathways and also may suggest a novel endogenous protective mechanism to limit Glu-induced excitotoxicity [279, 280].

Later, studies on the regulation of EAAC1 and GLT-1 involving mutagenesis and measurement of cell surface transporters identified a short amino acid motif in the carboxyl terminus of EAAC1 required for regulated trafficking. Additionally, it was shown that post-translational modification of GLT-1 by ubiquitin conjugation to lysine residues mediates its internalization and degradation. These studies demonstrate differential mechanisms whereby EAAC1 and GLT-1 are regulated and facilitate our knowledge of regulated trafficking of these transporters [281].

Additionally to drugs that inhibit the expression of EAATs, certain proteins may cause the same effect. For instance, one study demonstrated that exposure of astrocytes to high levels of *thrombin* (as may occur after a compromise of the blood–brain barrier after stroke and TBI), would reduce astrocyte Glu transporter levels. In isolated rat cortical astrocytes, thrombin induced a selective decrease in the expression of GLAST but not GLT1, with a corresponding decrease of Glu uptake by the astrocytes. Additionality, activation of the thrombin receptor PAR-1 induced a similar decrease in the expression of GLAST and compromised Glu uptake, suggesting that the GLAST downregulation occurs via the Rho kinase pathway. The study provides indirect evidence linking the compromise of the blood-brain barrier to thrombin-induced reduction in Glu transporter expression [282].

Moreover, *oncostatin M* (OSM), is a member of the interleukin-6 cytokine family. Elevated levels of OSM have been observed in HIV-1-associated neurocognitive disorders (HAND) and Alzheimer's disease. A recent study provided evidence that OSM treatment (10 ng/mL) time-dependently reduced GLAST and GLT-1 expression and inhibited Asp uptake in cultured astrocytes. This study also provided evidence that activation of JAK (Janus kinase)/signal transducers and activators of transcription 3 (STAT3) signaling by OSM inhibited Glu uptake and resulted in neuronal excitotoxicity in astrocytes, suggesting that targeting OSM receptor β signaling in astrocytes might alleviate HIV-1-associated excitotoxicity [283].

See Table 1 for a summary of the inhibitors of activity and expression described in Sects. 2.1–2.3.

3 Positive Modulators of EAATs

Many studies suggest that positive modulation of EAATs, especially EAAT2/GLT-1, provides a novel approach for the treatment of conditions in which extracellular concentration of Glu is elevated.

Table 1
Summary of inhibitors described that modulate activity or expression of EAATs

Class of compounds		Target	Selectivity and affinity	Basic research applications
Substrates (Fig. 3)	Glutamate Aspartate	Transporters and receptors	N.A.	N.A.
Competitive inhibitors (Fig. 3)	TBOA	EAAT1–3 (nonsubstrate)	More selective to EAAT2 and EAAT3, high selectivity for EAATs over Glu receptors	First nontransportable blocker for all subtypes of EAATs identified; facilitated the elucidation of several functions of the EAATs
	TFB-TBOA	EAAT1–3 (nonsubstrate)	Very potent, more selective to EAAT1 and EAAT2 over EAAT3 and over Glu receptors	Attenuates Glu-stimulated intracellular Na^+ elevation in vitro, induces severe convulsions in vivo
	DHK	EAAT2/GLT-1 (nonsubstrate)	Selective to EAAT2/GLT-1 over other transporter subtypes	Studies using DHK suggest that the bulk of Glu clearance in brain is mediated by EAAT1/GLAST and EAAT2/GLT-1
	WAY-213613	EAAT2/GLT-1 (nonsubstrate)	Selective to EAAT2/GLT-1 over other transporter subtypes and over Glu receptors	Facilitated the understanding of PKC involvement in the regulation of surface trafficking of GLT-1b and EAAC1 in glutamatergic cerebellar granule cells
	PDC	EAAT2/GLT-1 (substrate)	Selective to EAAT2/GLT-1 over other transporter subtypes	Triggers a neurodegenerative process mimicking several PD features
Noncompetitive inhibitors (Fig. 3)	HIP-A	EAATs (nontransportable)	Potent EAAT blocker, preferentially inhibits Glu release over uptake. No or low affinity for Glu receptors	Studies with HIP-A indicate a selective inhibition of the reverse transport under ischemic conditions, proposing a new strategy for neuroprotection
	HIP-B	EAATs (nontransportable)	Potent EAAT blocker, preferentially inhibits Glu release over uptake. No or low affinity for Glu receptors	More potent in inhibiting forward than reverse transport, suggesting regulatory transporter inhibition
	UCPH-101	EAAT1/GLAST (nonsubstrate)	Selective inhibitor of EAAT1 over EAAT2 and EAAT3, no inhibition of EAAT4 or EAAT5	Allosteric inhibition of UCPH-101 shows functional importance of trimerization domain of the EAAT

Expression inhibitors (Fig. 4)	MDPV	Several targets. Stimulant of the cathinone class, shown to inhibit NET/DAT and downregulate EAAT2/GLT-1 expression	N.A.		Studies with MDPV showed it downregulates EAAT2/GLT-1 expression and produce rewarding and locomotor-activating effects, associating a glutamatergic component to these behaviors
	Sevoflurane	Several targets, including the JAK/STAT pathway. Anesthetic shown to inhibit expression of EAAT1/GLAST and GFAP	N.A.		Studies with sevoflurane suggest that astrocytic dysfunction may contribute to neurotoxicity in the developing brain
	Amphetamine	Several targets, including DAT and EAAT3/EAAC1	N.A.		Studies demonstrate involvement of dynamin and Rho in stimulation of endocytosis of EAAT3/EAAC1 by amphetamine in dopamine neurons
	PMA	PKC	N.A.		Studies identified mechanism of EAAT2/GLT-1 rapid downregulated via lysosomal degradation, showed involvement of EAAT3/EAAC1 in the effects of antidepressants, revealed PKC involvement in EAAT3/EAAC1 activity and trafficking, and demonstrated motifs in transporters required for trafficking

Target, selectivity, affinity and applications for each is included. *NET* noradrenaline transporter, *DAT* dopamine transporter. *N.A.* not applicable

Other less selective inhibitors (discussed in the text), and proteins that decrease expression of EAATs (such as thrombin and oncostatin) are not shown in the table

This section will review the current research on the development of compounds with the ability to increase levels of EAAT2/GLT-1 protein expression (3.1) or compounds that interact directly or indirectly with EAAT2/GLT-1 enhancing its catalytic activity (3.2).

3.1 Compounds that Increase EAAT Expression

The expression of EAATs can be regulated in several ways, at both the transcriptional and translational level [22, 284]. This section will focus on exogenous drugs that can increase Glu transporter expression, and at the end briefly comment on proteins that also can upregulate EAAT expression. For a review on the mechanisms of regulation of Glu transporter function discussing transcriptional regulation and posttranscriptional regulation and modifications, see [11]. See Fig. 5 for a depiction of EAAT expression enhancers discussed in this chapter.

We start this section with the important discovery of *ceftriaxone*, an EAAT2/GLT-1 activator that has been extensively studied since the discovery of its mechanism. Although β-lactams have been historically used as antimicrobials, a notable ancillary effect in the host was identified by [231], with a blind screen of over 1000 FDA-approved drugs that discovered that ceftriaxone enhances the expression of EAAT2/GLT-1. This work was the foundation for exploring therapies using ceftriaxone as upregulator of EAAT2/GLT-1 expression for neuroprotection. Ceftriaxone treatment delayed neuronal death and muscle strength loss and increases survival in a mouse model of ALS [285], reduced neurodegeneration and motor deficits in rodent models of stroke [117, 286] and attenuated the damage observed in models of both acute and chronic neurodegenerative disorders [119, 287, 288]. In addition, ceftriaxone treatment before transient focal ischemia upregulates GLT-1 mRNA and protein and reduces infarct volume in rats [118].

Ceftriaxone was shown to have anti-nociceptive role in chronic neuropathic pain in rats [183]. It was also shown to reduce the level of brain Glu, brain edema, and neuronal death following lateral cortical impact injury in rats [289–291]. There is also evidence that ceftriaxone, by removing Glu from the synapse more efficiently through upregulation of GLT-1, resulted in a decrease in posttraumatic seizures after TBI [292]. Additionally, in the 6-OHDA -lesioned model of PD, ceftriaxone has been shown to increase striatal GLT-1 expression and attenuate the normally observed motor symptoms [293] and did not result in dyskinesia as a side effect [205]. Additionally, ceftriaxone was shown to have neuroprotective effects in Huntington's disease [174, 294] and to alleviate early brain injury after subarachnoid hemorrhage by increasing EAAT2/GLT-1 expression via the PI3K/Akt/NF-kappaB [295]. Recently ceftriaxone was shown to protect neurons against global brain ischemia via upregulation of EAAT2/GLT-1 expression and Glu uptake [296] and to preserve Glu transporters and prevent intermittent hypoxia-induced vulnerability to hypoxia in an in vitro on model of chronic intermittent hypoxia [297].

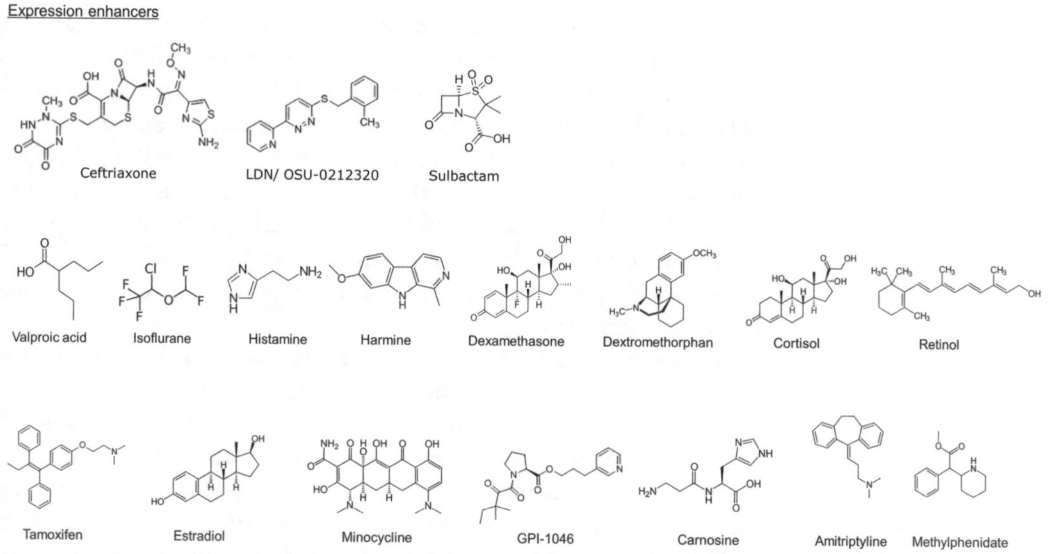

Expression enhancers

Ceftriaxone LDN/ OSU-0212320 Sulbactam

Valproic acid Isoflurane Histamine Harmine Dexamethasone Dextromethorphan Cortisol Retinol

Tamoxifen Estradiol Minocycline GPI-1046 Carnosine Amitriptyline Methylphenidate

Fig. 5 Structures of positive activators of EAATs expression. *Top row*: ceftriaxone ((6R,7R)-7-{[(2Z)-2-(2-amino-1,3-thiazol-4-yl)->2-(methoxyimino)acetyl]amino}-3-{[(2-methyl-5,6-dioxo-1,2,5,6-tetrahydro-1,2,4-triazin-3-yl)thio]methyl}-8-oxo-5-thia-1-azabicyclo[4.2.0]oct-2-ene-2-carboxylic acid), LDN/OSU-0212320 (3-[[(2-Methylphenyl)methyl]thio]-6-(2-pyridinyl)-pyridazine) and sulbactam ((2S,5R)-3,3-Dimethyl-7-oxo-4-thia-1-azabicyclo[3.2.0]heptane-2-carboxylic acid 4,4-dioxide). Not shown: β-lactam antibiotics ampicillin, cefazolin and cefoperazone. *Middle row*: valproic acid (2-propylpentanoic acid), isoflurane (RS)-2-chloro-2-(difluoromethoxy)-1,1,1-trifluoro-ethane), histamine (2-(1H-Imidazol-4-yl)ethanamine), harmine (7-MeO-1-Me-9H-pyrido[3,4-b]-indole), dexamethasone ((8S,9R,10S,11S,13S,14S,16R,17R)-9-Fluoro-11,17-dihydroxy-17-(2-hydroxyacetyl)-10,13,16-trimethyl-6,7,8,9,10,11,12,13,14,15,16,17-dodecahydro-3H-cyclopenta[a]phenanthren-3-one), dextromethorphan ((4bS,8aR,9S)-3-Methoxy-11-methyl-6,7,8,8a,9,10-hexahydro-5H-9,4b-(epiminoethano)phenanthrene), cortisol [(11β)-11,17,21-trihydroxypregn-4-ene-3,20-dione) and retinol ((2E,4E,6E,8E)-3,7-dimethyl-9-(2,6,6-trimethylcyclohex-1-enyl)nona-2,4,6,8-tetraen-1-ol)]. *Bottom row*: tamoxifen ((Z)-2-[4-(1,2-diphenylbut-1-enyl)phenoxy]-N,N-dimethylethanamine), estradiol ((8R,9S,13S,14S,17S)-13-methyl-6,7,8,9,11,12,14,15,16,17-decahydrocyclopenta[a]phenanthrene-3,17-diol), minocycline ((2E,4S,4aR,5aS,12aR)-2-(amino-hydroxy-methylidene)-4,7-bis(dimethylamino)-10,11,12a-trihydroxy-4a,5,5a,6-tetrahydro-4H-tetracene-1,3,12-trione), GPI-1046 (3-(3-pyridyl)-1-propyl (2S)-1-(3,3-dimethyl-1,2-dioxopentyl)-2-pyrrolidinecarboxylate), carnosine ((2S)-2-[(3-Amino-1-oxopropyl)amino]-3-(3H-imidazol-4-yl) propanoic acid), amitriptyline (3-(10,11-Dihydro-5H-dibenzo[a,d]cycloheptene-5-ylidene)-N,N-dimethyl-propan-1-amine) and methylphenidate (methyl 2-phenyl-2-(piperidin-2-yl)acetate). Not shown in the figure are proteins that activate EAAT expression: epidermal growth factor (EGF), pituitary adenylate cyclase-activating peptide (PACAP), mTOR (mammalian target of rapamycin), JAK2 (Janus kinase; JAK2) GSK3β (glycogen synthase kinase 3 beta) and hormones leptin and ghrelin

Regarding translational studies, an open trial of 108 ALS patients showed that ceftriaxone did not substantially improve muscle strength and disability scores [298]. Ceftriaxone has failed to show significant efficacy in clinical trials for stroke [299]. Nevertheless, a clinical trial of ceftriaxone treatment for ALS patients was conducted, which reported it to be a safe and tolerable drug for humans, in stages 1 and 2 [300]. Unfortunately, stage 3 was discontinued due to lack of efficacy in increasing length of patient survival or preventing a decline in function [301].

Recently, the effects of β-lactam antibiotics *ampicillin, cefazolin and cefoperazone* were investigated in a model of ethanol intake in alcohol-preferring rats. Chronic ethanol consumption is known to downregulate expression of GLT-1, which increases extracellular Glu levels in subregions of the mesocorticolimbic reward pathway. The compounds significantly reduced ethanol intake and significantly upregulated both GLT-1 and pAKT expressions in the nucleus accumbens and prefrontal cortex, demonstrating that these compounds appear to be potential therapeutic compounds for treating alcohol abuse and/or dependence [216].

However, not all of the studies have demonstrated increased Glu expression with β-lactams and neuroprotective effects. A study found that ceftriaxone did not increase GLT-1 promoter activity in rat hippocampal slices, although a neuroprotective effect was shown, suggesting that the neuroprotection was not associated to the level of GLT-1 protein expression [117]. Chronic administration of ceftriaxone in rats failed to increase GLT-1 mRNA and protein levels in different brain regions, however, intriguingly, it increased GLT-1 activity and conferred neuroprotection in a stroke model [286]. Additionally, no effect of ceftriaxone has been detected in various CNS regions in WT C57BL/6 mice, including the spinal cord, however, the drug attenuated disease course and severity in a model of autoimmune CNS inflammation [285]. Moreover, ceftriaxone slightly reduced GLT-1 levels and significantly decreased Glu uptake activity in mouse striatal astrocytes [302]. We believe there are several possible explanations for these discrepant results on GLT-1 regulation by β-lactam antibiotics. First, there seems to be indications that their protective effects are highly dependent on the experimental model. Second, the many ways that GLT-1 expression is measured can yield different results, if it is done indirectly (by measuring survival or amelioration of behavior defects), or directly (by quantifying transporter expression with immunoblots, measuring mRNA levels, or measuring uptake activity in tissue preparations such as synaptosomes). This is illustrated by the challenging finding that synaptosomal assays primarily detect the 5–10% of GLT-1 present in neurons rather than the 90–95% present in astrocytes [303]. If this is correct, one might expect that the uptake activity in synaptosomes could go up dramatically with minor changes to the total brain activity of GLT-1. This could have an impact on the interpretation of a number of studies, and this could also explain why some studies on EAAT2/GLT-1 activators demonstrate lack of upregulation of expression, despite the fact that neuroprotection is observed. Third, the preservation of tissue used in the assays should be considered, e.g. from human autopsy versus freshly obtained from animals, as these approaches can yield significant different levels of transport [304].

Further studies on ceftriaxone and other β-lactam antibiotics and derivatives are necessary to determine whether an analog compound with good brain penetrance, pharmacokinetic properties and with less risk of side effects and undesired toxicity can be developed into a therapeutic drug.

Thiopyridazine and pyridazine derivatives: an approach to identify translational activators of EAAT2/GLT-1 using a cell-based enzyme-linked immunosorbent assay identified 61 compounds that showed a dose-dependent increase in EAAT2/GLT-1 protein levels [232]. Later, they progressed the development of thiopyridazine and pyridazine derivatives, from the first group of hit compounds, which increase EAAT2/GLT-1 expression [305]. One of their lead compounds, *LDN/OSU-0212320*, a pyridazine derivative, was shown to increase EAAT2/GLT-1 expression through translational activation and was shown to protect cultured neurons from Glu-mediated excitotoxic injury. Moreover, it was demonstrated to delay motor function decline and extend lifespan in an animal model of ALS [234], and to restore EAAT2/GLT-1 function and improved cognitive functions, restore synaptic integrity, and reduce amyloid plaques in a transgenic model of Alzheimer disease [233].

A study investigated *sulbactam*, a β-lactamase inhibitor with a structure similar to ceftriaxone but with little antibacterial capability and presumably fewer side effects resulting from the antibacterial effect. Treatment with sulbactam protected pyramidal neurons against brain ischemia, and either antisense knockdown of GLT-1 expression or inhibition of the GLT-1 uptake activity with DHK, which inhibits selectively GLT-1, significantly blocked the neuronal protective effect of sulbactam, indicating that sulbactam has a neuronal protective effect though upregulation of GLT-1 [306].

Sodium valproate (or valproic acid), an anti-epileptic drug that has been shown to upregulate hippocampal Glu transport following chronic treatment [307]. Because the mechanism of action of valproate involves modulation of multiple mechanisms its beneficial effects cannot be contributed to only EAAT2/GLT-1 regulation. These other mechanisms include the regulation of γ-aminobutyric acid (GABA) and Glu neurotransmissions, activation of pro-survival protein kinases and inhibition of histone deacetylase. However, evidence for its neuroprotective properties is emerging. Recently, it was suggested that valproate reduces the development of chronic pain after nerve injury, in part by preventing downregulation of GLT-1 [308] and to prevent retinal degeneration in a murine model of normal tension glaucoma [309].

Isoflurane, a commonly used anesthetic, is suspected to be involved in alterations of glutamatergic system in prolonged isoflurane-induced learning/memory decline. One study has shown that isoflurane enhances the activity of EAAT3 by redistributing it to the plasma membrane. Additionally, it was shown that the phosphorylation of serine 465 in EAAT3 by PKC alpha mediates the increased transporter activity and redistribution to plasma membrane after isoflurane exposure [310]. Another study investigated whether Glu concentration and corresponding transporters or receptors display any alternations in aged rat suffering from isoflurane-induced learning/memory impairment. They found a continuous raised Glu and an upregulation of GLAST rather than GLT-1 and postsynaptic Glu receptors in the hippocampus of aged rats associated with isoflurane-induced learning/memory impairment [311]. A third study aimed to use isoflurane to determine how EAAT3/EAAC1 regulates learning and memory. They confirmed previous results that EAAT3 is upregulated with isoflurane but did not change levels of Glu receptors in the plasma membrane of mouse hippocampus. They observed that isoflurane increased protein phosphatase activity in both wild type (WT) and EAAT3/EAAC1(−/−) mice, and reduced GluR1 (a subunit of alpha-amino-3-hydroxy-5-methyl-4-isoxazole propionate (AMPA) receptor) expression in deficient mice. The phosphatase inhibitor okadaic acid attenuated these effects. Finally, isoflurane inhibited context-related fear conditioning in EAAT3/EAAC1 deficient mice, but not in WT. Thus, isoflurane may increase GluR1 trafficking to the plasma membrane via EAAT3/EAAC1 and inhibit GluR1 trafficking via protein phosphatase [312].

Histamine, histamine increased GLT-1 expression in pure cultured astrocytes, decreased the extracellular Glu content and alleviated neuronal cell death induced by exogenous Glu. Importantly, histamine was neuroprotective in brain slices after oxygen and glucose deprivation (OGD) and against ischemic injury in animal model of middle cerebral artery occlusion (MCAO) [313].

Harmine, a beta-carboline alkaloid, was identified through a screening of a library of 1040 FDA approved compounds and natural products in a fetal derived-human immortalized astroglial cells stably expressing a firefly luciferase reporter under the control of EAAT2/GLT-1 promoter. Harmine effectively increased GLT-1 expression and transporter activity in vivo and presented neuroprotection effects in a rat model of ALS, with the beneficial effects specifically due to upregulation of GLT-1 [314]. A recent study shows that harmine also provides neuroprotection in a global cerebral ischemia model [315].

Dexamethasone, a glucocorticoid, increases EAAT2/GLT-1 mRNA levels and subsequently upregulates EAAT2/GLT-1 protein expression and activity [316, 317]. A recent study shows that dexamethasone induces neuroprotection in hypoxic-ischemic brain injury in newborn rats, partly mediated via Akt activation [318]. Studies concluded that Akt induces the expression of EAAT2/GLT-1 through increased transcription and that Akt can regulate GLT-1 expression without increasing GLAST expression in astrocytes [319].

Dextromethorphan (DM), not to be confused with dexamethasone (above): is a non-narcotic antitussive drug that has been reported to have neuroprotective against cerebral ischemia, epilepsy, and acute brain injury. A study has explored the underlying mechanisms of the neuroprotective effects of DM in animals in the setting of TBI. It was identified that DM treatment following TBI significantly reduced brain edema and neurological deficits, as well as increased neuronal survival. These effects correlated with a decrease of tumor necrosis factor α, interleukin-1β (IL-1β) and IL-6 protein expression and an increase of GLAST and GLT-1 expression in the cortex. These results provided in vivo evidence that DM exerts neuroprotective effects via reducing inflammation and excitotoxicity induced following TBI, by increasing the expression of Glu transporters. [320]. However, it should be noted that this compound has also been reported to act as a NMDA receptor antagonist [321, 322], and in addition has actions as a nonselective serotonin reuptake inhibitor [323] and a sigma-1 receptor agonist [324], accounting for a complex mechanism that could be implicated in the neuroprotective effects observed.

Corticosterone and retinol are both able to increase the translation of EAAT2 transcripts and, moreover, many disease-associated insults affected the efficiency of the induced translation of the EAAT2 transcript by these compounds [325].

Tamoxifen, a selective estrogen receptor modulator that is used to treat breast cancer, was shown to enhance the expression and function of EAAT2/GLT-1 in rat astrocytes [326]. Tamoxifen regulates EAAT2/GLT-1 via the CREB (cAMP response element-binding protein) and NF-κB pathways [327].

Estradiol (17β-estradiol), one of the most active estrogen hormones possessing neuroprotective effects in both in vivo and in vitro models, has been shown to increase expression of both GLAST and GLT-1 mRNA and protein and Glu uptake in astrocytes [326], an effect that seems to be mediated by growth factors such as transforming growth factor-α (TGF-α). However, the adverse effects associated with long-term use of estradiol has hampered its clinical utility [33].

Minocycline, a broad-spectrum tetracycline antibiotic, was shown in neuropathic rats to maintain the expression of GLT-1 in the spinal dorsal horn and this was suggested to be involved in the attenuation of behavioral hypersensitivity [181]. This study supports the idea that EAAT2/GLT-1 may be a potential target for the development of analgesics.

GPI-1046, a synthetic neuroimmunophilin ligand that selectively increases GLT-1 expression and activity, was shown to protect motor neurons against excitotoxicity in vitro and increases the survival rate of transgenic ALS mice [328] and more recently was shown to have a role in reduction of ethanol consumption [218].

Carnosine (*beta*-alanyl-L-histidine): a dipeptide that was shown to upregulated the expression level of GLT-1 and to decrease neuronal cell death in cultured neuron/astrocyte exposed to OGD [329].

Amitriptyline, the tricyclic antidepressant was shown to reverse the downregulation of EAAT1/GLAST and EAAT2/GLT-1 and to induce upregulation of EAATs in an animal model of spared nerve injured [182], suggesting that this upregulation may be one of the therapeutic mechanisms of amitriptyline in the treatment of neuropathic pain.

Methylphenidate, is one of the most common drugs prescribed to treat autism spectrum disorders; and is known to increase dopamine extracellular levels. However, it was not established whether its sedative effects are related to regulation of the glutamatergic tone via the regulation of the Glu transport systems. A study using cultured chick cerebellum Bergmann glia cells showed that methylphenidate exposure resulted in increase in the activity and expression of GLAST. This study suggests that glial cells are targets for methylphenidate, and that could in turn regulate dopamine turnover [330].

Additionally to drugs that enhance EAATs expression, it is known that several proteins also have this effect. Early studies showed increased expression of GLT-1 protein in astroglial cultures by the treatment with epidermal growth factor (EGF) and this was thought to be mediated through intracellular signaling involving cAMP as dibutyryl-cAMP incubation had a similar effect [331]. Pituitary adenylate cyclase-activating peptide (PACAP) was also found to stimulate GLT-1 expression [332].

mTOR (a family of serine-threonine protein kinase mammalian target of rapamycin): is involved in the control of a wide variety of cellular processes, and there is evidence for the involvement of PI3K/Akt/mTOR signaling regulation of GLT-1 in astrocytes [333]. A study showed that treatment with PI3K or Akt inhibitors suppresses the phosphorylation of Akt and mTOR and decreased GLT-1 upregulation in astrocytes. Moreover, treatment with the mTOR inhibitor rapamycin decreased GLT-1 protein and mRNA levels and did not affect Akt phosphorylation, suggesting that mTOR is a downstream target of the PI3K/Akt pathway regulating GLT-1 expression [334]. Another study demonstrated upregulation of GLT-1 via mTOR-Akt-NF-κB (nuclear factor-κB) signaling cascade in a model of OGD, suggesting that this signaling cascade may work to promote Glu uptake in brain ischemia and neurodegenerative diseases [34]. However, even though mTOR inhibitors, such as rapamycin and its analogs may represent novel, rational therapies for a variety of neurological disorders, they have not been tested yet in patients with epilepsy, Alzheimer's, Huntington's, and Parkinson's disease. With continuing progress in basic and clinical research, there are reasons to be optimistic that the clinical indications of mTOR inhibitors for neurological disease will continue to expand in the future [335].

Targeting the *janus-activated kinase-2 (JAK2)* has also been demonstrated to be a potential therapeutic avenue for regulating Glu transporter function as it was described to be a powerful regulator of Glu transporters, by showing that co-expression of JAK2 with EAAT2 in *Xenopus* oocytes increased Glu induced current, and thus JAK2 was proposed to participate in the protection against excitotoxicity [336]. However, thus far EAAT2/GLT-1 has not been validated as target for JAK2 and its activation has not been investigated in animal models of excitotoxicity.

Another promising therapeutic target for regulating EAAT2/GLT-1 expression is *GSK3β* (glycogen synthase kinase 3 beta), a kinase involved in multiple cellular processes including neuronal development and synaptic plasticity, that was found through both co-expression and pharmacological studies to stimulate the expression of GLT-1 (while, interestingly, reduced that of GLAST) in COS-7 cells and *Xenopus* oocytes. The authors suggest that this differential modulation may be particularly relevant in pathological conditions such as Alzheimer's disease or ischemia in which Glu transporters and GSK3β are involved [337]. Recent evidence suggests that GSK3β regulation of GLT-1 expression in the spinal dorsal horn plays a role with neuropathic pain [180].

Leptin: a key regulator of appetite and metabolism, has previously been reported to influence glial structural proteins and morphology. A study demonstrated that metabolic status and leptin also modify GLAST and glucose transporters, indicating that metabolic signals influence synaptic efficacy and glucose uptake and, ultimately, neuronal function [338]. The same group extended their studies to g*hrelin*, a hormone that regulates appetite, adiposity, and glucose metabolism. In vivo and in vitro approaches demonstrate that acylated ghrelin rapidly stimulates GLAST expression and Glu uptake, and reduces glucose transporter levels and glucose uptake by astrocytes. The results indicate that acyl-ghrelin may mediate part of its metabolic actions through modulation of hypothalamic astrocytes. This effect may involve astrocyte-mediated changes in local glucose and Glu metabolism that alter the signals/nutrients reaching neighboring neurons [339].

In conclusion, increasing the in vivo expression levels of EAAT2/GLT-1 may be a promising therapy as exemplified by the many studies demonstrating neuroprotective effects. However, this strategy may also result in profound impact on glutamatergic neurotransmission throughout the brain with the intrinsic risk of inducing adverse effects, as exemplified in a study with ceftriaxone that resulted in disrupted hippocampal learning in rats [340]. Another potential issue is that many of the mechanisms targeted underlie the expression of numerous other genes, and thus transcriptional/translational modulators could potentially exert off-target effects outside the glutamatergic system. Interestingly, the kinetics and duration of the modulation exerted by at least

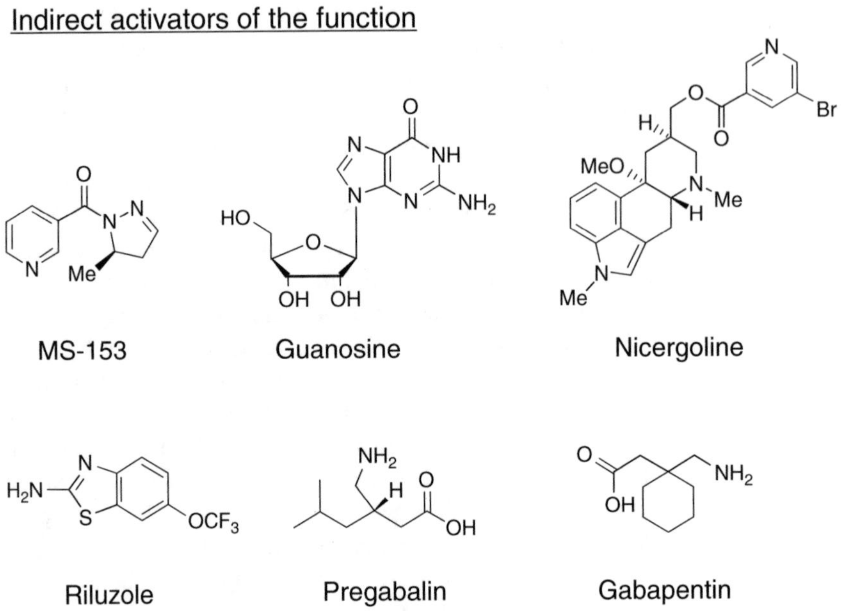

Indirect activators of the function

MS-153 Guanosine Nicergoline

Riluzole Pregabalin Gabapentin

Fig. 6 Structures of indirect activators of the function of EAATs: MS-153 ([R]-[−]-5-methyl-1-nicotinoyl-2-pyrazoline), guanosine (2-Amino-9-[3,4-dihydroxy-5-(hydroxymethyl)oxolan-2-yl]-3H–purin-6-one), nicergoline ([(8β)-10-methoxy-1,6-dimethylergolin-8-yl]methyl 5-bromopyridine-3-carboxylate), riluzole (6-(trifluoromethoxy)benzothiazol-2-amine), pregabalin ((S)-(+)-4-Amino-3-(2-methylpropyl)butanoic acid) and gapapentin (1-(aminomethyl)cyclohexaneacetic acid). Note that some studies propose gabapentin to induce glutamate release and to inhibit glutamate transport (Sect. 3.2). Not shown in the figure: direct activators Parawixin1 (unknown structure [385, 386]) and GT949, GT951 and GT939, novel allosteric modulators of EAAT2 (unpublished)

some of these modulators are likely to be significantly different from the properties of those targeting the EAAT protein, which could constitute an advantage or a problem depending on the specific use. Additional issues for EAAT2/GLT-1 translation activators lay in the effect that, although having neuroprotective properties, this class of compounds must be administered prophylactically to be neuroprotective; therefore they have low clinical relevance for acute conditions.

3.2 Compounds that Increase the Activity of EAATs

Figure 6 represents compounds that indirectly interact with EAAT2/GLT-1 augmenting its catalytic activity. Additionally, compounds that increase Glu uptake mediated by other transporter subtypes are discussed, and, finally, positive allosteric modulators, which represent a relatively novel approach to the treatment of conditions/diseases involving Glu excitotoxicity.

MS-153 ([R]-[−]-5-methyl-1-nicotinoyl-2-pyrazoline), characterized to enhance Glu uptake mediated by GLT-1 [341], was demonstrated to inhibit elevated brain Glu levels induced by MCAO [342], and to be neuroprotective in a model of focal ischemia in rats [343]. Additionally, studies showed that MS-153 inhibited high voltage-gated calcium channels through interactions with PKC, preventing massive release of Glu from nerve terminals in ischemic conditions [344]. MS-153 has also been shown to affect other conditions that are believed to involve excessive Glu signaling, including the inhibition of morphine tolerance and dependence [345], the attenuation of behavioral sensitization development to phencyclidine [346, 347], an anxiolytic effect in fear conditioning [348], and attenuation of the conditioned rewarding effects of morphine, methamphetamine, and cocaine [349]. MS-153 was also recently suggested to have potential as a therapeutic drug for the treatment of alcohol dependence [217, 350].

Our lab has studied the actions of MS-153 in the lateral fluid percussion model of TBI. We found that administration of MS-153 in the acute post-traumatic period provides acute and long-term neuroprotection for TBI, and that the neuroprotective actions of MS-153 can be explained on the basis of inhibition of calcium channels and interactions with PKC, rather than activation of Glu transport. We conclude that MS-153 displays promising potential for developing a much needed therapy for treatment of TBI [351].

Guanosine is a guanine-based purine that activates Glu uptake and exerts neurotrophic and neuroprotective effects in several models of CNS trauma [352, 353], including a model of neonatal hypoxia-ischemia [354], with recent evidence suggesting that guanosine is a promising therapeutic agent for the treatment of ischemic brain injury [355, 356].

Nicergoline, an ergot derivative, marketed under the trade name Sermion© for the treatment of cognitive, affective, and behavioral disorders of older people [357] has been shown to have neuroprotective actions through increase of Glu transport in rat cortical synaptosomes and cloned Glu transporters [358]. This compound was recently subject of a systematic review and meta-analysis that suggested a good safety profile for cerebrovascular disease and dementia [359].

Riluzole, a highly studied synthetic compound modulator of EAAT2/GLT-1, is a sodium channel-blocking benzothiazole anti-convulsant drug that was demonstrated to decrease excitotoxicity and neurodegeneration through many different mechanisms, including increased Glu uptake [360–364]. Riluzole significantly increased GLT-1-mediated uptake in transfected HEK293 [365], rat cortical astrocyte cultures [366] within striatal astrocytes [302] and cultured C6 astrogial cells [367]. Moreover, riluzole was shown to have neuroprotective effects on rat glial cells subjected to Glu excitotoxicity [368], to attenuate the behavioral effects of chronic unpredictable stress, a rodent model of depression [210] and to delay the onset of muscle weakness and extends lifespan by 4–6 weeks in animal models of ALS [369–371].

Currently riluzole is approved by the U.S. Food and Drug Administration for the treatment of ALS [372], albeit it prolongs patient life by only a few months [373]. Additionally, riluzole showed efficacy in preclinical models of spinal cord injury in reducing the extent of sodium- and Glu-mediated secondary injury [374]. Recently, riluzole has been suggested to be a potential therapy for Alzheimer's disease [375, 376].

Pregabalin (S)-3-aminomethyl-5-methyl hexanoic acid, (a derivative of GABA, marketed as Lyrica©)*:* is an analgesic and anticonvulsant, previously shown to be a ligand for the α2δ subunit component of voltage-gated calcium channels [377]. The effect of pregabalin on EAAT3/EAAC1 was measured using voltage patch clamping in *Xenopus* oocytes. The study demonstrated that pregabalin increased EAAT3 activity and PKC and PI3K were involved. These mechanisms could be involved in the analgesic effect of pregabalin in neuropathic pain [378].

Gabapentin, (a derivative of GABA, like pregabalin, marketed as Neurontin©), is used to treat epilepsy and neuropathic pain. Gabapentin is a α2/δ calcium channel modulator [379], and a study proposed that both riluzole and gabapentin activated descending inhibition in rats by increasing Glu receptor signaling in the locus coeruleus (LC) and hypothesized that these drugs share common mechanisms to enhance Glu release from astrocytes. In primary cultures of astrocytes, riluzole and gabapentin facilitated Glu-induced Glu release and significantly increased Glu uptake, the latter being completely blocked by TBOA [380]. Intriguingly, gabapentin was also shown to decrease the activity of EAAT3 in a concentration-dependent manner [381]. Moreover, a study that performed in vivo microdialysis in the LC in normal and spinal nerve ligated rats has observed basal extracellular concentration and tissue content of Glu in the LC to be greater in injured rats. Also, administration of gabapentin increased extracellular Glu concentration in the LC, and selective blockade by DHK or knockdown of GLT-1 abolished the gabapentin-induced Glu increase in the LC, whereas blockade of GABA-B receptors and depletion of noradrenalin did not. This suggests that gabapentin induces Glu release from astrocytes in the LC via GLT-1-dependent mechanisms to stimulate descending inhibition [382]. Additionally, gabapentin was also associated with downregulation of GLT-1 in the LC, which results in reduced spinal noradrenergic inhibition and contribute to impaired analgesic efficacy from gabapentin in chronic neuropathic pain [383].

Further studies should clarify the apparent discrepancies between the mechanisms of gabapentinoids pregabalin and gabapentin on glutamatergic systems.

Our lab has taken an approach to identify compounds that modulate Glu transport through the characterization of spider venoms, a valuable source of compounds bioactive on synaptic transmission [384–387]. This involved the purification of a compound from *Parawixia bistriata* spider venom, referred to as

Parawixin1, which was shown to stimulate Glu uptake in rat synaptosomes and to protect retinal tissue from ischemic damage [386] and further characterized to act directly and selectively on the Glu transporter EAAT2, specifically by facilitating conformational transitions involved in substrate translocation [385]. This spider venom provided proof of principle for the development of positive allosteric modulators of EAAT2/GLT-1 that could be useful to further understand the mechanisms involved in augmentation of the transporter function.

Additionally, indirect EAATs activators commonly have nonspecific effects that are associated with side effects that limit their utility or exhibit promiscuous pharmacological profiles. On the other hand, direct and selective EAAT2/GLT-1 allosteric positive modulators could become the next generation of compounds to be developed into safe and effective approaches to treating conditions involving excess Glu, devoid of numerous side effects.

Our lab has performed mutagenesis studies that identified a structural region within EAAT2 that is important for the transporter enhancing activity, comprised of TM domains 2, 5 and 8 [236] (see Fig. 7 for an illustration of putative allosteric and orthosteric sites within EAAT2). Subsequently, this unique structural information was employed in hybrid structure-based (HSB) virtual screening of a large library to identify novel EAAT2/GLT-1 activating compounds. This approach successfully identified hit compounds that enhance the activity of EAAT2 in cultured cells and that have neuroprotective properties in primary neuron cultures subjected to Glu excitotoxicity (Fontana, A.C.K. et al., not published). Further characterization of these novel allosteric modulators of EAAT2/GLT-1 using medicinal chemistry approaches, in vivo pharmacokinetic advanced profiling, in vitro and in vivo animal models, will determine whether this class of compounds could be developed into therapies for CNS disorders.

See Table 2 for a summary of the activators of expression and activity described in Sects. 3.1–3.2.

4 Conclusions and Perspectives

Modulation of extracellular Glu concentration with drugs that modulate Glu uptake systems has shown to be a very valuable approach for several reasons. Pharmacological inhibitors of Glu transport were essential to further our understanding of fundamental properties and function of Glu transporters and their role in shaping glutamatergic neurotransmission. On the other hand, the identification of positive modulators could be developed into therapies for CNS disorders that involve Glu dysfunction.

In conclusion, the discovery of selective inhibitors, the identification of structures of the bacterial homolog GlT_{Ph}, and computational modeling and experimental data gathered over the past

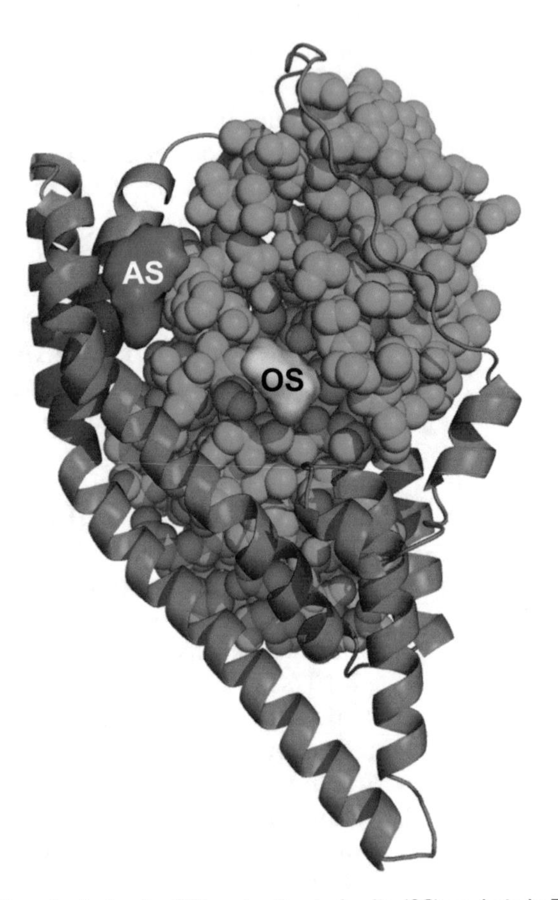

Fig. 7 Putative allosteric site (AS) and orthosteric site (OS) pockets in EAAT2. The rigid trimerization domain is visualized with ribbons in *red* and the transport domain is in *green color* in spheres

decade, were all critical to further our understanding of several aspects of Glu transporter function, including some of the molecular determinants of ligand interaction at orthosteric substrate and inhibitor binding pockets and allosteric pockets.

Nevertheless, small molecules designed to upregulate the activity of Glu transporters are still being developed, and these efforts may reveal intriguing and promising therapeutic avenues. It remains to be determined whether small molecules allosteric activators of EAAT2/GLT-1, as for many biologics, lend themselves to conventional pharmacokinetic analysis and are efficacious in animal models of CNS injury. Future research will reveal if such a class of compounds is effective in chronic conditions and in combination therapies.

Not until subtype-selective enhancers, inhibitors, and substrates for all subtypes of Glu transporters have been discovered one may fully understand how to fine tune the extracellular concentration of Glu in the brain. Collaboration between organic chemists and molecular pharmacologists is essential to develop

Table 2
Summary of activators described that modulate activity or expression of EAATs

Class of compounds		Mechanism, target, selectivity	Therapeutic potential
Activators of expression (Fig. 5)	Ceftriaxone	β-lactam antibiotic, transcriptional activator of EAAT2/GLT-1 expression, unknown selectivity	ALS, Alzheimer's disease, epilepsy, and several other CNS disorders
	LDN/OSU-0212320	Pyridazine derivative, translational activator of EAAT2/GLT-1, unknown selectivity	ALS, epilepsy, Alzheimer's, and possibly other CNS disorders
	Sulbactam	β-Lactamase inhibitor, transcriptional activator of EAAT2/GLT-1, unknown selectivity	Brain ischemia and possibly other CNS disorders
	Valproate	Multiple, including modulation of EAAT2/GLT-1, GABA transporters, kinases, and histone deacetylases. Shown to upregulate hippocampal Glu transport following chronic treatment	Epilepsy, bipolar disorder, neuropathic pain, migraines, glaucoma, and possibly other CNS disorders
	Isoflurane	General anesthetics, unknown exact mechanism, likely binds to GABA, Glu and glycine receptors with differential effects. Enhances the activity of EAAT3/EAAC1 by altering redistribution to plasma membrane	Reduces pain sensitivity (anesthesia)
	Histamine	Increases EAAT2/GLT-1 expression, activates G protein-coupled histamine receptors, activates ligand-gated chloride channels	Stroke and possibly other CNS disorders
	Harmine	Beta-carboline alkaloid with multiple mechanisms. Shown to increase EAAT2/GLT-1 expression and activity	ALS, cerebral ischemia, and possibly other CNS disorders
	Dexamethasone	Glucocorticoid, increases EAAT2/GLT-1 expression and activity, associated with neuroprotection partly mediated via Akt activation	Ischemic brain injury and possibly other CNS disorders
	Dextromethorphan	Antitussive, several targets and mechanisms: decreases interleukin 1β and IL 6 protein expression, increases EAAT1/GLAST and EAAT2/GLT-1 expression, NMDA receptor antagonist, nonselective serotonin uptake inhibitor and sigma-1 receptor agonist	Cerebral ischemia, epilepsy, TBI, and possibly other CNS disorders

(continued)

Table 2
(continued)

Class of compounds	Mechanism, target, selectivity	Therapeutic potential
Cortisol	Steroid hormone, increases EAAT2/GLT-1 translation	N.D.
Retinol	Vitamin A_1, increases EAAT2/GLT-1 translation	N.D.
Tamoxifen	Selective estrogen receptor modulator, shown to enhance EAAT2/GLT-1 expression, via CREB and NF-κB pathways	Bipolar disorders and possibly other CNS disorder, caution is needed due to possible severe side effects
Estradiol	Estrogen hormone, increases expression of both EAAT1/GLAST and EAAT2/GLT-1 protein, mediated by growth factors such as TGF-α	Neurodegenerative disorders, however, adverse effects associated with long-term use has hampered clinical utility
Minocycline	Broad-spectrum tetracycline antibiotic, maintains expression of EAAT2/GLT-1 in the spinal dorsal horn in neuropathy models	Neuropathic pain and possibly other CNS disorders
GPI-1046	Synthetic neuroimmunophilin ligand that selectively increases EAAT2/GLT-1 expression	Motor neuron disease, ALS, alcoholism, and possibly other CNS disorders
Carnosine	Dipeptide, upregulates EAAT2/GLT-1 expression	Stroke, Alzheimer, and possibly other CNS disorders
Amytrypline	Tricyclic antidepressant, reverses downregulation of EAAT1/GLAST and EAAT2/GLT-1 in neuropathic pain models	Neuropathic pain and possibly other CNS disorders
Methylphenidate	Shown to increase dopamine extracellular levels and the activity and expression of EAAT1/GLAST	Major depressive disorder, autism spectrum disorders and possibly other CNS disorders

Activators of activity (Fig. 6)	MS-153	Enhances activity of EAAT2/GLT-1, inhibits high voltage-gated calcium channels through PKC	Stroke, ischemia, anxiety, alcoholism, TBI and possibly other CNS disorders
	Guanosine	Activates Glu uptake through EAAT3/EAAC1, with indirect effects on oxidative stress pathways	Neonatal hypoxia-ischemia, ischemic brain injury and possibly other CNS disorders
	Nicergoline	Ergot derivative, shown to increase activity of EAAT2/GLT-1 and EAAT3/EAAC1	Cerebrovascular disease, dementia, and possibly other CNS disorders
	Riluzole	Several mechanisms, including increasing activity of EAAT2/GLT-1, block of sodium channel, and inhibition of kainate and NMDA receptors	Stroke, TBI, spinal cord injury, depression, ALS, Alzheimer and possibly other CNS disorders
	Pregabalin	Analgesic, $\alpha 2\delta$ subunit of voltage-gated calcium channels ligand, increases activity of EAAT3/EAAC1 through PKC and PI3K	Neuropathic pain, epilepsy and possibly other CNS disorders
	Gabapentin	Several mechanisms including induction of Glu release and increased Glu uptake through EAAT2/GLT-1 and decrease activity of to decrease the activity of EAAT3/EAAC1	Epilepsy, neuropathic pain and possibly other CNS disorders

Mechanism, target, selectivity and therapeutic potentials for each is included. *N.D.* not determined

Proteins that alter EAATs expression (such as mTOR, JAK2, GSK3β, leptin and ghrelin) and positive allosteric modulators of EAAT2/GLT-1 are not shown in table new compounds that target the EAATs and modulate extracellular concentration of Glu to develop novel tools and therapies for the CNS.

Acknowledgments

The author would like to thank Ole V. Mortensen (Drexel University College of Medicine) for critical reading of the manuscript and for help with selected figures, and Shaili Aggarwal (Drexel University College of Medicine) for assistance with selected figures.

References

1. McEntee WJ, Crook TH (1993) Glutamate: its role in learning, memory, and the aging brain. Psychopharmacology 111(4):391–401
2. Weiler IJ, Hawrylak N, Greenough WT (1995) Morphogenesis in memory formation: synaptic and cellular mechanisms. Behav Brain Res 66(1–2):1–6
3. Peng S, Zhang Y, Zhang J, Wang H, Ren B (2011) Glutamate receptors and signal transduction in learning and memory. Mol Biol Rep 38(1):453–460. doi:10.1007/s11033-010-0128-9
4. Lopez-Bayghen E, Ortega A (2011) Glial glutamate transporters: new actors in brain signaling. IUBMB Life 63(10):816–823. doi:10.1002/iub.536
5. Beart PM, O'Shea RD (2007) Transporters for L-glutamate: an update on their molecular pharmacology and pathological involvement. Br J Pharmacol 150(1):5–17. doi:10.1038/sj.bjp.0706949
6. Danbolt NC (2001) Glutamate uptake. Prog Neurobiol 65(1):1–105
7. Fontana AC (2015) Current approaches to enhance glutamate transporter function and expression. J Neurochem 134(6):982–1007. doi:10.1111/jnc.13200
8. Zhou Y, Danbolt NC (2014) Glutamate as a neurotransmitter in the healthy brain. J Neural Transm (Vienna) 121(8):799–817. doi:10.1007/s00702-014-1180-8
9. Zhou Y, Danbolt NC (2013) GABA and glutamate transporters in brain. Front Endocrinol 4:165. doi:10.3389/fendo.2013.00165
10. Vandenberg RJ, Ryan RM (2013) Mechanisms of glutamate transport. Physiol Rev 93(4):1621–1657. doi:10.1152/physrev.00007.2013
11. Grewer C, Gameiro A, Rauen T (2014) SLC1 glutamate transporters. Pflugers Arch 466(1):3–24. doi:10.1007/s00424-013-1397-7
12. Haugeto O, Ullensvang K, Levy LM, Chaudhry FA, Honore T, Nielsen M, Lehre KP, Danbolt NC (1996) Brain glutamate transporter proteins form homomultimers. J Biol Chem 271(44):27715–27722
13. Tanaka K, Watase K, Manabe T, Yamada K, Watanabe M, Takahashi K, Iwama H, Nishikawa T, Ichihara N, Kikuchi T, Okuyama S, Kawashima N, Hori S, Takimoto M, Wada K (1997) Epilepsy and exacerbation of brain injury in mice lacking the glutamate transporter GLT-1. Science 276(5319):1699–1702
14. Lehre KP, Danbolt NC (1998) The number of glutamate transporter subtype molecules at glutamatergic synapses: chemical and stereological quantification in young adult rat brain. J Neurosci 18(21):8751–8757
15. Suchak SK, Baloyianni NV, Perkinton MS, Williams RJ, Meldrum BS, Rattray M (2003) The 'glial' glutamate transporter, EAAT2 (Glt-1) accounts for high affinity glutamate uptake into adult rodent nerve endings. J Neurochem 84(3):522–532
16. Scofield MD, Kalivas PW (2014) Astrocytic dysfunction and addiction: consequences of impaired glutamate homeostasis. Neuroscientist 20(6):610–622. doi:10.1177/1073858413520347
17. Bergles DE, Jahr CE (1997) Synaptic activation of glutamate transporters in hippocampal astrocytes. Neuron 19(6):1297–1308
18. Diamond JS, Jahr CE (1997) Transporters buffer synaptically released glutamate on a submillisecond time scale. J Neurosci 17(12):4672–4687
19. Otis TS, Kavanaugh MP (2000) Isolation of current components and partial reaction cycles in the glial glutamate transporter EAAT2. J Neurosci 20(8):2749–2757

20. Herman MA, Jahr CE (2007) Extracellular glutamate concentration in hippocampal slice. J Neurosci 27(36):9736–9741. doi:10.1523/jneurosci.3009-07.2007
21. Maragakis NJ, Dykes-Hoberg M, Rothstein JD (2004) Altered expression of the glutamate transporter EAAT2b in neurological disease. Ann Neurol 55(4):469–477. doi:10.1002/ana.20003
22. Sheldon AL, Robinson MB (2007) The role of glutamate transporters in neurodegenerative diseases and potential opportunities for intervention. Neurochem Int 51(6–7):333–355. doi:10.1016/j.neuint.2007.03.012
23. Lauriat TL, McInnes LA (2007) EAAT2 regulation and splicing: relevance to psychiatric and neurological disorders. Mol Psychiatry 12(12):1065–1078. doi:10.1038/sj.mp.4002065
24. Kim K, Lee SG, Kegelman TP, Su ZZ, Das SK, Dash R, Dasgupta S, Barral PM, Hedvat M, Diaz P, Reed JC, Stebbins JL, Pellecchia M, Sarkar D, Fisher PB (2011) Role of excitatory amino acid transporter-2 (EAAT2) and glutamate in neurodegeneration: opportunities for developing novel therapeutics. J Cell Physiol 226(10):2484–2493. doi:10.1002/jcp.22609
25. Storck T, Schulte S, Hofmann K, Stoffel W (1992) Structure, expression, and functional analysis of a Na(+)-dependent glutamate/aspartate transporter from rat brain. Proc Natl Acad Sci U S A 89(22):10955–10959
26. Holmseth S, Dehnes Y, Huang YH, Follin-Arbelet VV, Grutle NJ, Mylonakou MN, Plachez C, Zhou Y, Furness DN, Bergles DE, Lehre KP, Danbolt NC (2012) The density of EAAC1 (EAAT3) glutamate transporters expressed by neurons in the mammalian CNS. J Neurosci 32(17):6000–6013. doi:10.1523/jneurosci.5347-11.2012
27. Bjorn-Yoshimoto WE, Underhill SM (2016) The importance of the excitatory amino acid transporter 3 (EAAT3). Neurochem Int. doi:10.1016/j.neuint.2016.05.007
28. Furuta A, Rothstein JD, Martin LJ (1997) Glutamate transporter protein subtypes are expressed differentially during rat CNS development. J Neurosci 17(21):8363–8375
29. Dehnes Y, Chaudhry FA, Ullensvang K, Lehre KP, Storm-Mathisen J, Danbolt NC (1998) The glutamate transporter EAAT4 in rat cerebellar Purkinje cells: a glutamate-gated chloride channel concentrated near the synapse in parts of the dendritic membrane facing astroglia. J Neurosci 18(10):3606–3619
30. Gincel D, Regan MR, Jin L, Watkins AM, Bergles DE, Rothstein JD (2007) Analysis of cerebellar Purkinje cells using EAAT4 glutamate transporter promoter reporter in mice generated via bacterial artificial chromosome-mediated transgenesis. Exp Neurol 203(1):205–212. doi:10.1016/j.expneurol.2006.08.016
31. Arriza JL, Eliasof S, Kavanaugh MP, Amara SG (1997) Excitatory amino acid transporter 5, a retinal glutamate transporter coupled to a chloride conductance. Proc Natl Acad Sci U S A 94(8):4155–4160
32. Sitcheran R, Gupta P, Fisher PB, Baldwin AS (2005) Positive and negative regulation of EAAT2 by NF-kappaB: a role for N-myc in TNFalpha-controlled repression. EMBO J 24(3):510–520. doi:10.1038/sj.emboj.7600555
33. Karki P, Smith K, Johnson J Jr, Lee E (2014) Astrocyte-derived growth factors and estrogen neuroprotection: role of transforming growth factor-alpha in estrogen-induced upregulation of glutamate transporters in astrocytes. Mol Cell Endocrinol 389(1–2):58–64. doi:10.1016/j.mce.2014.01.010
34. Ji YF, Zhou L, Xie YJ, Xu SM, Zhu J, Teng P, Shao CY, Wang Y, Luo JH, Shen Y (2013) Upregulation of glutamate transporter GLT-1 by mTOR-Akt-NF-small ka, CyrillicB cascade in astrocytic oxygen-glucose deprivation. Glia 61(12):1959–1975. doi:10.1002/glia.22566
35. Martinez-Lozada Z, Guillem AM, Robinson MB (2016) Transcriptional regulation of glutamate transporters: from extracellular signals to transcription factors. Adv Pharmacol 76:103–145. doi:10.1016/bs.apha.2016.01.004
36. Vornov JJ, Hollinger KR, Jackson PF, Wozniak KM, Farah MH, Majer P, Rais R, Slusher BS (2016) Still NAAG'ing after all these years: the continuing pursuit of GCPII inhibitors. Adv Pharmacol 76:215–255. doi:10.1016/bs.apha.2016.01.007
37. Stocchi F, Torti M (2016) Adjuvant therapies for Parkinson's disease: critical evaluation of safinamide. Drug Des Devel Ther 10:609–618. doi:10.2147/dddt.s77749
38. Trankner A, Sander C, Schonknecht P (2013) A critical review of the recent literature and selected therapy guidelines since 2006 on the use of lamotrigine in bipolar disorder. Neuropsychiatr Dis Treat 9:101–111. doi:10.2147/ndt.s37126
39. Olive MF, Cleva RM, Kalivas PW, Malcolm RJ (2012) Glutamatergic medications for the treatment of drug and behavioral addictions. Pharmacol Biochem Behav 100(4):801–810. doi:10.1016/j.pbb.2011.04.015
40. Pettorruso M, De Risio L, Martinotti G, Di Nicola M, Ruggeri F, Conte G, Di Giannantonio M, Janiri L (2014) Targeting

the glutamatergic system to treat pathological gambling: current evidence and future perspectives. Biomed Res Int 2014:109786. doi:10.1155/2014/109786

41. Castillo J, Loza MI, Mirelman D, Brea J, Blanco M, Sobrino T, Campos F (2016) A novel mechanism of neuroprotection: blood glutamate grabber. J Cereb Blood Flow Metab 36(2):292–301. doi:10.1177/0271678x15606721
42. Boyko M, Zlotnik A, Gruenbaum BF, Gruenbaum SE, Ohayon S, Kuts R, Melamed I, Regev A, Shapira Y, Teichberg VI (2011) Pyruvate's blood glutamate scavenging activity contributes to the spectrum of its neuroprotective mechanisms in a rat model of stroke. Eur J Neurosci 34(9):1432–1441. doi:10.1111/j.1460-9568.2011.07864.x
43. Wahle S, Stoffel W (1996) Membrane topology of the high-affinity L-glutamate transporter (GLAST-1) of the central nervous system. J Cell Biol 135(6 Pt 2):1867–1877
44. Seal RP, Amara SG (1998) A reentrant loop domain in the glutamate carrier EAAT1 participates in substrate binding and translocation. Neuron 21(6):1487–1498
45. Slotboom DJ, Sobczak I, Konings WN, Lolkema JS (1999) A conserved serine-rich stretch in the glutamate transporter family forms a substrate-sensitive reentrant loop. Proc Natl Acad Sci U S A 96(25):14282–14287
46. Grunewald M, Menaker D, Kanner BI (2002) Cysteine-scanning mutagenesis reveals a conformationally sensitive reentrant pore-loop in the glutamate transporter GLT-1. J Biol Chem 277(29):26074–26080. doi:10.1074/jbc.M202248200
47. Zarbiv R, Grunewald M, Kavanaugh MP, Kanner BI (1998) Cysteine scanning of the surroundings of an alkali-ion binding site of the glutamate transporter GLT-1 reveals a conformationally sensitive residue. J Biol Chem 273(23):14231–14237
48. Grunewald M, Kanner B (1995) Conformational changes monitored on the glutamate transporter GLT-1 indicate the existence of two neurotransmitter-bound states. J Biol Chem 270(28):17017–17024
49. Yernool D, Boudker O, Jin Y, Gouaux E (2004) Structure of a glutamate transporter homologue from Pyrococcus horikoshii. Nature 431(7010):811–818. doi:10.1038/nature03018
50. Boudker O, Ryan RM, Yernool D, Shimamoto K, Gouaux E (2007) Coupling substrate and ion binding to extracellular gate of a sodium-dependent aspartate transporter. Nature 445(7126):387–393. doi:10.1038/nature05455
51. Reyes N, Ginter C, Boudker O (2009) Transport mechanism of a bacterial homologue of glutamate transporters. Nature 462(7275):880–885. doi:10.1038/nature08616
52. Verdon G, Boudker O (2012) Crystal structure of an asymmetric trimer of a bacterial glutamate transporter homolog. Nat Struct Mol Biol 19(3):355–357. doi:10.1038/nsmb.2233
53. Gether U, Andersen PH, Larsson OM, Schousboe A (2006) Neurotransmitter transporters: molecular function of important drug targets. Trends Pharmacol Sci 27(7):375–383. doi:10.1016/j.tips.2006.05.003
54. Qu S, Kanner BI (2008) Substrates and nontransportable analogues induce structural rearrangements at the extracellular entrance of the glial glutamate transporter GLT-1/EAAT2. J Biol Chem 283(39):26391–26400. doi:10.1074/jbc.M802401200
55. Reyes N, Oh S, Boudker O (2013) Binding thermodynamics of a glutamate transporter homolog. Nat Struct Mol Biol 20(5):634–640. doi:10.1038/nsmb.2548
56. Kanner BI (1983) Bioenergetics of neurotransmitter transport. Biochim Biophys Acta 726(4):293–316
57. Kanner BI, Schuldiner S (1987) Mechanism of transport and storage of neurotransmitters. CRC Crit Rev Biochem 22(1):1–38
58. Kanner BI (1989) Ion-coupled neurotransmitter transport. Curr Opin Cell Biol 1(4):735–738
59. Szatkowski M, Attwell D (1994) Triggering and execution of neuronal death in brain ischaemia: two phases of glutamate release by different mechanisms. Trends Neurosci 17(9):359–365
60. Zerangue N, Kavanaugh MP (1996) Flux coupling in a neuronal glutamate transporter. Nature 383(6601):634–637. doi:10.1038/383634a0
61. Akyuz N, Altman RB, Blanchard SC, Boudker O (2013) Transport dynamics in a glutamate transporter homologue. Nature. doi:10.1038/nature12265
62. Akyuz N, Georgieva ER, Zhou Z, Stolzenberg S, Cuendet MA, Khelashvili G, Altman RB, Terry DS, Freed JH, Weinstein H, Boudker O, Blanchard SC (2015) Transport domain unlocking sets the uptake rate of an aspartate transporter. Nature 518(7537):68–73. doi:10.1038/nature14158
63. Drew D, Boudker O (2016) Shared molecular mechanisms of membrane transporters.

Annu Rev Biochem 85:543–572. doi:10.1146/annurev-biochem-060815-014520
64. Nilsson E, Alafuzoff I, Blennow K, Blomgren K, Hall CM, Janson I, Karlsson I, Wallin A, Gottfries CG, Karlsson JO (1990) Calpain and calpastatin in normal and Alzheimer-degenerated human brain tissue. Neurobiol Aging 11(4):425–431
65. Raghupathi R (2004) Cell death mechanisms following traumatic brain injury. Brain Pathol 14(2):215–222
66. Benveniste H, Drejer J, Schousboe A, Diemer NH (1984) Elevation of the extracellular concentrations of glutamate and aspartate in rat hippocampus during transient cerebral ischemia monitored by intracerebral microdialysis. J Neurochem 43(5):1369–1374
67. Brown JI, Baker AJ, Konasiewicz SJ, Moulton RJ (1998) Clinical significance of CSF glutamate concentrations following severe traumatic brain injury in humans. J Neurotrauma 15(4):253–263
68. Faden AI, Demediuk P, Panter SS, Vink R (1989) The role of excitatory amino acids and NMDA receptors in traumatic brain injury. Science 244(4906):798–800
69. Vespa P, Prins M, Ronne-Engstrom E, Caron M, Shalmon E, Hovda DA, Martin NA, Becker DP (1998) Increase in extracellular glutamate caused by reduced cerebral perfusion pressure and seizures after human traumatic brain injury: a microdialysis study. J Neurosurg 89(6):971–982. doi:10.3171/jns.1998.89.6.0971
70. Meldrum B, Garthwaite J (1990) Excitatory amino acid neurotoxicity and neurodegenerative disease. Trends Pharmacol Sci 11(9):379–387
71. Rosenberg PA, Amin S, Leitner M (1992) Glutamate uptake disguises neurotoxic potency of glutamate agonists in cerebral cortex in dissociated cell culture. J Neurosci 12(1):56–61
72. Clements J, Lester R, Tong G, Jahr C, Westbrook G (1992) The time course of glutamate in the synaptic cleft. Science 258(5087):1498–1501. doi:10.1126/science.1359647
73. Edwards RH (2015) Mobile binding sites regulate glutamate clearance. Nat Neurosci 18(2):166–168. doi:10.1038/nn.3931
74. Murphy-Royal C, Dupuis JP, Varela JA, Panatier A, Pinson B, Baufreton J, Groc L, Oliet SH (2015) Surface diffusion of astrocytic glutamate transporters shapes synaptic transmission. Nat Neurosci 18(2):219–226. doi:10.1038/nn.3901
75. Levy LM, Lehre KP, Walaas SI, Storm-Mathisen J, Danbolt NC (1995) Downregulation of glial glutamate transporters after glutamatergic denervation in the rat brain. Eur J Neurosci 7(10):2036–2041
76. Rothstein JD, Van Kammen M, Levey AI, Martin LJ, Kuncl RW (1995) Selective loss of glial glutamate transporter GLT-1 in amyotrophic lateral sclerosis. Ann Neurol 38(1):73–84. doi:10.1002/ana.410380114
77. Wilson JM, Khabazian I, Pow DV, Craig UK, Shaw CA (2003) Decrease in glial glutamate transporter variants and excitatory amino acid receptor downregulation in a murine model of ALS-PDC. Neuromol Med 3(2):105–118. doi:10.1385/nmm:3:2:105
78. Trotti D, Aoki M, Pasinelli P, Berger UV, Danbolt NC, Brown RH Jr, Hediger MA (2001) Amyotrophic lateral sclerosis-linked glutamate transporter mutant has impaired glutamate clearance capacity. J Biol Chem 276(1):576–582. doi:10.1074/jbc.M003779200
79. Li S, Mallory M, Alford M, Tanaka S, Masliah E (1997) Glutamate transporter alterations in Alzheimer disease are possibly associated with abnormal APP expression. J Neuropathol Exp Neurol 56(8):901–911
80. Rao VL, Baskaya MK, Dogan A, Rothstein JD, Dempsey RJ (1998) Traumatic brain injury downregulates glial glutamate transporter (GLT-1 and GLAST) proteins in rat brain. J Neurochem 70(5):2020–2027
81. Ikematsu K, Tsuda R, Kondo T, Nakasono I (2002) The expression of excitatory amino acid transporter 2 in traumatic brain injury. Forensic Sci Int 130(2–3):83–89
82. van Landeghem FK, Weiss T, Oehmichen M, von Deimling A (2006) Decreased expression of glutamate transporters in astrocytes after human traumatic brain injury. J Neurotrauma 23(10):1518–1528. doi:10.1089/neu.2006.23.1518
83. Yi JH, Hazell AS (2006) Excitotoxic mechanisms and the role of astrocytic glutamate transporters in traumatic brain injury. Neurochem Int 48(5):394–403. doi:10.1016/j.neuint.2005.12.001
84. Hakuba N, Koga K, Gyo K, Usami SI, Tanaka K (2000) Exacerbation of noise-induced hearing loss in mice lacking the glutamate transporter GLAST. J Neurosci 20(23):8750–8753
85. Katagiri H, Tanaka K, Manabe T (2001) Requirement of appropriate glutamate concentrations in the synaptic cleft for hippocampal LTP induction. Eur J Neurosci 14(3):547–553
86. Mitani A, Tanaka K (2003) Functional changes of glial glutamate transporter GLT-1 during

ischemia: an in vivo study in the hippocampal CA1 of normal mice and mutant mice lacking GLT-1. J Neurosci 23(18):7176–7182

87. Rothstein JD, Dykes-Hoberg M, Pardo CA, Bristol LA, Jin L, Kuncl RW, Kanai Y, Hediger MA, Wang Y, Schielke JP, Welty DF (1996) Knockout of glutamate transporters reveals a major role for astroglial transport in excitotoxicity and clearance of glutamate. Neuron 16(3):675–686
88. Rao VL, Dogan A, Bowen KK, Todd KG, Dempsey RJ (2001) Antisense knockdown of the glial glutamate transporter GLT-1 exacerbates hippocampal neuronal damage following traumatic injury to rat brain. Eur J Neurosci 13(1):119–128
89. Matsugami TR, Tanemura K, Mieda M, Nakatomi R, Yamada K, Kondo T, Ogawa M, Obata K, Watanabe M, Hashikawa T, Tanaka K (2006) From the cover: indispensability of the glutamate transporters GLAST and GLT1 to brain development. Proc Natl Acad Sci U S A 103(32):12161–12166. doi:10.1073/pnas.0509144103
90. Zhou Y, Waanders LF, Holmseth S, Guo C, Berger UV, Li Y, Lehre AC, Lehre KP, Danbolt NC (2014) Proteome analysis and conditional deletion of the EAAT2 glutamate transporter provide evidence against a role of EAAT2 in pancreatic insulin secretion in mice. J Biol Chem 289(3):1329–1344. doi:10.1074/jbc.M113.529065
91. Aoyama K, Suh SW, Hamby AM, Liu J, Chan WY, Chen Y, Swanson RA (2006) Neuronal glutathione deficiency and age-dependent neurodegeneration in the EAAC1 deficient mouse. Nat Neurosci 9(1):119–126. doi:10.1038/nn1609
92. Barnett NL, Pow DV (2000) Antisense knockdown of GLAST, a glial glutamate transporter, compromises retinal function. Invest Ophthalmol Vis Sci 41(2):585–591
93. Karlsson RM, Adermark L, Molander A, Perreau-Lenz S, Singley E, Solomon M, Holmes A, Tanaka K, Lovinger DM, Spanagel R, Heilig M (2012) Reduced alcohol intake and reward associated with impaired endocannabinoid signaling in mice with a deletion of the glutamate transporter GLAST. Neuropharmacology 63(2):181–189. doi:10.1016/j.neuropharm.2012.01.027
94. Szatkowski M, Barbour B, Attwell D (1990) Non-vesicular release of glutamate from glial cells by reversed electrogenic glutamate uptake. Nature 348(6300):443–446. doi:10.1038/348443a0
95. Billups B, Attwell D (1996) Modulation of non-vesicular glutamate release by pH. Nature 379(6561):171–174. doi:10.1038/379171a0
96. Jabaudon D, Scanziani M, Gahwiler BH, Gerber U (2000) Acute decrease in net glutamate uptake during energy deprivation. Proc Natl Acad Sci U S A 97(10):5610–5615
97. Levy LM, Warr O, Attwell D (1998) Stoichiometry of the glial glutamate transporter GLT-1 expressed inducibly in a Chinese hamster ovary cell line selected for low endogenous Na+-dependent glutamate uptake. J Neurosci 18(23):9620–9628
98. Katayama Y, Becker DP, Tamura T, Hovda DA (1990) Massive increases in extracellular potassium and the indiscriminate release of glutamate following concussive brain injury. J Neurosurg 73(6):889–900. doi:10.3171/jns.1990.73.6.0889
99. Soares HD, Thomas M, Cloherty K, McIntosh TK (1992) Development of prolonged focal cerebral edema and regional cation changes following experimental brain injury in the rat. J Neurochem 58(5):1845–1852
100. Phillis JW, O'Regan MH (1996) Mechanisms of glutamate and aspartate release in the ischemic rat cerebral cortex. Brain Res 730(1–2):150–164
101. Grewer C, Gameiro A, Zhang Z, Tao Z, Braams S, Rauen T (2008) Glutamate forward and reverse transport: from molecular mechanism to transporter-mediated release after ischemia. IUBMB Life 60(9):609–619. doi:10.1002/iub.98
102. Werner C, Engelhard K (2007) Pathophysiology of traumatic brain injury. Br J Anaesth 99(1):4–9. doi:10.1093/bja/aem131
103. Lima FD, Souza MA, Furian AF, Rambo LM, Ribeiro LR, Martignoni FV, Hoffmann MS, Fighera MR, Royes LF, Oliveira MS, de Mello CF (2008) Na+,K+-ATPase activity impairment after experimental traumatic brain injury: relationship to spatial learning deficits and oxidative stress. Behav Brain Res 193(2):306–310. doi:10.1016/j.bbr.2008.05.013
104. Kosugi T, Kawahara K (2006) Reversed actrocytic GLT-1 during ischemia is crucial to excitotoxic death of neurons, but contributes to the survival of astrocytes themselves. Neurochem Res 31(7):933–943. doi:10.1007/s11064-006-9099-6
105. Krzyzanowska W, Pomierny B, Filip M, Pera J (2014) Glutamate transporters in brain ischemia: to modulate or not? Acta Pharmacol Sin 35(4):444–462. doi:10.1038/aps.2014.1
106. Rothstein JD (1995) Excitotoxicity and neurodegeneration in amyotrophic lateral sclerosis. Clin Neurosci 3(6):348–359
107. Mark LP, Prost RW, Ulmer JL, Smith MM, Daniels DL, Strottmann JM, Brown WD, Hacein-Bey L (2001) Pictorial review of glu-

tamate excitotoxicity: fundamental concepts for neuroimaging. Am J Neuroradiol 22(10):1813–1824
108. Hazell AS (2007) Excitotoxic mechanisms in stroke: an update of concepts and treatment strategies. Neurochem Int 50(7–8):941–953. doi:10.1016/j.neuint.2007.04.026
109. Mehta A, Prabhakar M, Kumar P, Deshmukh R, Sharma PL (2013) Excitotoxicity: bridge to various triggers in neurodegenerative disorders. Eur J Pharmacol 698(1–3):6–18. doi:10.1016/j.ejphar.2012.10.032
110. Wang Y, Qin ZH (2010) Molecular and cellular mechanisms of excitotoxic neuronal death. Apoptosis 15(11):1382–1402. doi:10.1007/s10495-010-0481-0
111. Beresford IJ, Parsons AA, Hunter AJ (2003) Treatments for stroke. Expert Opin Emerg Drugs 8(1):103–122
112. Sisk DR, Kuwabara T (1985) Histologic changes in the inner retina of albino rats following intravitreal injection of monosodium L-glutamate. Graefes Arch Clin Exp Ophthalmol 223(5):250–258
113. Louzada-Junior P, Dias JJ, Santos WF, Lachat JJ, Bradford HF, Coutinho-Netto J (1992) Glutamate release in experimental ischaemia of the retina: an approach using microdialysis. J Neurochem 59(1):358–363
114. Taoufik E, Probert L (2008) Ischemic neuronal damage. Curr Pharm Des 14(33):3565–3573
115. Chao XD, Fei F, Fei Z (2010) The role of excitatory amino acid transporters in cerebral ischemia. Neurochem res 35(8):1224–1230. doi:10.1007/s11064-010-0178-3
116. Nishizawa Y (2001) Glutamate release and neuronal damage in ischemia. Life Sci 69(4):369–381
117. Lipski J, Wan CK, Bai JZ, Pi R, Li D, Donnelly D (2007) Neuroprotective potential of ceftriaxone in in vitro models of stroke. Neuroscience 146(2):617–629. doi:10.1016/j.neuroscience.2007.02.003
118. Chu K, Lee ST, Sinn DI, Ko SY, Kim EH, Kim JM, Kim SJ, Park DK, Jung KH, Song EC, Lee SK, Kim M, Roh JK (2007) Pharmacological induction of ischemic tolerance by glutamate transporter-1 (EAAT2) upregulation. Stroke 38(1):177–182. doi:10.1161/01.STR.0000252091.36912.65
119. Verma R, Mishra V, Sasmal D, Raghubir R (2010) Pharmacological evaluation of glutamate transporter 1 (GLT-1) mediated neuroprotection following cerebral ischemia/reperfusion injury. Eur J Pharmacol 638(1–3):65–71. doi:10.1016/j.ejphar.2010.04.021
120. Harvey BK, Airavaara M, Hinzman J, Wires EM, Chiocco MJ, Howard DB, Shen H, Gerhardt G, Hoffer BJ, Wang Y (2011) Targeted over-expression of glutamate transporter 1 (GLT-1) reduces ischemic brain injury in a rat model of stroke. PLoS One 6(8):e22135. doi:10.1371/journal.pone.0022135
121. Jabs R, Seifert G, Steinhauser C (2008) Astrocytic function and its alteration in the epileptic brain. Epilepsia 49(Suppl 2):3–12. doi:10.1111/j.1528-1167.2008.01488.x
122. Werner FM, Covenas R (2015) Review: Classical neurotransmitters and neuropeptides involved in generalized epilepsy in a multi-neurotransmitter system: how to improve the antiepileptic effect? Epilepsy Behav. doi:10.1016/j.yebeh.2015.01.038
123. Coutinho-Netto J, Abdul-Ghani AS, Collins JF, Bradford HF (1981) Is glutamate a trigger factor in epileptic hyperactivity? Epilepsia 22(3):289–296
124. Kaila K, Ruusuvuori E, Seja P, Voipio J, Puskarjov M (2014) GABA actions and ionic plasticity in epilepsy. Curr Opin Neurobiol 26:34–41. doi:10.1016/j.conb.2013.11.004
125. During MJ, Spencer DD (1993) Extracellular hippocampal glutamate and spontaneous seizure in the conscious human brain. Lancet 341(8861):1607–1610
126. Cavus I, Kasoff WS, Cassaday MP, Jacob R, Gueorguieva R, Sherwin RS, Krystal JH, Spencer DD, Abi-Saab WM (2005) Extracellular metabolites in the cortex and hippocampus of epileptic patients. Ann Neurol 57(2):226–235. doi:10.1002/ana.20380
127. van der Hel WS, Notenboom RG, Bos IW, van Rijen PC, van Veelen CW, de Graan PN (2005) Reduced glutamine synthetase in hippocampal areas with neuron loss in temporal lobe epilepsy. Neurology 64(2):326–333. doi:10.1212/01.wnl.0000149636.44660.99
128. Mathern GW, Mendoza D, Lozada A, Pretorius JK, Dehnes Y, Danbolt NC, Nelson N, Leite JP, Chimelli L, Born DE, Sakamoto AC, Assirati JA, Fried I, Peacock WJ, Ojemann GA, Adelson PD (1999) Hippocampal GABA and glutamate transporter immunoreactivity in patients with temporal lobe epilepsy. Neurology 52(3):453–472
129. Proper EA, Hoogland G, Kappen SM, Jansen GH, Rensen MG, Schrama LH, van Veelen CW, van Rijen PC, van Nieuwenhuizen O, Gispen WH, de Graan PN (2002) Distribution of glutamate transporters in the hippocampus of patients with pharmaco-resistant temporal lobe epilepsy. Brain 125(Pt 1):32–43
130. Lopes MW, Soares FM, de Mello N, Nunes JC, Cajado AG, de Brito D, de Cordova FM, da Cunha RM, Walz R, Leal RB (2013) Time-

dependent modulation of AMPA receptor phosphorylation and mRNA expression of NMDA receptors and glial glutamate transporters in the rat hippocampus and cerebral cortex in a pilocarpine model of epilepsy. Exp Brain Res 226(2):153–163. doi:10.1007/s00221-013-3421-8

131. David Y, Cacheaux LP, Ivens S, Lapilover E, Heinemann U, Kaufer D, Friedman A (2009) Astrocytic dysfunction in epileptogenesis: consequence of altered potassium and glutamate homeostasis? J Neurosci 29(34):10588–10599. doi:10.1523/jneurosci.2323-09.2009

132. Wong M, Ess KC, Uhlmann EJ, Jansen LA, Li W, Crino PB, Mennerick S, Yamada KA, Gutmann DH (2003) Impaired glial glutamate transport in a mouse tuberous sclerosis epilepsy model. Ann Neurol 54(2):251–256. doi:10.1002/ana.10648

133. Ueda Y, Doi T, Nagatomo K, Willmore LJ, Nakajima A (2007) Functional role for redox in the epileptogenesis: molecular regulation of glutamate in the hippocampus of FeCl3-induced limbic epilepsy model. Exp Brain Res 181(4):571–577. doi:10.1007/s00221-007-0954-8

134. Lu Z, Zhang W, Zhang N, Jiang J, Luo Q, Qiu Y (2008) The expression of glutamate transporters in chest compression-induced audiogenic epilepsy: a comparative study. Neurol Res 30(9):915–919. doi:10.1179/174313208x327964

135. Eid T, Thomas MJ, Spencer DD, Runden-Pran E, Lai JC, Malthankar GV, Kim JH, Danbolt NC, Ottersen OP, de Lanerolle NC (2004) Loss of glutamine synthetase in the human epileptogenic hippocampus: possible mechanism for raised extracellular glutamate in mesial temporal lobe epilepsy. Lancet 363(9402):28–37

136. Bjornsen LP, Eid T, Holmseth S, Danbolt NC, Spencer DD, de Lanerolle NC (2007) Changes in glial glutamate transporters in human epileptogenic hippocampus: inadequate explanation for high extracellular glutamate during seizures. Neurobiol Dis 25(2):319–330. doi:10.1016/j.nbd.2006.09.014

137. Akbar MT, Torp R, Danbolt NC, Levy LM, Meldrum BS, Ottersen OP (1997) Expression of glial glutamate transporters GLT-1 and GLAST is unchanged in the hippocampus in fully kindled rats. Neuroscience 78(2):351–359

138. Miller HP, Levey AI, Rothstein JD, Tzingounis AV, Conn PJ (1997) Alterations in glutamate transporter protein levels in kindling-induced epilepsy. J Neurochem 68(4):1564–1570

139. Simantov R, Crispino M, Hoe W, Broutman G, Tocco G, Rothstein JD, Baudry M (1999) Changes in expression of neuronal and glial glutamate transporters in rat hippocampus following kainate-induced seizure activity. Brain Res Mol Brain Res 65(1):112–123

140. Bough KJ, Paquet M, Pare JF, Hassel B, Smith Y, Hall RA, Dingledine R (2007) Evidence against enhanced glutamate transport in the anticonvulsant mechanism of the ketogenic diet. Epilepsy Res 74(2–3):232–236. doi:10.1016/j.eplepsyres.2007.03.002

141. Guo F, Sun F, Yu JL, Wang QH, Tu DY, Mao XY, Liu R, Wu KC, Xie N, Hao LY, Cai JQ (2010) Abnormal expressions of glutamate transporters and metabotropic glutamate receptor 1 in the spontaneously epileptic rat hippocampus. Brain Res Bull 81(4–5):510–516. doi:10.1016/j.brainresbull.2009.10.008

142. Kong Q, Takahashi K, Schulte D, Stouffer N, Lin Y, Lin CL (2012) Increased glial glutamate transporter EAAT2 expression reduces epileptogenic processes following pilocarpine-induced status epilepticus. Neurobiol Dis 47(2):145–154. doi:10.1016/j.nbd.2012.03.032

143. Jelenkovic AV, Jovanovic MD, Stanimirovic DD, Bokonjic DD, Ocic GG, Boskovic BS (2008) Beneficial effects of ceftriaxone against pentylenetetrazole-evoked convulsions. Exp Biol Med (Maywood) 233(11):1389–1394. doi:10.3181/0803-rm-83

144. Petr GT, Sun Y, Frederick NM, Zhou Y, Dhamne SC, Hameed MQ, Miranda C, Bedoya EA, Fischer KD, Armsen W, Wang J, Danbolt NC, Rotenberg A, Aoki CJ, Rosenberg PA (2015) Conditional deletion of the glutamate transporter GLT-1 reveals that astrocytic GLT-1 protects against fatal epilepsy while neuronal GLT-1 contributes significantly to glutamate uptake into synaptosomes. J Neurosci 35(13):5187–5201. doi:10.1523/jneurosci.4255-14.2015

145. Wetherington J, Serrano G, Dingledine R (2008) Astrocytes in the epileptic brain. Neuron 58(2):168–178. doi:10.1016/j.neuron.2008.04.002

146. Coulter DA, Eid T (2012) Astrocytic regulation of glutamate homeostasis in epilepsy. Glia 60(8):1215–1226. doi:10.1002/glia.22341

147. Crunelli V, Carmignoto G, Steinhauser C (2015) Novel astrocyte targets: new avenues for the therapeutic treatment of epilepsy. Neuroscientist 21(1):62–83. doi:10.1177/1073858414523320

148. Matute C, Domercq M, Sanchez-Gomez MV (2006) Glutamate-mediated glial injury: mechanisms and clinical importance. Glia 53(2):212–224. doi:10.1002/glia.20275

149. Guerriero RM, Giza CC, Rotenberg A (2015) Glutamate and GABA imbalance following traumatic brain injury. Curr Neurol Neurosci Rep 15(5):545. doi:10.1007/s11910-015-0545-1
150. Palmer AM, Marion DW, Botscheller ML, Swedlow PE, Styren SD, DeKosky ST (1993) Traumatic brain injury-induced excitotoxicity assessed in a controlled cortical impact model. J Neurochem 61(6):2015–2024
151. Baker AJ, Moulton RJ, MacMillan VH, Shedden PM (1993) Excitatory amino acids in cerebrospinal fluid following traumatic brain injury in humans. J Neurosurg 79(3):369–372. doi:10.3171/jns.1993.79.3.0369
152. Nilsson P, Hillered L, Ponten U, Ungerstedt U (1990) Changes in cortical extracellular levels of energy-related metabolites and amino acids following concussive brain injury in rats. J Cereb Blood Flow Metab 10(5):631–637. doi:10.1038/jcbfm.1990.115
153. Yamamoto T, Rossi S, Stiefel M, Doppenberg E, Zauner A, Bullock R, Marmarou A (1999) CSF and ECF glutamate concentrations in head injured patients. Acta Neurochir Suppl 75:17–19
154. Rothman SM, Olney JW (1995) Excitotoxicity and the NMDA receptor—still lethal after eight years. Trends Neurosci 18(2):57–58
155. Rothstein JD, Martin LJ, Kuncl RW (1992) Decreased glutamate transport by the brain and spinal cord in amyotrophic lateral sclerosis. N Engl J Med 326(22):1464–1468. doi:10.1056/NEJM199205283262204
156. Rothstein JD (1995) Excitotoxic mechanisms in the pathogenesis of amyotrophic lateral sclerosis. Adv Neurol 68:7–20. discussion 21–27
157. Bruijn LI, Becher MW, Lee MK, Anderson KL, Jenkins NA, Copeland NG, Sisodia SS, Rothstein JD, Borchelt DR, Price DL, Cleveland DW (1997) ALS-linked SOD1 mutant G85R mediates damage to astrocytes and promotes rapidly progressive disease with SOD1-containing inclusions. Neuron 18(2):327–338
158. Trotti D, Rolfs A, Danbolt NC, Brown RH Jr, Hediger MA (1999) SOD1 mutants linked to amyotrophic lateral sclerosis selectively inactivate a glial glutamate transporter. Nat Neurosci 2(5):427–433. doi:10.1038/8091
159. Bendotti C, Tortarolo M, Suchak SK, Calvaresi N, Carvelli L, Bastone A, Rizzi M, Rattray M, Mennini T (2001) Transgenic SOD1 G93A mice develop reduced GLT-1 in spinal cord without alterations in cerebrospinal fluid glutamate levels. J Neurochem 79(4):737–746
160. Van Den Bosch L, Van Damme P, Bogaert E, Robberecht W (2006) The role of excitotoxicity in the pathogenesis of amyotrophic lateral sclerosis. Biochim Biophys Acta 1762(11–12):1068–1082. doi:10.1016/j.bbadis.2006.05.002
161. Palmen SJ, van Engeland H, Hof PR, Schmitz C (2004) Neuropathological findings in autism. Brain 127(Pt 12):2572–2583. doi:10.1093/brain/awh287
162. Courchesne E (1997) Brainstem, cerebellar and limbic neuroanatomical abnormalities in autism. Curr Opin Neurobiol 7(2):269–278
163. Khodorov BI (2000) Mechanisms of destabilization of Ca2+-homeostasis of brain neurons caused by toxic glutamate challenge. Membr Cell Biol 14(2):149–162
164. Zeidan-Chulia F, Salmina AB, Malinovskaya NA, Noda M, Verkhratsky A, Moreira JC (2014) The glial perspective of autism spectrum disorders. Neurosci Biobehav Rev 38:160–172. doi:10.1016/j.neubiorev.2013.11.008
165. Ghanizadeh A, Berk M (2015) Beta-lactam antibiotics as a possible novel therapy for managing epilepsy and autism, a case report and review of literature. Iran J Child Neurol 9(1):99–102
166. Walker FO (2007) Huntington's disease. Lancet 369(9557):218–228. doi:10.1016/S0140-6736(07)60111-1
167. Bates G (2003) Huntingtin aggregation and toxicity in Huntington's disease. Lancet 361(9369):1642–1644. doi:10.1016/s0140-6736(03)13304-1
168. Browne SE, Beal MF (2006) Oxidative damage in Huntington's disease pathogenesis. Antioxid Redox Signal 8(11–12):2061–2073. doi:10.1089/ars.2006.8.2061
169. Massieu L, Garcia O (1998) The role of excitotoxicity and metabolic failure in the pathogenesis of neurological disorders. Neurobiology (Bp) 6(1):99–108
170. Shin JY, Fang ZH, ZX Y, Wang CE, Li SH, Li XJ (2005) Expression of mutant huntingtin in glial cells contributes to neuronal excitotoxicity. J Cell Biol 171(6):1001–1012. doi:10.1083/jcb.200508072
171. Estrada-Sanchez AM, Rebec GV (2012) Corticostriatal dysfunction and glutamate transporter 1 (GLT1) in Huntington's disease: interactions between neurons and astrocytes. Basal Ganglia 2(2):57–66. doi:10.1016/j.baga.2012.04.029
172. Arzberger T, Krampfl K, Leimgruber S, Weindl A (1997) Changes of NMDA receptor subunit (NR1, NR2B) and glutamate transporter (GLT1) mRNA expression in Huntington's disease—an in situ hybridization study. J Neuropathol Exp Neurol 56(4):440–454

173. Faideau M, Kim J, Cormier K, Gilmore R, Welch M, Auregan G, Dufour N, Guillermier M, Brouillet E, Hantraye P, Deglon N, Ferrante RJ, Bonvento G (2010) In vivo expression of polyglutamine-expanded huntingtin by mouse striatal astrocytes impairs glutamate transport: a correlation with Huntington's disease subjects. Hum Mol Genet 19(15):3053–3067. doi:10.1093/hmg/ddq212
174. Rebec GV (2013) Dysregulation of corticostriatal ascorbate release and glutamate uptake in transgenic models of Huntington's disease. Antioxid Redox Signal 19(17):2115–2128. doi:10.1089/ars.2013.5387
175. Estrada-Sanchez AM, Montiel T, Segovia J, Massieu L (2009) Glutamate toxicity in the striatum of the R6/2 Huntington's disease transgenic mice is age-dependent and correlates with decreased levels of glutamate transporters. Neurobiol dis 34(1):78–86. doi:10.1016/j.nbd.2008.12.017
176. Miller BR, Dorner JL, Bunner KD, Gaither TW, Klein EL, Barton SJ, Rebec GV (2012) Upregulation of GLT1 reverses the deficit in cortically evoked striatal ascorbate efflux in the R6/2 mouse model of Huntington's disease. J Neurochem 121(4):629–638. doi:10.1111/j.1471-4159.2012.07691.x
177. Petr GT, Schultheis LA, Hussey KC, Sun Y, Dubinsky JM, Aoki C, Rosenberg PA (2013) Decreased expression of GLT-1 in the R6/2 model of Huntington's disease does not worsen disease progression. Eur J Neurosci 38(3):2477–2490. doi:10.1111/ejn.12202
178. Tao YX, Gu J, Stephens RL Jr (2005) Role of spinal cord glutamate transporter during normal sensory transmission and pathological pain states. Mol Pain 1:30. doi:10.1186/1744-8069-1-30
179. Gegelashvili G, Bjerrum OJ (2014) High-affinity glutamate transporters in chronic pain: an emerging therapeutic target. J Neurochem 131(6):712–730. doi:10.1111/jnc.12957
180. Weng HR, Gao M, Maixner DW (2014) Glycogen synthase kinase 3 beta regulates glial glutamate transporter protein expression in the spinal dorsal horn in rats with neuropathic pain. Exp Neurol 252:18–27. doi:10.1016/j.expneurol.2013.11.018
181. Nie H, Zhang H, Weng HR (2010) Minocycline prevents impaired glial glutamate uptake in the spinal sensory synapses of neuropathic rats. Neuroscience 170(3):901–912. doi:10.1016/j.neuroscience.2010.07.049
182. Mao QX, Yang TD (2010) Amitriptyline upregulates EAAT1 and EAAT2 in neuropathic pain rats. Brain Res Bull 81(4–5):424–427. doi:10.1016/j.brainresbull.2009.09.006
183. Hu Y, Li W, Lu L, Cai J, Xian X, Zhang M, Li Q, Li L (2010) An anti-nociceptive role for ceftriaxone in chronic neuropathic pain in rats. Pain 148(2):284–301. doi:10.1016/j.pain.2009.11.014
184. Maeda S, Kawamoto A, Yatani Y, Shirakawa H, Nakagawa T, Kaneko S (2008) Gene transfer of GLT-1, a glial glutamate transporter, into the spinal cord by recombinant adenovirus attenuates inflammatory and neuropathic pain in rats. Mol Pain 4:65. doi:10.1186/1744-8069-4-65
185. Kumar A, Singh A, Ekavali (2015) A review on Alzheimer's disease pathophysiology and its management: an update. Pharmacol Rep 67(2):195–203. doi:10.1016/j.pharep.2014.09.004
186. Mattson MP, Chan SL (2003) Neuronal and glial calcium signaling in Alzheimer's disease. Cell Calcium 34(4–5):385–397
187. Hynd MR, Scott HL, Dodd PR (2004) Glutamate-mediated excitotoxicity and neurodegeneration in Alzheimer's disease. Neurochem Int 45(5):583–595. doi:10.1016/j.neuint.2004.03.007
188. Talantova M, Sanz-Blasco S, Zhang X, Xia P, Akhtar MW, Okamoto S, Dziewczapolski G, Nakamura T, Cao G, Pratt AE, Kang YJ, Tu S, Molokanova E, McKercher SR, Hires SA, Sason H, Stouffer DG, Buczynski MW, Solomon JP, Michael S, Powers ET, Kelly JW, Roberts A, Tong G, Fang-Newmeyer T, Parker J, Holland EA, Zhang D, Nakanishi N, Chen HS, Wolosker H, Wang Y, Parsons LH, Ambasudhan R, Masliah E, Heinemann SF, Pina-Crespo JC, Lipton SA (2013) Abeta induces astrocytic glutamate release, extrasynaptic NMDA receptor activation, and synaptic loss. Proc Natl Acad Sci U S A 110(27):E2518–E2527. doi:10.1073/pnas.1306832110
189. Hardy J, Cowburn R, Barton A, Reynolds G, Lofdahl E, O'Carroll AM, Wester P, Winblad B (1987) Region-specific loss of glutamate innervation in Alzheimer's disease. Neurosci Lett 73(1):77–80
190. Lauderback CM, Hackett JM, Huang FF, Keller JN, Szweda LI, Markesbery WR, Butterfield DA (2001) The glial glutamate transporter, GLT-1, is oxidatively modified by 4-hydroxy-2-nonenal in the Alzheimer's disease brain: the role of Abeta1-42. J Neurochem 78(2):413–416
191. Woltjer RL, Duerson K, Fullmer JM, Mookherjee P, Ryan AM, Montine TJ, Kaye JA, Quinn JF, Silbert L, Erten-Lyons D, Leverenz JB, Bird TD, Pow DV, Tanaka K, Watson GS, Cook DG (2010) Aberrant detergent-insoluble excitatory amino acid

transporter 2 accumulates in Alzheimer disease. J Neuropathol Exp Neurol 69(7):667–676. doi:10.1097/NEN.0b013e3181e24adb

192. Scott HA, Gebhardt FM, Mitrovic AD, Vandenberg RJ, Dodd PR (2011) Glutamate transporter variants reduce glutamate uptake in Alzheimer's disease. Neurobiology of aging 32(3):553.e1–553.11. doi:10.1016/j.neurobiolaging.2010.03.008
193. Mookherjee P, Green PS, Watson GS, Marques MA, Tanaka K, Meeker KD, Meabon JS, Li N, Zhu P, Olson VG, Cook DG (2011) GLT-1 loss accelerates cognitive deficit onset in an Alzheimer's disease animal model. J Alzheimers Dis 26(3):447–455. doi:10.3233/jad-2011-110503
194. Schallier A, Smolders I, Van Dam D, Loyens E, De Deyn PP, Michotte A, Michotte Y, Massie A (2011) Region- and age-specific changes in glutamate transport in the AbetaPP23 mouse model for Alzheimer's disease. J Alzheimers Dis 24(2):287–300. doi:10.3233/jad-2011-101005
195. Scimemi A, Meabon JS, Woltjer RL, Sullivan JM, Diamond JS, Cook DG (2013) Amyloid-beta1-42 slows clearance of synaptically released glutamate by mislocalizing astrocytic GLT-1. J Neurosci 33(12):5312–5318. doi:10.1523/jneurosci.5274-12.2013
196. Kulijewicz-Nawrot M, Sykova E, Chvatal A, Verkhratsky A, Rodriguez JJ (2013) Astrocytes and glutamate homoeostasis in Alzheimer's disease: a decrease in glutamine synthetase, but not in glutamate transporter-1, in the prefrontal cortex. ASN Neuro 5(4):273–282. doi:10.1042/an20130017
197. Bamford NS, Zhang H, Schmitz Y, Wu NP, Cepeda C, Levine MS, Schmauss C, Zakharenko SS, Zablow L, Sulzer D (2004) Heterosynaptic dopamine neurotransmission selects sets of corticostriatal terminals. Neuron 42(4):653–663
198. Ambrosi G, Cerri S, Blandini F (2014) A further update on the role of excitotoxicity in the pathogenesis of Parkinson's disease. J Neural Transm 121(8):849–859. doi:10.1007/s00702-013-1149-z
199. Rodriguez MC, Obeso JA, Olanow CW (1998) Subthalamic nucleus-mediated excitotoxicity in Parkinson's disease: a target for neuroprotection. Ann Neurol 44(3 Suppl 1):S175–S188
200. Deumens R, Blokland A, Prickaerts J (2002) Modeling Parkinson's disease in rats: an evaluation of 6-OHDA lesions of the nigrostriatal pathway. Exp Neurol 175(2):303–317. doi:10.1006/exnr.2002.7891
201. Oster S, Radad K, Scheller D, Hesse M, Balanzew W, Reichmann H, Gille G (2014) Rotigotine protects against glutamate toxicity in primary dopaminergic cell culture. Eur J Pharmacol 724:31–42. doi:10.1016/j.ejphar.2013.12.014
202. Massie A, Goursaud S, Schallier A, Vermoesen K, Meshul CK, Hermans E, Michotte Y (2010) Time-dependent changes in GLT-1 functioning in striatum of hemi-Parkinson rats. Neurochem Int 57(5):572–578. doi:10.1016/j.neuint.2010.07.004
203. Carbone M, Duty S, Rattray M (2012) Riluzole neuroprotection in a Parkinson's disease model involves suppression of reactive astrocytosis but not GLT-1 regulation. BMC Neurosci 13:38. doi:10.1186/1471-2202-13-38
204. Chotibut T, Davis RW, Arnold JC, Frenchek Z, Gurwara S, Bondada V, Geddes JW, Salvatore MF (2014) Ceftriaxone increases glutamate uptake and reduces striatal tyrosine hydroxylase loss in 6-OHDA Parkinson's model. Mol Neurobiol 49(3):1282–1292. doi:10.1007/s12035-013-8598-0
205. Kelsey JE, Neville C (2014) The effects of the beta-lactam antibiotic, ceftriaxone, on forepaw stepping and L-DOPA-induced dyskinesia in a rodent model of Parkinson's disease. Psychopharmacology 231(12):2405–2415. doi:10.1007/s00213-013-3400-6
206. Sabri F, Titanji K, De Milito A, Chiodi F (2003) Astrocyte activation and apoptosis: their roles in the neuropathology of HIV infection. Brain Pathol 13(1):84–94
207. Potter MC, Figuera-Losada M, Rojas C, Slusher BS (2013) Targeting the glutamatergic system for the treatment of HIV-associated neurocognitive disorders. J Neuroimmune Pharmacol 8(3):594–607. doi:10.1007/s11481-013-9442-z
208. Cisneros IE, Ghorpade A (2014) Methamphetamine and HIV-1-induced neurotoxicity: role of trace amine associated receptor 1 cAMP signaling in astrocytes. Neuropharmacology 85:499–507. doi:10.1016/j.neuropharm.2014.06.011
209. Rao VR, Ruiz AP, Prasad VR (2014) Viral and cellular factors underlying neuropathogenesis in HIV associated neurocognitive disorders (HAND). AIDS Res Ther 11:13–13. doi:10.1186/1742-6405-11-13
210. Banasr M, Chowdhury GM, Terwilliger R, Newton SS, Duman RS, Behar KL, Sanacora G (2010) Glial pathology in an animal model of depression: reversal of stress-induced cellular, metabolic and behavioral deficits by the glutamate-modulating drug riluzole.

Mol Psychiatry 15(5):501–511. doi:10.1038/mp.2008.106

211. Takahashi KK, Foster JB, Lin CL (2015) Glutamate transporter EAAT2: regulation, function, and potential as a therapeutic target for neurological and psychiatric disease. Cell Mol Life Sci 72(18):3489–3506. doi:10.1007/s00018-015-1937-8
212. Chen JX, Yao LH, BB X, Qian K, Wang HL, Liu ZC, Wang XP, Wang GH (2014) Glutamate transporter 1-mediated antidepressant-like effect in a rat model of chronic unpredictable stress. J Huazhong Univ Sci Technolog Med Sci 34(6):838–844. doi:10.1007/s11596-014-1362-5
213. Nakagawa T, Kaneko S (2013) SLC1 glutamate transporters and diseases: psychiatric diseases and pathological pain. Curr Mol Pharmacol 6(2):66–73
214. Gregg RA, Hicks C, Nayak SU, Tallarida CS, Nucero P, Smith GR, Reitz AB, Rawls SM (2016) Synthetic cathinone MDPV downregulates glutamate transporter subtype I (GLT-1) and produces rewarding and locomotor-activating effects that are reduced by a GLT-1 activator. Neuropharmacology 108:111–119. doi:10.1016/j.neuropharm.2016.04.014
215. Sari Y, Toalston JE, Rao PS, Bell RL (2016) Effects of ceftriaxone on ethanol, nicotine or sucrose intake by alcohol-preferring (P) rats and its association with GLT-1 expression. Neuroscience 326:117–125. doi:10.1016/j.neuroscience.2016.04.004
216. Rao PS, Goodwani S, Bell RL, Wei Y, Boddu SH, Sari Y (2015) Effects of ampicillin, cefazolin and cefoperazone treatments on GLT-1 expressions in the mesocorticolimbic system and ethanol intake in alcohol-preferring rats. Neuroscience 295:164–174. doi:10.1016/j.neuroscience.2015.03.038
217. Alhaddad H, Kim NT, Aal-Aaboda M, Althobaiti YS, Leighton J, Boddu SH, Wei Y, Sari Y (2014) Effects of MS-153 on chronic ethanol consumption and GLT1 modulation of glutamate levels in male alcohol-preferring rats. Front Behav Neurosci 8:366. doi:10.3389/fnbeh.2014.00366
218. Sari Y, Sreemantula SN (2012) Neuroimmunophilin GPI-1046 reduces ethanol consumption in part through activation of GLT1 in alcohol-preferring rats. Neuroscience 227:327–335. doi:10.1016/j.neuroscience.2012.10.007
219. Ayers-Ringler JR, Jia YF, Qiu YY, Choi DS (2016) Role of astrocytic glutamate transporter in alcohol use disorder. World J Psychiatry 6(1):31–42. doi:10.5498/wjp.v6.i1.31
220. Ozsoy S, Asdemir A, Karaaslan O, Akalin H, Ozkul Y, Esel E (2016) Expression of glutamate transporters in alcohol withdrawal. Pharmacopsychiatry 49(1):14–17. doi:10.1055/s-0035-1565134
221. McIntosh TK, Vink R, Soares H, Hayes R, Simon R (1990) Effect of noncompetitive blockade of N-methyl-D-aspartate receptors on the neurochemical sequelae of experimental brain injury. J Neurochem 55(4):1170–1179
222. Okiyama K, Smith DH, White WF, Richter K, McIntosh TK (1997) Effects of the novel NMDA antagonists CP-98,113, CP-101,581 and CP-101,606 on cognitive function and regional cerebral edema following experimental brain injury in the rat. J Neurotrauma 14(4):211–222
223. Okiyama K, Smith DH, White WF, McIntosh TK (1998) Effects of the NMDA antagonist CP-98,113 on regional cerebral edema and cardiovascular, cognitive, and neurobehavioral function following experimental brain injury in the rat. Brain Res 792(2):291–298
224. Ginsberg MD (2008) Neuroprotection for ischemic stroke: past, present and future. Neuropharmacology 55(3):363–389. doi:10.1016/j.neuropharm.2007.12.007
225. Ikonomidou C, Turski L (2002) Why did NMDA receptor antagonists fail clinical trials for stroke and traumatic brain injury? Lancet Neurol 1(6):383–386
226. Okiyama K, Smith DH, Thomas MJ, McIntosh TK (1992) Evaluation of a novel calcium channel blocker, (S)-emopamil, on regional cerebral edema and neurobehavioral function after experimental brain injury. J Neurosurg 77(4):607–615. doi:10.3171/jns.1992.77.4.0607
227. Cheney JA, Brown AL, Bareyre FM, Russ AB, Weisser JD, Ensinger HA, Leusch A, Raghupathi R, Saatman KE (2000) The novel compound LOE 908 attenuates acute neuromotor dysfunction but not cognitive impairment or cortical tissue loss following traumatic brain injury in rats. J Neurotrauma 17(1):83–91
228. Kristensen AS, Andersen J, Jorgensen TN, Sorensen L, Eriksen J, Loland CJ, Stromgaard K, Gether U (2011) SLC6 neurotransmitter transporters: structure, function, and regulation. Pharmacol Rev 63(3):585–640. doi:10.1124/pr.108.000869
229. Guo H, Lai L, Butchbach ME, Stockinger MP, Shan X, Bishop GA, Lin CL (2003)

Increased expression of the glial glutamate transporter EAAT2 modulates excitotoxicity and delays the onset but not the outcome of ALS in mice. Hum Mol Genet 12(19):2519–2532. doi:10.1093/hmg/ddg267

230. Bunch L, Erichsen MN, Jensen AA (2009) Excitatory amino acid transporters as potential drug targets. Expert Opin Ther Targets 13(6):719–731. doi:10.1517/14728220902926127

231. Rothstein JD, Patel S, Regan MR, Haenggeli C, Huang YH, Bergles DE, Jin L, Dykes Hoberg M, Vidensky S, Chung DS, Toan SV, Bruijn LI, Su ZZ, Gupta P, Fisher PB (2005) Beta-lactam antibiotics offer neuroprotection by increasing glutamate transporter expression. Nature 433(7021):73–77. doi:10.1038/nature03180

232. Colton CK, Kong Q, Lai L, Zhu MX, Seyb KI, Cuny GD, Xian J, Glicksman MA, Lin CL (2010) Identification of translational activators of glial glutamate transporter EAAT2 through cell-based high-throughput screening: an approach to prevent excitotoxicity. J Biomol Screen 15(6):653–662. doi:10.1177/1087057110370998

233. Takahashi K, Kong Q, Lin Y, Stouffer N, Schulte DA, Lai L, Liu Q, Chang LC, Dominguez S, Xing X, Cuny GD, Hodgetts KJ, Glicksman MA, Lin CL (2015) Restored glial glutamate transporter EAAT2 function as a potential therapeutic approach for Alzheimer's disease. J Exp Med 212(3):319–332. doi:10.1084/jem.20140413

234. Kong Q, Chang L-C, Takahashi K, Liu Q, Schulte DA, Lai L, Ibabao B, Lin Y, Stouffer N, Mukhopadhyay CD, Xing X, Seyb KI, Cuny GD, Glicksman MA, Lin C-LG (2014) Small-molecule activator of glutamate transporter EAAT2 translation provides neuroprotection. J Clin Invest 124(3):1255–1267. doi:10.1172/JCI66163

235. Mortensen OV, Kortagere S (2015) Designing modulators of monoamine transporters using virtual screening techniques. Front Pharmacol 6:223. doi:10.3389/fphar.2015.00223

236. Mortensen OV, Liberato JL, Coutinho-Netto J, Dos Santos WF, Fontana AC (2015) Molecular determinants of transport stimulation of EAAT2 are located at interface between the trimerization and substrate transport domains. J Neurochem 133(2):199–210. doi:10.1111/jnc.13047

237. Padmanabhan B (2013) Identification of novel modulators for ionotropic glutamate receptor, iGluA2 by in-silico screening. Theor Biol Med Model 10:46. doi:10.1186/1742-4682-10-46

238. Yuan H, Myers SJ, Wells G, Nicholson KL, Swanger SA, Lyuboslavsky P, Tahirovic YA, Menaldino DS, Ganesh T, Wilson LJ, Liotta DC, Snyder JP, Traynelis SF (2015) Context-dependent GluN2B-selective inhibitors of NMDA receptor function are neuroprotective with minimal side effects. Neuron 85(6):1305–1318. doi:10.1016/j.neuron.2015.02.008

239. Billups B, Szatkowski M, Rossi D, Attwell D (1998) Patch-clamp, ion-sensing, and glutamate-sensing techniques to study glutamate transport in isolated retinal glial cells. Methods Enzymol 296:617–632

240. Bozzo L, Chatton JY (2010) Inhibitory effects of (2S, 3S)-3-[3-[4-(trifluoromethyl)benzoylamino]benzyloxy]aspartate (TFB-TBOA) on the astrocytic sodium responses to glutamate. Brain Res 1316:27–34. doi:10.1016/j.brainres.2009.12.028

241. Innocenti B, Parpura V, Haydon PG (2000) Imaging extracellular waves of glutamate during calcium signaling in cultured astrocytes. J Neurosci 20(5):1800–1808

242. Hu Y, Mitchell KM, Albahadily FN, Michaelis EK, Wilson GS (1994) Direct measurement of glutamate release in the brain using a dual enzyme-based electrochemical sensor. Brain Res 659(1–2):117–125

243. Marvin JS, Borghuis BG, Tian L, Cichon J, Harnett MT, Akerboom J, Gordus A, Renninger SL, Chen TW, Bargmann CI, Orger MB, Schreiter ER, Demb JB, Gan WB, Hires SA, Looger LL (2013) An optimized fluorescent probe for visualizing glutamate neurotransmission. Nat Methods 10(2):162–170. doi:10.1038/nmeth.2333

244. Bridges RJ, Kavanaugh MP, Chamberlin AR (1999) A pharmacological review of competitive inhibitors and substrates of high-affinity, sodium-dependent glutamate transport in the central nervous system. Curr Pharm Des 5(5):363–379

245. Bridges RJ, Esslinger CS (2005) The excitatory amino acid transporters: pharmacological insights on substrate and inhibitor specificity of the EAAT subtypes. Pharmacol Ther 107(3):271–285. doi:10.1016/j.pharmthera.2005.01.002

246. Lebrun B, Sakaitani M, Shimamoto K, Yasuda-Kamatani Y, Nakajima T (1997) New beta-hydroxyaspartate derivatives are competitive blockers for the bovine glutamate/aspartate transporter. J Biol Chem 272(33):20336–20339

247. Shimamoto K (2008) Glutamate transporter blockers for elucidation of the function of

excitatory neurotransmission systems. Chem Rec 8(3):182–199. doi:10.1002/tcr.20145

248. Shigeri Y, Seal RP, Shimamoto K (2004) Molecular pharmacology of glutamate transporters, EAATs and VGLUTs. Brain Res Brain Res Rev 45(3):250–265. doi:10.1016/j.brainresrev.2004.04.004
249. Shimamoto K, Lebrun B, Yasuda-Kamatani Y, Sakaitani M, Shigeri Y, Yumoto N, Nakajima T (1998) DL-threo-beta-benzyloxyaspartate, a potent blocker of excitatory amino acid transporters. Mol Pharmacol 53(2):195–201
250. Abrahamsen B, Schneider N, Erichsen MN, Huynh TH, Fahlke C, Bunch L, Jensen AA (2013) Allosteric modulation of an excitatory amino acid transporter: the subtype-selective inhibitor UCPH-101 exerts sustained inhibition of EAAT1 through an intramonomeric site in the trimerization domain. J Neurosci 33(3):1068–1087. doi:10.1523/jneurosci.3396-12.2013
251. Arriza JL, Fairman WA, Wadiche JI, Murdoch GH, Kavanaugh MP, Amara SG (1994) Functional comparisons of three glutamate transporter subtypes cloned from human motor cortex. J Neurosci 14(9):5559–5569
252. Esslinger CS, Agarwal S, Gerdes J, Wilson PA, Davis ES, Awes AN, O'Brien E, Mavencamp T, Koch HP, Poulsen DJ, Rhoderick JF, Chamberlin AR, Kavanaugh MP, Bridges RJ (2005) The substituted aspartate analogue L-beta-threo-benzyl-aspartate preferentially inhibits the neuronal excitatory amino acid transporter EAAT3. Neuropharmacology 49(6):850–861. doi:10.1016/j.neuropharm.2005.08.009
253. O'Shea RD, Fodera MV, Aprico K, Dehnes Y, Danbolt NC, Crawford D, Beart PM (2002) Evaluation of drugs acting at glutamate transporters in organotypic hippocampal cultures: new evidence on substrates and blockers in excitotoxicity. Neurochem Res 27(1–2):5–13
254. Volterra A, Bezzi P, Rizzini BL, Trotti D, Ullensvang K, Danbolt NC, Racagni G (1996) The competitive transport inhibitor L-trans-pyrrolidine-2,4-dicarboxylate triggers excitotoxicity in rat cortical neuron-astrocyte cocultures via glutamate release rather than uptake inhibition. Eur J Neurosci 8(9):2019–2028
255. Izumi Y, Shimamoto K, Benz AM, Hammerman SB, Olney JW, Zorumski CF (2002) Glutamate transporters and retinal excitotoxicity. Glia 39(1):58–68. doi:10.1002/glia.10082
256. Bonde C, Noraberg J, Noer H, Zimmer J (2005) Ionotropic glutamate receptors and glutamate transporters are involved in necrotic neuronal cell death induced by oxygen-glucose deprivation of hippocampal slice cultures. Neuroscience 136(3):779–794. doi:10.1016/j.neuroscience.2005.07.020
257. Jabaudon D, Shimamoto K, Yasuda-Kamatani Y, Scanziani M, Gahwiler BH, Gerber U (1999) Inhibition of uptake unmasks rapid extracellular turnover of glutamate of nonvesicular origin. Proc Natl Acad Sci U S A 96(15):8733–8738
258. Shimamoto K, Sakai R, Takaoka K, Yumoto N, Nakajima T, Amara SG, Shigeri Y (2004) Characterization of novel L-threo-beta-benzyloxyaspartate derivatives, potent blockers of the glutamate transporters. Mol Pharmacol 65(4):1008–1015. doi:10.1124/mol.65.4.1008
259. Tsukada S, Iino M, Takayasu Y, Shimamoto K, Ozawa S (2005) Effects of a novel glutamate transporter blocker, (2S, 3S)-3-[3-[4-(trifluoromethyl)benzoylamino]benzyloxy]aspartate (TFB-TBOA), on activities of hippocampal neurons. Neuropharmacology 48(4):479–491. doi:10.1016/j.neuropharm.2004.11.006
260. Robinson MB, Dowd LA (1997) Heterogeneity and functional properties of subtypes of sodium-dependent glutamate transporters in the mammalian central nervous system. Adv Pharmacol 37:69–115
261. Dunlop J (2006) Glutamate-based therapeutic approaches: targeting the glutamate transport system. Curr Opin Pharmacol 6(1):103–107. doi:10.1016/j.coph.2005.09.004
262. Dunlop J, McIlvain HB, Carrick TA, Jow B, Lu Q, Kowal D, Lin S, Greenfield A, Grosanu C, Fan K, Petroski R, Williams J, Foster A, Butera J (2005) Characterization of novel aryl-ether, biaryl, and fluorene aspartic acid and diaminopropionic acid analogs as potent inhibitors of the high-affinity glutamate transporter EAAT2. Mol Pharmacol 68(4):974–982. doi:10.1124/mol.105.012005
263. Karatas-Wulf U, Koepsell H, Bergert M, Sönnekes S, Kugler P (2009) Protein kinase C-dependent trafficking of glutamate transporters excitatory amino acid carrier 1 and glutamate transporter 1b in cultured cerebellar granule cells. Neuroscience 161(3):794–805. doi:10.1016/j.neuroscience.2009.04.017
264. Nakamura Y, Kataoka K, Ishida M, Shinozaki H (1993) (2S,3S,4R)-2-(carboxycyclopropyl)glycine, a potent and competitive inhibitor of both glial and neuronal uptake of glutamate. Neuropharmacology 32(9):833–837
265. Zuiderwijk M, Veenstra E, Lopes da Silva FH, Ghijsen WE (1996) Effects of uptake carrier blockers SK & F 89976-A and

L-trans-PDC on in vivo release of amino acids in rat hippocampus. Eur J Pharmacol 307(3):275–282

266. Assous M, Had-Aissouni L, Gubellini P, Melon C, Nafia I, Salin P, Kerkerian-Le-Goff L, Kachidian P (2014) Progressive Parkinsonism by acute dysfunction of excitatory amino acid transporters in the rat substantia nigra. Neurobiol Dis 65:69–81. doi:10.1016/j.nbd.2014.01.011
267. Funicello M, Conti P, De Amici M, De Micheli C, Mennini T, Gobbi M (2004) Dissociation of [3H]L-glutamate uptake from L-glutamate-induced [3H]D-aspartate release by 3-hydroxy-4,5,6,6a-tetrahydro-3aH-pyrrolo[3,4-d]isoxazole-4-carboxylic acid and 3-hydroxy-4,5,6,6a-tetrahydro-3aH-pyrrolo[3,4-d]isoxazole-6-carboxylic acid, two conformationally constrained aspartate and glutamate analogs. Mol Pharmacol 66(3):522–529. doi:10.1124/mol.66.3
268. Colleoni S, Jensen AA, Landucci E, Fumagalli E, Conti P, Pinto A, De Amici M, Pellegrini-Giampietro DE, De Micheli C, Mennini T, Gobbi M (2008) Neuroprotective effects of the novel glutamate transporter inhibitor (−)-3-hydroxy-4,5,6,6a-tetrahydro-3aH-pyrrolo[3,4-d]-isoxazole-4-carboxylic acid, which preferentially inhibits reverse transport (glutamate release) compared with glutamate reuptake. J Pharmacol Exp Ther 326(2):646–656. doi:10.1124/jpet.107.135251
269. Callender R, Gameiro A, Pinto A, De Micheli C, Grewer C (2012) Mechanism of inhibition of the glutamate transporter EAAC1 by the conformationally constrained glutamate analogue (+)-HIP-B. Biochemistry 51(27):5486–5495. doi:10.1021/bi3006048
270. Jensen AA, Erichsen MN, Nielsen CW, Stensbol TB, Kehler J, Bunch L (2009) Discovery of the first selective inhibitor of excitatory amino acid transporter subtype 1. J Med Chem 52(4):912–915. doi:10.1021/jm8013458
271. Stensbol TB, Uhlmann P, Morel S, Eriksen BL, Felding J, Kromann H, Hermit MB, Greenwood JR, Brauner-Osborne H, Madsen U, Junager F, Krogsgaard-Larsen P, Begtrup M, Vedso P (2002) Novel 1-hydroxyazole bioisosteres of glutamic acid. Synthesis, protolytic properties, and pharmacology. J Med Chem 45(1):19–31
272. Conradt M, Stoffel W (1997) Inhibition of the high-affinity brain glutamate transporter GLAST-1 via direct phosphorylation. J Neurochem 68(3):1244–1251
273. Gegelashvili G, Dehnes Y, Danbolt NC, Schousboe A (2000) The high-affinity glutamate transporters GLT1, GLAST, and EAAT4 are regulated via different signalling mechanisms. Neurochem Int 37(2–3):163–170
274. Wang W, Lu R, Feng DY, Zhang H (2016) Sevoflurane inhibits glutamate-aspartate transporter and glial fibrillary acidic protein expression in hippocampal astrocytes of neonatal rats through the janus kinase/signal transducer and activator of transcription (JAK/STAT) pathway. Anesth Analg. doi:10.1213/ane.0000000000001238
275. Underhill SM, Wheeler DS, Li M, Watts SD, Ingram SL, Amara SG (2014) Amphetamine modulates glutamatergic neurotransmission through endocytosis of the excitatory amino acid transporter EAAT3 in dopamine neurons. Neuron 83(2):404–416. doi:10.1016/j.neuron.2014.05.043
276. Sheldon AL, Gonzalez MI, Krizman-Genda EN, Susarla BT, Robinson MB (2008) Ubiquitination-mediated internalization and degradation of the astroglial glutamate transporter, GLT-1. Neurochem Int 53(6–8):296–308. doi:10.1016/j.neuint.2008.07.010
277. Susarla BTS, Robinson MB (2008) Internalization and degradation of the glutamate transporter GLT-1 in response to phorbol ester. Neurochem Int 52(4–5):709–722. doi:10.1016/j.neuint.2007.08.020
278. Park HJ, Baik HJ, Kim DY, Lee GY, Woo JH, Zuo Z, Chung RK (2015) Doxepin and imipramine but not fluoxetine reduce the activity of the rat glutamate transporter EAAT3 expressed in Xenopus oocytes. BMC Anesthesiol 15:116. doi:10.1186/s12871-015-0098-5
279. Dowd LA, Robinson MB (1996) Rapid stimulation of EAAC1-mediated Na+-dependent L-glutamate transport activity in C6 glioma cells by phorbol ester. J Neurochem 67(2):508–516
280. Davis KE, Straff DJ, Weinstein EA, Bannerman PG, Correale DM, Rothstein JD, Robinson MB (1998) Multiple signaling pathways regulate cell surface expression and activity of the excitatory amino acid carrier 1 subtype of Glu transporter in C6 glioma. J Neurosci 18(7):2475–2485
281. Susarla BT, Seal RP, Zelenaia O, Watson DJ, Wolfe JH, Amara SG, Robinson MB (2004) Differential regulation of GLAST immunoreactivity and activity by protein kinase C: evidence for modification of amino and carboxyl termini. J Neurochem 91(5):1151–1163. doi:10.1111/j.1471-4159.2004.02791.x
282. Piao C, Ralay Ranaivo H, Rusie A, Wadhwani N, Koh S, Wainwright MS (2015) Thrombin

decreases expression of the glutamate transporter GLAST and inhibits glutamate uptake in primary cortical astrocytes via the Rho kinase pathway. Exp Neurol 273:288–300. doi:10.1016/j.expneurol.2015.09.009

283. Moidunny S, Matos M, Wesseling E, Banerjee S, Volsky DJ, Cunha RA, Agostinho P, Boddeke HW, Roy S (2016) Oncostatin M promotes excitotoxicity by inhibiting glutamate uptake in astrocytes: implications in HIV-associated neurotoxicity. J Neuroinflammation 13(1):144. doi:10.1186/s12974-016-0613-8

284. Anderson CM, Swanson RA (2000) Astrocyte glutamate transport: review of properties, regulation, and physiological functions. Glia 32(1):1–14

285. Melzer N, Meuth SG, Torres-Salazar D, Bittner S, Zozulya AL, Weidenfeller C, Kotsiari A, Stangel M, Fahlke C, Wiendl H (2008) A beta-lactam antibiotic dampens excitotoxic inflammatory CNS damage in a mouse model of multiple sclerosis. PLoS One 3(9):e3149. doi:10.1371/journal.pone.0003149

286. Thone-Reineke C, Neumann C, Namsolleck P, Schmerbach K, Krikov M, Schefe JH, Lucht K, Hortnagl H, Godes M, Muller S, Rumschussel K, Funke-Kaiser H, Villringer A, Steckelings UM, Unger T (2008) The beta-lactam antibiotic, ceftriaxone, dramatically improves survival, increases glutamate uptake and induces neurotrophins in stroke. J Hypertens 26(12):2426–2435. doi:10.1097/HJH.0b013e328313e403

287. Hota SK, Barhwal K, Ray K, Singh SB, Ilavazhagan G (2008) Ceftriaxone rescues hippocampal neurons from excitotoxicity and enhances memory retrieval in chronic hypobaric hypoxia. Neurobiol Learn Mem 89(4):522–532. doi:10.1016/j.nlm.2008.01.003

288. Lee SG, Su ZZ, Emdad L, Gupta P, Sarkar D, Borjabad A, Volsky DJ, Fisher PB (2008) Mechanism of ceftriaxone induction of excitatory amino acid transporter-2 expression and glutamate uptake in primary human astrocytes. J Biol Chem 283(19):13116–13123. doi:10.1074/jbc.M707697200

289. Pan XD, Wei J, Xiao GM (2011) Effects of beta-lactam antibiotics ceftriaxone on expression of glutamate in hippocampus after traumatic brain injury in rats. Zhejiang Da Xue Xue Bao Yi Xue Ban 40(5):522–526

290. Wei J, Pan X, Pei Z, Wang W, Qiu W, Shi Z, Xiao G (2012) The beta-lactam antibiotic, ceftriaxone, provides neuroprotective potential via anti-excitotoxicity and anti-inflammation response in a rat model of traumatic brain injury. J Trauma Acute Care Surg 73(3):654–660. doi:10.1097/TA.0b013e31825133c0

291. Cui C, Cui Y, Gao J, Sun L, Wang Y, Wang K, Li R, Tian Y, Song S, Cui J (2014) Neuroprotective effect of ceftriaxone in a rat model of traumatic brain injury. Neurol Sci 35(5):695–700. doi:10.1007/s10072-013-1585-4

292. Goodrich GS, Kabakov AY, Hameed MQ, Dhamne SC, Rosenberg PA, Rotenberg A (2013) Ceftriaxone treatment after traumatic brain injury restores expression of the glutamate transporter, GLT-1, reduces regional gliosis, and reduces post-traumatic seizures in the rat. J Neurotrauma 30(16):1434–1441. doi:10.1089/neu.2012.2712

293. Leung TC, Lui CN, Chen LW, Yung WH, Chan YS, Yung KK (2012) Ceftriaxone ameliorates motor deficits and protects dopaminergic neurons in 6-hydroxydopamine-lesioned rats. ACS Chem Neurosci 3(1):22–30. doi:10.1021/cn200072h

294. Miller BR, Dorner JL, Shou M, Sari Y, Barton SJ, Sengelaub DR, Kennedy RT, Rebec GV (2008) Upregulation of GLT1 expression increases glutamate uptake and attenuates the Huntington's disease phenotype in the R6/2 mouse. Neuroscience 153(1):329–337. doi:10.1016/j.neuroscience.2008.02.004

295. Feng D, Wang W, Dong Y, Wu L, Huang J, Ma Y, Zhang Z, Wu S, Gao G, Qin H (2014) Ceftriaxone alleviates early brain injury after subarachnoid hemorrhage by increasing excitatory amino acid transporter 2 expression via the PI3K/Akt/NF-kappaB signaling pathway. Neuroscience 268:21–32. doi:10.1016/j.neuroscience.2014.02.053

296. Hu YY, Xu J, Zhang M, Wang D, Li L, Li WB (2015) Ceftriaxone modulates uptake activity of glial glutamate transporter-1 against global brain ischemia in rats. J Neurochem 132(2):194–205. doi:10.1111/jnc.12958

297. Jagadapillai R, Mellen NM, Sachleben LR Jr, Gozal E (2014) Ceftriaxone preserves glutamate transporters and prevents intermittent hypoxia-induced vulnerability to brain excitotoxic injury. PLoS One 9(7):e100230. doi:10.1371/journal.pone.0100230

298. Beghi E, Bendotti C, Mennini T (2006) New ideas for therapy in ALS: critical considerations. Amyotroph Lateral Scler 7(2):126–127.; discussion 127. doi:10.1080/14660820510012040

299. Nederkoorn PJ, Westendorp WF, Hooijenga IJ, de Haan RJ, Dippel DW, Vermeij FH, Dijkgraaf MG, Prins JM, Spanjaard L, van de Beek D (2011) Preventive antibiotics in

stroke study: rationale and protocol for a randomised trial. Int J Stroke 6(2):159–163. doi:10.1111/j.1747-4949.2010.00555.x

300. Berry JD, Shefner JM, Conwit R, Schoenfeld D, Keroack M, Felsenstein D, Krivickas L, David WS, Vriesendorp F, Pestronk A, Caress JB, Katz J, Simpson E, Rosenfeld J, Pascuzzi R, Glass J, Rezania K, Rothstein JD, Greenblatt DJ, Cudkowicz ME (2013) Design and initial results of a multi-phase randomized trial of ceftriaxone in amyotrophic lateral sclerosis. PLoS One 8(4):e61177. doi:10.1371/journal.pone.0061177
301. Cudkowicz M, Shefner J, Consortium N (2013) STAGE 3 clinical trial of ceftriaxone in subjects with ALS (S36.001). Neurology 80:S36.001
302. Carbone M, Duty S, Rattray M (2012) Riluzole elevates GLT-1 activity and levels in striatal astrocytes. Neurochem Int 60(1):31–38. doi:10.1016/j.neuint.2011.10.017
303. Furness DN, Dehnes Y, Akhtar AQ, Rossi DJ, Hamann M, Grutle NJ, Gundersen V, Holmseth S, Lehre KP, Ullensvang K, Wojewodzic M, Zhou Y, Attwell D, Danbolt NC (2008) A quantitative assessment of glutamate uptake into hippocampal synaptic terminals and astrocytes: new insights into a neuronal role for excitatory amino acid transporter 2 (EAAT2). Neuroscience 157(1):80–94. doi:10.1016/j.neuroscience.2008.08.043
304. Li Y, Zhou Y, Danbolt NC (2012) The rates of postmortem proteolysis of glutamate transporters differ dramatically between cells and between transporter subtypes. J Histochem Cytochem 60(11):811–821. doi:10.1369/0022155412458589
305. Xing X, Chang LC, Kong Q, Colton CK, Lai L, Glicksman MA, Lin CL, Cuny GD (2011) Structure–activity relationship study of pyridazine derivatives as glutamate transporter EAAT2 activators. Bioorg med Chem Lett 21(19):5774–5777. doi:10.1016/j.bmcl.2011.08.009
306. Cui X, Li L, Hu YY, Ren S, Zhang M, Li WB (2015) Sulbactam plays neuronal protective effect against brain ischemia via upregulating GLT1 in rats. Mol Neurobiol 51(3):1322–1333. doi:10.1007/s12035-014-8809-3
307. Hassel B, Iversen EG, Gjerstad L, Tauboll E (2001) Upregulation of hippocampal glutamate transport during chronic treatment with sodium valproate. J Neurochem 77(5):1285–1292
308. Yoshizumi M, Eisenach JC, Hayashida K (2013) Valproate prevents dysregulation of spinal glutamate and reduces the development of hypersensitivity in rats after peripheral nerve injury. J Pain 14(11):1485–1491. doi:10.1016/j.jpain.2013.07.007
309. Kimura A, Guo X, Noro T, Harada C, Tanaka K, Namekata K, Harada T (2015) Valproic acid prevents retinal degeneration in a murine model of normal tension glaucoma. Neurosci Lett 588:108–113. doi:10.1016/j.neulet.2014.12.054
310. Huang Y, Feng X, Sando JJ, Zuo Z (2006) Critical role of serine 465 in isoflurane-induced increase of cell-surface redistribution and activity of glutamate transporter type 3. J Biol Chem 281(50):38133–38138. doi:10.1074/jbc.M603885200
311. Qu X, Xu C, Wang H, Xu J, Liu W, Wang Y, Jia X, Xie Z, Xu Z, Ji C, Wu A, Yue Y (2013) Hippocampal glutamate level and glutamate aspartate transporter (GLAST) are upregulated in senior rat associated with isoflurane-induced spatial learning/memory impairment. Neurochem Res 38(1):59–73. doi:10.1007/s11064-012-0889-8
312. Cao J, Wang Z, Mi W, Zuo Z (2014) Isoflurane unveils a critical role of glutamate transporter type 3 in regulating hippocampal GluR1 trafficking and context-related learning and memory in mice. Neuroscience 272:58–64. doi:10.1016/j.neuroscience.2014.04.049
313. Fang Q, Hu WW, Wang XF, Yang Y, Lou GD, Jin MM, Yan HJ, Zeng WZ, Shen Y, Zhang SH, Xu TL, Chen Z (2014) Histamine upregulates astrocytic glutamate transporter 1 and protects neurons against ischemic injury. Neuropharmacology 77:156–166. doi:10.1016/j.neuropharm.2013.06.012
314. Li Y, Sattler R, Yang EJ, Nunes A, Ayukawa Y, Akhtar S, Ji G, Zhang PW, Rothstein JD (2011) Harmine, a natural beta-carboline alkaloid, upregulates astroglial glutamate transporter expression. Neuropharmacology 60(7–8):1168–1175. doi:10.1016/j.neuropharm.2010.10.016
315. Sun P, Zhang S, Li Y, Wang L (2014) Harmine mediated neuroprotection via evaluation of glutamate transporter 1 in a rat model of global cerebral ischemia. Neurosci Lett 583c:32–36. doi:10.1016/j.neulet.2014.09.023
316. Zschocke J, Bayatti N, Clement AM, Witan H, Figiel M, Engele J, Behl C (2005) Differential promotion of glutamate transporter expression and function by glucocorticoids in astrocytes from various brain regions. J Biol Chem 280(41):34924–34932. doi:10.1074/jbc.M502581200
317. Wen ZH, Wu GJ, Chang YC, Wang JJ, Wong CS (2005) Dexamethasone modulates the

development of morphine tolerance and expression of glutamate transporters in rats. Neuroscience 133(3):807–817. doi:10.1016/j.neuroscience.2005.03.015

318. Feng Y, Lu S, Wang J, Kumar P, Zhang L, Bhatt AJ (2014) Dexamethasone-induced neuroprotection in hypoxic-ischemic brain injury in newborn rats is partly mediated via Akt activation. Brain Res 1589c:68–77. doi:10.1016/j.brainres.2014.09.073
319. Li LB, Toan SV, Zelenaia O, Watson DJ, Wolfe JH, Rothstein JD, Robinson MB (2006) Regulation of astrocytic glutamate transporter expression by Akt: evidence for a selective transcriptional effect on the GLT-1/EAAT2 subtype. J Neurochem 97(3):759–771. doi:10.1111/j.1471-4159.2006.03743.x
320. Pu B, Xue Y, Wang Q, Hua C, Li X (2015) Dextromethorphan provides neuroprotection via anti-inflammatory and anti-excitotoxicity effects in the cortex following traumatic brain injury. Mol Med Rep 12(3):3704–3710. doi:10.3892/mmr.2015.3830
321. Taylor CP, Traynelis SF, Siffert J, Pope LE, Matsumoto RR (2016) Pharmacology of dextromethorphan: relevance to dextromethorphan/quinidine (Nuedexta(R)) clinical use. Pharmacol Ther. doi:10.1016/j.pharmthera.2016.04.010
322. Reissig CJ, Carter LP, Johnson MW, Mintzer MZ, Klinedinst MA, Griffiths RR (2012) High doses of dextromethorphan, an NMDA antagonist, produce effects similar to classic hallucinogens. Psychopharmacology 223(1):1–15. doi:10.1007/s00213-012-2680-6
323. Schwartz AR, Pizon AF, Brooks DE (2008) Dextromethorphan-induced serotonin syndrome. Clin Toxicol (Phila) 46(8):771–773
324. Shin EJ, Nah SY, Chae JS, Bing G, Shin SW, Yen TP, Baek IH, Kim WK, Maurice T, Nabeshima T, Kim HC (2007) Dextromethorphan attenuates trimethyltin-induced neurotoxicity via sigma1 receptor activation in rats. Neurochem Int 50(6):791–799. doi:10.1016/j.neuint.2007.01.008
325. Tian G, Lai L, Guo H, Lin Y, Butchbach ME, Chang Y, Lin CL (2007) Translational control of glial glutamate transporter EAAT2 expression. J Biol Chem 282(3):1727–1737. doi:10.1074/jbc.M609822200
326. Lee E, Sidoryk-Wegrzynowicz M, Farina M, Rocha JB, Aschner M (2013) Estrogen attenuates manganese-induced glutamate transporter impairment in rat primary astrocytes. Neurotox Res 23(2):124–130. doi:10.1007/s12640-012-9347-2
327. Karki P, Webb A, Smith K, Lee K, Son DS, Aschner M, Lee E (2013) cAMP response element-binding protein (CREB) and nuclear factor kappaB mediate the tamoxifen-induced upregulation of glutamate transporter 1 (GLT-1) in rat astrocytes. J Biol Chem 288(40):28975–28986. doi:10.1074/jbc.M113.483826
328. Ganel R, Ho T, Maragakis NJ, Jackson M, Steiner JP, Rothstein JD (2006) Selective upregulation of the glial Na+-dependent glutamate transporter GLT1 by a neuroimmunophilin ligand results in neuroprotection. Neurobiol Dis 21(3):556–567. doi:10.1016/j.nbd.2005.08.014
329. Ouyang L, Tian Y, Bao Y, Xu H, Cheng J, Wang B, Shen Y, Chen Z, Lyu J (2016) Carnosine decreased neuronal cell death through targeting glutamate system and astrocyte mitochondrial bioenergetics in cultured neuron/astrocyte exposed to OGD/recovery. Brain Res Bull 124:76–84. doi:10.1016/j.brainresbull.2016.03.019
330. Guillem AMAM (2015) Methylphenidate increases glutamate uptake in Bergmann glial cells. Neurochem Res 40(11):2317–2324. doi:10.1007/s11064-015-1721-z
331. Zelenaia O, Schlag BD, Gochenauer GE, Ganel R, Song W, Beesley JS, Grinspan JB, Rothstein JD, Robinson MB (2000) Epidermal growth factor receptor agonists increase expression of glutamate transporter GLT-1 in astrocytes through pathways dependent on phosphatidylinositol 3-kinase and transcription factor NF-kappaB. Mol Pharmacol 57(4):667–678
332. Figiel M, Engele J (2000) Pituitary adenylate cyclase-activating polypeptide (PACAP), a neuron-derived peptide regulating glial glutamate transport and metabolism. J Neurosci 20(10):3596–3605
333. Pignataro G, Capone D, Polichetti G, Vinciguerra A, Gentile A, Di Renzo G, Annunziato L (2011) Neuroprotective, immunosuppressant and antineoplastic properties of mTOR inhibitors: current and emerging therapeutic options. Curr Opin Pharmacol 11(4):378–394. doi:10.1016/j.coph.2011.05.003
334. Wu X, Kihara T, Akaike A, Niidome T, Sugimoto H (2010) PI3K/Akt/mTOR signaling regulates glutamate transporter 1 in astrocytes. Biochem Biophys Res Commun 393(3):514–518. doi:10.1016/j.bbrc.2010.02.038
335. Wong M (2013) Mammalian target of rapamycin (mTOR) pathways in neurological diseases. Biomed J 36(2):10.4103/2319–14170.110365. doi:10.4103/2319-4170.110365

336. Hosseinzadeh Z, Bhavsar SK, Sopjani M, Alesutan I, Saxena A, Dermaku-Sopjani M, Lang F (2011) Regulation of the glutamate transporters by JAK2. Cell Physiol Biochem 28(4):693–702. doi:10.1159/000335763
337. Jimenez E, Nunez E, Ibanez I, Draffin JE, Zafra F, Gimenez C (2014) Differential regulation of the glutamate transporters GLT-1 and GLAST by GSK3beta. Neurochem Int 79:33–43. doi:10.1016/j.neuint.2014.10.003
338. Fuente-Martin E, Garcia-Caceres C, Granado M, de Ceballos ML, Sanchez-Garrido MA, Sarman B, Liu ZW, Dietrich MO, Tena-Sempere M, Argente-Arizon P, Diaz F, Argente J, Horvath TL, Chowen JA (2012) Leptin regulates glutamate and glucose transporters in hypothalamic astrocytes. J Clin Invest 122(11):3900–3913. doi:10.1172/JCI64102
339. Fuente-Martin E, Garcia-Caceres C, Argente-Arizon P, Diaz F, Granado M, Freire-Regatillo A, Castro-Gonzalez D, Ceballos ML, Frago LM, Dickson SL, Argente J, Chowen JA (2016) Ghrelin regulates glucose and glutamate transporters in hypothalamic astrocytes. Sci Rep 6:23673. doi:10.1038/srep23673
340. Matos-Ocasio F, Hernandez-Lopez A, Thompson KJ (2014) Ceftriaxone, a GLT-1 transporter activator, disrupts hippocampal learning in rats. Pharmacol Biochem Behav 122:118–121. doi:10.1016/j.pbb.2014.03.011
341. Shimada F, Shiga Y, Morikawa M, Kawazura H, Morikawa O, Matsuoka T, Nishizaki T, Saito N (1999) The neuroprotective agent MS-153 stimulates glutamate uptake. Eur J Pharmacol 386(2–3):263–270
342. Umemura K, Gemba T, Mizuno A, Nakashima M (1996) Inhibitory effect of MS-153 on elevated brain glutamate level induced by rat middle cerebral artery occlusion. Stroke 27(9):1624–1628
343. Kawazura H, Takahashi Y, Shiga Y, Shimada F, Ohto N, Tamura A (1997) Cerebroprotective effects of a novel pyrazoline derivative, MS-153, on focal ischemia in rats. Jpn J Pharmacol 73(4):317–324
344. Uenishi H, Huang CS, Song JH, Marszalec W, Narahashi T (1999) Ion channel modulation as the basis for neuroprotective action of MS-153. Ann N Y Acad Sci 890:385–399
345. Nakagawa T, Ozawa T, Shige K, Yamamoto R, Minami M, Satoh M (2001) Inhibition of morphine tolerance and dependence by MS-153, a glutamate transporter activator. Eur J Pharmacol 419(1):39–45
346. Abekawa T, Honda M, Ito K, Inoue T, Koyama T (2002) Effect of MS-153 on the development of behavioral sensitization to locomotion- and ataxia-inducing effects of phencyclidine. Psychopharmacology 160(2):122–131.doi:10.1007/s00213-001-0958-1
347. Abekawa T, Honda M, Ito K, Inoue T, Koyama T (2002) Effect of MS-153 on the development of behavioral sensitization to stereotypy-inducing effect of phencyclidine. Brain Res 926(1–2):176–180
348. Li X, Inouei T, Abekawai T, YiRui F, Koyama T (2004) Effect of MS-153 on the acquisition and expression of conditioned fear in rats. Eur J Pharmacol 505(1–3):145–149. doi:10.1016/j.ejphar.2004.10.041
349. Nakagawa T, Fujio M, Ozawa T, Minami M, Satoh M (2005) Effect of MS-153, a glutamate transporter activator, on the conditioned rewarding effects of morphine, methamphetamine and cocaine in mice. Behav Brain Res 156(2):233–239. doi:10.1016/j.bbr.2004.05.029
350. Aal-Aaboda M, Alhaddad H, Osowik F, Nauli SM, Sari Y (2015) Effects of (R)-(−)-5-methyl-1-nicotinoyl-2-pyrazoline on glutamate transporter 1 and cysteine/glutamate exchanger as well as ethanol drinking behavior in male, alcohol-preferring rats. J Neurosci Res 93(6):930–937. doi:10.1002/jnr.23554
351. Fontana AC, Fox DP, Zoubroulis A, Mortensen OV, Raghupathi R (2015) Neuroprotective effects of the glutamate transporter activator, MS-153, following traumatic brain injury. J Neurotrauma 33(11):1073–83. doi:10.1089/neu.2015.4079
352. Frizzo ME, Lara DR, Dahm KC, Prokopiuk AS, Swanson RA, Souza DO (2001) Activation of glutamate uptake by guanosine in primary astrocyte cultures. Neuroreport 12(4):879–881
353. Frizzo ME, Schwalm FD, Frizzo JK, Soares FA, Souza DO (2005) Guanosine enhances glutamate transport capacity in brain cortical slices. Cell Mol Neurobiol 25(5):913–921. doi:10.1007/s10571-005-4939-5
354. Moretto MB, Arteni NS, Lavinsky D, Netto CA, Rocha JB, Souza DO, Wofchuk S (2005) Hypoxic-ischemic insult decreases glutamate uptake by hippocampal slices from neonatal rats: prevention by guanosine. Exp Neurol 195(2):400–406. doi:10.1016/j.expneurol.2005.06.005
355. Hansel G, Tonon AC, Guella FL, Pettenuzzo LF, Duarte T, Duarte MM, Oses JP, Achaval M, Souza DO (2014) Guanosine protects

against cortical focal ischemia. Involvement of inflammatory response. Mol Neurobiol. doi:10.1007/s12035-014-8978-0

356. Hansel G, Ramos DB, Delgado CA, Souza DG, Almeida RF, Portela LV, Quincozes-Santos A, Souza DO (2014) The potential therapeutic effect of guanosine after cortical focal ischemia in rats. PLoS One 9(2):e90693. doi:10.1371/journal.pone.0090693
357. Fioravanti M, Flicker L (2001) Efficacy of nicergoline in dementia and other age associated forms of cognitive impairment. Cochrane Database Syst Rev (4):Cd003159. doi:10.1002/14651858.cd003159
358. Nishida A, Iwata H, Kudo Y, Kobayashi T, Matsuoka Y, Kanai Y, Endou H (2004) Nicergoline enhances glutamate uptake via glutamate transporters in rat cortical synaptosomes. Biol Pharm Bull 27(6):817–820
359. Fioravanti M, Nakashima T, Xu J, Garg A (2014) A systematic review and meta-analysis assessing adverse event profile and tolerability of nicergoline. BMJ Open 4(7):e005090. doi:10.1136/bmjopen-2014-005090
360. McIntosh TK, Smith DH, Voddi M, Perri BR, Stutzmann JM (1996) Riluzole, a novel neuroprotective agent, attenuates both neurologic motor and cognitive dysfunction following experimental brain injury in the rat. J Neurotrauma 13(12):767–780
361. Zhang C, Raghupathi R, Saatman KE, Smith DH, Stutzmann JM, Wahl F, McIntosh TK (1998) Riluzole attenuates cortical lesion size, but not hippocampal neuronal loss, following traumatic brain injury in the rat. J Neurosci Res 52(3):342–349
362. Azbill RD, Mu X, Springer JE (2000) Riluzole increases high-affinity glutamate uptake in rat spinal cord synaptosomes. Brain Res 871(2):175–180
363. Mu X, Azbill RD, Springer JE (2000) Riluzole and methylprednisolone combined treatment improves functional recovery in traumatic spinal cord injury. J Neurotrauma 17(9):773–780
364. Brothers HM, Bardou I, Hopp SC, Kaercher RM, Corona AW, Fenn AM, Godbout JP, Wenk GL (2013) Riluzole partially rescues age-associated, but not LPS-induced, loss of glutamate transporters and spatial memory. J Neuroimmune Pharmacol 8(5):1098–1105. doi:10.1007/s11481-013-9476-2
365. Fumagalli E, Funicello M, Rauen T, Gobbi M, Mennini T (2008) Riluzole enhances the activity of glutamate transporters GLAST, GLT1 and EAAC1. Eur J Pharmacol 578(2–3):171–176. doi:10.1016/j.ejphar.2007.10.023
366. Frizzo ME, Dall'Onder LP, Dalcin KB, Souza DO (2004) Riluzole enhances glutamate uptake in rat astrocyte cultures. Cell Mol Neurobiol 24(1):123–128
367. Dall'Igna OP, Bobermin LD, Souza DO, Quincozes-Santos A (2013) Riluzole increases glutamate uptake by cultured C6 astroglial cells. Int J Dev Neurosci 31(7):482–486. doi:10.1016/j.ijdevneu.2013.06.002
368. Dagci T, Yilmaz O, Taskiran D, Peker G (2007) Neuroprotective agents: is effective on toxicity in glial cells? Cell Mol Neurobiol 27(2):171–177. doi:10.1007/s10571-006-9082-4
369. Waibel S, Reuter A, Malessa S, Blaugrund E, Ludolph AC (2004) Rasagiline alone and in combination with riluzole prolongs survival in an ALS mouse model. J Neurol 251(9):1080–1084. doi:10.1007/s00415-004-0481-5
370. Gurney ME, Cutting FB, Zhai P, Doble A, Taylor CP, Andrus PK, Hall ED (1996) Benefit of vitamin E, riluzole, and gabapentin in a transgenic model of familial amyotrophic lateral sclerosis. Ann Neurol 39(2):147–157. doi:10.1002/ana.410390203
371. Del Signore SJ, Amante DJ, Kim J, Stack EC, Goodrich S, Cormier K, Smith K, Cudkowicz ME, Ferrante RJ (2009) Combined riluzole and sodium phenylbutyrate therapy in transgenic amyotrophic lateral sclerosis mice. Amyotrop Lateral Scler 10(2):85–94. doi:10.1080/17482960802226148
372. Traynor BJ, Bruijn L, Conwit R, Beal F, O'Neill G, Fagan SC, Cudkowicz ME (2006) Neuroprotective agents for clinical trials in ALS: a systematic assessment. Neurology 67(1):20–27. doi:10.1212/01.wnl.0000223353.34006.54
373. Miller RG, Mitchell JD, Moore DH (2012) Riluzole for amyotrophic lateral sclerosis (ALS)/motor neuron disease (MND). Cochrane Database Syst Rev 3:CD001447. doi:10.1002/14651858.CD001447.pub3
374. Wilson JR, Fehlings MG (2014) Riluzole for acute traumatic spinal cord injury: a promising neuroprotective treatment strategy. World Neurosurg 81(5–6):825–829. doi:10.1016/j.wneu.2013.01.001
375. Pereira AC, Gray JD, Kogan JF, Davidson RL, Rubin TG, Okamoto M, Morrison JH, McEwen BS (2016) Age and Alzheimer's disease gene expression profiles reversed by the glutamate modulator riluzole. Mol Psychiatry. doi:10.1038/mp.2016.33
376. Whitcomb DJ, Molnar E (2015) Is riluzole a new drug for Alzheimer's disease? J Neurochem 135(2):207–209. doi:10.1111/jnc.13260
377. Fink K, Dooley DJ, Meder WP, Suman-Chauhan N, Duffy S, Clusmann H, Göthert M (2002) Inhibition of neuronal Ca2+ influx

by gabapentin and pregabalin in the human neocortex. Neuropharmacology 42(2):229–236. doi:10.1016/S0028-3908(01)00172-1

378. Ryu JH, Lee PB, Kim JH, Do SH, Kim CS (2012) Effects of pregabalin on the activity of glutamate transporter type 3. Br J Anaesth 109(2):234–239. doi:10.1093/bja/aes120
379. Hendrich J, Van Minh AT, Heblich F, Nieto-Rostro M, Watschinger K, Striessnig J, Wratten J, Davies A, Dolphin AC (2008) Pharmacological disruption of calcium channel trafficking by the alpha2delta ligand gabapentin. Proc Natl Acad Sci U S A 105(9):3628–3633. doi:10.1073/pnas.0708930105
380. Yoshizumi M, Eisenach JC, Hayashida K (2012) Riluzole and gabapentinoids activate glutamate transporters to facilitate glutamate-induced glutamate release from cultured astrocytes. Eur J Pharmacol 677(1–3):87–92. doi:10.1016/j.ejphar.2011.12.015
381. Gil YS, Kim JH, Kim CH, Han JI, Zuo Z, Baik HJ (2015) Gabapentin inhibits the activity of the rat excitatory glutamate transporter 3 expressed in Xenopus oocytes. Eur J Pharmacol 762:112–117. doi:10.1016/j.ejphar.2015.05.038
382. Suto T, Severino AL, Eisenach JC, Hayashida K (2014) Gabapentin increases extracellular glutamatergic level in the locus coeruleus via astroglial glutamate transporter-dependent mechanisms. Neuropharmacology 81:95–100. doi:10.1016/j.neuropharm.2014.01.040
383. Kimura M, Eisenach JC, Hayashida KI (2016) Gabapentin loses efficacy over time after nerve injury in rats: role of glutamate transporter-1 in the locus coeruleus. Pain. doi:10.1097/j.pain.0000000000000608
384. de O Beleboni R, Pizzo AB, Fontana AC, Carolino RO, Coutinho-Netto J, Dos Santos WF (2004) Spider and wasp neurotoxins: pharmacological and biochemical aspects. Eur J Pharmacol 493(1–3):1–17. doi:10.1016/j.ejphar.2004.03.049
385. Fontana AC, de Oliveira Beleboni R, Wojewodzic MW, Ferreira Dos Santos W, Coutinho-Netto J, Grutle NJ, Watts SD, Danbolt NC, Amara SG (2007) Enhancing glutamate transport: mechanism of action of Parawixin1, a neuroprotective compound from Parawixia bistriata spider venom. Mol Pharmacol 72(5):1228–1237. doi:10.1124/mol.107.037127
386. Fontana AC, Guizzo R, de Oliveira Beleboni R, Meirelles ESAR, Coimbra NC, Amara SG, dos Santos WF, Coutinho-Netto J (2003) Purification of a neuroprotective component of Parawixia bistriata spider venom that enhances glutamate uptake. Br J Pharmacol 139(7):1297–1309. doi:10.1038/sj.bjp.0705352
387. Estrada G, Villegas E, Corzo G (2007) Spider venoms: a rich source of acylpolyamines and peptides as new leads for CNS drugs. Nat Prod Rep 24(1):145–161. doi:10.1039/b603083c

Chapter 8

Drugs to Tune Up Glutamatergic Systems: Modulators of Glutamate Metabotropic Receptors

Kathy Sengmany and Karen J. Gregory

Abstract

The metabotropic glutamate (mGlu) receptors are Class C G protein-coupled receptors that offer promising potential therapeutic targets for multiple central nervous system (CNS) disorders. Dysfunction or dysregulation of the glutamatergic system, the main excitatory neurotransmitter system in the CNS, is thought to be associated with multiple CNS disorders including schizophrenia, depression, anxiety, Fragile X syndrome, Parkinson's disease, and refractory chronic pain. Here, we describe drug discovery and development approaches for targeting mGlu receptors, with particular focus on allosteric modulation and biased agonism. Binding to allosteric sites that are topographically distinct from the endogenous binding site, allows for increased selectivity between receptor subtypes, and maintaining tonal and temporal glutamatergic responses. Biased agonism allows for further specialization, with the potential to design drugs that target desired receptor responses at the exclusion of those leading to adverse effects. Collectively, these newer paradigms of drug action offer the promise for discovery of therapeutics with favorable outcomes and minimized adverse effects within the delicate CNS environment.

Key words Metabotropic glutamate receptor, Allosteric modulator, Biased agonism

1 Introduction

Glutamate (Glu), the main excitatory neurotransmitter in the central nervous system (CNS), acts via three ionotropic, *N*-methyl-D-aspartate (NMDA), α-amino-3-hydroxy-5-methyl-4-isoxazoleproprionic acid (AMPA) and kainate, and eight metabotropic Glu (mGlu) receptors [1, 2]. The eight mGlu receptors ($mGlu_{1-8}$) are further divided into three groups based on pharmacology, signal transduction mechanisms, and sequence homology. Receptors within groups display 70% sequence identity, while between groups, homology decreases to approximately 45% [3]. Group I mGlu receptors ($mGlu_1$ and $mGlu_5$) preferentially activate phospholipase C (PLC) via $G\alpha_{q/11}$ coupling leading to intracellular Ca^{2+} (iCa^{2+}) mobilization. On the other hand, group II ($mGlu_2$ and $mGlu_3$) and III ($mGlu_4$, $mGlu_6$, $mGlu_7$ and $mGlu_8$) receptors

Sandrine Parrot and Luc Denoroy (eds.), *Biochemical Approaches for Glutamatergic Neurotransmission*, Neuromethods, vol. 130,
DOI 10.1007/978-1-4939-7228-9_8,

preferentially couple to $G_{i/o}$, resulting in inhibition of adenylate cyclase and subsequently decreased cAMP production [3, 4]. With the exception of $mGlu_6$, mostly found within the retina, all mGlu receptors are expressed ubiquitously throughout the CNS and are crucial for neuronal plasticity and function.

While the mGlu receptors have been classified based on canonical coupling to their respective G proteins and signaling pathways, it is becoming increasingly clear that these receptors, like many G protein-coupled receptors (GPCRs), are promiscuously coupled. In addition to canonical $G\alpha_{q/11}$ coupling that leads to modulation of ion channels and phosphorylation of protein kinase C (PKC), mitogen-activated protein kinases (MAPK) and phospholipase A_2 (PLA_2) [3, 5], the group I mGlu receptors can also mediate extracellular signal-regulated kinase 1/2 (ERK1/2) phosphorylation independent of PKC, phosphoinositide 3-kinase (PI3K) and iCa^{2+} mobilization [6, 7]. Evidence for convergence/divergence of intracellular signaling pathways is significant given that both the ERK-MAPK and PI3K-mTOR cascades [8, 9] are involved in the balance between long-term depression and long-term potentiation [10]. Further, in the post-synaptic density, group I receptors are also physically linked to ionotropic Glu receptors mediated by scaffolding proteins such as Homer, SHANK3 [11–13]. In recombinant cell lines, both $mGlu_1$ and $mGlu_5$ couple to alternative G proteins ($G_{i/o}$ and/or G_s) [14–17], however, cellular context can influence whether or not coupling to G proteins other than $G_{q/11}$ is evident [18, 19]. Importantly, pleiotropic coupling is not limited to the group I receptors. For example, group III receptors are generally found presynaptically and modulate neurotransmitter release via interactions with ion channels [20, 21], namely high-voltage activated Ca^{2+} channel inhibition [22], K^+ channel activation [23–25], in a manner that is generally independent of changes in cAMP, although this may not always be the case [26]. Modulation of vesicular cycling and release by group III receptors can be dependent on cAMP/PKA cascade inhibition [27]. It is clear that activation of different mGlu subtypes results in a multitude of coordinated converging and divergent cellular responses. The mGlu receptors are critical for maintaining the balance between excitatory and inhibitory activity within the brain. As such, glutamatergic system dysfunction has been associated with a wide variety of CNS disorders including schizophrenia, anxiety, depression, autism, and chronic pain [28–31]. Here, we describe several strategies to pharmacologically target the glutamatergic system, with particular focus on allosteric modulation of metabotropic Glu receptors, and the newer paradigm of biased agonism. We review therapeutic indications for mGlu receptor allosteric modulators in CNS disorders associated with Glu dysfunction, encompassing both preclinical and clinical studies.

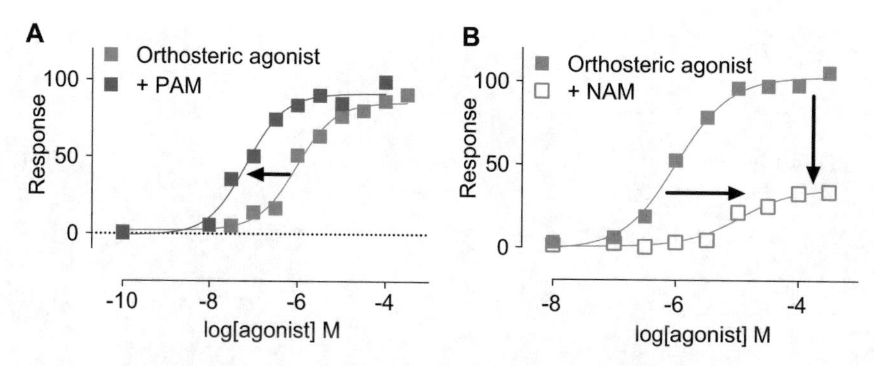

Fig. 1 Simulations of allosteric interactions detected using a functional response to orthosteric agonist. (**a**) In the presence of a positive allosteric modulator (PAM) orthosteric agonist potency is increased, shifting to the left ($\alpha\beta > 1$). (**b**) The presence a negative allosteric modulator (NAM) decreases the potency of orthosteric agonist ($0 < \alpha\beta < 1$), and may also diminish the agonist maximal response

2 Allosteric Modulation of mGlu Receptors

Traditional drug development has largely focused on understanding the endogenous (orthosteric) ligand binding pocket and designing small molecules to bind competitively to either activate or block the desired receptor. However, a greater understanding of GPCR structure and conformational dynamism has allowed a paradigm shift toward designing drugs that bind to alternate pockets distinct from the orthosteric binding site [32]. These topographically distinct pockets are referred to as allosteric sites. Simultaneous receptor occupation with an allosteric ligand can modulate the binding of, or response to, an endogenous ligand, a property referred to as cooperativity [33]. Positive allosteric modulators (PAMs) enhance, while negative allosteric modulators (NAMs) diminish orthosteric agonist binding/efficacy (Fig. 1). Neutral allosteric ligands (NALs) bind to allosteric sites without altering orthosteric agonist activity/binding. In addition, allosteric modulators may also possess intrinsic efficacy (either inverse or positive) in their own right, such compounds are classified first on the basis of the allosteric effect followed by intrinsic efficacy, for e.g. PAM-agonists, NAM-inverse agonists [34]. Discovery efforts for mGlu receptor allosteric modulators have been particularly fruitful over the past two decades. A wealth of different chemotypes comprising the full spectrum of allosteric ligand pharmacology is now available (Table 1).

Advantages of allosteric ligands are threefold; first, allosteric sites generally show lower sequence conservation than orthosteric pockets, allowing for greater selectivity between receptor subtypes. Second, as pure allosteric modulators require orthosteric agonist to have an effect there is an inherent potential to fine-tune receptor

Table 1
Molecular pharmacology and therapeutic potential of select mGlu ligands

Compound	Chemical name	Target	Activity	Therapeutic/biological significance and refs
Nonselective mGlu ligands				
L-Glutamate	(*S*)-1-Aminopropane-1,3-dicarboxylic acid	$mGlu_{1-3,5} > mGlu_{4,6-8}$	Orthosteric agonist	Pharmacological tool [3, 4]
(1*S*,3*R*)-ACPD	(1*S*,3*R*)-1-Aminocyclopentane-1,3-dicarboxylic acid	$mGlu_{1-3,5} > mGlu_{4,6-8}$	Orthosteric agonist	Pharmacological tool [3, 4]
Group I mGlu ligands				
(*S*)-3,5-DHPG	(*S*)-3,5-Dihydroxyphenylglycine	$mGlu_{1/5}$	Orthosteric agonist	Pharmacological tool [3, 4]
L-Quisqualate	(2*S*)-2-Amino-3-(3,5-dioxo-1,2,4-oxadiazolidin-2-yl)propanoic acid	$mGlu_{1/5}$ #	Orthosteric agonist	Pharmacological tool [3, 4]
FITM	4-Fluoro-*N*-(4-(6-(isopropylamino) pyrimidin-4-yl) thiazol-2-yl)-*N*-methylbenzamide	$mGlu_1$	NAM	Pharmacological tool [35]
HTL14242	(3-Chloro-5-[6-(5-fluoropyridin-2-yl)pyrimidin-4-yl]benzonitrile)	$mGlu_5$	NAM	Lead compound [36]
CPPHA	*N*-[4-Chloro-2-[(1,3-dihydro-1,3-dioxo-2*H*-isoindol-2-yl)methyl] phenyl]-2-hydroxybenzamide	$mGlu_5 > mGlu_1 > mGlu_{4/8}$	PAM: IP_1, iCa^{2+} Agonist: pERK1/2	Pharmacological tool, second site PAM [37]
CPCCOEt	7-(Hydroxyimino)cyclopropa[*b*] chromen-1a-carboxylate ethyl ester	$mGlu_1$	NAM	Pain [38, 39]
MPEP	2-Methyl-6-(phenylethynyl)pyridine hydrochloride	$mGlu_5 > mGlu_4$	NAM: iCa^{2+}, pERK1/2 Inverse agonist: IP_1	Anxiety, depression, pain, addiction, PD [40–44]
JNJ16259685	3,4-Dihydro-2H-pyrano [2, 3]b quinolin-7-yl)(cis-4-methoxycyclohexyl) methanone	$mGlu_1$	NAM	Pain [45]

CDPPB	3-Cyano-*N*-(1,3-diphenyl-1*H*-pyrazol-5-yl)benzamide	$mGlu_5$	Agonism: IP_1 > pERK1/2 ≫ iCa^{2+} [a]	Psychosis [46]
VU0360172	*N*-Cyclobutyl-6-((3-fluorophenyl) ethynyl)nicotinamide	$mGlu_5$	Agonism: IP_1 > pERK1/2[a] PAM: iCa^{2+}	Psychosis [46]
DPFE	1-(4-(2,4-Difluorophenyl)piperazin-1-yl)-2-((4-fluorobenzyl)oxy) ethanone	$mGlu_5$	Agonism: IP_1 ≫ iCa^{2+}; biphasic pERK1/2[a]	Psychosis, cognition [46]
VU0424465	5-[2-(3-Fluorophenyl)ethynyl]-*N*-[(2R)-3-hydroxy-3-methylbutan-2-yl]pyridine-2-carboxamide	$mGlu_5$	Agonism: IP_1 > pERK1/2 ≫ iCa^{2+} [a]	Seizures [46, 47]
VU0403602	*N*-Cyclobutyl-5-((3-fluorophenyl) ethynyl)picolinamide	$mGlu_5$	Agonism: IP_1 > pERK1/2 ≫ iCa^{2+} [a]	Metabolite causes seizures [46, 48]
Fenobam	*N*-(3-Chlorophenyl)-*N'*-(4,5-dihydro-1-methyl-4-oxo-1*H*-imidazol-2-yl)urea	$mGlu_5$ #	NAM	Anxiety, depression, FXS, pain, PD [49–56]
Basimglurant	(2-Chloro-4-[1-(4-fluoro-phenyl)-2,5-dimethyl-1H-imidazol-4-ylethynyl]-pyridine)	$mGlu_5$	NAM	Major depressive disorder [57]
Mavoglurant	Methyl (3aR,4S,7aR)-4-hydroxy-4-[2-(3-methylphenyl) ethynyl]-3,3a,5,6,7,7a–hexahydro-2H-indole-1-carboxylate	$mGlu_5$	NAM	FXS, PD, Huntington's disease [58–61]
CTEP	2-Chloro-4-((2,5-dimethyl-1-(4-(trifluoromethoxy)phenyl)-1H–imidazol-4-yl)ethynyl)pyridine	$mGlu_5$	NAM	FXS [62–64]

(continued)

Table 1 (continued)

Compound	Chemical name	Target	Activity	Therapeutic/biological significance and refs
Nonselective mGlu ligands				
VU0477573	*N*,*N*-Diethyl-5-((3-fluorophenyl) ethynyl)picolinamide	$mGlu_5$	Limited NAM	Anxiety [65]
Group II mGlu ligands				
LY354740	(1S,2S,5R,6S)-2-aminobicyclo[3.1.0] hexane-2,6-dicarboxylate monohydrate	$mGlu_{2/3}$	Orthosteric agonist	Psychosis, anxiety, addiction [66–68]
LY379268	(1*R*,4*R*,5*S*,6*R*)-4-Amino-2-oxabicyclo[3.1.0]hexane-4,6-dicarboxylic acid	$mGlu_{2/3}$	Orthosteric agonist	Psychosis [66]
LY341495	(2*S*)-2-Amino-2-[(1*S*,2*S*)-2-carboxycycloprop-1-yl]-3-(xanth-9-yl) propanoic acid	$mGlu_{2/3}$	Orthosteric antagonist	Psychosis [66]
BINA	3′-[[(2-Cyclopentyl-2,3-dihydro-6,7-dimethyl-1-oxo-1*H*-inden-5-yl) oxy]methyl]-[1,1′-biphenyl]-4-carboxylic acid	$mGlu_2$	PAM	Psychosis, anxiety [28, 69]
LY487379	*N*-(4-(2-Methoxyphenoxy) phenyl)-N-(2,2,2-trifluoroethylsulfonyl)-pyrid-3-ylmethylamine	$mGlu_2$	PAM	Psychosis [28]
Group III mGlu ligands				
ACPT-I	(1*S*,3*R*,4*S*)-1-Aminocyclopentane-1,3,4-tricarboxylic acid	Group III	Orthosteric agonist	Pain [70]

LSP4-2022	(2*S*)-2-Amino-4-({[4-(carboxymethoxy)phenyl](hydroxy)methyl}-(hydroxy)phosphoryl)butanoic acid	$mGlu_4$	Orthosteric agonist	Pain [70–72]
PHCCC	*N*-Phenyl-7-(hydroxyimino)cyclopropa[*b*]chromen-1a-carboxamide	$mGlu_4 > mGlu_1$	PAM: iCa^{2+} > GIRK[b]	Pain, PD [71–75]
VU0155041	*cis*-2-[[(3,5-Dichlorophenyl)amino]carbonyl]cyclohexane carboxylic acid	$mGlu_4$	PAM	PD [76–78]
ADX88178	4-Methyl-*N*-[5-methyl-4-(1H–pyrazol-4-yl)-1,3-thiazol-2-yl]pyrimidin-2-amine	$mGlu_4$	PAM	PD [76–78]
AF21934	(1S,2R)-2-[(aminooxy)methyl]-N-(3,4-dichlorophenyl)cyclohexane-1-carboxamide	$mGlu_4$	PAM	Anxiety, PD [76–78]

has activity at non-mGlu receptors (ionotropic Glu receptors for L-quisqualate, A_3 adenosine receptors for fenobam)
[a]Bias relative to DHPG
[b]Bias relative to glutamate

activity in a temporal and spatially refined manner, that simply cannot be achieved by a synthetic orthosteric ligand [79]. However, pure allosteric modulators that have no intrinsic activity and require endogenous neurotransmitter may also present as a disadvantage. For example, in disease states that have a diminished endogenous neurotransmitter levels, such as dopamine loss in Parkinson's disease [80], pure allosteric modulators may not have efficacy. Third, the cooperativity between allosteric modulators and orthosteric ligands is saturable, and thus there is reduced risk in the case of overdose. Overall, targeting allosteric binding pockets offers a promising and viable option to develop CNS therapeutics with desirable outcomes and minimal adverse effects.

2.1 Quantifying Allosteric Interactions

As alluded to earlier, allosteric modulators can influence both orthosteric ligand affinity and efficacy, which has the capacity to result in complex pharmacological profiles. Multiple models have emerged to facilitate quantification of GPCR allosteric interactions. The simplest model to describe the influence of an allosteric modulator on the binding on an orthosteric ligand to its receptor is the allosteric ternary complex model (ATCM) [81–83]. The simultaneous occupation of a receptor by two ligands influences affinity, denoted by the cooperativity factor α, and is governed by ligand concentrations, equilibrium dissociation constants K_A and K_B of orthosteric (A) and allosteric (B) ligands respectively [84, 85] (Fig. 2). Affinity modulation is reciprocal, the binding of an allosteric ligand stabilizes receptor conformations that increase or decrease orthosteric ligand affinity and vice versa. Increasingly, it has become apparent that allosteric interactions may also influence GPCR activation states, in addition to, or exclusive of, effects on affinity. In order to accommodate allosteric modulation of efficacy the simple ATCM was extended to an allosteric two-state model, and the now widely applied operational model of allosterism [81, 85]. The operational model of allosterism amalgamates the ATCM and Black and Leff's operational model of agonism [86]. Within this framework the effect of an allosteric ligand on both orthosteric agonist affinity (α) and efficacy (β) is accommodated [85] (Fig. 2). PAMs are defined as having a cooperativity ($\alpha\beta$) value greater than 1, NAMs, a cooperativity value between 0 and 1, and NALs are defined as having neutral cooperativity (α and/or β = 1). It is important to note that an allosteric modulator may influence affinity and efficacy to different degrees and in different directions [87]. Furthermore, there is the potential that cooperativity drives subtype selectivity rather than affinity. For example, MPEP is an $mGlu_5$ NAM of Glu efficacy for iCa^{2+} mobilization and ERK1/2 phosphorylation, and DHPG/quisqualate mediated IP_1 accumulation [40–43]. On the other hand, MPEP is an $mGlu_4$ PAM-agonist of L-AP4 mediated iCa^{2+} mobilization [44]. The operational model of allosterism allows for

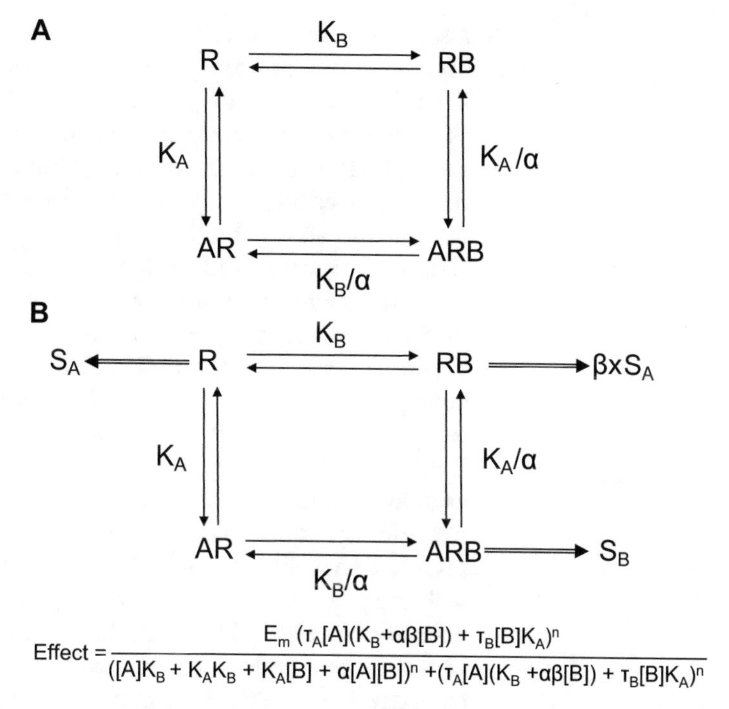

Fig. 2 Ternary complex models for G protein-coupled receptors. (**a**) The simplest framework to describe allosteric interactions is the ATCM, where an orthosteric ligand (*A*) and allosteric modulator (*B*) can simultaneously bind to a receptor (*R*) influencing the respective equilibrium dissociation constants (K_A and K_B), as described by the cooperativity factor α. (**b**) To describe effects on efficacy, the ATCM was incorporated in to Black and Leff's operational model of agonism. The change in stimulus response coupling (S_A) in the presence of allosteric modulator is described by the scaling factor β. Intrinsic allosteric agonist efficacy is also accommodated

delineation of effects on affinity versus efficacy. Further, intrinsic allosteric agonist efficacy is also accommodated as denoted by τ_B [34, 79]. For the mGlu receptors this is an important consideration since for multiple chemotypes and at multiple subtypes, allosteric agonist activity is evident [42, 43, 88]. Thus rigorous quantification of allosteric ligands remains pertinent to accurately define pharmacology of mGlu allosteric ligands and facilitate comparisons between chemotypes and targets [34].

2.2 Validation and Detection of Allosteric Modulators

In addition to application of analytical models to quantify allosteric interactions, robust experimental approaches are required to define and validate allosteric interactions. Radioligand binding assays are commonly used to verify allosteric agonist binding of a studied ligand. This is readily done for $mGlu_1$, $mGlu_2$, and $mGlu_5$, where allosteric radioligands have been developed [89–92], such that novel allosteric ligands can be validated via inhibition binding assays. Allosteric ligand affinity can be directly determined from inhibition binding assays via application of the standard Cheng-Prusoff equation [93]. For many GPCRs the gold-standard approach to definitively demonstrate an allosteric mechanism is through demonstration of affinity modulation between an unlabeled allosteric ligand and radiolabelled orthosteric ligand using dissociation kinetic binding assays [94]. A change in ligand dissociation rate from the receptor when simultaneously occupied is unambiguously demonstrative of an allosteric interaction. For mGlu receptors the prevailing approach has been to determine changes in orthosteric radioligand affinity (e.g. [^{3}H]quisqualate for Group I mGlu receptors, and [^{3}H]LY354740 for Group II mGlu receptors) rather than kinetics [89, 95, 96]. The mGlu$_5$ PAMs, DFB (or [(3-Fluorophenyl)methylene]hydrazone-3-fluorobenzaldehyde) and CDPPB, however, enhance mGlu$_5$ orthosteric agonist affinity (quisqualate) in rat cortical astrocytes [42], while mGlu$_2$ PAMs also enhance affinity of several orthosteric ligands including Glu and LY354740 [97]. Affinity modulation has also been shown with [^{3}H]quisqualate and [^{3}H]DCG-IV (i.e. [^{3}H]-2-(2,3-dicarboxycyclopropyl)glycine) by mGlu$_1$ and mGlu$_2$ PAMs respectively [95, 98]. Furthermore, several mGlu allosteric ligands are thought to be efficacy-only modulators, including the mGlu$_5$ PAM CDPPB (with Glu), the mGlu$_5$ NAM MPEP, and mGlu$_1$ NAM CPCCOEt [99–102]. Functional assays are required to detect and validate allosteric modulation of efficacy.

Allosteric efficacy modulation can be determined via cell-based functional assays [103]. Rapid screening of small allosteric molecules commonly involves administration of allosteric ligand at varying concentrations, with either an EC_{20} orthosteric agonist concentration for PAMs, or an EC_{80} (submaximal) orthosteric agonist concentration for NAMs [103]. Orthosteric agonist concentrations are chosen to allow for clear determination of modulatory activity, as enhancement of a small response, or reduction of a large response allows for ease of classification as either a PAM or NAM respectively. A triple add paradigm has more recently surpassed the single add protocol, as it allows for simultaneous screening for agonist, PAM, and NAM activity for receptor measures that are detected in real-time [103]. In this protocol, compounds are first added, followed by a second addition of orthosteric agonist EC_{20}, and a third addition of orthosteric agonist. Agonist activity is determined through the first addition, and PAM or NAM activity is determined through the second and third addition of orthosteric ligand respectively [103]. These experimental paradigms provide a quick snapshot of prospective modulators via a potency determination. It is important to consider that modulator potency is a composite measure of allosteric modulator affinity, affinity and efficacy cooperativity, and intrinsic agonist activity. Furthermore, modulator potency is also dependent upon the orthosteric agonist concentration [43, 104]. In order to delineate allosteric ligand affinity, cooperativity, and efficacy, interaction studies with full dose response curves of orthosteric agonist in the presence of a range of allosteric ligand concentrations is required. PAM activity may manifest as an increase in agonist potency that approaches a limit as defined by cooperativity and/or an increase in the maximal agonist response [4, 82]. Conversely, NAM activity will manifest as a progressive right shift in agonist potency that reaches a limit and/or a depression in the agonist maximal response [4, 82]. The magnitude and direction of the shifts in agonist potency correspond to cooperativity whereas the concentrations over which these effects happen are related to modulator affinity. In addition to the saturable nature of allosteric ligand interactions, determination of allosteric binding sites provides a second level of validation for an allosteric mechanism of action.

2.3 Structural Basis of mGlu Allosteric Modulation

In addition to sharing the common 7TM typical of all GPCRs, Class C receptors have a large extracellular N-terminal domain, widely known as the Venus flytrap (VFT) domain. Early homology modeling against bacterial periplasmic amino acid binding proteins, such as the leucine/isoleucine/valine binding protein (LIVBP), suggested the N-terminal domain comprised two distinct lobes that closed around Glu upon ligand binding [105, 106]. Subsequently, several mGlu VFT domain crystal structures have definitively localized the Glu orthosteric binding site within this domain [107, 108]. Of note, receptors with only the soluble N-terminal region retain their ability to bind mGlu agonists [109–112], highlighting the presence of the orthosteric binding pocket within the VFT. Consistent with these findings, modeling suggests Glu binding to only one VFT is required for conformational changes to induce a receptor activation [113]. Linking the N-terminal region to the 7TM is the cysteine rich domain (CRD) with four disulfide bridges [108, 114, 115]. The CRD is crucial in signal transduction from the VFT to the 7TM upon orthosteric ligand binding [108, 110]. To date, the majority of mGlu allosteric ligands bind to the 7TM domain (discussed below).

2.3.1 Common Allosteric Site Within 7TM Domain

Early mGlu receptor chimeras where the VFT of different subtypes were interchanged, provided invaluable information on localization of allosteric ligand binding sites. CPCCOEt was validated as an allosteric modulator via $mGlu_1$ and calcium sensing receptor chimera [38], as well as chimeras with other mGlu subtypes [39]. Numerous mGlu chimeras were consequently used to characterize the binding of a diverse range of mGlu allosteric ligands to the 7TM domain [73, 98, 100, 112]. In addition, truncated receptor constructs also offered insight into allosteric binding locations, with the use of "headless" mGlu receptors—that is, mGlu receptors lacking the extracellular N-terminus that retain a functional

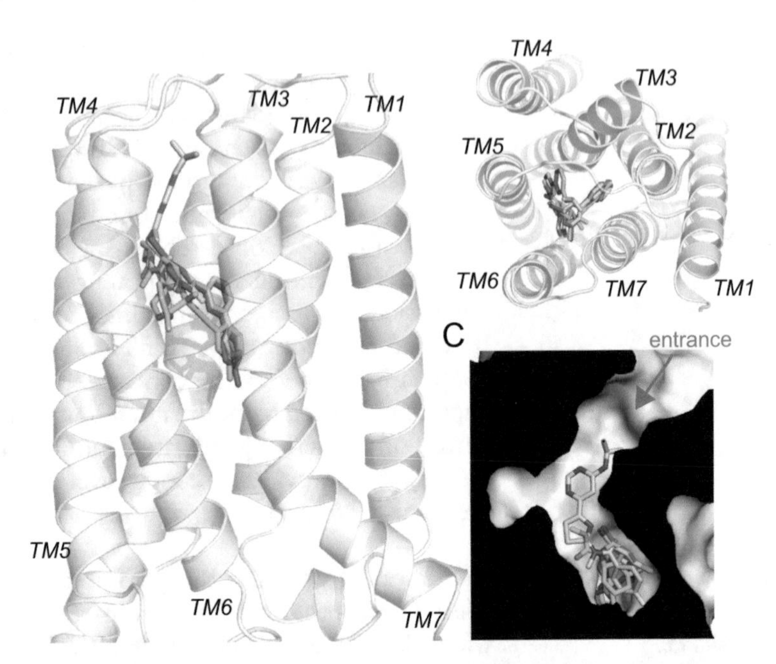

Fig. 3 Recent crystal structures provide new insights into allosteric modulator binding within mGlu receptors. (**a**) Alignment of $mGlu_1$ (PDB ID: 4OR2) and $mGlu_5$ 7TM crystal structures (PDB ID's: 4OO9, 5CGC, 5CGD) reveals the overlapping binding cavity. For clarity the backbone of 4OO9 only is shown in cartoon, ligands are depicted in sticks: *green* is FITM; *cyan* is mavoglurant; *yellow* (5CGC) and *magenta* (5CGD) are HTL14242. (**b**) Top-down view of the binding pocket (*panel A* rotated 90°). (**c**) Slice through of the protein surface reveals the narrow entrance

7TM and C-terminus capable of coupling to intracellular effectors [111]. In these modified $mGlu_{1/5}$ receptors, PAMs have agonist activity, and NAMs were inverse agonists [111, 112], highlighting the fact that these ligands bind within the 7TM, rather than VFT. Subsequently, multiple site-directed mutagenesis-based studies primarily focused on group I receptors demonstrated that residues within the top third of TM3, TM5, TM6 and TM7 were critical for mGlu allosteric ligand interactions within a proposed common allosteric site across the entire Class C GPCR family [116–118].

Recently, X-ray crystal structures of 7TM domains of human $mGlu_1$ and $mGlu_5$ receptor, in complex with the $mGlu_1$ NAM FITM [35] and $mGlu_5$ NAMs mavoglurant [58] and HTL14242 [36] confirmed the location of the proposed mGlu common allosteric binding pocket, otherwise known as the MPEP binding pocket (Fig. 3a and b). The $mGlu_5$ mavoglurant/HTL14242 allosteric binding pocket resides between TM2, TM3, TM5, TM6 and TM7, and has a restricted ~7 Å entrance into the helical bundle due to anchoring of extracellular loop 2 across the top of the 7TMs

[58] (Fig. 3c). The FITM binding pocket in $mGlu_1$ overlaps with that observed in $mGlu_5$, but is located higher within the 7TMs (Fig. 3a), these differences likely allow subtype selective targeting between group I mGlu receptors [35]. Further, these studies revealed new structural insights into the overall architecture of Class C GPCR 7TM region and activation mechanisms. For example, water was crystallized in the bottom of the $mGlu_5$/mavoglurant and $mGlu_5$/HTL14242 structures coordinated with residues previously shown to engender switches in allosteric modulator pharmacology [36, 58, 119]. These NAM co-crystal structures have opened up the possibility of rational discovery approaches informed by structural information.

2.3.2 Secondary Allosteric Sites

At least one other secondary allosteric binding pocket within $mGlu_5$, distinct from the well-characterized MPEP site, was proposed with the discovery of the $mGlu_5$ PAM CPPHA, and its derivative NCFP (or N-(4-chloro-2-[(1,3-dioxoisoindolin-2-yl) methyl)phenyl)picolinamide]), that have no effect or an allosteric interaction with the common site $mGlu_5$ allosteric radioligand [^{3}H]methoxyPEPy [120, 121]. Mutagenesis studies supported a lack of interaction with the common MPEP site, and identified a single residue in TM1 that influenced CPPHA potentiation [37] as well as the related compound NCFP [121]. Further, high throughput screening resulted in the discovery of another class of $mGlu_5$ with a benzamine scaffold with subsequent optimization leading to the $mGlu_5$ PAM VU0357121 and the first non-MPEP site NAL VU0365396 [122]. Radioligand binding studies also suggest a binding site distinct from that of MPEP, whereas mutagenesis studies were inconsistent with CPPHA site, but suggested overlap with the common allosteric site [122]. A third $mGlu_5$ PAM, VU0400100, is also proposed to bind to a site distinct from the common/MPEP site [123]. Recently, a fourth ligand, XAP044 (or 7-Hydroxy-3-(4-iodophenoxy)-4H-1-benzopyran-4-one), an $mGlu_7$ selective NAM, was discovered that binds allosterically within the VFT [124]. The presence of different 'druggable' allosteric pockets opens up the possibility of engendering unique pharmacological profiles through stabilization of different receptor conformations.

2.4 Complexities of Allosteric Modulation

Complexities in allosteric modulation arise from translating findings from a controlled in vitro environment to physiologically relevant systems, preclinical models, and to patients. First, mGlu receptors are obligate homodimers, however, may also exist as heteromers, or higher order oligomers, within native systems [3]. Heteromers and oligomers are complexes of nonidentical monomers and often have vastly different receptor pharmacology to the individual monomers [125]. Functional heteromers have been shown with a combination of group II and III mGlu receptors

[126–129], between group I mGlu receptors, $mGlu_5$/calcium-sensing receptor [130], $mGlu_5$/μ-opioid receptor [131] and $mGlu_5$/adenosine A_{2A} [132, 133]. The presence of functional heteromers can perturb allosteric modulator pharmacology. For example, PHCCC, a prototypical $mGlu_4$ PAM, potentiates Glu activity at $mGlu_4$/$mGlu_4$ homomers but not $mGlu_{2/4}$ heteromers, whereas the structurally distinct $mGlu_4$ PAM, VU0155041, modulates both $mGlu_4$ homomers and $mGlu_{2/4}$ heteromers [127]. As a result, PHCCC does not modulate $mGlu_4$ activity at cortico-striatal synapses due to the presence of $mGlu_2$/$mGlu_4$ heteromers [127]. Small molecule drug discovery programs often screen initial drug hits at a single target, therefore the presence of heteromers in native systems may result in unanticipated effects. If the relationship between mGlu and other receptors can be understood, the intricacies between individual receptor "subunits" may offer a unique avenue of drug development. Indeed, this concept has been manipulated with the $mGlu_5$ and μ-opioid receptor, where application of the $mGlu_5$ NAM, MPEP, reduced μ-opioid receptor internalization, phosphorylation, and desensitization in response to agonist [131]. Targeting $mGlu_5$ and μ-opioid receptors in tandem may provide a means to maximize efficacy of current opioid therapies with reduced tolerance, an adverse effect commonly seen with morphine and its derivatives [134]. However, receptor heteromerization, and indeed receptor cross-talk, remains a variable in drug design that must be accounted for, as drugs are taken from in vitro studies to the complex physiological environment.

Localization of ligand binding, whether on intracellular or on plasma membrane-bound receptors provides another potential therapeutic target in drug development. Increasing evidence has shown the presence of $mGlu_5$ on intracellular membranes [135–137], thus leading to the notion of location bias, where differential activation of signaling cascades may be observed in discrete regions of the cell. $mGlu_5$ receptors are in fact, mostly found intracellularly, within inner and outer nuclear membranes, as well as within the endoplasmic reticulum [137, 138]. Since intracellular $mGlu_5$ receptors are oriented (N-terminus is within the lumen of the organelle, and C-terminus located within the cytoplasm) such that identical second messenger proteins are available to these receptors [139]. The difference, however, lies in receptor activation. Only agonists that are able to diffuse, or be actively transported across the plasma membrane are able to activate intracellular receptors [140]. Moreover, diverse physiological outcomes may be location dependent, for example, intracellular $mGlu_5$ receptors mediate sustained elevations in iCa^{2+} whereas cell surface $mGlu_5$ mediates a rapid and transient elevation in iCa^{2+} [136]. In the striatum, intracellular but not plasma membrane $mGlu_5$ mediates activation of ERK1/2, Elk-1 and Arc that are critical for synaptic plasticity [136, 141]. Hippocampal $mGlu_5$-dependent synaptic plasticity can be influenced by receptor compartmentalization, where intracellular $mGlu_5$ mediates long-term depression (LTD) and plasma membrane $mGlu_5$ mediates both LTD and long-term potentiation [135, 140]. Thus, in targeting intracellular receptors, drug design may also need to consider formulations that allow delivery to intracellular compartments.

At the level of the ligand, complexities arise from the steep structure activity relationships (SAR) commonly observed with mGlu allosteric ligands [142]. "Molecular switches" have been observed where small changes in chemical structure result in dramatic changes in cooperativity, e.g. a PAM arising from a NAM scaffold or vice versa [142]. In phenylethynyl pyrimidine chemotypes, different methyl substitutions converted an $mGlu_5$ partial antagonist to either a NAM or PAM respectively [143] or a NAM to NAL [144], while modifications of $mGlu_5$ PAM ADX-472373 resulted in NAMs, PAMs, and partial agonists [145]. Moreover, small changes in chemical structure can result in changes in mGlu subtype selectivity [142, 146]. Thus, the complexities of mGlu allosteric modulator SAR remain a challenge in medicinal chemistry efforts to optimize lead compounds and highlight the need to delineate affinity and cooperativity rather than potency to best understand allosteric modulator SAR.

Ultimately, the key difference between currently marketed central nervous system (CNS) drugs, which act largely via orthosteric agonism or antagonism, and allosteric modulators, is the potential to maintain receptor physiology, rather than be completely silenced or switched on [33, 85, 103]. Therefore allosteric modulators may maintain favorable neurotransmitter tone within the CNS environment. A further level of specificity as well as complexity is presented with the emerging drug action paradigm of biased agonism, where agonist and/or allosteric modulators dictate different functional outcomes through interactions at the same receptor.

3 Biased Agonism and Modulation

Biased agonism describes the phenomenon where the binding of either orthosteric or allosteric ligands results in different effector coupling and signaling profiles in a ligand dependent fashion [147, 148] (Fig. 4). Bias can be manifested as ligand-dependent alterations in archetypal G protein coupling and second-messenger activation, to internalization, desensitization, compartmentalization, and even receptor oligomerization [147, 148]. The underlying molecular basis for biased agonism is thought to be ligand-induced stabilization of distinct subsets of receptor conformations resulting in different signaling fingerprints [149, 150]. As mentioned earlier, pleiotropic coupling in mGlu receptors is

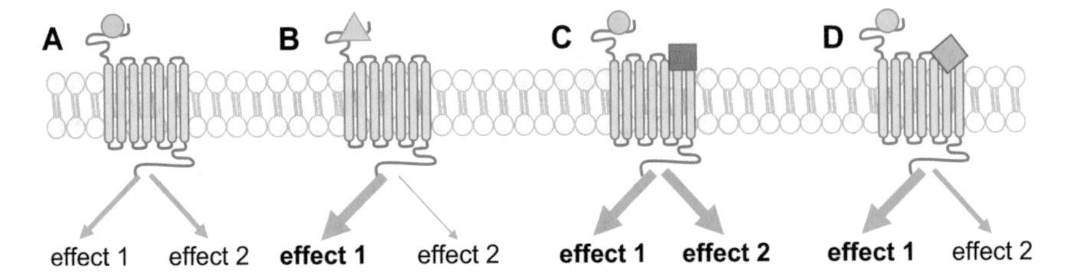

Fig. 4 Biased agonism and modulation of receptor responses. Receptor activation by different agonists (**a**, **b**) can result in different pharmacological profiles. (**c**) Binding of an allosteric modulator may modulate coupling to all pathways to a similar extent. (**d**) Binding of a biased modulator will differentially modulate different signaling pathways

gaining greater appreciation. It is clear that assessment of ligand activity in a single functional assay is insufficient to understand the full scope of drug pharmacology. Indeed, biased agonism is operative for orthosteric ligands of $mGlu_1$ $mGlu_4$, $mGlu_7$ and $mGlu_8$ receptors [70, 151, 152]. Interestingly, comparison of different endogenous orthosteric $mGlu_1$ agonists revealed glutaric acid and succinic acid were biased toward sustained ERK1/2 phosphorylation and cytoprotection versus phosphoinositide (PI) hydrolysis relative to Glu, which coupled more strongly to PI hydrolysis over pERK1/2 and cytoprotection [151]. The concept of biased endogenous ligands may explain apparent biological redundancies in activating mGlu receptors. Indeed, this notion has been explored with the existence of multiple somatostatin and opioid endogenous ligands [153], with each endogenous ligand potentially stabilizing distinct receptor conformations to engender different physiological effects.

Biased agonism is not limited to orthosteric ligands. Interestingly, $mGlu_5$ PAMs or PAM-agonists (classified based on potentiation of Glu-stimulated $mGlu_5$-iCa^{2+} mobilization in recombinant cells) are biased agonists (relative to DHPG) between second messengers traditionally considered to be interdependent, namely IP_1 accumulation and iCa^{2+} mobilization [46]. Biased agonism (relative to DHPG) for $mGlu_5$ allosteric agonists was also evident in both recombinant and primary neuron cultures between ERK1/2 phosphorylation, IP_1 accumulation, and intracellular versus extracellular sources for iCa^{2+} mobilization [46]. Of note, the most divergent biased agonism fingerprints were associated with $mGlu_5$ PAM-agonists with distinct in vivo behavioral effects. VU0424465 has seizure liability mediated via $mGlu_5$ [47] and was associated with the largest degrees of bias between pathways [46]. In contrast, DPFE and VU0409551 were unbiased (relative to DHPG) in recombinant cells and had distinct bias profiles between IP_1 and pERK1/2 in cortical neuron cultures [46]. The ability of DPFE and VU0409551 to engender different biased agonism fingerprints relative to both DHPG and VU0424465 may underlie in vivo efficacy in antipsychotic and cognition models coupled with improved safety profiles [154, 155]. Indeed, VU0409551 first presented as a biased ligand due to its lack of efficacy in modulating NMDA receptor currents or NMDA receptor-dependent synaptic plasticity in in vitro electrophysiological preparations [154]. It remains to be determined whether biased allosteric agonists will translate to novel therapeutics with improved therapeutic efficacy and safety profiles.

Direct activation of receptors may not be desirable within the CNS environment. Related to biased agonism, there is the potential for allosteric ligands to differentially modulate different signaling pathways stimulated by an orthosteric agonist, a phenomenon referred to as biased modulation. Biased modulation may manifest as different degrees of cooperativity with the same orthosteric agonist, or a different apparent affinity of an allosteric modulator for a receptor, depending on response measured [156]. For $mGlu_5$, biased modulation is evident between iCa^{2+} and ERK1/2 phosphorylation in recombinant cells and cortical astrocyte and neuron cultures [43, 46, 157]. Furthermore, multiple $mGlu_5$ PAMs show different degrees of cooperativity with the same orthosteric agonist between iCa^{2+} and IP_1 accumulation [46]. Biased modulation has the potential to result in distinct physiological effects. For example, the second-site $mGlu_5$ PAM NCFP did not potentiate hippocampal synaptic plasticity, but did potentiate DHPG-induced depolarization in subthalamic nucleus neurons [121]. Beyond $mGlu_5$, a pan-group III mGlu potentiator (VU0422288) was recently disclosed that has different degrees of cooperativity with Glu at $mGlu_7$ in an assay dependent manner [70]. The prevalence and therapeutic potential of biased agonism and/or modulation is only beginning to be realized. If signaling fingerprints can be linked to a desired therapeutic outcome, in the future it may be possible to rationally design mGlu allosteric ligands that tailor receptor activity toward therapeutic effects and avoid adverse effects. However, the possibility of biased agonism and/or modulation also raises additional complexity when investigating mGlu allosteric modulators.

Allosteric interactions by their nature are sensitive to the two ligands that simultaneously occupy the receptor, a phenomenon referred to as probe dependence. Probe dependence describes observations that the magnitude and direction of cooperativity can change depending upon which orthosteric ligand is used to detect an allosteric interaction [158, 159]. Probe dependence is a key consideration when moving from a recombinant system, where Glu may be readily used as an orthosteric ligand due to the controlled nature of receptor expression, toward a native system, where the presence of various mGlu receptors may confound results if Glu is used. Probe dependence is operative between Glu and the surrogate group I mGlu agonist DHPG with several $mGlu_5$ PAMs [46]. Further, two recent pan-group III mGlu PAMs

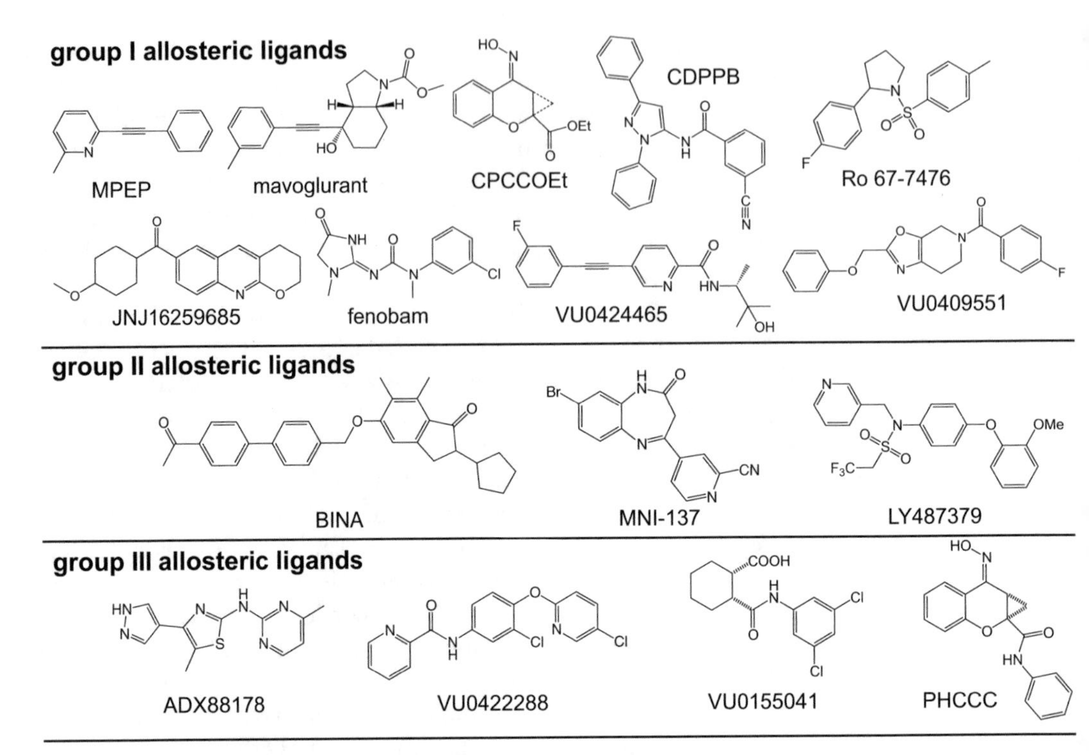

Fig. 5 Structures of select mGlu allosteric modulators

showed differing degrees of affinity and efficacy cooperativity depending on the orthosteric agonist utilized [70]. Despite the complexity and potential pitfalls, biased agonism and/or modulation offer the potential to fine-tune glutamatergic signaling to the level of intracellular effectors that may be altered in disease.

4 CNS Disorders Linked to an Altered Glutamatergic System

As Glu is the main excitatory neurotransmitter in the brain, it is unsurprising that changes in either CNS Glu levels [160–163], or the function of Glu receptors [164–170] are associated with several CNS disorders. A plethora of mGlu allosteric modulators show promise as novel therapeutics for a variety of CNS disorders (Fig. 5). With greater understanding of the full scope of drug action, allosteric drug development has the potential to move toward greater specificity and selectivity for desired therapeutic receptor endpoints, minimizing both off- and on-target adverse effects.

4.1 Schizophrenia

Schizophrenia is a debilitating disease comprising of three symptom classes: positive (hallucinations, delusions, paranoia), negative (depression, anhedonia) and cognitive (working memory deficits, inability to plan or anticipate outcomes). Current therapeutic options are generally dopamine receptor antagonists, however, these drugs offer minimal relief of negative and cognitive symptoms, and can be associated with severe extrapyramidal side effects [171]. The glutamatergic system is implicated in the undertreated pathophysiology of schizophrenia. For example, all three symptom clusters are evident in rodent models treated with the nonspecific NMDA receptor antagonist phencyclidine (PCP) [172]. Further, another NMDA receptor antagonist, ketamine, produced psychosis in healthy human volunteers, closely resembling the thought disorders observed in symptomatic schizophrenia [173]. Glutamatergic dysfunction in schizophrenia is postulated to arise from NMDA receptor hypofunction on GABAergic interneurons within the cortical and subcortical circuits within the brain [174] and as such, enhancing glutamatergic tone is a potential therapeutic option [175]. While the aforementioned evidence suggests NMDA receptor function enhancement is a viable objective in schizophrenia treatment, this receptor's fast-acting ionotropic properties may reduce its potential as a therapeutic target due to the risk of excitotoxicity [176]. Metabotropic Glu receptors are slower acting and modulate activity of ionotropic Glu receptors [177], therefore offering an attractive therapeutic strategy.

4.1.1 $mGlu_{2/3}$ Agonists and PAMs in the Treatment of Schizophrenia

Group II mGlu receptors are implicated as potential therapeutic targets due to autoreceptor properties and localization within the limbic and forebrain regions associated with schizophrenia [178]. Preclinical studies showed normalization of a PCP-induced rat model of hyperlocomotion, working memory, and stereotypy by the $mGlu_{2/3}$ agonist LY354740, with a favorable adverse effect profile and minimal effects on spontaneous activity or dopamine neurotransmission [179]. The related $mGlu_{2/3}$ agonist LY379268 also ameliorated amphetamine-induced hyperlocomotion evoked by PCP, with minimal adverse effects, which was reversed by the selective $mGlu_{2/3}$ antagonist LY341495 [66]. While agonism may provide therapeutic effect, group II mGlu receptor PAMs may provide a greater safety profile due to fine-tuning, rather than direct receptor activation. Indeed, the $mGlu_2$ selective PAMs, LY487379 and BINA, have behavioral effects similar to $mGlu_{2/3}$ agonists in PCP- and amphetamine-induced mouse models of hyperlocomotion and prepulse inhibition [28] Thus, the preclinical studies of Group II mGlu agonists and PAMs show great promise in the treatment of this multifaceted disease.

4.1.2 $mGlu_5$ PAMs in the Treatment of Schizophrenia

$mGlu_5$ activation positively modulates the NMDA receptor through increasing open channel probability, via PKC-dependent Src signaling, and stabilization of the Homer-Shank protein anchor [12, 180]. Interestingly, NMDA receptor activation is suggested to be involved in early phase synaptic plasticity, with rapid calcium influx leading to modulation of AMPA receptor trafficking [181]. This relationship between ionotropic and metabotropic Glu receptors has resulted in extensive efforts in designing therapies targeting $mGlu_5$ to treat CNS disorders related to NMDA receptor dysfunction. Knockout rodent models have highlighted the close correlation between $mGlu_5$ and schizophrenia-like symptoms or psychosis, with $mGlu_5$ knock out mice [182] show consistent deficits in prepulse inhibition relative to wild-type controls (reviewed in [175].

Promisingly, selective $mGlu_5$ PAMs have efficacy in preclinical model of antipsychotic-like effects. 3-Cyano-*N*-(1,3-diphenyl-1*H*-pyrazol-5-yl)benzamide (CDPPB) was the first selective $mGlu_5$ modulator to reverse the multidimensional aspects associated with psychosis: amphetamine induced hyperlocomotion, prepulse inhibition, and cognitive deficits [99, 183]. Several other selective $mGlu_5$ PAMs have since been shown to be efficacious in rodent models of psychosis [154, 155, 184]. In addition to promising preclinical efficacy, select $mGlu_5$ PAMs and PAM-agonists have been associated with adverse effect liability. VU0424465, a robust PAM-agonist, did not produce any antipsychotic effects, but rather resulted in dose-dependent seizures [47], while 4-fluorophenyl((2R,5S)-5-[5-(5-fluoropyridin-2-yl)-1,2,4-oxadiazol-3-yl]-2-methylpiperidin-1-yl)methanone (5PAM523), an $mGlu_5$ PAM with little/no agonism, had anti-psychotic-like effects, however, 5PAM523 also caused neurotoxicity in rats [185]. Interestingly, another $mGlu_5$ PAM agonist, VU0403602, dose-dependently reversed amphetamine-induced hyperlocomotion, however, produced time-dependent seizures and forelimb asymmetry [48]. VU0403602 adverse effects were attributed to an active metabolite, thereby highlighting the importance of understanding not only drug pharmacodynamics, but also pharmacokinetic profile within the living system [48]. Recently, the $mGlu_5$ PAM VU0409551 was shown to be a biased ligand, differentially activating/modulating different signaling pathways and lacked the ability to potentiate $mGlu_5$ modulation of NMDA receptor currents [46, 154]. Despite the absence of NMDA receptor modulation, VU0409551 had anti-psychotic-like and cognition-enhancing effects in preclinical rodent models, potentially negating the role of NMDA receptors in $mGlu_5$ PAM efficacy [154]. Nonetheless, targeting $mGlu_5$ receptors may prove fruitful in designing therapies for schizophrenia that treat both psychosis and cognition impairments.

4.2 Anxiety and Depression

Generalized anxiety and major depressive disorder remain a large burden, not only on the individual, but on society and the economy as a whole [186]. Current therapies include benzodiazepines, tricyclic antidepressants, and serotonergic agents; however, these therapies are associated with adverse effects such as dry mouth, constipation, diarrhea, and dizziness [187]. These adverse effects may be attributed to the general "dirty" nature of current drugs [188, 189], with many of the adverse effects linked to unwanted cholinergic activation [187]. Thus, there remains a clear need for more efficacious therapeutics, with greater selectivity and minimal off-target effects. Altered Glu levels within the brains of patients with anxiety and depression [162, 163] have implicated mGlu receptors as potential targets in the treatment of these disorders [190, 191].

4.2.1 $mGlu_{2/3}$ PAMs in the Treatment of Anxiety

Group II mGlu agonists display anxiolytic effects in preclinical rodent models [192], with LY354740 increasing open-arm time in a mouse behavioral model of anxiety without the adverse effect of sedation associated with benzodiazepines [67]. LY354740 also showed clinical efficacy in the treatment of panic attacks and generalized anxiety disorder with a favorable adverse effect profile [67]. Despite its promise in the preclinical setting, LY354740 activates both $mGlu_2$ and $mGlu_3$ receptors, and hence does not allow differentiation of subtype activity [68]. The design of subtype selective PAMs has paved the way for greater selectivity of pharmacological interventions [68]. Positive allosteric modulators derived from BINA and LY487379 [97, 193, 194], with increased selectivity toward $mGlu_2$, exhibit anxiolytic effects [194]. Therefore, $mGlu_2$ selective PAMs are a novel treatment strategy for the treatment of anxiety.

4.2.2 $mGlu_5$ NAMs in the Treatment of Anxiety and Depression

As NMDA receptor overactivity is implicated in anxiety and depression [195], $mGlu_5$ presents as an attractive drug target due to its ability to modulate NMDA receptor activity. In support of this hypothesis, $mGlu_5$ knockout mice have reduced depressive symptoms relative to their wild-type counterparts, despite a paradoxical increase in anxiety [196, 197]. Inta and colleagues suggest this may be the result of general ablation of $mGlu_5$, resulting in altered neurogenesis [196]. Nevertheless, the prototypical $mGlu_5$ NAM MPEP is both anxiolytic and antidepressive in preclinical studies [197–200]. Furthermore, MPEP improved the antidepressive effects of imipramine, a currently marketed tricyclic antidepressant, suggesting synergistic efficacy in targeting multiple CNS neurotransmitters [197]. $mGlu_5$ NAMs from multiple chemical scaffolds also show efficacy in decreasing mouse marble burying, a behavioral model of anxiolytic drug activity [184, 201, 202]. Fenobam is an $mGlu_5$ NAM, which was originally discovered as a non-benzodiazepine anxiolytic, with clinical efficacy in a small double-blind, placebo-controlled study with fewer adverse effects reported relative to diazepam control [49–51]. Safety and efficacy of basimglurant, an $mGlu_5$ NAM, as an adjunct treatment of major depressive disorder was recently assessed in a phase IIb trial. Basimglurant had anti-depressive effects in all secondary outcomes, with good tolerability, however, the primary outcome measure of clinician-reported changes in depression were not met [57]. While promising, it should be noted that select $mGlu_5$ NAMs, such as fenobam and MPEP, are associated with cognitive impairments and psychotomimetic-like effects [88]. It has been postulated that one means to overcome $mGlu_5$ NAM adverse effect liability is via the development of "partial" NAMs, which have limited cooperativity, such that even at full receptor occupancy by the NAM, a degree of Glu agonist activity is retained [65, 182]. A number of partial $mGlu_5$ NAMs have now been discovered that have anxiolytic and anti-addiction efficacy [65, 182], and promisingly are devoid of psychotomimetic-like effects or cognitive deficits [182]. Therefore, development of "partial" or NAMs with limited cooperativity may provide greater control in modulating glutamatergic neurotransmission to yield a broader therapeutic index.

4.3 Addiction

While dopamine, within the reward neurocircuitry of the brain, is implicated in addiction, Glu may play a pivotal role in both development and maintenance of addiction [203]. The glutamatergic system is highly integrated and thus able to modulate the dopaminergic reward system, with glutamatergic projections observed in key dopaminergic brain areas such as the ventral tegmental area and nucleus accumbens [204, 205]. Knockout of $mGlu_5$ ameliorates preclinical models of addiction, such as operant sensation seeking, extinction, and reinstatement [206–211]. In preclinical models, fenobam attenuated drug-seeking behavior in rats and reduced cocaine-induced behavioral sensitization [52, 53]. $mGlu_5$ NAMs show preclinical efficacy in relapse and reinstatement of amphetamine, cocaine, nicotine, and alcohol addictions [212], and thus provide attractive pharmacological options in targeting the multiple facets of addiction. In addition, Group II mGlu receptors are expressed within brain regions associated with reward and addiction [213]. Nicotine self-administration down-regulates Group II mGlu receptor expression, thus, up-regulation of these receptors may offer relief from addiction [213]. Indeed, $mGlu_{2/3}$ agonists LY354740 and LY379268 decreased nicotine self-administration and cue-reinstatement, although tolerance to LY379268 quickly developed, highlighting the malleable and highly adaptable CNS environment [213, 214]. Administration of the $mGlu_2$ PAM AZD8529 in rodent models of abstinence, reduced relapse of methamphetamine administration and cue-seeking behavior [215], while similar effects were observed with the $mGlu_2$ PAM BINA and cocaine addiction [69]. Thus, there is much promise in fine-tuning glutamatergic neurotransmission for the treatment of addictive disorders.

4.4 Pain and Inflammation

Chronic or neuropathic pain has been associated with dysfunction of several mGlu receptors [216]. Group I mGlu receptors are located within dorsal root ganglia, the spinal cord, and brain regions associated with pain sensation and transmission [217–219], with group I knockout mice displaying decreased pain responses [220, 221]. The $mGlu_1$ selective NAM JNJ16259685 produced analgesia in a rodent model of formalin-induced hyperalgesia [45], while an $mGlu_1$ antagonist produced analgesia in a spinal nerve ligation assay [222]. Fenobam also produced similar effects, with favorable activity in mouse models of pain and importantly, did not induce tolerance after chronic administration [54, 55]. Interestingly, while the analgesia is thought to arise from central pathways, peripherally injected MPEP, into an inflamed rat hind paw, was more effective than intracerebroventricular or intrathecal administration to attenuate hyperalgesia [218]. Thus, targeting peripheral mGlu receptors represent yet another avenue in which to design drugs that act directly to the site of injury or inflammation. In addition to group I receptors, $mGlu_4$ is an attractive target for pain and inflammation. $mGlu_4$ is located within spinal neuronal terminals and unmyelinated C-fibers, and $mGlu_4$ deletion produced hypersensitivity to noxious stimuli [71]. Group III agonists including the $mGlu_4$ selective agonist LSP4-2022, as well as the $mGlu_4$ PAM PHCCC reduced hyperalgesia without affecting healthy controls [71, 72]. Current analgesics are suboptimal and associated with tolerance [223], there remains a need for better pain therapeutics, which may be addressed through targeting mGlu receptors.

4.5 Autism Spectrum Disorders

Fragile X Syndrome (FXS) is a genetic disorder where mutations silence the *Fmr1* gene, which encodes fragile X mental retardation protein (FMRP), and is one of the most common inherited causes of autism spectrum disorders (ASD) and mental retardation [224]. FMRP is an RNA binding protein involved in mRNA translation, its absence leads to impaired synaptic plasticity [224]. Phenotypically, patients present with autistic-like behaviors and cognitive deficits ranging from poor working memory to impaired executive function and social skills [224]. The glutamatergic system has been heavily implicated in this disorder, with rodent models of FXS showing an imbalance in long-term depression and long-term activation, resulting in disturbed glutamatergic synaptic plasticity and abnormal neuronal growth [225–228]. $mGlu_5$ hyperactivity is suggested to play a significant role in FXS pathophysiology, and hence $mGlu_5$ negative modulation may be a therapeutic option [226]. Indeed, $mGlu_5$ NAMs MPEP and CTEP rescue multiple aspects of the FXS phenotype in numerous rodent models [62–64]. These promising preclinical studies led to recent clinical trials. Fenobam was tested in an open-label pilot study for the treatment of FXS, where no clinically significant adverse effects were observed [56]. Another $mGlu_5$ NAM, mavoglurant, improved behavioral symptoms in FXS in two phase IIb trials, however, primary outcome measures were not met [59]. Of note, primary outcomes were parental observations of patients, and hence were potentially subjective [60]. Despite this, mavoglurant has a predictable and tolerable adverse effect profile [59, 60], and may instruct further development of $mGlu_5$ NAMs to remedy glutamatergic hyperactivity in FXS.

Tuberous sclerosis complex (TSC) is a genetic disorder affecting multiple organ systems, with up to 95% of patients experiencing CNS involvement including epilepsy, intellectual disability, and autism spectrum disorders [229]. Malfunction of the *TSC1* and *TSC2* genes results in the loss of the hamartin–tuberin complex, and subsequent aberration in mTOR signaling, protein synthesis, and neuronal development [230, 231]. Unlike FXS, TSC rodent models display deficiencies in mGlu mediated LTD [232, 233], thus leading to enhancement of mGlu receptors being a tractable therapy in this disease. Pretreatment of hippocampal slices with the $mGlu_5$ PAM CDPPB in TSC mouse models restored LTD to wild type levels, and normalized protein synthesis [233]. CDPPB also reversed the cognitive and behavioral deficits in the TSC mouse models [233]. Thus, targeting $mGlu_5$ receptors, through inhibition or enhancement informed by the underlying pathophysiology, provides potential therapies in the treatment of both FXS and TSC respectively.

4.6 Parkinson's Disease

Parkinson's disease (PD) is characterized by bradykinesia, rigidity, and tremors arising from the aggregation of Lewy bodies and loss of dopaminergic innervation within the substantia nigra [80]. Current treatment for PD involves replacement of dopaminergic input through administration of the dopamine precursor L-3,4-dihydroxyphenylalanine (L-DOPA), however, with long term therapy up to 80% of patients experience L-DOPA induced dyskinesias (LID) [234]. Evidence suggests LID may arise from deranged synaptic plasticity, or 'pathological' long-term potentiation, between dopaminergic and glutamatergic inputs within the nigrostriatal pathway [235, 236], and as such, current treatments for LID involve NMDA antagonism by amantadine [237]. There remains a need for more efficacious treatments for both PD and LID, without the adverse motor and cognitive effects seen with L-DOPA and amantadine.

4.6.1 $mGlu_4$ PAMs in the Treatment of PD and LID

$mGlu_4$ receptors are widely expressed throughout brain regions involved in the pathophysiology of PD, including the basal ganglia, hippocampus, and cerebellum [167, 238]. Administration of the group III selective mGlu agonists into the substantia nigra reversed akinesia in rodent models of PD, while also providing neuroprotection against 6-hydroxydopamine lesions [239, 240]. The selective $mGlu_4$ PAM PHCCC also reversed reserpine-induced akinesia and was protective against neuronal degeneration in rodents [74, 75]. Haloperidol-induced catalepsy was also reversible by the

$mGlu_4$ PAMs VU0155041, ADX88178, AF21934 with the absence of LID adverse effects [76–78]. Interestingly, both $mGlu_4$ PAMs ADX88178 and AF21934 showed synergistic effects with administration of L-DOPA, such that the therapeutic dose of L-DOPA may be reduced, thereby reducing the risk of LID [77, 78]. Thus, $mGlu_4$ PAMs in conjunction with current therapies may provide a novel therapeutic strategy to reduce adverse effects while maximizing therapeutic effects.

4.6.2 $mGlu_5$ NAMs in the Treatment of PD and LID

Elevated $mGlu_5$ receptor expression has been reported in the striatum of patients of PD and in 1-methyl-4-phenyl-1,2,3,6-tetrahydropyridine (MPTP)-lesioned nonhuman primates with dyskinesias [241]. Treatment with MPEP in conjunction with L-DOPA produced an L-DOPA sparing effect, reducing the overall incidence of LID [242, 243]. Further, fenobam treatment in rodent and nonhuman primate models of PD reduced peak-dose dyskinesias [244]. The immense potential of $mGlu_5$ targeted PD and LID therapies have been summarized in [245]. Mavoglurant was assessed in the treatment of LID in two phase II trials, however, failed to meet the primary outcome of antidyskinetic activity, with a greater adverse effect profile relative to placebo [61]. Despite lack of clinical trial success to date, $mGlu_5$ allosteric ligands remain viable options in the treatment of PD and LID, and may find a role in synergistic therapies to reduce L-DOPA doses.

5 Concluding Remarks

In various preclinical models of CNS disorders, mGlu allosteric ligands have demonstrated promising efficacy profiles. Unfortunately, the limited examples of mGlu allosteric ligands that have progressed into phase II clinical trials have yet to demonstrate efficacy. To address this disconnect between animal models and human trials, there remains a need to better understand the full scope of mGlu allosteric modulator activity. In order to harness the newer paradigms of biased agonism and modulation to fine-tune glutamatergic neurotransmission, a deeper understanding of the physiology and pathophysiology of the different mGlu receptors is sorely needed. In particular, by understanding how different mGlu-stimulated signaling pathways are linked to pathology and therapeutic efficacy, it may be possible to develop biased mGlu modulators that normalize defective signaling cascades and avoid target-mediated adverse effects. It is clear that modulators of glutamatergic neurotransmission show immense potential in a wide range of CNS disorders, not only as prospective therapies, but also pharmacological tools in understanding the complexities of metabotropic Glu receptor activity in health and disease.

References

1. Houamed KM et al (1991) Cloning, expression, and gene structure of a G protein-coupled glutamate receptor from rat brain. Science 252(5010):1318–1321
2. Masu M et al (1991) Sequence and expression of a metabotropic glutamate receptor. Nature 349(6312):760–765
3. Conn PJ, Pin JP (1997) Pharmacology and functions of metabotropic glutamate receptors. Annu Rev Pharmacol Toxicol 37: 205–237
4. Niswender CM, Conn PJ (2010) Metabotropic glutamate receptors: physiology, pharmacology, and disease. Annu Rev Pharmacol Toxicol 50:295–322
5. Hermans E, Challiss RA (2001) Structural, signalling and regulatory properties of the group I metabotropic glutamate receptors: prototypic family C G-protein-coupled receptors. Biochem J 359(Pt 3):465–484
6. Peavy RD, Conn PJ (1998) Phosphorylation of mitogen-activated protein kinase in cultured rat cortical glia by stimulation of metabotropic glutamate receptors. J Neurochem 71(2):603–612
7. Thandi S et al (2002) Group-I metabotropic glutamate receptors, mGlu1a and mGlu5a, couple to extracellular signal-regulated kinase (ERK) activation via distinct, but overlapping, signalling pathways. J Neurochem 83(5):1139–1153
8. Hou L, Klann E (2004) Activation of the phosphoinositide 3-kinase-Akt-mammalian target of rapamycin signaling pathway is required for metabotropic glutamate receptor-dependent long-term depression. J Neurosci 24(28):6352–6361
9. Chan TO et al (1999) AKT/PKB and other D3 phosphoinositide-regulated kinases: kinase activation by phosphoinositide-dependent phosphorylation. Annu Rev Biochem 68: 965–1014
10. Waung MW, Huber KM (2009) Protein translation in synaptic plasticity: mGluR-LTD, Fragile X. Curr Opin Neurobiol 19(3):319–326
11. Mao L et al (2005) The scaffold protein Homer1b/c links metabotropic glutamate receptor 5 to extracellular signal-regulated protein kinase cascades in neurons. J Neurosci 25(10):2741–2752
12. Tu JC et al (1999) Coupling of mGluR/Homer and PSD-95 complexes by the Shank family of postsynaptic density proteins. Neuron 23(3):583–592
13. Grabrucker AM et al (2011) Postsynaptic ProSAP/Shank scaffolds in the cross-hair of synaptopathies. Trends Cell Biol 21(10): 594–603
14. Aramori I, Nakanishi S (1992) Signal transduction and pharmacological characteristics of a metabotropic glutamate receptor, mGluR1, in transfected CHO cells. Neuron 8(4):757–765
15. Hermans E et al (2000) Complex involvement of pertussis toxin-sensitive G proteins in the regulation of type 1alpha metabotropic glutamate receptor signaling in baby hamster kidney cells. Mol Pharmacol 58(2):352–360
16. Francesconi A, Duvoisin RM (1998) Role of the second and third intracellular loops of metabotropic glutamate receptors in mediating dual signal transduction activation. J Biol Chem 273(10):5615–5624
17. Joly C et al (1995) Molecular, functional, and pharmacological characterization of the metabotropic glutamate receptor type 5 splice variants: comparison with mGluR1. J Neurosci 15(5 Pt 2):3970–3981
18. Abe T et al (1992) Molecular characterization of a novel metabotropic glutamate receptor mGluR5 coupled to inositol phosphate/Ca2+ signal transduction. J Biol Chem 267(19):13361–13368
19. Balazs R et al (1997) Metabotropic glutamate receptor mGluR5 in astrocytes: pharmacological properties and agonist regulation. J Neurochem 69(1):151–163
20. Cochilla AJ, Alford S (1998) Metabotropic glutamate receptor-mediated control of neurotransmitter release. Neuron 20(5): 1007–1016
21. Takahashi T et al (1996) Presynaptic calcium current modulation by a metabotropic glutamate receptor. Science 274(5287):594–597
22. Stefani A et al (1998) Group III metabotropic glutamate receptor agonists modulate high voltage-activated Ca2+ currents in pyramidal neurons of the adult rat. Exp Brain Res 119(2):237–244
23. Daniel H, Crepel F (2001) Control of Ca(2+) influx by cannabinoid and metabotropic glutamate receptors in rat cerebellar cortex requires K(+) channels. J Physiol 537(Pt 3): 793–800
24. Cain SM et al (2008) mGlu4 potentiation of K(2P)2.1 is dependant on C-terminal dephosphorylation. Mol Cell Neurosci 37(1):32–39
25. Saugstad JA et al (1996) Metabotropic glutamate receptors activate G-protein-coupled

inwardly rectifying potassium channels in Xenopus oocytes. J Neurosci 16(19): 5979–5985

26. Martin R et al (2007) mGluR7 inhibits glutamate release through a PKC-independent decrease in the activity of P/Q-type Ca2+ channels and by diminishing cAMP in hippocampal nerve terminals. Eur J Neurosci 26(2):312–322
27. Chavis P et al (1998) Visualization of cyclic AMP-regulated presynaptic activity at cerebellar granule cells. Neuron 20(4):773–781
28. Muguruza C et al (2016) Group II metabotropic glutamate receptors as targets for novel antipsychotic drugs. Front Pharmacol 7:130
29. Bruno V et al (2016) The impact of metabotropic glutamate receptors into active neurodegenerative processes: a "dark side" in the development of new symptomatic treatments for neurologic and psychiatric disorders. Neuropharmacology. doi:10.1016/j.neuropharm.2016.04.044
30. Senter RK et al (2016) The role of mGlu receptors in hippocampal plasticity deficits in neurological and psychiatric disorders: implications for allosteric modulators as novel therapeutic strategies. Curr Neuropharmacol 14(5):455–473
31. Koga K et al (2016) Metabotropic glutamate receptor dependent cortical plasticity in chronic pain. Curr Neuropharmacol 14(5):427–434
32. Lagerstrom MC, Schioth HB (2008) Structural diversity of G protein-coupled receptors and significance for drug discovery. Nat Rev Drug Discov 7(4):339–357
33. Christopoulos A, Kenakin T (2002) G protein-coupled receptor allosterism and complexing. Pharmacol Rev 54(2):323–374
34. Christopoulos A et al (2014) International Union of Basic and Clinical Pharmacology. XC. Multisite pharmacology: recommendations for the nomenclature of receptor allosterism and allosteric ligands. Pharmacol Rev 66(4):918–947
35. Wu H et al (2014) Structure of a class C GPCR metabotropic glutamate receptor 1 bound to an allosteric modulator. Science 344(6179):58–64
36. Christopher JA et al (2015) Fragment and structure-based drug discovery for a class C GPCR: discovery of the mGlu5 negative allosteric modulator HTL14242 (3-chloro-5-[6-(5-fluoropyridin-2-yl)pyrimidin-4-yl]benzonitrile). J Med Chem 58(16): 6653–6664
37. Chen Y et al (2008) N-{4-Chloro-2-[(1,3-dioxo-1,3-dihydro-2H-isoindol-2-yl)methyl] phenyl}-2-hydroxybenzamide (CPPHA) acts through a novel site as a positive allosteric modulator of group 1 metabotropic glutamate receptors. Mol Pharmacol 73(3): 909–918
38. Brauner-Osborne H et al (1999) Interaction of CPCCOEt with a chimeric mGlu1b and calcium sensing receptor. Neuroreport 10(18): 3923–3925
39. Gasparini F et al (2001) Discovery and characterization of non-competitive antagonists of group I metabotropic glutamate receptors. Farmaco 56(1–2):95–99
40. Gasparini F et al (1999) 2-Methyl-6-(phenylethynyl)-pyridine (MPEP), a potent, selective and systemically active mGlu5 receptor antagonist. Neuropharmacology 38(10): 1493–1503
41. Chen Y et al (2007) Interaction of novel positive allosteric modulators of metabotropic glutamate receptor 5 with the negative allosteric antagonist site is required for potentiation of receptor responses. Mol Pharmacol 71(5):1389–1398
42. Bradley SJ et al (2011) Quantitative analysis reveals multiple mechanisms of allosteric modulation of the mGlu5 receptor in rat astroglia. Mol Pharmacol 79(5):874–885
43. Gregory KJ et al (2012) Investigating metabotropic glutamate receptor 5 allosteric modulator cooperativity, affinity, and agonism: enriching structure-function studies and structure-activity relationships. Mol Pharmacol 82(5):860–875
44. Rovira X et al (2015) Overlapping binding sites drive allosteric agonism and positive cooperativity in type 4 metabotropic glutamate receptors. FASEB J 29(1):116–130
45. Mabire D et al (2005) Synthesis, structure-activity relationship, and receptor pharmacology of a new series of quinoline derivatives acting as selective, noncompetitive mGlu1 antagonists. J Med Chem 48(6):2134–2153
46. Sengmany K et al (2016) Biased allosteric agonism and modulation of metabotropic glutamate receptor 5: implications for optimizing preclinical neuroscience drug discovery. Neuropharmacology. doi:10.1016/j.neuropharm.2016.07.001
47. Rook JM et al (2013) Unique signaling profiles of positive allosteric modulators of metabotropic glutamate receptor subtype 5 determine differences in in vivo activity. Biol Psychiatry 73(6):501–509

48. Bridges TM et al (2013) Biotransformation of a novel positive allosteric modulator of metabotropic glutamate receptor subtype 5 contributes to seizure-like adverse events in rats involving a receptor agonism-dependent mechanism. Drug Metab Dispos 41(9): 1703–1714
49. Pecknold JC et al (1982) Treatment of anxiety using fenobam (a nonbenzodiazepine) in a double-blind standard (diazepam) placebo-controlled study. J Clin Psychopharmacol 2(2):129–133
50. Porter RH et al (2005) Fenobam: a clinically validated nonbenzodiazepine anxiolytic is a potent, selective, and noncompetitive mGlu5 receptor antagonist with inverse agonist activity. J Pharmacol Exp Ther 315(2):711–721
51. Wieronska JM, Pilc A (2013) Glutamate-based anxiolytic ligands in clinical trials. Expert Opin Investig Drugs 22(8):1007–1022
52. Watterson LR et al (2013) Attenuation of reinstatement of methamphetamine-, sucrose-, and food-seeking behavior in rats by fenobam, a metabotropic glutamate receptor 5 negative allosteric modulator. Psychopharmacology 225(1):151–159
53. Huang CC et al (2015) Cocaine withdrawal impairs mGluR5-dependent long-term depression in nucleus accumbens shell neurons of both direct and indirect pathways. Mol Neurobiol 52(3):1223–1233
54. Jacob W et al (2009) The anxiolytic and analgesic properties of fenobam, a potent mGlu5 receptor antagonist, in relation to the impairment of learning. Neuropharmacology 57(2):97–108
55. Montana MC et al (2011) Metabotropic glutamate receptor 5 antagonism with fenobam: examination of analgesic tolerance and side effect profile in mice. Anesthesiology 115(6):1239–1250
56. Berry-Kravis E et al (2009) A pilot open label, single dose trial of fenobam in adults with fragile X syndrome. J Med Genet 46(4): 266–271
57. Quiroz JA et al (2016) Efficacy and safety of basimglurant as adjunctive therapy for major depression: a randomized clinical trial. JAMA Psychiatry. doi:10.1001/jamapsychiatry.2016.0838
58. Dore AS et al (2014) Structure of class C GPCR metabotropic glutamate receptor 5 transmembrane domain. Nature 511(7511): 557–562
59. Berry-Kravis E et al (2016) Mavoglurant in fragile X syndrome: results of two randomized, double-blind, placebo-controlled trials. Sci Transl Med 8(321):321ra5
60. Bailey DB Jr et al (2016) Mavoglurant in adolescents with fragile X syndrome: analysis of clinical global impression-improvement source data from a double-blind therapeutic study followed by an open-label, long-term extension study. J Neurodev Disord 8:1
61. Trenkwalder C et al (2016) Mavoglurant in Parkinson's patients with l-dopa-induced dyskinesias: two randomized phase 2 studies. Mov Disord. doi:10.1002/mds.26585
62. Yan QJ et al (2005) Suppression of two major Fragile X Syndrome mouse model phenotypes by the mGluR5 antagonist MPEP. Neuropharmacology 49(7):1053–1066
63. Dolen G et al (2007) Correction of fragile X syndrome in mice. Neuron 56(6):955–962
64. Michalon A et al (2012) Chronic pharmacological mGlu5 inhibition corrects fragile X in adult mice. Neuron 74(1):49–56
65. Nickols HH et al (2016) VU0477573: partial negative allosteric modulator of the subtype 5 metabotropic glutamate receptor with in vivo efficacy. J Pharmacol Exp Ther 356(1): 123–136
66. Cartmell J et al (1999) The metabotropic glutamate 2/3 receptor agonists LY354740 and LY379268 selectively attenuate phencyclidine versus d-amphetamine motor behaviors in rats. J Pharmacol Exp Ther 291(1): 161–170
67. Schoepp DD et al (2003) LY354740, an mGlu2/3 receptor agonist as a novel approach to treat anxiety/stress. Stress 6(3):189–197
68. Conn PJ, Jones CK (2009) Promise of mGluR2/3 activators in psychiatry. Neuropsychopharmacology 34(1):248–249
69. Jin X et al (2010) The mGluR2 positive allosteric modulator BINA decreases cocaine self-administration and cue-induced cocaine-seeking and counteracts cocaine-induced enhancement of brain reward function in rats. Neuropsychopharmacology 35(10):2021–2036
70. Jalan-Sakrikar N et al (2014) Identification of positive allosteric modulators VU0155094 (ML397) and VU0422288 (ML396) reveals new insights into the biology of metabotropic glutamate receptor 7. ACS Chem Neurosc 5(12):1221–1237
71. Vilar B et al (2013) Alleviating pain hypersensitivity through activation of type 4 metabotropic glutamate receptor. J Neurosci 33(48) 18951–18965
72. Goudet C et al (2008) Group III metabotropic glutamate receptors inhibit hyperalgesia in animal models of inflammation and neuropathic pain. Pain 137(1):112–124
73. Maj M et al (2003) (−)-PHCCC, a positive allosteric modulator of mGluR4: characterization

mechanism of action, and neuroprotection. Neuropharmacology 45(7):895–906

74. Marino MJ et al (2003) Allosteric modulation of group III metabotropic glutamate receptor 4: a potential approach to Parkinson's disease treatment. Proc Natl Acad Sci U S A 100(23): 13668–13673
75. Battaglia G et al (2006) Pharmacological activation of mGlu4 metabotropic glutamate receptors reduces nigrostriatal degeneration in mice treated with 1-methyl-4-phenyl-1,2, 3,6-tetrahydropyridine. J Neurosci 26(27): 7222–7229
76. Jones CK et al (2011) Discovery, synthesis, and structure-activity relationship development of a series of N-4-(2,5-dioxopyrrolidin-1-yl)phenylpicolinamides (VU0400195, ML182): characterization of a novel positive allosteric modulator of the metabotropic glutamate receptor 4 (mGlu(4)) with oral efficacy in an antiparkinsonian animal model. J Med Chem 54(21):7639–7647
77. Le Poul E et al (2012) A potent and selective metabotropic glutamate receptor 4 positive allosteric modulator improves movement in rodent models of Parkinson's disease. J Pharmacol Exp Ther 343(1):167–177
78. Bennouar KE et al (2013) Synergy between L-DOPA and a novel positive allosteric modulator of metabotropic glutamate receptor 4: implications for Parkinson's disease treatment and dyskinesia. Neuropharmacology 66: 158–169
79. Keov P et al (2011) Allosteric modulation of G protein-coupled receptors: a pharmacological perspective. Neuropharmacology 60(1): 24–35
80. Kalia LV, Lang AE (2015) Parkinson's disease. Lancet 386(9996):896–912
81. Hall DA (2000) Modeling the functional effects of allosteric modulators at pharmacological receptors: an extension of the two-state model of receptor activation. Mol Pharmacol 58(6):1412–1423
82. Kenakin T (2004) Principles: receptor theory in pharmacology. Trends Pharmacol Sci 25(4):186–192
83. Kenakin T (2009) Quantifying biological activity in chemical terms: a pharmacology primer to describe drug effect. ACS Chem Biol 4(4):249–260
84. Ehlert FJ (1988) Estimation of the affinities of allosteric ligands using radioligand binding and pharmacological null methods. Mol Pharmacol 33(2):187–194
85. Leach K et al (2007) Allosteric GPCR modulators: taking advantage of permissive receptor pharmacology. Trends Pharmacol Sci 28(8):382–389
86. Black JW, Leff P (1983) Operational models of pharmacological agonism. Proc R Soc Lond B Biol Sci 220(1219):141–162
87. Price MR et al (2005) Allosteric modulation of the cannabinoid CB1 receptor. Mol Pharmacol 68(5):1484–1495
88. Gregory KJ et al (2013) Pharmacology of metabotropic glutamate receptor allosteric modulators: structural basis and therapeutic potential for CNS disorders. Prog Mol Biol Transl Sci 115:61–121
89. Lavreysen H et al (2003) [3H]R214127: a novel high-affinity radioligand for the mGlu1 receptor reveals a common binding site shared by multiple allosteric antagonists. Mol Pharmacol 63(5):1082–1093
90. Kohara A et al (2005) Radioligand binding properties and pharmacological characterization of 6-amino-N-cyclohexyl-N,3-dimethylthiazolo[3,2-a]benzimidazole-2-carboxamide (YM-298198), a high-affinity, selective, and noncompetitive antagonist of metabotropic glutamate receptor type 1. J Pharmacol Exp Ther 315(1):163–169
91. Gasparini F et al (2002) [(3)H]-M-MPEP, a potent, subtype-selective radioligand for the metabotropic glutamate receptor subtype 5. Bioorg Med Chem Lett 12(3):407–409
92. Cosford ND et al (2003) [3H]-methoxymethyl-MTEP and [3H]-methoxy-PEPy: potent and selective radioligands for the metabotropic glutamate subtype 5 (mGlu5) receptor. Bioorg Med Chem Lett 13(3): 351–354
93. Cheng Y, Prusoff WH (1973) Relationship between the inhibition constant (K1) and the concentration of inhibitor which causes 50 per cent inhibition (I50) of an enzymatic reaction. Biochem Pharmacol 22(23):3099–3108
94. Gregory KJ et al (2010) Overview of receptor allosterism. Curr Protoc Pharmacol. Chapter 1:Unit 1.21
95. Schaffhauser H et al (2003) Pharmacological characterization and identification of amino acids involved in the positive modulation of metabotropic glutamate receptor subtype 2. Mol Pharmacol 64(4):798–810
96. Lundstrom L et al (2011) Structural determinants of allosteric antagonism at metabotropic glutamate receptor 2: mechanistic studies with new potent negative allosteric modulators. Br J Pharmacol 164(2b): 521–537
97. Johnson MP et al (2005) Metabotropic glutamate 2 receptor potentiators: receptor

modulation, frequency-dependent synaptic activity, and efficacy in preclinical anxiety and psychosis model(s). Psychopharmacology 179(1):271–283

98. Knoflach F et al (2001) Positive allosteric modulators of metabotropic glutamate 1 receptor: characterization, mechanism of action, and binding site. Proc Natl Acad Sci U S A 98(23):13402–13407

99. Kinney GG et al (2005) A novel selective positive allosteric modulator of metabotropic glutamate receptor subtype 5 has in vivo activity and antipsychotic-like effects in rat behavioral models. J Pharmacol Exp Ther 313(1):199–206

100. Pagano A et al (2000) The non-competitive antagonists 2-methyl-6-(phenylethynyl)pyridine and 7-hydroxyiminocyclopropan[b] chromen-1a-carboxylic acid ethyl ester interact with overlapping binding pockets in the transmembrane region of group I metabotropic glutamate receptors. J Biol Chem 275(43):33750–33758

101. Litschig S et al (1999) CPCCOEt, a noncompetitive metabotropic glutamate receptor 1 antagonist, inhibits receptor signaling without affecting glutamate binding. Mol Pharmacol 55(3):453–461

102. Hemstapat K et al (2006) A novel class of positive allosteric modulators of metabotropic glutamate receptor subtype 1 interact with a site distinct from that of negative allosteric modulators. Mol Pharmacol 70(2):616–626

103. Melancon BJ et al (2012) Allosteric modulation of seven transmembrane spanning receptors: theory, practice, and opportunities for central nervous system drug discovery. J Med Chem 55(4):1445–1464

104. Lindsley CW et al (2016) Practical strategies and concepts in GPCR allosteric modulator discovery: recent advances with metabotropic glutamate receptors. Chem Rev 116(11): 6707–6741

105. O'Hara PJ et al (1993) The ligand-binding domain in metabotropic glutamate receptors is related to bacterial periplasmic binding proteins. Neuron 11(1):41–52

106. Costantino G, Pellicciari R (1996) Homology modeling of metabotropic glutamate receptors. (mGluRs) structural motifs affecting binding modes and pharmacological profile of mGluR1 agonists and competitive antagonists. J Med Chem 39(20):3998–4006

107. Kunishima N et al (2000) Structural basis of glutamate recognition by a dimeric metabotropic glutamate receptor. Nature 407(6807): 971–977

108. Muto T et al (2007) Structures of the extracellular regions of the group II/III metabotropic glutamate receptors. Proc Natl Acad Sci U S A 104(10):3759–3764

109. Okamoto T et al (1998) Expression and purification of the extracellular ligand binding region of metabotropic glutamate receptor subtype 1. J Biol Chem 273(21): 13089–13096

110. Rondard P et al (2006) Coupling of agonist binding to effector domain activation in metabotropic glutamate-like receptors. J Biol Chem 281(34):24653–24661

111. Goudet C et al (2004) Heptahelical domain of metabotropic glutamate receptor 5 behaves like rhodopsin-like receptors. Proc Natl Acad Sci U S A 101(1):378–383

112. Suzuki G et al (2007) Pharmacological characterization of a new, orally active and potent allosteric metabotropic glutamate receptor 1 antagonist, 4-[1-(2-fluoropyridin-3-yl)-5-methyl-1H-1,2,3-triazol-4-yl]-N-isopropyl-N-methyl- 3,6-dihydropyridine-1(2H)-carboxamide (FTIDC). J Pharmacol Exp Ther 321(3):1144–1153

113. Pin J, Duvoisin R (1995) The metabotropic glutamate receptors: structure and functions. Neuropharmacology 34:1–26

114. Romano C et al (2001) Covalent and noncovalent interactions mediate metabotropic glutamate receptor mGlu5 dimerization. Mol Pharmacol 59(1):46–53

115. Romano C et al (1996) Metabotropic glutamate receptor 5 is a disulfide-linked dimer. J Biol Chem 271(45):28612–28616

116. Malherbe P et al (2006) Comparison of the binding pockets of two chemically unrelated allosteric antagonists of the mGlu5 receptor and identification of crucial residues involved in the inverse agonism of MPEP. J Neurochem 98(2):601–615

117. Malherbe P et al (2003) Mutational analysis and molecular modeling of the binding pocket of the metabotropic glutamate 5 receptor negative modulator 2-methyl-6-(phenylethynyl)-pyridine. Mol Pharmacol 64(4):823–832

118. Sheffler DJ et al (2011) Allosteric modulation of metabotropic glutamate receptors. Adv Pharmacol 62:37–77

119. Gregory KJ et al (2013) Probing the metabotropic glutamate receptor 5 (mGlu5) positive allosteric modulator (PAM) binding pocket: discovery of point mutations that engender a "molecular switch" in PAM pharmacology. Mol Pharmacol 83(5):991–1006

120. O'Brien JA et al (2004) A novel selective allosteric modulator potentiates the activity of native metabotropic glutamate receptor subtype 5 in rat forebrain. J Pharmacol Exp Ther 309(2):568–577
121. Noetzel MJ et al (2013) A novel metabotropic glutamate receptor 5 positive allosteric modulator acts at a unique site and confers stimulus bias to mGlu5 signaling [Erratum appears in Mol Pharmacol. 2013;84(4):654]. Mol Pharmacol 83(4):835–847
122. Hammond AS et al (2010) Discovery of a novel chemical class of mGlu(5) allosteric ligands with distinct modes of pharmacology. ACS Chem Neurosci 1(10):702–716
123. Rodriguez AL et al (2010) Identification of a glycine sulfonamide based non-MPEP site positive allosteric potentiator (PAM) of mGlu5. In: Probe reports from the NIH molecular libraries program. National Center for Biotechnology Information (US), Bethesda, MD
124. Gee CE et al (2014) Blocking metabotropic glutamate receptor subtype 7 (mGlu7) via the Venus flytrap domain (VFTD) inhibits amygdala plasticity, stress, and anxiety-related behavior. J Biol Chem 289(16):10975–10987
125. Prinster SC et al (2005) Heterodimerization of g protein-coupled receptors: specificity and functional significance. Pharmacol Rev 57(3): 289–298
126. Doumazane E et al (2011) A new approach to analyze cell surface protein complexes reveals specific heterodimeric metabotropic glutamate receptors. FASEB J 25(1):66–77
127. Yin S et al (2014) Selective actions of novel allosteric modulators reveal functional heteromers of metabotropic glutamate receptors in the CNS. J Neurosci 34(1):79–94
128. Kammermeier PJ (2012) Functional and pharmacological characteristics of metabotropic glutamate receptors 2/4 heterodimers. Mol Pharmacol 82(3):438–447
129. Sevastyanova TN, Kammermeier PJ (2014) Cooperative signaling between homodimers of metabotropic glutamate receptors 1 and 5. Mol Pharmacol 86(5):492–504
130. Gama L et al (2001) Heterodimerization of calcium sensing receptors with metabotropic glutamate receptors in neurons. J Biol Chem 276(42):39053–39059
131. Schroder H et al (2009) Allosteric modulation of metabotropic glutamate receptor 5 affects phosphorylation, internalization, and desensitization of the micro-opioid receptor. Neuropharmacology 56(4):768–778
132. Rodrigues RJ et al (2005) Co-localization and functional interaction between adenosine A(2A) and metabotropic group 5 receptors in glutamatergic nerve terminals of the rat striatum. J Neurochem 92(3):433–441
133. Ferre S et al (2002) Synergistic interaction between adenosine A2A and glutamate mGlu5 receptors: implications for striatal neuronal function. Proc Natl Acad Sci U S A 99(18):11940–11945
134. Williams JT et al (2013) Regulation of mu-opioid receptors: desensitization, phosphorylation, internalization, and tolerance. Pharmacol Rev 65(1):223–254
135. Purgert CA et al (2014) Intracellular mGluR5 can mediate synaptic plasticity in the hippocampus. J Neurosci 34(13):4589–4598
136. Jong YJ et al (2009) Intracellular metabotropic glutamate receptor 5 (mGluR5) activates signaling cascades distinct from cell surface counterparts. J Biol Chem 284(51): 35827–35838
137. Kumar V et al (2008) Activated nuclear metabotropic glutamate receptor mGlu5 couples to nuclear Gq/11 proteins to generate inositol 1,4,5-trisphosphate-mediated nuclear Ca2+ release. J Biol Chem 283(20): 14072–14083
138. Hubert GW et al (2001) Differential subcellular localization of mGluR1a and mGluR5 in the rat and monkey Substantia nigra. J Neurosci 21(6):1838–1847
139. Jong YJ et al (2014) Location-dependent signaling of the group 1 metabotropic glutamate receptor mGlu5. Mol Pharmacol 86(6): 774–785
140. Jong YJ et al (2005) Functional metabotropic glutamate receptors on nuclei from brain and primary cultured striatal neurons. Role of transporters in delivering ligand. J Biol Chem 280(34):30469–30480
141. Kumar V et al (2012) Activation of intracellular metabotropic glutamate receptor 5 in striatal neurons leads to up-regulation of genes associated with sustained synaptic transmission including Arc/Arg3.1 protein. J Biol Chem 287(8):5412–5425
142. Wood MR et al (2011) "Molecular switches" on mGluR allosteric ligands that modulate modes of pharmacology. Biochemist 50(13): 2403–2410
143. Sharma S et al (2009) Discovery of molecular switches that modulate modes of metabotropic glutamate receptor subtype 5 (mGlu5) pharmacology in vitro and in vivo within a series of functionalized, regioisomeric 2- and

5-(phenylethynyl)pyrimidines. J Med Chem 52(14):4103–4106
144. Rodriguez AL et al (2005) A close structural analog of 2-methyl-6-(phenylethynyl)-pyridine acts as a neutral allosteric site ligand on metabotropic glutamate receptor subtype 5 and blocks the effects of multiple allosteric modulators. Mol Pharmacol 68(6):1793–1802
145. Lamb JP et al (2011) Discovery of molecular switches within the ADX-47273 mGlu5 PAM scaffold that modulate modes of pharmacology to afford potent mGlu5 NAMs, PAMs and partial antagonists. Bioorg Med Chem Lett 21(9):2711–2714
146. Sheffler DJ et al (2012) Development of a novel, CNS-penetrant, metabotropic glutamate receptor 3 (mGlu3) NAM probe (ML289) derived from a closely related mGlu5 PAM. Bioorg Med Chem Lett 22(12):3921–3925
147. Galandrin S et al (2016) Delineating biased ligand efficacy at 7TM receptors from an experimental perspective. Int J Biochem Cell Biol. doi:10.1016/j.biocel.2016.04.009
148. Luttrell LM (2014) Minireview: More than just a hammer: ligand "bias" and pharmaceutical discovery. Mol Endocrinol 28(3): 281–294
149. Kenakin T et al (2012) A simple method for quantifying functional selectivity and agonist bias. ACS Chem Neurosci 3(3):193–203
150. Kenakin T, Christopoulos A (2013) Signalling bias in new drug discovery: detection, quantification and therapeutic impact. Nat Rev Drug Discov 12(3):205–216
151. Emery AC et al (2012) Ligand bias at metabotropic glutamate 1a receptors: molecular determinants that distinguish beta-arrestin-mediated from G protein-mediated signaling. Mol Pharmacol 82(2):291–301
152. Hathaway HA et al (2015) Pharmacological characterization of mGlu1 receptors in cerebellar granule cells reveals biased agonism. Neuropharmacology. doi:10.1016/j.neuropharm.2015.02.007
153. Thompson GL et al (2014) Biological redundancy of endogenous GPCR ligands in the gut and the potential for endogenous functional selectivity. Front Pharmacol 5:262
154. Rook JM et al (2015) Biased mGlu5-positive allosteric modulators provide in vivo efficacy without potentiating mGlu5 modulation of NMDAR currents. Neuron 86(4): 1029–1040
155. Gregory KJ et al (2013) N-aryl piperazine metabotropic glutamate receptor 5 positive allosteric modulators possess efficacy in preclinical models of NMDA hypofunction and cognitive enhancement. J Pharmacol Exp Ther 347(2):438–457
156. Cook AE et al (2015) Biased allosteric modulation at the CaS receptor engendered by structurally diverse calcimimetics. Br J Pharmacol 172(1):185–200
157. Zhang Y et al (2005) Allosteric potentiators of metabotropic glutamate receptor subtype 5 have differential effects on different signaling pathways in cortical astrocytes. J Pharmacol Exp Ther 315(3):1212–1219
158. Suratman S et al (2011) Impact of species variability and 'probe-dependence' on the detection and in vivo validation of allosteric modulation at the M4 muscarinic acetylcholine receptor. Br J Pharmacol 162(7): 1659–1670
159. Valant C et al (2012) Probe dependence in the allosteric modulation of a G protein-coupled receptor: implications for detection and validation of allosteric ligand effects. Mol Pharmacol 81(1):41–52
160. Sanacora G et al (2012) Towards a glutamate hypothesis of depression: an emerging frontier of neuropsychopharmacology for mood disorders. Neuropharmacology 62(1): 63–77
161. Kaiser LG et al (2005) Age-related glutamate and glutamine concentration changes in normal human brain: (1)H MR spectroscopy study at 4 T. Neurobiol Aging 26(5): 665–672
162. Auer DP et al (2000) Reduced glutamate in the anterior cingulate cortex in depression: an in vivo proton magnetic resonance spectroscopy study. Biol Psychiatry 47(4):305–313
163. Hashimoto K et al (2007) Increased levels of glutamate in brains from patients with mood disorders. Biol Psychiatry 62(11):1310–1316
164. Amalric M (2015) Targeting metabotropic glutamate receptors (mGluRs) in Parkinson's disease. Curr Opin Pharmacol 20:29–34
165. Bruno V et al (2001) Metabotropic glutamate receptor subtypes as targets for neuroprotective drugs. J Cereb Blood Flow Metab 21(9):1013–1033
166. Conn PJ (2003) Physiological roles and therapeutic potential of metabotropic glutamate receptors. Ann N Y Acad Sci 1003:12–21
167. Nicoletti F et al (2011) Metabotropic glutamate receptors: from the workbench to the bedside. Neuropharmacology 60(7–8): 1017–1041
168. Nicoletti F et al (2015) Metabotropic glutamate receptors as drug targets: what's new? Curr Opin Pharmacol 20:89–94

169. Spooren W et al (2003) Insight into the function of Group I and Group II metabotropic glutamate (mGlu) receptors: behavioural characterization and implications for the treatment of CNS disorders. Behav Pharmacol 14(4):257–277
170. Spooren WP et al (2001) Novel allosteric antagonists shed light on mglu(5) receptors and CNS disorders. Trends Pharmacol Sci 22(7):331–337
171. Noetzel MJ et al (2012) Emerging approaches for treatment of schizophrenia: modulation of glutamatergic signaling. Discov Med 14(78): 335–343
172. Morris BJ et al (2005) PCP: from pharmacology to modelling schizophrenia. Curr Opin Pharmacol 5(1):101–106
173. Adler CM et al (1999) Comparison of ketamine-induced thought disorder in healthy volunteers and thought disorder in schizophrenia. Am J Psychiatry 156(10):1646–1649
174. Marek GJ et al (2010) Glutamatergic (N-methyl-D-aspartate receptor) hypofrontality in schizophrenia: too little juice or a miswired brain? Mol Pharmacol 77(3): 317–326
175. Wieronska JM et al (2016) Metabotropic glutamate receptors as targets for new antipsychotic drugs: historical perspective and critical comparative assessment. Pharmacol Ther 157:10–27
176. Serafini G et al (2013) Pharmacological properties of glutamatergic drugs targeting NMDA receptors and their application in major depression. Curr Pharm Des 19(10): 1898–1922
177. Matosin N, Newell KA (2013) Metabotropic glutamate receptor 5 in the pathology and treatment of schizophrenia. Neurosci Biobehav Rev 37(3):256–268
178. Wright RA et al (2013) CNS distribution of metabotropic glutamate 2 and 3 receptors: transgenic mice and [(3)H]LY459477 autoradiography. Neuropharmacology 66:89–98
179. Moghaddam B, Adams BW (1998) Reversal of phencyclidine effects by a group II metabotropic glutamate receptor agonist in rats. Science 281(5381):1349–1352
180. Lu WY et al (1999) G-protein-coupled receptors act via protein kinase C and Src to regulate NMDA receptors. Nat Neurosci 2(4):331–338
181. Borgdorff AJ, Choquet D (2002) Regulation of AMPA receptor lateral movements. Nature 417(6889):649–653
182. Gould RW et al (2016) Partial mGlu(5) negative allosteric modulators attenuate cocaine-mediated behaviors and lack psychotomimetic-like effects. Neuropsychopharmacology 41(4):1166–1178
183. Horio M et al (2013) Therapeutic effects of metabotropic glutamate receptor 5 positive allosteric modulator CDPPB on phencyclidine-induced cognitive deficits in mice. Fundam Clin Pharmacol 27(5):483–488
184. Rodriguez AL et al (2010) Discovery of novel allosteric modulators of metabotropic glutamate receptor subtype 5 reveals chemical and functional diversity and in vivo activity in rat behavioral models of anxiolytic and antipsychotic activity. Mol Pharmacol 78(6): 1105–1123
185. Parmentier-Batteur S et al (2014) Mechanism based neurotoxicity of mGlu5 positive allosteric modulators—development challenges for a promising novel antipsychotic target. Neuropharmacology 82:161–173
186. Kessler RC et al (2009) The global burden of mental disorders: an update from the WHO World Mental Health (WMH) surveys. Epidemiol Psichiatr Soc 18(1):23–33
187. Tham A et al (2016) Efficacy and tolerability of antidepressants in people aged 65 years or older with major depressive disorder – a systematic review and a meta-analysis. J Affect Disord 205:1–12
188. Woods JH et al (1992) Benzodiazepines: use, abuse, and consequences. Pharmacol Rev 44(2):151–347
189. Galling B et al (2015) Safety and tolerability of antidepressant co-treatment in acute major depressive disorder: results from a systematic review and exploratory meta-analysis. Expert Opin Drug Saf 14(10):1587–1608
190. Chojnacka-Wojcik E et al (2001) Glutamate receptor ligands as anxiolytics. Curr Opin Investig Drugs 2(8):1112–1119
191. Chaki S et al (2013) mGlu2/3 and mGlu5 receptors: potential targets for novel antidepressants. Neuropharmacology 66:40–52
192. Swanson CJ et al (2005) Metabotropic glutamate receptors as novel targets for anxiety and stress disorders. Nat Rev Drug Discov 4(2): 131–144
193. Galici R et al (2005) A selective allosteric potentiator of metabotropic glutamate (mGlu) 2 receptors has effects similar to an orthosteric mGlu2/3 receptor agonist in mouse models predictive of antipsychotic activity. J Pharmacol Exp Ther 315(3): 1181–1187
194. Galici R et al (2006) Biphenyl-indanone A, a positive allosteric modulator of the metabotropic glutamate receptor subtype 2, has antipsychotic- and anxiolytic-like effects in mice. J Pharmacol Exp Ther 318(1):173–185

195. Valenti O et al (2002) Distinct physiological roles of the Gq-coupled metabotropic glutamate receptors co-expressed in the same neuronal populations. J Cell Physiol 191(2): 125–137
196. Inta D et al (2013) Significant increase in anxiety during aging in mGlu5 receptor knockout mice. Behav Brain Res 241:27–31
197. Li X et al (2006) Metabotropic glutamate 5 receptor antagonism is associated with antidepressant-like effects in mice. J Pharmacol Exp Ther 319(1):254–259
198. Spooren WP et al (2000) Anxiolytic-like effects of the prototypical metabotropic glutamate receptor 5 antagonist 2-methyl-6-(phenylethynyl)pyridine in rodents. J Pharmacol Exp Ther 295(3):1267–1275
199. Schulz B et al (2001) The metabotropic glutamate receptor antagonist 2-methyl-6-(phenylethynyl)-pyridine (MPEP) blocks fear conditioning in rats. Neuropharmacology 41(1):1–7
200. Belozertseva IV et al (2007) Antidepressant-like effects of mGluR1 and mGluR5 antagonists in the rat forced swim and the mouse tail suspension tests. Eur Neuropsychopharmacol 17(3):172–179
201. Felts AS et al (2013) Discovery of VU0409106: a negative allosteric modulator of mGlu5 with activity in a mouse model of anxiety. Bioorg Med Chem Lett 23(21): 5779–5785
202. Mueller R et al (2012) Discovery of 2-(2-benzoxazoyl amino)-4-aryl-5-cyanopyrimidine as negative allosteric modulators (NAMs) of metabotropic glutamate receptor 5 (mGlu(5)): from an artificial neural network virtual screen to an in vivo tool compound. ChemMedChem 7(3):406–414
203. Tzschentke TM, Schmidt WJ (2003) Glutamatergic mechanisms in addiction. Mol Psychiatry 8(4):373–382
204. Christie MJ et al (1987) Excitatory amino acid projections to the nucleus accumbens septi in the rat: a retrograde transport study utilizing D[3H]aspartate and [3H] GABA. Neuroscience 22(2):425–439
205. Gorelova N, Yang CR (1997) The course of neural projection from the prefrontal cortex to the nucleus accumbens in the rat. Neuroscience 76(3):689–706
206. Eiler WJ 2nd et al (2011) mGlu5 receptor deletion reduces relapse to food-seeking and prevents the anti-relapse effects of mGlu5 receptor blockade in mice. Life Sci 89 (23–24):862–867
207. Bird MK et al (2010) Cocaine-mediated synaptic potentiation is absent in VTA neurons from mGlu5-deficient mice. Int J Neuropsychopharmacol 13(2):133–141
208. Olsen CM et al (2010) Operant sensation seeking requires metabotropic glutamate receptor 5 (mGluR5). PLoS One 5(11): e15085
209. Stoker AK et al (2012) Involvement of metabotropic glutamate receptor 5 in brain reward deficits associated with cocaine and nicotine withdrawal and somatic signs of nicotine withdrawal. Psychopharmacology 221(2):317–327
210. Chesworth R et al (2013) The metabotropic glutamate 5 receptor modulates extinction and reinstatement of methamphetamine-seeking in mice. PLoS One 8(7):e68371
211. Chiamulera C et al (2001) Reinforcing and locomotor stimulant effects of cocaine are absent in mGluR5 null mutant mice. Nat Neurosci 4(9):873–874
212. Olive M (2009) Metabotropic glutamate receptor ligands as potential therapeutics for addiction. Curr Drug Abuse Rev 2:83–98
213. Liechti ME et al (2007) Metabotropic glutamate 2/3 receptors in the ventral tegmental area and the nucleus accumbens shell are involved in behaviors relating to nicotine dependence. J Neurosci 27(34):9077–9085
214. Helton DR et al (1997) LY354740: a metabotropic glutamate receptor agonist which ameliorates symptoms of nicotine withdrawal in rats. Neuropharmacology 36(11–12):1511–1516
215. Caprioli D et al (2015) Effect of the novel positive allosteric modulator of metabotropic glutamate receptor 2 AZD8529 on incubation of methamphetamine craving after prolonged voluntary abstinence in a rat model. Biol Psychiatry 78(7):463–473
216. Woolf CJ, Salter MW (2000) Neuronal plasticity: increasing the gain in pain. Science 288(5472):1765–1769
217. Martin LJ et al (1992) Cellular localization of a metabotropic glutamate receptor in rat brain. Neuron 9(2):259–270
218. Walker K et al (2001) mGlu5 receptors and nociceptive function: II. mGlu5 receptors functionally expressed on peripheral sensory neurones mediate inflammatory hyperalgesia. Neuropharmacology 40(1):10–19
219. Crawford JH et al (2000) Mobilisation of intracellular Ca2+ by mGluR5 metabotropic glutamate receptor activation in neonatal rat cultured dorsal root ganglia neurones. Neuropharmacology 39(4):621–630
220. Galik J et al (2008) Involvement of group I metabotropic glutamate receptors and glutamate transporters in the slow excitatory

synaptic transmission in the spinal cord dorsal horn. Neuroscience 154(4):1372–1387

221. Kolber BJ et al (2010) Activation of metabotropic glutamate receptor 5 in the amygdala modulates pain-like behavior. J Neurosci 30(24):8203–8213
222. Bennett CE et al (2012) Fused tricyclic mGluR1 antagonists for the treatment of neuropathic pain. Bioorg Med Chem Lett 22(4):1575–1578
223. Jamison RN, Mao J (2015) Opioid analgesics. Mayo Clin Proc 90(7):957–968
224. Garber KB et al (2008) Fragile X syndrome. Eur J Hum Genet 16(6):666–672
225. Huber KM et al (2002) Altered synaptic plasticity in a mouse model of fragile X mental retardation. Proc Natl Acad Sci U S A 99(11): 7746–7750
226. Bear MF et al (2004) The mGluR theory of fragile X mental retardation. Trends Neurosci 27(7):370–377
227. Irwin SA et al (2002) Dendritic spine and dendritic field characteristics of layer V pyramidal neurons in the visual cortex of fragile-X knockout mice. Am J Med Genet 111(2): 140–146
228. Nimchinsky EA et al (2001) Abnormal development of dendritic spines in FMR1 knockout mice. J Neurosci 21(14):5139–5146
229. Sahin M (2012) Targeted treatment trials for tuberous sclerosis and autism: no longer a dream. Curr Opin Neurobiol 22(5):895–901
230. Kwiatkowski DJ, Manning BD (2005) Tuberous sclerosis: a GAP at the crossroads of multiple signaling pathways. Hum Mol Genet 14(Spec No 2):R251–R258
231. Wullschleger S et al (2006) TOR signaling in growth and metabolism. Cell 124(3):471–484
232. Chevere-Torres I et al (2012) Metabotropic glutamate receptor-dependent long-term depression is impaired due to elevated ERK signaling in the DeltaRG mouse model of tuberous sclerosis complex. Neurobiol Dis 45(3):1101–1110
233. Auerbach BD et al (2011) Mutations causing syndromic autism define an axis of synaptic pathophysiology. Nature 480(7375):63–68
234. Bastide MF et al (2015) Pathophysiology of L-dopa-induced motor and non-motor complications in Parkinson's disease. Prog Neurobiol 132:96–168
235. Picconi B et al (2012) Synaptic dysfunction in Parkinson's disease. Adv Exp Med Biol 970: 553–572
236. Calabresi P et al (2000) Levodopa-induced dyskinesia: a pathological form of striatal synaptic plasticity? Ann Neurol 47(4 Suppl 1): S60–S68. discussion S68–S69
237. Fox SH et al (2011) The movement disorder society evidence-based medicine review update: treatments for the motor symptoms of Parkinson's disease. Mov Disord 26(S3): S2–S41
238. Duty S (2010) Therapeutic potential of targeting group III metabotropic glutamate receptors in the treatment of Parkinson's disease. Br J Pharmacol 161(2):271–287
239. Austin PJ et al (2010) Symptomatic and neuroprotective effects following activation of nigral group III metabotropic glutamate receptors in rodent models of Parkinson's disease. Br J Pharmacol 160(7):1741–1753
240. Lopez S et al (2012) Antiparkinsonian action of a selective group III mGlu receptor agonist is associated with reversal of subthalamonigral overactivity. Neurobiol Dis 46(1):69–77
241. Ouattara B et al (2011) Metabotropic glutamate receptor type 5 in levodopa-induced motor complications. Neurobiol Aging 32(7):1286–1295
242. Morin N et al (2013) MPEP, an mGlu5 receptor antagonist, reduces the development of L-DOPA-induced motor complications in de novo parkinsonian monkeys: biochemical correlates. Neuropharmacology 66:355–364
243. Morin N et al (2013) Chronic treatment with MPEP, an mGlu5 receptor antagonist, normalizes basal ganglia glutamate neurotransmission in L-DOPA-treated parkinsonian monkeys. Neuropharmacology 73:216–231
244. Rylander D et al (2010) A mGluR5 antagonist under clinical development improves L-DOPA-induced dyskinesia in parkinsonian rats and monkeys. Neurobiol Dis 39(3):352–361
245. Litim N. et al. (2016) Metabotropic glutamate receptors as therapeutic targets in Parkinson's disease: an update from the last 5 years of research. Neuropharmacology. doi:10.1016/j.neuropharm.2016.03.036

Chapter 9

Glutamatergic Synthesis, Recycling, and Receptor Pharmacology at *Drosophila* and Crustacean Neuromuscular Junctions

Joshua S. Titlow and Robin L. Cooper

Abstract

Invertebrate glutamatergic synapses have been at the forefront of major discoveries into the mechanisms of neurotransmission. In this chapter we recount many of the neurophysiological advances that have been made using invertebrate model organisms, from receptor pharmacology to synaptic plasticity and glutamate recycling. We then direct your attention to the crayfish and fruit fly larva neuromuscular junctions, glutamatergic synapses that have been extraordinarily insightful, the crayfish because of its experimental tractability and *Drosophila* because of its extensive genetic and molecular resources. Detailed protocols with schematics and representative images are provided for both preparations, along with references to more advanced techniques that have been developed in these systems. The chapter concludes with a discussion of unresolved questions and future directions for which invertebrate neuromuscular junction preparations would be particularly well suited.

Key words Neuromuscular junction, Glutamatergic synapse, Invertebrate, Crayfish, *Drosophila*

1 Overview of Glutamate Activity at Neuronal Synapses

Glutamate (Glu) is one of the most common neurotransmitters in animals as it is known to be present in some of the most primitive animal species [1–3] and is one of the most abundant transmitters in the central nervous system (CNS) of vertebrates [4]. Various receptor subtypes have evolved to provide a wide range of responses to Glu, from fast acting ion channels (ionotropic) to slow acting second messenger cascades (metabotropic), and excitatory as well as inhibitory responses. The types of receptors show a wide diversity across the animal kingdom [3, 5] and are even present in roots of some plants to respond to environmental Glu [6]. Classically, receptors have been defined by their pharmacological profile with agonist and antagonist binding affinities [7]. More recently receptors have been taxonomically defined by gene and protein sequence homology. On the presynaptic side, neurons employ various

Sandrine Parrot and Luc Denoroy (eds.), *Biochemical Approaches for Glutamatergic Neurotransmission*, Neuromethods, vol. 130,
DOI 10.1007/978-1-4939-7228-9_9,

mechanisms to incorporate Glu and organize its release. Glutamate can be taken into cells by plasma membrane transporters (GluT or excitatory amino acid transporters—EAAT) or indirectly by a transporter for other amino acids such as glutamine [8].

Through intracellular biochemical reactions amino acids and intermediate compounds can be converted to Glu. Intracellular Glu is packaged into synaptic vesicles against a concentration gradient through vesicular transporters (vGluT; [9, 10]). The process of Glu release, re-uptake, and repackaging to be released again is dependent on many molecular functions. The recycling process can be estimated by kinetic rates; however, there are various pathways depending on the synaptic circuit. In the CNS of vertebrates, Glu recycling occurs directly through GluT and indirectly through glial glutamate–glutamine–glutamate pathways, making it difficult to discreetly measure the various rates in intact systems. Glutamate can also be taken up into neurons that use γ-amino butyric acid (GABA) as a transmitter since Glu is converted to GABA in GABA-ergic neurons [11, 12].

Invertebrate neuromuscular junction (NMJ) preparations have played an important role in fostering our understanding of neurotransmission at glutamatergic synapses. The aim of this chapter is to consolidate the knowledge of invertebrate NMJs and discuss the experimental potential of invertebrate synapses going forward. In doing so we highlight the important similarities and differences in the molecular mechanisms underlying invertebrate NMJ and mammalian glutamatergic transmission, including pharmacological and physiological characteristics of Glu receptors. We then provide a brief description of protocols for the crayfish and fruit fly larva neuromuscular junctions and conclude with some ideas for future research directions with these systems.

2 Glutamatergic Transmission at Invertebrate Neuromuscular Junctions

Various invertebrate models have been used to investigate the regulation and developmental aspects of Glu receptors and their action on cells [13–18]. Likely due to the ease of experimental setup and long viability in a minimal saline, invertebrate neuromuscular junctions (NMJs) of insects and crustaceans lead the way in obtaining pharmacological profiles with a battery of compounds that would later be screened on isolated neural preparations of vertebrates to address similar physiological questions [19–25]. Thus, early on, due to the simplicity of NMJs for physiological recordings and observation these specimens served as models for understanding potential actions for vertebrate systems. Invertebrate NMJs were not necessarily a model for vertebrate NMJs, as acetylcholine (Ach) had already been touted as a transmitter for the heart [26] and NMJs in frog and mammals [27]; however in an assay to demonstrate Ach was the active substance for vertebrate NMJs, the leech skeletal muscle preparation was used [28]. Likely a need to replicate findings from the frog NMJ for Ach drove similar questions about glutamate action on the crustacean and locust NMJs, such as quantal responses [29, 30] and desensitization with prolonged application. Since Ach did not have an action at the crustacean NMJs, other potential transmitters known in the vertebrate CNS were tried from homogenized CNS samples of dog and guinea-pig on NMJs of the limbs as well as the hindgut of crayfish. This lead to further studies into the specific compounds that activated or inhibited transmission at crustacean NMJs on the limbs and gut [31]. Rapid progress followed in primarily crayfish preparations to determine the specific compounds that resulted in muscle contraction and inhibition. Ach and Ach antagonists were shown not to have a direct effect on NMJ preparations and would not block the actions of L-Glu [31–35].

An historical review detailing the discovery of GABA [36] walks one through the intriguing science from a compound termed "inhibitory factor," which was isolated from homogenized bovine brain tissue, to the observed effects and postulation that GABA was an active synaptic compound [32, 37, 38]. It was shown by Kuffler and Edwards [39], Boistel and Fatt [40], and later proved by Kravitz [41–44] that indeed GABA does exhibit inhibitory action as a neurotransmitter released from lobster motor neurons on the opener muscle of the walking leg. The discovery that GABA is released from nerves at the crustacean NMJs was of interest since GABA could block the response of Glu. It was later shown that GABA not only had reception on the contractile muscle directly but presynaptically on the excitatory motor neuron which released Glu [45].

After the initial discoveries demonstrated that amino acids were the compounds released from the motor neurons innervating crustacean muscle, a focus then turned to examining which various amino acids could have an effect on the NMJ responses in various crustacean and insect preparations. Past reviews by Usherwood [1, 2] mention various species used for investigating glutamatergic NMJs. Of crustaceans the crab [46–48], lobster [49], shrimp [50, 51], and heart of the isopod [52] were some of the preparations used. As for insects the cockroach [46], locust [53], moth [54], cabbage looper caterpillar [55], and blowfly [56] have been used for physiological studies. Other invertebrates such as an acorn barnacle [57] and snails [46] were also used.

Various agonist and antagonist as well as modulators of transmission were uncovered using invertebrates as experimental organisms over the years. The rationale to focus on crustaceans was most likely due to accessibility of the animal, viability, and ease to examine the responses from nerve stimulation, which was occurring even before the neurotransmitters were identified. In addition, there is a long history of anatomical characterization for these preparations going as far back as the 1880s [58] with observations that nerve stimulation could cause muscle contractions that lead to facilitation in force development ([59, 60], see review on the history of experimentation using the opener muscle of crayfish: [61]). When one considers that Sidney Ringer [62, 63] had only developed a saline for maintaining the viability of the frog heart preparations around the same time crayfish were being used to demonstrate muscle contraction from stimulating nerves over long periods of time in isolation, the crayfish offered further hope in addressing the properties of synaptic transmission. It was not until Van Harreveld [64] developed a saline for crayfish that prolonged physiological studies were practical. Synaptic physiology and dissection of the pharmacology and function of Glu and GABA receptors grew steadily afterwards using the crayfish and other crustaceans [65–67].

Using various stimulation paradigms of the motor nerve, short-term facilitation (STF) [68] and long-term facilitation (LTF) was first demonstrated at crustacean NMJs [69] and later long-term potentiation (LTP) was shown to be present in mammalian CNS preparations [70]. These findings directed investigations to determine if the mechanisms were due to purely presynaptic or postsynaptic modifications in the receptor density or receptor subtypes to account for the effects. Pharmacological profiling of crustacean NMJs continued in the early days [19, 71–76] providing assays to determine mechanisms for modulation of the response to Glu with a wide variety of compounds. Shank and Freeman [77] demonstrated that aspartate produced a cooperative effect with Glu at lobster NMJs. This was also confirmed to occur at NMJs in a Hermit crab [78]. L-Proline was shown to act as a Glu antagonist [79] which is surprising as proline increases in the hemolymph with cold stress in insects [80]; thus, it would appear to further limit NMJ function in response to cold. The effects of other compounds such as piperidine dicarboxylates [81], 5-methyl-1-phenyl-2-(3-piperidinopropylamino)-hexane-1-ol (MLV-5860) [82], chlorisondamine and TI-233 [83, 84], spermidine [85], streptomycin, and similar antibiotics [86], quisqualic acid [87], stizolobic acid [88], α-amino-3-hydroxy-5-methyl-4-isoxazole-propionic acid (AMPA), *N*-methyl-D-aspartate (NMDA) and (1S,3R)-1-aminocyclo-pentane-1.3-dicarboxylic acid (t-ACPD) [89] were also discovered. The Glu receptor subtype on the body wall muscles of the crayfish and many crustaceans is primarily classified as quisqualate sensitive (~100 times increased responsiveness than Glu; [90]) and ionotropic [19, 91] with Na^+ being the predominate ion, in addition to some Ca^{2+} influx and K^+ efflux when opened at resting membrane levels [92].

During synaptic transmission Glu induces a rapid current influx that produces a rapid depolarization of the muscle membrane followed by a much slower decay in the synaptic potential. The amplitudes of the excitatory synaptic responses varies greatly at crustacean NMJs as there are a variety of synaptic responses from spiking muscles to graded excitatory postsynaptic potentials (EPSPs) that can arise from high- and low-output synapses [93–97]. The non-spiking EPSPs show a slow decay which is partly due to desensitization of the receptors [92, 98–101] and if the muscle is bathed in Glu the receptors will fully desensitize, blocking transmission [98, 102]. The presence of high extracellular calcium ions is known to decrease the rate of desensitization by Glu [103, 104] and concanavalin A (a plant lectin; [105]) can not only partially decrease desensitization on its own but it can also block the effect of Ca^{2+} on the receptors [103]. Thus, the desensitization effect of Ca^{2+} is extracellular on the receptors or membrane. As far as we are aware this has not been addressed in insect NMJs.

The potential for presynaptic glutamatergic autoreceptors has also been investigated at the crustacean and insect NMJs. Since presynaptic glutamatergic receptors occur in the mammalian CNS [106] it would not be surprising to also predict they might occur at NMJs in the invertebrates. The use of a metabotropic agonist t-ACPD on NMJs of the crayfish provided confounding results with some preparations being enhanced and others depressed [89]. Since some preparations showed an effect there may well be presynaptic autoreceptors for Glu in the crustacean preparations [89]. It would be of interest to examine high-output as well as low-output NMJs for differences in effects to t-ACPD as well as other potential metabotropic agonists and antagonists.

While the crustaceans were being examined for glutamatergic actions at the NMJs and pharmacological profiling, the NMJ of locust legs served as an insect counterpart. This preparation was used likely due to accessibility and being a relatively large insect preparation for physiologists at the time. There is a rich history of physiology and pharmacology using the locust preparation [21, 22, 24, 25]. Similarly, the locust NMJ paralleled the crayfish NMJ in physiology and pharmacological profiling as well as in desensitization with Glu. A literature search in PUBMED.GOV using the key words "Insect glutamate neuromuscular junction" returned 319 hits. The first 142 references and most following ones focused exclusively on *Drosophila* which indicates the recent research focus among the vast array of insect species present. As with the crayfish and other crustacean preparations, the locust model fell short in being able to genetically manipulate the expression of Glu receptor subunits and proteins involved with synaptic transmission. Though these model systems are still valuable for addressing particular physiological questions, the era of molecular biology has given way to the more genetically amenable *Drosophila melanogaster* as a model for synaptic studies using the neuromuscular junction.

3 Glutamate Receptors in the *Drosophila* Neuromuscular Junction

Sophisticated gene manipulation, extensive collections of mutant lines, and relatively simple, inexpensive maintenance make *Drosophila melanogaster* an excellent experimental system for neurobiology. The *Drosophila* larva NMJ in particular has been steadily revealing the physiological mechanisms of synaptic transmission for over 40 years [23]. It is a rare system where individual synapses from identified neurons can easily be manipulated in the context of development or plasticity in vivo. In partially dissected preparations the glutamatergic synapses lie directly on the muscle cell surface, providing uninhibited optical access for single molecule localization, super resolution, and other advanced microscopy techniques in combination with electrophysiology. Physiologically relevant salines that allow prolonged viability have been a breakthrough for physiologists in the *Drosophila* field [107–109]. Optogenetic stimulation and calcium imaging are also well established in this system [110, 111]. The purpose of this section is to describe what is known about glutamatergic neurotransmission at the *Drosophila* NMJ, while pointing out essential similarities and differences between it and mammalian neural synapses. We then discuss recent discoveries in Glu receptor pharmacology and synaptic plasticity at the *Drosophila* NMJ, and finish the section with an overview of molecular mechanisms that are required for proper Glu receptor localization. For detailed information on experimental paradigms and other molecular factors that have been described in the larva NMJ there is an entire book and several comprehensive review articles [112–114].

3.1 Pharmacological Properties of Glutamate Receptors in the Drosophila Larva NMJ

Pharmacological and genetic analyses have provided a clear picture of the ionotropic Glu receptor (iGluR) subtypes present at the *Drosophila* larva NMJ. The field unanimously asserts that the iGluRs present at the post synaptic density are heterotetramers composed of three common subunits (GluRIIC, GluRIID, and GluRIIE) and an interchangeable forth subunit (either GluRIIA or GluRIIB) [15, 115–117]. Though these iGluR subunits most closely resemble vertebrate AMPA and kainate receptors at the amino acid sequence level, *Drosophila* iGluRs exhibit distinct differences in their pharmacological profile. Most notable is that AMPA type iGluRs expressed at the *Drosophila* NMJ are not especially sensitive to AMPA, kainate, or NMDA, but respond to quisqualate [118, 119]. The molecular difference underlying species-specific agonist activity may have been detected in a recent study that reported the crystal structure of GluRIIB bound to Glu. Though the volume of the GluRIIB ligand binding cavity is similar to the vertebrate ligand binding cavity, the presence of Tyr481, through interactions with Asp509 and Arg429, appears to prevent binding of the common ligands [120]. A similar finding was later made for the GluRIIA Glu complex, which also exhibits a pharmacological profile that diverges from the vertebrate iGluR [7]. Importantly, heterologous expression approaches were achieved in both studies that enable functional reconstitution of the iGluR complex, providing the opportunity to test different gene products with single channel resolution *in vitro*, quickly transfer those gene products into the organism with *Drosophila* gene editing [121], and verify the hypotheses in vivo at the larva NMJ.

Another difference in *Drosophila* larva NMJ receptor pharmacology is sensitivity to toxins. Lobster and cricket NMJs as well as at mammalian hippocampal pyramidal neurons are blocked by the Joro spider toxin (JSTX) [122–124], but to our knowledge JSTX does not block *Drosophila* Glu receptors. Philanthotoxin-433 (PhTx), however, a noncompetitive open channel Glu receptor blocker derived from wasp venom, has proven to be a powerful pharmacological tool for investigating glutamatergic transmission at the *Drosophila* larva NMJ. When injected into the larva or applied directly to the exposed NMJ, PhTx induces presynaptic compensation within ~10 min [125]. This form of synaptic plasticity, referred to as homeostatic plasticity [126], is achieved through an increase in quantal content, and is also observed in GluRIIa mutants [127]. Not only is this form of plasticity observed at hippocampal glutamatergic synapses, some of the key molecular components are conserved, including presynaptic calcium channels [128, 129] and postsynaptic mTOR signaling [130, 131]. A notable mechanistic aspect of homeostatic plasticity at the larval NMJ is that it requires retrograde signaling from the postsynaptic muscle cell to the motor neuron. Retrograde signaling appears to be a widespread mechanism that has emerged throughout nervous system evolution to regulate various forms of synaptic plasticity. Cell-specific control of gene expression in pre- and post-synaptic compartments has made the *Drosophila* NMJ a convenient system to address the location of action for many molecules.

Two metabotropic Glu receptors are found in the *Drosophila* genome though only one was found to be functional (mGluR; [132, 133]). *Drosophila* mGluR has 45% and 43% amino acid sequence homology with its mammalian homologs, mGluR3 and mGluR2 respectively, and it was responsive to several mammalian mGluR agonists and antagonists in a mammalian heterologous expression system, showing negative coupling to the adenylate cyclase pathway [132]. At the larval NMJ, mGluR is expressed predominantly in the presynaptic compartment where it has a role in activity-dependent plasticity [134]. mGluR mutants exhibited normal baseline synaptic transmission but significantly enlarged bouton size and reduced bouton number. A relatively limited panel of pharmacological agents has been tested in this system in vivo, and it is also not yet known whether these receptors have a role in rapid activity-dependent structural modifications at the NMJ.

3.2 Physiological Properties of Glutamatergic Neurotransmission at the Drosophila Larva NMJ

Simple electrophysiological accessibility to a genetically specified synapse is a valuable feature of the *Drosophila* larva NMJ. In the larva filet preparation, motor synapses on the dorsoventral longitudinal muscles lie directly on the cell surface. These muscles are large (~100 μm × 300 μm), isopotential, and do not exhibit active membrane properties under normal culturing conditions. Muscles 6 and 7 are the most often used and best characterized [135], but they do exhibit an important drawback, which is that they are each innervated by two separate motor neurons. The larval muscles also receive innervation from aminergic neurons [136].

Ionic currents in the larva muscle have been well characterized through genetic and pharmacological analysis. Iontophoresis of L-Glu at the synaptic termini was used to determine that the excitatory transmitter at the larva NMJ is Glu ([137], though for inhibitory effects of L-Glu see: [138]). iGluRs in the larva NMJ rapidly desensitize in the presence of excessive extracellular Glu [137, 139]. Synaptic potentials can be investigated by electrical stimulation of the segmental nerves innervating dorsoventral longitudinal muscles. A nonspecific cation synaptic current can be recorded intracellularly throughout the muscle in response to nerve stimulation or as a result of endogenous activity if the nerves are not severed from the brain. Single quantal events can also be observed in intracellular recordings from the muscle. Kinetics of the evoked potentials have been analyzed using ion exchange, common reagents for blocking ion channels, and through analysis of ion channel mutants that were isolated from genetic screens. As described in the synaptic plasticity section below, transmission at the NMJ is extremely sensitive to extracellular Ca^{2+} levels [140]. Passive membrane properties of the muscles are well characterized. An inward Ca^{2+} current and outward K^{+} current are readily observable under two-electrode voltage clamp. The K^{+} current is sensitive to tetraethyl ammonium and is significantly reduced in *either-a-go-go* and *shaker* mutants, which code for potassium channels known to be responsible for the inward rectifying and transient A current respectively [141]. Peron et al. [142] provide a detailed overview of the other ion channel genes expressed at the larva NMJ.

3.3 Synaptic Plasticity at the Drosophila Larva NMJ

Activity-dependent synaptic plasticity has been extensively studied at the *Drosophila* larva NMJ. Throughout larva development the muscle size increases exponentially and requires equivalent expansion of the synaptic field. Through genetic analysis it was determined that synaptic expansion during NMJ development is an activity-dependent process that also requires trophic factors associated with tissue development (reviewed in [113]). The mature NMJ synapse at the last stage of larva development has also been shown to exhibit several forms of short- and long-term synaptic plasticity resembling long-term potentiation (LTP) and long-term depression (LTD) that are investigated in mammalian brain preparations. Here we describe the characteristics of acutely inducible forms of synaptic plasticity at the larva NMJ.

Different forms of activity-dependent synaptic plasticity can be assessed at the larva NMJ simply by adjusting the stimulus parameters. Similar to long-term facilitation (LTF) in crustaceans and long-term potentiation (LTP) in mammals, nerve evoked synaptic potentials in the *Drosophila* larva NMJ can be enhanced by trains of high-frequency stimuli, referred to as post-tetanic potentiation (PTP, [143]). This form of activity-dependent plasticity is evoked by stimulus frequencies between 5–20 Hz, low extracellular Ca^{2+} (0.2 mmol/L), is cAMP-dependent, and lasts on average for 158 s [143]. Lower stimulus frequencies (0.1–1 Hz) induce a form of depression called low-frequency short-term depression [144], whereas higher frequencies (40–60 Hz) induce short-term depression in low extracellular calcium conditions (<1 mmol/L). Paired pulse facilitation is another form of short-term plasticity that provides a robust readout of synaptic physiology at the larva NMJ [140]. Currently there is no widely accepted example of nerve-induced stimulation that induces long-term synaptic changes resembling LTP. Given that NMDA-like iGluRs have not been identified at the larva NMJ it is unlikely that a strictly homologous LTP phenomenon exists. However activity-dependent synaptic phenomena resembling the cellular changes in LTP have been identified. Increasing synaptic activity through elevated temperature or induced crawling can cause NMJ growth and potentiated transmitter release [145]. Spaced potassium depolarization, in dissected but intact NMJ preparations, also induces synapse formation and potentiated transmitter release [146]. Both phenomena require translation and the latter process requires transcription. Taken together, the physiological changes that are consolidated with new structures that require changes in gene expression, activity-dependent plasticity at the larva NMJ is a legitimate experimental system for investigating the molecular mechanisms of long-term information storage in glutamatergic synapses. Wnt signaling, BMP signaling, miRNAs, and CamKII have already been implicated in long-term facilitation at the NMJ [146–149], and others are sure to follow.

3.4 Molecular Mechanism of Glutamate Receptor Localization at the Drosophila Larva NMJ

The *Drosophila* larva NMJ has been especially valuable for determining how Glu receptors are localized to synaptic sites. Success in this field is due in large part to ease of imaging the larva filet preparation and reliable, commercially available antibodies for labeling Glu receptors and other synaptic markers [150]. Live imaging of fluorescent protein tagged Glu receptors in the *Drosophila* larva has provided unparalleled insight into the dynamics of Glu receptor assembly in vivo [17, 151]. Genetic analysis has also provided extensive insight into glutamatergic synapse formation. We identified at least 41 separate studies that reported a change in larva

NMJ Glu receptor level as a result of loss of function gene mutation (Table 1). A more recent reverse genetic screen, which focused specifically on PDZ containing genes, determined that null mutations in 42.8% (48 of the 112 nonlethal mutations) of the PDZ containing genes disrupt GluRIIa localization in vivo [152]. These results indicate that Glu receptor localization is an amazingly complex process that is regulated by several convergent molecular pathways. Physiological state of the NMJ is also important, as spontaneous neurotransmission is required for proper iGluR localization [192, 193]. Localization of the presynaptic active zones and the postsynaptic receptor array are tightly correlated; however, it does appear that spontaneous vesicle fusion events and evoked events do not always use the same synaptic sites [194]. Thus, synaptic sites have varying probabilities of transmission which may also have to do with differences in synaptic complexity [194, 195].

The mechanism for iGluR localization appears to be post translational, as RNA fluorescence in situ hybridization (FISH) shows very little overall enrichment for iGluR mRNA at the post synaptic density [196], though studies have shown that mRNA and RNA binding proteins are present in close proximity to the synapse [161, 197]. With recent application of single molecule FISH to the larva NMJ [198] it will be possible to determine how different aspects of glutamatergic transmission are locally regulated at the mRNA level.

4 Glutamate Recycling in *Drosophila* and Crayfish NMJs

It is apparent that the glutamatergic synapses at the invertebrate NMJs function similarly to other chemical synapses, although some of the synaptic ultrastructure may differ [95–97]. Generally, transmitter is packaged into clear core synaptic vesicles within the presynaptic nerve terminal via vesicular transporters (vGluT) [199, 200] and vesicles exist in various states, from being docked and readily releasable to being sequestered in reserve pools [199, 201–204]. The vesicle pools are dynamic with stimulation dependent recruitment [18, 205] and can be depressed with repetitive stimulation [206, 207]. As with other synaptic preparations of NMJs in vertebrates and invertebrates there are low- and high-output type synapses and differing muscle phenotypes (slow, intermediate, and fast; [208, 209]). The general characteristics are that the low-output synapses have few docked vesicles but can show dramatic facilitation due to reserve vesicles and recruitment to active zone sites on synapses, whereas the high-output synapses fuse many vesicles and produce large EPSPs but fatigue quickly due to a limited reserve pool [207].

The process of Glu uptake through the presynaptic plasma membrane transporter (GluT/ EAAT) and repackaging in the vesicles (vGluT) [9, 10] in *Drosophila* and crayfish models serve as

Table 1
Factors affecting GluR localisation at the *Drosophila* larva NMJ

Gene name	GluR subtype	LoF effect on iGluR levels	Reference
PDZ containing genes	GluRIIA	42 Decrease, 6 increase	[152]
Filamin	GluRIIA	Decrease	[153]
Diablo	GluRIIA	Increase	[154]
Lk6 kinase	GluRIIA	Decrease	[155]
Monensin sensitivity 1	GluRIIA	Increase	[156]
Neuropilin and toll-like protein	GluRIIA	Increase	[157]
Kismet	GluRIIA	Decrease	[158]
Activin	GluRIIA/B	Decrease	[159]
Neurologin 3	GluRIIA	Decrease	[160]
Staufen	GluRIIA	Decrease	[161]
Wingless	GluRIIA	Increase	[162]
Reverse polarity	GluRIIA	Increase	[162]
Mgat1	GluRIIB	Decrease	[163]
Slowpoke	GluRIIA/B	Decrease	[164]
Tbc1d15-17	GluRIIA	Decrease	[165]
Akt1	GluRIIA	Decrease	[166]
Lethal giant larvae	GluRIIB	Increase	[167]
Longitudinal lacking	GluRIIA/B, III	Decrease	[168]
Tor and eIF2a	GluRIIA	Decrease	[133]
Calcium/calmodulin-dependent serine protein kinase	GluRIIA	Decrease	[169]
Neuroligin 2	GluRIIB, III	Decrease	[170]
Neurexin	GluRIIA	Decrease	[171]
Neuroligin 1	GluRIID	Decrease	[172]
Metro	GluRIID	Decrease	[173]
Twinfilin	GluRIIA	Decrease	[174]
Rho GTPase activating protein at 100F	GluRIIA	Increase	[175]
CamKII	GluRIIA	Increase	[176]
Nanos	GluRIIA/B	Decrease/increase	[177]

(continued)

Table 1 (continued)

Gene name	GluR subtype	LoF effect on iGluR levels	Reference
Dystroglycan	GluRIIA	Decrease	[178]
Protein-O-mannosyl transferase 1	GluRIIB	Decrease	[179]
Dorsal	GluRIIA	Decrease	[180]
Mind the gap	GluRIIC/D	Mislocalised	[181]
β2 and β6 proteasome	GluRIIB	Increase	[182]
Coracle	GluRIIA	Decrease	[183]
Discs large 1	GluRIIB	Decrease	[184]
Pumilio	GluRIIA	Increase	[185]
p21-activated kinase	GluRIIA	Decrease	[186]
Survival of motor neuron	GluRIIA	Decrease	[187]
G protein α s subunit	GluRIIA	Decrease	[188]
Rho-type guanine nucleotide exchange factor	GluRIIA	Decrease	[189]
Actin 57b	GluRIII	Decrease	[190]
Nesprin	GluRIIA	Decrease	[191]

models for vertebrate glutamatergic synapses as they are pharmacologically similar. TBOA blocks reuptake via GluT [18, 210–212] and Bafilomycin A1(B1793) blocks vacuolar ATPase which drives vGluT [18, 200].

Although novel proteins and functional significance associated with vesicle and Glu receptor dynamics are continuously being discovered at *Drosophila* NMJs, homologs are sought in analogous glutamatergic synaptic sites in vertebrates and in synapses which are not glutamatergic [153, 155, 213, 214]. The ability to examine the effect of overexpression or knock down is rapidly able to be addressed using the *Drosophila* model. Temporally regulated expression with Gal80 has promoted this model over others to separate acute molecular mechanisms from developmental issues in synaptogenesis. Diseases inflicting glutamatergic synapses in humans are also modeled at the *Drosophila* NMJ [215]. The glutamatergic synapses at the *Drosophila* NMJ show effects of aging and disuse with a loss of presynaptic vesicles and prolonged recovery due to stressors of activity [216], which are similar to those shown in crustaceans [217] and mammals [218–220].

5 Interesting Side Notes

An interesting phenomenon that occurs at crayfish and *Drosophila* NMJs is that CO_2 blocks Glu receptors directly, independent of decreased intracellular or extracellular pH induced by CO_2 exposure [221–223]. CO_2 also rapidly paralyzes honeybees [224]. Lower extracellular and intracellular pH to 5.0 still allows synaptic transmission to occur but in the presence of CO_2 the synaptic transmission is rapidly blocked and removing CO_2, even though intracellular pH is still reduced, reverses the receptor block. Hypoxia or displacement of O_2 with N_2 does not mimic the rapid effect of CO_2. Interestingly CO_2 was used as an anesthetic for human and animal surgeries in early medicine [225]. Vertebrate NMDA receptors on cerebellar neurons are inhibited by protons even within a physiological pH range [226]. The open channel blocker MK-801 decreases its affinity in low pH suggesting that possible low Ca^{2+} flux with low pH results in causes inward currents through the NMDA receptors to decrease [227, 228]. The effect of protons on the NMDA receptors may be extracellular [229] but such detail as to the potential mechanism of action on the quisqualate Glu receptors of the invertebrates has not been investigated.

Invertebrate models, with the exception of *Drosophila*, have not previously been genetically amenable to investigate molecular mechanisms of synaptic transmission at NMJs. Some clever alternative approaches have enabled successful molecular investigation in these systems. Because many crustacean motor axons are large enough for pressure injection or iontophoresis, chemical compounds, small proteins, siRNA, or mRNA have been injected directly into the presynaptic terminal to examine functionality [230]. Gene transfection in primary cell culture has also been an effective approach to study molecular mechanisms of synaptic transmission in invertebrates [231]. With CRISPR/Cas9 mutagenesis, it is now possible to perform genetic analysis in crustacean organisms that are not typical genetic model organisms [232], making it possible that we will see a resurrection in the use of crustacean NMJs systems for glutamatergic synapse biology.

6 Protocols for the Crayfish and *Drosophila* Larva Neuromuscular Junction Preparations

To help visualize the invertebrate neuromuscular junction preparations we provide a brief overview and pertinent references for the crayfish leg and fruit fly larva neuromuscular junction protocols, instead of giving step-by-step protocols for each preparation which are referenced in the sections below. The aim here is to introduce the experimental procedure for accessing these synapses and highlight some of the key advantages and limitations.

6.1 Crayfish Neuromuscular Junction Preparations

The crayfish and lobster offer several types of NMJ preparations, many of which have been described in numerous publications for teaching modules or detailed research-based protocols. The crayfish NMJ preparations tend to have a better viability than lobster or crab models over longer periods in a defined saline for teaching labs and for addressing experimental questions. The muscle phenotypes and innervation profiles of the commonly used crayfish and lobster muscles have been described [96, 209, 233, 234]. Depending on the synaptic responses to be investigated in the crayfish model one can readily choose a low-output tonic-like NMJ or a high-output phasic-like NMJ on a single muscle fiber that is dually innervated (i.e., the walking leg extensor muscle), or muscle fibers that are mostly innervated by one type of innervation profile (tonic-like or phasic-like) that also correlates with the muscle phenotype. Abdominal muscles which are tonic-like are as follows: superficial extensor lateral, superficial extensor medial, and superficial flexor muscles. The abdominal muscles which are phasic-like are as follows: deep extensor lateral, deep extensor medial, and deep flexor muscles. The anatomical arrangement of the abdominal muscles are highlighted in Sohn et al. [235, 236], and dissection procedures are shown in video format in Baierlein et al. [237]. The opener muscle of the walking leg contains regional variation in the innervation and muscle fiber profiles even though the muscle is innervated by a single excitatory motor neuron. However, the innervation and muscle phenotype generally displays a tonic-like profile. The dissection and physiological procedures for the opener muscle is shown schematically in Fig. 1a and video format [61]. The dissection and recording procedures for the walking leg extensor with the dually innervated muscle fibers of high- and low-output synapses is also shown in video format [238]. An advantage to using the walking legs is that the animal will autotomize the leg when pinched at the base so four or more preparations can be obtained from one animal and if the animal is left alive for some time the legs will regenerate.

The tonic-like innervation profile is one that will show smaller EPSP amplitudes but will rapidly facilitate in amplitude in a stimulation frequency dependent manner. The synaptic responses are fatigue resistance and generally contain larger varicosities than the thin filiform like nerve terminals of the phasic-like innervation. The high-output phasic innervation usually produces large amplitude EPSPs and will fatigue relatively quickly compared to the tonic innervation. Intracellular recordings in the axons of the motor neurons are a key asset to the crustacean preparations. Substances such as peptide fragments, ionic indicators, and direct measures of the action potential shape to address presynaptic contributions to synaptic physiology have been conducted in the crayfish opener preparation [230, 239–241]. The caveat in working with NMJs is the fact the muscle can contract. If one is only examining the presynaptic terminal then the Glu receptors on the muscle fibers can be desensitized by adding Glu (1–10 mmol/L) to the bath. However, in measuring functional synaptic responses the intracellular electrode in a contracting muscle fiber may be dislodged. This can usually be prevented by maintaining the muscle in a taut position when pinning the preparation in a recording dish.

Synaptic responses can readily be measured with standard intracellular recording techniques. However, due to the large size of some muscle fibers two electrode voltage clamp is not as feasible compared to larval *Drosophila* muscles due to space clamp issues. In addition, in larger muscles the minis can be difficult to detect, which is in part due to lower input resistance but also a decrement in the electrical responses due to cable properties of the muscle membrane. To directly measure quantal responses from select regions of a motor nerve, focal macropatch recordings offer excellent resolution of single quantal events. The single quantal responses can be used to address synaptic efficacy and shapes of the synaptic responses related to Glu receptor function [242].

In investigating proteins involved in synaptic structure the crayfish preparations offers tissue with sufficient material for Western blots and in situ staining. The crayfish nerve terminals have shown to be immunocytochemically similar to *Drosophila* in terms of some antibody staining (i.e., synaptotagmin staining, [243]) but not for horseradish peroxidase (HRP) antibody staining. The vesicular uptake of FM1-43 is similar in crayfish and *Drosophila* NMJs; however, the vital fluorescent dye, 4-14-(diethylamino) styryll-*N*-methylpyridiniumio dide (4-Di-2-Asp; [244]), obtained from Molecular Probes (Eugene, OR) works extremely well for crayfish NMJs but not for *Drosophila* (Fig. 1b; [245]). In addition, a dilute methylene blue stain made in crayfish saline can also be used to highlight the innervation of the muscle.

Unlike rodent brain slices or cultured rodent neurons the crayfish and *Drosophila* preparations function well at room temperature without any special considerations of an incubator and gas mixtures for maintaining the pH of the media. In addition, the preparations function well within a temperature range of 18–22 °C. The buffers added to the salines used for the *Drosophila* and crayfish are stable. Even though the crayfish NMJs can last several hours in minimal saline, attempts to culture intact NMJs for days have not been successful with the crayfish NMJ preparations.

6.2 The Drosophila Larva Filet Preparation

The *Drosophila* larva filet preparation can be used for electrophysiology, optogenetics, live or fixed imaging. To perform the procedure one needs very fine insect pins, a petri dish filled partially with a solid elastomer, a simple physiological saline [107], forceps, micro dissection scissors, and a dissecting microscope. Prior to dissection a larva is rinsed, dried, and pinned dorsal side up in the head and tail region, as shown in Fig. 2a. After submerging the

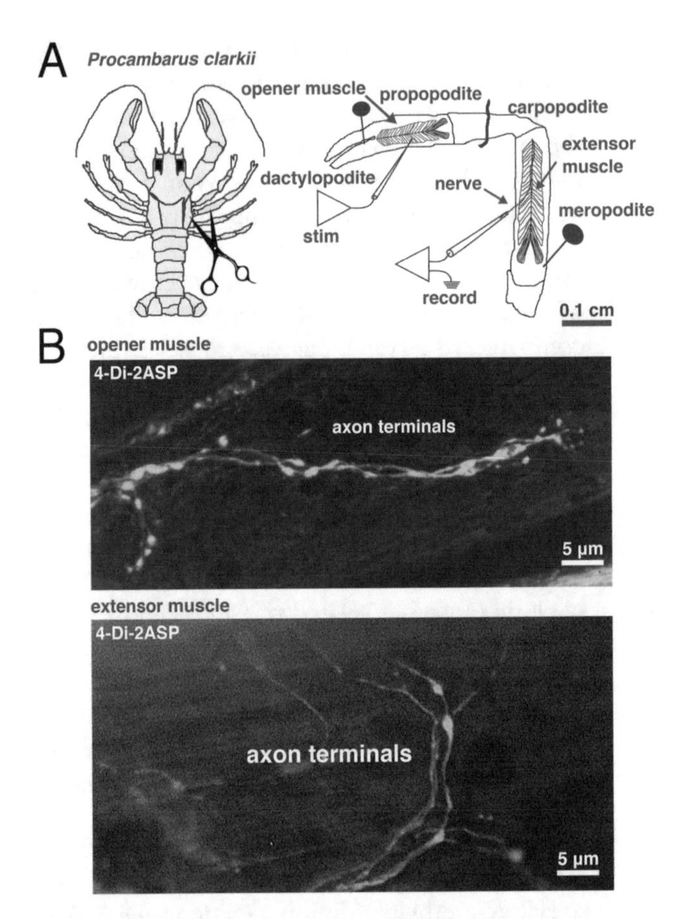

Fig. 1 Overview of the crayfish walking leg preparation. (**a**) The walking leg is readily obtained by pinching at the base of the leg above the autotomy plane to minimize bleeding when the leg is removed. After pinning the leg down, the ventral aspect of the opener preparation is accessible by removing the closer muscle and if one would like to stimulate the excitatory nerve to the opener this can be accomplished by finding the branch in the meropodite region (see video of dissection Cooper and Cooper (2009) [61]). (**b**) Staining of the living motor nerve terminals with 4-Di-2ASP highlights the innervation, making it easier to place a focal macropatch electrode over a desired region of the terminal. The *top panel* illustrates innervation on the opener muscle with the two motor nerve terminals. One is the excitor and one is the inhibitor, but the two cannot be differentiated with the 4-Di-2ASP staining. However they can be selectively stimulated by separating the nerve in the meropodite region. The *lower panel* illustrates innervation on the leg extensor muscle with the large varicosities of the tonic excitatory motor nerve and the thin filiform terminals of the phasic terminal

larva in saline, a shallow incision is made along the dorsal midline of the larva and the internal organs are carefully removed. The internal organs can be removed in a single step by first cutting the trachea attachments to the bodywall along each segment. Additional pins can then be placed in the four corners to gently spread the carcass, as shown in the second panel of Fig. 2a. Alternatively, the preparation can be dissected on a glass slide fitted with magnetic tape and insect pins attached to paper clips that can be easily

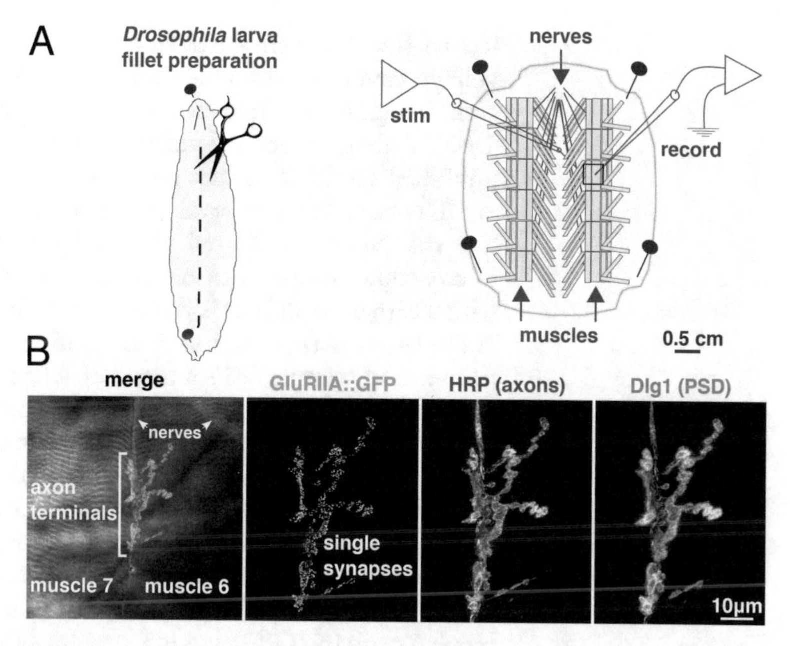

Fig. 2 Overview of the *Drosophila* larva filet preparation. (**a**) Schematics for dissecting the larva for direct access to the neuromuscular junction (NMJ). After pinning the larva in an elastomer-lined dish with saline, a shallow incision is made along the midline, then viscera and central nervous system are removed, making the nerves and bodywall muscles easily accessible to electrodes or a fluorescence light source for microscopy. (**b**) Anatomy of a fixed larva NMJ as visualized through a 60× objective. Fluorescence markers are superimposed onto a DIC brightfield image in the Merge panel to show the muscle ultrastructure. Nerve fibers that innervate the bodywall muscles are also seen in this image. Glutamate receptors are observed by an endogenous GFP-tagged GlurIIA construct. Commercially available reagents are available for labeling the axons (anti horse-radish peroxidase (HRP)) and the post synaptic density (PSD—discs large antibody (Dlg1))

maneuvered [246, 247]. At this point the specimen could be fixed for immunohistochemistry, imaged on an upright fluorescence microscope with water dipping objective, or prepared for electrophysiology. The schematic in Fig. 2a shows a basic configuration to evoke excitatory junction potentials (eEJPs). A small glass capillary suction electrode is placed on a severed nerve and a sharp glass capillary intracellular electrode is placed into a muscle fiber of the same segment. Resting membrane potential of the muscle should be larger than −60 mV. Frequent (>1 Hz) miniature excitatory junction potentials (mEJPs) should be observed with amplitudes larger than 1 mV. Supra threshold electrical stimulation of the nerve should evoke excitatory junction potentials larger than 20 mV. Once this fundamental procedure can be reliably performed, one can embark on the more exotic techniques that have led to the discoveries described in Sect. 3, e.g., two-electrode voltage clamp, paired pulse and high-frequency stimulation, calcium imaging, optogenetic or thermogenetic activation, and FM 1-43 labeling. Several detailed protocols and videos have been published on the larva filet preparation [248–250].

The larva filet preparation can be a powerful tool in laboratories that aren't equipped with electrophysiology or advanced microscopy equipment. A standard epifluorescent microscope with a camera is all that is required to assess synaptic morphology at the larva NMJ. Immunohistochemistry protocols for the NMJ are relatively simple (<24 h total, <1 h hands on time) and involve standard reagents (phosphate-buffered saline, Triton-X, formaldehyde, glycerol). Antibody markers for the larva NMJ are also inexpensive and robust. Antibodies raised against the horseradish peroxidase (HRP) enzyme, which are commercially available with a wide selection of conjugated fluorophores, specifically label the axon terminals. An antibody against the discs large protein (Dlg1) reliably marks the post synaptic density. Endogenous green fluorescent protein (GFP)-tagged proteins are publicly available for thousands of *Drosophila* genes. An example microscopy image of the larva NMJ with each of the markers and a GFP-tagged GluRIIA is shown in Fig. 2b. The specimen was prepared using the technique described above and standard immunohistochemistry procedures [251, 252]. Quantitative analysis of axon terminal and glutamatergic synapse morphology in various mutant backgrounds [253–255] has provided a wealth of understanding about the molecular mechanisms of synapse development and plasticity.

There are some caveats for the larva filet preparation. It does not involve exotic culturing techniques but the dissection does require a fair bit of skill, especially for physiology and live imaging experiments. Not only does the tissue have to remain healthy and be consistent from preparation to preparation, but it should be well restrained to minimize movement from muscle contractions. Supplemented media can maintain the NMJ preparation in culture for over 24 h without substantial physiological changes if the cells are not disturbed. However prolonged recording, stimulation, or imaging can cause the preparation to run down over time, e.g., decreased resting membrane potential and decreased mEJP and eEJP amplitudes. A source of variability in NMJ phenotypes that must be carefully accounted for is developmental plasticity. The structure and function of the NMJ is very plastic and subtle changes in the environment can have significant physiological effects, therefore culturing conditions must be rigorously controlled when using *Drosophila* larva.

7 Summary and Future Directions for Invertebrate and Vertebrate Glutamate Synapses

There are various topics we feel are worth continuing as well as novel directions where using the invertebrate glutamatergic synapse preparations could have implications for mammalian glutamatergic synapses or even chemical transmission in general. The effects of pH and the idea that molecular CO_2 may block the pore of the Glu receptor could have direct implications for pH sensing and regulation throughout the animal kingdom. Details of potential mechanisms of action for putative presynaptic glutamatergic auto-receptors in influencing synaptic transmission still need to be determined, as well as the possibility that such presynaptic receptors reside on non-glutamatergic presynaptic neurons as a means of detecting volume transmission. Differences in the postsynaptic array of Glu receptors have not been addressed in the context of synaptic output or the rate of spontaneous events. There are many accessory proteins now known to be present pre- and post-synaptically but their functional roles will take some time to determine and how they are regulated. The influence of an animal's diet and metabolism on Glu receptor function is an area that was prominent earlier in pharmacological studies but today there are so many herbal supplements containing plant and algae extracts known to have an action on Glu receptors yet careful monitoring of long-term consequences in low-level concentrations have not been addressed. The common monosodium glutamate (MSG) added to food as a supplement, domoic acid from red algae and kainic acid from seaweed are a few of the compounds that are well known to have consequences in humans and other animals. On a clinical note, one treatment for epileptic seizures involves manipulation of vGluT through the action of acetoacetate, a metabolite of fat, which competes with Cl^- for the binding site on vGluT and hinders Glu transport [256]. The natural body metabolite homocysteine, which can act as an agonist and an antagonist on Glu receptors, is now gaining attention. *Drosophila* and possibly other crustacean systems could provide a well-defined system to investigate the physiological effects of disorders related to glutamatergic transmission or outstanding questions about chemical transmission in general.

Acknowledgments

We thank Dr. J. Troy Littleton (Massachusetts Institute of Technology, Cambridge, MA, USA) for editorial comments and suggestions on improving this chapter. J.S.T. is supported by a Wellcome Trust Senior Basic Biomedical Research Fellowship (096144) awarded to Professor Ilan Davis.

References

1. Usherwood PN (1977) Glutamatergic synapses in invertebrates [proceedings]. Biochem Soc Trans **5**(4):845–849
2. Duce IR (1988) Glutamate. In: Lunt GG, Olsen RW (eds) Comparative invertebrate neurochemistry. Springer US, Boston, MA, pp 42–89
3. Greer JB, Khuri S, Fieber LA (2017) Phylogenetic analysis of ionotropic L-glutamate receptor genes in the Bilateria, with special notes on Aplysia californica. BMC Evol Biol **17**(1):11
4. Petroff OA (2002) GABA and glutamate in the human brain. Neuroscientist **8**(6):562–573
5. Traynelis SF et al (2010) Glutamate receptor ion channels: structure, regulation, and function. Pharmacol Rev **62**(3):405–496
6. Price MB, Jelesko J, Okumoto S (2012) Glutamate receptor homologs in plants: functions and evolutionary origins. Front Plant Sci **3**:235
7. Li Y et al (2016) Novel functional properties of Drosophila CNS glutamate receptors. Neuron **92**(5):1036–1048
8. Hawkins RA, Vina JR (2016) How glutamate is managed by the blood-brain barrier. Biology (Basel) **5**(4)
9. Anne C, Gasnier B (2014) Vesicular neurotransmitter transporters: mechanistic aspects. Curr Top Membr **73**:149–174
10. Ziegler AB et al (2016) The amino acid transporter JhI-21 coevolves with glutamate receptors, impacts NMJ physiology, and influences locomotor activity in Drosophila larvae. Sci Rep **6**:19692
11. Ishibashi H et al (2013) Dynamic regulation of glycine-GABA co-transmission at spinal inhibitory synapses by neuronal glutamate transporter. J Physiol **591**(16):3821–3832
12. Vandenberg RJ, Ryan RM (2013) Mechanisms of glutamate transport. Physiol Rev **93**(4):1621–1657
13. Littleton JT, Ganetzky B (2000) Ion channels and synaptic organization: analysis of the Drosophila genome. Neuron **26**(1):35–43
14. Pawlu C, DiAntonio A, Heckmann M (2004) Postfusional control of quantal current shape. Neuron **42**(4):607–618
15. Featherstone DE et al (2005) An essential Drosophila glutamate receptor subunit that functions in both central neuropil and neuromuscular junction. J Neurosci **25**(12): 3199–3208
16. Guerrero G et al (2005) Heterogeneity in synaptic transmission along a Drosophila larval motor axon. Nat Neurosci **8**(9): 1188–1196
17. Rasse TM et al (2005) Glutamate receptor dynamics organizing synapse formation in vivo. Nat Neurosci **8**(7):898–905
18. Logsdon S et al (2006) Regulation of synaptic vesicles pools within motor nerve terminals during short-term facilitation and neuromodulation. J Appl Physiol (1985) **100**(2): 662–671
19. Shinozaki H, Shibuya I (1974) A new potent excitant, quisqualic acid: effects on crayfish neuromuscular junction. Neuropharmacology **13**(7):665–672
20. Shinozaki H, Ishida M (1981) An attempt at an analysis of the factors determining the time course of the glutamate response in the crayfish neuromuscular junction. J Pharmacobiodyn **4**(7):483–489
21. Anderson CR, Cull-Candy SG, Miledi R (1976) Glutamate and quisqualate noise in voltage-clamped locust muscle fibres. Nature **261**(5556):151–153
22. Cull-Candy SG, Parker I (1983) Experimental approaches used to examine single glutamate-receptor ion channels in locust muscle fibers. In: Sakmann B, Neher E (eds) Single-channel recording. Springer US, Boston, MA, pp 389–400
23. Jan LY, Jan YN (1976) Properties of the larval neuromuscular junction in Drosophila melanogaster. J Physiol **262**(1):189–214
24. Patlak JB, Gration KA, Usherwood PN (1979) Single glutamate-activated channels in locust muscle. Nature **278**(5705):643–645
25. Gration KA et al (1981) Agonist potency determination by patch clamp analysis of single glutamate receptors. Brain Res **230**(1-2): 400–405
26. Loewi O (1921) Über humorale Übertragbarkeit der Herznervenwirkung i.v. Mitteilung. Pflügers Arch Ges Physiol **189**:239–242
27. Dale HH, Feldberg W, Vogt M (1936) Release of acetylcholine at voluntary motor nerve endings. J Physiol **86**(4):353–380
28. Minz B (1932) Pharmakologische Untersuchungen am Blutegelpräparat,zugleich eine Methode zum biologischen Nachweis von Acetylcholin bei Anwesenheit anderer pharmakologisch wirksamer körpereigener Stoffe. Arch Exp Pharmal **168**:292–304

29. Dudel J, Kuffler SW (1961) The quantal nature of transmission and spontaneous miniature potentials at the crayfish neuromuscular junction. J Physiol **155**:514–529
30. Cull-Candy SG (1984) Inhibitory synaptic currents in voltage-clamped locust muscle fibres desensitized to their excitatory transmitter. Proc R Soc Lond B Biol Sci **221**(1224):375–383
31. Jones HC (1962) The action of L-glutamic acid and of structurally related compounds on the hind gut of the crayfish. J Physiol **164**:295–300
32. Florey E (1954) An inhibitory and an excitatory factor of mammalian central nervous system, and their action of a single sensory neuron. Arch Int Physiol **62**(1):33–53
33. Robbins J (1959) The excitation and inhibition of crustacean muscle by amino acids. J Physiol **148**:39–50
34. Van Harreveld A (1959) Compounds in brain extracts causing spreading depression of cerebral cortical activity and contraction of crustacean muscle. J Neurochem **3**(4):300–315
35. Van Harreveld A, Mendelson M (1959) Glutamate-induced contractions in crustacean muscle. J Cell Comp Physiol **54**:85–94
36. Harris-Warrick R (2005) Synaptic chemistry in single neurons: GABA is identified as an inhibitory neurotransmitter. J Neurophysiol **93**(6):3029–3031
37. Florey E (1961) A new test preparation for bio-assay of factor I and gamma-aminobutyric acid. J Physiol **156**:1–7
38. Bazemore A, Elliott KA, Florey E (1956) Factor I and gamma-aminobutyric acid. Nature **178**(4541):1052–1053
39. Kuffler SW, Edwards C (1958) Mechanism of gamma aminobutyric acid (GABA) action and its relation to synaptic inhibition. J Neurophysiol **21**(6):589–610
40. Boistel J, Fatt P (1958) Membrane permeability change during inhibitory transmitter action in crustacean muscle. J Physiol **144**(1): 176–191
41. Kravitz EA (1962) Enzymic formation of gamma-aminobutyric acid in the peripheral and central nervous system of lobsters. J Neurochem **9**:363–370
42. Kravitz EA, Kuffler SW, Potter DD (1963) Gamma-aminobutyric acid and other blocking compounds in crustacea: III. Their relative concentrations in separated motor and inhibitory axons. J Neurophysiol **26**: 739–751
43. Kravitz EA et al (1963) Gamma-aminobutyric acid and other blocking compounds in crustacea: II. Peripheral nervous system. J Neurophysiol **26**:729–738
44. Kravitz EA, Potter DD (1965) A further study of the distribution of gamma-aminobutyric acid between excitatory and inhibitory axons of the lobster. J Neurochem **12**:323–328
45. Dudel J, Kuffler SW (1961) Presynaptic inhibition at the crayfish neuromuscular junction. J Physiol **155**:543–562
46. Kerkut GA et al (1965) The presence of glutamate in nerve-muscle perfusates of Helix, Carcinus and Periplaneta. Comp Biochem Physiol **15**(4):485–502
47. Crawford AC, McBurney RN (1976) Proceedings: The time course of action of L-glutamate at the excitatory neuromuscular junction in Maia squinado. J Physiol **254**(1): 47P–48P
48. King AE, Wheal HV (1984) The excitatory actions of kainic acid and some derivatives at the crab neuromuscular junction. Eur J Pharmacol **102**(1):129–134
49. Gray SR et al (1991) Solubilization and purification of a putative quisqualate-sensitive glutamate receptor from crustacean muscle. Biochem J 273(Pt 1):165–171
50. Fiszer de Plazas S, De Robertis E (1974) Isolation of hydrophobic proteins binding neurotransmitter aminoacids. Glutamate receptor of the shrimp muscle. J Neurochem **23**(6):1115–1120
51. Chiba C, Tazaki K (1992) Glutamatergic motoneurons in the stomatogastric ganglion of the mantis shrimp Squilla oratoria. J Comp Physiol A **170**(6):773–786
52. Sakurai A, Yamagishi H (2000) Graded neuromuscular transmission in the heart of the isopod crustacean Ligia exotica. J Exp Biol **203**(Pt 9):1447–1457
53. Lunt GG (1973) Hydrophobic proteins from locust (Shistocerca gregaria) muscle with glutamate receptor properties. Comp Gen Pharmacol **4**(13):75–79
54. Issberner JP et al (2002) Combined imaging and chemical sensing of L-glutamate release from the foregut plexus of the lepidopteran, Manduca sexta. J Neurosci Methods **120**(1): 1–10
55. Gardiner RB et al (2002) Cellular distribution of a high-affinity glutamate transporter in the nervous system of the cabbage looper Trichoplusia ni. J Exp Biol **205**(Pt 17): 2605–2613
56. Grigor'ev VV, Ragulin VV (1988) Action of phosphorus-containing aminocarboxylic acids on neuromuscular transmission in the blowfly. Neirofiziologiia **20**(2):256–258

57. Gallus L et al (2010) NMDA R1 receptor distribution in the cyprid of Balanus amphitrite (=Amphibalanus amphitrite) (Cirripedia, Crustacea). Neurosci Lett **485**(3):183–188
58. Huxley TH (1880) The crayfish: an introduction to the study of zoology. By T.H. Huxley. With eighty-two illustrations. D. Appleton and Company, New York
59. Richet C (1879) Contribution a la physiologic des centres nerveux et des muscles de l'ecrevisse. Arch Physiol **6**(262–299): 522–576
60. Richet C (1881) Physiologie des muscles et des nerfs. Le ons prof sees la Facult de m decine en 1881, par Charles Richet. G. Bailli re, Paris
61. Cooper AS, Cooper RL (2009) Historical view and physiology demonstration at the NMJ of the crayfish opener muscle. J Vis Exp (33):1595
62. Ringer S (1882) Concerning the influence exerted by each of the constituents of the blood on the contraction of the ventricle. J Physiol **3**(5–6):380–393
63. Ringer S (1882) Regarding the action of hydrate of soda, hydrate of ammonia, and hydrate of potash on the ventricle of the frog's heart. J Physiol **3**(3–4):195–202.6
64. Van Harreveld A (1936) A physiological solution for freshwater crustaceans. Proc Soc Exp Biol Med **34**(4):428–432
65. Katz B, Kuffler SW (1946) Excitation of the nerve-muscle system in Crustacea. Proc R Soc Lond B Biol Sci **133**:374–389
66. Katz B (1949) Neuro-muscular transmission in invertebrates. Biol Rev Camb Philos Soc **24**(1):1–20
67. Wiersma CA (1949) Synaptic facilitation in the crayfish. J Neurophysiol **12**(4):267–275
68. Fatt P, Katz B (1953) The effect of inhibitory nerve impulses on a crustacean muscle fibre. J Physiol **121**(2):374–389
69. Sherman RG, Atwood HL (1971) Synaptic facilitation: long-term neuromuscular facilitation in crustaceans. Science **171**(3977): 1248–1250
70. Bliss TV, Lomo T (1973) Long-lasting potentiation of synaptic transmission in the dentate area of the anaesthetized rabbit following stimulation of the perforant path. J Physiol **232**(2):331–356
71. Grundfest H, Reuben JP, Rickles WH Jr (1959) The electrophysiology and pharmacology of lobster neuromuscular synapses. J Gen Physiol **42**(6):1301–1323
72. Takeuchi A, Takeuchi N (1964) Iontophoretic application of gamma-aminobutyric acid crayfish muscle. Nature **203**:1074–1075
73. Takeuchi A, Takeuchi N (1964) The effect on crayfish muscle of iontophoretically applied glutamate. J Physiol **170**:296–317
74. Takeuchi A, Takeuchi N (1965) Localized action of gamma-aminobutyric acid on the crayfish muscle. J Physiol **177**(2):225–238
75. Takeuchi A, Takeuchi N (1966) A study of the inhibitory action of gamma-aminobutyric acid on neuromuscular transmission in the crayfish. J Physiol **183**(2):418–432
76. Barker JL (1975) Divalent cations: effects on post-synaptic pharmacology of invertebrate synapses. Brain Res **92**(2):307–323
77. Shank RP, Freeman AR (1975) Cooperative interaction of glutamate and aspartate with receptors in the neuromuscular excitatory membrane in walking limbs of the lobster. J Neurobiol **6**(3):289–303
78. McBain AE, Wheal HV, Collins JF (1984) The pharmacology of the piperidine dicarboxylates on the crustacean neuromuscular junction. Neuropharmacology **23**(1):23–30
79. Van Harreveld A (1980) L-proline as a glutamate antagonist at a crustacean neuromuscular junction. J Neurobiol **11**(6):519–529
80. Olsson T et al (2016) Hemolymph metabolites and osmolality are tightly linked to cold tolerance of Drosophila species: a comparative study. J Exp Biol **219**(Pt 16): 2504–2513
81. McBain AE, Wheal HV (1984) Further structure activity studies on the excitatory amino acid receptors of the crustacean neuromuscular junction. Comp Biochem Physiol C 77(2): 357–362
82. Shinozaki H, Ishida M (1986) A new potent channel blocker: effects on glutamate responses at the crayfish neuromuscular junction. Brain Res **372**(2):260–268
83. Shinozaki H, Ishida M, Mizuta T (1982) Glutamate inhibitors in the crayfish neuromuscular junction. Comp Biochem Physiol C **72**(2):249–255
84. Shinozaki H, Ishida M (1983) Excitatory junctional responses and glutamate responses at the crayfish neuromuscular junction in the presence of chlorisondamine. Brain Res **273**(2):325–333
85. Klose MK, Atkinson JK, Mercier AJ (2002) Effects of a hydroxy-cinnamoyl conjugate of spermidine on arthropod neuromuscular junctions. J Comp Physiol A Neuroethol Sens Neural Behav Physiol **187**(12):945–952
86. Onodera K, Takeuchi A (1977) Inhibitory effect of streptomycin and related antibiotics on the glutamate receptor of the crayfish neuromuscular junction. Neuropharmacology **16**(3):171–177

87. Shinozaki H, Ishida M (1981) Electrophysiological studies of kainate, quisqualate, and ibotenate action on the crayfish neuromuscular junction. Adv Biochem Psychopharmacol **27**:327–336
88. Shinozaki H, Ishida M (1988) Stizolobic acid, a competitive antagonist of the quisqualate-type receptor at the crayfish neuromuscular junction. Brain Res **451**(1–2):353–356
89. Schramm M, Dudel J (1997) Metabotropic glutamate autoreceptors on nerve terminals of crayfish muscle depress or facilitate release. Neurosci Lett **234**(1):31–34
90. Stettmeier H, Finger W (1983) Excitatory postsynaptic channels operated by quisqualate in crayfish muscle. Pflugers Arch **397**(3): 237–242
91. Shinozaki H, Ishida M (1981) Quisqualate action on the crayfish neuromuscular junction. J Pharmacobiodyn **4**(1):42–48
92. Dudel J, Franke C, Hatt H (1992) Rapid activation and desensitization of transmitter-liganded receptor channels by pulses of agonists. Ion Channels **3**:207–260
93. Atwood HL (1982) 3—Synapses and neurotransmitters. In: The biology of crustacea. Academic Press, San Diego, pp 105–150
94. Atwood HL, Cooper RL, Wojtowicz JM (1994) Nonuniformity and plasticity of quantal release at crustacean motor nerve terminals. Adv Second Messenger Phosphoprotein Res **29**:363–382
95. Atwood HL, Cooper RL (1995) Functional and structural parallels in crustacean and Drosophila neuromuscular systems. Am Zool **35**(6):556–565
96. Atwood HL, Cooper RL (1996) Assessing ultrastructure of crustacean and insect neuromuscular junctions. J Neurosci Methods **69**(1):51–58
97. Atwood HL, Cooper RL (1996) Synaptic diversity and differentiation: crustacean neuromuscular junctions. Invert Neurosci **1**(4): 291–307
98. Dudel J, Franke C, Luboldt W (1993) Reaction scheme for the glutamate-ergic, quisqualate type, completely desensitizing channels on crayfish muscle. Neurosci Lett **158**(2):177–180
99. Tour O, Parnas H, Parnas I (1995) The double-ticker: an improved fast drug-application system reveals desensitization of the glutamate channel from a closed state. Eur J Neurosci 7(10):2093–2100
100. Tour O, Parnas H, Parnas I (1998) Depolarization increases the single-channel conductance and the open probability of crayfish glutamate channels. Biophys J **74**(4): 1767–1778
101. Tour O, Parnas H, Parnas I (2000) On the mechanism of desensitization in quisqualate-type glutamate channels. J Neurophysiol **84**(1):1–10
102. Shinozaki H, Ishida M (1981) The recovery from desensitization of the glutamate receptor in the crayfish neuromuscular junction. Neurosci Lett **21**(3):293–296
103. Thieffry M (1984) The effect of calcium ions on the glutamate response and its desensitization in crayfish muscle fibres. J Physiol **355**:119–135
104. Hatt H, Franke C, Dudel J (1988) Calcium dependent gating of the L-glutamate activated, excitatory synaptic channel on crayfish muscle. Pflugers Arch **411**(1):17–26
105. Stettmeier H, Finger W, Dudel J (1983) Effects of concanavalin A on glutamate operated postsynaptic channels in crayfish muscle. Pflugers Arch **397**(1):20–24
106. Park H, Popescu A, Poo MM (2014) Essential role of presynaptic NMDA receptors in activity-dependent BDNF secretion and corticostriatal LTP. Neuron **84**(5):1009–1022
107. Stewart BA et al (1994) Improved stability of Drosophila larval neuromuscular preparations in haemolymph-like physiological solutions. J Comp Physiol A **175**(2):179–191
108. Macleod GT et al (2002) Fast calcium signals in Drosophila motor neuron terminals. J Neurophysiol **88**(5):2659–2663
109. de Castro C et al (2014) Analysis of various physiological salines for heart rate, CNS function, and synaptic transmission at neuromuscular junctions in Drosophila melanogaster larvae. J Comp Physiol A Neuroethol Sens Neural Behav Physiol **200**(1):83–92
110. Pulver SR et al (2009) Temporal dynamics of neuronal activation by Channelrhodopsin-2 and TRPA1 determine behavioral output in Drosophila larvae. J Neurophysiol **101**(6): 3075–3088
111. Macleod GT (**2012**) Calcium imaging at the Drosophila larval neuromuscular junction. Cold Spring Harb Protoc 2012(7):758–766
112. Budnik V, Ruiz-Canada C (2006) The fly neuromuscular junction: structure and function, 2nd edn. Elsevier Science, San Diego, CA
113. Menon KP, Carrillo RA, Zinn K (2013) Development and plasticity of the Drosophila larval neuromuscular junction. Wiley Interdiscip Rev Dev Biol **2**(5):647–670
114. Harris KP, Littleton JT (2015) Transmission, development, and plasticity of synapses. Genetics **201**(2):345–375

115. Marrus SB et al (2004) Differential localization of glutamate receptor subunits at the Drosophila neuromuscular junction. J Neurosci **24**(6):1406–1415
116. Qin G et al (2005) Four different subunits are essential for expressing the synaptic glutamate receptor at neuromuscular junctions of Drosophila. J Neurosci **25**(12):3209–3218
117. DiAntonio A (2006) Glutamate receptors at the Drosophila neuromuscular junction. Int Rev Neurobiol **75**:165–179
118. Bhatt D, Cooper RL (2005) The pharmacological and physiological profile of glutamate receptors at the Drosophila larval neuromuscular junction. Physiol Entomol **30**(2): 205–210
119. Lee JY et al (2009) Furthering pharmacological and physiological assessment of the glutamatergic receptors at the Drosophila neuromuscular junction. Comp Biochem Physiol C Toxicol Pharmacol **150**(4):546–557
120. Han TH et al (2015) Functional reconstitution of Drosophila melanogaster NMJ glutamate receptors. Proc Natl Acad Sci U S A **112**(19):6182–6187
121. Gratz SJ et al (2013) Genome engineering of Drosophila with the CRISPR RNA-guided Cas9 nuclease. Genetics **194**(4):1029–1035
122. Abe T, Kawai N, Miwa A (1983) Effects of a spider toxin on the glutaminergic synapse of lobster muscle. J Physiol **339**:243–252
123. Kawai N et al (1984) Spider toxin (JSTX) on the glutamate synapse. J Physiol (Paris) **79**(4):228–231
124. Kawasaki F, Kita H (1996) Physiological and immunocytochemical determination of the neurotransmitter at cricket neuromuscular junctions. Zool Sci **13**(4):503–507
125. Frank CA et al (2006) Mechanisms underlying the rapid induction and sustained expression of synaptic homeostasis. Neuron **52**(4): 663–677
126. Frank CA (2014) Homeostatic plasticity at the Drosophila neuromuscular junction. Neuropharmacology **78**:63–74
127. DiAntonio A et al (1999) Glutamate receptor expression regulates quantal size and quantal content at the Drosophila neuromuscular junction. J Neurosci **19**(8):3023–3032
128. Frank CA, Pielage J, Davis GW (2009) A presynaptic homeostatic signaling system composed of the Eph receptor, ephexin, Cdc42, and CaV2.1 calcium channels. Neuron **61**(4):556–569
129. Jakawich SK et al (2010) Local presynaptic activity gates homeostatic changes in presynaptic function driven by dendritic BDNF synthesis. Neuron **68**(6):1143–1158
130. Henry FE et al (2012) Retrograde changes in presynaptic function driven by dendritic mTORC1. J Neurosci **32**(48):17128–17142
131. Penney J et al (2012) TOR is required for the retrograde regulation of synaptic homeostasis at the Drosophila neuromuscular junction. Neuron **74**(1):166–178
132. Parmentier ML et al (1996) Cloning and functional expression of a Drosophila metabotropic glutamate receptor expressed in the embryonic CNS. J Neurosci **16**(21):6687–6694
133. Mitri C et al (2004) Divergent evolution in metabotropic glutamate receptors. A new receptor activated by an endogenous ligand different from glutamate in insects. J Biol Chem **279**(10):9313–9320
134. Bogdanik L et al (2004) The Drosophila metabotropic glutamate receptor DmGluRA regulates activity-dependent synaptic facilitation and fine synaptic morphology. J Neurosci **24**(41):9105–9116
135. Kurdyak P et al (1994) Differential physiology and morphology of motor axons to ventral longitudinal muscles in larval Drosophila. J Comp Neurol **350**(3):463–472
136. Ormerod KG et al (2013) Action of octopamine and tyramine on muscles of Drosophila melanogaster larvae. J Neurophysiol **110**(8): 1984–1996
137. Jan LY, Jan YN (1976) L-Glutamate as an excitatory transmitter at the Drosophila larval neuromuscular junction. J Physiol **262**(1): 215–236
138. Delgado R et al (1989) L-Glutamate activates excitatory and inhibitory channels in Drosophila larval muscle. FEBS Lett **243**(2): 337–342
139. Chen K, Augustin H, Featherstone DE (2009) Effect of ambient extracellular glutamate on Drosophila glutamate receptor trafficking and function. J Comp Physiol A Neuroethol Sens Neural Behav Physiol **195**(1):21–29
140. Zucker RS, Regehr WG (2002) Short-term synaptic plasticity. Annu Rev Physiol **64**: 355–405
141. Wu CF et al (1983) Potassium currents in Drosophila: different components affected by mutations of two genes. Science **220**(4601): 1076–1078
142. Peron S et al (2009) From action potential to contraction: neural control and excitation-contraction coupling in larval muscles of Drosophila. Comp Biochem Physiol A Mol Integr Physiol **154**(2):173–183

143. Zhong Y, Wu CF (1991) Altered synaptic plasticity in Drosophila memory mutants with a defective cyclic AMP cascade. Science **251**(4990):198–201
144. Wu Y, Kawasaki F, Ordway RW (2005) Properties of short-term synaptic depression at larval neuromuscular synapses in wild-type and temperature-sensitive paralytic mutants of Drosophila. J Neurophysiol **93**(5): 2396–2405
145. Sigrist SJ et al (2003) Experience-dependent strengthening of Drosophila neuromuscular junctions. J Neurosci **23**(16):6546–6556
146. Ataman B et al (2008) Rapid activity-dependent modifications in synaptic structure and function require bidirectional Wnt signaling. Neuron **57**(5):705–718
147. Nesler KR et al (2013) The miRNA pathway controls rapid changes in activity-dependent synaptic structure at the Drosophila melanogaster neuromuscular junction. PLoS One **8**(7):e68385
148. Nesler KR et al (2016) Presynaptic CamKII regulates activity-dependent axon terminal growth. Mol Cell Neurosci **76**:33–41
149. Piccioli ZD, Littleton JT (2014) Retrograde BMP signaling modulates rapid activity-dependent synaptic growth via presynaptic LIM kinase regulation of cofilin. J Neurosci **34**(12):4371–4381
150. Budnik V, Gorczyca M, Prokop A (2006) Selected methods for the anatomical study of Drosophila embryonic and larval neuromuscular junctions. Int Rev Neurobiol **75**: 323–365
151. Fuger P et al (2007) Live imaging of synapse development and measuring protein dynamics using two-color fluorescence recovery after photo-bleaching at Drosophila synapses. Nat Protoc **2**(12):3285–3298
152. Sturgeon M et al (2016) The Notch ligand E3 ligase, Mind Bomb1, regulates glutamate receptor localization in Drosophila. Mol Cell Neurosci **70**:11–21
153. Lee G, Schwarz TL (2016) Filamin, a synaptic organizer in Drosophila, determines glutamate receptor composition and membrane growth. Elife 5. doi:10.7554/eLife.19991
154. Wang M et al (2016) Dbo/Henji modulates synaptic dPAK to gate glutamate receptor abundance and postsynaptic response. PLoS Genet **12**(10):e1006362
155. Hussein NA et al (2016) The extracellular-regulated kinase effector Lk6 is required for glutamate receptor localization at the Drosophila neuromuscular junction. J Exp Neurosci **10**:77–91
156. Deivasigamani S et al (2015) A presynaptic regulatory system acts transsynaptically via Mon1 to regulate glutamate receptor levels in Drosophila. Genetics **201**(2):651–664
157. Ramos CI et al (2015) Neto-mediated intracellular interactions shape postsynaptic composition at the Drosophila neuromuscular junction. PLoS Genet **11**(4):e1005191
158. Ghosh R et al (2014) Kismet positively regulates glutamate receptor localization and synaptic transmission at the Drosophila neuromuscular junction. PLoS One **9**(11):e113494
159. Kim MJ, O'Connor MB (2014) Anterograde Activin signaling regulates postsynaptic membrane potential and GluRIIA/B abundance at the Drosophila neuromuscular junction. PLoS One **9**(9):e107443
160. Xing G et al (2014) Drosophila neuroligin3 regulates neuromuscular junction development and synaptic differentiation. J Biol Chem **289**(46):31867–31877
161. Gardiol A, St Johnston D (2014) Staufen targets coracle mRNA to Drosophila neuromuscular junctions and regulates GluRIIA synaptic accumulation and bouton number. Dev Biol **392**(2):153–167
162. Kerr KS et al (2014) Glial wingless/Wnt regulates glutamate receptor clustering and synaptic physiology at the Drosophila neuromuscular junction. J Neurosci **34**(8): 2910–2920
163. Parkinson W et al (2013) N-glycosylation requirements in neuromuscular synaptogenesis. Development **140**(24):4970–4981
164. Lee J, Ueda A, Wu CF (2014) Distinct roles of Drosophila cacophony and Dmca1D Ca(2+) channels in synaptic homeostasis: genetic interactions with slowpoke Ca(2+)-activated BK channels in presynaptic excitability and postsynaptic response. Dev Neurobiol **74**(1):1–15
165. Lee MJ et al (2013) Tbc1d15-17 regulates synaptic development at the Drosophila neuromuscular junction. Mol Cells **36**(2): 163–168
166. Lee HG et al (2013) Akt regulates glutamate receptor trafficking and postsynaptic membrane elaboration at the Drosophila neuromuscular junction. Dev Neurobiol **73**(10): 723–743
167. Staples J, Broadie K (2013) The cell polarity scaffold Lethal Giant Larvae regulates synapse morphology and function. J Cell Sci **126**(Pt 9):1992–2003
168. Fukui A et al (2012) Lola regulates glutamate receptor expression at the Drosophila neuromuscular junction. Biol Open **1**(4):362–375

169. Chen K, Featherstone DE (2011) Pre and postsynaptic roles for Drosophila CASK. Mol Cell Neurosci **48**(2):171–182
170. Sun M et al (2007) Presynaptic contributions of chordin to hippocampal plasticity and spatial learning. J Neurosci **27**(29): 7740–7750
171. Chen K et al (2010) Neurexin in embryonic Drosophila neuromuscular junctions. PLoS One **5**(6):e11115
172. Banovic D et al (2010) Drosophila neuroligin 1 promotes growth and postsynaptic differentiation at glutamatergic neuromuscular junctions. Neuron **66**(5):724–738
173. Bachmann A et al (2010) A perisynaptic menage a trois between Dlg, DLin-7, and Metro controls proper organization of Drosophila synaptic junctions. J Neurosci **30**(17):5811–5824
174. Wang D et al (2010) Drosophila twinfilin is required for cell migration and synaptic endocytosis. J Cell Sci **123**(Pt 9):1546–1556
175. Owald D et al (2010) A Syd-1 homologue regulates pre- and postsynaptic maturation in Drosophila. J Cell Biol **188**(4):565–579
176. Morimoto T et al (2010) Subunit-specific and homeostatic regulation of glutamate receptor localization by CaMKII in Drosophila neuromuscular junctions. Neuroscience **165**(4): 1284–1292
177. Menon KP et al (2009) The translational repressors Nanos and Pumilio have divergent effects on presynaptic terminal growth and postsynaptic glutamate receptor subunit composition. J Neurosci **29**(17):5558–5572
178. Bogdanik L et al (2008) Muscle dystroglycan organizes the postsynapse and regulates presynaptic neurotransmitter release at the Drosophila neuromuscular junction. PLoS One **3**(4):e2084
179. Wairkar YP et al (2008) Synaptic defects in a Drosophila model of congenital muscular dystrophy. J Neurosci **28**(14):3781–3789
180. Heckscher ES et al (2007) NF-kappaB, IkappaB, and IRAK control glutamate receptor density at the Drosophila NMJ. Neuron **55**(6):859–873
181. Rohrbough J et al (2007) Presynaptic establishment of the synaptic cleft extracellular matrix is required for post-synaptic differentiation. Genes Dev **21**(20):2607–2628
182. Haas KF et al (2007) The ubiquitin-proteasome system postsynaptically regulates glutamatergic synaptic function. Mol Cell Neurosci **35**(1):64–75
183. Chen K et al (2005) The 4.1 protein coracle mediates subunit-selective anchoring of Drosophila glutamate receptors to the postsynaptic actin cytoskeleton. J Neurosci **25**(28): 6667–6675
184. Chen K, Featherstone DE (2005) Discs-large (DLG) is clustered by presynaptic innervation and regulates postsynaptic glutamate receptor subunit composition in Drosophila. BMC Biol **3**:1
185. Menon KP et al (2004) The translational repressor Pumilio regulates presynaptic morphology and controls postsynaptic accumulation of translation factor eIF-4E. Neuron **44**(4):663–676
186. Albin SD, Davis GW (2004) Coordinating structural and functional synapse development: postsynaptic p21-activated kinase independently specifies glutamate receptor abundance and postsynaptic morphology. J Neurosci **24**(31):6871–6879
187. Chan YB et al (2003) Neuromuscular defects in a Drosophila survival motor neuron gene mutant. Hum Mol Genet **12**(12):1367–1376
188. Renden RB, Broadie K (2003) Mutation and activation of Galpha s similarly alters pre- and postsynaptic mechanisms modulating neurotransmission. J Neurophysiol **89**(5): 2620–2638
189. Parnas D et al (2001) Regulation of postsynaptic structure and protein localization by the Rho-type guanine nucleotide exchange factor dPix. Neuron **32**(3):415–424
190. Blunk AD et al (2014) Postsynaptic actin regulates active zone spacing and glutamate receptor apposition at the Drosophila neuromuscular junction. Mol Cell Neurosci **61**: 241–254
191. Morel V et al (2014) Drosophila Nesprin-1 controls glutamate receptor density at neuromuscular junctions. Cell Mol Life Sci **71**(17): 3363–3379
192. Saitoe M et al (2001) Absence of junctional glutamate receptor clusters in Drosophila mutants lacking spontaneous transmitter release. Science **293**(5529):514–517
193. Verstreken P, Bellen HJ (2002) Meaningless minis? Mechanisms of neurotransmitter-receptor clustering. Trends Neurosci **25**(8): 383–385
194. Melom JE et al (2013) Spontaneous and evoked release are independently regulated at individual active zones. J Neurosci **33**(44): 17253–17263
195. Deitcher DL et al (1998) Distinct requirements for evoked and spontaneous release of neurotransmitter are revealed by mutations in the Drosophila gene neuronal-synaptobrevin. J Neurosci **18**(6):2028–2039

196. Ganesan S, Karr JE, Featherstone DE (2011) Drosophila glutamate receptor mRNA expression and mRNP particles. RNA Biol **8**(5):771–781
197. Sigrist SJ et al (2000) Postsynaptic translation affects the efficacy and morphology of neuromuscular junctions. Nature **405**(6790): 1062–1065
198. Titlow JS et al. Super-resolution single molecule FISH at the Drosophila neuromsucular junction. Methods Mol Biol, in press
199. Sudhof TC (2004) The synaptic vesicle cycle. Annu Rev Neurosci **27**:509–547
200. Wu WH, Cooper RL (2012) The regulation and packaging of synaptic vesicles as related to recruitment within glutamatergic synapses. Neuroscience **225**:185–198
201. Rizzoli SO, Betz WJ (2005) Synaptic vesicle pools. Nat Rev Neurosci **6**(1):57–69
202. Fredj NB, Burrone J (2009) A resting pool of vesicles is responsible for spontaneous vesicle fusion at the synapse. Nat Neurosci **12**(6): 751–758
203. Denker A et al (2011) A small pool of vesicles maintains synaptic activity in vivo. Proc Natl Acad Sci U S A **108**(41):17177–17182
204. Denker A et al (2011) The reserve pool of synaptic vesicles acts as a buffer for proteins involved in synaptic vesicle recycling. Proc Natl Acad Sci U S A **108**(41): 17183–17188
205. Akbergenova Y, Bykhovskaia M (2009) Stimulation-induced formation of the reserve pool of vesicles in Drosophila motor boutons. J Neurophysiol **101**(5):2423–2433
206. Johnstone AFM, Viele K, Cooper RL (2011) Structure/function assessment of synapses at motor nerve terminals. Synapse **65**(4): 287–299
207. Wu WH, Cooper RL (2013) Physiological separation of vesicle pools in low- and high-output nerve terminals. Neurosci Res **75**(4): 275–282
208. Walrond JP, Govind CK, Huestis SE (1993) Two structural adaptations for regulating transmitter release at lobster neuromuscular synapses. J Neurosci **13**(11):4831–4845
209. Mykles DL et al (2002) Myofibrillar protein isoform expression is correlated with synaptic efficacy in slow fibres of the claw and leg opener muscles of crayfish and lobster. J Exp Biol **205**(Pt 4):513–522
210. Dudel J, Schramm M (2003) A receptor for presynaptic glutamatergic autoinhibition is a glutamate transporter. Eur J Neurosci **18**(4): 902–910
211. Pinard A et al (2003) Glutamatergic modulation of synaptic plasticity at a PNS vertebrate cholinergic synapse. Eur J Neurosci **18**(12): 3241–3250
212. Kim WM et al (2016) The role of inversely operating glutamate transporter in the paradoxical analgesia produced by glutamate transporter inhibitors. Eur J Pharmacol **793**: 112–118
213. Koles K et al (2012) Mechanism of evenness interrupted (Evi)-exosome release at synaptic boutons. J Biol Chem **287**(20):16820–16834
214. Spring AM, Brusich DJ, Frank CA (2016) C-terminal Src kinase gates homeostatic synaptic plasticity and regulates fasciclin II expression at the Drosophila neuromuscular junction. PLoS Genet **12**(2):e1005886
215. Deshpande M, Rodal AA (2016) The crossroads of synaptic growth signaling, membrane traffic and neurological disease: insights from Drosophila. Traffic **17**(2):87–101
216. Petralia RS, Mattson MP, Yao PJ (2014) Communication breakdown: the impact of ageing on synapse structure. Ageing Res Rev **14**:31–42
217. Cooper AS, Johnstone AFM, Cooper RL (2013) Motor nerve terminal morphology with unloading and reloading of muscle in Procambarus clarkii. Journal of Crustacean Biology **33**(6):818–827
218. Morrison JH, Baxter MG (2012) The ageing cortical synapse: hallmarks and implications for cognitive decline. Nat Rev Neurosci **13**(4):240–250
219. Picconi B, Piccoli G, Calabresi P (2012) Synaptic dysfunction in Parkinson's disease. Adv Exp Med Biol **970**:553–572
220. Yeoman M, Scutt G, Faragher R (2012) Insights into CNS ageing from animal models of senescence. Nat Rev Neurosci **13**(6): 435–445
221. Badre NH, Martin ME, Cooper RL (2005) The physiological and behavioral effects of carbon dioxide on Drosophila melanogaster larvae. Comp Biochem Physiol A Mol Integr Physiol **140**(3):363–376
222. Bierbower SM, Cooper RL (2010) The effects of acute carbon dioxide on behavior and physiology in Procambarus clarkii. J Exp Zool A Ecol Genet Physiol **313**(8):484–497
223. Bierbower SM, Cooper RL (2013) The mechanistic action of carbon dioxide on a neural circuit and NMJ communication. J Exp Zool A Ecol Genet Physiol **319**(6):340–354
224. Sugahara M, Sakamoto F (2009) Heat and carbon dioxide generated by honeybees

jointly act to kill hornets. Naturwissenschaften **96**(9):1133–1136

225. Eisele JH, Eger EI 2nd, Muallem M (1967) Narcotic properties of carbon dioxide in the dog. Anesthesiology **28**(5):856–865
226. Traynelis SF, Cull-Candy SG (1990) Proton inhibition of N-methyl-D-aspartate receptors in cerebellar neurons. Nature **345**(6273): 347–350
227. Tong CK, Chesler M (2000) Modulation of spreading depression by changes in extracellular pH. J Neurophysiol **84**(5):2449–2457
228. Williams K et al (2003) Pharmacology of delta2 glutamate receptors: effects of pentamidine and protons. J Pharmacol Exp Ther **305**(2):740–748
229. Low CM et al (2003) Molecular determinants of proton-sensitive N-methyl-D-aspartate receptor gating. Mol Pharmacol **63**(6):1212–1222
230. He P et al (1999) Role of alpha-SNAP in promoting efficient neurotransmission at the crayfish neuromuscular junction. J Neurophysiol **82**(6):3406–3416
231. Kandel ER (2012) The molecular biology of memory: cAMP, PKA, CRE, CREB-1, CREB-2, and CPEB. Mol Brain **5**:14
232. Martin A et al (2016) CRISPR/Cas9 mutagenesis reveals versatile roles of Hox genes in crustacean limb specification and evolution. Curr Biol **26**(1):14–26
233. LaFramboise WA et al (2000) Muscle type-specific myosin isoforms in crustacean muscles. J Exp Zool **286**(1):36–48
234. Griffis B, Moffett SB, Cooper RL (2001) Muscle phenotype remains unaltered after limb autotomy and unloading. J Exp Zool **289**(1):10–22
235. Sohn J, Mykles DL, Cooper RL (2000) Characterization of muscles associated with the articular membrane in the dorsal surface of the crayfish abdomen. J Exp Zool **287**(5): 353–377
236. Strawn JR, Neckameyer WS, Cooper RL (2000) The effects of 5-HT on sensory, central and motor neurons driving the abdominal superficial flexor muscles in the crayfish. Comp Biochem Physiol B Biochem Mol Biol **127**(4):533–550
237. Baierlein B et al (2011) Membrane potentials, synaptic responses, neuronal circuitry, neuromodulation and muscle histology using the crayfish: student laboratory exercises. J Vis Exp (47)
238. Wu WH, Cooper RL (2010) Physiological recordings of high- and low-output NMJs on the crayfish leg extensor muscle. J Vis Exp (45)
239. Cooper RL et al (1996) Synaptic structural complexity as a factor enhancing probability of calcium-mediated transmitter release. J Neurophysiol **75**(6):2451–2466
240. Winslow JL, Cooper RL, Atwood HL (2002) Intracellular ionic concentration by calibration from fluorescence indicator emission spectra, its relationship to the Kd, Fmin, Fmax formula, and use with Na-Green for presynaptic sodium. J Neurosci Methods **118**(2):163–175
241. Majeed ZR et al (2015) New insights into the acute actions from a high dosage of fluoxetine on neuronal and cardiac function: Drosophila, crayfish and rodent models. Comp Biochem Physiol C Toxicol Pharmacol 176–177: 52–61
242. Cooper RL et al (1995) Quantal measurement and analysis methods compared for crayfish and Drosophila neuromuscular junctions, and rat hippocampus. J Neurosci Methods **61**(1-2):67–78
243. Cooper RL, Hampson DR, Atwood HL (1995) Synaptotagmin-like expression in the motor nerve terminals of crayfish. Brain Res **703**(1–2):214–216
244. Magrassi L, Purves D, Lichtman JW (1987) Fluorescent probes that stain living nerve terminals. J Neurosci 7(4):1207–1214
245. Cooper RL, Marin L, Atwood HL (1995) Synaptic differentiation of a single motor neuron: conjoint definition of transmitter release, presynaptic calcium signals, and ultrastructure. J Neurosci **15**(6):4209–4222
246. Muller KJ, Nicholls JG, Stent GS (1981) Neurobiology of the leech. Cold Spring Harbor Laboratory, New York
247. Cooper AS et al (2009) Monitoring heart function in larval Drosophila melanogaster for physiological studies. J Vis Exp (33). doi:10.3791/1596
248. Verstreken P, Ohyama T, Bellen HJ (2008) FM 1-43 labeling of synaptic vesicle pools at the Drosophila neuromuscular junction. Methods Mol Biol **440**:349–369
249. Imlach W, McCabe BD (2009) Electrophysiological methods for recording synaptic potentials from the NMJ of Drosophila larvae. J Vis Exp (24). doi:10.3791/1109
250. Zhang B, Stewart B (2010) Electrophysiological recording from Drosophila larval

body-wall muscles. Cold Spring Harb Protoc 2010(9):pdb prot5487

251. Brent J, Werner K, McCabe BD (2009) Drosophila larval NMJ immunohistochemistry. J Vis Exp (25)
252. Ramachandran P, Budnik V (2010) Immunocytochemical staining of Drosophila larval body-wall muscles. Cold Spring Harb Protoc 2010(8):pdb prot5470
253. Andlauer TF, Sigrist SJ (2012) Quantitative analysis of Drosophila larval neuromuscular junction morphology. Cold Spring Harb Protoc 2012(4):490–493
254. Nijhof B et al (2016) A new Fiji-based algorithm that systematically quantifies nine synaptic parameters provides insights into Drosophila NMJ morphometry. PLoS Comput Biol **12**(3):e1004823
255. Sanhueza M, Kubasik-Thayil A, Pennetta G (2016) Why quantification matters: characterization of phenotypes at the Drosophila larval neuromuscular junction. J Vis Exp (111). doi:10.3791/53821
256. Juge N et al (2010) Metabolic control of vesicular glutamate transport and release. Neuron **68**(1):99–112

Chapter 10

Optical Control of Glutamate Receptors of the NMDA-Kind in Mammalian Neurons, with the Use of Photoswitchable Ligands

Shai Berlin and Ehud Y. Isacoff

Abstract

N-Methyl-D-aspartate receptors (NMDAR) are members of the glutamate binding ligand-gated receptors. They are primarily found at excitatory synapses, essential for some of the most prominent forms of synaptic plasticity pertaining to learning and memory (Nabavi et al., 2014; Bliss and Lomo, 1973) and their dysfunction underlies diverse diseases (Newcomer et al., 2000; Burnashev and Szepetowski, 2015). By combining genetic manipulations of NMDAR subunits and synthetic chemical photoswitches, we have recently developed a family of light-gated, or photoswitchable, NMDA receptors to gate plasticity in vitro and in vivo. This approach—synthetic optogenetics (Berlin and Isacoff, 2017)—enables to confer remote, rapid, and reversible optical modulation of NMDA receptors of a particular subunit composition. This chapter describes the use of azobenzene-based tethered photoswitches and engineered NMDAR subunits to engender the NMDA-receptor light-sensitive; in cultured hippocampal neurons.

Key words Glutamate receptor, NMDA receptors, Synthetic optogenetics, Light-controlled receptors, Azobenzene, LiGluN

1 Introduction

1.1 Light-Activated Proteins

Optical control of membrane proteins has an advantage over other existing methodologies (e.g., electrical or pharmacological) owing to light noninvasive and orthogonal nature towards most biological systems, and its inherent spatial resolution [1, 2]. To engage and apply light on most cells, unlike in the visual system, the addition of light-sensitive proteins (i.e., photoreceptors) is required. Photoreceptors are specialized proteins, naturally harboring light-sensitive chromophores. Chromophores contain unique electronic and photochemical properties that set in motion rapid, light-dependent reactions such as isomerization (e.g., retinal and tetrapyrrole), proton coupled electron transfer (flavin and tryptophan), or chemical redox reactions (flavin) [3, 4]. Thus, the interaction of light with different recipes of chromophores and

Sandrine Parrot and Luc Denoroy (eds.), *Biochemical Approaches for Glutamatergic Neurotransmission*, Neuromethods, vol. 130, DOI 10.1007/978-1-4939-7228-9_10,

photoreceptors results in the modulation of a wide variety of cellular activities. This realization promoted the harnessing and repurposing of photoreceptors into valuable light-operated optogenetic tools [5–7]. Today, photo-control over cellular activities can be obtained by a large variety of photoreceptor-dependent tools, such as light-gated channels and pumps (the opsin superfamily [7–9]), light-activated G-protein Coupled Receptors [10, 11], light-operated enzymes [12, 13], light-dependent dimerizing molecules [14], and light-sensitive CRISPR/Cas9 systems for controlling transcription [15–17] or genomic editing [16], to name but a few! Thus, the expanded palette of light-gated tools and improved genetic-targeting methods [18, 19], in combination with state-of-the-art light manipulations [20, 21], have accelerated the use of optogenetic tools in almost every field of biology [22, 23], neuroscience in particular [7, 24].

Another class of optogenetics, already introduced in the early 1970s [25–27], consists of non-photoreceptor-dependent tools. These tools are obtained by tagging *blind* proteins (i.e., insensitive to light) or molecules (e.g., DNA [28]) with light-sensitive synthetic chromophores to, consequently, render the latter receptive to light. We dub this methodology *synthetic optogenetics* [29] owing to the nonbiological (i.e., synthetic) origin of the engineered light-gated protein. Synthetic chromophores—or photoswitches [30], bear similarity to their natural counterparts (e.g., retinal) in that they too absorb light and, in turn, undergo light-dependent reactions. Of the various possible structural changes of various light-sensing synthetic moieties used in the development of photoswitches, the reversible *trans*-to-*cis* isomerization of azobenzenes is the most commonly used [30], and rightly so as azobenzenes undergo the largest geometrical changes when the molecule swops from –*trans* to –*cis* following near-UV or green light absorption, respectively (Fig. 1a and b) [35–37]. This change of geometry can be employed, for example, to present or withdraw an active chemical headgroup from the receptor to modulate its activity (e.g., Fig. 2, and see below).

Soluble azobenzene photoswitches; or *p*hoto*c*hromic *l*igands (PCL), are soluble, nontethered switches that modulate a protein activity upon their isomerization; by steric or allosteric interactions. This approach has been applied by several groups to photomanipulate enzymes [40], G protein-coupled receptors (GPCRs) [41] and ion channels [42–45]. For example, a derivatized PCL, bearing tetraethylammonium (TEA) as a functional chemical headgroup (i.e., pharmacophore) at one end of the molecule, has been engaged to act as a light-dependent pore blocker of a voltage-gated K^+-channel [45]. Conversely, PCLs may encompass ligands (e.g., glutamate) to act as diffusible photo-agonists of ligand-gated ion channels [26, 43, 44], akin to caged compounds [46]. Though the

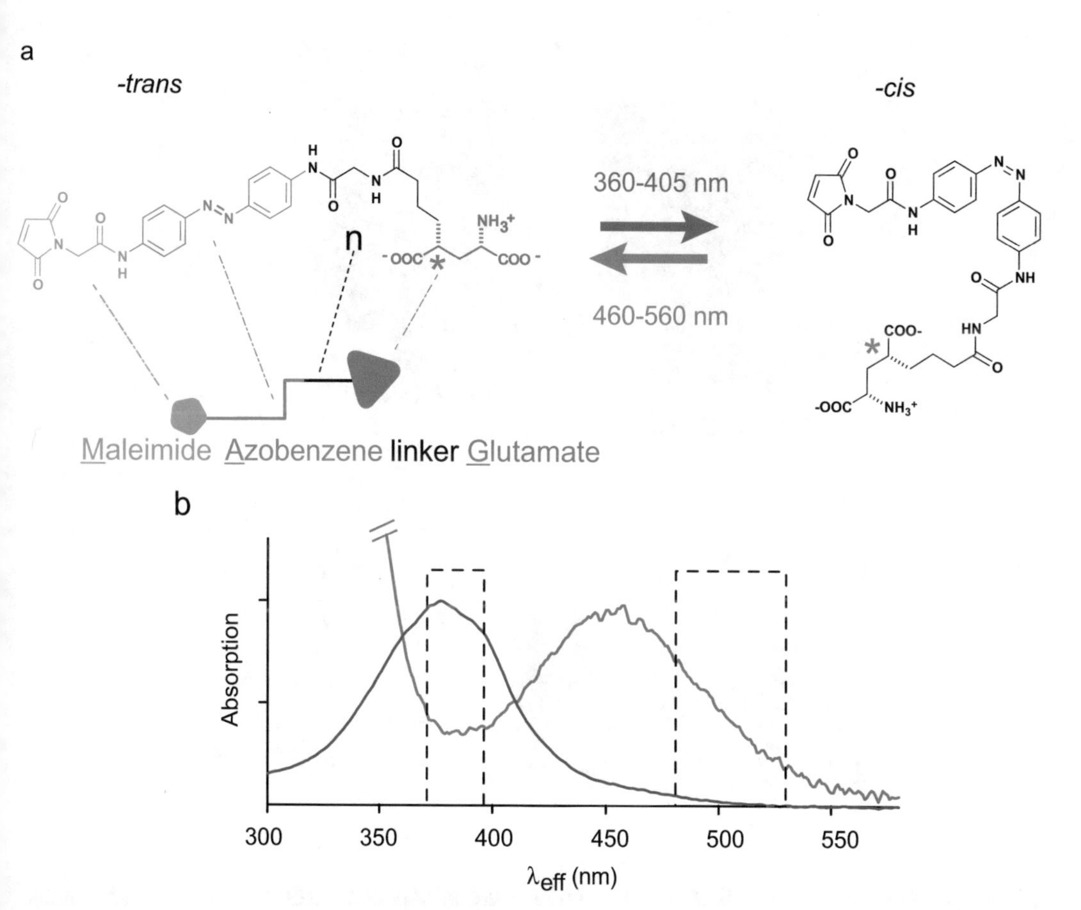

Fig. 1 Isomerization of an azobenze-based phototethered ligand, denoted MAG. (**a**) Chemical structure of the MAG photoswitch. MAG0 and MAG1 differ in length ($n = 0, 1$) [31], whereas L-MAG and D-MAG [32] differ in stereochemistry (*red asterisk*). The azobenzene moiety (*red*) enables –*trans* to –*cis* isomerization; *the* maleimide group (*cyan*) allows conjugation with a cysteine residue and the glutamate headgroup (*magenta*) serves as an agonist or antagonist. (**b**) Normalized absorption spectra of the –*trans* (*purple*) and –*cis* (*green*) states of L-MAG0. *Boxed regions* denote wavelength ranges that facilitate high percentages of photostationary states. Panels (**a**) was adapted from Berlin S. and Szobota S., et al. [33]. Panel (**b**) was adapted from Carroll E., et al. [34]

use of PCLs is highly advantageous as it allows targeting endogenous unmodified receptors, this method is limited by the specificity of the pharmacophore. In addition, PCLs cannot be used to genetically target a particular cellular population, nor spatially confined.

The other strategy, on which we will expand in this chapter, consists of *p*hotoswitchable *t*ethered *l*igands (PTLs) [47]. PTLs are conceptually similar to PCLs, but contain an additional reactive group at the end(s) of the photoswitch to allow its site-specific conjugation onto a protein. PTLs can be tethered to one- or two-sites of the protein. Two-site conjugating (i.e., bridge) PTLs do not contain a pharmacophore, and therefore have been employed

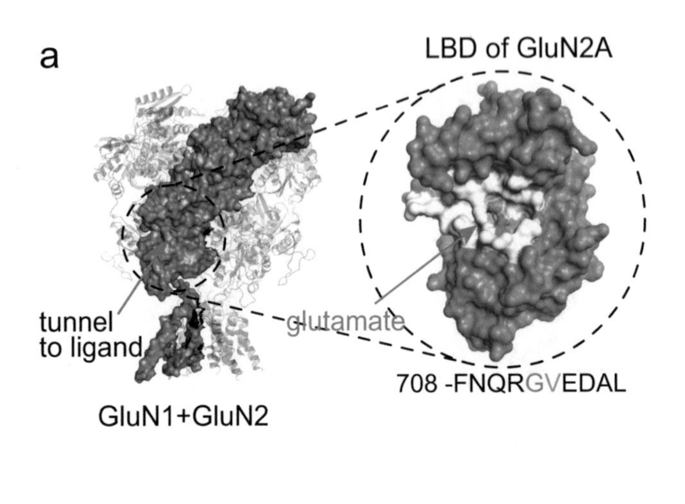

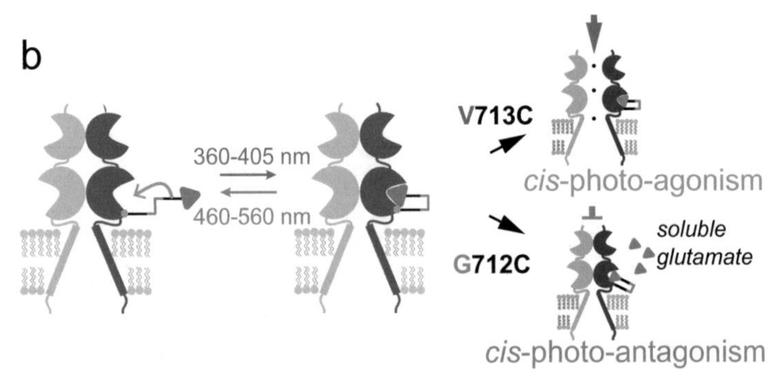

Fig. 2 Light-gated GluN receptor (LiGluN). (**a**) Space filled, crystal structure of a GluN1a/GluN2B heterotetramer (PDB-4PE5 [38]) and GluN2A' Ligand Binding Domain (*inset*, PDB-2A5S [39]) showing the access tunnel (*green arrow*), the glutamate ligand (*magenta*) and residues mutated to cysteines and tested for photo-responses (*white surfaces*). Two positions that yield the largest photo-responses—agonism and antagonism, are highlighted (*green* and *red surfaces*, respectively; bottom sequence). (**b**) Cartoon depiction of photo-agonism and antagonism. For simplicity, one LiGluN1a-*wt* subunit (*light grey*, glycine-devoid) co-assembling with one engineered LiGluN2A subunits (*dark grey*) are shown (instead of tetramer). The MAG photoswitch is *color-coded* as in Fig. 1a. When bound to residue V713C, near-UV illumination isomerizes MAG to *–cis,* correctly positioning the glutamate headgroup in the glutamate binding pocket, inducing LBD closure and channel opening (*cis*-photo-agonism; *top scheme*). Conversely, when MAG is bound to position G712C, in *–cis* the glutamate headgroup does not properly dock within the LBD, obstructing LBD closure, out-competing soluble glutamate, to antagonize and close the channel

to yield forceps-like motion to alter protein structure and, thereby, function [48, 49]. However, the more common use of PTLs is when tethered to a single site on the protein. In this instance, the PTL is designed to include two distinct chemical groups spanning the light-reactive azobenzene: one end of the photoswitch tethers it to the protein (Fig. 1a, cyan), whereas the other end contains the functional pharmacophore which interacts with the protein (Fig. 1a, magenta and see Sect. 1.3).

In an attempt to manipulate NMDA receptors (NMDARs or GluNRs), we employed an azobenzene-based PTL bearing a maleimide (to irreversibly conjugate cysteines) and a glutamate (Glu) headgroup as the ligand; denoted MAG (Maleimide-Azobenzene-Glutamate). We modified specific residues on several different GluN- subunits that, when bound to MAG, exhibit photo-agonism (Fig. 2b, top scheme) or antagonism (Fig. 2b, bottom) following illumination. In the first sections of the chapter, we discuss GluNRs and current methods used to study them in their native environment (Sect. 1.2). This section will highlight several advantages and disadvantages, some of which have prompted us to engineer light-gated GluNRs (LiGluNRs) [33]. In subsequent sections, we will describe in length the MAG PTL (Sect. 1.3.1), the necessary protein modification used to tether it to GluN subunits (Sect. 1.3.2) and several screening methods that can be utilized to assess light-dependent channel responses (Sect. 1.3.3). Section 1.4 will entail photoswitching LiGluNs in neurons, highlighting expected outcomes. Last sections include detailed protocols and discussion.

1.2 Studying Glutamate Receptors in Neurons

GluNRs are heterotetrameric ligand-gated ion channels (Fig. 2a). Functional GluNRs at the plasma membrane (PM) typically consist of two obligatory glycine-binding GluN1 subunits that multimerize with two Glu-binding GluN2 and/or glycine-binding GluN3 subunits [50, 51]. In particular, the Glu-binding GluN2 subunits are encoded by four discrete genes (GluN2A-D), and the resulting subunits exhibit different expression patterns throughout the brain and during development [52, 53]. GluNRs are primarily found at excitatory synapses, at synaptic and extrasynaptic locations [54], where they play major roles in synaptic plasticity [55, 56], owing to several unique features. First, GluNRs are coincidence detectors. To open, they require the binding of Glu (and glycine), but also local depolarization to relieve their intrinsic Mg^{2+}-block. Thus, their activation depends on the coinciding activities of the pre- and the postsynapse. Secondly, though all ionotropic Glu receptors (iGluRs) are nonspecific cation channels (E_{rev} ~ 0 mV [57, 58]), GluNRs conduct the largest fraction of Ca^{2+} (11–14% compared to 1–3% in α-amino-3-hydroxy-5-methyl-4-isoxazolepropionic acid (AMPA) receptors [50, 51, 59]), a key component in synaptic plasticity [60].

GluNRs have been extensively studied over the years. In neurons, where multiple classes of Glu receptors and GluNRs of different compositions can be found [51], the activity of subtype-specific GluNRs has largely been addressed by conventional pharmacology [51]. Specific activation of *all* GluNRs can be obtained by the GluNR-specific agonist *N*-methyl-D-aspartic acid (NMDA), whereas specific antagonism or channel block of *all* GluNRs is obtained by AP-5 [61] or the activity-dependent pore blocker

MK-801 [62], respectively. Notably, whereas no subunit-specific agonists have been developed, there are few successful subunit-specific blockers: Ifenprodil (and prodil-derivatives [63, 64]) and Ro 25-6981 [65] specifically target GluN2B-containing receptors, whereas NVP-AAM077 [66] and TCN-201/213 [67] target GluN2A-containing receptors. Despite their utility, NVP-AAM077 poorly discriminates between GluN2A and B-containing receptors, and there is no concentration that allows full inhibition of GluN2A-containing receptors, without affecting GluN2B-containing receptors [68]. Compounds from the TCN-family are poorly soluble (<10 μmol/L) [69] and their effect is highly glycine-dependent [69, 70]. This complicates their usage in preparations with high or unknown glycine concentrations (e.g., the brain [71, 72]). Lastly, Ro 25-6981 (and other prodils) contain several off-targets [73, 74]. Notably, to date there are no subunit-specific antagonists to differentiate between GluN2C and –D containing receptors [75–77] or between GluN3A vs. –B containing receptors [50].

Genetic manipulations of specific GluN- subunits have been extremely valuable, as can be obtained by gene-knockout [78, 79], RNA-interference [80], conditional Cre-dependent ablations [81], or more recently CRISPR/Cas9-induced deletions [82, 83]. Despite the absolute subunit-specificity, these too present several shortcomings. First, as the genetic manipulations are typically done in the germlines, genetic deletions are typically global and often lethal [84–86]. These broad effects may lead to compensation and gross cellular changes [87, 88]. Secondly, these methods are very hard to localize to specific brain regions or cell types [89] and as off-target can occur, these are hard to predict and expose [90–92].

Several optical methods have been developed to probe GluNRs activity. Caged compounds are chemically trapped drugs, and those can only be released, and become functional, only following illumination. These include caged-glutamate, NMDA, and the pore blocker MK-801. These are extremely powerful tools, easily combined with other recording or imaging methods and enable high temporal and spatial resolution. Importantly, these are readily photoactivated by two-photon excitation [46]. However, as with conventional pharmacology and PCLs, these too are limited by the specificity of the pharmacophore they harbor. In addition, as these reagents are typically applied at high concentrations (~mmol/L), they have been found to affect other targets [93]. Another recent optical development involves the usage of light-reactive unnatural amino acids (UAA) incorporated within the protein. UAA are incorporated within the protein at designated stop codons, so that only channels that have had the unnatural amino acid incorporated are fully translated and reach the PM. Thus, this system offers the genetic targeting so desperately sought-after, but the technique is still not very widespread and only few light-gated ion channels have been successfully engineered: a potassium channel [94], an AMPA receptor [95] and a single GluN- subunit; a photo-inactivated GluN2B [96]. This likely stems from the method complexity, requirements of multiple steps and numerous components, not to mention irreversibility.

To try and address some of these concerns, we employed the approach of synthetic optogenetics to render GluNRs amenable by light. The approach enables reversible gating of specific GluN-subunits, expressed at specific cellular populations, with unmet spatiotemporal resolution. These will be further explored in following sections.

1.3 Experimental Strategy

1.3.1 Glutamate-Based Photoswitch

The azobenzene-based photoswitch used for photoswitching LiGluNRs is composed of three parts: a cysteine-reactive *M*aleimide, an *A*zobenzene, and a *G*lutamate headgroup (MAG, Fig. 1a). As noted above, azobenzenes have several key features making them particularly suitable for use in biological systems. They undergo reversible photochemistry so that many rounds of *–trans/–cis* (inactive/active, irrespectively) states can be obtained by alternating light illumination (Fig. 1b). The *–trans* state is energetically favored, and in the dark more than 95% of the molecules will occupy this state [97]. Though the absorption spectra of the *–trans* and *–cis* isomers overlap substantially (Fig. 1b) [34], irradiation at optimal wavelengths of 380 and 500 nm can produce photostationary states with a maximum of ~80% *–cis* or ~90% *–trans*, respectively [31]. Though thermal *cis*-to-*trans* relaxation of the azobenzene occurs, it does so in tens of minutes (~20 min at RT) [31, 35] and from a biological point of view this slow rate of relaxation makes this molecule virtually *bistable* (e.g., Fig. 3b). This is particularly desirable, because short light exposures are sufficient to obtain long durations of activity, thus reducing the need of frequent and repetitive illumination and light-induced toxicity.

There are several MAG variants differing in stereochemistry, length, and absorption properties. Though it is difficult to predict in advance which of these will perform best with the subunit of choice, structural data may be helpful (see below). In the case of GluNRs, we found that MAG molecules bearing a L-stereochemistry (L-MAG), as opposed to D-MAGs [32, 34] (Fig. 1a, red asterisk), of different lengths (via glycine linkers, Fig. 1a, *n*; linker) perform best [33].

1.3.2 Protein Modification of GluN Subunits to Include Reactive Handle for PTL Attachment

The attachment of the MAG to the GluN- subunit is obtained by site-specific cysteine mutagenesis at the ligand-binding domain (LBD) of the subunit. Before rationally introducing the mutation, it is first necessary to identify an appropriate cysteine-attachment site. This can be obtained by studying the structure of the Glu binding pocket (Fig. 2a), in order to identify potential residues located at the surface of the LBD and at a reasonable distance away from the pocket. Luckily, ample structural data exists for several

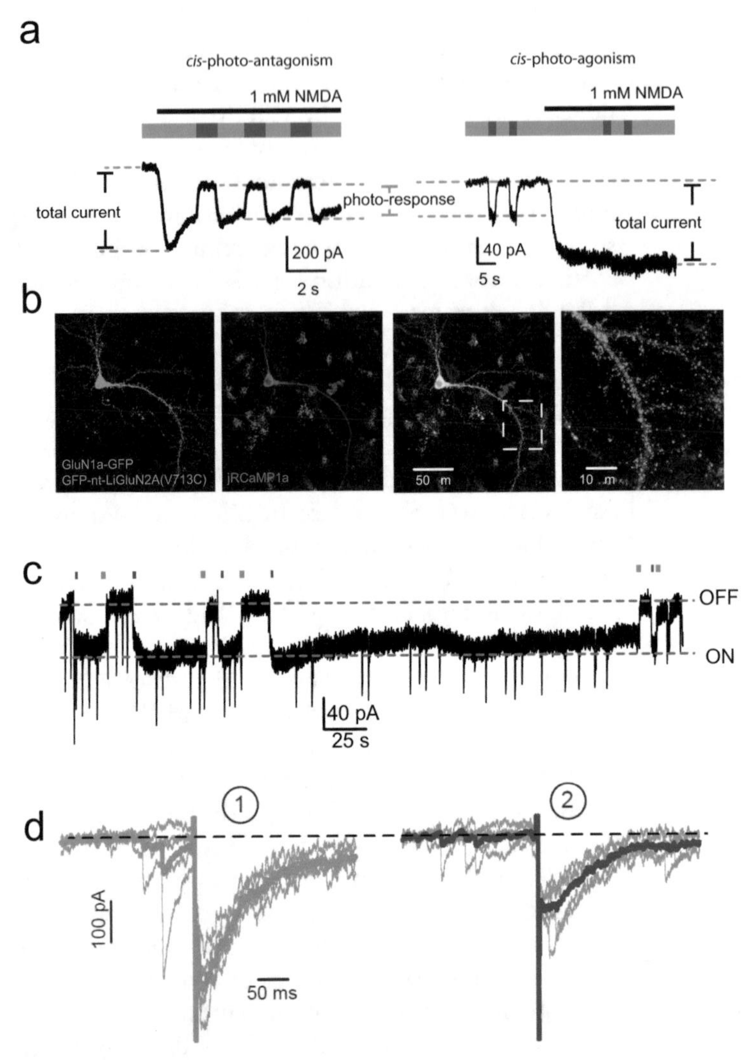

Fig. 3 Photo-agonism and -antagonism of LiGluNs in hippocampal neurons. (**a**) Traces of *cis*-photo-antagonism (*left*) and *cis*-photo-agonism (*right*), displaying the relative sizes of the NMDA-saturated currents and the photo-responses (extent of inhibition and activation, respectively). (**b**) Overexpression of amino-terminal GFP-tagged GluN subunits (GFP-nt-LiGluN) enables detection of fluorescence in internal membranes (*green*, likely endoplasmic reticulum), but also in dendritic structures, i.e., spines. Cytoplasmic fill was obtained by jRCaMP1a (*magenta*). (**c**) Trace showing the bistability of the MAG photoswitch in a cultured hippocampal neuron transfected with LiGluN2A(V713C). The -*trans* and -*cis* states of MAG are photo-stable. In the dark, or after green-light illumination (*green bars*; 488 nm laser, Zeiss confocal system), the channel is closed (*black trace*; OFF). However, brief scanning over the dendrites by near-UV illumination (*purple bars*; 405 nm) induces an inward current (*black trace*; ON). The current persists in the dark without visible decay for minutes. Spontaneous EPSCs are observed in the background of the photo-current. (**d**) Autaptic NMDA-dependent evoked EPSCs ($eEPSC_{NMDA}$) under light control during voltage clamp. (*1*) $eEPSC_{NMDA}$ from neurons expressing GluN2A(G712C) and GluN1a(E406C), evoked by brief (~2–4 ms) intracellular depolarizations; during green light illumination (510 nm). (*2*) Near-UV light (370 nm) antagonizes the synaptic receptors, seen as reduction in $eEPSC_{NMDA}$' amplitude. Individual (*grey traces*) are superimposed with the average of five consecutive EPSCs recordings (*green* for 510 nm light; *violet* for 370 nm light). Panels (**a**), (**c**), and (**d**) were adapted from Berlin S. and Szobota S., et al. [33]

GluN- subunits, such as isolated LBDs (example Fig. 2a, inset; PDB: 2A5S [39]), intact heterotetrameric receptors (Fig. 2a PDB: 4PE5 [38, 98]), or structures of sister-receptors from the Glu-binding receptor superfamily (e.g., [99]), as iGluRs share a high degree of homology at the LBD. Notably, there is even an isolated Kainate receptor LBD bound to an azobenzene PCL; *gluazo* [100]. This structure may prove even more valuable than GluNR structures for the purpose of examining tethering positions for the PTL. Other methods for determining potential residues can include dynamic simulations [32] or, more easily, by systematic mutagenesis of external, water-accessible residues located on the surface of the LBD, as we have performed (Fig. 2a, white surface) [33]. Notably, when studying the structure of the receptor, we emphasize the need to identify a potential entry tunnel so as to allow the linker of MAG to protrude and reach its attachment site on the surface of the protein, without obstructing the clamshell from closing over the agonist (Fig. 2a, green arrow). Luckily, most LBD structures of iGluRs are in the closed conformation that enable examination of this entry tunnel.

In the case of the LiGluNs, the first round of cysteine scanning around the LBD resulted in ~20 single-cysteine variants [33]. We found one residue at the LBD (V713) that when bound to L-MAG1 via a cysteine (i.e., V713C mutation) enabled light-dependent channel opening, without obstructing clamshell closure (Fig. 2b, top scheme). We also noted that most cysteine mutants when bound to MAG and following illumination, yielded photo-antagonism. We hypothesized that this is due to the more stringent structural constraints required for channel opening, rather than for antagonism. Thus, any position that did not enable Glu to properly dock within the LBD or that did not allow the clamshell to close properly antagonized the receptor and closed the channel. We used this to create a photo-antagonistic LiGluN (Fig. 2b, bottom scheme).

1.3.3 Optical Screening Methods for Viable Cysteine Mutants

Prior testing whether a modified GluN-subunit labeled with MAGx can produce photo-currents, it is critical to consider the following. First, it is important to consider that all GluNRs require the obligatory GluN1-*wt* subunit for membrane trafficking, otherwise retained in the endoplasmic reticulum (ER) [55, 101, 102]. Thus, depending on the expression system (particularly in non-neuronal cells such as HEK cells, *Xenopus* oocytes, etc.), it may be necessary to co-express the GluN1 subunit. Second, it is important to provide and saturate GluN1 with its ligand glycine (or D-serine). Third, it is crucial to remove Mg^{2+} from the extracellular solution, as it readily blocks all NMDARs and no photocurrents will be observed, unless the cell is actively depolarized.

To quickly and efficiently screen for photo-responses of engineered cysteine variants bound to MAG, an orthogonal system for measuring responses is warranted. As noted above, because of the high Ca^{2+}-permeability of GluNRs, Ca^{2+}-imaging is particularly suited for this as it can allow rapid monitoring of many different clones and PTL combinations [31], noninvasively. However, careful considerations should be taken when choosing a fluorescent Ca^{2+}-probe. It is best to avoid probes that require excitation wavelengths that may also excite and actively switch the azobenzene molecule from *–trans* to *–cis* during imaging. To this end, red-shifted probes are better suited than green probes (see absorption spectrum Fig. 1b). Bright fluorescent probes are especially recommended, as these require lower imaging intensities (~$\mu W/mm^2$) [103]. In this case, even when the excitation spectra of *cis*-MAG and the fluorescent probe may slightly overlap, the lower intensities ensure less effect on MAG [104], which could then be efficiently photoswitched by intermittent, higher intensity near-UV or green light illumination. Red Genetically Encoded Ca^{2+}-Indicators (red GECIs), such as R-GECO [105] and RCaMP [106], are particularly advantageous since their peak absorption (~550 nm) is almost completely orthogonal to the *–cis* absorption spectrum of most MAG photoswitches [31, 34]. Thus, during the imaging of the red GECIs, MAG would not be isomerized. However, red GECIs are prone to UV-induced artifacts [106, 107]. These can be circumvented by lowering the near-UV illumination intensity applied for photoswitching to *–cis* or by combining multiple illuminations wavelengths [33, 107]. Luckily, newer red GECIs have been reported to be less prone to UV-excitation artifacts [108]. Additional means to avoid artifacts is to physically separate the imaging wavelengths from those used for photoswitching by spatial confinement of the different wavelengths to different cellular regions (or planes), such as near-membrane near-UV illumination to gate the receptors, and imaging of the GECI in the cytosol from above [109]. In all cases, additional controls are highly recommended to discriminate between *bona-fide* photo-responses from light-induced artifacts, such as testing transfected cells that were not incubated with MAG, or cells expressing the fluorescent sensors without expressing LiGluNs, etc.

1.3.4 Electrophysiological Recordings of Photo-Currents

Clones that yielded positive effects with the GECIs should be further characterized by electrophysiology, as this method is the direct readout of channel activity. Notably, this technique is mostly insensitive to light illuminations and offers an extremely high degree of sensitivity and temporal resolution. Recordings can be done using Two-Electrode Voltage Clamp (TEVC) in *Xenopus* oocytes for screening [44] and pharmacological characterization [33], or in mammalian cells by the patch clamp method [110]. Of note, when recording from *Xenopus* oocytes, it is highly recommended to remove all Ca^{2+} from the extracellular solutions, as activation of NMDARs and entry of Ca^{2+} into the oocyte will induce the activation of Ca^{2+}-dependent Cl^- channels (see [33] for *Xenopus* recording solutions).

There are different types of activation schemes that we recommend considering prior recordings. For instance, if the receptors are activated by the *–cis* form of MAG (denoted *cis*-agonism), increases in current will be observed only during near-UV illumination (~380 nm), whereas cessation of the photo-currents will be obtained by MAG in *–trans*, following green light illumination (~510 nm) (Fig. 2b and Fig. 3a, right trace, purple and green bars, respectively). *Trans*-agonism is not as easily discernable. When MAG acts as a *trans*-agonist, the channel is thereby activated by the relaxed form of the photoswitch. This signifies that the channel will be activated and opened during the labeling process, in the dark, if not persistently illuminated by near-UV light [103, 111] or if the receptors are not blocked by soluble antagonists. In this instance, near-UV illumination (without the presence of soluble Glu) will remove the Glu headgroup from the LBD, to close the channel. Then, illumination with green light (~510 nm) will toggle MAG back into *–trans* to open the channel. For photo-antagonism (Fig. 2b, bottom scheme), akin to soluble antagonists, photo-effects require the channel to be open beforehand to exert their effect (Fig. 3a, left trace). Without soluble Glu, neither *cis*- or *trans-antagonists* will display any apparent effects. To test for antagonism, this requires the application of soluble Glu (and glycine) to first open the channel to then allow to test whether near-UV or green light can antagonize the receptor by expelling the Glu from the LBD (Fig. 2b, bottom scheme and Fig. 3a, left trace).

When testing for agonism, it is also highly recommended to end the test with the perfusion of soluble Glu. This application enables to assess the degree (or extent) of activation by light (photo-response (pA)/total current (pA) * 100; Fig. 3a). MAG typically behaves as a full agonist [33], however this protocol also enables to test whether MAG has any potentiating effect, i.e., further increase of the maximal glutamate-current. A detailed recording protocol is found in Sect. 3.5.

1.4 LiGluNRs in Hippocampal Neurons

In neurons, the endogenous pool of GluN1 subunit makes it possible to express a single GluN2 or -3 subunit in order to obtain PM expression. This method of expression is more physiologically relevant as it can yield moderate expression levels of GluNRs containing the modified subunit [33, 112, 113]. This may also ensure that GluNRs remain uniquely localized to synapses as opposed to pan-cellular expression in dendrites, soma and axon. At physiological levels, GluNRs are found in small numbers at the synapse (~60 molecules/synapse) [114], making it hard to detect green fluorescent protein (GFP)-tagged receptors by regular fluorescent methods (see Sect. 2). This is particularly true when expressing a GFP-tagged subunit compared to expressing soluble GFP that fills the entire cytosol. Overexpression of two tagged-subunits (e.g., GluN1a-GFP and GluN2A-GFP) enhances detection at single synapses (Fig. 3b), likely at the cost of overexpression. Nonetheless, for screening and characterization purposes overexpression may be justified, as overexpression will yield higher PM levels of light-ready channels and, importantly, larger photo-currents [33]. Notably, if dual expression is not desired, there are other means to potentially increase expression at the synapse. This may involve overexpression of other proteins, such as PSD-95; known to potentiate synaptic currents, increase synapse size, and reduce turnover of postsynaptic components [115].

To test for *cis*-photoagonism in neurons, LiGluN2A(V713C) or LiGluN2B(V714C) should be transfected. As noted above, these will complex with the native GluN1 subunits and reach the PM within a couple of days. Then, prior recordings, neurons will be incubated with L-MAG1 (50-300 μmol/L) for ~40–45 min to allow the conjugation of MAG with the cysteine-modified subunits found in receptor complexes at the PM. Then, during perfusion of extracellular media containing glycine but without Mg^{2+}, broad and brief (~1 s) cellular illumination with near-UV light (~380 nm) will induce inward NMDA-dependent currents. These currents will be sustained in the dark until illuminated with 510 nm light (Fig. 3c). Notably, this current will appear on top of the spontaneous synaptic activity of the neuron, if TTX (1–2 μmol/L) isn't applied. Following recordings of photocurrents, application of saturating (~mmol/L) soluble NMDA should be applied to compare the size of the photocurrents obtained to the total NMDA-dependent current. We recommend using NMDA as opposed to Glu as the ligand because this bypasses the need to use a cocktail of inhibitors/blockers required to block all other metabotropic and ionotropic GluRs; explicitly mGluRs, GluKRs, and GluARs. In addition, when testing for GluN2A-containing receptors, all other GluN2-receptors should be blocked to isolate the GluN2A-current (for example ifenprodil, see above).

The same process should be applied when using the photo-antagonistic variants; GluN2A(G712C) or GluN1a(E406C) can be expressed alone or together (to obtain stronger block). These require the shorter L-MAG photoswitch; L-MAG0 (linker = 0 and 1, see Table 1) [33]. Here, light illumination before the application of the soluble ligand does not typically produce any effect, though valuable to test. Then, during NMDA perfusion and appropriate blockers, illumination will cause the reduction of the current.

To test for the functional incorporation of the receptors into the synapses (aside imaging GFP of the tagged receptors), there are several optical and electrophysiological methods available. Optically, light could be targeted to individual spines, where Ca^{2+},

Table 1
LiGluN variants and their appropriate L-MAG photoswitches. List of the four LiGluN subunits, noting the incorporated cysteine mutation, the best MAG photoswitch combination, and the light-dependent responses (antagonism or agonism)

Light-gated GluN subunits	MAG attachment site (cysteine point mutation)	MAG derivative	Photo-effect
LiGluN1a	E406C	L-MAG0	Photo-antagonism
LiGluN2A	G712C	L-MAG0	Photo-antagonism
LiGluN2A	V713C	L-MAG1	Photo-agonism
LiGluN2B	V714C	L-MAG1	Photo-agonism

voltage or structural changes, namely spine expansion, could be imaged and assessed. With the use of electrophysiology, we have used the autaptic system. In this system, the neuron synapses onto itself permitting the recording of both pre- and postsynaptic responses from the same neuron, via the same patch pipette. We used the *cis*-photo-antagonistic variants to block synaptic NMDARs, emerging as reduction in the size of the NMDA-dependent evoked excitatory post synaptic currents ($eEPSC_{NMDA}$) (Fig. 3d) [33].

Lastly, we have also used these tools to control cellular excitability. To this end, broad cellular illumination of neurons expressing either agonistic or antagonistic variants allowed to induce (even in the presence of Mg^{2+} if neurons are slightly depolarized) or inhibit action potential firing, respectively (Fig. 4a, b). Induction of firing is likely obtained by the depolarizing synaptic NMDAR currents, in particular their large Ca^{2+}-plateaus, as obtained by *wt* GluNRs [116, 117]. Conversely, inhibition of firing is achieved from the reduction of the synaptic inputs, as shown with the autaptic responses (see Fig. 3d).

2 Materials

2.1 DNA Constructs

2.1.1 Constructing Light-Activated, Fluorescently Tagged and Untagged GluN Subunits

1. For mammalian cells expression, receptor variants need to be cloned into pcDNA-like vectors (e.g., Fig. 4, Invitrogen Cat. No.: V79520 or similar variants such as pNICE, pRK5, or viral vectors—pAAV and pCSC; for Adeno-Associated Virus and Lentivirus production, respectively) (**Note 1**). Commonly, expression vectors contain promoters such as the cytomegalovirus (CMV) promoter or its modified version; CAG (**Note 2**) [118, 119] to yield strong and robust expression in many mammalian heterologous expression systems (e.g., HEK293 cells) [120], as well as in primary cultured neurons. For exclusive neuronal expression, the human synapsin promoter (hSyn) [121] may be used.

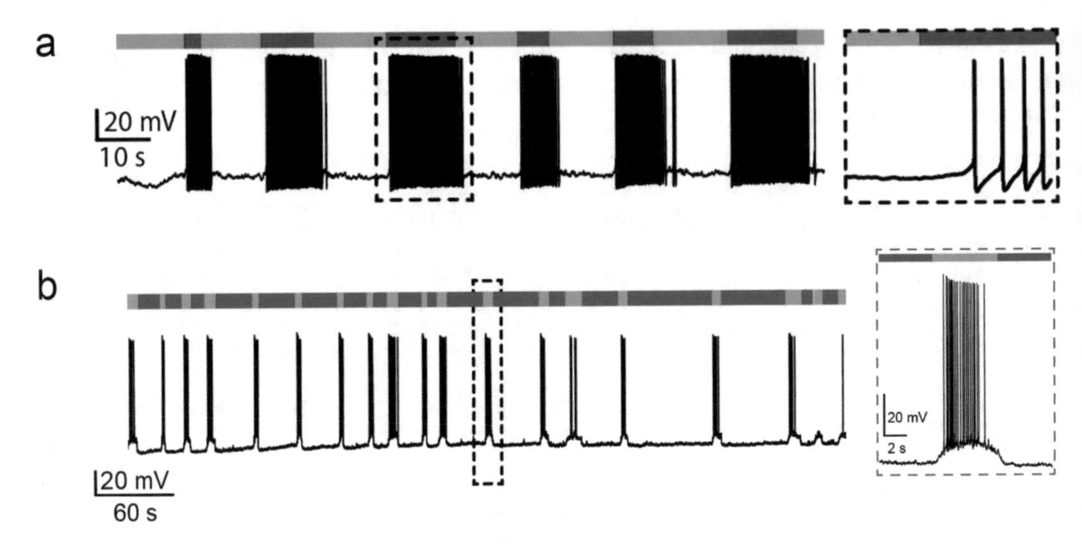

Fig. 4 Modulating neuronal excitability and synaptic activity via GluNR photoactivation or photoblock. (**a**, **b**) Representative traces (current clamp) of hippocampal neurons transfected with (**a**) LiGluN2A(V713C) labeled with L-MAG1 or (**b**) LiGluN1a(E406C) and LiGluN2A(G712C) labeled with L-MAG0, and illuminated with ~380 nm (*violet bars*) or 510 nm (*green bars*) light. Neurons were held at −45 to −50 mV. During photoactivation (**a**) or block (**b**) of LiGluNs, neurons display significant increase or decrease in firing frequency (*insets*). Panels (**a**) and (**d**) were adapted from Berlin S. and Szobota S., et al. [33]

2. The transgene should be inserted at the multiple cloning site (MCS) of the expression vector following a Kozak translation initiation sequence (Fig. 5; (**G/A**)NNATG**G**) [122], with the ATG initiation codon for correct initiation of translation.
3. Single cysteine point mutations are introduced near, but outside of the Glu-binding site of the Glu receptor subunits (see Fig. 1c and Table 1), using site-directed mutagenesis. PCR-mutated transgenes need to be verified that no additional mutations were mistakenly obtained by full sequencing.
4. Fluorescent-tagging of LiGluN subunits can be obtained by the insertion of fluorescent proteins at the amino (NT) or carboxy termini (CT) of the gene. For NT-fusion, the open reading frame of the fluorescent protein (e.g., enhanced GFP or eGFP) should follow the signal peptide sequence of the GluN-subunit (Fig. 5 and **Note 3**). If placed at the CT, it should be placed before the stop codon. In both cases, GFP will be localized to the membrane of the cell (but green fluorescence will also be observed intracellularly, in internal membranes, e.g., ER, Fig. 3b, left image, green cytosolic fill).
5. Fluorescent cytosolic markers could be used to better identity transfected cells instead of fused receptors. For independent cytoplasmic fill, the fluorescent protein can be expressed independently of the LiGluN subunit(s), by co-transfection of two expression vectors. Another expression method includes

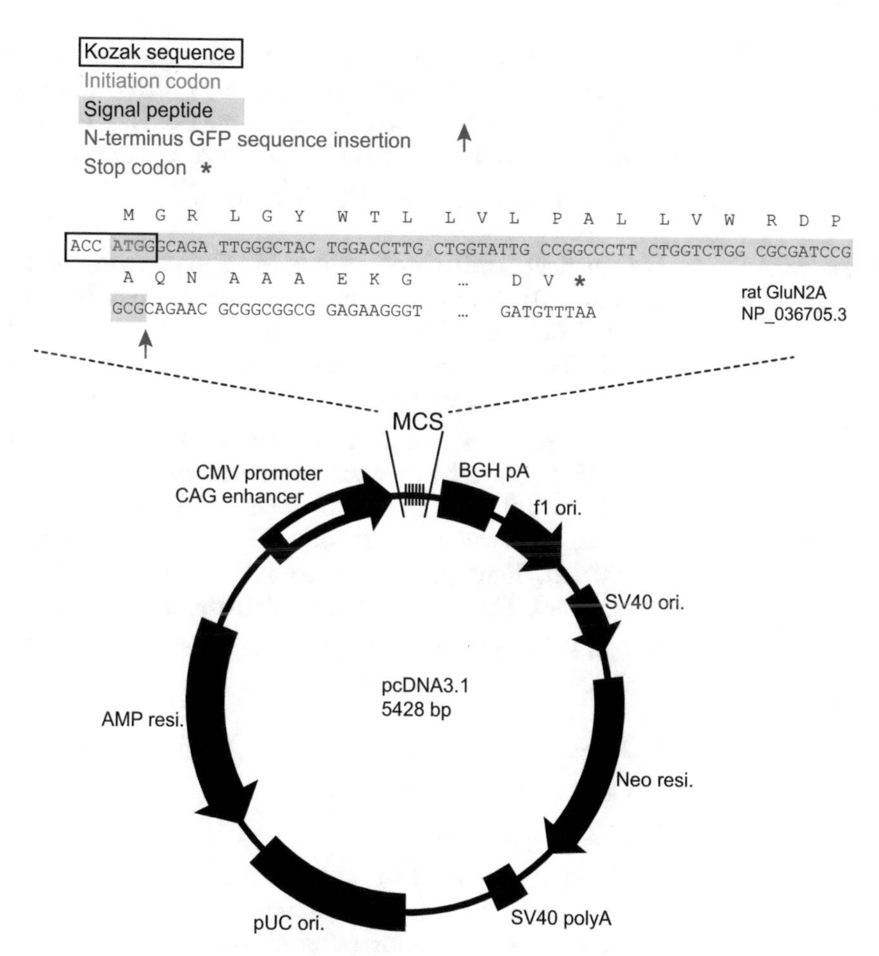

Fig. 5 Molecular engineering and mammalian expression of LiGluNs in hippocampal neurons. To express LiGluN in mammalian cells (e.g., neurons), a pcDNA-like vector is required containing an appropriate promoter. Inclusion of an upstream Kozak sequence and initiation codon are required. The fusion of eGFP at the amino-terminal of LiGluN should be introduced after the signal peptide, to enable proper insertion of the receptor into the membrane during translation. Carboxy-terminal tagging with GFP should precede the stop codon

dicistronic vectors that allows the translation of two independent (i.e., not fused) transgenes. This expression system typically includes an $IRES_{GTX}$ (Internal Ribosome Entry Site) promoter between the two transgenes [123, 124]. For stoichiometric expression of the transgenes, self-cleaving peptides can be employed. In this manner, a single polypeptide is transcribed from a single transgene, followed by a self, co-translational, intraribosomal cleavage, yielding multiple discrete translation products (**Note 4**) [125].

6. For expression in *Xenopus laevis* oocytes [126], the LiGluNs should be cloned into vectors that allow in vitro transcription of RNA, such as pGEM-HE. These vectors are suited for the oocyte expression system because they include both 5′-and 3′-untranslated regions (UTRs) from the *Xenopus* β-globin

gene that boosts expression of the transgenes following RNA injection [127]. This will require in vitro transcription of capped RNA (e.g., mMESSAGE mMACHINE transcription kit, ThermoFisher Scientific, Cat. #: AM1344).

7. DNA purifications kits (see Sect. 3).

2.1.2 Photoswitches

1. Chemical synthesis of various MAG derivatives (L/D-MAG0, 1, 2, MAG_{460}) can be done as previously described [103, 128, 129]. Several MAG photoswitches are now commercially available (**Note 5**) and others can be obtained from the Isacoff lab (UC Berkeley) by request.
2. MAG photoswitches, in powder form, are dissolved in dry DMSO (**Note 6**) to stock concentration of 50–100 mmol/L (Mw ~ 600 g/mol), aliquoted (in clear PCR tubes at 0.3 μL/tube) and stored at −20 or −80 °C in light protected tubes (50 mL light protection conical tubes, Cat. #: 227 280, Greiner). These are preferably desiccated (**Note 7**).
3. L-MAG0 and L-MAG1 are in use with different the different GluN subunits (see Table 1) [33].

2.1.3 Cell Culture

1. Tissue culture hood (laminar flow/sterile hood).
2. CO_2 cell-culture incubator.
3. Heated water bath.
4. Inverted live-cell tissue culture microscope; including bright-field, phase contrast, and fluorescence for cell inspection and fluorescence evaluation (after transfection).
5. Round cover glasses, 12 mm, thickness ~0.1 mm (e.g., Carolina Biologicals Cover glasses, Item #: 633029).
6. Polystyrene flat bottom 24-well plates, sterile (for 12 mm cover glasses).
7. Poly-L-lysine hydrobromide (e.g., Sigma P2636) solution is prepared as a 10 g/mL stock solution (20×) in distilled water (filtered for sterilization) and stored as 1 mL stock solution aliquots at −20°C.
8. Borate Buffer (60 mmol/L, pH 8.5). Dissolve 1.5 g boric acid, 2.3 g sodium borate in 500 mL distilled water. Filter to sterilize and store at 4 °C.
9. 60 and 80 mm Petri dishes.
10. 15 and 50 mL conical tubes.
11. 70% EtOH.
12. Dissection tools—(scissors, large forceps, two sharpened forceps, scalpel).
13. Dissection scope.

14. Cell strainer (40 μm, BD Biosciences, Cat. No. 352340).
15. Glass Pasteur pipettes (flame polished).
16. Minimum essential medium (MEM), glutamine-free (e.g., Gibco, Cat. No. 11090-081).
17. Neuronal growth medium: 2.5 mL fetal bovine serum (FBS, Sigma F6178), 1 mL B27, 0.5 mL Glutamax (100×, Invitrogen), 1 mL 1 mol/L D-glucose, 50 μL serum extender (BD Biosciences, Cat. No. 355006), fill MEM to final volume 50 mL Filter to sterilize and store at 4 °C. Add 4 μmol/L cytosine arabinosine (Ara-C, Sigma, C6645).
18. NMEM-B27 transfection medium (in mmol/L): MEM, 1 Sodium Pyruvate, 15 HEPES, 2 L-glutamine (GlutaMax, Gibco Cat No. 35050), 10 D-glucose and 1× B-27 (Gibco, Cat. No. 17504044). Filter to sterilize and store at 4 °C.
19. Wash-HBSS buffer (in mmol/L): 135 NaCl, 20 HEPES, 4 KCl, 1 Na_2HPO_4, 2 $CaCl_2$, 1 $MgCl_2$, 10 D-glucose, pH 7.3. Filter to sterilize and store at 4 °C.
20. 2× BBS (in mmol/L): 50 BES, 1.5 Na_2HPO_4, 280 NaCl. Make several different pH solutions ranging from pH 6.9 to 7.15 (in 0.05 steps). Filter to sterilize and store at 4 °C.
21. Dissection/Dissociation medium for stock 500 mL solution (in mmol/L): HBSS (Ca^{+2}- and Mg^{2+}-free), 10 HEPES (5 mL from Gibco, Cat. #: 315630-080 or 1.19 g HEPES, Sigma), 20 Glucose, pH 7.4 (NaOH). Filter to sterilize and store at 4 °C.
22. 2.5 mol/L $CaCl_2$ solution (in water).
23. 2.5% trypsin (10×, no phenol red, Gibco, Cat. #: 15090046).
24. P0–P5 rat neonates. Four–five pups yield 24 cover slips (12 mm) with a density of ~80K cells/well.

2.1.4 Labeling and Recording Solutions for Patching Neurons

1. Recording solution (nominally Mg^{2+}-free and high external Ca^{2+}) (in mmol/L): 138 NaCl, 1.5 KCl, 10 D-glucose, 2.5–5 $CaCl_2$, 5 HEPES, pH 7.4 (NaOH) and 0.01–0.05 glycine (**Note 8**).
2. Labeling solution (in mmol/L): 150 *N*-methyl-D-glucamine (NMDG)-HCl, 3 KCl, 0.5 $CaCl_2$, 5 $MgCl_2$, 10 HEPES, 5 D-glucose, pH 7.4 (KOH).
3. Intracellular patch pipette solution: (in mmol/L): 135 K-gluconate, 10 NaCl, 10 HEPES, 2 $MgCl_2$, 2 MgATP, 1 EGTA, pH 7.4 (~300 mOsm), pH 7.2–7.3 (NaOH) [130].
4. L-Glutamic acid (e.g., Sigma 49449) or *N*-Methyl-D-Aspartic acid (NMDA, Sigma, M3262) solutions. Prepare a 100 mmol/L stock in extracellular solution. Readjust to pH 7.3 (NaOH). Filter and store aliquots at −20 °C.

5. Neurons can be, optionally, pre-treated with reducing agents, such as 1 mmol/L DTT for 5 min, rinsed for 10 min with extracellular solution followed by MAG incubation.
6. During MAG incubation (45 min at 37 °C, 5% CO_2) neurons can be incubated in the presence of GluNR blockers such as MK-801 maleate (10 μmol/L, Tocris, Cat. #: 0924), to prevent excessive activation of the receptors by MAG [31] and possible cytotoxicity (**Note 9**).

2.1.5 Imaging and Photoswitching Illumination Sources

1. Inverted microscope equipped with a brightfield condenser and epifluorescence light path or confocal laser scanning microscope. These are typically placed on anti-vibration tables (and surrounded by a Faraday cage when in use with electrophysiology systems).
2. Filter set for imaging GFP or another fluorescent expression marker (**Note 10**), including a suitable excitation filters, dichroic mirror, and emission filters. These filters are commonly combined into filter cubes and typically mounted in a turret under the objectives. Most advanced imaging systems (e.g., confocal microscopes) are typically equipped with dichroics and filters sets required for blue (CFP/DAPI), green (GFP) and red (RFP/DsRED) imaging.
3. Optional—a CCD camera can be used for better fluorescence detection and can also be used for guiding the patch pipette, as opposed to the eyepieces (oculars).
4. An intense light source for photoswitching in the desired wavelength ranges (e.g., for –*trans* to –*cis* isomerization of L-MAG0/1: ~375–395 nm and ~488–532 nm for the inverse [31, 34]; Fig. 1b and see **Note 11**). This light source system should have the capacity to rapidly switch between wavelengths, and likely a dark position (i.e., no light emission) to test for bistability. Commonly, Xe-arc lamps (e.g., DG-4 with band-pass filters, Sutter Instruments) or monochromators (e.g., Polychrome, Till Photonics) are used. Other light sources may include LED systems (e.g., pE-4000, CoolLED), confocal microscopes equipped with different wavelength lasers (e.g., Diode laser 405 nm at 30 mW and a multiline Argon laser, 458/488/514 nm at 25 mW, Zeiss) (**Note 12**).
5. Optional, an optical power and energy meter with a photodiode sensor to measure light intensities (e.g., PM100D equipped with a C-series photodiode, λ_{range}: 200–1100 nm and $Power_{range}$: 50 nW–50 mW, Thorlabs).

2.1.6 Electrophysiology Setup [131] and Accessories for Cultured Primary Hippocampal Neuron

1. Patch-clamp amplifier, headstage, A/D converter and recording software (e.g., Axopatch 200B, Digidata 1440A and pClamp software, Molecular Devices).
2. Plated (chlorided) Ag/AgCl wires (e.g., item # 64-1317, Warner Instruments) for headstage and reference electrodes (**Note 13**).
3. 3-axis micromanipulator with amplifier headstage mount (e.g., item # MP-285, Sutter Instruments). The manipulator should be mounted on a rotating base, either on the microscope stage or on a separate stand.
4. Glass micropipette puller (e.g., P-97, Sutter Instruments).
5. Pipette glass capillaries, borosilicate, thin wall (e.g., G150TF-3, Warner Instruments).
6. Syringe tips for filling micropipettes (e.g., Microfil 28 AWG, MF28G67-5, World Precision Instruments). Alternatively, long, soft plastic pipette tips used for microcapillaries and microinjections can be used (**Note 14**). A filter (0.2 μm, PES, 13 mm) can be placed between the syringe and tip to remove any precipitates, as well as dust particles from the intracellular solution.
7. Perfusion chamber for mounting cover glasses (e.g., RC-25 F/PM3, Warner Instruments). Vacuum grease can be used for sealing (e.g., High Vacuum Grease, Dow Corning) the chamber to prevent leakage.
8. Salt agar bridges. These are glass capillaries bent into a U-shape over a flame. The salt bridge solution, typically 3 mol/L NaCl in 1% agarose (electrophoresis grade), is prepared by microwave heating. After cooling down (<60 °C), a pipette tip is used to fill the bent glass capillary. Following polymerization (hardening) of the agar in the capillary bridges, store in 3 mol/L NaCl at room temperature or at 4 °C; for longer durations.
9. Perfusion system and vacuum line. A gravity-driven perfusion system (manual or automatic, e.g., ALA-VC3 8SG, ALA scientific) is sufficient for slow perfusion experiments. These systems typically include reservoirs of solution (~60 mL), mounted above the stage and flow is controlled by valves (manual or electronic). Flow speed can be further adjusted by flow regulators. When using several solutions, multiple lines can be combined via a manifold (e.g., MMF MiniManifold, ALA scientific) before connecting to the inlet of the recording chamber. The outlet of the perfusion chamber should be connected to a vacuum line; encompassing a collection bottle and a trap (for overflow). Suction flow rate can be regulated by a stopcock.

3 Methods

3.1 DNA Preparation

1. Transform *E. coli* (e.g., DH5α-Z competent cells, Zymo research) with the expression vector(s) encoding the LiGluN subunits and grow on agar plates with appropriate resistance (e.g., Ampicillin).
2. For amplification, pick a single colony and grow in liquid culture.
3. For plasmid purification, use Mini or Midi kits. Midi preparations are preferable as they tend to yield higher concentrations of DNA at larger volumes and virtually free of endotoxins (<0.05 endotoxin units/mg plasmid DNA). New pyrogen- or endotoxin-free plastic ware and pipette tips should be used. For glassware, heat overnight at 180 °C to destroy endotoxins (**Note 15**).
4. Elute with sterile elution buffer or distilled water (*see* **Note 16**).
5. Determine the concentration of the eluted DNA using a spectrophotometer (e.g., Nanodrop). This procedure will also give an indication regarding the purity of the sample. Nucleotides, RNA, and DNA absorb at 260 nm, whereas other molecules (e.g., proteins) absorb at 280. The ratio between 260/280 is therefore used to determine sample purity. A 260/280 ratio of ~1.8 represents a pure DNA sample a ratio of ~2.0 for a RNA sample. If the ratio is significantly lower, it may indicate contamination of proteins, phenols or other 280 nm-absorbing contaminants. A 260/230 may also be used to determine purity and should range between 2.0–2.2 (for examples see technical notes—Nanodrop, ThermoScientific). DNA concentration typically yield ≥0.4 μg/μL.

3.2 Coverslip Preparation

1. Distribute coverslips into individual wells of a 24-well plate.
2. Add 70–100% EtOH for 10 min.
3. Wash 3× with sterile water (1 min incubation/wash).
4. Add poly-L-lysine (70–130K MW; 0.5 mg/mL PLL in borate buffer).
5. Wait overnight (leave plate in tissue-culture hood or in cell-culture incubator). PLL-coated coverslips can be stored in 4 °C.
6. Wash 3–5× with sterile water (5–10 min incubation/wash).
7. **When preparing cultured hippocampal neurons—allow coverslips to soak in sterile water for ~1 h (remove water during the hippocampi trypsin digestion—see Sect. 3.3) and then allow coverslips to dry in the tissue-culture hood.

3.3 Isolation of Postnatal Hippocampal Neurons [132]

3.3.1 Pups

1. Time pregnant female rats should be ordered.
2. Euthanize 4–5 neonate pups (P0–P4; rat) using an approved method.

3.3.2 Dissection Procedure

1. Warm 100 mL of dissociation/dissection medium in 37 °C water bath (prepare two 50 mL conical tubes).
2. Warm neuronal growth media in cell culture incubator. For 50 coverslips, evenly distributing 50 mL in a six well plate (~1 mL/coverslip).
3. Fill two 60 mm Petri dishes with warm dissociation/dissection medium.
4. Thaw trypsin (0.5 mL aliquot of 2.5% trypsin) at RT.
5. In a laminar flow hood, decapitate one pup at a time pup and place head in dry, 60 mm Petri dishes. Dispose of body into biohazard bags.
6. Hold the sides of the head by hands (or forceps) and with fine scissors cut the skin from back to front (neck to nose). Peal skin away to expose the skull.
7. With the same scissors, cut the skull open (midline) by making an incision from back to front. Be careful not to cut through the brain tissue when removing skull.
8. Peel back skull and transfer (slide) the brain into the 60 mm dishes containing the warm dissociation/dissection medium using large forceps (or spatula).Repeat procedure for all 4–5 pups.
9. Move brains into new Petri dish containing clean dissociation/dissection medium.
10. Under a dissecting microscope, separate hemispheres using a fine scalpel. This could be done by impaling the brain stem (or olfactory) with fine forceps, while using a scalpel to separate the hemispheres (mind the hippocampus so that it is not damaged in the process).
11. Under a dissecting microscope, peel the meninges away from the hippocampus using sharpened forceps by gently grabbing loose ends (see **Note 17**).
12. Separate and remove the hippocampus by using a scalpel.
13. **Discard hippocampus if it is not removed in one piece or if it is damaged.
14. Transfer the hippocampus to a 15 mL conical tube filled with dissociation/dissection medium. Repeat procedure for all hemispheres.

15. Remove all but 4.5 mL of dissociation/dissection medium from the tube containing the hippocampi.
16. Add 0.5 mL trypsin (2.5%).
17. Mix by gently inverting several times.
18. Place in a 37 °C water bath for 15 min.
19. During this waiting period, finish washing the PLL-coated coverslips.
20. Add ~10 mL dissociation/dissection medium to hippocampi to dilute trypsin.
21. Gently invert the tube to wash hippocampi.
22. Remove solution from hippocampi using a flame-polished serological pipet. Careful not to draw the hippocampi.
23. Add ~10 mL of dissociation/dissection medium to wash hippocampi and repeat 5–6 times.
24. At the last wash, remove as much medium as possible, without touching and damaging the hippocampi.
25. Add 1 mL of neuronal growth medium to hippocampi.
26. Triturate six times with a flame-polished Pasteur pipette with a large opening at the tip.
27. Allow large tissue debris to precipitate (~1 min).
28. Place a cell-strainer onto a new 50 mL conical tube. Wet the strainer with 1 mL of neuronal growth medium.
29. Transfer the supernatant (containing the suspended cells) onto the cell-strainer.
30. Add 1 mL of neuronal growth media to the precipitated debris.
31. Triturate 6 times with the smaller opening Pasteur pipette.
32. Repeats step 29.
33. Discard remaining hippocampal debris.
34. Wash cell-strainer with 1 mL of neuronal growth medium.

3.3.3 Cell Dilution and Plating

1. With the use of a hemocytometer, count cells and dilute in neuronal growth medium to ~200K cells/mL.
2. Transfer 0.5 mL to each well to obtain ~100K cells/well.
3. Place in CO_2-incubator and following 15 min aspirate and replace medium with 0.5 mL of new neuronal growth medium.
4. Five days after plating add 0.5 mL of neuronal growth medium supplemented with 8 μmol/L Ara-C (to a final concentration of 4 μmol/L) to the existing ~0.5 mL.

3.3.4 Calcium-Phosphate Transfection

1. Warm Transfection medium (~1.5 mL/coverslip), wash-HBSS medium (0.5 mL/coverslip) and neuronal growth medium supplemented with 4 μmol/L Ara-C (0.5 mL/coverslip); in a six-well plate and place in the cell-culture incubator.

2. Bring 2.5 mol/L $CaCl_2$ and 2×BBS to RT.
3. Prepare 15 μL/coverslip transfection DNA-mixtures by adding (in order): 1.5 μL of $CaCl_2$, water (subtract the volume that will contain the DNA) and DNA volumes from midi preparations to contain 0.5–1 μg LiGluN and (optional) 0.3 μg eGFP DNA. Set premixes aside.
4. Bring the 24-well plate containing the neurons as well as the six-well plate containing the heated media.
5. Aspirate and save the old neuronal growth medium and concomitantly wash the coverslips by adding 0.5 mL/coverslip of transfection medium (do not let coverslips dry).
6. Aspirate and add 0.5 mL/coverslip of transfection medium.
7. Add to the DNA mixtures 15 μL/coverslip 2×BBS. Mix thoroughly.
8. Add 30 μL/coverslip of the DNA-2xBBS mixtures onto the cells immediately.
9. Incubate for 45–60 min. Use the shortest amount of time required to generate a fine layer of noticeable precipitates (discernable by brightfield).
10. Aspirate transfection medium containing the DNA mixtures and replace with wash-HBSS.
11. Incubate for 10–15 min, during which all precipitates will have been dissolved.
12. Aspirate and wash with 0.5 mL/coverslip of transfection medium.
13. Aspirate and replace with 0.5 mL/coverslip of supplemented neuronal growth medium.
14. Add 0.5 mL/coverslip from the neuronal growth medium that was initially collected and saved.

3.4 Preparing Stock Solution of L-MAG and Labeling Neurons

1. Dissolve MAG powder aliquot in anhydrous DMSO (~15 μL/mg). The solution will quickly appear as dark orange. If precipitations remain, vortex or add DMSO. Do not heat the solution.
2. (Optional) To determine stock concentration take an aliquot (step 1) and dilute in extracellular recording solution by 1:2000. Use a spectrophotometer to determine concentration (using equation of $A = \varepsilon cl$, where A—absorbance at 360 nm, ε—molar extinction coefficient; $\varepsilon_{360\text{-MAG}} = 0.024$ L/μmol cm and L—light path).
3. Prepare aliquots of 0.3 μL of 50–100 mmol/LM. Store at −20 or −80 °C in the dark, preferably over desiccant. To note, these small aliquots will suffice for labeling a single coverslip. Following labeling, discard the solution as it should not be reused to label additional coverslips.

4. For labeling, take a single aliquot and dilute in 300 μL of NMDG-labeling solution (to a final 50–100 μmol/L final concentration).
5. 3–5 days after transfection (i.e. 13–15 d.i.v.) neurons are ready for recording. Remove media from neurons and immediately replace with the NMDG-labeling solution.
6. Replace 24-well plate in the incubator for 45 min incubation in the dark at 37 °C. For the first couple of trials, consistently check for neuronal health by examining cellular morphology (by brightfield or fluorescence). If massive cell death appears, try less concentrated MAG solution, incubation with receptor blockers or lower extracellular Ca^{2+} concentrations.
7. Wash coverslip thoroughly with extracellular recording solution (3×).

3.5 Patch Clamping Neurons

1. Prepare patch pipettes with resistance of 8–12 MΩ when filled with intracellular solution (see **Note 18**).
2. Thaw intracellular solution on ice.
3. Warm extracellular solution to RT.
4. Place the labeled coverslip in a glass-bottom recording chamber onto the electrophysiology setup. The reference electrode should be placed in a NaCl reservoir and connected to the recording chamber via a salt agar bridge.
5. Set the gravity-driven perfusion system by filling two of its reservoirs with- extracellular solution and extracellular solution with 1–5 mmol/L Glutamate (NMDA).
6. Adjust the perfusion speed so as to have a slow and constant level of medium above the coverslip to avoid vibrations.
7. Select a transfected cell while using the microscope GFP settings (illumination wavelengths, excitation, and emission filters).
8. Fill the patch pipette with the syringe tips for filling micropipettes with thawed intracellular solution from the back. To remove bubbles from the tip (front end) of the pipette, tapping or flick against the side of the pipette. Another method of getting rid of air bubbles trapped at the finest tip at the front of the patch pipette is by backfilling. This requires placing the tip of the patch pipette into intracellular solution. This will draw solution into the very finest tip of the pipette.
9. Place the patch pipette in the electrode holder so that the silver wire is in contact with the solution in the pipette.
10. Before moving the headstage, apply negative pressure (gently suck) to remove any remaining bubbles from the tip of the pipette.

11. Move the headstage so that the patch pipette is situated above the recording chamber filled with extracellular solution.
12. Before inserting the pipette in the bath, apply positive pressure to keep the tip of the patch pipette clear of any particles once in the bath.
13. Insert the pipette into the bath and observe its resistance (~8–12 MΩ).
14. Using the micromanipulator, place the pipette tip directly above the transfected cell.
15. Adjust the offset while using a test pulse of 5–10 mV, to monitor for pipette resistance and capacitance.
16. Approach the cell until a seal is formed. In the interim, change to the optical settings (e.g., rotate turret to appropriate position) to enable photoswitching (~380 and 500 nm) and turn off the brightfield light. During the gigaseal formation (gradual increases in resistance to >1 Ω), gradually adjust to negative holding potentials to promote seal formation.

4 Notes

1. For a more comprehensive list of expression vectors, visit plasmid repository databases, such as Addgene: https://www.addgene.org/vector-database/.
2. CAG: CMV immediate early enhancer/chicken β-actin promoter/rabbit β-globin intron composite promoter.
3. Peptide databases, including detailed description of topology: http://www.uniprot.org/ and http://signalpeptide.de/index.php.
4. For a large list of self-cleaving peptides sequences visit: http://www.st-andrews.ac.uk/ryanlab/2A_2Alike.pdf.
5. Commercially available photoswitches are sold by Aspira Scientific http://www.aspirasci.com/neuroscience-probes.
6. The use of ultrapure DMSO is recommended for stock solution and since DMSO is hygroscopic, it is preferable to store it in small aliquots in desiccants, to prevent moisture.
7. Contamination of MAG (in powder form), DMSO or stock solutions with water should be avoided, as the maleimide is easily hydrolysable to products that will no longer allow cysteine conjugation. Once the MAG solution has been diluted in the labeling buffer, it should be used as quickly as possible.
8. Mg^{2+} is removed from the extracellular milieu as GluNR are easily blocked by magnesium [133].

9. D-AP5 (competitive antagonist, 100 μmol/L, Tocris, Cat. No. 0106) can also be used during MAG incubation, however it has been suggested that it may interfere with the insertion of recombinant GluN2A-containing receptors into the synapses [112].
10. It is preferable to use green or red fluorescent proteins (rather than blue shifted variants, e.g. CFP), as expression markers, to minimize activation of the receptors (by isomerization f MAG into *–cis*), during the search for positively transfected neurons on the patching rig.
11. The light intensity (in mW/mm^2) will determine how fast the population of MAG labeled receptors can be isomerized at a given wavelength [31, 34, 104, 128]. The higher the intensity the faster the photoswitching. For example, moderate light illumination (~mW/mm_2) yield activation kinetics on the order of tens of milliseconds ($\tau_{\text{photo on}}$ ~ 40 ms), whereas high power illumination (e.g., Diode laser, W/mm^2) allows sub-millisecond photoswitching ($t_{10\text{–}90\%} < 50\text{–}200$ μs) [104].
12. Laser scanning systems (e.g., Laser Scanning confocal Microscopes) can provide very high intensities, but one has to take into consideration that the entire region of interest does not get illuminated concurrently, rather each point gets illuminated sequentially (though quickly, typically in μs/pixel). For large regions, this may require longer illumination times.
13. Unused Ag electrodes are not plated (i.e., coated) and appear as bright silver. Chloriding the wire is achieved by making it positive relative to a NaCl (0.9%) solution it is immersed in and passing a current through the wire. The color of a plated wire should appear as light- to dark-gray. A simpler method of chloriding is to place the wire for prolonged durations (>30 min) in liquid bleach (NaClO).
14. Filling-syringes, with fine long tips, can be fabricated manually by carefully holding a plastic 5 mL syringe with forceps from both ends and placing the central part of the syringe above a flame (e.g., Bunsen burner). As the plastic begins to heat, pulling the ends of the syringe apart stretches the plastic to create a very thin and long tip.
15. Autoclaving does not inactivate endotoxins.
16. Eluting and storing of DNA in elution buffer (Tris EDTA or TE buffer) is possible, though TE buffers usually contain EDTA that can affect downstream applications, such as enzymatic linearization of DNA required for in vitro RNA synthesis.
17. The meninges are sticky and can be difficult to remove. However, it is important to remove as much as possible,

so that they do not contribute cells to the culture (such as microglia).

18. Sutter Instruments- Pipette Cookbook (http://www.sutter.com/PDFs/pipette_cookbook.pdf).

5 Conclusion

In this chapter, we have described the general approach of synthetic optogenetics and how to apply it to photoswitch NMDARs found at the PM of mammalian neurons. We show that a single cysteine mutation is both necessary and sufficient to tether a unique, light-sensitive photoswitch to the subunit LBD. In turn, this renders the subunit, and thereby the receptor that incorporates the subunit, light-ready (i.e., amenable by light). Application of the cysteine-reactive photoswitch onto the preparation does not seem to affect cell health or affect other membrane proteins since, fortuitously, solvent accessible and unconjugated cysteines are rare in proteins (see above). Thus, only modified receptors bioconjugated to the photoswitch become light-ready. Therefore, even though all cells and membrane proteins in the preparation are exposed to the photoswitch during the labeling period, only the genetically modified NMDAR subunit(s), expressed at select cells, can be light-gated.

LiGluNs, as synthetic optogenetic tools, enable to explore NMDARs with unprecedented spatiotemporal control: intense light enables to open or close the receptors with kinetics resembling those observed during native Glu transmission or intrinsic channel closure. Localized illumination can activate receptors at unique cellular regions. Importantly, this approach is completely reversible, enabling to scrutinize the contribution of specific NMDAR- subunits to processes pertaining to synaptic plasticity.

References

1. Forne I, Ludwigsen J, Imhof A, Becker PB, Mueller-Planitz F (2012) Probing the conformation of the ISWI ATPase domain with genetically encoded photoreactive cross-linkers and mass spectrometry. Mol Cell Proteomics 11(4):M111.012088
2. Schmid F-X (2001) Biological macromolecules: UV-visible spectrophotometry—eLS. John Wiley & Sons, Ltd, Hoboken, NJ
3. McBee JK, Kuksa V, Alvarez R, de Lera AR, Prezhdo O, Haeseleer F, Sokal I, Palczewski K (2000) Isomerization of all-trans-retinol to cis-retinols in bovine retinal pigment epithelial cells: dependence on the specificity of retinoid-binding proteins. Biochemistry 39(37):11370–11380
4. Kennis JT, Mathes T (2013) Molecular eyes: proteins that transform light into biological information. Interface Focus 3(5):20130005
5. Nagel G, Ollig D, Fuhrmann M, Kateriya S, Musti AM, Bamberg E, Hegemann P (2002) Channelrhodopsin-1: a light-gated proton channel in green algae. Science 296(5577): 2395–2398
6. Nagel G, Szellas T, Huhn W, Kateriya S, Adeishvili N, Berthold P, Ollig D, Hegemann P, Bamberg E (2003) Channelrhodopsin-2, a directly light-gated cation-selective membrane

channel. Proc Natl Acad Sci U S A 100(24): 13940–13945
7. Deisseroth K (2015) Optogenetics: 10 years of microbial opsins in neuroscience. Nat Neurosci 18(9):1213–1225
8. Boyden ES, Zhang F, Bamberg E, Nagel G, Deisseroth K (2005) Millisecond-timescale, genetically targeted optical control of neural activity. Nat Neurosci 8(9):1263–1268
9. Hegemann P, Nagel G (2013) From channelrhodopsins to optogenetics. EMBO Mol Med 5(2):173–176
10. Li X, Gutierrez DV, Hanson MG, Han J, Mark MD, Chiel H, Hegemann P, Landmesser LT, Herlitze S (2005) Fast noninvasive activation and inhibition of neural and network activity by vertebrate rhodopsin and green algae channelrhodopsin. Proc Natl Acad Sci U S A 102(49):17816–17821
11. Airan RD, Thompson KR, Fenno LE, Bernstein H, Deisseroth K (2009) Temporally precise in vivo control of intracellular signalling. Nature 458(7241):1025–1029
12. Scheib U, Stehfest K, Gee CE, Korschen HG, Fudim R, Oertner TG, Hegemann P (2015) The rhodopsin-guanylyl cyclase of the aquatic fungus Blastocladiella emersonii enables fast optical control of cGMP signaling. Sci Signal 8(389):rs8
13. Schroder-Lang S, Schwarzel M, Seifert R, Strunker T, Kateriya S, Looser J, Watanabe M, Kaupp UB, Hegemann P, Nagel G (2007) Fast manipulation of cellular cAMP level by light in vivo. Nat Methods 4(1):39–42
14. Levskaya A, Weiner OD, Lim WA, Voigt CA (2009) Spatiotemporal control of cell signalling using a light-switchable protein interaction. Nature 461(7266):997–1001
15. Polstein LR, Gersbach CA (2015) A light-inducible CRISPR-Cas9 system for control of endogenous gene activation. Nat Chem Biol 11(3):198–200
16. Nihongaki Y, Kawano F, Nakajima T, Sato M (2015) Photoactivatable CRISPR-Cas9 for optogenetic genome editing. Nat Biotechnol 33(7):755–760
17. Konermann S, Brigham MD, Trevino AE, Hsu PD, Heidenreich M, Cong L, Platt RJ, Scott DA, Church GM, Zhang F (2013) Optical control of mammalian endogenous transcription and epigenetic states. Nature 500(7463):472–476
18. Luo L, Callaway EM, Svoboda K (2008) Genetic dissection of neural circuits. Neuron 57(5):634–660
19. Fenno LE, Mattis J, Ramakrishnan C, Hyun M, Lee SY, He M, Tucciarone J, Selimbeyoglu A, Berndt A, Grosenick L, Zalocusky KA, Bernstein H, Swanson H, Perry C, Diester I, Boyce FM, Bass CE, Neve R, Huang ZJ, Deisseroth K (2014) Targeting cells with single vectors using multiple-feature Boolean logic. Nat Methods 11(7):763–772
20. Papagiakoumou E, Anselmi F, Begue A, de Sars V, Gluckstad J, Isacoff EY, Emiliani V (2010) Scanless two-photon excitation of channelrhodopsin-2. Nat Methods 7(10): 848–854
21. Kim CK, Yang SJ, Pichamoorthy N, Young NP, Kauvar I, Jennings JH, Lerner TN, Berndt A, Lee SY, Ramakrishnan C, Davidson TJ, Inoue M, Bito H, Deisseroth K (2016) Simultaneous fast measurement of circuit dynamics at multiple sites across the mammalian brain. Nat Methods 13(4):325–328
22. Adamantidis A, Arber S, Bains JS, Bamberg E, Bonci A, Buzsaki G, Cardin JA, Costa RM, Dan Y, Goda Y, Graybiel AM, Hausser M, Hegemann P, Huguenard JR, Insel TR, Janak PH, Johnston D, Josselyn SA, Koch C, Kreitzer AC, Luscher C, Malenka RC, Miesenbock G, Nagel G, Roska B, Schnitzer MJ, Shenoy KV, Soltesz I, Sternson SM, Tsien RW, Tsien RY, Turrigiano GG, Tye KM, Wilson RI (2015) Optogenetics: 10 years after ChR2 in neurons—views from the community. Nat Neurosci 18(9): 1202–1212
23. Editorial (2011) Method of the year 2010. Nat Methods 8(1):1
24. Boyden ES (2015) Optogenetics and the future of neuroscience. Nat Neurosci 18(9): 1200–1201
25. Sheridan RE, Lester HA (1982) Functional stoichiometry at the nicotinic receptor. The photon cross section for phase 1 corresponds to two bis-Q molecules per channel. J Gen Physiol 80(4):499–515
26. Lester HA, Krouse ME, Nass MM, Wassermann NH, Erlanger BF (1980) A covalently bound photoisomerizable agonist: comparison with reversibly bound agonists at Electrophorus electroplaques. J Gen Physiol 75(2):207–232
27. Bartels E, Wassermann NH, Erlanger BF (1971) Photochromic activators of the acetylcholine receptor. Proc Natl Acad Sci U S A 68(8):1820–1823
28. Biswas M, Burghardt I (2014) Azobenzene photoisomerization-induced destabilization of B-DNA. Biophys J 107(4):932–940
29. Berlin S, Isacoff EY (2017) Synapses in the spotlight with synthetic optogenetics. EMBO Rep 18(5):677–692. doi:10.15252/embr.201744010

30. Szymanski W, Beierle JM, Kistemaker HA, Velema WA, Feringa BL (2013) Reversible photocontrol of biological systems by the incorporation of molecular photoswitches. Chem Rev 113(8):6114–6178
31. Gorostiza P, Volgraf M, Numano R, Szobota S, Trauner D, Isacoff EY (2007) Mechanisms of photoswitch conjugation and light activation of an ionotropic glutamate receptor. Proc Natl Acad Sci U S A 104(26): 10865–10870
32. Levitz J, Pantoja C, Gaub B, Janovjak H, Reiner A, Hoagland A, Schoppik D, Kane B, Stawski P, Schier AF, Trauner D, Isacoff EY (2013) Optical control of metabotropic glutamate receptors. Nat Neurosci 16(4):507–516
33. Berlin S, Szobota S, Reiner A, Carroll EC, Kienzler MA, Guyon A, Xiao T, Tauner D, Isacoff EY (2016) A family of photoswitchable NMDA receptors. Elife 5
34. Carroll EC, Berlin S, Levitz J, Kienzler MA, Yuan Z, Madsen D, Larsen DS, Isacoff EY (2015) Two-photon brightness of azobenzene photoswitches designed for glutamate receptor optogenetics. Proc Natl Acad Sci U S A 112(7):E776–E785
35. Beharry AA, Woolley GA (2011) Azobenzene photoswitches for biomolecules. Chem Soc Rev 40(8):4422–4437
36. De Poli M, Zawodny W, Quinonero O, Lorch M, Webb SJ, Clayden J (2016) Conformational photoswitching of a synthetic peptide foldamer bound within a phospholipid bilayer. Science 352(6285):575–580
37. Renner C, Moroder L (2006) Azobenzene as conformational switch in model peptides. Chembiochem 7(6):868–878
38. Karakas E, Furukawa H (2014) Crystal structure of a heterotetrameric NMDA receptor ion channel. Science 344(6187):992–997
39. Furukawa H, Singh SK, Mancusso R, Gouaux E (2005) Subunit arrangement and function in NMDA receptors. Nature 438(7065): 185–192
40. Kaufman H, Vratsanos SM, Erlanger BF (1968) Photoregulation of an enzymic process by means of a light-sensitive ligand. Science 162(3861):1487–1489
41. Nargeot J, Lester HA, Birdsall NJ, Stockton J, Wassermann NH, Erlanger BF (1982) A photoisomerizable muscarinic antagonist. Studies of binding and of conductance relaxations in frog heart. J Gen Physiol 79(4): 657–678
42. Deal WJ, Erlanger BF, Nachmansohn D (1969) Photoregulation of biological activity by photochromic reagents. 3. Photoregulation of bioelectricity by acetylcholine receptor inhibitors. Proc Natl Acad Sci U S A 64(4): 1230–1234
43. Volgraf M, Gorostiza P, Szobota S, Helix MR, Isacoff EY, Trauner D (2007) Reversibly caged glutamate: a photochromic agonist of ionotropic glutamate receptors. J Am Chem Soc 129(2):260–261
44. Laprell L, Repak E, Franckevicius V, Hartrampf F, Terhag J, Hollmann M, Sumser M, Rebola N, DiGregorio DA, Trauner D (2015) Optical control of NMDA receptors with a diffusible photoswitch. Nat Commun 6:8076
45. Banghart MR, Mourot A, Fortin DL, Yao JZ, Kramer RH, Trauner D (2009) Photochromic blockers of voltage-gated potassium channels. Angew Chem Int Ed Engl 48(48): 9097–9101
46. Ellis-Davies GC (2007) Caged compounds: photorelease technology for control of cellular chemistry and physiology. Nat Methods 4(8):619–628
47. Reiner A, Levitz J, Isacoff EY (2015) Controlling ionotropic and metabotropic glutamate receptors with light: principles and potential. Curr Opin Pharmacol 20:135–143
48. Woolley GA (2005) Photocontrolling peptide alpha helices. Acc Chem Res 38(6): 486–493
49. Browne LE, Nunes JP, Sim JA, Chudasama V, Bragg L, Caddick S, North RA (2014) Optical control of trimeric P2X receptors and acid-sensing ion channels. Proc Natl Acad Sci U S A 111(1):521–526
50. Paoletti P, Bellone C, Zhou Q (2013) NMDA receptor subunit diversity: impact on receptor properties, synaptic plasticity and disease. Nat Rev Neurosci 14(6):383–400
51. Traynelis SF, Wollmuth LP, McBain CJ, Menniti FS, Vance KM, Ogden KK, Hansen KB, Yuan H, Myers SJ, Dingledine R (2010) Glutamate receptor ion channels: structure, regulation, and function. Pharmacol Rev 62(3):405–496
52. Akazawa C, Shigemoto R, Bessho Y, Nakanishi S, Mizuno N (1994) Differential expression of five N-methyl-D-aspartate receptor subunit mRNAs in the cerebellum of developing and adult rats. J Comp Neurol 347(1):150–160
53. Sheng M, Cummings J, Roldan LA, Jan YN, Jan LY (1994) Changing subunit composition of heteromeric NMDA receptors during development of rat cortex. Nature 368(6467): 144–147
54. Hardingham GE, Bading H (2010) Synaptic versus extrasynaptic NMDA receptor signalling: implications for neurodegenerative disorders. Nat Rev Neurosci 11(10):682–696

55. Dingledine R, Borges K, Bowie D, Traynelis SF (1999) The glutamate receptor ion channels. Pharmacol Rev 51(1):7–61
56. Hunt DL, Castillo PE (2012) Synaptic plasticity of NMDA receptors: mechanisms and functional implications. Curr Opin Neurobiol 22(3):496–508
57. Yu XM, Salter MW (1998) Gain control of NMDA-receptor currents by intracellular sodium. Nature 396(6710):469–474
58. Jahr CE, Stevens CF (1993) Calcium permeability of the N-methyl-D-aspartate receptor channel in hippocampal neurons in culture. Proc Natl Acad Sci U S A 90(24): 11573–11577
59. Burnashev N, Zhou Z, Neher E, Sakmann B (1995) Fractional calcium currents through recombinant GluR channels of the NMDA, AMPA and kainate receptor subtypes. J Physiol 485(Pt 2):403–418
60. Zucker RS (1999) Calcium- and activity-dependent synaptic plasticity. Curr Opin Neurobiol 9(3):305–313
61. Davies J, Francis AA, Jones AW, Watkins JC (1981) 2-Amino-5-phosphonovalerate (2APV), a potent and selective antagonist of amino acid-induced and synaptic excitation. Neurosci Lett 21(1):77–81
62. Wong EH, Kemp JA, Priestley T, Knight AR, Woodruff GN, Iversen LL (1986) The anticonvulsant MK-801 is a potent N-methyl-D-aspartate antagonist. Proc Natl Acad Sci U S A 83(18):7104–7108
63. Reynolds IJ, Miller RJ (1989) Ifenprodil is a novel type of N-methyl-D-aspartate receptor antagonist: interaction with polyamines. Mol Pharmacol 36(5):758–765
64. Williams K (1993) Ifenprodil discriminates subtypes of the N-methyl-D-aspartate receptor: selectivity and mechanisms at recombinant heteromeric receptors. Mol Pharmacol 44(4):851–859
65. Fischer G, Mutel V, Trube G, Malherbe P, Kew JN, Mohacsi E, Heitz MP, Kemp JA (1997) Ro 25-6981, a highly potent and selective blocker of N-methyl-D-aspartate receptors containing the NR2B subunit. Characterization in vitro. J Pharmacol Exp Ther 283(3):1285–1292
66. Auberson YP, Allgeier H, Bischoff S, Lingenhoehl K, Moretti R, Schmutz M (2002) 5-Phosphonomethylquinoxalinediones as competitive NMDA receptor antagonists with a preference for the human 1A/2A, rather than 1A/2B receptor composition. Bioorg Med Chem Lett 12(7):1099–1102
67. Bettini E, Sava A, Griffante C, Carignani C, Buson A, Capelli AM, Negri M, Andreetta F, Senar-Sancho SA, Guiral L, Cardullo F (2010) Identification and characterization of novel NMDA receptor antagonists selective for NR2A- over NR2B-containing receptors. J Pharmacol Exp Ther 335(3):636–644
68. Neyton J, Paoletti P (2006) Relating NMDA receptor function to receptor subunit composition: limitations of the pharmacological approach. J Neurosci 26(5):1331–1333
69. Edman S, McKay S, Macdonald LJ, Samadi M, Livesey MR, Hardingham GE, Wyllie DJ (2012) TCN 201 selectively blocks GluN2A-containing NMDARs in a GluN1 co-agonist dependent but non-competitive manner. Neuropharmacology 63(3):441–449
70. McKay S, Griffiths NH, Butters PA, Thubron EB, Hardingham GE, Wyllie DJ (2012) Direct pharmacological monitoring of the developmental switch in NMDA receptor subunit composition using TCN 213, a GluN2A-selective, glycine-dependent antagonist. Br J Pharmacol 166(3):924–937
71. Ferraro TN, Hare TA (1985) Free and conjugated amino acids in human CSF: influence of age and sex. Brain Res 338(1):53–60
72. Johnson JW, Ascher P (1987) Glycine potentiates the NMDA response in cultured mouse brain neurons. Nature 325(6104):529–531
73. Pinard E, Alanine A, Bourson A, Buttelmann B, Gill R, Heitz M, Jaeschke G, Mutel V, Trube G, Wyler R (2001) Discovery of (R)-1-[2-hydroxy-3-(4-hydroxy-phenyl)-propyl]-4-(4-methyl-benzyl)-piperidin-4-ol: a novel NR1/2B subtype selective NMDA receptor antagonist. Bioorg Med Chem Lett 11(16): 2173–2176
74. Mony L, Kew JN, Gunthorpe MJ, Paoletti P (2009) Allosteric modulators of NR2B-containing NMDA receptors: molecular mechanisms and therapeutic potential. Br J Pharmacol 157(8):1301–1317
75. Hansen KB, Traynelis SF (2011) Structural and mechanistic determinants of a novel site for noncompetitive inhibition of GluN2D-containing NMDA receptors. J Neurosci 31(10):3650–3661
76. Acker TM, Yuan H, Hansen KB, Vance KM, Ogden KK, Jensen HS, Burger PB, Mullasseril P, Snyder JP, Liotta DC, Traynelis SF (2011) Mechanism for noncompetitive inhibition by novel GluN2C/D N-methyl-D-aspartate receptor subunit-selective modulators. Mol Pharmacol 80(5):782–795
77. Mullasseril P, Hansen KB, Vance KM, Ogden KK, Yuan H, Kurtkaya NL, Santangelo R, Orr AG, Le P, Vellano KM, Liotta DC, Traynelis SF (2010) A subunit-selective potentiator of NR2C- and NR2D-containing NMDA receptors. Nat Commun 1:90

78. Sakimura K, Kutsuwada T, Ito I, Manabe T, Takayama C, Kushiya E, Yagi T, Aizawa S, Inoue Y, Sugiyama H et al (1995) Reduced hippocampal LTP and spatial learning in mice lacking NMDA receptor epsilon 1 subunit. Nature 373(6510):151–155
79. Bannerman DM, Bus T, Taylor A, Sanderson DJ, Schwarz I, Jensen V, Hvalby O, Rawlins JN, Seeburg PH, Sprengel R (2012) Dissecting spatial knowledge from spatial choice by hippocampal NMDA receptor deletion. Nat Neurosci 15(8):1153–1159
80. Fire A, Xu S, Montgomery MK, Kostas SA, Driver SE, Mello CC (1998) Potent and specific genetic interference by double-stranded RNA in Caenorhabditis elegans. Nature 391(6669):806–811
81. Tsien JZ, Chen DF, Gerber D, Tom C, Mercer EH, Anderson DJ, Mayford M, Kandel ER, Tonegawa S (1996) Subregion- and cell type-restricted gene knockout in mouse brain. Cell 87(7):1317–1326
82. Straub C, Granger AJ, Saulnier JL, Sabatini BL (2014) CRISPR/Cas9-mediated gene knock-down in post-mitotic neurons. PLoS One 9(8):e105584
83. Incontro S, Asensio CS, Edwards RH, Nicoll RA (2014) Efficient, complete deletion of synaptic proteins using CRISPR. Neuron 83(5):1051–1057
84. Forrest D, Yuzaki M, Soares HD, Ng L, Luk DC, Sheng M, Stewart CL, Morgan JI, Connor JA, Curran T (1994) Targeted disruption of NMDA receptor 1 gene abolishes NMDA response and results in neonatal death. Neuron 13(2):325–338
85. Li Y, Erzurumlu RS, Chen C, Jhaveri S, Tonegawa S (1994) Whisker-related neuronal patterns fail to develop in the trigeminal brainstem nuclei of NMDAR1 knockout mice. Cell 76(3):427–437
86. Kutsuwada T, Sakimura K, Manabe T, Takayama C, Katakura N, Kushiya E, Natsume R, Watanabe M, Inoue Y, Yagi T, Aizawa S, Arakawa M, Takahashi T, Nakamura Y, Mori H, Mishina M (1996) Impairment of suckling response, trigeminal neuronal pattern formation, and hippocampal LTD in NMDA receptor epsilon 2 subunit mutant mice. Neuron 16(2):333–344
87. Nakazawa K, McHugh TJ, Wilson MA, Tonegawa S (2004) NMDA receptors, place cells and hippocampal spatial memory. Nat Rev Neurosci 5(5):361–372
88. Rossi A, Kontarakis Z, Gerri C, Nolte H, Holper S, Kruger M, Stainier DY (2015) Genetic compensation induced by deleterious mutations but not gene knockdowns. Nature 524(7564):230–233
89. Aronoff R, Petersen CC (2006) Controlled and localized genetic manipulation in the brain. J Cell Mol Med 10(2):333–352
90. Persengiev SP, Zhu X, Green MR (2004) Nonspecific, concentration-dependent stimulation and repression of mammalian gene expression by small interfering RNAs (siRNAs). RNA 10(1):12–18
91. Alvarez VA, Ridenour DA, Sabatini BL (2006) Retraction of synapses and dendritic spines induced by off-target effects of RNA interference. J Neurosci 26(30):7820–7825
92. Cho SW, Kim S, Kim Y, Kweon J, Kim HS, Bae S, Kim JS (2014) Analysis of off-target effects of CRISPR/Cas-derived RNA-guided endonucleases and nickases. Genome Res 24(1):132–141
93. Kramer RH, Mourot A, Adesnik H (2013) Optogenetic pharmacology for control of native neuronal signaling proteins. Nat Neurosci 16(7):816–823
94. Kang JY, Kawaguchi D, Coin I, Xiang Z, O'Leary DD, Slesinger PA, Wang L (2013) In vivo expression of a light-activatable potassium channel using unnatural amino acids. Neuron 80(2):358–370
95. Klippenstein V, Ghisi V, Wietstruk M, Plested AJ (2014) Photoinactivation of glutamate receptors by genetically encoded unnatural amino acids. J Neurosci 34(3):980–991
96. Zhu S, Riou M, Yao CA, Carvalho S, Rodriguez PC, Bensaude O, Paoletti P, Ye S (2014) Genetically encoding a light switch in an ionotropic glutamate receptor reveals subunit-specific interfaces. Proc Natl Acad Sci U S A 111(16):6081–6086
97. Rau H (1973) Spectroscopic properties of organic azo compounds. Angew Chem Int Ed Engl 12(3):224–235
98. Lee CH, Lu W, Michel JC, Goehring A, Du J, Song X, Gouaux E (2014) NMDA receptor structures reveal subunit arrangement and pore architecture. Nature 511(7508):191–197
99. Sobolevsky AI, Rosconi MP, Gouaux E (2009) X-ray structure, symmetry and mechanism of an AMPA-subtype glutamate receptor. Nature 462(7274):745–756
100. Reiter A, Skerra A, Trauner D, Schiefner A (2013) A photoswitchable neurotransmitter analogue bound to its receptor. Biochemistry 52(50):8972–8974
101. Scott DB, Blanpied TA, Swanson GT, Zhang C, Ehlers MD (2001) An NMDA receptor ER retention signal regulated by phosphorylation

and alternative splicing. J Neurosci 21(9): 3063–3072

102. McIlhinney RA, Le Bourdelles B, Molnar E, Tricaud N, Streit P, Whiting PJ (1998) Assembly intracellular targeting and cell surface expression of the human N-methyl-D-aspartate receptor subunits NR1a and NR2A in transfected cells. Neuropharmacology 37(10-11):1355–1367

103. Numano R, Szobota S, Lau AY, Gorostiza P, Volgraf M, Roux B, Trauner D, Isacoff EY (2009) Nanosculpting reversed wavelength sensitivity into a photoswitchable iGluR. Proc Natl Acad Sci U S A 106(16): 6814–6819

104. Reiner A, Isacoff EY (2014) Tethered ligands reveal glutamate receptor desensitization depends on subunit occupancy. Nat Chem Biol 10(4):273–280

105. Zhao Y, Araki S, Wu J, Teramoto T, Chang YF, Nakano M, Abdelfattah AS, Fujiwara M, Ishihara T, Nagai T, Campbell RE (2011) An expanded palette of genetically encoded Ca(2)(+) indicators. Science 333(6051): 1888–1891

106. Akerboom J, Carreras Calderon N, Tian L, Wabnig S, Prigge M, Tolo J, Gordus A, Orger MB, Severi KE, Macklin JJ, Patel R, Pulver SR, Wardill TJ, Fischer E, Schuler C, Chen TW, Sarkisyan KS, Marvin JS, Bargmann CI, Kim DS, Kugler S, Lagnado L, Hegemann P, Gottschalk A, Schreiter ER, Looger LL (2013) Genetically encoded calcium indicators for multi-color neural activity imaging and combination with optogenetics. Front Mol Neurosci 6:2

107. Wu J, Liu L, Matsuda T, Zhao Y, Rebane A, Drobizhev M, Chang YF, Araki S, Arai Y, March K, Hughes TE, Sagou K, Miyata T, Nagai T, Li WH, Campbell RE (2013) Improved orange and red Ca(2)+/- indicators and photophysical considerations for optogenetic applications. ACS Chem Neurosci 4(6):963–972

108. Dana H, Mohar B, Sun Y, Narayan S, Gordus A, Hasseman JP, Tsegaye G, Holt GT, Hu A, Walpita D, Patel R, Macklin JJ, Bargmann CI, Ahrens MB, Schreiter ER, Jayaraman V, Looger LL, Svoboda K, Kim DS (2016) Sensitive red protein calcium indicators for imaging neural activity. Elife 5

109. Li D, Herault K, Isacoff EY, Oheim M, Ropert N (2012) Optogenetic activation of LiGluR-expressing astrocytes evokes anion channel-mediated glutamate release. J Physiol 590(4):855–873

110. Neher E, Sakmann B (1976) Single-channel currents recorded from membrane of denervated frog muscle fibres. Nature 260(5554): 799–802

111. Lemoine D, Habermacher C, Martz A, Mery PF, Bouquier N, Diverchy F, Taly A, Rassendren F, Specht A, Grutter T (2013) Optical control of an ion channel gate. Proc Natl Acad Sci U S A 110(51):20813–20818

112. Barria A, Malinow R (2002) Subunit-specific NMDA receptor trafficking to synapses. Neuron 35(2):345–353

113. Prybylowski K, Fu Z, Losi G, Hawkins LM, Luo J, Chang K, Wenthold RJ, Vicini S (2002) Relationship between availability of NMDA receptor subunits and their expression at the synapse. J Neurosci 22(20): 8902–8910

114. Sheng M, Kim E (2011, 2011) The postsynaptic organization of synapses. Cold Spring Harb Perspect Biol 3(12). doi:10.1101/cshperspect.a005678

115. Kim MJ, Futai K, Jo J, Hayashi Y, Cho K, Sheng M (2007) Synaptic accumulation of PSD-95 and synaptic function regulated by phosphorylation of serine-295 of PSD-95. Neuron 56(3):488–502

116. Grienberger C, Chen X, Konnerth A (2014) NMDA receptor-dependent multidendrite Ca(2+) spikes required for hippocampal burst firing in vivo. Neuron 81(6):1274–1281

117. Schiller J, Schiller Y (2001) NMDA receptor-mediated dendritic spikes and coincident signal amplification. Curr Opin Neurobiol 11(3):343–348

118. Niwa H, Yamamura K, Miyazaki J (1991) Efficient selection for high-expression transfectants with a novel eukaryotic vector. Gene 108(2):193–199

119. Miyazaki J, Takaki S, Araki K, Tashiro F, Tominaga A, Takatsu K, Yamamura K (1989) Expression vector system based on the chicken beta-actin promoter directs efficient production of interleukin-5. Gene 79(2): 269–277

120. Qin JY, Zhang L, Clift KL, Hulur I, Xiang AP, Ren BZ, Lahn BT (2010) Systematic comparison of constitutive promoters and the doxycycline-inducible promoter. PLoS One 5(5):e10611

121. Kugler S, Kilic E, Bahr M (2003) Human synapsin 1 gene promoter confers highly neuron-specific long-term transgene expression from an adenoviral vector in the adult rat brain depending on the transduced area. Gene Ther 10(4):337–347

122. Kozak M (1987) An analysis of 5′-noncoding sequences from 699 vertebrate messenger RNAs. Nucleic Acids Res 15(20):8125–8148

123. Hartenbach S, Fussenegger M (2006) A novel synthetic mammalian promoter derived from an internal ribosome entry site. Biotechnol Bioeng 95(4):547–559
124. Chappell SA, Edelman GM, Mauro VP (2000) A 9-nt segment of a cellular mRNA can function as an internal ribosome entry site (IRES) and when present in linked multiple copies greatly enhances IRES activity. Proc Natl Acad Sci U S A 97(4):1536–1541
125. de Felipe P, Luke GA, Hughes LE, Gani D, Halpin C, Ryan MD (2006) E unum pluribus: multiple proteins from a self-processing polyprotein. Trends Biotechnol 24(2):68–75
126. Gurdon JB, Lane CD, Woodland HR, Marbaix G (1971) Use of frog eggs and oocytes for the study of messenger RNA and its translation in living cells. Nature 233(5316):177–182
127. Liman ER, Tytgat J, Hess P (1992) Subunit stoichiometry of a mammalian K+ channel determined by construction of multimeric cDNAs. Neuron 9(5):861–871
128. Volgraf M, Gorostiza P, Numano R, Kramer RH, Isacoff EY, Trauner D (2006) Allosteric control of an ionotropic glutamate receptor with an optical switch. Nat Chem Biol 2(1): 47–52
129. Kienzler MA, Reiner A, Trautman E, Yoo S, Trauner D, Isacoff EY (2013) A red-shifted, fast-relaxing azobenzene photoswitch for visible light control of an ionotropic glutamate receptor. J Am Chem Soc 135(47):17683–17686
130. Szobota S, Gorostiza P, Del Bene F, Wyart C, Fortin DL, Kolstad KD, Tulyathan O, Volgraf M, Numano R, Aaron HL, Scott EK, Kramer RH, Flannery J, Baier H, Trauner D, Isacoff EY (2007) Remote control of neuronal activity with a light-gated glutamate receptor. Neuron 54(4):535–545
131. Molleman A (2003) Patch clamping: an introductory guide to patch clamp electrophysiology, 1st edn. Wiley, Chichester, UK
132. Beaudoin GM 3rd, Lee SH, Singh D, Yuan Y, Ng YG, Reichardt LF, Arikkath J (2012) Culturing pyramidal neurons from the early postnatal mouse hippocampus and cortex. Nat Protoc 7(9):1741–1754
133. Mayer ML, Westbrook GL (1987) Permeation and block of N-methyl-D-aspartic acid receptor channels by divalent cations in mouse cultured central neurones. J Physiol 394: 501–527
134. Nabavi S, Fox R, Proulx CD, Lin JY, Tsien RY, Malinow R (2014) Engineering a memory with LTD and LTP. Nature 511(7509): 348–352
135. Bliss TV, Lomo T (1973) Long-lasting potentiation of synaptic transmission in the dentate area of the anaesthetized rabbit following stimulation of the perforant path. J Physiol 232(2):331–356
136. Newcomer JW, Farber NB, Olney JW (2000) NMDA receptor function, memory, and brain aging. Dialogues Clin Neurosci 2(3): 219–232
137. Burnashev N, Szepetowski P (2015) NMDA receptor subunit mutations in neurodevelopmental disorders. Curr Opin Pharmacol 20: 73–82

Chapter 11

In Vivo Electrochemical Studies of Optogenetic Control of Glutamate Signaling Measured Using Enzyme-Based Ceramic Microelectrode Arrays

Jason J. Burmeister, Francois Pomerleau, Jorge E. Quintero, Peter Huettl, Yi Ai, Johan Jakobsson, Martin Lundblad, Andreas Heuer, John T. Slevin, and Greg A. Gerhardt

Abstract

Direct electrochemical measurements of glutamate release in vivo were combined with optogenetics in order to examine light-induced control of glutamate neurotransmission in the rodent brain. Self-referenced recordings of glutamate using ceramic-based microelectrode arrays (MEAs) in hippocampus and frontal cortex demonstrated precise optical control of light-induced glutamate release through channelrhodopsin (ChR2) expression in both rat hippocampus and frontal cortex. Although the virus was only injected unilaterally, bilateral and rostro-caudal expression was observed in slice imaging, indicating diffusion and active transport of the viral particles. Methodology for the optogenetic control of glutamate signaling in the rat brain is thoroughly explained with special attention paid to MEA enzyme coating and cleaning for the benefit of other investigators. These data support that optogenetic control of glutamate signaling is robust with certain advantages as compared to other methods to modulate the in vivo control of glutamate signaling.

Key words Glutamate, Optogenetics, Electrochemistry, Microelectrode, Array, Amperometry, Glutamate oxidase, Neurotransmitter, Biosensor

1 Introduction: Electrochemistry and Optogenetics

Clinicians and scientists have been recording neuronal electrical activity of the brain since 1930s. However, it was not until 1970s that researchers began direct chemical measurements of neurotransmitters, which makes up approximately 90% of brain communication [1–3]. Since that time the field of microelectrode technology has greatly advanced the direct chemical measurements of neurotransmitters in the extracellular space of the brain including electroactive and non-electroactive neurotransmitters [3–10]. Finer temporal and spatial resolution and less tissue damage compared to

Sandrine Parrot and Luc Denoroy (eds.), *Biochemical Approaches for Glutamatergic Neurotransmission*, Neuromethods, vol. 130, DOI 10.1007/978-1-4939-7228-9_11, © Springer Science+Business Media LLC 2018

microdialysis approaches make microelectrodes desirable for in vivo neurochemical measurements for drug effects and behavioral studies. Handmade carbon fiber-based microelectrodes are the most widely used and have been employed for primarily measuring rapid changes in central nervous system (CNS) levels of dopamine and some other electroactive molecules in anesthetized and awake, behaving animals [11, 12]. However, carbon fibers have shown little success in monitoring non-electroactive neurotransmitters such as glutamate (Glu) and acetylcholine due to poor temporal resolution and detection limits and lack of self-referencing capabilities. Biosensors capable of measuring these non-electroactive neurotransmitters have been developed using handmade platinum (Pt) or platinum/iridium (Pt/Ir) wire microelectrodes [5, 13, 14]. The catalytic Pt surface offers improved sensitivity and temporal resolution for direct oxidation of H_2O_2 produced from the oxidase-based enzymes. To overcome the limitations and inconsistencies of handmade microelectrodes, our group has developed a more universal microelectrode array (MEA) technology platform for second-by-second measurements of neurotransmitters, neuromodulators, and markers of brain energy/metabolism. Enzyme-based ceramic MEAs have been used for chronic neurotransmitter measurements in rodents and nonhuman primates [15–17]. Although, a variety of neurochemicals can be measured using the MEA technology in conjunction with constant voltage amperometry coupled with the Fast Analytical Sensing Technology (FAST16mkIII; Quanteon, LLC, Nicholasville, KY) recording system, we have the most extensive experience measuring the major excitatory neurotransmitter in the CNS, L-Glu [15, 16, 18].

More specific, targeted control of neural systems to establish causality between neurotransmitter release events and behavior has remained difficult to achieve, but has been progressing [16, 18, 19]. Electrical stimulation and behavioral paradigms have been useful for studying release and regulation of dopamine neuronal systems [20, 21]. Our group has also extensively employed the use of locally applied potassium to induce membrane depolarization and cause focal release of neurotransmitters around microelectrodes. This allows for reproducible and "dose-controlled" studies for mapping release and uptake of neurotransmitters [17–19]. However, more specific and dose-dependent approaches have been needed to more thoroughly understand in vivo regulation and control of neurotransmitter systems.

Recent advancements in optogenetics, which introduces light-sensitive proteins (opsins) into neurons that regulate transmembrane ion conductance shows great promise for neuronal control [22–25]. Electrophysiological studies have shown that optical excitation or inhibition of neuronal activity is correlated with behavior, but to date, very few studies have examined neurotransmitter release combined with optical stimulation [26]. In this chapter, we discuss combining our expertise of direct electrochemical measurements of in vivo neurotransmitter release, specifically Glu, in order to examine optogenetic control of Glu neurotransmission in the rodent brain. The basic methodology we have developed, which combines the technology of optogenetics in vivo with our MEA technology is explained. In addition, we report some of our recent data on optogenetic-induced release of Glu in the rat hippocampus and frontal cortex. Finally, we present detailed methodology for use of our MEA technology for self-referencing measures of Glu combined with optogenetics.

2 Materials

2.1 Ceramic-Based MEAs

Silicon was the first substrate material used to form MEAs using photolithographic techniques [27]. We later introduced the use of ceramic substrates (e.g., Al_2O_3), which has physical, chemical, and electronic properties that make it useful as a substrate for MEAs including strength, relative inertness, and its property as a good insulator [4]. Other relatively "biologically inert" materials that have been FDA approved for implantable devices including polyimide for encapsulation and Pt recording (electrophysiological or chemical) sites and Pt/Ir stimulation sites make up the MEA designs employed by our group. Mass fabrication using photolithographic techniques along with computer-controlled diamond saw cut-outs result in MEAs with reproducible surface areas and recording characteristics. Hundreds of MEAs can be simultaneously patterned on a single ceramic wafer with different or identical designs. This reduces cost and allows for "conformal" MEA designs based on the anatomy and experimental goals. In many cases, MEAs may be cleaned and reused for more than one experiment. See Sect. 4 for MEA cleaning procedure. Biocompatibility based on the ability to record single-unit activity of neurons, electrochemical signals of a test molecule such as peroxide, post-experimentation integrity, and postmortem histopathology have been demonstrated 6 months after implantation in vivo in awake rats [28].

The fabrication procedure of the MEAs has been previously described [4, 29, 30], briefly to fabricate the MEAs, non-conducting substrates composed of high purity Al_2O_3 (99.9%, Coors, CoorsTek® Golden, CO, up to 3 in. diameter wafers) are polished to the desired thickness (125 μm). A photomask is used to pattern Pt (0.25 μm) recording/stimulation sites, connecting lines and bonding pads on one or both sides of the MEAs. Polyimide (1.25 μm) is patterned using a second mask to insulate the connecting lines. Additional Pt or Pt/Ir (0.25 μm) can be sputtered onto the recording or stimulation sites to increase surface area or to alter stimulation properties. MEAs have been fabricated with 4–8 recording sites per side ranging from 100 to

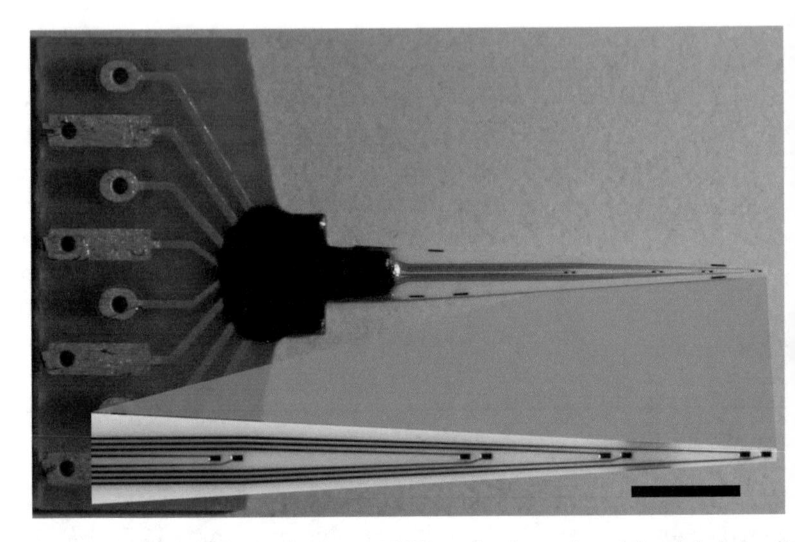

Fig. 1 8-track single-sided 8 channel MEA on freely moving stub printed circuit board (PCB). Magnified inset of ceramic tip shows each recording site is 50 μm × 100 μm with recording site pairs having 100 μm separation. Starting from the tip, spacing between first three site pairs is 1 mm. Spacing between third and fourth site pairs is 2 mm. Black scale bar is 1 mm

7500 μm^2 in area (Fig. 1). Conformal biomorphic arrays with various recording site sizes and arrangements have been designed to record from specific brain layers [28, 29, 31]. Patterning both the front and back of the ceramic MEA substrate increases the recording site density and creates isolated front and back site pairs. In addition, site pairs can be used as redundant or "fail safe" recording sites in the event of recording site failure or noise induction. MEAs can be obtained from the Center for Microelectrode Technology (Lexington, KY, www.ukycenmet.com). Ag/AgCl reference electrodes are used for all recordings as well as the FAST-16mkIII recording system and software (rev. 2, 2014; Quanteon, USA).

2.2 Animals

Male Sprague-Dawley rats (Harlan, Indianapolis, IN), aged 2–3 months, were used for all experiments ($n = 17$). Animals were individually housed and maintained on a 12 h light:12 h dark cycle (lights on at 06:00 h). Animals had access to food and water *ad libitum* and were monitored by the Department of Laboratory Animal Resources at the University of Kentucky. All protocols used in the experiments were approved by the Institutional Animal Care and Use Committee.

2.3 MEA and Optical Fiber Configuration

For our initial study, we stimulated the transfected channelrodopsins (ChR2) locally by affixing a guide cannula (Plastics One, Roanoke, VA) to our MEA through which a fiber optic cannula (Ø200 μm Core, 0.39 NA; Thorlab, Newton, NJ) was inserted with the tip of

Fig. 2 Photographs of the MEA coupled with waveguide and micropipette for in vivo electrochemical recordings of optogenetically controlled release of glutamate

the fiber positioned ~200 μm from the surface of the MEA in between all 4 Pt sites (see Fig. 2). The laser source (OBIS LS 488 nm, 100 mW, Coherent, Santa Clara, CA) was controlled through its analog port using a Master-8 pulse generator (A.M.P.I., Jerusalem, Israel). Various pulse trains of 2.5 s duration or constant stimulation of various durations were delivered locally through the fiber optic cannula (waveguide). Photovoltaic effects at higher pulse wattages are effectively removed by self-referencing subtraction. Figure 2 shows an image of a laser pulse at the tip of the MEA. We also affixed a glass micropipette for local application of drugs (e.g., Glu transporter inhibitors).

3 Methods

3.1 AAV Packaging

The viral vectors used in this study were pseudotyped $AAV_{2/5}$ vectors, where the transgene of interest is flanked by inverted terminal repeats (ITR) of the AAV2 packaged in an AAV5 capsid. The AAV vectors were produced using a double-transfection method with the plasmid pAAV-hSyn-hChR2(H134R)-Enhanced Yellow

Fluorescent Protein (EYFP) and the appropriate helper plasmid containing the essential adenoviral packaging genes, as previously described [32]. Vectors were purified by iodixanol step gradients and Sepharose Q column chromatography. The purified viral vector suspension was titrated with TaqMan quantitative PCR and primers targeting the WPRE sequence. The final titers of the injected AAV vector suspensions were between $2.4e^{+14}$ and $2.8e^{+14}$ genome copies/mL.

3.2 AAV Intracranial Injections

Animals were anesthetized with isoflurane (1–3%) and placed in a stereotaxic frame. The head was shaved and cleaned with alternating proviodine and alcohol solution. A midline incision was made on the scalp to expose the skull. We targeted two brain areas, a burr hole was made above the left infralimbic cortex (from bregma: AP: +3.2 mm; ML: +0.8 mm) and another was made above the dentate gyrus (DG) of the hippocampus (from bregma: AP: −4.4 mm; ML: −2.2 mm). A Hamilton syringe was loaded with 2 μL of the AAV2/5-hSyn-hChR2 (H134R)-EYFP suspension in phosphate buffer saline (PBS) and was lowered to the infralimbic cortex (DV: −5.5) where 1 μL of the suspension was slowly injected over 4 min. The syringe was then moved to the DG (DV: −4.0) where the remaining 1 μL was injected. The tissue was allowed to rest for 5 min before the needle was slowly retracted. A minimum of 5 weeks was allowed before experimentation was started.

3.3 In Vivo Amperometric Measures of Glutamate Using MEAs

Before in vivo use, an in vitro calibration, as described previously [30], was used to calculate selectivity (Glu vs. ascorbic acid), sensitivity (slope, in nAmps/μmol/L), and limit of detection (in μmol/L based on a signal to noise ratio of 3). Briefly, the enzyme-coated MEA (see notes for coating procedure) was calibrated by adding aliquots of stock solutions to a stirred, 37 °C PBS solution to achieve the following concentrations: 250 μmol/L ascorbate, 20 μmol/L Glu, 40 μmol/L Glu, 60 μmol/L Glu, and 8.8 μmol/L H_2O_2.

Animals were anesthetized with isoflurane (1–3%) and placed in a stereotaxic frame. The head was shaved and a midline incision was made on the scalp to expose the skull. We targeted the same brain areas where the vector was injected. Burr holes were made above left infralimbic cortex (from bregma: AP: +3.2 mm; ML: +0.8 mm) and above hippocampus dentate gyrus (DG) (from bregma: AP: −4.4 mm; ML: −2.2 mm). A small burr hole was also made away from the recording areas for placement of the Ag/AgCl reference electrode.

The MEA/waveguide/pipette configured for self-referencing measure of Glu as described above (Fig. 2) was lowered in the targeted areas and recording started. As explained in details below (Sect. 3.2), this coating configuration allows for background measures and subtraction to obtain a self-referenced Glu signal furthermore any photovoltaic effects are effectively removed by self-referencing subtraction.

Amperometric recordings, (100 Hz displayed at a frequency of 10 Hz), were performed using the FAST16mkIII recording system.

3.4 Self-Referencing Recordings

Photolithography combined with mass fabrication processes allow the MEAs to be precisely fabricated with control of the geometric layout of multiple recording sites. Matched pairs of recording sites within close proximity allow for background subtraction with a technique we call self-referencing [33]. The complexity of biological systems complicates in vivo electrochemical measurements. Electrochemical recordings have historically focused on ways to minimize the "non-Faradaic" or often non-neurochemical/neurotransmitter specific background signals with the goal of maximizing the signal of the analyte of interest. Non-Faradaic signals include solvent dipole reorientation, adsorption, desorption, and movement of electrolyte ions at the recording site surfaces. In biological systems, the background, or non-Faradaic response, can also include pH shifts, changes in Ca^{2+} and other ions. Many effects of chemical interference can be removed by self-referencing including subtraction of noise from the analyte signal, which is present on both the analyte and control sites. In addition, self-referencing with multisite MEAs makes it possible to measure both tonic levels and phasic changes of the neurochemical, rather than only the changes from a baseline or transient neurotransmitter changes [33]. This powerful feature is not possible using single microelectrodes such as the classic carbon fiber microelectrode as the non-Faradaic signals cannot be easily removed from the recorded signal. Differentially coating recording sites on a single MEA results in discrete signals for control and analyte sites that can be subtracted to yield more specifically the analyte signal. Achieving this coating on control and analyte-sensing sites in close, defined proximity is an experimental challenge. A micro-coating apparatus capable of differentially coating side-by-side pairs has been under development and shows great promise for future generations of MEAs that will have faster recording properties, improved sensitivity, and more optimized self-referencing recording capabilities. Two-sided MEAs with identical recording sites on both the front and back of the microelectrode may also simplify site coating with enzymes and other layers. One side can be coated for Glu detection while the reverse side is coated as a control for self-referencing. MEAs with sites on both sides can be coated by the micro-coating apparatus or by hand methods. In addition, for optogenetic studies, photovoltaic effects produced by photon/metal interactions from the laser excitation can also be removed by subtraction. Glutamate oxidase (GluOx) immobilized onto the MEA surface by entrapment in a protein-matrix of bovine serum albumin (BSA) cross-linked with glutaraldehyde retains high enzyme-selectivity and activity [30]. (See Sect. 4 for details on enzyme coating.)

The immobilized GluOx converts local Glu molecules that physically touch the recoding site surface into the reporter molecule, H_2O_2, plus α-ketoglutarate. Glutamate oxidase is extremely selective for L-Glu over other amino acids [34]. This H_2O_2 is directly proportional to local Glu in a given brain region. It is quantified at the Pt recording sites by amperometric oxidation to yield a current that is proportional to local Glu concentration. The addition of a selective layer improves MEA performance for Glu detection even though interfering species can be removed by self-referencing subtraction. Nafion® is often used to repel the anions such as uric acid, ascorbic acid, and 3,4-dihydroxyphenylacetic acid (DOPAC) by charge repulsion from the sulfonated ionomer. However, interfering chemicals that are positively charged including dopamine, serotonin, and norepinephrine remain detectable when using Nafion®. Alternatively, size exclusion may be used to block interfering chemicals. Poly-(meta-phenylenediamine) (mPD) has been an effective size exclusion layer on the MEAs [35, 36]. Once electro-polymerized onto the recording sites, mPD limits molecules larger than H_2O_2 regardless of charge, which virtually eliminates dopamine, serotonin, and norepinephrine as well as uric acid, ascorbic acid, and DOPAC. Note, small electrochemically active molecules like nitric oxide can still be detected when using mPD as an exclusion layer. This may be a problem if using only a single microelectrode. However, when self-referencing subtraction with two electrodes or an MEA is used these interfering signals can be detected and removed along with the non-Faradaic background current. To perform self-referencing subtraction for Glu measures, one recording site is capable of recording Glu while an adjacent site is insensitive to Glu (see Fig. 3) ([15], see Figs. 1.2 and 1.5 and [33, 37]). When the control site response (blue area in Fig. 3) is subtracted from the Glu site response (green area in Fig. 3), non-Faradaic and other interfering signals are removed resulting in the subtracted more specific Glu concentration in vivo (see Figs. 5 and 7) (e.g., see Fig. 5; [16, 37]). This methodology allows Glu concentrations as low as 0.1 μmol/L ($S/N = 3$, $n = 20$) to be measured with a spatial resolutions of about 50 μm × 50 μm. Another metric to confirm the identity of a signal is the correlation between responses at the Glu and sentinel sites. A custom software package for in vivo electrochemical signaling analysis allows large data sets to be analyzed for spontaneous, transient Glu signals as well as potassium-induced phasic changes of Glu [31, 38]. This software has aided in analysis of many studies of chronic recordings of Glu using MEAs implanted in rats and mice for up to 17 days [15, 16, 18, 39].

3.5 Histology

The animal was euthanized by transcardial perfusion with cold PBS (pH 7.4) followed by 4% paraformaldehyde in 0.1 mol/L phosphate buffer (PB), pH 7.4. The brain was immediately removed and postfixed in the same fixative for 24 h at 4 °C and subsequently

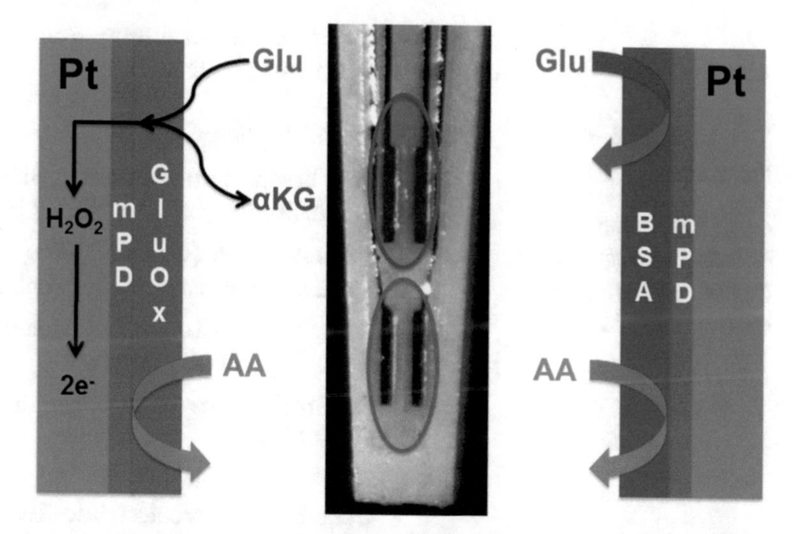

Fig. 3 Basic schematic of the MEA used to record tonic and phasic changes in Glu. The left and right figures show the glutamate oxidase, BSA and mPD layers on the recording sites of the MEA and their function for detection of Glu (*left*) and subtraction/detection of interference and noise (*right*). *Glu* glutamate, *α-KG* alpha ketoglutarate, *AA* ascorbic acid, *GluOx* glutamate oxidase, *BSA* bovine serum albumin, *Glut* glutaraldehyde

transferred into 15% and 30% buffered sucrose. The brain was coronal cut into 40 μm-thick sections on a frozen sliding microtome from the frontal cortex to the medial pontine nuclei. All the sections were serially collected into a cryoprotectant solution and stored at −20 °C.

A series of one in every 12 sections were rinsed with 0.1 mol/L PB (pH 7.4) and mounted on a gelatin-coated slide. After air dried, the slide was coverslipped with mounting medium containing 4′,6-diamidino-2-phenylindole (DAPI). EYFP expression in the brain was observed at low and high magnifications under a Nikon E50 fluorescence microscope equipped with a Nikon DS-Ril digital camera.

4 Notes

Perhaps the most difficult aspect of the MEA technology is the proper coating of the sentinel and Glu recording sites with the crosslinking layers and enzymes. Therefore, we have included below a detailed "step-by-step" protocol for the preparation of the Glu recording MEAs (Note 4.1). In addition, cleaning and reuse of the MEAs is recommended and economically driven because the MEAs are still rather expensive to manufacture due to the current fabrication scale necessitating a custom process for the microelectrode design specifications. We have therefore included our current cleaning procedure for maximum utilization of the MEAs for multiple experiments (Note 4.2).

4.1 MEA Coating Procedures

4.1.1 MEA Coating Procedures for Glutamate Oxidase Layer and Inactive Matrix for Sentinel Sites

1. Materials and Equipment
   - 1.1 BSA (bovine serum albumin, Sigma-Aldrich, 99%, catalog #A-3059) store at 4 °C, replace monthly.
   - 1.2 Glutaraldehyde (Grade I, 25%, Sigma-Aldrich Catalog #G-6257), store at −20 °C, replace yearly.
   - 1.3 Lyophilized L-Glutamate Oxidase (US Biological G400-01), 4 °C, replace yearly.
   - 1.4 Hamilton microsyringe (Gastight 80383, 701SN 10 μL (22 s/2″/3)).
   - 1.5 MEA—ceramic-based multisite microelectrode array (Center for Microelectrode Technology, www.ukycenmet.com).
   - 1.6 Microcentrifuge tube (1.5 mL).
   - 1.7 Microcentrifuge tube (500 μL).
   - 1.8 Microscope (dissecting).
   - 1.9 mPD (m-Phenylenediamine dihydrochloride, Acros, 130560250) store desiccated at RT, replace yearly.
   - 1.10 Pipettes (1000 and 10 μL).
2. Procedures

   Enzyme application to MEAs.

   Enzyme layer application is performed *before* mPD coating. MEAs should be cleaned and peroxide tested to acceptable standards before enzyme layer application.

   - 2.1 Solution Preparation—vials containing all proteins and enzymes should only be opened and used after reaching room temperature.
     - 2.1.1 L-glutamate oxidase preparation

       L-glutamate oxidase stock solution is prepared by adding deionized water (diH_2O) to the lyophilized, purified enzyme. 50 μL diH_2O is added to 25 units of GluOx to yield a 0.5 U/μL.

       Note: Stock solution is aliquoted (1 μL) and the aliquots are stored in appropriately sized containers (250 μL microcentrifuge tubes) at −20 °C. Aliquots may be stored for up to 6 months under these conditions.

       The following steps are performed immediately before MEA coating (i.e., the day of coating).

2.1.2 Glutaraldehyde/BSA coating solution preparation (protein-matrix)

2.1.2.1 BSA (10 mg) is transferred to a 1.5 mL microcentrifuge tube.

2.1.2.2 985 μL of diH_2O is added to the microcentrifuge tube containing BSA using a using 1000 μL pipette.

2.1.2.3 The solution is mixed by manual agitation (re-pipetting using 1000 μL pipette, until BSA is dissolved).

Note: Do not use vortex to mix solution as doing so may introduce air bubbles into the solution. Also, set pipette to volume < 1000 μL (e.g., 700 μL) to avoid introducing air bubbles. The solution may be placed in a microcentrifuge if necessary.

2.1.2.4 Once BSA solution is dissolved, 5 μL of glutaraldehyde (25%) solution is added. Caution: glutaraldehyde is a sensitizer.

2.1.2.5 The solution is mixed by inverting the closed tube 3–5 times.

Note: The resulting solution is 1% BSA, 0.125% glutaraldehyde.

2.1.2.6 The protein-matrix solution is set aside for 5 min. Solution will appear light yellow in color.

2.1.3 Glutamate oxidase coating solution preparation—performed just before MEA coating.

2.1.3.1 4 μL of the BSA/glutaraldehyde solution (step 2.1.2) is added to the microcentrifuge tube containing GluOx (see step 2.1.1).

2.1.3.2 The solution is mixed by manual agitation (re-pipetting with 10 μL pipette (pipette should be set to volume < 3 μL).

Note: Resulting solution is 1% BSA, 0.125% glutaraldehyde, and 0.1 units/μL GluOx. Solution should be used for coating immediately (proceed to step 2.2.2).

2.1.3.3 Set timer for 15 min.

Note: Glutamate oxidase coating solution may only be used within 15 min. Although several MEAs may be coated with one prepared solution, it is recommended not to exceed the useable time window for GluOx. The number of MEAs that can be coated within 15 min vary. Generally, 4–6 MEAs can be coated at a time.

2.2 Coating MEAs with Enzyme or Protein-matrix

Typically, a pair or a single recording pad on an MEA is coated with GluOx enzyme and a different pair or single recording pad ("sentinel" site(s)) is coated with the inactive protein-matrix.

2.2.1 Hamilton microsyringes (10 μL) are used to coat MEAs with L-GluOx (step 2.2.2) or inactive protein-matrix solution (BSA + glutaraldehyde (step 2.2.2)). The procedures used for enzyme or inactive protein-matrix coatings are the same. Clean syringes before and after use, i.e., three rinses with methanol, clean the tip with a wipe to clear any build up on the tip, three more rinses with methanol, three rinses with diH_2O.

Note: Dedicated syringes are used for applying coats of enzyme (e.g., GluOx) or protein-matrix. Do not use the same syringe for different solution.

Note: It is recommended to perform protein-matrix coatings before enzyme coatings, i.e., steps 2.2.2–2.2.6 will be repeated three times (total of three coats) for protein-matrix coatings and then three times for enzyme.

2.2.2 The solution (GluOx or protein-matrix) is drawn up (~1 μL using a 10 μL Hamilton Syringe).

2.2.3 The plunger is gently depressed to dispense a small bead of solution at the syringe tip (visualized using a dissecting microscope).

2.2.4 Using the dissecting microscope to target the MEA recording sites, the solution is applied to the appropriate recording sites by briefly contacting the recording sites with the solution droplet/bead starting from the opposite end from the center of the contacts and drawing the bead of solution to the edge of the contacts that are being coated.

2.2.5 The solution droplet is raised straight up and off the recordings sites, i.e., the syringe tip should not scrape across the MEA surface.

2.2.6 Set timer for 1 min. This is the minimum time requirement between coating applications to the recording site(s).

2.2.7 Enzyme-coated MEAs should cure a minimum of 48 h but we recommend 5 to 7 days before use (e.g., mPD plating, calibration or experimentation). It is recommended that enzyme-coated MEAs be used within 2 months of coating for most applications.

4.2 MEA Cleaning Protocol

4.2.1 Cleaning Materials and Equipment

1. Methanol
2. Small cotton balls or pellets
3. Small tweezers

4.2.2 Procedures

1. Find small dental or surgical cotton pills or balls. (If using pills, find the hole and expand it with tweezers.) Soak the cotton in methanol.
2. Take the cotton and place it on the tip of the MEA to soak. (With the cotton pill place the hole over the tip of the electrode to help keep it in place).
3. Use a swab wetted with methanol to keep the cotton wet. Gently rubbing the top of the MEA every ~2 min.
4. After 5 min of soaking, remove cotton pill or ball.
5. Hold the back end of the MEA so it does not move when cleaning.
6. Use the swab to gently rub the top of the MEA from the base of the dried enzyme to the tip of the electrode.
7. If the enzyme does not seem to be dislodging, repeat steps 2–6.
8. The whole MEA does not need to be clear of residue but the recording sites need to be clear. Ensure that the platinum recording sites are a uniform color and appearance and the polyimide insulation has not been damaged.

5 Discussion and Conclusions

5.1 Basic Recording Capabilities of the MEAs

Ceramic-based conformal MEAs have been applied to a myriad of applications including many experiments in rats, mice, and nonhuman primates. These experiments include chronic recordings of multiple single-unit neural activity, electrochemical recordings of evoked and stimulated release of neurotransmitters in awake rats, nonhuman primates, and possibly future applications in human neurosurgery [17, 37, 39–49]. In freely moving rats, single-unit neuronal activity has been recorded for up to 6 months in the hippocampus. Stimulation and recordings from hippocampal pyramidal cells in the rat have been performed. Specially designed deep recording electrodes for the nonhuman primate brain have been used for single-unit recordings from nonhuman primate frontal cortex and hippocampus in awake animals. Neurochemical Glu signaling recordings in the motor and frontal cortex of anesthetized monkeys as well as Glu concentration levels in the prefrontal cortex, hippocampus, and putamen of awake rats and nonhuman primates have been performed. Pt/Ir has been applied to the MEA recording site surfaces to improve electrical stimulation and increase recording site surface area. Flexible shank electrodes have been developed in conjunction with Ad-Tech® Medical Instrument Corporation for future studies in nonhuman primates and patients with epilepsy, brain tumors, Parkinson's disease, deep brain stimulation (DBS) surgery, or traumatic brain injury (TBI). A multi-shank electrode with up to 40 active sites for recording and stimulation in the hippocampus of awake primates is under development for electrochemical and electrophysiological studies.

5.2 AAV Distribution

It was previously shown that a similar viral construct took 35 days before expression of the ChR2 protein started to plateau [50]. We therefore waited a minimum of 5 weeks after viral injection before starting experimentation. For our initial study, we decided not to dilute the AAV solution. As seen in Fig. 4, robust expression of the ChR2-EYFP seen 5 weeks post injection was not limited to the initial DG injection site (see Sect. 2 for coordinates). Clear bilateral and rostro-caudal expression of the virus can be observed in the images from the unilateral injections of the virus into the DG. These studies clearly show that the virus can be transported by the neuronal projections. Furthermore, DAPI staining of DNA shows that expression of the AAV is located in the synapses (no apparent co-staining in cell body regions) as predicted with the use of the synapsin promoter.

Similarly, the histological analysis of the distribution of the ChR2 protein in the rat frontal cortex 6 weeks following a single frontal cortex injection into the prelimbic frontal cortex (Prl) showed robust expression of the ChR2 protein throughout the frontal cortex regions and adjacent brain areas (Fig. 8). Similar to data collected in the rat hippocampus, AAV5-hSyn-hChR2(H134R)-EYFP into the infralimbic cortex was not limited to the injection site and was widespread through the cortex and other adjacent regions such as caudate putamen (CPu) and nucleus accumbens (Acb). Taken together with the data from the ChR2 virus injections into the rat DG, the data support the hypothesis that the opsin expression extends far beyond the likely diffusion-controlled delivery and support active transport of the viral particles to other regions of the CNS.

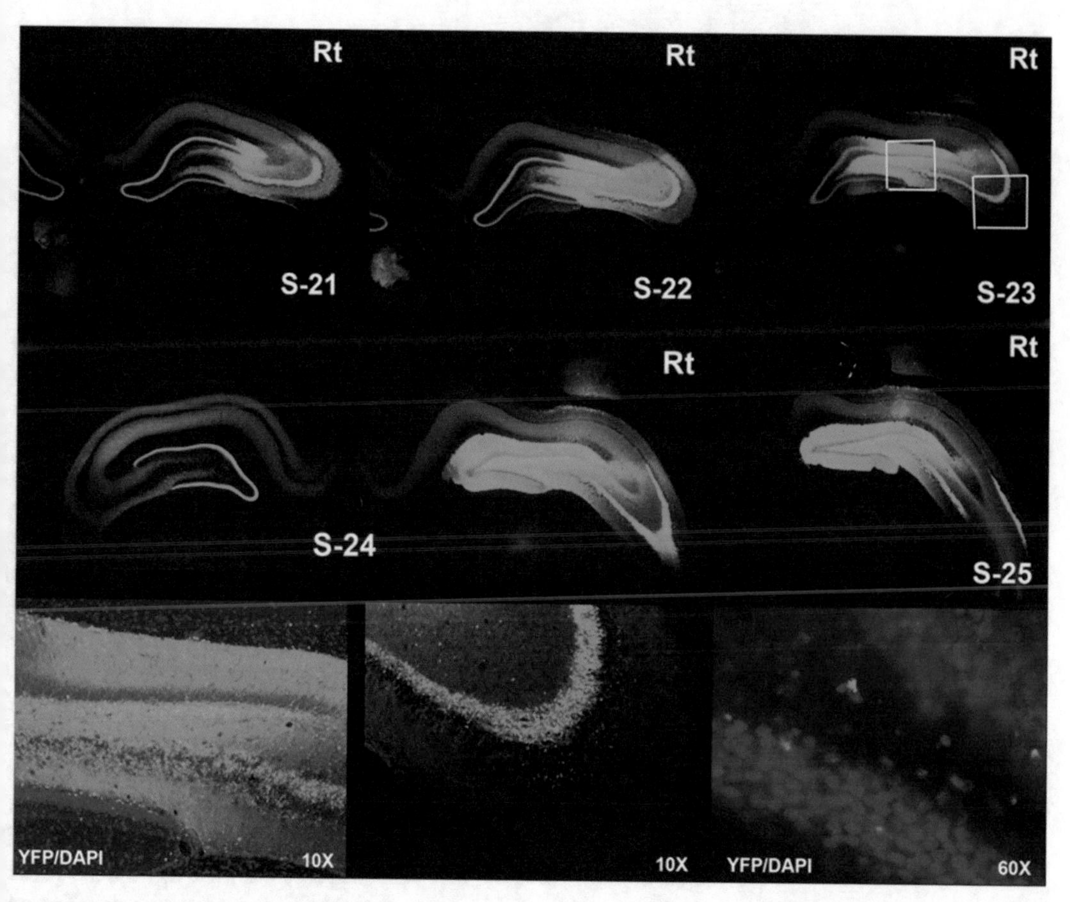

Fig. 4 AAV$_{2/5}$-hSyn-hChR2(H134R)-EYFP distribution throughout the right and left hippocampus 6 weeks post injection of the virus into the right DG. Each section was taken 360 μm apart. *Bottom row* shows the magnified section of the right hippocampus in combination with DAPI staining. "Rt" indicates the right hemisphere

5.3 Light-Induced Glutamate Release Observed in the Dentate Gyrus and Frontal Cortex

The Deisseroth lab showed early on in their work that ChR2, a rapidly gated light-sensitive cation channel, could be reliably used for millisecond timescale optical control of neuronal spikes and neural activity [23]. In our studies, we wanted to observe if the increase in spiking actually involved release of neurotransmitters such as Glu. Our approach used MEAs configured for self-referenced measures of both tonic and phasic Glu release. The light waveguide was affixed ~200 μm from the Pt surface of the MEA, which we believed would stimulate local synapses expressing ChR2 to release Glu and possibly other neurotransmitters.

As a control experiment, the MEA/waveguide was lowered in hippocampus of naïve animals and we examined the effects of the light on the MEA and surrounding tissue. The results showed that when the power reached 20 mW, some photovoltaic effects were measured by the MEAs but no Glu was released as evidenced through the self-referencing MEA measures (Unpublished data).

Because the MEA is configured for self-referenced Glu measures, we were able to subtract any interfering photovoltaic effects from our Glu measures. Thus, the photovoltaic effects, while predicted by theory and observed in this study, can be easily subtracted from the MEA recordings of Glu and other neurochemicals.

Studies were then carried out involving lowering the MEA/waveguide into the right DG of animals that had previously been injected with AAV into the right DG 5-weeks prior to in vivo recordings. Light activation through the waveguide was seen to produce light power-dependent release of Glu. Figure 5d shows precise optical control of Glu release starting from a very low power light pulse train (1 ms, 40 Hz, ~0.5 mW for 2.5 s) and reaching a plateau of Glu release using constant light stimulation at 50 mW (for 2–3 s). From our previous experience with evoked release of Glu, the observed signal plateau is likely due to two major factors: the density of synapses close to the MEA and the maximal amount of light dispersion in the brain, which for blue light (488 nm) is estimated at no more than 1 mm [51, 52]. Interestingly, initial studies show that light pulse trains (20 ms, 40 Hz, ~4 mW) or constant light stimulation (DC, 5 mW) at equivalent power generated relatively the same amount of Glu release. Figure 5e is a plot of the measured Glu release amplitude versus applied laser energy. There was clear indication that the amplitude of the Glu release was dose-dependently related to the duration of the light pulses and the power of the light pulses. Furthermore, the temporal dynamics of Glu release produced by the viral expression and light activation were similar to what we have previously reported using other forms of stimulation (high potassium, drug induced, or behavior) [45, 53–56]. The dose–response of the Glu signals vs. light power seen from light activation had a wider dynamic range and precision than we have seen from other methods involving micropipette methods, electrical stimulation or behavioral activation [18, 29, 43]. These studies support that the ChR2-control of Glu release by the virus was robust and different from our previous studies of Glu neuronal systems.

Histological evaluation of the viral vector distribution showed that it was not limited to the injection site and was also distributed bilaterally into the left hippocampus. We wanted to determine if the ChR2 located in the contralateral hemisphere was capable of producing light-evoked release of Glu so we lowered the MEA/waveguide into the left DG area. Figure 6 shows that both hemispheres exhibited robust release of Glu using similar light parameters supporting that the viral vector has an effective distribution away from the injection site in the rat hippocampus. Based on EYFP fluorescence even though we observe less fluorescence on the contralateral side of injection, it appears that the synapse density with transfected ChR2 is similar to the ipsilateral injection site. For investigations where a more focal transfected area is needed or

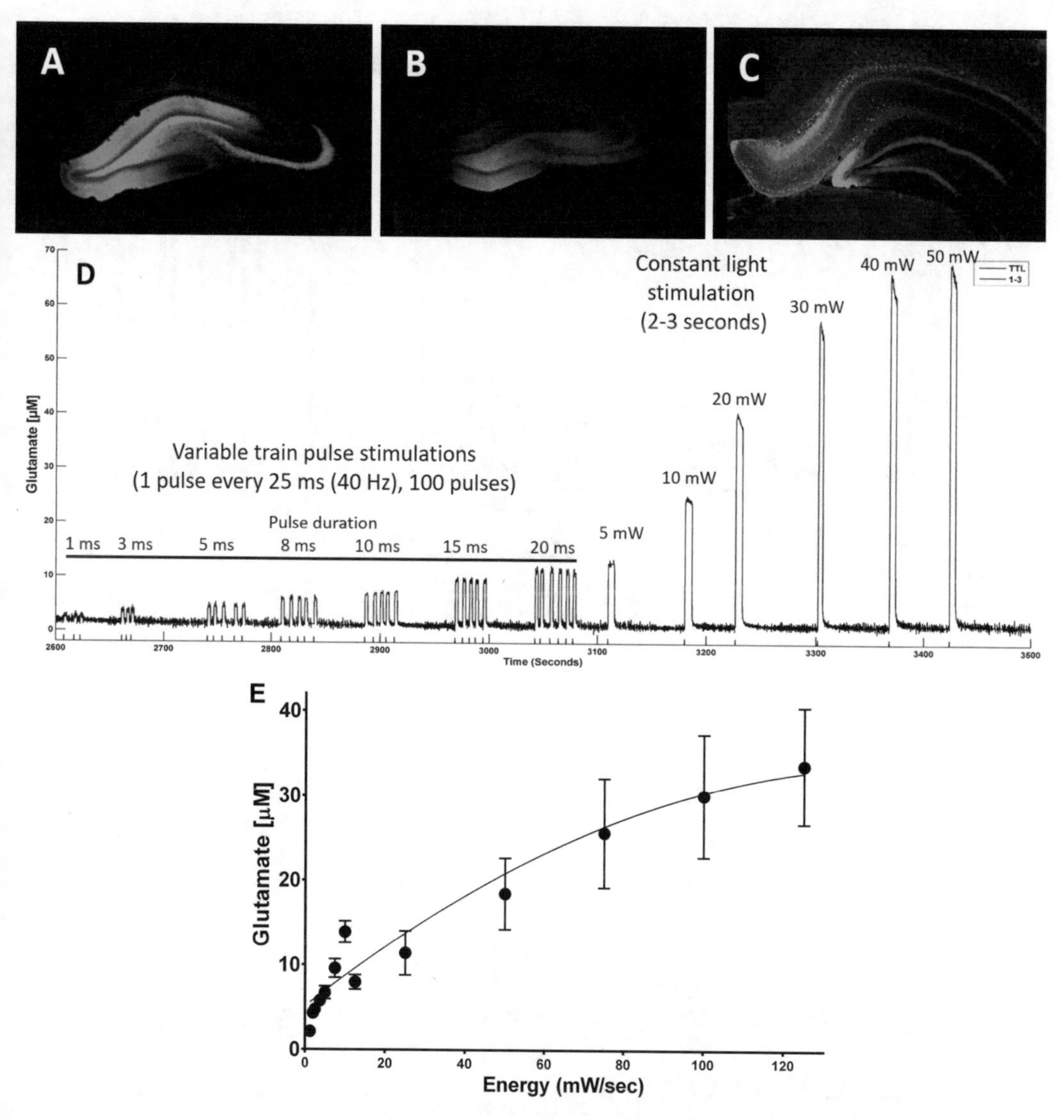

Fig. 5 Images (**a–c**) of the enhanced yellow fluorescent protein (EYFP) expression showing distribution of $AAV_{2/5}$-ChR2-Syn-EYFP 6 weeks after a single injection (1 μL) into the DG. Channelrodopsin distribution was not limited to the injection site DG (**a**) and can be observed in the CA1–CA3 (**b**) and the EC (**c**). (**c**) shows that there is no co-staining between DAPI and EYFP. (**d**) Variable train laser pulse stimulations between 1 and 20 ms (power equivalent from 0.5 mW to ~4 mW) that were delivered at 40 Hz (every 25 ms) followed by increasing power using constant light stimulation. (**e**) Plot of measured Glu amplitude concentration versus applied laser energy. The *solid line* is a curve fit of the data

desirable, we recommend using a lower titer of viral vector. However, we have not tested this hypothesis to prove that a lower amount of light-evoked Glu release is observed.

We have begun to carry out pharmacological testing of the optogenetic control of Glu signaling. Our preliminary studies demonstrate that local application of the Glu uptake inhibitor,

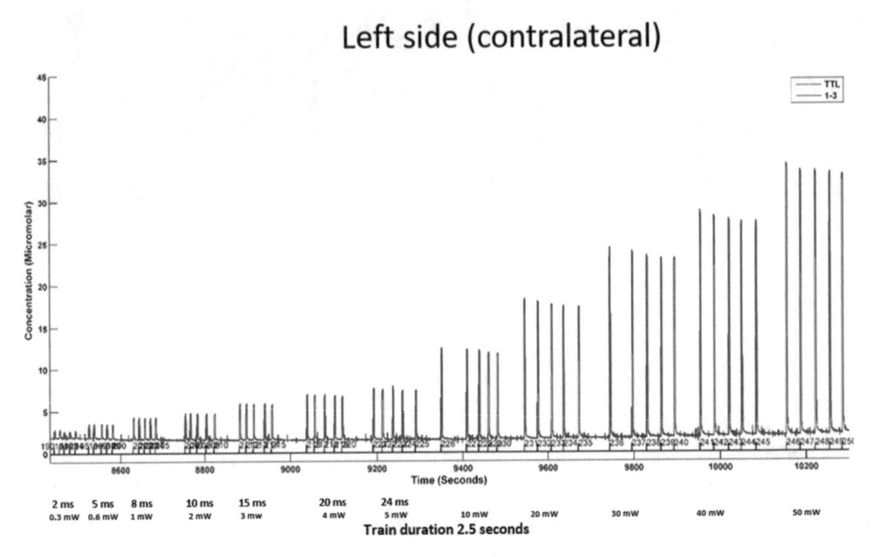

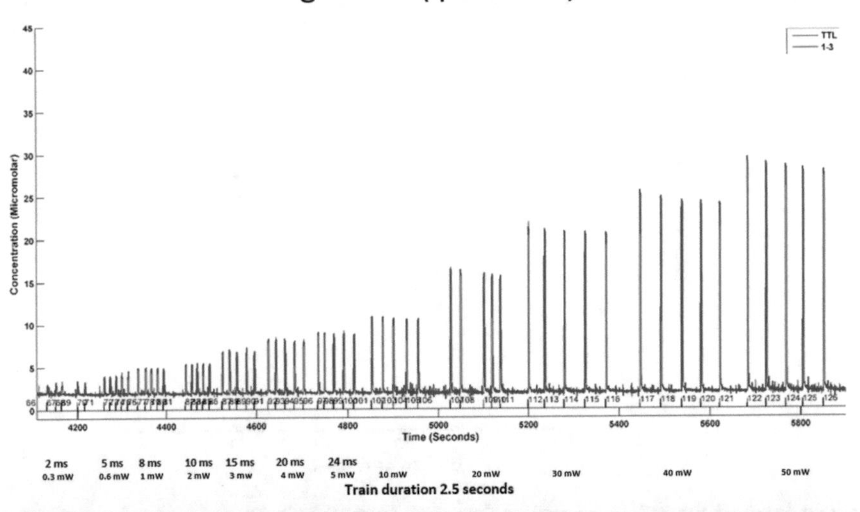

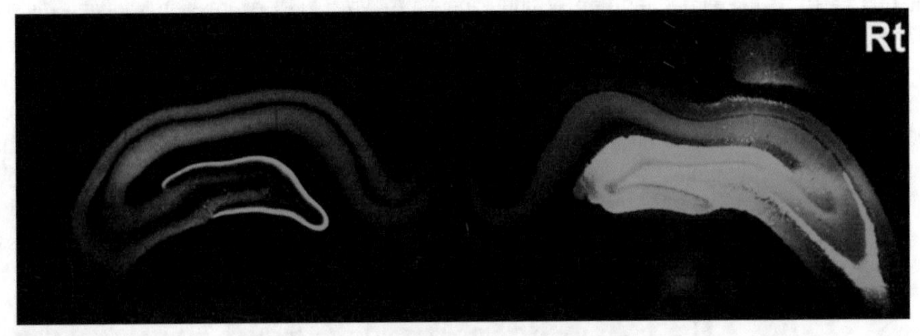

Fig. 6 Comparison of power-dependent light-induced glutamate release in the left hippocampus (contralateral) (**a**) vs. the right hippocampus (ipsilateral) (**b**). (**c**) showing EYFP (Enhanced Yellow Fluorescent Protein) bilateral distribution 6 weeks after a single injection of AAV5- ChR2-Syn-EYFP (1 μL) into the right DG

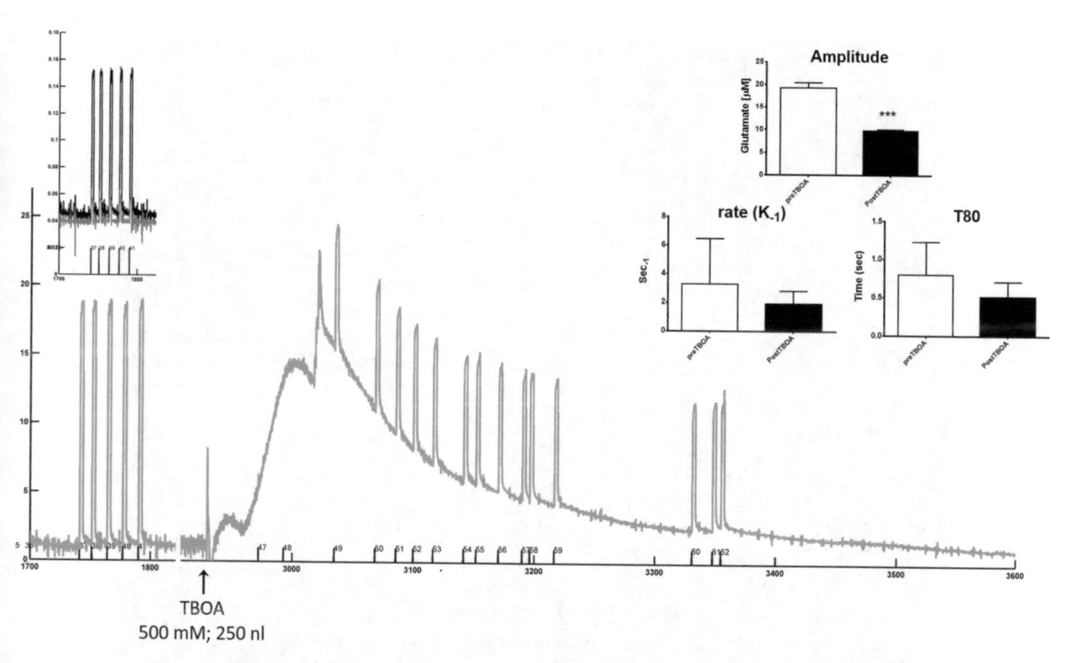

Fig. 7 Self-referenced glutamate trace (*green*) showing that the uptake inhibitor TBOA caused an increase in tonic (basal) Glu and a decrease in light-induced (5 mW) Glu release. This concentration of TBOA did not significantly affect the uptake rates of the extracellular Glu signals produced by light stimulation of ChR2. *Left inset* showing the raw signal of the light-induced Glu release (pre-TBOA) showing both the Glu measuring channel (*black trace*) and the self-referencing channel (*red trace*). The self-referencing or sentinel channel shows a small amount of photovoltaic effect on the MEAs that is removed by the self-referencing recording method. The three inset bar graphs indicate amplitude, first-order uptake rate, and the time it takes for the Glu concentration to decrease by 80% of the max (T80) for pre- and post-TBOA applications

DL-threo ß-Benzyloxyaspartate (TBOA), a potent inhibitor of all subtypes of the excitatory amino acid transporters, can greatly affect both tonic and phasic release of Glu neurotransmission produced by ChR2 expression. As seen in Fig. 7, we locally applied TBOA from a micropipette next to the MEA/waveguide tip. Our prior studies have demonstrated that locally applied TBOA produces robust extracellular changes in Glu in the frontal cortex and striatum of anesthetized or awake rats [18, 38, 57]. Our current studies show that TBOA produced an increase in tonic (basal) Glu and a decrease in light-induced Glu release (Fig. 8). These data are consistent with the direct effects of TBOA on extracellular Glu and most likely the increased extracellular levels of Glu produced an effect on presynaptic mGluR2/3 receptors to inhibit light-induced Glu release. We have seen that direct application of selective mGluR2/3 agonists and antagonists produce decreases and increases in extracellular Glu, respectively [18, 58]. These data support that the Glu system is responding to pharmacological modulation and further support that the optogenetic control of Glu signaling is similar to pharmacological modulation of glutamatergic systems.

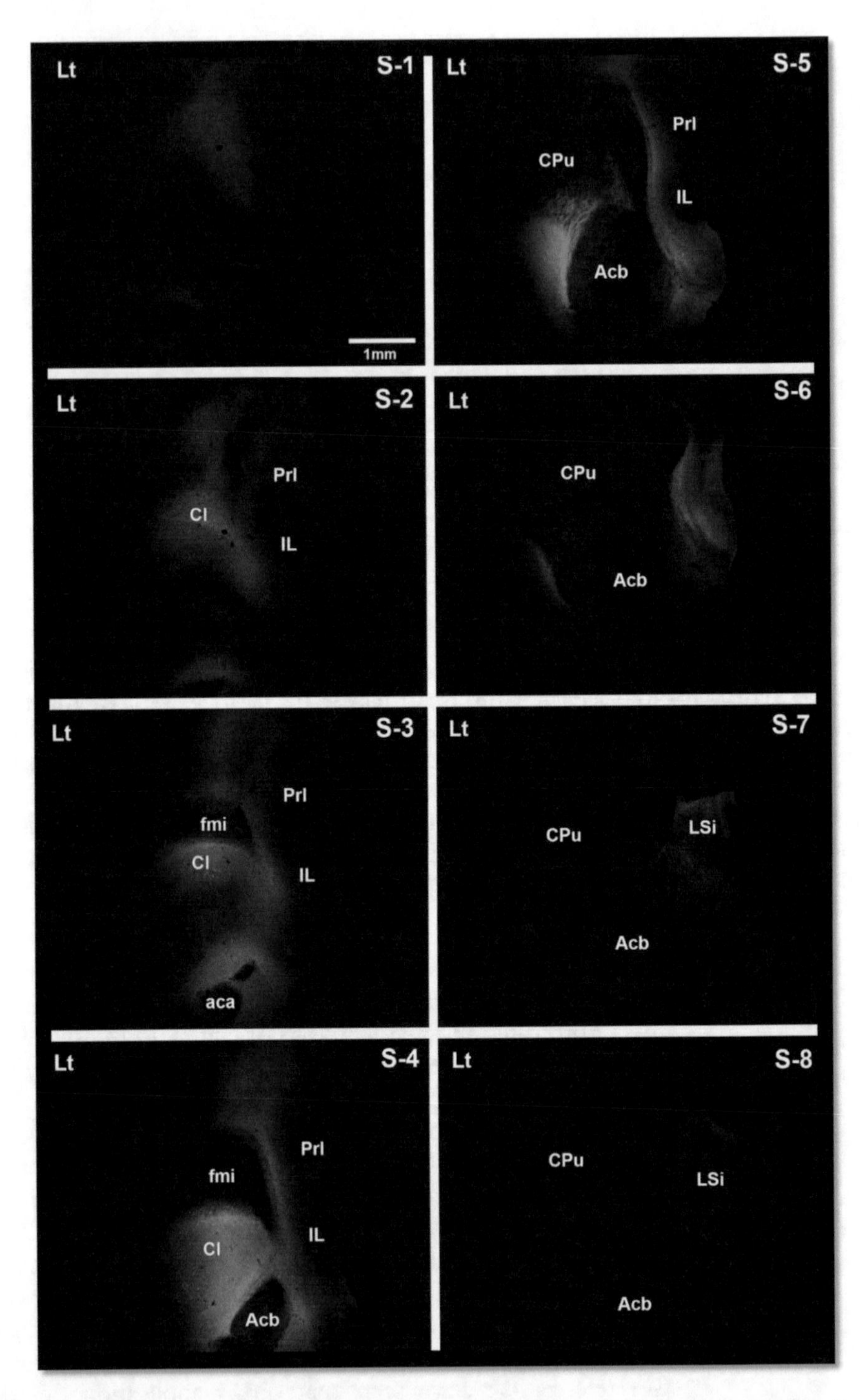

Fig. 8 EYFP expression showing distribution of the virus throughout the rat brain after 6 weeks after a single injection (1 μL) of AAV5-hSyn-hChR2(H134R)-EYFP in the infralimbic frontal cortex. Abbreviations: *Lt* left hemisphere, *CPu* striatum, *Prl* IL-infralimbic, *aca* anterior part of anterior commissure, *Acb* nucleus accumbens, *Cl* central lateral nucleus, *fmi* anterior forceps of corpus callosum, *LSi* lateral septal nucleus

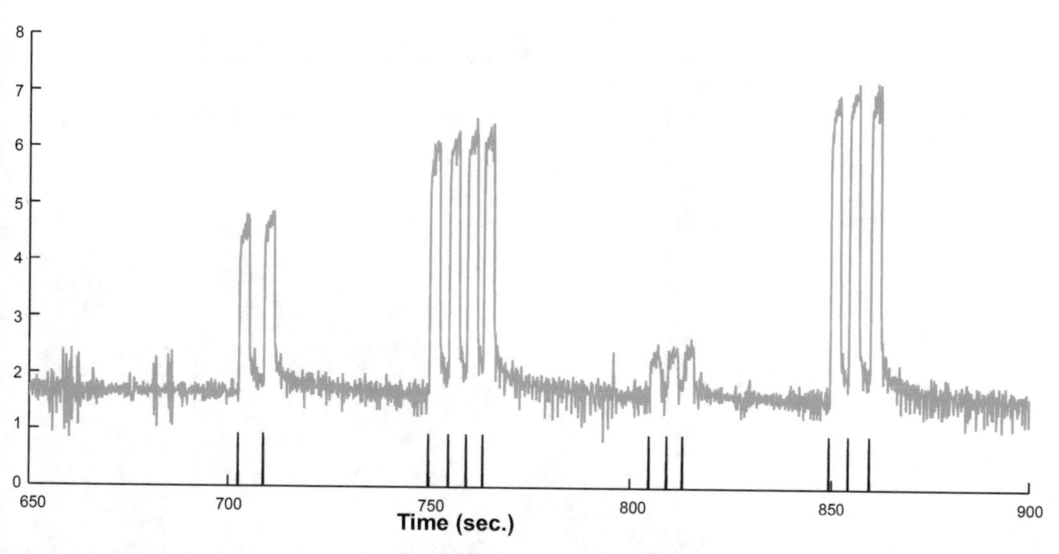

Fig. 9 Self-referenced trace showing light-induced Glu release from the infralimbic cortex is light power dependent and very reproducible. Randomized light train pulses (40 Hz) for 2.5 s (5 ms ~ 1 mW; 8 ms ~ 2 mW; 1 ms, ~0.5 mW; 10 ms, ~3 mW) are shown. Bars along time axis indicate light pulses

Similar to the hippocampal studies, waveguide/MEA arrays were used to investigate light-evoked release of Glu in the rat frontal cortex. Robust and light dose-dependent Glu signals were recorded. As seen in the Fig. 9, we observed power-dependent light-induced Glu release and very reproducible, which is similar to what we observed in the hippocampus. However, when we compare similar signals from both brain regions (Fig. 10) we observed significantly more release in the hippocampus than in the cortex suggesting a higher fiber density in the hippocampus and/or greater expression of the ChR2 protein. Additional experiments are needed to investigate this finding to determine if this is a brain region-specific effect or inherent variability of viral expression.

5.4 Conclusions

The present studies support that optogenetics can be used as a methodology to precisely control the release of Glu in the rat hippocampus and frontal cortex. The precision of light-evoked release of Glu was exceptional, exceeding our previous experiences with potassium-evoked, electrically evoked and behaviorally induced release of Glu in vivo. Interestingly, in our studies, AAV5-hSyn-hChR2(iH134R)-EYFP distribution was not limited to the injection site suggesting that a low titer may be required for a more precise distribution and/or there is active transport of the viral particles. Light-induced Glu release was independent of light pulses vs. constant light. Also, measured Glu release was power dependent up to a limit around 50 mW. We observed more Glu release in the DG suggesting a higher synapse density than the infralimbic cortex or a higher titer of the hCR2 protein. Finally, Glu dynamics obtained with optogenetics were in the same range

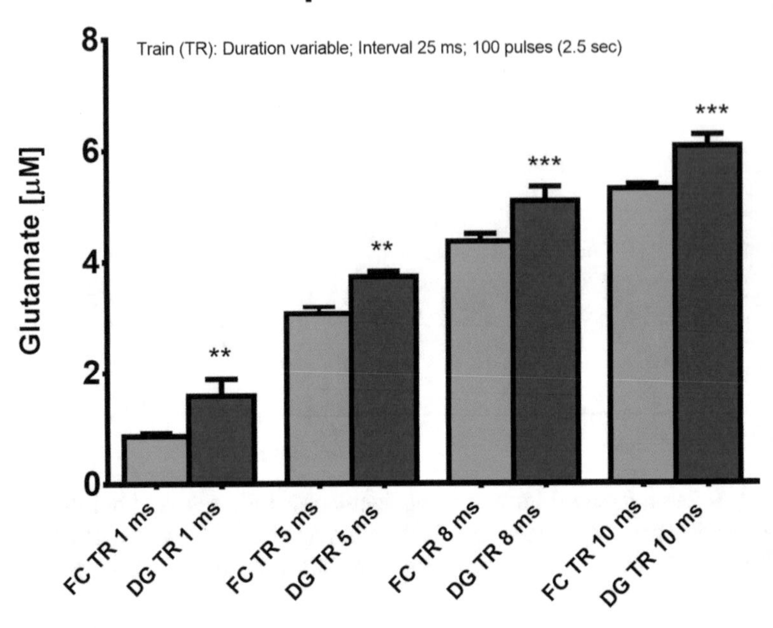

Fig. 10 Comparisons of light-induced glutamate release between IL (*green*) and DG (*blue*) in the rat brain. Student's *t*-test; $^{**}P < 0.01$, $^{***}P < 0.001$

as we have previously reported using other forms of stimulation such as high potassium, drug induced, or behavior [45, 53–56]. These results demonstrate the feasibility of directly measuring Glu release in vivo using optogenetics.

Future directions of our optogenetic studies include demonstrating combined MEA/optogenetics use in freely moving animals. Also, chemical concentration information obtained from the Glu-sensing MEA can be used to modulate in vivo Glu concentration using feedback to the laser, which is being explored for studies of epileptogenesis and possible control of seizures in animals that are models of epilepsy. Other neurochemicals and neurochemical combinations may be measured including but not limited to glucose, acetylcholine, choline, lactate, dopamine, as well as O_2 using MEAs and possibly modulated with optogenetics. The combination of optogenetics with the MEA technology promises to be a powerful tool for studying the CNS and continues to show capabilities that may have clinical applications in the future.

Acknowledgments

Supported by NIDA; R21DA033796-01, DARPA; N66001-09-C-2080 and NIH; CTSA 1 UL1RR033173-01. The content is solely the responsibility of the authors and does not necessarily represent the official views of the NIH.

Disclosure of competing interest: G.A.G. is principal owner of Quanteon LLC. J.E.Q., P.H., and F.P. have served as consultants to Quanteon LLC.

References

1. Kissinger PT, Hart JB, Adams RN (1973) Voltammetry in brain tissue—a new neurophysiological measurement. Brain Res 55(1):209–213
2. McCreery RL, Dreiling R, Adams RN (1974) Voltammetry in brain tissue: the fate of injected 6-hydroxydopamine. Brain Res 73(1):15–21
3. Wightman RM et al (1976) Monitoring of transmitter metabolites by voltammetry in cerebrospinal fluid following neural pathway stimulation. Nature 262(5564):145–146
4. Burmeister JJ, Moxon K, Gerhardt GA (2000) Ceramic-based multisite microelectrodes for electrochemical recordings. Anal Chem 72(1):187–192
5. Hu Y et al (1994) Direct measurement of glutamate release in the brain using a dual enzyme-based electrochemical sensor. Brain Res 659(1–2):117–125
6. Dale N, Pearson T, Frenguelli BG (2000) Direct measurement of adenosine release during hypoxia in the CA1 region of the rat hippocampal slice. J Physiol 526(1):143–155. doi:10.1111/j.1469-7793.2000.00143.x
7. Oldenziel WH et al (2006) Evaluation of hydrogel-coated glutamate microsensors. Anal Chem 78(10):3366–3378. doi:10.1021/ac052146s
8. Zesiewicz T et al (2013) Update on treatment of essential tremor. Curr Treat Options Neurol 15(4):410–423. doi:10.1007/s11940-013-0239-4
9. Wassum KM et al (2008) Silicon wafer-based platinum microelectrode array biosensor for near real-time measurement of glutamate in vivo. Sensors Basel Sensors 8(8):5023–5036. doi:10.3390/s8085023
10. Lowry JP et al (1998) An amperometric glucose-oxidase/poly(o-phenylenediamine) biosensor for monitoring brain extracellular glucose: in vivo characterisation in the striatum of freely-moving rats. J Neurosci Methods 79(1):65–74
11. Westerink RH (2004) Exocytosis: using amperometry to study presynaptic mechanisms of neurotoxicity. Neurotoxicology 25(3):461–470. doi:10.1016/j.neuro.2003.10.006
12. Suaud-Chagny MF et al (1993) High sensitivity measurement of brain catechols and indoles in vivo using electrochemically treated carbon-fiber electrodes. J Neurosci Methods 48(3):241–250
13. Oldenziel WH et al (2006) In vivo monitoring of extracellular glutamate in the brain with a microsensor. Brain Res 1118(1):34–42. doi:10.1016/j.brainres.2006.08.015. S0006-8993(06)02370-5 [pii]
14. Tian FM et al (2009) A microelectrode biosensor for real time monitoring of L-glutamate release. Anal Chim Acta 645(1–2):86–91. doi:10.1016/j.aca.2009.04.048
15. Rutherford EC et al (2007) Chronic second-by-second measures of L-glutamate in the central nervous system of freely moving rats. J Neurochem 102(3):712–722
16. Hascup KN et al (2008) Second-by-second measures of L-glutamate in the prefrontal cortex and striatum of freely moving mice. J Pharmacol Exp Ther 324(2):725–731. doi:10.1124/jpet.107.131698
17. Stephens ML et al (2010) Real-time glutamate measurements in the putamen of awake rhesus monkeys using an enzyme-based human microelectrode array prototype. J Neurosci Methods 185(2):264–272. doi:10.1016/j.jneumeth.2009.10.008. S0165-0270(09)00564-0 [pii]
18. Hascup ER et al (2010) Rapid microelectrode measurements and the origin and regulation of extracellular glutamate in rat prefrontal cortex. J Neurochem 115(6):1608–1620. doi:10.1111/j.1471-4159.2010.07066.x
19. Wassum KM, Phillips PE (2015) Probing the neurochemical correlates of motivation and decision making. ACS Chem Neurosci 6(1):11–13. doi:10.1021/cn500322y
20. Phillips PE et al (2003) Subsecond dopamine release promotes cocaine seeking. Nature 422(6932):614–618
21. Park J, Takmakov P, Wightman RM (2011) In vivo comparison of norepinephrine and dopamine release in rat brain by simultaneous measurements with fast-scan cyclic voltammetry. J Neurochem 119(5):932–944. doi:10.1111/j.1471-4159.2011.07494.x
22. Gritton HJ et al (2016) Cortical cholinergic signaling controls the detection of cues. Proc Natl Acad Sci U S A 113(8):E1089–E1097. doi:10.1073/pnas.1516134113

23. Boyden ES et al (2005) Millisecond-timescale, genetically targeted optical control of neural activity. Nat Neurosci 8(9):1263–1268. doi:10.1038/nn1525
24. Schweizer N et al (2014) Limiting glutamate transmission in a Vglut2-expressing subpopulation of the subthalamic nucleus is sufficient to cause hyperlocomotion. Proc Natl Acad Sci U S A 111(21):7837–7842. doi:10.1073/pnas.1323499111
25. Yau HJ et al (2016) Pontomesencephalic tegmental afferents to VTA non-dopamine neurons are necessary for appetitive Pavlovian learning. Cell Rep. doi:10.1016/j.celrep.2016.08.007
26. Parrot S et al (2015) Why optogenetics needs in vivo neurochemistry. ACS Chem Neurosci 6(7):948–950. doi:10.1021/acschemneuro.5b00003
27. Wise WS, Loh SE (1976) Equilibria and origin of minerals in system Al_2O_3-$AlPO_4$-H_2O. Am Mineral 61(5–6):409–413
28. Opris I et al (2011) Neural activity in frontal cortical cell layers: evidence for columnar sensorimotor processing. J Cogn Neurosci 23(6):1507–1521. doi:10.1162/jocn.2010.21534
29. Burmeister J et al (2002) Improved ceramic-based multisite microelectrode for rapid measurements of L-glutamate in the CNS. J Neurosci Methods 119(2):163–171
30. Burmeister JJ et al (2013) Glutaraldehyde cross-linked glutamate oxidase coated microelectrode arrays: selectivity and resting levels of glutamate in the CNS. ACS Chem Neurosci 4(5):721–728. doi:10.1021/cn4000555
31. Hascup KN et al (2011) Resting glutamate levels and rapid glutamate transients in the prefrontal cortex of the flinders sensitive line rat: a genetic rodent model of depression. Neuropsychopharmacology 36(8):1769–1777. doi:10.1038/npp.2011.60. npp201160 [pii]
32. Ulusoy A et al (2009) Dose optimization for long-term rAAV-mediated RNA interference in the nigrostriatal projection neurons. Mol Ther 17(9):1574–1584. doi:10.1038/mt.2009.142
33. Burmeister JJ, Gerhardt GA (2001) Self referencing ceramic based multisite microelectrodes for the detection and elimination of interferences from the measurement of L-glutamate and other analytes. Anal Chem 73(5):1037–1042
34. Kusakabe H et al (1983) Purification and properties of a new enzyme, L-glutamate oxidase, from *streptomyces* SP X-199-6 grown on wheat bran. Agric Biol Chem 47(6):1323–1328
35. Hinzman JM et al (2010) Diffuse brain injury elevates tonic glutamate levels and potassium-evoked glutamate release in discrete brain regions at two days post-injury: an enzyme-based microelectrode array study. J Neurotrauma 27(5):889–899. doi:10.1089/neu.2009.1238
36. Stephens ML et al (2009) Age-related changes in glutamate release in the CA3 and dentate gyrus of the rat hippocampus. Neurobiol Aging. doi:10.1016/j.neurobiolaging.2009.05.009. S0197-4580(09)00172-9 [pii]
37. Day BK et al (2006) Microelectrode array studies of basal and potassium-evoked release of L-glutamate in the anesthetized rat brain. J Neurochem 96(6):1626–1635
38. Hinzman JM et al (2012) Disruptions in the regulation of extracellular glutamate by neurons and glia in the rat striatum two days after diffuse brain injury. J Neurotrauma 29(6):1197–1208. doi:10.1089/neu.2011.2261
39. Dash MB et al (2009) Long-term homeostasis of extracellular glutamate in the rat cerebral cortex across sleep and waking states. J Neurosci 29(3):620–629. doi:10.1523/JNEUROSCI.5486-08.2009. 29/3/620 [pii]
40. Hampson RE et al (2012) Facilitation and restoration of cognitive function in primate prefrontal cortex by a neuroprosthesis that utilizes minicolumn-specific neural firing. J Neural Eng 9(5):056012. doi:10.1088/1741-2560/9/5/056012
41. Opris I et al (2012) Closing the loop in primate prefrontal cortex: inter-laminar processing. Front Neural Circuits 6:88. doi:10.3389/fncir.2012.00088
42. Opris I et al (2012) Columnar processing in primate pFC: evidence for executive control microcircuits. J Cogn Neurosci. doi:10.1162/jocn_a_00307
43. Onifer SM, Quintero JE, Gerhardt GA (2012) Cutaneous and electrically evoked glutamate signaling in the adult rat somatosensory system. J Neurosci Methods 208(2):146–154. doi:10.1016/j.jneumeth.2012.05.013
44. Zhang H, Lin SC, Nicolelis MA (2009) Acquiring local field potential information from amperometric neurochemical recordings. J Neurosci Methods 179(2):191–200. doi:10.1016/j.jneumeth.2009.01.023. S0165-0270(09)00052-1 [pii]
45. Quintero JE et al (2007) Amperometric measures of age-related changes in glutamate regulation in the cortex of rhesus monkeys. Exp Neurol 208(2):238–246
46. Howe WM et al (2013) Prefrontal cholinergic mechanisms instigating shifts from monitoring for cues to cue-guided performance: converging electrochemical and fMRI evidence from rats and humans. J Neurosci 33(20):8742–8752. doi:10.1523/jneurosci.5809-12.2013

47. Konradsson-Geuken A et al (2010) Cortical kynurenic acid bi-directionally modulates prefrontal glutamate levels as assessed by microdialysis and rapid electrochemistry. Neuroscience 169(4):1848–1859. doi:10.1016/j.neuroscience.2010.05.052. S0306-4522(10)00792-X [pii]
48. Zhou N et al (2013) Regenerative glutamate release by presynaptic NMDA receptors contributes to spreading depression. J Cereb Blood Flow Metab 33(10):1582–1594. doi:10.1038/jcbfm.2013.113
49. Choi Hyun B et al (2012) Metabolic communication between astrocytes and neurons via bicarbonate-responsive soluble adenylyl cyclase. Neuron 75(6):1094–1104. doi:10.1016/j.neuron.2012.08.032
50. Zhang F et al (2010) Optogenetic interrogation of neural circuits: technology for probing mammalian brain structures. Nat Protoc 5(3):439–456. doi:10.1038/nprot.2009.226
51. Yizhar O et al (2011) Optogenetics in neural systems. Neuron 71(1):9–34. doi:10.1016/j.neuron.2011.06.004
52. Bernstein JG et al (2008) Prosthetic systems for therapeutic optical activation and silencing of genetically-targeted neurons. Proc SPIE Int Soc Opt Eng 6854:68540h. doi:10.1117/12.768798
53. Miller EM et al (2015) Simultaneous glutamate recordings in the frontal cortex network with multisite biomorphic microelectrodes: new tools for ADHD research. J Neurosci Methods. doi:10.1016/j.jneumeth.2015.01.018
54. Stephens ML et al (2014) Tonic glutamate in CA1 of aging rats correlates with phasic glutamate dysregulation during seizure. Epilepsia 55(11):1817–1825. doi:10.1111/epi.12797
55. Hunsberger HC et al (2015) P301L tau expression affects glutamate release and clearance in the hippocampal trisynaptic pathway. J Neurochem 132(2):169–182. doi:10.1111/jnc.12967
56. Thomas TC et al (2012) Hypersensitive glutamate signaling correlates with the development of late-onset behavioral morbidity in diffuse brain-injured circuitry. J Neurotrauma 29(2):187–200. doi:10.1089/neu.2011.2091
57. Hinzman JM et al (2015) Spreading depolarizations mediate excitotoxicity in the development of acute cortical lesions. Exp Neurol 267:243–253. doi:10.1016/j.expneurol.2015.03.014
58. Quintero JE et al (2011) Methodology for rapid measures of glutamate release in rat brain slices using ceramic-based microelectrode arrays: basic characterization and drug pharmacology. Brain Res 1401:1–9. doi:10.1016/j.brainres.2011.05.025

Chapter 12

Separation-Based Sensors Using Microchip Electrophoresis with Microdialysis for Monitoring Glutamate and Other Bioactive Amines

Nathan Oborny, Michael Hogard, and Susan M. Lunte

Abstract

Large increases in the release of glutamate (Glu) into the extracellular space of the brain have been shown to produce excitotoxicity, which can lead to long-term neuronal damage seen in many neurological disorders, including traumatic brain injury (TBI) and stroke. Therefore, methods for continuous monitoring of extracellular Glu concentrations in the brain can be very useful to clinicians for determining the best timing for pharmacological intervention. Microdialysis (MD) is an in vivo sampling method that, when coupled with a separation-based analytical method, makes it possible to monitor multiple analytes simultaneously with high temporal resolution. Very low flow rates are associated with this sampling method (typically between 0.1 and 1 μL/min), leading to very small sample volumes per unit time. To maximize the analysis of microdialysis samples with good temporal resolution, several methods for integrating MD with microchip electrophoresis (ME) have been described. Microdialysis coupled to microchip electrophoresis (MD-ME) can provide rapid analysis of complex samples using sample injection volumes less than 100 nL. However, the design and construction of MD-ME devices requires a high degree of planning and knowledge of microfabrication. The processes outlined in this chapter detail what we have found to be the best practices of design, construction, and operation of MD-ME devices.

Key words Microchip electrophoresis, Fluorescence, Microdialysis, Glutamate, Derivatization

1 Introduction

Changes in extracellular glutamate (Glu) concentrations in the brain following neurological insults, such as traumatic brain injury and hypoxia, play a critical role in secondary injury, and subsequently affect long-term patient outcome [1]. The currently accepted hypothesis is that damaged neurons release Glu into the extracellular spaces at a concentration significantly above basal levels following the primary injury. The large excess of Glu overwhelms the Glu transporters that normally maintain low micromolar (1–3 μmol/L) [2] concentrations in the extracellular space,

Sandrine Parrot and Luc Denoroy (eds.), *Biochemical Approaches for Glutamatergic Neurotransmission*, Neuromethods, vol. 130, DOI 10.1007/978-1-4939-7228-9_12, © Springer Science+Business Media LLC 2018

resulting in rapid uptake by surrounding healthy neurons. This leads to an increased uptake of calcium (Ca^{2+}) [3–6] into the cells, causing a cascade of events and culminating in apoptosis. The whole process is generally referred to as excitotoxicity [7].

Since elevated Glu concentrations are the proximal cause of this excitotoxicity, the desire for methods to monitor extracellular Glu in a clinical setting has grown in recent years. The hope is that by monitoring changes in Glu concentration in the brain, rapid intervention that would prevent continued excitotoxicity and tissue damage may be possible. With this in mind, two primary methods for monitoring Glu in vivo have been developed: electrochemical biosensors and microchip electrophoresis (ME).

1.1 Biosensors for the Glutamate Detection

Biosensors can be broadly defined as devices that convert the recognition of a biochemical species into an electrical or optical signal [8]. For this to be useful, not only must the recognition be specific to the analyte of interest, but the sensor must be able to detect it at biologically relevant concentrations. Several classes of biosensors exist, including electrochemical sensors that are described in another chapter of this book.

The most commonly used Glu biosensors in a clinical setting are based on amperometric detection [6, 9]. These biosensors use an enzyme specific to Glu to generate electroactive species. These biosensors commonly incorporate glutamate oxidase [6] or glutamate dehydrogenase, which generate hydrogen peroxide or NADPH in the presence of Glu [9].

These enzyme-based biosensors have many advantages for monitoring Glu in the brain. The biggest of these is their high temporal resolution. Biosensors can respond to changes in Glu concentration on time scales from 5–10 s. Additionally, due to their small diameter (10–100 μm), they can exhibit excellent spatial resolution in heterogenous tissue such as the brain [10]. The excellent temporal and spatial resolution offered by Glu biosensors makes them very useful for monitoring rapid, localized changes such as Glu release into the synaptic cleft [11]. However, these biosensors have the disadvantage that they can usually monitor only a single analyte at a time. This means that multiple biosensors with different biorecognition elements must be used to detect multiple species in a specific region of the brain.

1.2 Analysis of Complex Samples via Microchip Electrophoresis

Microchip electrophoresis (ME) is based on the same principles as capillary electrophoresis and, thus, separates compounds based on the ratio of their size and charge. This technique is therefore well suited for the analysis of small, charged molecules such as Glu. In contrast to biosensors, microchip electrophoresis permits the simultaneous analysis of multiple analytes in a single run.

Detection in ME can be accomplished using a variety of methods, the most common being electrochemistry, mass spectrometry,

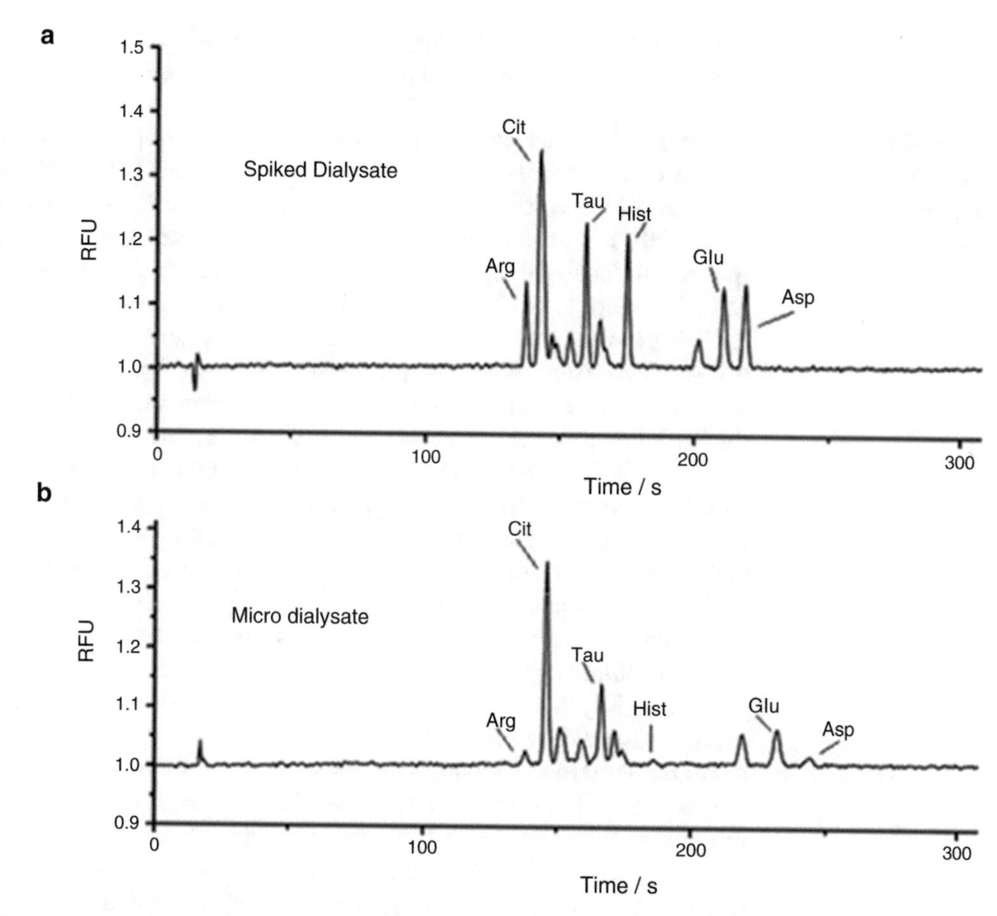

Fig. 1 Electropherograms of neuroactive NDA-derivatized amino acids obtained by microchip electrophoresis. (**a**) An electropherogram of an NDA/CN-derivatized MD sample spiked with 5 μmol/L arginine (Arg), citrulline (Cit), taurine (Tau), histamine (Hist), glutamate (Glu), and aspartate (Asp) for peak identification. Analysis performed offline using a 15-cm ME device. (**b**) An electropherogram of an NDA/CN-derivatized MD sample from the striatum of a male Sprague Dawley rat. Sample was collected at 1 μL/min and derivatized offline prior to separation. Reproduced with permission from Ref. 12

and fluorescence. Glutamate is most commonly measured using fluorescence detection following derivatization with a fluorescent (or fluorogenic) reagent. Examples of these reagents are discussed at length later in this chapter. Using microchip electrophoresis, it is possible to separate Glu from other important biogenic amines in a brain microdialysate sample. In particular, the detection of Glu in conjunction with aspartate (Asp) and γ-aminobutyric acid (GABA) can provide particular insight into the status of the brain tissue being sampled. Arginine and citrulline are indicators of nitric oxide synthase activity in the brain and can be measured under the same conditions. As Fig. 1 demonstrates, sample collection via microdialysis, derivatization, and separation using ME (all of which are described in detail later in this chapter) can provide a powerful view of multiple analytes within minutes of sample acquisition [12].

1.3 Sampling Considerations for Microchip Electrophoresis

Microdialysis sampling (MD) allows continuous sampling of both in vivo environments, such as the extracellular fluid (ECF) of the brain, and in vitro tissue cultures. Sampling is accomplished using a 100–500 μm probe containing a semi-porous membrane with a specific molecular weight cutoff that is inserted at the biological site of interest. A perfusate solution, similar in composition to the ECF, is pumped through the probe at flow rates generally between 0.1 and 1 μL/min. Since the perfusate lacks any of the analytes of interest, a concentration-based diffusion gradient is created across the probe membrane. Analytes smaller than the pore size, typically on the order of 20–60 kDa [13], diffuse through the membrane based on their concentration gradient into the perfusate, now referred to as the dialysate. Slower flow rates allow longer equilibration times and higher analyte recovery through the probe, but lead to much smaller sample volumes. Finally, the dialysate leaves the probe via the exit tubing [13]. Dialysate samples can then be analyzed offline or online for the compounds of interest.

Microdialysis is an ideal sampling method to couple to ME for several reasons. The first of these is that the sampling process exhibits no net fluid loss, allowing samples to be taken for extended periods of time [5]. Second, strategic selection of pore size can prevent interfering proteins and other large molecules from entering the perfusate, effectively eliminating the need for further sample preparation steps. Finally, as sampling is due entirely to analyte diffusion across a semipermeable membrane, virtually any small molecule can be sampled using the technique. However, this last point can also be problematic. Because sampling is diffusion limited, the ultimate concentration of analyte collected is dependent primarily on the flow rate of the perfusate [14]. While flow rates of around 1 μL/min are commonly used, only 30–40% of the analyte is recovered with such a high flow rate. Thus, to recover 100% of a target analyte, flow rates of approximately 100 nL/min are required [2].

Practically speaking, the trade-off between flow rate of the perfusate and analyte concentration in the dialysate, shown in Fig. 2, dictates that any subsequent assay must have a very low volume requirement, a very high mass sensitivity, or a low requirement for temporal resolution for analytes found at low concentrations in the ECF. An example of how perfusate flow rate can influence an assay can be envisioned by imagining a hypothetical assay requiring 4 μL of sample. If the microdialysis flow rate used to collect this sample is 1 μL/min, we can expect to recover only 30–40% of any target analyte present in the ECF due to the limited time available for analyte diffusion. If an analyte of interest—extracellular Glu at a concentration of 1 μmol/L for instance—is collected at this flow rate, we can therefore expect to only recover 300–400 nmol/L. Additionally, the

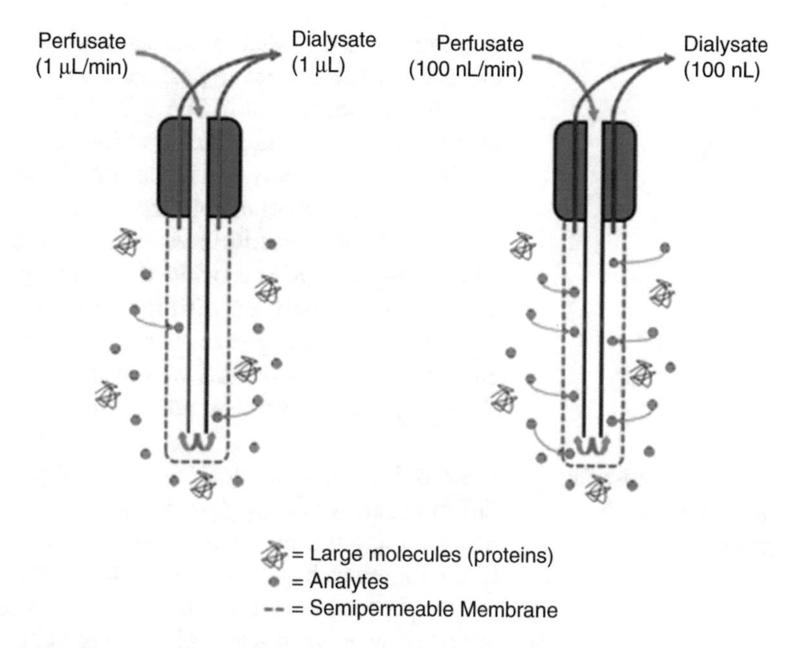

Fig. 2 Effect of microdialysis flow rate on analyte recovery. Schematic of a microdialysis brain probe with regard to flow-dependent recovery of an analyte. (**a**) At faster flow rates, more sample volume can be collected, but fewer analytes will diffuse across the semipermeable membrane. (**b**) At slower flow rates, more analytes can diffuse across the membrane, but at the cost of collecting much less sample. In both cases, compounds above the molecular weight cutoff of the membrane (such as proteins) are not collected

assay could only be performed 15 times per hour. Therefore, this flow rate has dictated both the temporal resolution and the minimum limits of detection our assay must have. Decreasing our flow rate to 100 nL/min would allow us to have higher limits of detection (with full recovery of 1 μmol/L), but at the expense of worsening our temporal resolution to 40 min per assay [15].

The continuous sampling nature of MD makes it an extremely useful tool to evaluate the health of patients following traumatic brain injury or other neurotrauma. An example of this can be found in the CMA Cerebral Tissue Monitoring System (CMA Microdialysis AB, Kista Sweden). The CMA system uses microdialysis to collect perfusate in microvials for offline analysis in a "microdialysis analyzer" such as the CMA 600, ISCUS, or ISCUS^{Flex} model [16]. These microdialysis analyzers are capable of monitoring multiple analytes including glucose, lactate, pyruvate, glycerol, urea, and Glu from multiple patients in an offline batch mode [17]. By collecting and analyzing samples in this manner, the ISCUS^{Flex} is able to make 30 discrete measurements per hour using sample volumes between 200 nL and 2 μL [18]. However, offline processing of samples in this manner can be problematic, as evaporation from such small sample volumes can skew analysis, even when the system is designed to minimize such evaporation. The direct coupling of MD to ME (MD-ME) avoids both of these problems by enabling rapid, continuous sample analysis.

The combination of MD and ME for continuous monitoring of drugs and neurotransmitters in vivo has recently been reviewed [19]. By taking continuous, online sampling into consideration when designing a microchip-based analysis system, nanoliters of sample can be analyzed with high mass sensitivity, following rapid separations and analysis of complex samples. This chapter will discuss the principles of ME with a focus on design considerations, along with the fabrication and use of ME devices.

1.4 An Introduction to Electrophoretic Separations

Electrophoresis is a separation technique whereby analytes with differing ratios of charge-to-hydrodynamic radius can be separated using an applied external electric field. The general technique of electrophoresis has been in use for many years [20–22]. Briefly, an electrophoretic separation begins with the application of an electric potential to a conductive electrolyte solution containing the analytes of interest. The application of this potential results in a force on all charged analytes toward the electrode of opposite charge, with positive analytes migrating toward the negatively charged cathode and negative analytes migrating toward the positively charged anode. The degree that an individual analyte responds to the electric field is known as its electrophoretic mobility.

A traditional application of electrophoresis is the separation of DNA following amplification using PCR. In this example, negatively charged DNA migrates toward a positive anode through a gel that acts as a sieve, slowing the migration of larger fragments. This sieve results in a characteristic *ladder* distribution [23]. However, the migration of charged analytes toward opposite electrodes can be problematic in cases where both positive and negatively charged analytes are of interest. In addition, low separation efficiencies and inadequate resolution make the use of slab gel electrophoresis impractical for small molecules.

1.5 Capillary Electrophoresis

Like gel electrophoresis, capillary electrophoresis (CE) functions via the application of an electric potential to a conductive background electrolyte (BGE). In this case, the BGE is contained within a long capillary tube, typically made of silica glass and optionally coated with a polyimide coating that provides additional robustness. The capillary can range in length from 20 cm to over 100 cm, with an internal diameter between 10 and 100 μm. The outer diameter of the capillary generally ranges from 200 to 375 μm. These dimensions have important repercussions that have resulted in vastly improved performance of this system over gel-based electrophoresis for small molecules. First, the high ratio of surface area to volume allows substantial amounts of heat to be

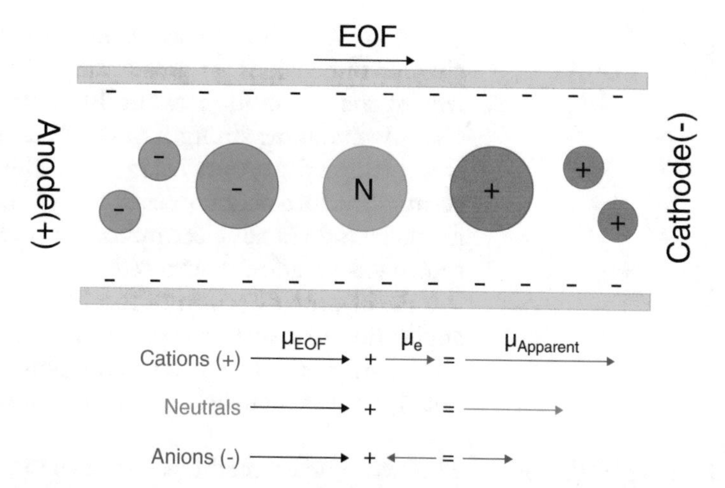

Fig. 3 Electroosmotic flow. A negative charge is generated on the surface of silica capillaries due to the use of a BGE with pH of 3 or greater. As an external voltage is applied to the channel, a bulk flow (μ_{EOF}) is created from anode (+) to cathode (−). Simultaneously, analytes begin to migrate toward their respective opposite charge (negatively charged cations toward the anode, positively charged anions toward the cathode). This electrophoretic mobility (μ_e) of each analyte is governed by its individual charge-to-mass ratio. The overall apparent mobility of each analyte is the resulting sum of μ_{EOF} and μ_e

dissipated. This in turn allows the use of higher potentials (typically on the order of several hundred volts per centimeter), thereby increasing separation efficiency. Secondly, the small internal dimensions limit sample volume requirements. Injection volumes for CE are typically on the order of 100 nL, although they can be larger if desired. Finally, the charged groups on the inner surface of the capillary can create an electrical double layer and cause the production of electroosmotic flow (EOF) when an electric potential is applied. This EOF makes it possible to detect all compounds, regardless of charge, at one end of the capillary.

1.6 Electroosmotic Flow

Electroosmotic flow (EOF), shown in Fig. 3, is generated in the capillary as a result of an electrical double layer that is formed on the surface of silica capillaries when using a BGE of pH 3 or greater. Above this pH, silanol groups on the surface of the capillary become ionized, resulting in a net negative surface charge. In response to the formation of this surface charge, a layer of cations from the BGE forms a compact layer of positive charges near the surface. This layer becomes more diffuse as the distance from the negative capillary surface increases. When a separation potential is applied to the capillary, these diffuse cations migrate toward the cathode, dragging the solvent with them. If the EOF is stronger than the attraction of the analytes of interest to the anode (electrophoretic mobility), then all analytes, regardless of charge, will migrate toward the cathode where the detector is placed.

Many variables can affect the formation of an EOF, including channel dimensions (at greater than ~200 μm, viscous forces overwhelm the net motion of the BGE and Joule heating increases), viscosity and ionic strength of the BGE, pH, and applied potential. It is through the proper balancing of these variables and those concerning the interaction of each analyte with the BGE and resulting in their individual electrophoretic mobilities (μ_e) that a well-resolved separation is achieved.

Finally, the direction of the EOF can be reversed through the application of a surface coating of positive charge and a reversal of the polarity of the applied electric potential. Operating in reversed polarity can be beneficial when separating some charged species.

1.7 Sample Injection

The two primary methods by which samples are injected into a CE or ME system are electrokinetic and hydrodynamic injections. Electrokinetic injection uses control of the EOF and electroosmotic mobility generated by an applied potential difference to inject sample into the capillary. To accomplish this, the capillary is placed into a sample and a potential is applied for a prescribed amount of time. Analytes are introduced into the capillary using the same principles of migration that are used during a separation. Due to the fact that analytes migrate at varying rates depending on their electrophoretic mobility, electrokinetic injection can therefore be biased toward smaller analytes and those having charges opposite to the polarity of the separation voltage. In the extreme case where an EOF is not present, this bias will result in only analytes of a single charge being injected.

By contrast, hydrodynamic injections exhibit no such bias. In this injection scheme, a defined volume of sample is injected into the capillary by first placing the capillary in the sample and then applying a pressure differential to the sample itself. The result is fluid displacement into the capillary. Following this injection, a separation potential is applied and electrophoretic separation proceeds normally.

1.8 Microchip Electrophoresis

Like CE, microchip electrophoresis (ME) has many benefits over more common separation methods such as liquid chromatography, including the ability to analyze extremely small sample volumes, high separation efficiencies, and rapid analysis times. In addition to these advantages, ME has several added benefits. The first of these is that ME devices have a much smaller footprint than existing CE systems. The decrease in length from a 75-cm channel in CE to a 5-cm channel, along with the smaller dimensions of the channels (15 μm deep by 50 μm wide), allows a fraction of the voltages to be used while preserving the 333 V/cm separation potential. The dimensions of the channels in a typical ME setup can be seen in Fig. 4. The decrease in voltage needed as well as in the associated instrumentation can lead to a much smaller overall system. The separation efficiency, which is

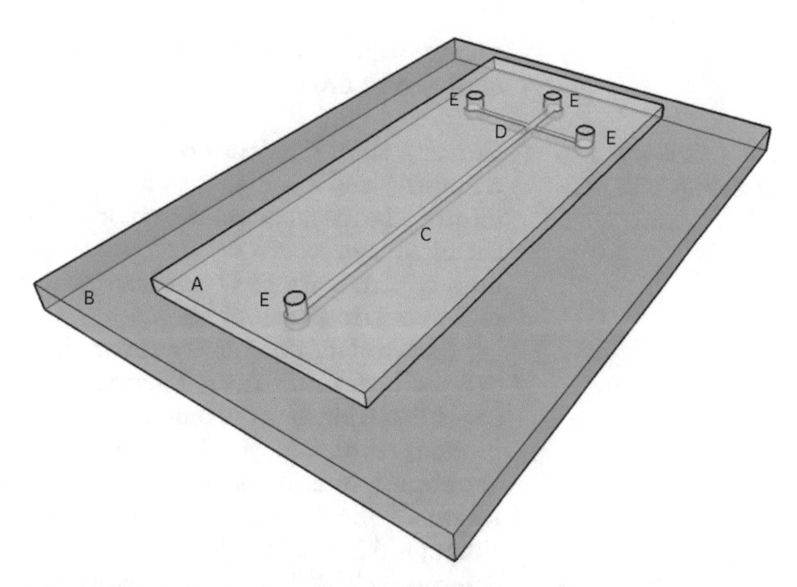

Fig. 4 Schematic of a microchip electrophoresis device. Schematic of a typical ME device with a simple-T design. The entire system is usually no bigger than 3 cm × 10 cm × 5 mm thick. (**a**) Channel substrate. The channel's dimensions can vary with application, but are usually on the order of 50 μm wide × 15 μm deep. (**b**) Base substrate. The channel substrate can be reversibly or irreversibly bound to the base, as discussed in Sect. 3. (**c**) Separation channel, which can vary in length from 2.5 to 15 cm depending on the device design. (**d**) Electrokinetic gate. (**e**) Sample and buffer reservoirs. The exact conditions to perform an electrokinetic injection on an ME device are discussed further in Fig. 9

a function of field strength, is preserved, resulting in decreased analysis times as analytes travel shorter distances to the detection point.

Another advantage of ME is that modern micro-manufacturing techniques such as photolithography, hot embossing, injection molding, and casting make it possible to integrate multiple complex features into a single device [24]. These features, which can include gating for the introduction of sample to the separation channel, tapered geometries to improve separation efficiencies, the addition of electrodes for electrochemical detection, and micromixers to facilitate on-chip derivatization of samples, add greatly to the utility of microchip systems. By integrating these functions directly into the microchip, it is also possible to radically decrease sample and reagent volume requirements. Finally, these techniques allow microchip designers to select materials based on properties such as optical clarity, surface chemistry, material expense, and ease of integration.

In light of the limitations placed on any analytical method by a sampling technique such as microdialysis, as mentioned in the previous section, ME represents an ideal method for analyzing small sample volumes and enabling high temporal resolution. To take full advantage of the ability to handle small volumes, a detection strategy with sufficiently low limits of detection (mass sensitivity) must also be used.

1.9 Microchip Detection Strategies

Virtually, all of the detection strategies used in modern analytical chemistry have been successfully integrated with the microchip format. The most common methods include fluorescence, electrochemical, and contactless conductivity detection. Additionally, mass spectrometry has been interfaced with ME using methods such as electrospray ionization.

As with CE, the most common detection methods used with ME are optical in nature. However, unlike CE, in which absorbance detection is very common, the thick planar substrates used in the construction of microchips can make it difficult to obtain absorbance measurements. Additionally, the short optical path length of the separation channel leads to a lack of sensitivity, although this can be mitigated by the use of a specially designed detection cell [25]. Electrochemical and conductivity detection are popular methods of detection in the microchip format due to the ease of integrating electrodes directly into the microchip substrate. However, fluorescence detection is more often used due to its generally low limits of detection and the fact that it does not suffer from limitations due to pathlength, and that it exhibits a lower background than other detection methods. The additional ease of focusing an excitation light source such as a laser on the ME channel results in it being commonly used with ME. However, because the majority of compounds and, specifically, Glu are not inherently fluorescent, it is often necessary to first derivatize a sample with a fluorophore prior to separation using ME.

In this chapter and, in particular, the next section, we will discuss the types of materials used in the construction of an ME device, the methods used, and the relative pros and cons of each. Due to the advantages mentioned above, our detection method of choice for Glu is fluorescence. While the methods discussed reflect this preference, the design and construction principles are broadly applicable to other detection strategies.

1.10 Derivatization for Fluorescence Detection

As mentioned previously, many analytes, such as Glu, are not natively fluorescent and, consequently, require a derivatization reaction to produce a fluorescently active product that can be detected. When choosing a fluorogenic compound (or fluorophore, more generally), several parameters must be taken into consideration. The first is the selectivity of the reagent for specific functional groups on the analyte of interest. In the case of amino acid analysis, the most commonly used reagents are selective for the primary amine group. A reagent containing a fluorophore can complicate analysis and require additional work to separate the fluorescent reagent from the analytes of interest. Ideally then, a reagent will be fluorescently active only following derivatization of

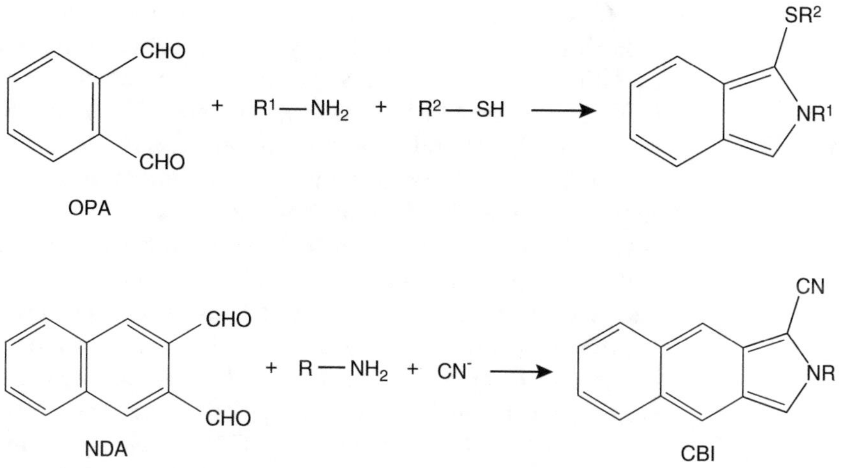

Fig. 5 NDA and OPA derivatization reactions

the target analyte. Another consideration when selecting a reagent is that of high quantum efficiency of the reaction product at the desired excitation wavelength. Finally, when performing online separations using ME or CE, the rate and yield of the derivatization reaction, and its reproducibility, are of utmost importance.

With these considerations in mind, two fluorogenic compounds that are typically used for the determination of amino acids are naphthalene-2,3-dicarboxaldehyde (NDA) and *o*-phthaldialdehyde (OPA). This is due to their rapid reaction times, selectivity for primary amine groups, and compatibility with commercially available lasers. These reagents are non-fluorescent prior to reacting with a primary amine, eliminating additional fluorescent species that could complicate analysis [26]. The NDA reaction occurs in the presence of CN and takes 120–240 s to complete, while the OPA reaction occurs in the presence of 2-mercaptoethanol (2-ME) and takes 10–30 s to complete [26, 27]. These reactions are shown in Fig. 5. It should be noted that NDA can also react with primary amines in the presence of thiols, such as 2-ME, and does so at a faster rate than it reacts with CN. However, the fluorescent products created via this reaction are less stable and have a lower fluorescence quantum efficiency than those created via the reaction with CN, undermining any gains to be had via the increased rate [28].

In the case of an online assay, where the sample is derivatized between the microdialysis probe and the electrophoretic separation, the derivatization reaction does not always go to completion. Due to their rapid reaction rates, both NDA and OPA can be used for online derivatization. However, the instability of OPA can complicate the separation. Consequently, in further discussions here, we focus on the use of NDA.

Care should be taken when using NDA (or OPA) for derivatization, as both CN and 2-ME are highly toxic. NDA is optimally dissolved in 100% acetonitrile (ACN) to the desired stock concentration [29]. NaCN should be dissolved in water. NDA and NaCN are both photolabile and should be stored in a refrigerator and protected from light and atmospheric conditions when not in use. Experience has shown that fresh stock solutions of NDA and NaCN should be made weekly. 2-ME is both flammable and toxic and should be handled accordingly. In addition, it is easily oxidized and should be stored in a sealed container. Finally, 2-ME is extremely noisome and should be used in a fume hood.

When using NDA (and OPA), the optimal derivatization for amino acids normally occurs at a pH that corresponds to the pK_a for the analyte amine group [26] . This is typically around pH 9.2 for amino acid analytes. Whether performing offline or online derivatization, the reagents used should be prepared in 10- to 100-fold excess of the estimated concentration of primary amines in the sample. An example derivatization procedure can be found in Sects. 3.8 and 3.9 of this chapter.

2 Materials

The construction and operation of ME devices, whether they are operated offline or online coupled to MD, is a multistep process involving photolithography, chemical etching, and high temperature bonding. These procedures are best performed in a cleanroom environment for the highest level of reproducibility, although it is possible to construct these devices in a laboratory setting if care is taken to maintain a clean environment. The materials, facilities, and instrumentation listed below describe what is used in our laboratory for the fabrication of glass and polydimethylsiloxane (PDMS) microfluidic devices. A detailed explanation of their use in the fabrication process is found in the methods section following this section.

2.1 Cleanroom Facilities and Instrumentation

ABM UV Flood Source and Mask Alignment System (ABM, Scotts Valley, CA).

Brewer Science Cee 200CBX Spin Coater (Brewer Science, Rolla, MO).

Harrick PDC-32G Plasma Cleaner (Harrick Plasma, Ithaca, NY).

BD-20 AC Laboratory Corona Treater (Electro-Technic Products, Chicago, IL).

Tencor Alpha-Step 200 Profilometer (KLA Tencor, Milpitas, CA).

Fisher Scientific Muffle Furnace (Fisher, Waltham, MA).

2.2 Derivatization Reagents

2-Mercaptoethanol (2-ME) (Fisher, Waltham, MA).

Sodium Cyanide (NaCN) (Fisher, Waltham, MA).

o-phthaldialdehyde (OPA) (Fisher, Waltham, MA).
Naphthalene-2,3-dicarboxaldehyde (NDA) (Fisher, Waltham, MA).

2.3 Online MD Coupling

Syringe Pump, Model CMA 100 (CMA Microdialysis, Holliston, MA).
1 mL Syringes (CMA Microdialysis, Holliston, MA).
Polyethylene Tubing, I.D. 0.39 mm, O.D. 1.09 mm (Fisher Scientific, Waltham, MA).
Bonded Port Connector C360-400 (Labsmith, Livermore, CA).

2.4 Glass-Glass Microchip Fabrication Materials

AZ1518 pre-coated glass substrates such as borosilicate (borofloat) or soda lime glass (Telic, Valencia, CA).
AZ1518 (AZ Electronic Materials Corp, Somerville, NJ).
MIF 300 AZ1518 developer (Emd Performance Materials, Somerville, NJ).
Acetone (Fisher, Waltham, MA).
Water.
Chrome Etchant, CR-75 (Cyantek, Fremont, CA).
HF Etchant solution consisting of: 20% HF, Nitric Acid, Water.
Ultra Pure Calcium Acetate (MP Biomolecules, Solon, OH).
Alconox™ Powdered Detergent (Fisher, Waltham, MA).
Diamond Drill Bits (Ukam Industrial Superhard Tools, Valencia, CA).
Programmable High Temperature Oven (Fisher, Waltham, MA).

2.5 PDMS Microchip Fabrication Materials

Silicon wafer (100-mm thickness) (University Wafer, Boston MA).
SU-8 10 photoresist (Microchem Corp, Flanders, MA).
SU-8 developer (Microchem Corp, Flanders, MA).
Isopropyl alcohol (Fisher, Waltham, MA).
Sylgard™ 184 Silicone Elastomer Base (Dow Corning, Auburn, MI).
Sylgard™ 184 Silicone Elastomer Curing Agent (Dow Corning, Auburn, MI).
Parafilm™ (Fisher, Waltham, MA).

3 Methods

3.1 Device Design

The development of a microfluidic devices begins with the design process. In designing an ME device, it is helpful to first determine what requirements will be placed on the function of the device itself. Four initial questions that must be asked:

1. Will the ME device interface directly to a microdialysis probe in an online manner or will samples be added to the device offline?
2. Will sample derivatization be performed on-chip or off?
3. How will sample be injected into the separation channel?
4. What length of separation channel will be required?

If sample acquisition is to be performed via online interface to microdialysis, pressure at the device inlet must be taken into account and channel widths adjusted accordingly. If derivatization will occur on-chip as well, multiple laminar flows must be mixed, either actively or passively, and the reactants must be given sufficient time following mixing for derivatization to occur. Sample injection, whether via pressure or performed electrokinetically, requires channel geometries that allow flow in one direction (down the separation channel) to be prevented.

Pressure injection is accomplished in ME devices by using additional PDMS membrane layers and the application of external positive and/or negative pressures to prevent fluid flow through channels. Unlike electrokinetic injection, which can result in biased injections of sample ions depending on their charges, pressure injection of sample is unbiased. However, due to the requirement of a flexible layer of PDMS, this approach is not typically used with all-glass ME devices. Providing a source of negative or positive pressure also further complicates the system design. For this reason, we focus primarily on the use of electrokinetic injection for all three types of ME devices described here.

When using electrokinetic injection, potentials are applied to each channel in order to establish a "gate," with longer channels requiring higher potentials. Additionally, asymmetrical geometries (such as having channels of varying lengths) can necessitate multiple high voltage supplies.

Finally, the length and geometry of the separation channel itself will directly determine many performance characteristics of the ME device, including peak resolution, EOF magnitude, and the separation voltage required. Separation channel widths greater than 200 μm should be avoided as viscous fluid properties begin to overwhelm those of the EOF beyond this point and Joule heating increases substantially. When comparing channel lengths, longer separation channels will result in greater separation efficiencies, assuming a comparable applied potential (measured in volts per centimeter). However, care should be taken when designing channels longer than 10 cm, since these necessitate serpentine channels that can result in a "racetrack" effect. This effect can be mitigated by tapering turns. Tapered turns themselves can be problematic, however, as the localized electric field strengths are increased. As the field strength increases, the current passing through that region of the channel also increases. This leads, in turn, to a resistive heating of the BGE termed Joule heating, which can bring about poor resolution. Despite the potential for this effect, optimal channel dimensions can greatly improve resolution when properly implemented for longer channel lengths if Joule heating is prevented [30, 31].

Modeling of ME devices using software such as Comsol® prior to any production steps can be very advantageous. However, modeling entire geometries can be computationally intensive and time consuming. Unless it is necessary to model all parts of the chip simultaneously, the best practice is to model specific regions, such as the electrokinetic gate and turns, in order to optimize those particular geometries. The design of mixing and derivatization geometries can also benefit from modeling the device prior to construction to ensure that the chosen combination of mixing geometry and sample flow rate will result in a fully derivatized sample prior to injection and separation. Designs must typically be exported into a computer-aided drafting (CAD) format prior to being made into a photomask, a process that may require additional review to ensure that the conversion of formats itself did not introduce errors.

Finally, printed photomasks must be of an extremely high resolution, and these are not attainable without the use of a specialized photomask printer. For that reason, professional services such as that provided by Infinite Graphics (Minneapolis, MN) are typically used. It is more cost-effective both in terms of printing photomasks and ultimately in terms of substrate usage to have multiple copies of devices on a single photomask. For reasons that will be explained in the following sections, these copies should be laid out in such a way that they can easily be separated into individual devices (i.e., easy to cut the glass substrate without scratching a neighboring device) while fitting the maximum number of devices onto a single glass substrate. It is important that the photoresist is removed along a line between the devices, due to the fact that glass cutters will not cut through photoresist or chrome layers. Finally, if holes are to be drilled in the device following development, the diameter of the drill bit must be taken into account by providing an additional margin to prevent the destruction of neighboring features.

3.2 Material Selection

Selecting a material with which to construct an ME device must begin with consideration of the detection method used. In the case of optical detection, the optical clarity of the material, including any autofluorescence of the material itself at the target wavelength, is of paramount concern. This can be of particular importance with plastic substrates. When using electrochemical or conductivity detection, the ability to integrate an electrode into the substrate must be considered. The surface chemistry of materials must also be taken into account, not only when considering the EOF generated by the electric field but also when determining whether or not analyte adsorption is likely to be a problem.

With this in mind, two materials have been widely evaluated and are commonly employed for ME: glass and PDMS. A third option, which will also be discussed, is to combine these two materials into a hybrid device in an effort to take advantage of aspects of each. These three options are by no means an exhaustive list of substrates for ME construction. Many other polymers, including poly(methyl methacrylate) (PMMA) and cyclic olefin copolymer (COC), are commonly used in device manufacture, each with their individual pros and cons. This chapter is focused on the use of glass, PDMS, and PDMS-glass hybrid devices because these substrates represent ideal material characteristics (glass) and ideal ease of use (PDMS). Their individual properties are described in the following sections.

3.3 Glass ME Devices

In many ways, glass is an ideal substrate for an ME device. Its optical clarity is superb over a range of wavelengths, its surface chemistry is stable over a range of pH values resulting in high reproducibility, and it has the ability to withstand high voltages (and subsequent Joule heating) associated with electrophoretic separations [32]. Additionally, because CE capillaries are typically glass, separation conditions can easily be transferred from CE to ME and vice versa.

The construction of glass-glass ME devices begins with the use of photolithography to transfer a device pattern to a borosilicate glass substrate pre-coated with chrome and photoresist. While it is possible to coat glass substrates manually, pre-coated substrates purchased from a company such as Telic (Valencia, CA) are inexpensive and consistent in performance. Following the transfer of the pattern, a multistep process to remove sections of the pattern followed by etching with a hydrofluoric acid solution is used to create the channels in the bottom glass substrate. Finally, holes are drilled, the glass is cut to size, and a two-step bonding process is used to permanently bond a blank glass piece to the etched glass. The result is an ME device that can withstand both the pressures of a microdialysis flow as well as the high voltages necessary for electrophoresis.

The following quick reference guide is a step-by-step explanation of the steps necessary to construct a glass-glass ME device. A schematic of this process is also shown in Fig. 6.

3.3.1 Quick Reference to Glass-Glass ME Device Construction

Photolithography (Fig. 6a and b)

1. Pre-bake glass substrate, photoresist side up, on a hotplate at 100 °C for 2 min.

   Note: The prebake time will vary according to the photoresist type and thickness. Consult the photoresist documentation for more information.

2. Allow glass to cool to room temperature before proceeding.

3. Place photomask on glass substrate such that the printed side of the photomask is in contact with the photoresist. Apply a vacuum to ensure minimal distance between mask and photoresist.

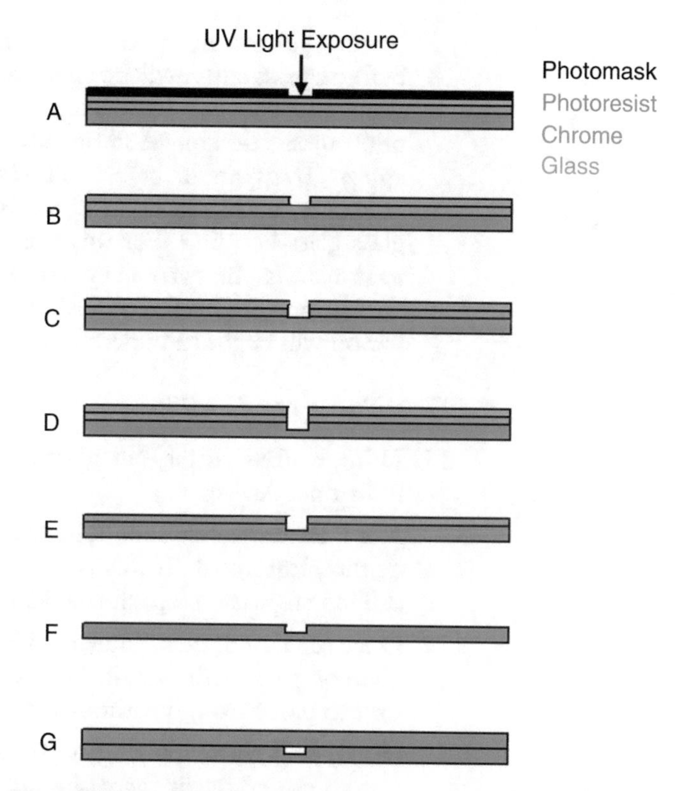

Fig. 6 Construction of an all glass microchip electrophoresis device. Schematic outline of glass microchip fabrication. (**a**) Align photomask on coated glass and expose to UV light. (**b**) Remove photomask and soak in developer to remove exposed photoresist. (**c**) Remove exposed chrome using chrome etchant. (**d**) Etch exposed glass with hydrofluoric acid, then confirm channel depth with profilometer. (**e**) Remove remaining photoresist with acetone. (**f**) Remove remaining chrome with chrome etchant. (**g**) Bond to another piece of unmodified glass to form complete chip

4. While maintaining a vacuum, expose the photoresist to UV through the photomask. Duration of exposure will depend on resist type, thickness, and UV intensity.
5. Place exposed glass substrate in developer to remove exposed area.
6. Wash with water and dry gently with compressed nitrogen or air.
7. Place on hotplate at 100 °C for 10 min to fully evaporate developer.

Chrome Removal (Fig. 6c)

8. Place wafer in a Pyrex™ dish of chrome etchant.
9. Agitate gently until chrome regions are clear +45 s.
10. Rinse with water and dry using compressed nitrogen or air.

Note: Chrome and photoresist must be removed from glass before a glass cutter will be effective. If making multiple devices simultaneously, it is important to include these cut lines in the photomask design. Additionally, chrome-coated glass will appear transparent before all chrome has been removed. However, a thin layer of chrome that remains will prevent glass-glass bonding. For this reason, allow the glass substrate to remain in the chrome etchant for 45 s *after* it appears that the chrome has been removed. Additional etching beyond this period will begin to undercut features.

Glass Cutting and Drilling

11. Using a glass cutter, cut glass along exposed paths between individual devices.

    Note: Extreme care should be taken not to scratch the surface of the photoresist. If this is a problem, the glass cutting and drilling steps can be performed after the HF wet etching.

12. Once separated, place individual ME devices in approximately 1 cm of water with a sacrificial layer under the chip to allow the drill to pass through without being damaged.

13. Slowly drill glass, applying constant pressure. It may be necessary to use gradually increasing drill bit sizes.

    Note: More ME devices are destroyed in this step than any other. Using new diamond-coated bits can help prevent glass breakage. When choosing drill bit sizes, two things to consider are whether or not MD access ports will be glued into place over the holes and evaporation during operation. If the holes drilled in the glass to create sample and buffer wells are too small, these solutions will evaporate rapidly during operation. As this evaporation proceeds, ionic strength will change. This in turn will cause additional current draw and, consequently, alter migration times. If MD access ports are to be placed over the holes, the holes must of course be smaller than the ports themselves.

HF Wet Etching (Fig. 6d)

14. Place glass with photoresist/chrome side up in HF etchant and agitate for duration of etching process. Etching time will depend on desired depth as well as glass type.

15. Remove from HF etchant, rinse with a solution of calcium acetate to deactivate the HF etchant. Following this, rinse with water and dry using compressed nitrogen or air.

16. Measure channel depth using a profilometer. If the desired depth has not been achieved, calculate the etch rate and place the device back in the HF for the necessary amount of time.

Note: Due to the extreme toxicity of HF, the best practice is to minimize the number of times the device is taken in and out of the HF etchant if at all possible. When handling the substrate, the HF remaining on the device should first be deactivated by rinsing with a slurry of calcium acetate in water, followed by rinsing with water.

Removal of Remaining Photoresist and Chrome (Fig. 6e and f)

17. Remove all remaining photoresist by rinsing with acetone.
18. Rinse with water and dry with compressed nitrogen or air.
19. Place the glass substrate in chrome etchant, agitating gently, until all remaining chrome has been removed +1 min.

    Note: The removal of all photoresist and chrome is necessary to facilitate a glass-glass bond.

Bonding Glass-Glass ME Devices (Fig. 6g)

20. Cut blank glass to the necessary size.

    Note: To avoid breakage due to differences in thermal expansion, blanks should be of the same glass type as the etched device.
21. Clean both glass pieces using Alconox™ mixed to the manufacturer's specifications.
22. Rinse both sides well with deionized water (DI) water.
23. Incubate the glass surfaces to be bonded using a solution of 0.5% (w/v) calcium acetate, approximately 30 mmol/L, and 0.5% (w/v) Alconox™ in DI water for no less than 30 s.
24. Rinse thoroughly with DI water.
25. Under running DI water, bring glass pieces together from bottom to top.

    Note: Force out any remaining bubbles but allow a layer of water to remain.
26. Place the device in an oven for 1 h at 60 °C.

    Note: Do NOT force all water from between the layers. The evaporation from the oven will force the glass together.
27. At the 1-h mark, inspect the device for Newton rings indicating incomplete bonding of the surfaces. If found, separate using a razor blade and repeat steps 21–26.
28. Place binding clips around the device to apply additional pressure. Increase the temperature to 105 °C for 2 h.
29. Inspect device. If Newton rings have formed, separate using a razor blade and repeat steps 21–28.
30. If no Newton rings are found, execute the following program, slowly increasing the oven temperature to 630 °C and holding

Table 1
Oven program to bond a glass-glass microchip device

Elapsed time (h)	Time for step (h)	Oven temp (°C)	Rate (°C/min)
0.00	0.00	25	0.0
2.97	2.97	560	3.0
3.30	0.33	630	3.5
5.80	2.50	630	0.0
6.16	0.36	565	−3.0
6.77	0.61	510	−1.5
7.27	0.50	510	0.0
8.77	1.50	465	−0.5
10.24	1.47	25	−5.0

before cooling gradually to room temperature. Steps shown in Table 1.

31. If the ME device is meant to be used online (MD-ME), access ports must be glued to the chip to allow the flow from the syringe pump to enter the device. Polyethelene tubing connecting the syringe pump can be connected directly to ports glued to the surface of the device. To glue ports in place, a strong UV source such as the ELC-450 (Electrolite Corp, Bethel, CT) should be used. UV glue (Fisher Scientific, Waltham MA) should be applied carefully to the ports while holding them in place.

    Note: Should glue enter the channels, flush with water and methanol until the glue has been cleared. Avoid UV exposure during this period if possible.

3.4 Polydimethylsiloxane (PMDS) ME Devices

Glass microchips permit highly reproducible separations and can last for extended periods of time if properly cared for. However, the fabrication process is lengthy and difficult and requires the use of dangerous and expensive materials. As these chips are irreversibly bonded, if there are any issues with the construction of the separation channel or solution reservoirs (such as clogging or scratches), they must be disposed of and replaced. This makes all-glass microchips a poor choice for prototyping methods, in which multiple chip designs are studied before settling on a final pattern.

Polymeric substrates are often used for device development to avoid these complications and allow faster prototyping. A wide variety of polymeric substrates are used for microchip electrophoresis; however, PDMS is by far the most common and will be the

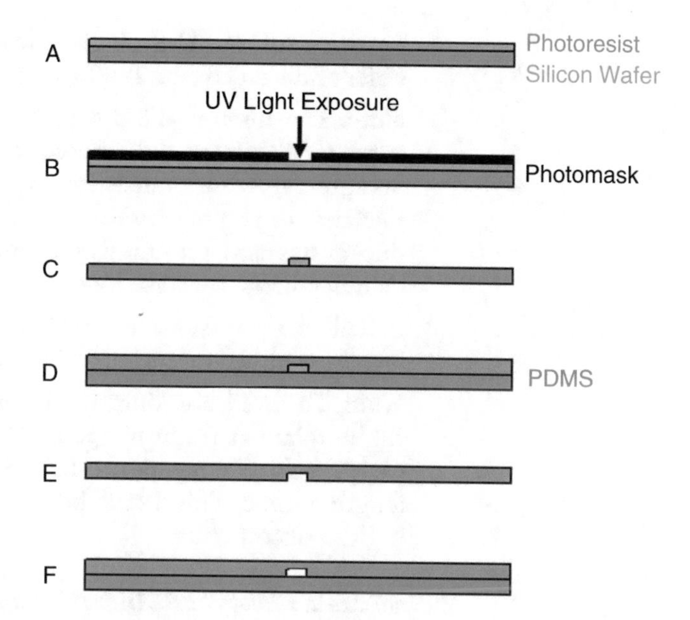

Fig. 7 Construction of a PDMS-PDMS microchip electrophoresis device. Schematic of PDMS-PDMS or hybrid microchip fabrication. (**a**) Spin-coat silicon wafer with negative photoresist. (**b**) Align photomask on coated wafer and expose to UV light. (**c**) Use developer to remove unexposed photoresist, then post-bake and confirm microfeature dimensions with profilometer. (**d**) Pour PDMS over master and cure in oven for at least 2 h. (**e**) Peel PDMS off of master and cover with Parafilm™ if not bonding immediately. (**f**) Punch holes for buffer reservoirs in PDMS, remove Parafilm™, if necessary, and bond to substrate of choice (PDMS or glass)

only polymer described here. Its elastomeric properties (including the ability to seal to surfaces without distorting), optical transparency, low cost, and ease of fabrication make it an excellent candidate for microchip fabrication [33]. After a master mold with the desired microfeatures is made, multiple PDMS chips can be prepared using the same master for extended periods. A schematic of this process is shown in Fig. 7.

PDMS has only a very small surface charge, which means an EOF will not occur in PDMS channels. To achieve a separation, additives that will adhere to the walls and lend it their charge must be included in the BGE. The additive used depends on whether a normal or reverse polarity is applied. A common additive for normal polarity is sodium dodecyl sulfate (SDS), while a common additive for reverse polarity is tetradecyltrimethylammonium bromide (TTAB).

3.4.1 Quick Reference to PDMS-PDMS ME Device Construction

Silicon Master Spin-Coating (Fig. 7a)

1. Use nitrogen flow to clean surface of silicon wafer.
2. Center silicon wafer on spin coater platform.

3. Pipette 4 mL of SU-8 10 photoresist onto the exact center of wafer and spin coat at 2000 rpm until fully coated.

   Note: The amount of SU-8 10 photoresist used directly influences its thickness and, subsequently, the final depth of the microfeatures. The values stated in these steps are for a final thickness of 15 μm with SU-8 10. Values associated with other thicknesses and photoresists should be calculated based on manufacturers' specifications.

4. Prebake silicon wafer at 65 °C for 2 min on hotplate, then increase temperature to 95 °C for 5 min.

   Note: To avoid breaking the silicon wafer, it is better to start the hotplate at room temperature and gradually ramp up to each specified temperature before heating it for the desired length of time. This should be done in any step requiring heating the silicon wafer.

Silicon Master Photomask Alignment and UV Exposure (Fig. 7b)

5. Align negative photomask on top of wafer on photolithography platform; use vacuum to hold substrate and mask in place.

   Note: When using a printed photomask, make sure that the printed side is facing down (flush with the substrate). This will prevent any diffusion of light across the width of the transparency and avoid unwanted enlargement of the features. Make sure the silicon wafer has cooled to room temperature before aligning the photomask.

6. Expose to UV light for 16 s.

   Note: Full UV exposure of SU-8 is dependent on the substrate thickness and light source intensity. In this case, a flood-source with output intensity of 21 mW/cm^2 is used.

Silicon Master Post-Exposure Processing and Profiling (Fig. 7c)

7. Post-bake silicon wafer at 65 °C for 1 min on hotplate, then increase temperature to 95 °C for 2 min.
8. Place developed wafer in SU-8 developer; gently swirl for 2 min to remove undeveloped photoresist.
9. Remove master from developer; wash with isopropyl alcohol and dry with nitrogen flow.

   Note: Acetone will cause developed SU-8 to dissolve. Only isopropyl alcohol should be used for cleaning silicon masters using this photoresist.

10. Hardbake at 200 °C for 2 h.
11. Use a profilometer to confirm the dimensions of the microfeatures on the master.

PDMS Curing (Fig. 7d and e)

12. Measure 12 g total of a 9:1 ratio of Sylgard™ 184 elastomer (10.4 g) and curing agent (1.6 g) into a small cup; mix thoroughly.

    Note: This amount of polymer was chosen because the surface tension will be sufficient to hold the mixture on the master's surface. Greater amounts can be used, but a mold will be necessary to hold the polymer on the wafer's surface. Using 12 g will result in a final thickness of approximately 1.2 mm.

    Note: Different ratios of elastomer to curing agent can be used in order to modify the elasticity of the PDMS. In general, the more curing agent used, the less elastic is the final polymeric substrate.

13. Place cup inside vacuum desiccator and remove all gas bubbles via repeated application of vacuum, being careful not to spill polymer over the edges of the container.
14. Pour mixture onto master and allow it to settle to a uniform thickness; place in oven and heat at 70 °C for at least 2 h.

    Note: Make sure that the master is level once it is placed in the oven. If it is placed at an angle, the PDMS will cure unevenly and potentially be unusable.

15. Use a razor blade to scrape the PDMS off of the edges of the master, then slowly peel off the polymer layer, being careful not to stretch or rip the substrate.
16. Either bond immediately (see next section) or place the PDMS on a piece of Parafilm™, channel-side down; fold Parafilm™ over the top of PDMS to fully seal it.

    Note: If protected from air exposure, PDMS can safely be sealed within Parafilm™ for up to 1 week. If cracks are visible in the microchannels when examined using a microscope, the polymer has dried out and should not be used.

PDMS Hole Punching and Bonding (Fig. 7f)

17. Use a PDMS hole puncher to remove PDMS at reservoirs and create wells, using a Kimwipe to clean the puncher surface between each use.
18. Peel Parafilm™ off PDMS.
19. Place the PDMS with the channels facing up next to another piece of PDMS.

    Note: Although the parafilm should keep any dust or other detritus from reaching the PDMS surface, it can still happen. This is most easily avoided by bonding the PDMS in a clean-room environment, but a piece of clear tape can be used to remove any dust from the surface immediately before step 4.

20. Run a handheld plasma oxidizer over the surface with the electrode a centimeter above the polymer surface, spending around 30 s on each piece of PDMS [34].

    Note: A conventional plasma oxidizer can also be used.

21. Firmly press the substrates together, taking care to avoid touching the oxidized surface; run fingers over the PDMS to force out any air bubbles that form between the layers.

22. Apply pressure with fingertips around the edges of the microchip until sealed, usually within 1 or 2 min.

    Note: Applying pressure directly over the channels can result in forcing them closed and ruining the device.

23. Examine device using a microscope to guarantee that the microfeatures are preserved and that no clogs are present.

3.5 Hybrid ME Devices

Despite the ease of fabricating PDMS microchips, they suffer in terms of longevity and reproducibility. A PDMS chip can often be used for only a 6–8 h period before a new one must be made. This is especially true if using high field strengths, which can cause the PDMS polymer to degrade or burn. This results in many different chips being required for even the simplest of studies, which can produce slight changes in separation efficiency and migration times for the analytes of interest. This variability in chips is further exacerbated because PDMS is not natively charged, which means additives must be added to the buffer in order to establish an EOF.

One solution to these reproducibility issues is the fabrication of PDMS-glass hybrid microchips. In this design, the microfeatures are prepared in either PDMS or glass; they can then be bonded to the opposite substrate to create the completed device. This method combines the strengths of glass and polymeric devices. The inclusion of glass as a substrate leads to more reproducible EOFs, while the inclusion of PDMS still allows rapid fabrication and easy cleaning of clogs [35].

When making a hybrid device, the main consideration is which substrate will contain the separation channel. This depends on the analytical needs of your method. If a reproducible EOF is required or higher field strengths are a necessity, having three walls of the channel made from glass will result in better separations. However, if there are no available facilities for etching and drilling glass or rapid prototyping of the channel design is required, a device with three walls of PDMS can be used.

The previous sections outlined methods of preparing microchannels in both glass (Sect. 3.3.) and PDMS (Sect. 3.4). These methods are unchanged; the only difference in constructing a hybrid device is the method of bonding. For most purposes, reversible bonding will suffice, as it is much faster and is compatible with most separations. Irreversible bonding is required only when there are concerns regarding the seal of the substrates, such as when incorporating a pressure-driven flow as the sample source.

Reversible Bonding

1. Wipe down the glass substrate with a Kimwipe soaked in 50% isopropyl alcohol (IPA), then dry with lint-free paper and nitrogen flow.
2. If necessary, use a PDMS puncher to remove PDMS at reservoirs and create wells, using a Kimwipe to clean the puncher surface between each use.

   Note: This step is not needed if wells have been drilled in the glass substrate.
3. Peel Parafilm™ off PDMS.

   Note: If bonding in a cleanroom facility, there should be no dust or other detritus on the substrate surface. A piece of clear tape can be used to remove any particulates on both glass and PDMS if necessary.
4. Slowly place PDMS on glass substrate, channels facing inwards; run fingers over the PDMS to force out any air bubbles that form between the layers.
5. Examine device using a microscope to guarantee the microfeatures are preserved and no clogs are present.

   Note: If there is particulate in the channels or improper sealing, the PDMS can be peeled off the glass and reapplied after using tape to clean both substrate surfaces.

Irreversible Bonding

1. Wipe down the glass substrate with a Kimwipe soaked in 50% IPA, then dry with lint-free paper and nitrogen flow.
2. If necessary, use a PDMS puncher to remove PDMS at reservoirs and create wells, using a Kimwipe to clean the puncher surface between each use.

   Note: This step is not needed if wells have been drilled in the glass substrate.
3. Peel Parafilm™ off PDMS.

   Note: If bonding in a cleanroom facility, there should be no dust or other detritus on the substrate surface. A piece of clear tape can be used to remove any particulates on both glass and PDMS if necessary.
4. Run a handheld plasma oxidizer (BD-20 AC, Electro-Technic Products, Chicago IL) over the surface with the electrode a centimeter above the polymer surface, spending around 30 s on each piece of PDMS.

Note: A conventional plasma oxidizer (Harrick Scientific, Ithaca NY) can also be used. It is not recommended to oxidize PDMS for more than 2 min, as this can result in degradation of the surface and potential damage to any microfeatures or improper bonding.

5. Firmly press the substrates together, taking care to avoid touching the oxidized surface; run fingers over the PDMS to force out any air bubbles that form between the layers.
6. Apply pressure with fingertips around the edges of the microchip until sealed, usually within one or 2 min.

   Note: Applying pressure directly over the channels can result in forcing them closed and ruining the device.
7. Examine device using a microscope to guarantee that the microfeatures are preserved and no clogs are present.

   Note: If there is particulate in the channels or improper sealing, the PDMS must be removed by scraping it off with a razor blade. The glass substrate should be washed with IPA and acetone to clean away any residual PDMS before attempting a second bonding.

3.6 Device Use

As mentioned in the previous sections, the detection of Glu using ME can be performed either in an offline or online fashion. These two methods of operation differ primarily in terms of sample acquisition, derivatization and, in the case of online, the added complexity of integrating fluid handling into the ME device. Consequently, the following sections are divided into offline sample preparation and handling, online sample preparation and handling, and, finally, the sample injection, detection, and separation steps common to both. In either case however, as noted previously, ME devices have exceedingly small internal dimensions, on the order of 15 μm deep by 50 μm wide, making it easy for particles to block channels. This is of less concern for online microdialysis samples as the process of microdialysis itself prevents larger molecules from entering the dialysate. Samples, stock solutions, and any other solution used in the device including cleaning solutions should be filtered prior to use using a 0.22 μm filter. Additionally, care should be taken to prevent dust from collecting in the channel inlets of glass-glass devices between use. Two common methods for preventing dust accumulation are wrapping the chip in Parafilm™ and storing the chip under water. The latter is a popular method used to maintain the chip overnight while the former is used for long-term storage. New PDMS-PDMS and PDMS/glass hybrid devices should be made between uses, preventing this problem.

3.7 Cleaning and Conditioning ME Devices

3.7.1 Glass-Glass ME Devices

Prior to each use of a glass-glass ME device, the channels must be cleaned and conditioned (referred to from this point onward simply as conditioning) using a sequence of steps designed to remove any of a series of surface contaminants, finishing with the application of sodium hydroxide (NaOH) to leave the separation channel with a negatively charged surface. The following sequence should be applied for each full conditioning. For each step, the solution should be added to all wells but one, allowing the application of negative pressure at that well to pull the solutions through. Negative pressure can be generated using a pump or a vacuum aspirator. The sequence for full conditioning:

1. Deionized, Millipore™ filtered water for >5 min
2. 0.1 mol/L HCl for >5 min
3. Deionized, Millipore™ filtered water for >5 min
4. 0.1 mol/L NaOH for >5 min
5. Deionized, Millipore™ filtered water for >5 min
6. Background electrolyte (BGE) until channels are filled completely

Between individual uses of the device, it may be possible to skip steps 2 and 3. Care should be taken not to add acids or bases to the device without first cleaning with water as rapid changes in pH can result in precipitation within the channel. Additional information regarding chip cleaning can be found in Sect. 4.5 at the end of this chapter.

3.7.2 PDMS-PDMS and PDMS/Glass Hybrid ME Devices

Conditioning a PDMS-PDMS or PDMS/glass hybrid device requires a different procedure, primarily because using strong acids can lead to delamination of the PDMS. Therefore, a full conditioning for these devices consists of only the following sequence of flushing the chip:

1. Deionized, Millipore™ filtered water for >5 min
2. 0.1 mol/L NaOH for >5 min
3. Deionized, Millipore™ filtered water for >5 min
4. Background electrolyte (BGE) (fill channels with BGE)

3.8 Offline Sample Preparation

When using fluorescence detection, offline samples must first be derivatized before separation and detection using ME. As mentioned in the derivatization section, a variety of fluorophores and fluorogenic compounds are available for this purpose. While NDA/CN and OPA/2-ME are popular due to their rapid reaction rate and ease of use for online derivatization, offline sample preparation allows the researcher more latitude with regard to derivatization. Selectivity and quantum efficiency should be the chief concerns in this case, and selection of a fluorophore/fluorogenic compound should be made to maximize both. That said, instability of the products of OPA derivatization can complicate offline analysis. For this reason, we recommend using NDA for offline derivatization.

NDA stock should be prepared in ACN on a weekly basis and stored in a refrigerator, protected from light exposure and atmospheric oxidation. NaCN should be made in water on a weekly basis and stored under the same conditions as NDA.

Offline sample derivatization should be performed with an excess of both NDA and NaCN compared to the analyte(s) of interest. A feature of using these compounds is that they are fluorogenic, that is, not fluorescently active prior to reacting with a primary amine. Fluorescent reagents that are not fluorogenic require better separations for the parent compound and side products that complicate analysis. However, care should be taken not to use too high a concentration of fluorogenic reagents since fluorescent side products can form. Typical reaction conditions for analyte concentrations ranging from 1 to 100 μmol/L include equal volumes of NDA and NaCN at stock concentrations of 5 and 10 mmol/L, respectively. An example of the derivatization of a solution containing 100 μmol/L Glu and 100 μmol/L Asp might be as follows:

Stock Concentrations:

Glu stock concentration: 2 mmol/L in H_2O.

Asp stock concentration: 2 mmol/L in H_2O.

NDA stock concentration (in 100% ACN): 5 mmol/L.

NaCN stock concentration (in 100% Water): 10 mmol/L.

Tetraborate buffer (BGE) concentration at pH 9.2: 15 mmol/L.

Volumes Used for Final Volume of 200 μL:

Glu: 10 μL for final concentration of 100 μmol/L.

Asp: 10 μL for final concentration of 100 μmol/L.

NDA: 2× the volume of all amino acids or 20 μL for a final concentration of 500 μmol/L.

NaCN: again, 2× the volume of all amino acids or 20 μL. Final concentration is 1 mmol/L.

BGE volume: 140 μL.

Therefore, the limiting reagent in this hypothetical sample is the amino acids, *not* the derivatization agent. As mentioned above, high concentrations of NDA and NaCN alone can result in the formation of fluorescent side products. However, these are usually very low in concentration and do not typically complicate the assay.

One thing to note regarding this derivatization procedure is the final concentration of ACN in the sample. A total of 20 μL of ACN was added, resulting in a sample that is 10% ACN v/v. Experience has shown that, due to the hydrophobicity of NDA, injecting a sample containing NDA dissolved in ACN into a channel containing only borate buffer (for instance) will result in NDA precipitating out of solution and forming a clog. To prevent this from occurring, ACN or another similar hydrophobic solvent must be added to the BGE at a percentage greater than approximately 3%. Should a clog form, refer to Sect. 4.5 for more information.

3.9 Online Derivatization

The same principles are maintained for online derivatization as in offline. The differences between the two arise from differences in fluid mixing at the microscale and in the fact that reactions typically do not proceed to completion. Mixing of laminar flows commonly found in microfluidic devices is beyond the scope of this chapter. However, broadly speaking, the design of the MD-ME interface should take into account mixing architectures (passive or active) to ensure full mixing of sample inlet and derivatization agents.

As mentioned, NDA/CN and OPA/2-ME are used for many MD-ME systems because of their rapid derivatization reactions and the fact that they are fluorogenic. However, even these reagents can require several minutes to fully derivatize a sample. Consequently, when performing derivatization online, much lower levels of fluorescent product are likely to be created prior to detection, leading to relatively high limits of detection.

3.10 System Setup

Just as ME designs vary, the orientation of the supporting equipment (power supplies, detection equipment, associated optics, etc.) is often specific to the application or laboratory. Therefore, the description we offer here represents only one possible implementation. We have found this implementation useful for a variety of ME devices and applications, including both online and offline analysis. A schematic diagram of the general system setup can be seen in Fig. 8.

3.10.1 Materials List

Nikon Eclipse Ti Microscope.
Stanford Research Labs Model SR570 Preamplifier.
Hamamatsu PMT Model R1527 with Pacific Instruments Model 227 PMT Power Supply.
National Instruments USB-6229.
Ultravolt Model HVRack-4-250 High Voltage Power Supply.
Omicron PhoxX 445 nm Diode Laser.

3.10.2 Optical System

Briefly, a Nikon Eclipse Ti microscope is used as a stage upon which an ME device is placed. A 445 nm laser, specific for NDA excitation, is fired into a fiber optic cable, which is subsequently interfaced to the Nikon Eclipse Ti microscope. The incoming laser

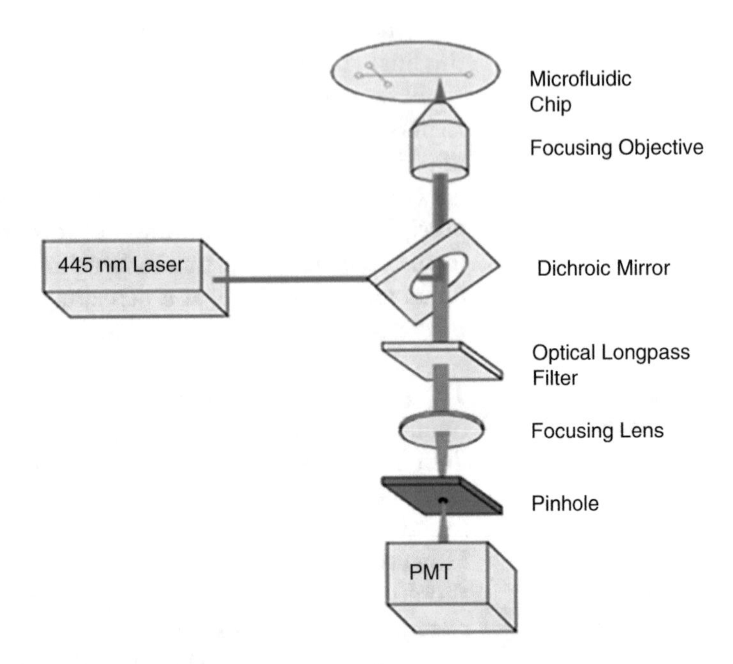

Fig. 8 Schematic overview of a microchip electrophoresis system with laser-induced fluorescence detection. An excitation light source, here at 445 nm laser, is focused such that its output is reflected upward by a dichroic mirror mounted at 45°. The ME device is mounted so that the excitation light intersects the separation channel near the end of the channel. As fluorescent analytes pass through the beam, emitted light passes through the dichroic mirror due to its longer wavelength. Continuing along this path, an optical longpass filter is used to remove residual excitation light before the remaining light is focused onto a photomultiplier tube (PMT) for amplification and conversion to an electrical signal

light passes through an excitation optical filter, attenuating any wavelengths of light other than 445 nm before being reflected upward using a dichroic mirror mounted at 45°. This dichroic mirror is chosen specifically to reflect wavelengths lower than that of the emitted light from the target fluorophore. The result is laser light reflected upward through a focusing objective lens, and intersecting the end of the ME device's separation channel, which has been placed on the microscope stage. As NDA-tagged analytes, such as Glu, migrate toward the end of the separation channel, they pass through the beam of 445 nm laser light. Encountering this results in fluorescent excitation and emission of light at approximately 490 nm. The emission of this light occurs in all directions, including back down through the focusing optics. As this emitted light continues along its path, it passes through the dichroic mirror due to its longer wavelength. A final emission filter is used to remove any light having a wavelength less than a cutoff of ~490 nm (chosen to prevent excitation laser light from striking the detector).

3.10.3 Signal Detection and Amplification

A photomultiplier tube (PMT) with an applied bias voltage of 1000 V was used to detect the light emitted by the analytes, converting the light into an electrical current. The current produced by the PMT during this process is typically on the order of 1 nA to 1 μA and requires subsequent amplification before being converted to a digital signal for analysis. This amplification step is performed by a current-to-voltage amplifier (Model SR570). This particular model of amplifier allows the user to specify a low noise, high gain setting, which we typically set to a gain of 200 μA/V, and a low pass filter set to attenuate high frequency noise over 3 Hz.

Finally, following amplification, a National Instruments data acquisition system, model USB-6229, was used to convert the analog output of the amplifier into a digital signal for display and analysis using custom Labview-based software.

3.10.4 Injection and Separation

As mentioned in the previous sections, electrokinetic injection is our preferred method of sample injection due to the relative simplicity of the system. That said, the coordinated use of multiple high voltage (HV) potentials in a small area while maintaining optical alignment of microscale features can be nontrivial. Safety is a chief concern when using HV power supplies. Additionally, a HV supply should be chosen to have high stability over time, digital control allowing potentials to be toggled on and off quickly, short circuit protection to protect both user and supply, and feedback to the user regarding the actual voltage and current output of the system. Feedback to the user is incredibly useful in diagnosing problems and is discussed in Sect. 4. With these requirements in mind, we typically use Ultravolt power supplies such as the HVRack-4-250. Finally, for the purposes of this example, we will assume a "simple T" architecture such as the one in Fig. 9 that uses two HV ports and two ground ports to create a gate and separation potential for offline sample analysis.

To begin, the chip should be mounted on the microscope platform such that the excitation light is focused on the gate of the chip to first determine the correct ratio of HV potentials to use to establish a gate. For the simple T with 2.5 cm side channels and a separation channel 5 cm in length such as the design in Fig. 9, a ratio 3:4 is typically sufficient, meaning that applying 2000 V to the "Buffer" port as a separation voltage and *(2000*.75)* = 1500 V to the "Sample" port will create a gate. This ratio will differ depending on the chip geometry used. Once established, this gate can be seen through the microscope by adjusting the position of the chip such that the excitation light is focused on the gate region of the chip, a fact that can be useful for diagnosis. Wires should be attached and secured with HV applied to the sample and buffer ports and ground wires in the buffer waste and sample waste ports. Alligator clips should be soldered to the ends of both the HV and ground wires, and platinum wires should be placed in the wells

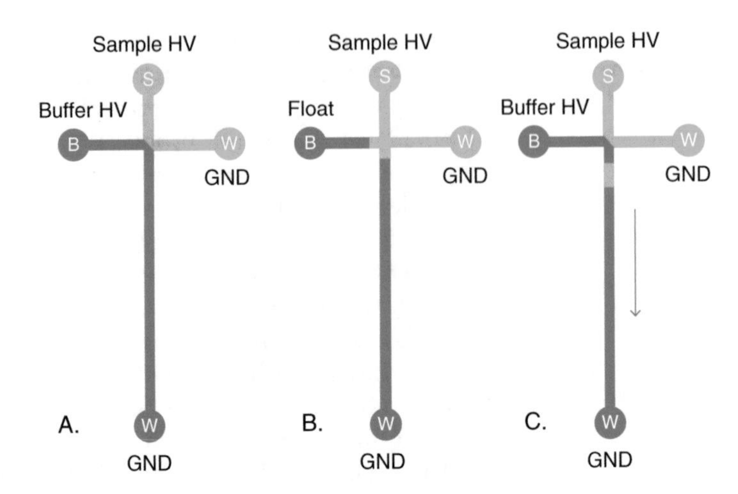

Fig. 9 Electrokinetic injection in a simple T chip. (**a**) The microchip is loaded with sample (S) and buffer (B). A high voltage potential is applied, resulting in electroosmotic flow toward the waste ports (W). (**b**) Buffer voltage is floated, allowing sample to be injected. (**c**) Buffer voltage is reestablished and analyte separation proceeds

themselves. Once the presence of a gate has been established, the detection point should be moved to the end of the separation channel. To inject a sample, the "Buffer" voltage will need to be turned off momentarily, allowing the potential to float. This is not to be confused with grounding the potential, which would result in sample flowing into the buffer reservoir as the EOF reversed direction.

After sample has been allowed to enter the separation channel (a 1 s injection time is typically more than sufficient), the voltage applied to the "Buffer" port is reapplied. This has the effect of both reestablishing the electrokinetic gate, preventing additional sample from entering the channel, and applying a potential to the separation channel. It is at this point in the process of analysis that analytes will begin to migrate with the EOF, separating according to their individual electrophoretic mobilities, toward the end of the separation channel.

3.10.5 Online MD-ME

The operation of an integrated MD-ME system is very similar in function to the offline example described above. As with offline analysis, the ME device is mounted over the detection system with the excitation light focused on the end of the separation channel. Integrating an MD flow to an ME device, however, requires that ports first be bonded to the surface of the device. As mentioned previously, UV glue is typically used in conjunction with bonded

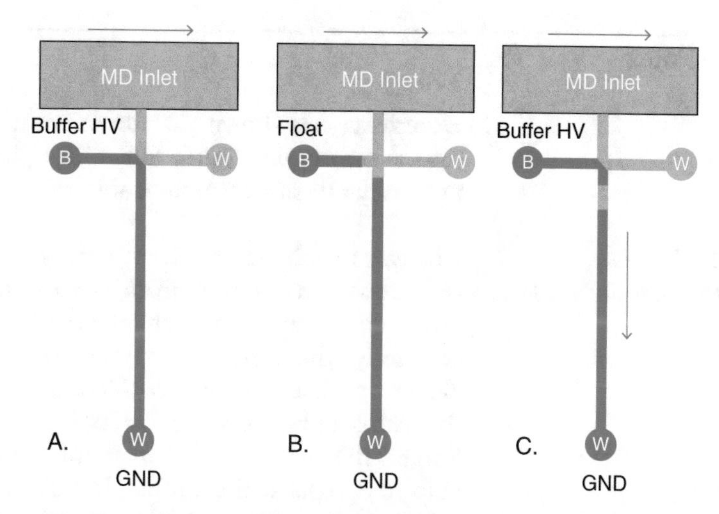

Fig. 10 Simple T chip with MD flow combined with electrokinetic injection. (**a**) Similar to the T chip shown in Fig. 4, the MD-ME interface shown here has sample applied to the top port via integration with an MD probe and run buffer added to the remaining three ports. A potential is applied to the run buffer reservoir (B), resulting in electroosmotic flow toward the waste ports (W), while pressure from the MD interface results in sample flow to sample waste. The MD flow must be grounded between the microdialysis probe and the ME device (not show in this diagram) to prevent electrocution. However, the distance to this ground must be longer than the distance to the ground potentials on the ME device itself to prevent a counter EOF flow. (**b**) Buffer voltage is floated, allowing sample to be injected. (**c**) Buffer voltage is reestablished, analyte separation proceeds

port connectors to create an interface capable of withstanding the pressure produced by the microdialysis flow.

Once the ports have been attached, polyethylene tubing can be used to connect the syringes containing NDA in ACN, NaCN, and H_2O the output of the microdialysis probe to the ME device. The fluids from these three inlets must be given sufficient time to mix on-chip, allowing analytes in the dialysate to form fluorescent products.

As in offline analysis, an electrokinetic gate is established at the inlet to the separation channel to prevent sample from continuously entering the channel. However, in the case of a coupled MD-ME system, the MD interface must be grounded in order to prevent damage to equipment or test subjects. If, however, the port is grounded near the gate, a counter-flowing EOF can be created. Therefore, the MD flow should be grounded as far from the separation channel as possible. Once a grounded pressure flow has been established, a separation voltage can be applied. An example of this can be seen in Fig. 10.

4 Notes

Experience has shown that there are many ways in which an ME separation can encounter difficulties. Below are the steps to diagnosing the most common problems.

4.1 Unstable Separation Voltage or Current

Changes in voltage or current during a separation can be the result of gradual evaporation from the sample and buffer wells. This evaporation can, in turn, change the ionic concentration and, consequently, the resistance of the ME system. However, sudden changes such as a spike or oscillation in current are indicative of the formation of bubbles (due to Joule heating) or a clogged channel. Should this occur once, it may only be necessary to clean the ME chip and replace the buffer. If bubbles are forming repeatedly, using a buffer of lower ionic strength may be necessary.

4.2 Spikes in Signal Output

Sudden signal spikes that are either too narrow to be genuine peaks or substantially higher in signal than the concentration would warrant can be caused by bubbles and/or particles in the channels. Both of these conditions can be quickly diagnosed by inspecting the channel under a microscope. Should particles be found during offline use, filtering the sample more thoroughly prior to injection using a 0.22 μm syringe filter is likely the best solution, as sample volume is typically not a limiting factor. During online use, samples are filtered through the porous membrane of the microdialysis probe. Particles found in online samples could indicate that a smaller pore size membrane should be used. Despite this, particles will sometimes form in the channel itself due to, for instance, NDA precipitating from solution. This condition can sometimes be prevented by adding a small amount of ACN to the BGE.

Bubbles are typically indicative of one of two conditions. The first is that the ME device was not properly prepared before use. To ensure that no bubbles remain in the channel, continue to apply negative pressure to the waste port while adding the BGE. Visually inspect prior to use to verify that all bubbles have been removed. Should bubbles form during the run, it is likely due to Joule heating boiling the BGE inside the channel itself. Decreasing the applied potential or decreasing the ionic strength of the BGE (and thereby lowering the current draw) will prevent further bubble formation.

4.3 Gradual Increases in Baseline, No Peaks Detected

Gradual increases in the baseline are indicative of an unstable or nonexistent gate, therefore allowing fluorescent product to gradually enter the separation channel and accumulate. This can be diagnosed by visually inspecting the gate using a microscope and corrected by adjusting the ratio of HV potentials. Additionally, a parametric search for appropriate combinations of voltages using Comsol modeling can save considerable time when first using an ME device.

4.4 Short Circuit During Application of HV Potentials

Often, this is due to a thin layer of conductive liquid between wells on the chip. Should this be the case, carefully drying the surface using a Kimwipe® is often sufficient to resolve the problem. However, if a surfactant such as SDS is present in the buffer, a conductive monolayer can still be present. This monolayer can be removed using methanol. Finally, it is possible for high voltage arcing to occur within the chip itself if bonding of the two layers was incomplete.

4.5 Clogged or Blocked Channels

The micrometer channel dimensions in ME devices make them highly susceptible to clogging. If any particulates get into the channels or crystals form in the buffer, there can be a variety of effects on the separation. If no peaks are detected or if there are inconsistencies in the separation current and none of the previous issues are present, the channels should be examined with a microscope to see if any clogs are visible. Clogs are especially common near the sample reservoir inlets and at the electrokinetic gate, although they can occur anywhere within the microchip.

The easiest solution for clogs is to avoid them in the first place. Using lint-free paper any time it is necessary to physically wipe down the chip will help to avoid getting fibers in the wells. Additionally, any solution put into the device should be filtered to remove any particulate in the liquid. If the device is not in use for an extended length of time, the channels should be completely filled with water and then stored within a beaker; this will prevent any evaporation within the channels leading to clogs.

4.5.1 Removing Clogs from Glass-Glass Microchips

Due to the length of time required to fabricate new all-glass microchips, cleaning a clog from one of these devices is often preferable to making a new one. Various solutions can be flowed through the channels via the application of negative pressure in order to reduce or completely dissolve clogs. We have previously used hydrochloric acid, nitric acid, sodium hydroxide, ACN, IPA, and DMSO in order to clean clogs from chips. Unfortunately, there is no one specific solution that works more often than any other, so this process is mostly trial and error.

Sonicating the chip has also successfully removed clogs. In these cases, the channels were filled with water before being placed in a full beaker and placed in a sonicator bath. The vibrations can break down any blockages to the point at which they can be washed out. We have also had limited success using high temperatures to eliminate clogs. Putting a clogged glass-glass device in the kiln cycle for bonding has degraded blockages to the point where they can be easily washed out.

4.5.2 Removing Clogs from PDMS Microchips

It is often easier and faster to simply make a new PDMS device than to spend time clearing out clogs. If the PDMS is reversibly bound, it can be peeled off of the substrate and cleaned using a piece of clear tape. The substrate should also be wiped down with IPA to remove any detritus before attempting to reapply the PDMS.

If the PDMS is irreversibly bound, the easiest way to remove a clog is briefly flowing IPA through the channels via the application of a negative pressure. IPA will slowly dissolve PDMS, so any particulate stuck on the channel walls can be washed away. However, this also means that repeated use can damage device features.

5 Applications

The first report of the use of microchip electrophoresis to detect Glu came from the Robert Kennedy group, who published a study investigating amino acids with OPA derivatization [36]. These experiments had limits of detection near 200 nmol/L and achieved a separation in 95 s using an all-glass microchip. By using a different chip design and high potential field, they also managed to achieve the separation in under 20 s. They continued this work using a segmented flow PDMS-glass microchip design and the NDA/CN derivatization reaction [37]. This separation also occurred in under 20 s and had improved limits of detection. This method was used to track the in vivo change of Asp and Glu following the introduction of a Glu transport inhibitor and microdialysis sampling. More recent work from this group has further improved separation speeds and detection limits using "water-in-oil" detection schemes that can provide excellent temporal resolution by derivatizing samples in discrete plugs of 8–10 nL prior to injection and separation [37].

The Susan Lunte group has also used microchip electrophoresis for Glu detection, furthering their earlier work using capillary electrophoresis to analyze amino acid neurotransmitters [38] and microchip electrophoresis to detect fluorescein in the brain [39]. This led to the development of a PDMS-glass hybrid microchip that allowed continuous, online, in vivo monitoring with microdialysis sampling, as shown in Fig. 11 [40].

This study used the NDA/CN reaction to detect amino acid neurotransmitters, using a 20-cm separation channel with on-chip mixing to facilitate derivatization. This method was used to study the change in concentration of fluorescein in rat brain dialysate over time, but was also capable of separating and detecting Glu. These results were expanded upon in an additional study, which improved the micromixer chip design and used a full-PDMS microchip [41]. This allowed near real-time in vivo monitoring of amino acid neurotransmitters, including Glu, with a 10-min lag time. Further work has been done to design a robust and portable

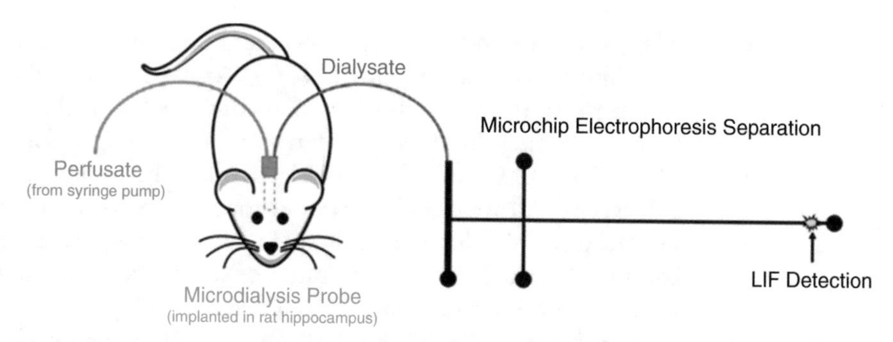

Fig. 11 In vivo microdialysis-microchip electrophoresis system. Schematic of a basic in vivo microdialysis-microchip electrophoresis system. In a brain experiment, the perfusate is normally artificial cerebrospinal fluid (aCSF) pumped at a flow rate of 1 μL/min. The derivatization reaction normally occurs between the collection of dialysate and its injection on the microchip system (not pictured)

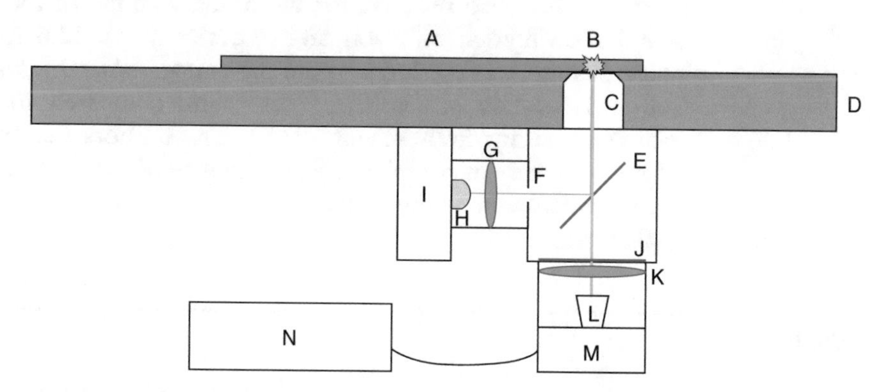

Fig. 12 Portable LED excitation and fluorescence detection system. Schematic representation of portable LED excitation and fluorescence detection system. Given wavelengths are for a separation using the NDA/CN derivatization reaction. (*A*) Microchip. (*B*) In-channel detection point. (*C*) Focusing objective. (*D*) *X*–*Y* positioner. (*E*) Dichroic mirror. (*F*) Pinhole. (*G*) Collimating lens. (*H*) LED at 445 nm. (*I*) Heat sink. (*J*) Long pass filter at 470 nm. (*K*) Focusing lens. (*L*) Avalanche photodiode (APD). (*M*) APD power supply. (*N*) Lock-in amplifier and data acquisition software

detection system with a light-emitting diode (LED) excitation source, shown in Fig. 12. This will allow in vivo microdialysis experiments to be carried out in a medical setting, where the use of a laser excitation source is impractical. Using this system, Glu was separated from multiple other amines and detected in brain microdialysis samples derivatized using the NDA/CN reaction [12].

While the bulk of published studies involving Glu detection using microchip electrophoresis have used either NDA or OPA for derivatization, several other fluorophores have been explored. Among these, an ME-based chiral separation of D/L Glu as well as D/L Asp in rat and human cerebral spinal fluid samples was developed by Huang et al. [42]. The separation of each enantiomer was accomplished through the addition of 12 mmol/L γ-cyclodextrin to a run buffer containing 30 mmol/L sodium dodecylsulfate

(SDS) and 25 mmol/L sodium borate. Pre-column derivatization with fluorescein isothiocyante (FITC) was used to label the amino acids prior to the separation on a 7.6 cm simple-T microchip, ultimately resulting in detection limits of 40 nmol/L for D-Glu. FITC, like OPA and NDA, is selective for the amine moiety of amino acids. However, FITC is a fluorophore and is therefore fluorescent prior to derivatization, which can complicate analysis. Additionally, the long reaction times of several hours necessary for derivatization limit its utility for rapid assays [43].

Another example of Glu detection via ME is the development of a system for in situ analysis of organic matter in Martian soil [44]. This method was first tested as a proof-of-concept study using soil samples from the Atacama Desert. Amino acids were first extracted and then derivatized using Pacific Blue succinimidyl ester, which is also selective for the primary amine of the amino acid. Following derivatization, an ME device with a 22.6-cm separation channel was used to separate the sample. With this system, a 200-fold increase in sensitivity was found compared to earlier techniques using fluorescamine [45]. The authors felt that the minimal increase in complexity from using an ME device was more than compensated by the smaller size, mass, and reagent use in this application.

6 Conclusions

Microchip electrophoresis provides an ideal platform for rapid separation of Glu as well as other neuroactive amines in neurological samples. Through either offline or online methods of sample acquisition and fluorescence derivatization, endogenous levels of Glu can be monitored in near real-time and for extended periods while simultaneously monitoring other analytes of interest, providing an important view of brain health.

Future research will focus on the development of techniques that will allow for better throughput, automation, and a wider analysis of neurobiological samples. The direct coupling of ME to continuous sampling methods still has room for improvement, as many current devices suffer from irreproducibility and the need for intensive fabrication procedures. One goal is to reduce the lag time between sampling and detection, allowing for closer near real-time analysis of neurobiological samples by minimizing the connection volumes between the dialysis probe and the ME separation. There are already research efforts in this direction, such as the incorporation of microdialysis sampling into a microchip designed to be used on-animal [46]. A method for continuous near real-time monitoring of multiple neuroactive amines could also be potentially useful for understanding of neurological diseases such as stroke and traumatic brain injury. Ultimately, it could be incorporated into a clinical instrument that could be used in the Intensive Care Unit to monitor treatment of patients with traumatic brain injury.

Furthermore, expanding the scope of analytes monitored using these separation-based sensors is important. Although the primary focus of this chapter is the detection of Glu, it is not the only neuroactive amine important in assessing brain health. Since ME is primarily a separation technique, it has the potential to monitor multiple analytes simultaneously with sufficient resolution for quantitation. Although there are other amines present in the brain that could be derivatized via the methods discussed in this chapter, a more universal method of measurement would allow for a fuller picture of neurobiological methods. Studies coupling electrochemical sensors with ME will allow for this, but introduce their own complications regarding electrode sensitivity and robustness (as discussed in another chapter of this book). Other detection methods, such as conductivity and mass spectrometry, can also allow for a wider range of analytes to be measured in a single ME analysis.

Acknowledgements

This research was financially supported by NIH R01NS042929, NIH R01NS066466, and COBRE P20GM103638. The authors would like to thank Richard de Campos for his assistance in producing graphics for this chapter. The authors would also like to thank Nancy Harmony for her editing skills as well as the Ralph N. Adams Institute Cleanroom Facilities.

References

1. Maas AI, Stocchetti N, Bullock R (2008) Moderate and severe traumatic brain injury in adults. Lancet Neurol 7(8):728–741
2. Benveniste H (1989) Brain microdialysis. J Neurochem 52(6):1667–1679
3. Sattler R, Tymianski M (2000) Molecular mechanisms of calcium-dependent excitotoxicity. J Mol Med 78(1):3–13
4. Kristián T, Siesjö BK (1998) Calcium in ischemic cell death. Stroke 29(3):705–718
5. Chamoun R, Suki D, Gopinath SP, Goodman JC, Robertson C (2010) Role of extracellular glutamate measured by cerebral microdialysis in severe traumatic brain injury. J Neurosurg 113(3):564
6. Soldatkin O, Nazarova A, Krisanova N, Borysov A, Kucherenko D, Kucherenko I, Pozdnyakova N, Soldatkin A, Borisova T (2015) Monitoring of the velocity of high-affinity glutamate uptake by isolated brain nerve terminals using amperometric glutamate biosensor. Talanta 135:67–74
7. Szydlowska K, Tymianski M (2010) Calcium, ischemia and excitotoxicity. Cell Calcium 47(2):122–129
8. Dale N, Hatz S, Tian F, Llaudet E (2005) Listening to the brain: microelectrode biosensors for neurochemicals. Trends Biotechnol 23(8):420–428
9. Zhang M, Mao L (2005) Enzyme-based amperometric biosensors for continuous and on-line monitoring of cerebral extracellular microdialysate. Front Biosci 10:345–352
10. van der Zeyden M, Oldenziel WH, Rea K, Cremers TI, Westerink BH (2008) Microdialysis of GABA and glutamate: analysis, interpretation and comparison with microsensors. Pharmacol Biochem Behav 90(2):135–147
11. Khan AS, Michael AC (2003) Invasive consequences of using micro-electrodes and microdialysis probes in the brain. TrAC Trends Anal Chem 22(8):503–508

12. Oborny NJ, Costa EEM, Suntornsuk L, Abreu FC, Lunte SM (2016) Evaluation of a portable microchip electrophoresis fluorescence detection system for the analysis of amino acid neurotransmitters in brain dialysis samples. Anal Sci 32(1):35–40
13. Nandi P, Lunte SM (2009) Recent trends in microdialysis sampling integrated with conventional and microanalytical systems for monitoring biological events: a review. Anal Chim Acta 651(1):1–14
14. Stenken JA (1999) Methods and issues in microdialysis calibration. Anal Chim Acta 379(3):337–358
15. Georganopoulou DG, Carley R, Jones DA, Boutelle MG (2000) Development and comparison of biosensors for in-vivo applications. Faraday Discuss 116:291–303
16. Cecil S, Chen PM, Callaway SE, Rowland SM, Adler DE, Chen JW (2011) Traumatic brain injury advanced multimodal neuromonitoring from theory to clinical practice. Crit Care Nurse 31(2):25–37
17. 510(k) Summary for CMA Cerebral Tissue Monitoring System (2002) Food and Drug Administration, accessdata.fda.gov
18. AB MD (2015) ISCUSflex Microdialysis Analyzer. http://www.mdialysis.com/analyzers/iscusflex-for-point-of-care
19. Saylor RA, Lunte SM (2015) A review of microdialysis coupled to microchip electrophoresis for monitoring biological events. J Chromatogr A 1382:48–64
20. Smithies O (1955) Zone electrophoresis in starch gels: group variations in the serum proteins of normal human adults. Biochem J 61(4):629
21. Li SFY (1992) Capillary electrophoresis: principles, practice and applications, vol 52. Elsevier, Amsterdam
22. Weinberger R (2000) Practical capillary electrophoresis. Academic Press, New York
23. Lee PY, Costumbrado J, Hsu C-Y, Kim YH (2012) Agarose gel electrophoresis for the separation of DNA fragments. J Vis Exp 62:e3923
24. Landers JP (2007) Handbook of capillary and microchip electrophoresis and associated microtechniques. CRC press, Boca Raton, FL
25. Ohlsson PD, Ordeig O, Mogensen KB, Kutter JP (2009) Electrophoresis microchip with integrated waveguides for simultaneous native UV fluorescence and absorbance detection. Electrophoresis 30(24):4172–4178
26. De Montigny P, Stobaugh JF, Givens RS, Carlson RG, Srinivasachar K, Sternson LA, Higuchi T (1987) Naphthalene-2,3-dicarboxyaldehyde/cyanide ion: a rationally designed fluorogenic reagent for primary amines. Anal Chem 59(8):1096–1101
27. Chen RF, Scott C, Trepman E (1979) Fluorescence properties of o-phthaldialdehyde derivatives of amino acids. Biochim Biophys Acta 576(2):440–455
28. Matuszewski BK, Givens RS, Srinivasachar K, Carlson RG, Higuchi T (1987) N-substituted 1-cyanobenz [f] isoindole: evaluation of fluorescence efficiencies of a new fluorogenic label for primary amines and amino acids. Anal Chem 59(8):1102–1105
29. Bantan-Polak T, Kassai M, Grant KB (2001) A comparison of fluorescamine and naphthalene-2,3-dicarboxaldehyde fluorogenic reagents for microplate-based detection of amino acids. Anal Biochem 297(2):128–136
30. Molho JI, Herr AE, Mosier BP, Santiago JG, Kenny TW, Brennen RA, Gordon GB, Mohammadi B (2001) Optimization of turn geometries for microchip electrophoresis. Anal Chem 73(6):1350–1360
31. Paegel BM, Hutt LD, Simpson PC, Mathies RA (2000) Turn geometry for minimizing band broadening in microfabricated capillary electrophoresis channels. Anal Chem 72(14):3030–3037
32. Jacobson SC, Moore AW, Ramsey JM (1995) Fused quartz substrates for microchip electrophoresis. Anal Chem 67(13):2059–2063
33. Duffy DC, McDonald JC, Schueller OJ, Whitesides GM (1998) Rapid prototyping of microfluidic systems in poly (dimethylsiloxane). Anal Chem 70(23):4974–4984
34. Haubert K, Drier T, Beebe D (2006) PDMS bonding by means of a portable, low-cost corona system. Lab Chip 6(12):1548–1549
35. Huynh BH, Fogarty BA, Nandi P, Lunte SM (2006) A microchip electrophoresis device with on-line microdialysis sampling and on-chip sample derivatization by naphthalene 2,3-dicarboxaldehyde/2-mercaptoethanol for amino acid and peptide analysis. J Pharm Biomed Anal 42(5):529–534
36. Sandlin ZD, Shou M, Shackman JG, Kennedy RT (2005) Microfluidic electrophoresis chip coupled to microdialysis for in vivo monitoring of amino acid neurotransmitters. Anal Chem 77(23):7702–7708
37. Wang M, Roman GT, Perry ML, Kennedy RT (2009) Microfluidic chip for high efficiency electrophoretic analysis of segmented flow from a microdialysis probe and in vivo chemical monitoring. Anal Chem 81(21):9072–9078
38. Zhou SY, Zuo H, Stobaugh JF, Lunte CE, Lunte SM (1995) Continuous in vivo monitoring of amino acid neurotransmitters by

microdialysis sampling with online derivatization and capillary electrophoresis separation. Anal Chem 67(3):594–599

39. Huynh BH, Fogarty BA, Martin RS, Lunte SM (2004) On-line coupling of microdialysis sampling with microchip-based capillary electrophoresis. Anal Chem 76(21):6440–6447
40. Nandi P, Desai DP, Lunte SM (2010) Development of a PDMS-based microchip electrophoresis device for continuous online in vivo monitoring of microdialysis samples. Electrophoresis 31(8):1414–1422
41. Nandi P, Scott DE, Desai D, Lunte SM (2013) Development and optimization of an integrated PDMS based-microdialysis microchip electrophoresis device with on-chip derivatization for continuous monitoring of primary amines. Electrophoresis 34(6):895–902
42. Huang Y, Shi M, Zhao S (2009) Quantification of D-Asp and D-Glu in rat brain and human cerebrospinal fluid by microchip electrophoresis. J Sep Sci 32(17):3001–3006
43. Maeda H, Ishida N, Kawauchi H, Tuzimura K (1969) Reaction of fluorescein-isothiocyanate with proteins and amino acids. The. J Biochem 65(5):777–783
44. Chiesl TN, Chu WK, Stockton AM, Amashukeli X, Grunthaner F, Mathies RA (2009) Enhanced amine and amino acid analysis using Pacific Blue and the Mars Organic Analyzer microchip capillary electrophoresis system. Anal Chem 81(7):2537–2544
45. Skelley AM, Aubrey AD, Willis PA, Amashukeli X, Ehrenfreund P, Bada JL, Grunthaner FJ, Mathies RA (2007) Organic amine biomarker detection in the Yungay region of the Atacama Desert with the Urey instrument. J Geophys Res Biogeosci 112(G4)
46. Scott DE, Willis SD, Gabbert S, Johnson D, Naylor E, Janle EM, Krichevsky JE, Lunte CE, Lunte SM (2015) Development of an on-animal separation-based sensor for monitoring drug metabolism in freely roaming sheep. Analyst 140(11):3820–3829

Chapter 13

Brain Glutamate Monitoring by Microdialysis and Separation Methods with a Special Focus on Capillary Electrophoresis with Laser-Induced Fluorescence Detection

Luc Denoroy and Sandrine Parrot

Abstract

The in vivo monitoring of extracellular brain glutamate, i.e., of the glutamate fraction potentially acting on receptors, is mostly performed by brain microdialysis coupled to a separation micro-method, such an approach offering high chemical selectivity combined to good anatomical and temporal resolution. Among the two separation micro-methods used to determine glutamate in brain microdialysates, namely liquid chromatography and capillary electrophoresis, only the latter one, especially when coupled to laser-induced fluorescence detection, can allow to routinely analyze glutamate in low-volume (<1 μL) microdialysates for high temporal resolution (≤ 1 min) monitoring of fast neurobiological events.

This chapter will present detailed capillary electrophoresis with laser-induced fluorescence detection protocols for the monitoring of glutamate in brain microdialysate. Various aspects of the technology, including sample derivatization, injection procedures, modes of separation and their effects on analytical performance such as sensitivity, temporal resolution in glutamate monitoring and potential to simultaneously determine other neurotransmitters will be reviewed and discussed. Finally, in order to give a well-balanced opinion regarding the analysis of glutamate in brain microdialysate, the high performance liquid chromatography technology will be also reviewed.

Key words Glutamate, Brain, Microdialysis, Capillary electrophoresis, Liquid chromatography

1 Introduction

Glutamate (Glu) is a major excitatory amino acid in the central nervous system and its involvement in various neurophysiological processes can be studied by monitoring changes of its extracellular concentration. This is commonly performed by in vivo microdialysis, a technique allowing sampling chemical compounds from the extracellular fluid in targeted brain areas, in combination with a separation method.

Sandrine Parrot and Luc Denoroy (eds.), *Biochemical Approaches for Glutamatergic Neurotransmission*, Neuromethods, vol. 130, DOI 10.1007/978-1-4939-7228-9_13, © Springer Science+Business Media LLC 2018

The microdialysis is a sampling technique which is detailed in Chaps12 and 14 and which is based on diffusion across a dialysis membrane. Most of the dialysis probes consist of two concentric tubes with the distal part covered by a semipermeable hollow fiber. After implantation into a brain area, the probe is perfused with artificial cerebrospinal fluid, the fluid being pushed through the outer tube and returning through the inner tube. Molecules diffuse down their concentration gradient across the dialysis membrane, from the extracellular environment into the probe. Once extracted, the chemicals can be directly analyzed, since the samples collected (called microdialysates) are aqueous and relatively "clean."

Capillary electrophoresis (CE) in conjunction with laser-induced fluorescence (LIF) detection has become one of the most powerful analytical tools for the determination of Glu in brain microdialysates. CE is an electrokinetic separation technique performed in a fused silica capillary having an internal diameter smaller than 100 μm. A schematic representation of a CE instrument is shown in Fig. 1. When using that approach, the separation of Glu is readily achieved by free solution capillary electrophoresis (FSCE), a CE separation mode in which the background electrolyte is a simple electrolyte solution such as borate buffer. The electrophoretic migration and the electroosmotic flow (EOF) allow the separation of molecules. These forces, as well as the characteristics of the capillary (e.g., length and internal diameter), characterize the separation system. Indeed, due to its plug-like flow and minimal diffusion, CE possesses enormous resolving power and large peak capacities, offering the advantage of rapidity, high resolution, and sensitivity, while requiring very small sample sizes, i.e., from a few μL to sub-μL volume [1]. Furthermore, the suitability of very small volume of dialysate makes possible short collection times at low flow-rate, thereby leading to an increase in microdialysis temporal resolution and to a better respect of the neuronal environment. With this approach, sampling times as short as 5–30 s have been reported [2–4], allowing the monitoring of rapid changes associated with neuronal events. Although many of these studies have been performed using homemade CE instruments which cannot easily be set up in most biology laboratories, such short sampling times can also be achieved by coupling microdialysis with commercially available CE-LIF detection systems [5].

As Glu is not fluorescent at available laser wavelengths, microdialysis samples must be derivatized with a tagging agent. The fluorogenic reagent naphthalene-2,3-dicarboxyaldehyde (NDA) reacts rapidly with the primary amine function of amino acids in the presence of CN^{-}, to give stable fluorescent derivatives (cyanobenzo[f]isoindol or CBI products) which can be detected following excitation with the 442-nm HeCd or the 410-nm diode laser (Fig. 2). The figures of merit of the quantitative determina-

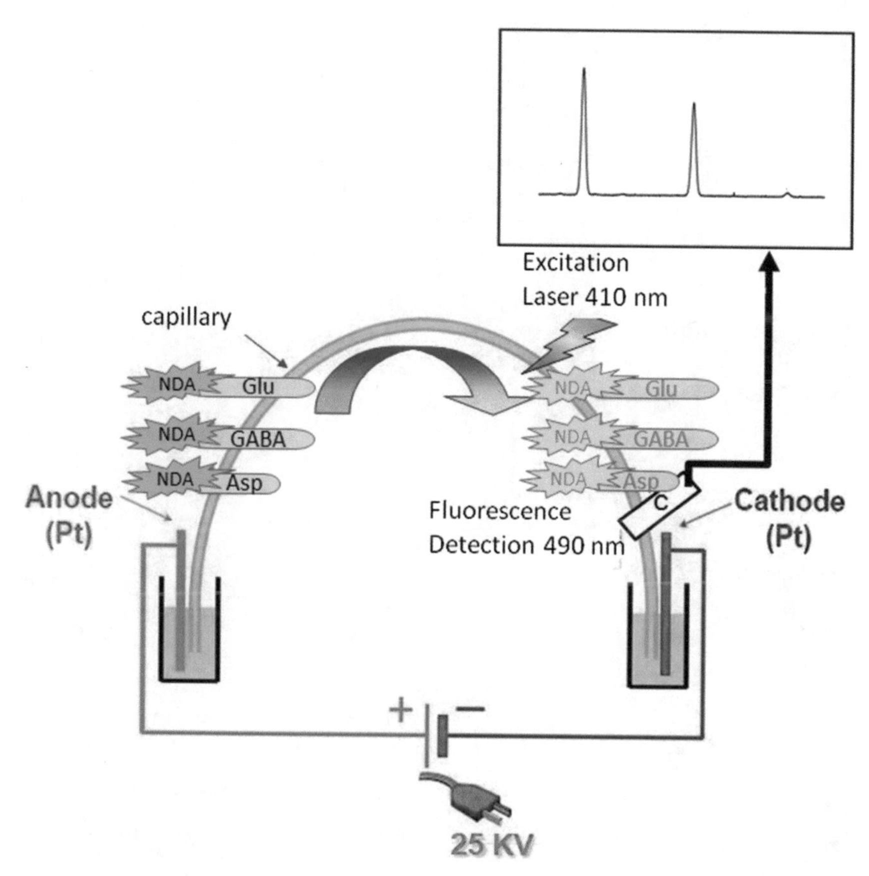

Fig. 1 Schematic representation of a capillary electrophoresis instrument with laser-induced fluorescence detection. Separation of derivatized analytes occurs within a thermostated fused silica capillary (internal diameter 10–75 μm; length 20–100 cm) filled with an electrolyte buffer. The capillary extremities and the two electrodes are plunged in reservoirs filled with the same electrolyte buffer. Hydrodynamic (i.e., pressure assisted) or electrokinetic (i.e., voltage assisted) injections of the sample are performed after a transient reservoir removal at the anode side of the capillary. Afterwards, a high voltage (5–30 kV) is applied between the two electrodes plunged in electrolyte reservoirs. Molecules migrate from the anode inlet to the cathode outlet of the capillary. Separated compounds are detected in situ through a window in the external polyimide coating of the capillary

tion of NDA derivatives of amino acids by CE-LIF have been previously reported, showing good repeatability, linearity, and accuracy associated with a low limit of detection, i.e., in the nmol/L range.

The major procedures involved in the free solution capillary electrophoresis (FSCE)-LIF analysis of Glu from brain microdialysates, i.e., NDA derivatization and separation, will be described in the following parts of this chapter. In addition, other electrophoretic protocols based on micellar electrokinetic chromatography (MEKC) and allowing the simultaneous determination of Glu with other amino acid neurotransmitters will be also outlined.

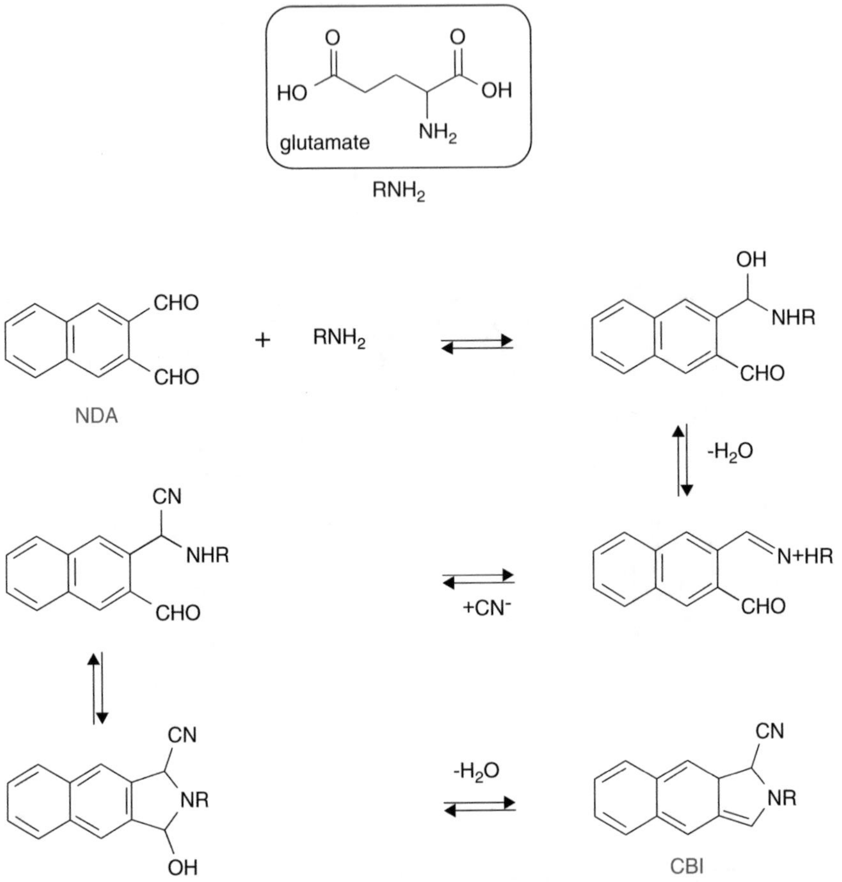

Fig. 2 Successive steps of the derivatization reaction of glutamate. Naphthalene-2,3-dicarboxaldehyde (NDA) reacts with the primary amine moiety of Glu in the presence of cyanide ions (CN^-) as the nucleophilic agent leading to the formation of a highly fluorescent cyanobenzo[f]-isoindole product (CBI)

2 Materials

2.1 Reagent for Microdialysis

Artificial cerebrospinal fluid (aCSF):

1.2 mmol/L $CaCl_2$, 1.0 mmol/L $MgCl_2$, 2.8 mmol/L KCl 149 mmol/L NaCl. The pH is adjusted to 7.35–7.40 with 222 mmol/L phosphate buffer pH 7.4. The solution is stored at +4 °C for up to 4 weeks and should be filtered through a 0.2-μm filter before use.

2.2 Reagents for FSCE of Glutamate

2.2.1 Sample Derivatization

Derivatization buffer:

500 mmol/L sodium borate buffer, pH 8.7 (made from boric acid and borax). Store at ambient temperature to avoid crystallization.

Derivatization reagents:

87 mmol/L NaCN (4.264 mg/mL of water). Store up to 1 week at 4 °C in a capped glass container.

2.925 mmol/L NDA (0.539 mg/mL of acetonitrile-water, 50:50). For dissolving, add acetonitrile first, then water. Store up to 1 week at 4 °C in a foil-wrapped, capped glass container. The solution should be crystal clear and transparent. Discard and replace the solution if oxidized (pink-colored).

NaCN is a highly toxic chemical and must be detoxified to harmless waste before disposal. For that purpose, the waste should be treated with glucose (ten times more concentrated than CN^-) in borate buffer or with alkaline sodium hypochlorite (<1 g/L) in excess. NDA solution could be disposed with organic solvents wastes.

Standards:

Amino acids: 1 mmol/L Glu in 100 mmol/L HCl, 1 mmol/L aspartate (Asp) in 100 mmol/L HCl.

Internal standard: 1 mmol/L α-amino adipic acid (AAD) in 100 mmol/L HCl.

Dispense into 1.5-ml microcentrifuge tubes. Store up to 6 months at −20 °C.

2.2.2 Sample Separation

Background electrolyte: 100 mmol/L sodium borate buffer, pH 9.2. Store at +4 °C for 1 month. Filter on cellulose acetate (0.2 μm) before use.

Autoclaved ultrapure water.

1% (w/v) NaOH.

2.3 Reagents for MEKC of Glutamate

2.3.1 Sample Derivatization

Same derivatization reagents and buffer as for FSCE of Glu.

Standards:

Amino acids: 1 mmol/L Glu in 100 mmol/L HCl, 1 mmol/L Asp in 100 mmol/L HCl, 1 mmol/L γ-aminobutyric acid (GABA) in 100 mmol/L HCl.

Internal standard: 1 mmol/L L-cysteic acid in 100 mmol/L HCl.

2.3.2 Sample Separation

Background electrolyte: 75 mmol/L borate buffer, pH 9.2, containing 10 mmol/L hydroxypropyl-β-cyclodextrin and 70 mmol/L sodium dodecyl sulfate (SDS). Store at ambient temperature (to prevent precipitation due to the presence of high concentration of SDS) for 1 month. Filter on cellulose acetate (0.2 μm) before use. Discard and replace the solution if cloudy.

Autoclaved ultrapure water.

1% (w/v) NaOH.

2.4 CE-LIF System

CE instrument with 1–30 kV voltage capabilities, refrigerated sampler (Beckman-MDQ or Agilent 3D).

Fused silica capillary (50-μm i.d., 375-μm o.d.; Polymicro Technology).

LIF detector (Picometrics Zetalif) equipped with 410-nm diode laser (Melles Griot) or 442-nm HeCd laser (Melles Griot), the emission wavelength being set to 490 nm.

2.4.1 General Requirements

Because primary amine contaminants can affect detection, special care must be taken to avoid any dust or bacterial contamination. Ultrapure water, pipet tips, and microcentrifuge polymerase chain reaction (PCR) tubes should be autoclaved. Clean gloves must be worn during manipulation of samples and reagents.

Filtration of solutions through 0.22-μm cellulose acetate filter is important for proper functioning of CE since particulate matter can clog the capillary lumen leading to baseline disturbances and current failure.

3 Methods

3.1 Manual Derivatization of Microdialysates with NDA/NaCN

1. Prepare NaCN-borate solution by adding 111 μL of 87 mmol/L NaCN to 556 μL of 500 mmol/L sodium borate buffer. Store in a sealed 1-mL glass vial up to 1 day at 4 °C.
2. In a PCR tube containing 10 μL of microdialysis sample, add in the following order:

   1 μL 10 μM AAD (internal standard).

   2 μL NaCN-borate solution.

   1 μL 2.925 mmol/L NDA.

   If volume of sample is >10 μL, increase volume of each reagent proportionately.

   The aim of the internal standard is to take into account any inter-run variability in derivatization and separation. It allows the ease of peak identification and a better quantification, signal being calculated as the ratio between the peak area of analyte and the peak area of internal standard.
3. If volume of microdialysis sample is <10 μL, prepare the following stock reagent mixture immediately prior to use:

   10 μL 10 μM AAD (internal standard).

   20 μL NaCN-borate solution.

   10 μL 2.925 mmol/L NDA.

   Add 0.8 μL of the reagent mixture to each tube containing 2 μL of microdialysis sample. If the sample volume is >2 μL, increase volume of each reagent accordingly.
4. Vortex and then let the reaction develop 15 min at room temperature. Store up to 6 h at 4 °C prior to injection for analysis.

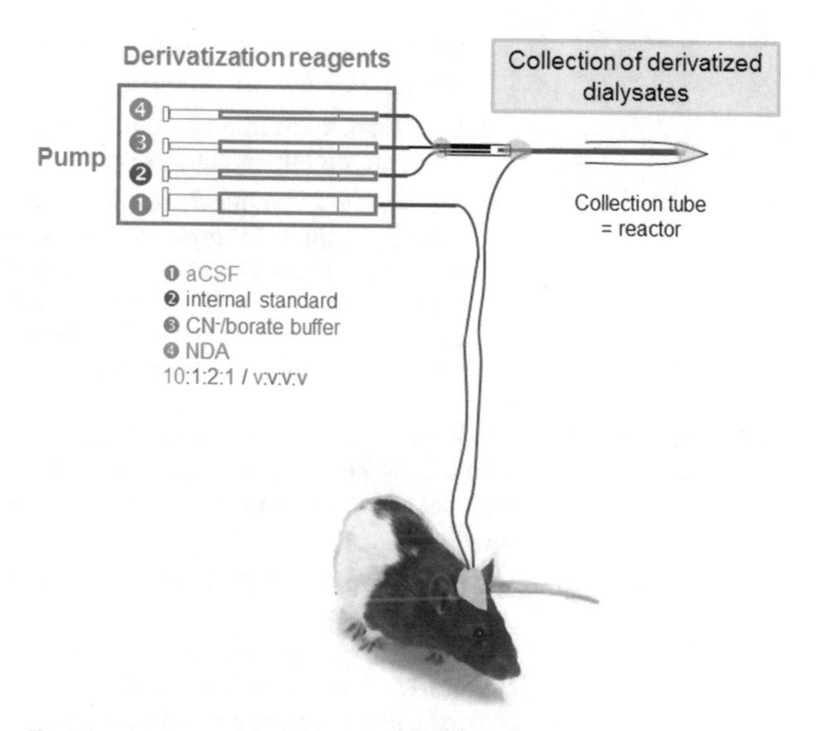

Fig. 3 Setup for online derivatization of brain microdialysates. The outlet capillary of the microdialysis probe was glued lengthwise to a capillary bringing the derivatization reagents (NaCN/borate buffer and NDA) together with internal standard directly into the collection tube. In this setup, the dead volume was only due to the tubing between the probe and collection tube

3.2 Online Derivatization of Microdialysates with NDA/NaCN

Manual derivatization of sample volumes lower than 2 μL cannot be accomplished with good repeatability and should be performed within a continuous flow derivatization system placed at the outlet of the microdialysis probe. This procedure allows to continuously adding very small volumes of reagents to the microdialysate with good reproducibility, while preventing evaporation and sample loss. A setup for online derivatization of brain microdialysates is shown in Fig. 3.

3.2.1 Reactor for Online Derivatization

1. Adapt a 40-μm i.d., 105-μm o.d. capillary to the outlet of the microdialysis probe.
2. Introduce and fix with epoxy glue three 75-μm i.d., 150-μm o.d. capillaries into one side of a 2.5-cm long, 300-μm i.d. polyethylene tubing.

   These capillaries will bring reagents and the internal standard into the lumen of the polyethylene tubing; the solutions contained within the capillaries will mix by passive diffusion inside this polyethylene tubing to produce the reaction mixture.
3. Introduce and fix with epoxy glue another 75-μm i.d. capillary onto the other side of the polyethylene tubing.

The reaction mixture made inside the polyethylene tubing will flow out through this capillary.

4. Glue capillary from probe outlet to the capillary from polyethylene tubing outlet lengthwise.

   This system for online derivatization must have a minimal dead volume to limit diffusion of solutes. The use of tubing with very small inner diameters and very slow perfusion rates of reagents allows sub-microliter volumes to be derivatized and collected with good yield and repeatability, avoiding any fall in sensitivity as compared to manual derivatization.

3.2.2 Protocol for Online Derivatization

Place four microsyringes (two 50-μL, one 100-μL, and one 500-μL) on the rack of a microsyringe pump. Adjust the flow rate so that the 500-μL syringe delivers aCSF at a flow rate of 0.5 to 2 μL/min.

(a) The 500-μL syringe is connected to the inlet of the microdialysis probe.

(b) A 50-μL syringe contains the internal standard (1 mmol/L AAD for FSCE of Glu or 1 mmol/L cysteic acid for the MEKC determination of amino acids) and is connected to capillary no. 2.

(c) The 100-μL syringe contains 14.66 mmol/L NaCN in 0.5 M borate buffer, pH 8.7 and is connected to capillary no. 3.

(d) The other 50-μL syringe contains 2.925 mmol/L NDA in acetonitrile-water (50:50, v/v) and is connected to capillary no. 4.

With this system, the microdialysate coming from the outlet of the probe (capillary no. 1) is delivered together with the derivatization mixture into collecting vials in which the derivatization reaction takes place.

3.3 FSCE Separation of Glutamate

The values of time and pressure for rinsing and injection steps are given for a 63-cm long 50-μm i.d. capillary. They have to be altered when using capillary of different size and the CE Expert software from Beckman could be of great help for calculating the new values (https://sciex.com/ce-features-and-benefits/ce-expert-lite).

1. In the inlet buffer rack of the CE instrument place one vial containing filtered, autoclaved water, three vials containing filtered background electrolyte (one for washing the capillary, one for rinsing the inlet electrode and the outside of capillary inlet, one for the separation), and one vial containing 1% (w/v) NaOH.
2. Place one vial containing filtered background electrolyte in the outlet buffer rack of the CE instrument.
3. Perform washing and conditioning of the capillary (once at the beginning of each working day) by flushing it with the following order of reagents: 1% NaOH (20 psi, 10 min), ultrapure water (20 psi, 10 min), and background electrolyte

(20 psi,10 min). Check capillary and background electrolyte by applying a positive voltage of 25 kV between inlet and outlet electrodes. Note value of current (this value will depend on the total length and the inner diameter of the capillary and may slightly vary according to batch of separation buffer).

The current should be ~30 μA when the total length of a 50-μm i.d. capillary is 63 cm. If the current is lower than expected, this may indicate a clogged capillary, resulting usually from salt deposits at the capillary inlet. Trimming the capillary inlet may rectify the problem. If not, the capillary should be changed. If the current is higher than expected, this may indicate a contaminated background electrolyte. A new washing procedure or the change of the electrolyte solution in vial should rectify the problem.

4. Turn on the laser 5 min (for 410-nm diode laser) or 20 min (for 442-nm HeCd laser) before the beginning of the first analysis.
5. Load derivatized samples in the sample rack of the CE instrument.

   Sample can be loaded in 200-μL polyethylene PCR tubes, plastic, or glass microtubes. Make sure that samples lie at the bottom of each tube; a brief centrifugation may help to ensure that small volume samples are deposited to the bottom of the tubes. Moreover, one must check that sample volume is sufficient so that the inlet of the capillary can plunge into the sample, at least 2 μL for PCR tubes for instance.
6. Briefly wash the capillary with background electrolyte (20 psi, 5 min). Inject sample (0.2 psi, 10 s). After sample injection, briefly (0.01 min) dip the inlet electrode and capillary inlet into a second vial of background electrolyte (for rinsing the outside of capillary inlet and not contaminating the separation buffer). Plunge the inlet electrode and the inlet of the capillary into a third vial containing background electrolyte and apply voltage (25 kV) for 6–15 min (the run time is depending of capillary length as it adjusted to be slightly higher than the migration time of the last peak of interest). Repeat this sequence for each sample. A typical FSCE separation of Glu in a brain microdialysate is shown in Fig. 4.
7. The inter-run capillary washing step is crucial to obtain a reproducible inner surface of the capillary and hence stable EOF and migration times between runs.

3.4 Data Analysis

Peak areas of Glu and Asp standards (ranging from 10 nmol/L to 10 μmol/L) as well as of AAD (internal standard, constant concentration 10 μmol/L) are measured using the software of the CE instrument.

Relative peak areas of standards (i.e., relative to internal standard AAD) are calculated and used to draw a calibration curve.

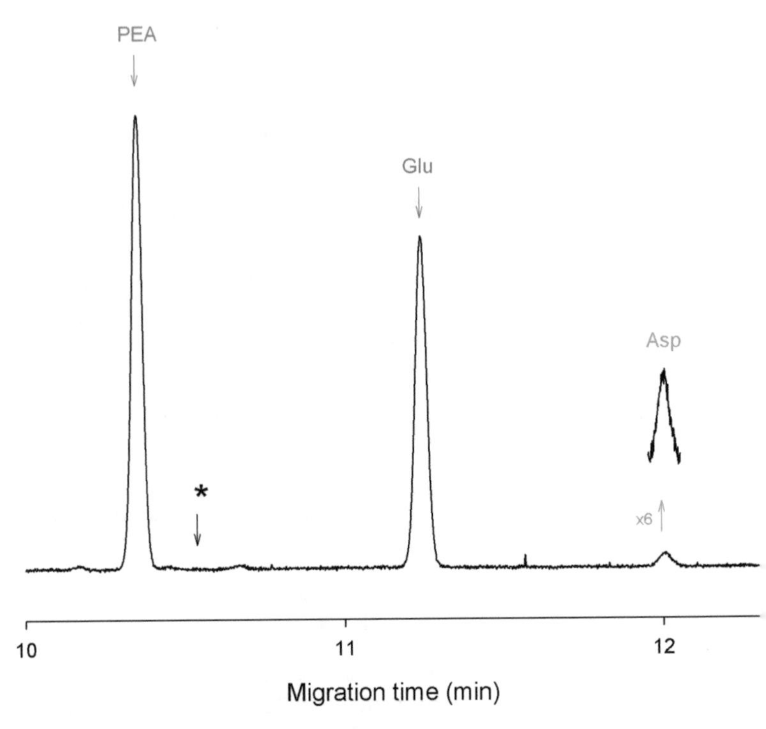

Fig. 4 Free solution capillary electrophoresis separation of a microdialysate fron rat striatum. *PEA* phenylethanolamine (a marker of membrane integrity), *Glu* glu tamate, *Asp* aspartate. The *arrow* with an *asterisk* indicates the migration time o the internal standard α-amino adipic acid and shows that this compound, whe added to the sample, does not co-migrate with any endogenous compound

Linear regression analysis is performed to test the significance o linearity of this curve and to obtain the linear regressio equation.

Areas of microdialysate Glu and Asp peaks, as well as of AAI added to microdialysates, are measured. Relative peak areas o microdialysate Glu and Asp are calculated.

The actual concentrations of Glu and Asp in the microdialy sates are determined using the linear regression equation of th calibration curve.

As the basal microdialysate Glu level can vary between animals concentrations are normalized for each animal, i.e., expressed as percentage of their own basal values. For each group of experimen tal animals, data are given as mean ± SEM expressed as a percent age of the basal values.

An example of representative data is shown in Fig. 5.

The determination of Asp allows ensuring the optimum resolu tion since Glu and Asp are closely related molecules. Moreover, it wid ens the biological interest of the analysis, since Asp is an excitator neurotransmitter, although less studied than Glu.

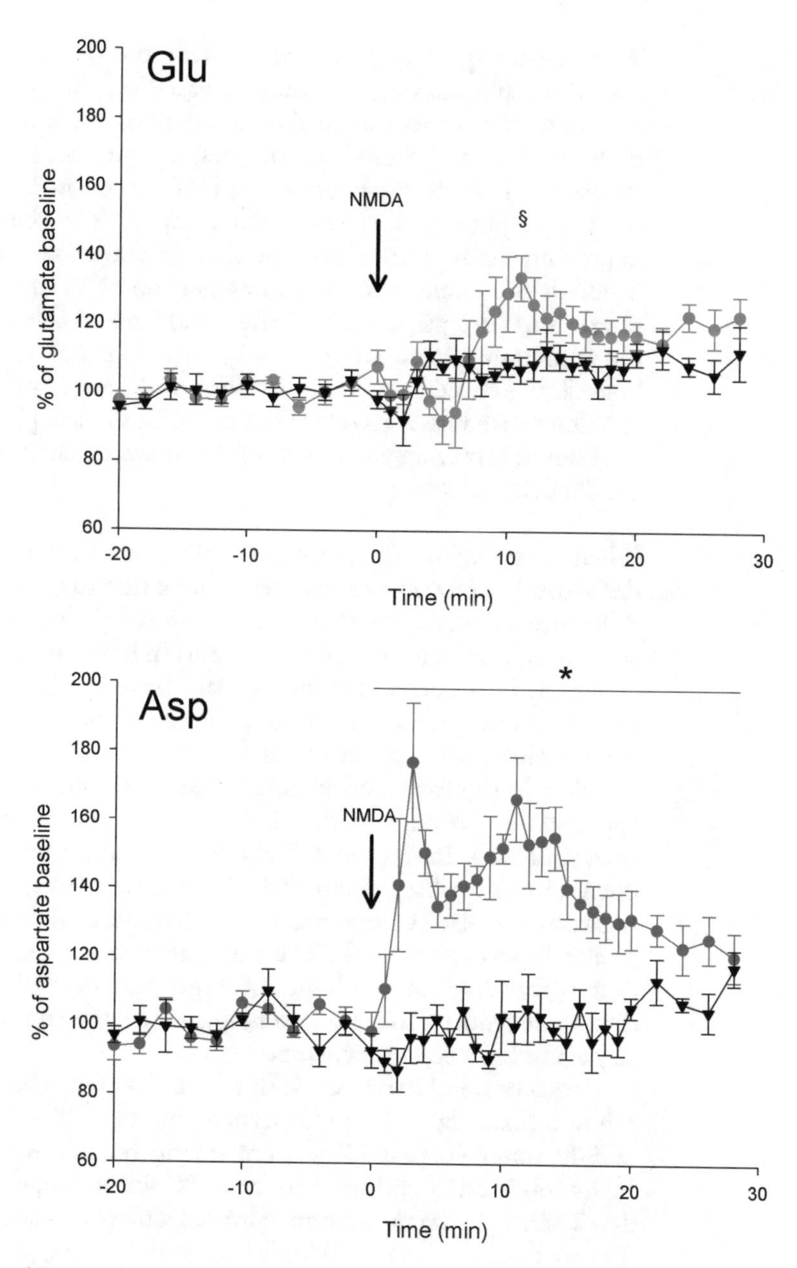

Fig. 5 Effects of the administration of NMDA (0.5 μL, 5 mmol/L) by pneumatic ejection on microdialysate Glu (*top*) and Asp (*bottom*) levels from striatum of urethane-anesthetized rats ($n = 6$). NMDA was administered during a 1-min period (*black arrow*). Control animals received the vehicle (artificial CSF, $n = 5$). The NMDA-induced changes in Glu and Asp levels are expressed as percentages of the mean of their respective basal values. The basal concentrations of Glu and Asp in the microdialysates were 10.89 ± 1.55 μmol/L and 128.0 ± 13.6 nmol/L, respectively ($n = 11$). $^{*}p < 0.05$, ANOVA with repeated measures, followed by PLSD Fisher's test, $^{\S}p < 0.01$

3.5 Variations and Improvements on the Same Technique

3.5.1 Stacking

When injecting sample, an in-capillary preconcentration (also named sample stacking) procedure, based on the change in the analyte migration velocity at the boundary of discontinuous solutions system, can be used to compress analytes bands within the capillary [1]. By decreasing the length of the analyte bandwidth in the sample plug, such stacking procedure leads to sharper peaks, improving the separation and the limit of detection. In this case, injection of sample is followed by injection of 0.1 mol/L phosphoric acid (0.2 psi, 5 s), inducing a sudden and brief increase in the migration velocity of analytes by changing their charge at the boundary between solutions exhibiting different pH. Another stacking procedure is to dilute the microdialysis sample two times with water, achieving an electrical field enhancement in such a low conductivity sample.

3.5.2 In-Capillary Derivatization

When considering the manual, online or in CE-instrument derivatization procedures, one can remark that there is still a large difference between the total volume of sample and the volume actually injected into the CE instrument. In both cases, the volume effectively introduced into the capillary (several nL) is one thousandth to one hundredth of the solution of the derivatives. The rest remains unused or discarded.

The in-capillary derivatization has been introduced as an approach to eliminate the limitations of these precapillary derivatizations. Indeed, as CE allows free solution analysis in an open tubular capillary, a part of the inner space of a capillary can be regarded as a place where reaction derivatization could occur prior to electrophoretic analysis. The major advantage of such a procedure is that only small volumes of reagent are needed and sample dilution is reduced to a minimum, since the front end of the capillary is used as a reaction chamber.

Reagent solution was made by mixing 20 μL of a borate-NaCN solution (made by adding 111 μL of 87 mmol/L NaCN to 556 μL of 500 mmol/L pH 8.7 sodium borate buffer) and 10 μL of 10 mmol/L NDA and was kept at +4 °C for no longer than half a day. Thereafter, water, reagent solution, amino acid standard solution in aCSF, or brain microdialysis sample were put into four respective 200 μL PCR tubes which were placed in the sample tray of the CE instrument (+4 °C). To carry out derivatization at the inlet of the capillary filled with background electrolyte, the following steps were performed sequentially: injection of reagent solution (5 kV, 5 s), injection of sample (5 kV, 2.5 s), and a second injection of reagent solution (5 kV, 5 s). After each of the three consecutive injections, the inlet of the capillary was briefly dipped into the water tube to avoid a cross-contamination. After the second injection of the reagent solution, the inlet of the capillary was moved to a vial containing the background electrolyte and a 4 kV separation voltage was applied for 30 sec, followed by a 1-min step without any potential (for derivatives amplification) prior to the application of the 25 kV separation voltage. Figure 6 shows the separation of a microdialysate from rat hippocampus (together with blank and standards) obtained after in-capillary derivatization.

Although the advantages of NDA as a fluorogenic agent have been detailed [7, 8], *several previous papers have mentioned that it could lead to interfering by-products if its concentration is too high in the reaction medium. The optimal concentration of NDA that we found for in-capillary derivatization was higher than the one generally used in manual* [8–10] *or online batchwise derivatization* [5, 11]. *We suggest that there was less chance of forming by-products of CBI-amino acids when using the in-capillary derivatization mode because they were separated quite rapidly from the excess of reagent upon applying the separation voltage.*

3.5.3 Simultaneous Analysis of Glutamate and GABA

Neuroscientists may wish to assess both excitatory and inhibitory neurotransmissions in some experimental paradigms. For that purpose, they may perform simultaneous determination of Glu and GABA in brain microdialysates by CE-LIF after derivatization of sample with NDA. However, as the NDA derivative of GABA is neutral, it cannot be separated by FSCE, since in this mode of separation neutral compounds migrate all together with the EOF. NDA derivative of GABA must be separated by MEKC, a mode of separation using a running buffer containing micelles of surfactant and allowing to separate compounds according to both charge and hydrophobicity.

NDA derivatization protocols already described for the FSCE separation of Glu (e.g., manual and online derivatization of microdialysates with NDA/NaCN) can be used prior to the MEKC separation of Glu and GABA. In contrast, the in-capillary derivatization protocol is not suitable, due to the binding of NDA on micelles of surfactant. An alternative is the in-sample vial automated derivatization described further below.

MEKC Separation of Glutamate and GABA

The procedure for MEKC-LIF analysis of Glu and GABA is similar to the protocol above for FSCE analysis of Glu, with the following important modifications:

(a) The background electrolyte is a 75 mmol/L borate buffer, pH 9.2, containing 10 mmol/L hydroxypropy-β-cyclodextrine (HP-β-CD, Sigma) and 70 mmol/L SDS (can be stored up to 2 weeks at room temperature).

(b) AAD cannot be used as an internal standard since it co-migrates with several endogenous compounds contained in brain microdialysate samples. Cysteic acid is used instead since this acidic derivative of cysteine exhibits a high migration time and is therefore well separated from analytes of interest and other compounds present in the microdialysates.

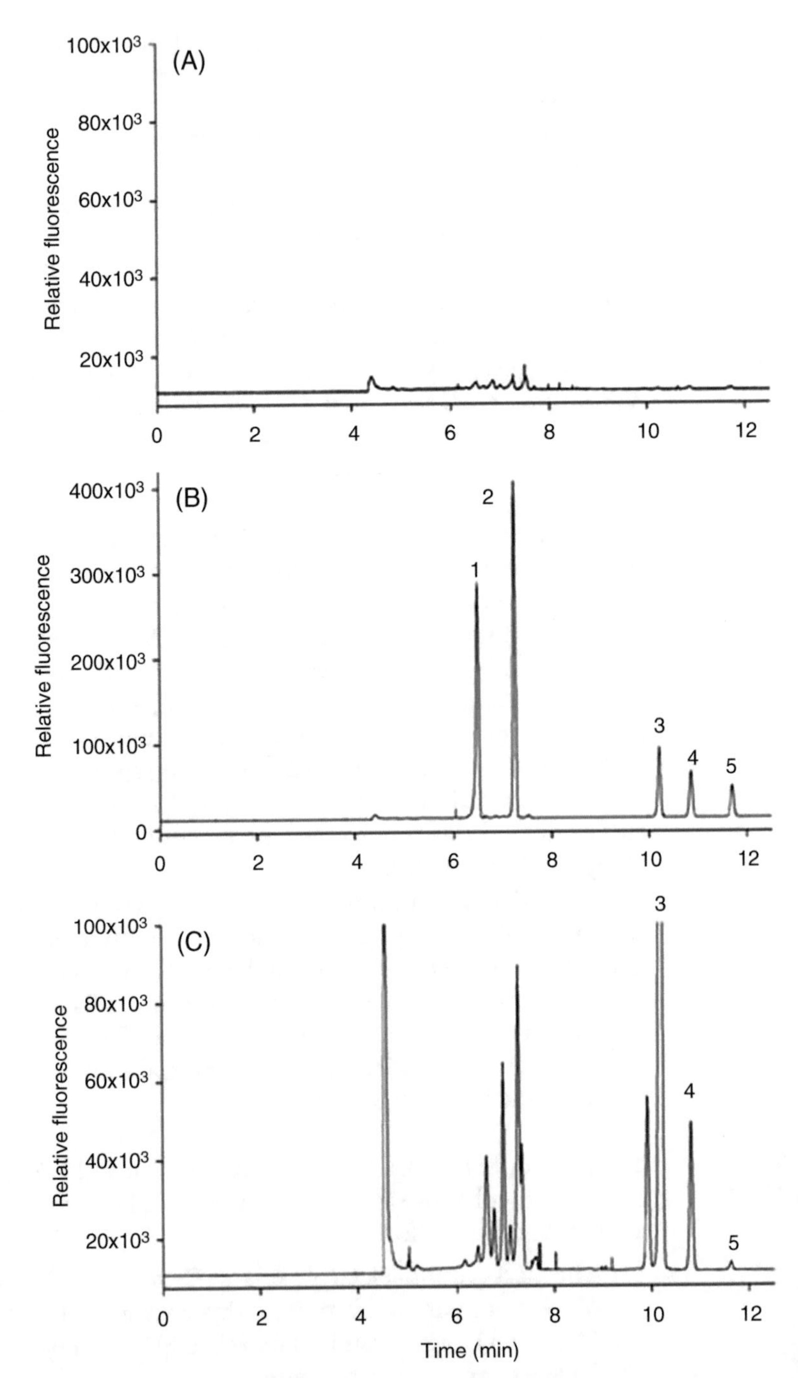

Fig. 6 Electropherograms of (**a**) blank (aCSF), (**b**) 10 μmol/L leucine (*1*), glycine (*2*), α-amino adipic acid (*3*), Glu (*4*) and Asp (*5*) standards, and (**c**) a basal rat hippocampal microdialysate sample showing the peaks of Glu (3.64 μmol/L) and Asp (0.31 μmol/L) after in-capillary derivatization (from [6], with permission from Elsevier)

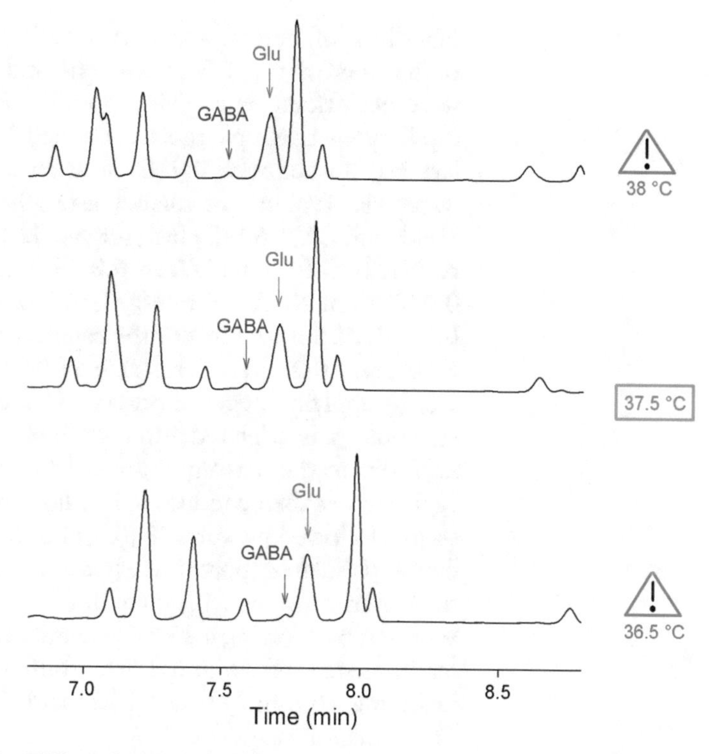

Fig. 7 Effect of capillary temperature on the separation between NDA-derivatized Glu and GABA. In this example, best separation is obtained at 37.5 °C. If the temperature is too low (36.5 °C), GABA and Glu partly overlap. If the temperature is too high (38 °C), Glu is not well separated from the following peak

(c) Between each run, the capillary is washed sequentially with 1% NaOH (20 psi, 1 min), ultrapure water (20 psi, 2 min), and background electrolyte (20 psi, 5 min).

(d) Hydrodynamic injection of sample is 0.6 psi, 10 s.

Capillary temperature is crucial for the separation of GABA in microdialysates, as shown in Fig. 7. The optimal temperature is 36–38 °C. When using a new batch of background electrolyte, an adjustment of the separation temperature with a 0.5 °C precision is required. This is made by trial and error while analyzing a few microdialysis samples (such an adjustment cannot be made with standards). If the temperature is too low, GABA and Glu co-elute. Any increase in temperature increases the migration time of Glu and allows the resolution of GABA from Glu. If the temperature is too high, Glu may be poorly resolved from the following peak.

3.5.4 In-Sample Vial Automated Derivatization

This automated derivatization procedure suitable for unattended derivatization and injection of the samples by the P/ACE MDQ system (Beckman) is described using a 50-μm i.d. fused silica capillary, 62-cm long (46 cm from injection to detection window).

Handling of buffers and derivatization solutions located in the buffer tray of the CE instrument and hydrodynamic injections were performed by applying positive or negative pressure at the capillary inlet. Sample tubes contained 2 μL of dialysate. The capillary was flushed with 0.9 μL of water by applying pressure in the water vial [volume calculated using Expert software (Beckman, Fullerton, CA, USA)]. Next, the capillary was loaded with 0.020 μL of NaCN (300 mmol/L in 0.5 M borate buffer, pH 10.5) and 0.040 μL of NDA (15 mmol/L in 75% dimethylsulfoxide). Then, 0.33 μL of the contents of the capillary (including the NaCN, the NDA, and 0.27 μL of water) were delivered into the microdialysate by applying negative pressure in the sample vial. A brief pressure pulse was delivered into the sample vial in order to push all the solutions to the bottom of the vial and ensure proper mixing. The capillary was then conditioned by flushing with 0.1 mol/L NaOH (4 μL) followed by water (3 μL), and then filled with running buffer (3 μL). After approximately 5 min of derivatization, 0.015 μL of the mixture was injected into the capillary. Separation was achieved by applying a 24 kV potential at 33 °C. The running buffer consisted of sodium borate buffer (75 mmol/L, pH 9.2) including 10 mmol/L HP-β-CD and 70 mmol/L SDS to which 5% methanol was added daily.

4 Notes

4.1 The Handling of Small Volume Samples

In CE, the small inner diameter of the capillary tube is well adapted to the analysis of extremely small sample volumes: samples ranging from a few microliters down to sub-microliter volumes can routinely be injected.

When microdialysates are manually collected and analyzed offline, the minimal sample volume required for the CE analysis must be determined. Sample injections are currently made at one side of the capillary by applying pressure. The volume of the sample must be sufficient (1) to allow the capillary to plunge into it, preventing the injection of air microbubbles with the sample, and (2) to avoid any significant loss by evaporation when a series of microdialysates (i.e., at least 30 samples) is placed in the CE sample rack before being injected. The authors have found that 940 nL is the minimal volume required for the analysis of a large series of sample with a good repeatability, provided the use of sample vials with a conic-shaped bottom, instead of PCR tubes with a flat bottom. With a shorter series of samples, determination of amino acid neurotransmitter content can be performed using 500 nL of derivatized microdialysates.

4.2 The Lowering of Contaminant Peaks

The presence of contaminant peaks can dramatically hamper the separation and quantification of Glu or other amino acid neurotransmitters. A huge Asp peak or a large increase in the Asp to Glu ratio may indicate such a contamination. To keep any contaminants as low as possible, special care must be taken to avoid any dust, fingerprint, or bacterial contamination. Ultrapure water, pipet tips, and microcentrifuge (PCR) tubes should be autoclaved. Clean gloves must be worn during manipulation of samples and reagents.

5 Discussion and Conclusion

5.1 CE-LIF: Derivatization Issues

5.1.1 General Consideration on Tagging Agents and Derivatization

Several tagging agents have been used for the CE-LIF determination of Glu in brain microdialysate and most of them are commercially available. Some of them are fluorophore, i.e., being themselves fluorescent, while others are fluorogenic, i.e., being nonfluorescent until they react with a primary amine, usually in the presence of a nucleophile. Fluorogenic agents lead to less interfering peaks and background than fluorophore agents and may be considered as very valuable. However, quantum yield, rapidity of the reaction, and stability of the derivatized products should also be considered when choosing the optimal tagging reagent. Another requirement may be the capability to derivatize low concentrations of Glu. That is one reason why NDA is used in our laboratory for determining Glu in brain microdialysates.

On the other hand, one must keep in mind that the choice of the tagging agent will affect the separation conditions. Indeed, the addition of a tagging agent to an analyte will change its charge to mass ratio and hydrophobicity, i.e., its electrophoretic characteristics, and such a change will be different for each tagging agent. Therefore, the separation conditions which have been optimized for one given fluorogene (or fluorophore) will not be valid when using another tagging agent.

5.1.2 Drawbacks When Derivatizing low Concentrations

A drawback of derivatization is that tagging agents do not fully react with their target compounds when the latter are not well concentrated. A likely reason for such a decline in the efficiency of reaction at low analyte concentration is the occurrence of a significant competition between hydrolysis and labeling reaction. The consequence of the incomplete derivatization of highly dilute analyte solution is the loss of linear correlation between the concentration and peak area.

Wagner et al. [12] have compared the performance of three fluorescence labels for the CE-LIF analysis of microdialysate Glu, NBD-F, FITC, and CFSE. If similar limit of quantification (LOQ) of 0.1 μmol/L were obtained, limit of detection (LOD) values were different: 9.8 nmol/L for NBD-F, 3.5 nmol/L for FITC, and 1.5 nmol/L for CFSE derivatives. The considerable difference between LOD and LOQ is due to the unreliable derivatization reaction at low sample concentration. In order to overcome that problem, a high excess of tagging agent with respect to the estimated concentration of analyte may lead to a quantitative labeling, but at the expenses of side reaction and/or detection of labeling agent, especially if a fluorophore is used.

A mean for derivatizing low concentrations of Glu (as of other amino acid neurotransmitters) is to choose a fluorogenic agent that may lead to less detectable side products than a fluorophore and to carefully optimize the derivatization protocol in order to enhance the reaction yield and to minimize any interferents and contaminants. It is thus possible to derivatize down to nanomolar concentrations of Glu or amine neurotransmitters, as we have performed using NDA as a tagging agent [10, 13, 14].

5.1.3 Advantages and Limitations of Most Used Tagging Agents

Aldehyde Tagging Agents

OPA

O-phthaldialdehyde (OPA) reacts rapidly with primary amines in the presence of β-mercaptoethanol to produce fluorescent isoindole products. In addition, it is itself not highly fluorescent. These characteristics made OPA an attractive tagging agent for online derivatization and online analysis approaches where short reaction time and low interferences are desired [15]. However, the isoindole derivatives of OPA are chemically unstable. Moreover, their absorbance maximum is in the near UV, at ~350 nm and, for fluorescence excitation, this will require a 325-nm HeCd laser, which may be poorly reliable. Furthermore, the absorptivity and quantum yield of the OPA-tagged amines may be lower than those of derivatives obtained with other labeling agents, leading to a limited sensitivity. To achieve low LODs of OPA derivatives, a sheath-flow cuvette system for detection that is not commercially available may be required [16, 17].

NDA

Naphthalene-2,3-dicarboxyaldehyde (NDA) is a fluorogenic tagging agent which can replace OPA, providing isoindole derivatives which are more stable than OPA derivatives. Although the derivatization reaction is slower than with OPA, it is completed after a few minutes, making NDA suitable for online derivatization and online analysis approaches. Moreover, the NDA-derivatized compounds can be excited at either 442 or 410 nm to get the optimum sensitivity, the 410-nm diode laser being well reliable. Thus, due to its numerous advantages, i.e., stability and high quantum yield of derivatives, fast reaction, suitability to manual, in-capillary and online derivatization, suitability to reliable laser, NDA has been chosen and is routinely used in our laboratory since 20 years. Its sole drawback is to handle the toxic CN^- when performing the derivatization reaction and to follow a procedure when disposing of this waste (see Sect. 2).

CBQCA

3-(4-Carboxybenzoyl)-2-quinolinecarboxaldehyde (CBQCA) may be an interesting dye because it is fluorogenic, providing a good fluorescence yield only when it has reacted with a primary amine function. Interestingly, it can be used with both HeCd (442 nm) and argon-ion lasers (488 nm) which are available with commercial CE-LIF instruments [18]. However, as for NDA, the derivatization, reaction of CBQCA with amino acids involves the using of the toxic ion CN^-.

FQ

5-Furoylquinoline-3-carboxaldehyde (FQ) is a fluorogenic tagging agent leading to fluorescent products which are excited by the 488-nm line of the argon-ion laser. However, the FQ derivatives are not soluble enough in aqueous buffers to be compatible with an online reaction [19].

Fluorescein-Based Dyes

The tagging agents related to fluorescein are fluorophore; after reaction with amine-containing compounds, they lead to highly fluorescent derivatives.

FITC

Fluorescein isothiocyanate (FITC) is a widely used fluorescent derivatizing agent for the amine groups, primarily due to its high quantum yield and its compatibility with the commercial 488 nm argon-ion laser. However, FITC requires a long reaction time (12–16 h) for derivatization and yields numerous degradation and side products which may hamper the separation and prevent from detecting very low concentration of amino acids in a mixture.

Nevertheless, FITC derivatives may be considered as stable during a few hours [12].

CFSE

5-CarboxyfluoresceinN-succinimidyl ester (CFSE) is another fluorescein-based dye which fits the excitation wavelength of the 488-nm argon-ion laser. It is a fluorophore which gives interfering peaks, although it has been claimed to form less hydrolysis products than FITC. However, CFSE derivatives exhibit inferior stability compared to FITC derivatives and interfering peaks increase over time [12, 20].

SIFA

N-Hydroxysuccinimidyl fluorescein-O-acetate (SIFA) is a labeling reagent whose derivatization reaction is usually performed in aqueous phase. In this case, competition occurs between the derivatization of SIFA with analytes and the hydrolysis of SIFA itself, which may give rise to a large interfering peak [21].

DTAF

5-(4,6-Dichloro-s-triazin-2-ylamino) fluorescein (DTAF) is a fluorescent fluorescein analog used in CE-LIF analysis and exhibiting good stability. The excitation/emission wavelengths of DTAF are 492/520 nm, so a commercially available argon-ion laser could be used.

Benzoxadiazole-Based Dye

NBD-F

4-Fluoro-7-nitro-2,1,3-benzoxadiazole (NBD-F) is an interesting tagging agent which addresses some of the limitations of the other labeling reagents. Due to its fluorogenic nature, less interfering peaks are expected and the optimum excitation wavelength for NBD-F derivatives is ~470 nm, compatible with argon-ion lasers or newer diode pumped solid state lasers. Moreover, NBD-F is a fluorogenic reagent that efficiently reacts with primary and secondary amines, allowing detecting proline, if needed, in addition to amino acid neurotransmitters. However, NBD-F derivatives could exhibit a limited stability: standing in the refrigerated autosampler for 2 h, peak areas of the analytes have been claimed to decrease by about 50% [12, 16].

5.2 LIF Detection: Laser Issues

To detect NDA-tagged amino acids, the 442-nm line of the HeCd laser was first used, but its main drawback is its limited lifetime (~2000–3000 h). An alternative is the 410-nm diode laser which offers a longer lifetime at a lower cost while providing comparable results in terms of sensitivity [22]. Furthermore, cheaper alternative are light-emitting diodes (LEDs) which are attractive due to their long lifetime and small size. However, due to weaker intensities and the broader emission light, LED-induced fluorescence detection (LEDIF) is intrinsically less sensitive for analytes than LIF detection [23]. However, by optimizing the detector setup (see below) or performing in-capillary sample concentration, level of sensitivity sufficient for analyzing microdialysates may be obtained.

5.3 Sensitivity Issues

5.3.1 Needs for Reaching a Low Limit of Detection

In microdialysis-based Glu monitoring, high sensitivity of the analytical method is required due to the low concentration of Glu in the extracellular space and the even lower concentration of Glu in the microdialysate resulting from the moderate recovery of microdialysis probe (usually ~12% at ambient temperature for a 3-mm active dialysis length membrane). This will be worsened if the microdialysis probe is implanted into a small-sized brain area (as currently encountered when using mice), as the probes suitable for such structure (1.5 mm) will exhibit lower recovery.

5.3.2 Relative Significance of the Limit of Detection

One may emphasize the difficulties for comparing LODs between studies, since according to publications LOD may be expressed from standards analysis (usually performed when using off-line analysis) [10, 22, 24] or expressed as the concentration at the microdialysis probe for amino acid neurotransmitters (usually performed when CE is online coupled to microdialysis and derivatization) [2, 16]. Furthermore, in the former case, derivatization is sometimes carried out at high analyte concentration followed by dilution, leading to impressive LOD values. However, these data are mostly informative about the sensitivity of the LIF detector and are far lower than achievable in real sample analysis. To evaluate the analytical potential of a CE-LIF method, the LOD must be determined from a calibration curve constructed with several derivatized concentrations of standards and not from the dilutions of a derivatized high concentration. Nevertheless, LOD expressed as the concentration at the microdialysis probe best reflects the sensitivity which may be expected in real sample analysis [5]. In this latter case, however, a good LOD is not only aided by a high-sensitivity LIF detector, but also by the relative recovery of the microdialysis step, which depends on the aCSF flow rate used, and of the surface of the dialysis membrane.

5.3.3 Factors Affecting the Sensitivity Independently of Detection System

Pre-analytical Issues

Change with time of level of amino acid neurotransmitters as a result of storage and sample preparation has been documented for CSF and to a lesser extent for microdialysis samples. Spontaneous synthesis of GABA from homocarnosine has been reported in human CSF at room temperature; however, this was not reported in rat CSF [25]. Changes in human CSF Glu level during storage and treatment have also been reported [26]. One must be aware that a number of different factors can alter the recoveries of amino acid neurotransmitters in CSF and microdialysis samples. Thus, uniform sample storage and sample preparation are critical. The best solution to alleviate this problem could eventually be online derivatization and online microdialysis CE system.

In-Capillary Preconcentration Step (Stacking)

To lower the limit of detection, a stacking technique can be applied. Such approach, which can dramatically increase the level of sensitivity, has been thoroughly studied and applied to various types of samples [1, 27–29]. However, it has not been widely used in the CE analysis of amino acid neurotransmitters from microdialysates or CSF samples, mainly for two reasons: (a) most of the stacking techniques are efficient only on low-salt samples and are not suitable to aCSF matrix and (b) the stacking can only be performed in the off-line microdialysis CE setup. Nevertheless, some stacking protocols can be successfully applied to brain microdialysates, as shown in the few examples below.

Indeed, a stacking technique using poly(ethylene oxide) (PEO) as a high-viscosity barrier can handle high-salt samples and has been applied to the CE-LEDIF analysis of NDA-tagged Glu, GABA, and glycine (Gly) in human CSF. In this technique, the amino acid derivatives (anions) migrating against EOF are stacked at the boundary between sample zone and PEO (neutral) solutions as a result of increases in viscosity and possible interactions with PEO molecules (hydrogen bonding) [23]. Another stacking approach related to large-volume sample stacking injection was used to increase the sensitivity in the analysis of Glu, GABA, and Gly from periaqueductal gray matter microdialysates, by CE with contactless conductivity detection (C4D). Although the intrinsic sensitivity of such detector is lower than LIF, the stacking approach performed on underivatized amino acids led to the same level of sensitivity than LIF [30]. Similarly, the LODs for underivatized amino acids from CSF were improved, reaching 20 nmol/L range, when performing in-capillary preconcentration based on pH-mediated stacking in CE-mass spectrometry (MS) analysis [31].

5.3.4 Factor Affecting the Sensitivity When Using CE-LIF

Tagging Agents

The fluorescence of the derivatized amino acid neurotransmitters will depend on the quantum efficiency of the dye used. Nevertheless, if some tagging agents may exhibit higher quantum yield than others (FITC may be superior to OPA), the molecular structure which is tagged and the physicochemical characteristics of the medium can also affect the fluorescence of the derivatized amino acid neurotransmitters.

Molecular Structure of Analyte

CE-LIF electropherograms corresponding to the separation of several amino acids clearly show that the peak intensities may be different for each amino acid, despite they were labeled at the same concentration and by the same tag. This is clearly shown in Fig. 6b, in which peaks from derivatives of Glu and Asp are smaller than those from derivatives of Gly and leucine. These differences could be related to variations in absorbance and fluorescence properties of each derivative, as well as to differences in their kinetics of labeling and/or of degradation [18, 22].

Electrolyte Properties

The peak intensities of the derivatized amino acids are also affected by the background electrolyte [22]. The fluorescence intensities of compounds labeled with FITC or with NDA are commonly higher at basic conditions [23, 24]. Besides pH, the chemical composition of the background electrolyte can also affect fluorescence. For example, when used as separation buffer additives, cyclodextrins are capable of increasing the fluorescence signal of NDA-derivatized acidic amino acids [32]. Moreover, when performing MEKC separation of NDA-derivatized amino acids, the residence time of analytes in the hydrophobic core of micelles is more important when using LDS instead of SDS and narrower and higher peak could be obtained [22].

Type of LIF Detector

Besides the commercially available LIF detectors which offer on-capillary detection of analytes through an excitation light from outside the capillary, improved sensitivity may be obtained in some homemade detectors. Highest sensitivity may be offered by the sheath-flow cuvette detector; indeed, this arrangement reduces the background signal caused by laser light scattering and background fluorescence, which limits sensitivity when performing on-capillary detection. Moreover, gain in sensitivity made through the use of the sheath-flow cuvette allows a reduction in injection volume and hence an improvement in separation efficiency [2].

In another home-build setup allowing a reduction of background and improved fluorescence, the excitation light is brought by an optical fiber inserted inside the separation capillary. Indeed, when the excitation light excites analytes from outside the capillary at the detection window, light reflecting and scattering on capillary surface could result in background noise and low detection sensitivity. In contrast, by inserting an optical fiber directly into the outlet end of the separation capillary and settling it at the detection window, the excitation light could completely excite the analytes while avoiding the light reflecting and scattering on the capillary surface [33, 34].

5.4 Other Detections than LIF

5.4.1 EC

Amino acids are not electroactive enough to be detected by electrochemistry (EC) at low concentration. Therefore, a precolumn derivatization is needed to obtain electroactive derivatives. This could be performed using NDA and the approach is sensitive enough for the monitoring of Glu in brain microdialysates, but an electrochemical pretreatment of the carbon fiber microelectrode was found to be necessary between each CE run, as the amino acid derivatives appeared to foul the microelectrode surface of the home-build detector [35]. However, as no EC detector for CE instrument is commercially available, this type of detection is not widely used for the determination of the amino acid neurotransmitters after CE separation.

5.4.2 MS

Despite its high selectivity, mass spectrometry (MS) is not yet widely used for the determination of amino acid neurotransmitters in brain microdialysate or CSF. Its feasibility has been shown for determining neurotransmitters in brain tissue after solid-phase extraction, with an LOD (nmol/L range) which could be suitable for microdialysate analysis [36]. The high cost of the instrument and the expertise needed to master this type of detection do not favor its widespread use in neurobiology laboratories. Nevertheless, metabolomics studies have revealed the power and interest of CE-MS in the profiling and determination of CSF amino acids [31].

5.4.3 C4D

The sensitivity of the contactless conductivity detection (C4D) is not particularly high, but the optimization of the BGE and the use of online stacking procedures allow attaining submicromolar LOD values, i.e., in the 10 nmol/L range, in the CE analysis of underivatized amino acid neurotransmitters. The sensitivity is thus fully comparable with the use of LIF and permits monitoring of neurotransmitters in brain microdialysates [30].

5.5 Advantages of Simultaneous Determination of Other Neurotransmitters than Glutamate

The high-speed multianalyte capability of CE offers opportunities to simultaneously separate several neurotransmitters in addition to Glu and therefore to monitor interactions between neurotransmitter systems in a physiological or pharmacological experiment. Several CE-LIF methods have been developed in order to separate microdialysate excitatory amino acids (Glu, Asp), inhibitory amino acids (GABA, Gly), modulator of NMDA receptor (D-serine), and even catecholamines (dopamine) within the same run [22, 24, 37, 38].

5.6 Separation Modes: FSCE vs. MEKC

Many derivatized amino acids tend to have close electrophoretic mobilities and therefore cannot be separated by CE when using simple inorganic buffers as background electrolytes, as in most of the FSCE separations. However, this is less encountered when analyzing the sole Glu because it bears two carboxylic acid functions which make its migration time longer than those of other amino acids and amine compounds. Thus, the FSCE analysis of the sole Glu is one of the simplest and easiest to run CE separation. When separating other neurotransmitters in addition to Glu, the peak capacity can be improved nevertheless by adding to the BGE complexing agents such as cyclodextrins (β-cyclodextrin or HP-β-cyclodextrin) which may differentially alter the mobility of each analyte, while keeping a FSCE mode of separation [2].

However, the number of compounds that could be resolved and detected will be more significantly increased by using MEKC as the separation mode, this mode allowing separating compounds according to both their charge to mass ratio and hydrophobicity. In MEKC, the concentration of surfactant has a significant effect on the separation because it modifies the selectivity by adjusting the partition of analytes between micellar pseudophase and aqueous phase. The analyte separation clearly improves with increasing concentrations of SDS, probably due to increased micelle analyte interaction. Moreover, the addition of organic solvent has been shown to improve the resolution in MEKC, as it can increase the hydrophobicity of the aqueous phase, easing the interactions between analytes and micelles and changing the distribution of analytes between the aqueous phase and the micellar pseudophase [18, 21]. A drawback using MEKC may be longer migration times.

5.7 CE Determination: Other Factors Affecting the Quality of the Analysis

In order to ensure the repeatability of the CE method, stability of EOF from run to run is mandatory, i.e., the properties of the inner surface of the separation capillary must remain constant. For that purpose, a rinsing sequence consisting of NaOH, water, and running buffer should be performed between each run to effectively eliminate the adsorption of analytes and/or matrix constituents onto the capillary wall, as shown in Fig. 8. A simple rinse with the running buffer is not sufficient to get a proper analysis of GABA/Glu with our MEKC-LIFD protocol.

Moreover, some publications have suggested that the value of EOF may affect the quality of the quantification. Indeed, Lu et al. [23] have reported that, despite longer migration time and broader peaks, the sensitivity of CE-LIF was better in the absence of EOF as a result of a stable baseline and lower fluorescence background. Similarly, Tuma et al. [39] have noted that the addition of polyethyleneglycol to the running buffer decreases and stabilizes the EOF, thus improving the repeatability of the migration times and peak areas.

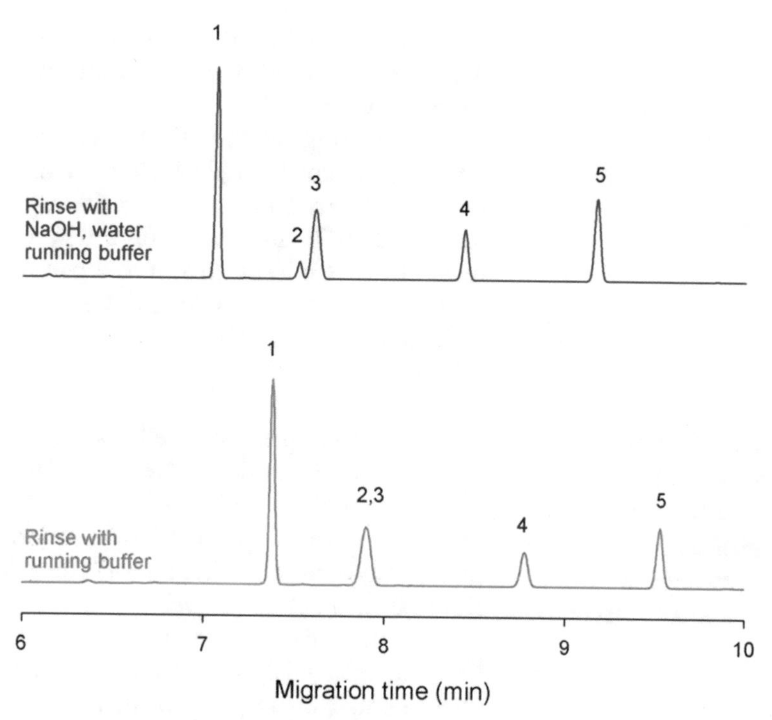

Fig. 8 Electropherograms obtained using the protocol for GABA/Glu quantification. The quality of the separation is depending on the rinses before sample injection: rinse with the running buffer (30 s) versus rinse using NaOH (30 s), water (30 s), then running buffer (30 s). The standard sample contained glycine (*1*), Glu (*3*), L-Asp (*4*) at 5 μmol/L, and GABA at 0.5 μmol/L was derivatized manually using cysteic acid as internal standard (*5*). Note that the separation is warranted only with a preconditioning with NaOH, followed by water and buffer rinses. The rinses have also an impact on migration times of all the molecules considered here

Besides, when working with samples from different sources, matrix effect may change. For example, the migration times for Glu are different for various CSF samples, resulting from matrix effects. Thus, it may be important to apply a standard addition method to minimize the matrix interference in compound identification and quantification [23].

5.8 Coating of Capillary

In most of the CE separation of amino acid neurotransmitters from brain microdialysates, bare fused silica capillaries are used. In such capillaries however, Glu possesses the least apparent mobility and migrates well after the other amino acids, making the analysis time relatively long, i.e., usually in the 3–20 min depending on the CE instrument setup and separation conditions (capillary length, temperature, buffer concentration, etc.…). However, using polyacrylamide-coated capillaries exhibiting very low EOF and applying reversed polarity for separation, the migration order will be reversed, leading to a reduction in the migration time of Glu. This may be useful when large number of microdialysis samples is to be analyzed [12].

When using bare fused silica capillaries for the CE analysis of CSF with minimal sample pretreatment, the adsorption of proteins or other matrix components to the capillary wall may cause irreproducible EOF and migration times. This problem may be minimized by dynamically coating the bare fused silica capillaries with charged polymers. Such noncovalently coated capillaries are proven to be successful for the CE-MS analysis of underivatized amino acids from CSF [31].

5.9 Comparison of Online Versus Off-Line Analysis after Online Derivatization of Microdialysates

Derivatized samples obtained just after the derivatization device connected to the outlet of the microdialysis probe, can be analyzed on line through an analytical interface allowing the direct coupling of the microdialysis probe/derivatization device to the CE instrument, or off-line, i.e., after sample collection in vials, which may be stored frozen before analysis.

As microdialysis samples are protein-free, online coupling between microdialysis/derivatization and custom-made CE-LIF instruments can be performed, provided that derivatization and separation are performed rapidly. This setup prevents sample loss and evaporation, removing the requirement of collecting, storing, and analyzing large numbers of nanoliter-volume microdialysate fractions and giving results in real time. Using various interfaces connected to a continuous flow derivatization device, 20-s to 3-min sampling rates has been reported for the in vivo monitoring of excitatory amino acids [2, 40]. With the online microdialysis/CE system, however the CE voltage is always on and the samples are injected electrokinetically into the capillary. Therefore, in order to get the best injection precision, it is necessary to use a run buffer of ionic strength higher than that of the sample. With online methods, however, temporal resolution becomes limited by the speed of the analytical method used. In addition, for an accurate monitoring of fast events, one must take into account the total delay in the response of the online system to a biochemical event.

The analysis of low-volume microdialysates can also be performed off-line using a commercially available CE-LIF system, provided that online derivatization of sample is carried out [5]. Consequently, this technique can be set up in neuroscience laboratories that have no access to a specialized workshop for making custom-made CE instruments. Furthermore, one advantage of the off-line approach is to uncouple microdialysis sampling from the CE analysis, allowing optimal separation and derivatization conditions to be utilized. Off-line analysis can be useful since, if any breakdown of the CE-LIF system occurs, microdialysates can be stored up to 3 days prior to analysis. In contrast, if online analysis is used, samples cannot be saved and data are lost. Thus, off-line analysis offers more flexibility for planning experiments.

5.10 Online Analysis and High Temporal Resolution

The online microdialysis/derivatization/CE analysis approach allows the monitoring of fast neurobiological events with a high temporal resolution, which is characterized by a delay time and a response time. The delay time is defined as the time required for the change in neurotransmitter concentration to be first detected. It is the sum of: (1) the time needed for the sample to be collected and transferred to the derivatization device, (2) the time spent into the derivatization device, (3) the time needed for the sample to be moved from the derivatization device to the injection interface, and (4) the CE separation time. The response time is defined as the time required for an increase to reach its maximum after the increase was first detected. It is limited by the broadening of a concentration pulse by diffusion and parabolic flow when it is pumped through the microdialysis probe/derivatization device.

Temporal resolutions on the order of a few seconds have become feasible using the online microdialysis CE approach, providing that short separation capillary is used, since the separation time is a limiting factor. Temporal resolution can be further improved by performing overlapping injections if no peaks are detected at the beginning of the electropherogram. For example, if no peak is detected during the first 6 s of a 17 s separation, the separation is stopped after ~6 s and a second injection is performed. Analytes from the first injection that have not migrated out of the capillary pass through the detector before the first peak of the subsequent separation. Thus the total time for separation becomes ~10 s [41].

5.11 Liquid Chromatography vs. CE-LIF for the Determination of Glutamate in Brain Microdialysis Samples

Besides CE, high performance liquid chromatography (HPLC) methods, coupled to fluorescence (i.e., lamp-induced fluorescence), EC or MS detection, are commonly employed for determining Glu in brain microdialysates and therefore need to be considered.

5.11.1 HPLC with Fluorescence Detection

In HPLC-fluorescence methods, samples are derivatized before to be injected into a reversed phase column. For example, OPA/beta-mercaptoethanol derivatization was used to determine Glu in 20 μL (10-min collection) rat brain microdialysates and 20 μL of derivatized samples were injected onto a C18 250 × 4.6 mm column. LOD was 37 nmol/L [42]. The same derivatization was used to determine Glu in 5 μL rat brain microdialysate; the separation was performed on an 80 × 4.6 mm 3-μm packing reversed phase column and the fluorimetry allowed detecting Glu with an LOD of 20 nmol/L [43]. A similar method has been used [44, 45] to analyze Glu in 4 μL (2-min collection time) OPA-derivatized human brain microdialysates. A ODS 100 × 3.2 mm 3-μm column was employed and the LOQ was ~33 nmol/L. A lower LOD (2 nmol/L) has been reported for another method, in which a

C18 150 × 4.6 mm 5-μm column was used for separating 10-min (20 μL) mice brain microdialysates [46]. In an earlier paper, Kehr et al. [47] reported a comprehensive comparison of five different systems HPLC system with fluorescence detection for the fast determination of OPA-derivatized Glu and Asp in microdialysis sample. It was shown that the fluorescence detection allowed LOD of 5 nmol/L using conventional columns (60 × 4 mm Nucleosil 5 C18) and of about 2–3 nmol/L for the microbore system (100 × 1 mm column, C18 silica, 5-μm particle size).

The precolumn NDA/CN^- derivatization has been also employed for the chromatographic separation and fluorescence detection of Glu from brain microdialysates and offers the same level of sensitivity than the OPA/beta-mercaptoethanol derivatization. NDA/CN^- derivatization has been used [48] for determining Glu in 7 μL of rat brain microdialysate, the separation being performed on an ODS 100 × 4.6 mm 3-μm particle column. The same derivatization has been used for the determination of eight amino acids, including Glu in 4-min rat brain microdialysates. Samples were separated on a 250 × 4.6 mm C18 5-μm column before fluorescence detection which gave LOD around 6 nmol/L [49].

5.11.2 HPLC with Electrochemical Detection

As Glu is not easily electroactive, samples need to be derivatized before to be injected into a reversed phase column and detected by electrochemistry. OPA/beta-mercaptoethanol derivatization and HPLC-EC have been used to analyze cat CSF (25-μL sample volume). In the derivatization process, excess thiol reagent was removed with iodoacetamide in order to improve amperometric detection. The separation was performed using C18 25 × 4.6 mm 5-μm column and allowed to determine eight amino acids, including Glu with an LOD in the μmol/L range [50]. The need for a second derivatization step with iodoacetamide to remove excess of a thiol moiety was also emphasized [47]. It allows to suppress the frontal peak and improves the baseline for a better resolution of Glu and Asp from brain microdialysates. LOD was in the 40 nmol/L range, using amperometric detection. Other thiols have been investigated to counter the relative instability of derivatives when using the OPA/beta-mercaptoethanol derivatization. In that respect, OPA/tert-butylthiol derivatization has been performed on 20 μL of pull-push perfusates from rat brain. The separation on a C18 3-μm column (size not indicated) and amperometric detection allowed the determination of six amino acids, including Glu with LODs in the 50–250 nmol/L range [51]. Lower LOD has been reported when using microbore-HPLC with amperometric detection. Indeed, in the determination of Glu and Asp in 15 μL of OPA/tert-butylthiol derivatized cat brain microdialysate using a C8 100 × 1 mm 3-μm column, the LOD was found equal to 1 nmol/L [52]. Sulfite has been proposed as an alternative to thiols for the OPA derivatization. It has been used in a method for the determination of Glu in OPA/sulfite-derivatized rat brain microdialysates (20 μL sample volume). Using a C18 25 cm 5-μm column and amperometric detection, LOD was around 0.5 nmol/L [53].

Coulometry is another mode of electrochemical detection which has been used in the HPLC analysis of OPA/beta-mercaptoethanol-derivatized microdialysate, as in the work by Donzanti and Yamamoto [54] in which samples were separated on a C18 8 × 4.3 mm 3-μm column. Seven amino acids including Glu were measured in 25-μL (2.5 μL/min perfusion flow rate) rat brain microdialysate and LOD was in the 50–100 nmol/L range. Using the same technique, a similar column has been used for detecting Glu in 2-min (2 μL) microdialysates [55]. More recently, HPLC with coulometric detection has also been used for determining Glu in mice brain microdialysate after precolumn derivatization with OPA. Twenty μL samples were injected into a C18 50 × 3.2 mm column and LOD as low as 0.01 fg/20 μL (3.4 fmole/L) has been claimed [56].

Besides the direct electrochemical detection of OPA-derivatized Glu, this neurotransmitter could also be separated as native and electrochemically detected thanks to a postcolumn reactor containing the enzyme glutamate oxidase. The hydrogen peroxide arising from the enzymatic reaction is detected by amperometry on a platinum working electrode. This has been used for monitoring Glu in 20 μL (20-min collection) mice brain microdialysate [57].

5.11.3 HPLC with Mass Spectrometry Detection

For the determination of Glu in brain microdialysate, HPLC coupled to MS detection has emerged as a suitable alternative to HPLC with fluorescence or electrochemical detection, thanks to the enhanced chemical selectivity brought by the MS detection and to the possibility to simultaneously quantify other neuroactive compounds. For example, 5 μL of rat brain microdialysate (20-min collection time) were separated by ultra-high performance liquid chromatography (UHPLC) using a BEH C18 100 × 2.1 mm, 1.7-μm column and analytes were detected and quantified by a triple quadrupole (TQ) MS operated in multiple reaction monitoring (MRM) mode. Nine compounds including Glu were detected and the LOD for Glu was ~136 nmol/L [58]. As native (i.e., underivatized) neuroactive compounds such as amino acids are hydrophilic, they can be more easily separated by hydrophilic interaction liquid chromatography (HILIC) than by reversed phase chromatography. In that respect, capillary HILIC-MS, using a homemade 30 cm × 200 μm column packed with 5-μm polyhydroxyethylaspartamide particles, has been used for separating extracellular brain fluid from rhesus monkey brain obtained by the push-pull technique, instead of microdialysis sampling. Detection was performed by ion-trap MS operated in MRM mode, allowing the determination of six neurotransmitters, including Glu with an LOD equal to

20 nmol/L [59]. In another study, 20-μL microdialysates from rat brain were separated by HILIC on a 20 × 2.1 mm 3.5-μm column and eluates were detected on a TQ MS. Using the MRM mode, several fragment ions were generated for the identification of analytes and the LOQ for Glu was 10 nmol/L [60].

The derivatization is another way to improve the separation of amino acids and other hydrophilic neuroactive compounds on reversed phase column. For example, prior derivatization of microdialysate with benzoyl chloride increases the hydrophobicity of compounds so that they can be more easily separated by reversed phase liquid chromatography. Such a derivatization also increases the sensitivity of MS detection. It has been performed on 5-μL rat brain microdialysates prior to their separation by UHPLC on a BEH C18 100 × 1 mm 1.7-μm column and their detection by TQ MS operated in MRM mode. Seven neuroactive compounds were detected, including Glu with an LOD equal to 5 nmol/L [61]. In another work, 10-min rat brain microdialysates (30 μL) were collected and were derivatized with Symdaq (Symmetrical DiAldehydes Quaternary ions) reagent. Forty-five microliters of the derivatized sample (corresponding to 8.3 μL of underivatized microdialysate) were injected on a C12 100 × 3.0 mmol/L 2.5-μm column. Analytes were detected on a TQ MS and acquisition was run on the MRM mode. LOD on column was 20 nmol/L, i.e., five times lower than for underivatized sample [62].

5.12 To Conclude: What to Choice for Glutamate Monitoring, HPLC or CE

The above short reviews clearly show that both HPLC methods and CE-LIF could be suitable for determining Glu in brain microdialysis samples, as they can offer similar concentration sensitivity. Moreover, HPLC-fluo and HPLC-EC need about the same level of expertise in separation science as CE-LIF. In contrast, HPLC-MS needs special expertise and could be mastered only by well-trained scientists.

The major advantage of CE-LIF over HPLC, in its classical or microchip (see Chap. 12) format, is its suitability for analyzing low-volume microdialysates (i.e., <1 μL) for high-temporal resolution monitoring. HPLC methods cannot handle such low-volume samples arising from experiments aimed to monitor rapid neurobiological events. If μL range samples can be analyzed using microbore chromatography, one must be aware that this technology is less straightforward to run that classical HPLC, needing special equipment and a constant care to minimize extra-column volumes in order to keep a sufficient separation resolution. Another major advantage of CE-LIF is its lower cost to run, mainly related to the lower price of fused silica capillaries as compared to packed HPLC columns and to the lower volumes of solvent and buffer used to perform the separation.

Finally, one must always keep in mind that CE-LIF will give its best only when performed by well-trained scientists fully aware of

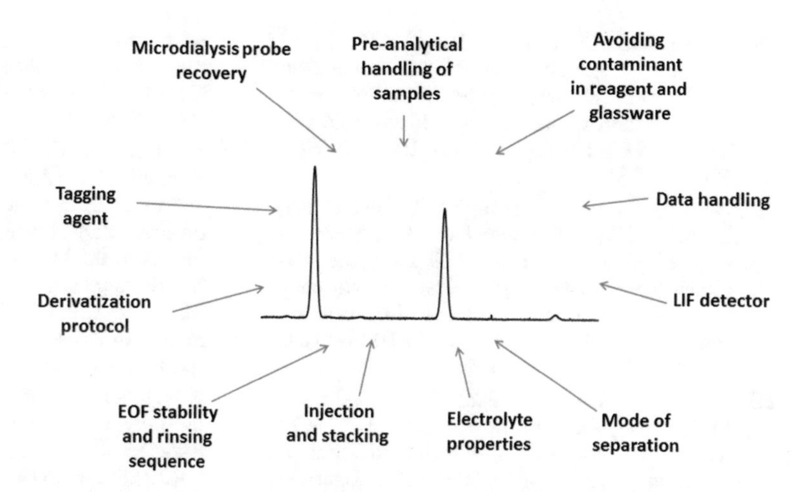

Fig. 9 Synthetic representation of factors influencing the performances of CE-LIF determination of glutamate in brain microdialysates

the numerous factors influencing the performances of CE-LIF determination of Glu in brain microdialysates, which are detailed in this chapter and summarized on Fig. 9.

Acknowledgments

The authors thank the Institut National de la Santé et de la Recherche Médicale (INSERM), the Centre National de la Recherche Scientifique (CNRS), and the Université Claude Bernard Lyon 1. The authors warmly thank Dr. Chloé Hegoburu for her help for the Figs. 1 and 3.

References

1. Denoroy L, Parrot S (2017) Analysis of amino acids and related compounds by capillary electrophoresis. Sep Purif Rev 46(2):108–151
2. Bowser MT, Kennedy RT (2001) In vivo monitoring of amine neurotransmitters using microdialysis with on-line capillary electrophoresis. Electrophoresis 22(17):3668–3676. doi:10.1002/1522-2683(200109)22:17<3668::AID-ELPS3668>3.0.CO;2-M
3. Kennedy RT, Watson CJ, Haskins WE, Powell DH, Strecker RE (2002) In vivo neurochemical monitoring by microdialysis and capillary separations. Curr Opin Chem Biol 6(5):659–665
4. Powell PR, Ewing AG (2005) Recent advances in the application of capillary electrophoresis to neuroscience. Anal Bioanal Chem 382(3):581–591. doi:10.1007/s00216-005-3075-x
5. Parrot S, Sauvinet V, Riban V, Depaulis A, Renaud B, Denoroy L (2004) High temporal resolution for in vivo monitoring of neurotransmitters in awake epileptic rats using brain microdialysis and capillary electrophoresis with laser-induced fluorescence detection. J Neurosci Methods 140(1–2):29–38. doi:10.1016/j.jneumeth.2004.03.025
6. Denoroy L, Parrot S, Renaud L, Renaud B, Zimmer L (2008) In-capillary derivatization and capillary electrophoresis separation of amino acid neurotransmitters from brain microdialysis samples. J Chromatogr A 1205(1–2):144–149. doi:10.1016/j.chroma.2008.07.043
7. Bardelmeijer HA, Lingeman H, de Ruiter C, Underberg WJ (1998) Derivatization in capillary electrophoresis. J Chromatogr A 807(1):3–26

8. Chang PL, Chiu TC, Chang HT (2006) Stacking, derivatization, and separation by capillary electrophoresis of amino acids from cerebrospinal fluids. Electrophoresis 27(10):1922–1931. doi:10.1002/elps.200500496
9. De Montigny P, Stobaugh JF, Givens RS, Carlson RG, Srinivasachar K, Sternson LA, Higuchi T (1987) Naphthalene-2,3-dicarboxyaldehyde/cyanide ion: a rationally designed fluorogenic reagent for primary amines. Anal Chem 59(8):1096–1101. doi:10.1021/ac00135a007
10. Sauvinet V, Parrot S, Benturquia N, Bravo-Moraton E, Renaud B, Denoroy L (2003) In vivo simultaneous monitoring of gamma-aminobutyric acid, glutamate, and L-aspartate using brain microdialysis and capillary electrophoresis with laser-induced fluorescence detection: analytical developments and in vitro/in vivo validations. Electrophoresis 24(18):3187–3196. doi:10.1002/elps.200305565
11. Zhou SY, Zuo H, Stobaugh JF, Lunte CE, Lunte SM (1995) Continuous in vivo monitoring of amino acid neurotransmitters by microdialysis sampling with on-line derivatization and capillary electrophoresis separation. Anal Chem 67(3):594–599
12. Wagner Z, Tabi T, Zachar G, Csillag A, Szoko E (2011) Comparison of quantitative performance of three fluorescence labels in CE/LIF analysis of aspartate and glutamate in brain microdialysate. Electrophoresis 32(20):2816–2822. doi:10.1002/elps.201100032
13. Bert L, Robert F, Denoroy L, Renaud B (1996) High-speed separation of subnanomolar concentrations of noradrenaline and dopamine using capillary zone electrophoresis with laser-induced fluorescence detection. Electrophoresis 17(3):523–525. doi:10.1002/elps.1150170318
14. Bert L, Robert F, Denoroy L, Stoppini L, Renaud B (1996) Enhanced temporal resolution for the microdialysis monitoring of catecholamines and excitatory amino acids using capillary electrophoresis with laser-induced fluorescence detection. Analytical developments and in vitro validations. J Chromatogr A 755(1):99–111
15. Lada MW, Kennedy RT (1996) Quantitative in vivo monitoring of primary amines in rat caudate nucleus using microdialysis coupled by a flow-gated interface to capillary electrophoresis with laser-induced fluorescence detection. Anal Chem 68(17):2790–2797
16. Klinker CC, Bowser MT (2007) 4-Fluoro-7-nitro-2,1,3-benzoxadiazole as a fluorogenic labeling reagent for the in vivo analysis of amino acid neurotransmitters using online microdialysis-capillary electrophoresis. Anal Chem 79(22):8747–8754. doi:10.1021/ac071433o
17. Shou M, Smith AD, Shackman JG, Peris J, Kennedy RT (2004) In vivo monitoring of amino acids by microdialysis sampling with on-line derivatization by naphthalene-2,3-dicarboxyaldehyde and rapid micellar electrokinetic capillary chromatography. J Neurosci Methods 138(1–2):189–197. doi:10.1016/j.jneumeth.2004.04.006
18. Bergquist J, Gilman SD, Ewing AG, Ekman R (1994) Analysis of human cerebrospinal fluid by capillary electrophoresis with laser-induced fluorescence detection. Anal Chem 66(20):3512–3518
19. Wu J, Chen Z, Dovichi NJ (2000) Reaction rate, activation energy, and detection limit for the reaction of 5-furoylquinoline-3-carboxaldehyde with neurotransmitters in artificial cerebrospinal fluid. J Chromatogr B Biomed Sci Appl 741(1):85–88
20. Chen HL, Zhang XJ, Qi SD, Xu HX, Sung JJ, Bian ZX (2009) Simultaneous determination of glutamate and aspartate in rat periaqueductal gray matter microdialysates by capillary electrophoresis with laser-induced fluorescence. J Chromatogr B Anal Technol Biomed Life Sci 877(27):3248–3252. doi:10.1016/j.jchromb.2009.08.006
21. Deng YH, Wang H, Zhang HS (2008) Determination of amino acid neurotransmitters in human cerebrospinal fluid and saliva by capillary electrophoresis with laser-induced fluorescence detection. J Sep Sci 31(16–17):3088–3097. doi:10.1002/jssc.200800339
22. Siri N, Lacroix M, Garrigues JC, Poinsot V, Couderc F (2006) HPLC-fluorescence detection and MEKC-LIF detection for the study of amino acids and catecholamines labelled with naphthalene-2,3-dicarboxyaldehyde. Electrophoresis 27(22):4446–4455. doi:10.1002/elps.200600165
23. Lu MJ, Chiu TC, Chang P, Chang HT (2005) Determination of glycine, glutamine, glutamate, and γ-aminobutyric acid in cerebrospinal fluids by capillary electrophoresis with light emitting diode-induced fluorescence detection. Anal Chim Acta 538(1–2):143–150. doi:10.1016/j.aca.2005.02.041
24. Li H, Li C, Yan ZY, Yang J, Chen H (2010) Simultaneous monitoring multiple neurotransmitters and neuromodulators during cerebral ischemia/reperfusion in rat by microdialysis and capillary electrophoresis. J Neurosci Methods 189(2):162–168. doi:10.1016/j.jneumeth.2010.03.022

25. Grove J, Schechter PJ, Tell G, Rumbach L, Marescaux C, Warter JM, Koch-Weser J (1982) Artifactual increases in the concentration of free GABA in samples of human cerebrospinal fluid are due to degradation of homocarnosine. J Neurochem 39(4):1061–1065
26. Ferrarese C, Pecora N, Frigo M, Appollonio I, Frattola L (1993) Assessment of reliability and biological significance of glutamate levels in cerebrospinal fluid. Ann Neurol 33(3):316–319. doi:10.1002/ana.410330316
27. Osbourn DM, Weiss DJ, Lunte CE (2000) On-line preconcentration methods for capillary electrophoresis. Electrophoresis 21(14):2768–2779. doi:10.1002/1522-2683(20000801)21:14<2768::AID-ELPS2768>3.0.CO;2-P
28. Simpson SL Jr, Quirino JP, Terabe S (2008) On-line sample preconcentration in capillary electrophoresis. Fundamentals and applications. J Chromatogr A 1184(1–2):504–541. doi:10.1016/j.chroma.2007.11.001
29. Slampova A, Mala Z, Pantuckova P, Gebauer P, Bocek P (2013) Contemporary sample stacking in analytical electrophoresis. Electrophoresis 34(1):3–18. doi:10.1002/elps.201200346
30. Tuma P, Sustkova-Fiserova M, Opekar F, Pavlicek V, Malkova K (2013) Large-volume sample stacking for in vivo monitoring of trace levels of gamma-aminobutyric acid, glycine and glutamate in microdialysates of periaqueductal gray matter by capillary electrophoresis with contactless conductivity detection. J Chromatogr A 1303:94–99. doi:10.1016/j.chroma.2013.06.019
31. Ramautar R, Mayboroda OA, Deelder AM, Somsen GW, de Jong GJ (2008) Metabolic analysis of body fluids by capillary electrophoresis using noncovalently coated capillaries. J Chromatogr B Anal Technol Biomed Life Sci 871(2):370–374. doi:10.1016/j.jchromb.2008.06.004
32. Church WH, Lee CS, Dranchak KM (1997) Capillary electrophoresis of glutamate and aspartate in rat brain dialysate. Improvements in detection and analysis time using cyclodextrins. J Chromatogr B Biomed Sci Appl 700(1–2):67–75
33. Wang C, Zhao S, Yuan H, Xiao D (2006) Determination of excitatory amino acids in biological fluids by capillary electrophoresis with optical fiber light-emitting diode induced fluorescence detection. J Chromatogr B Anal Technol Biomed Life Sci 833(2):129–134. doi:10.1016/j.jchromb.2006.01.013
34. Zhao S, Yuan H, Xiao D (2006) Optical fiber light-emitting diode-induced fluorescence detection for capillary electrophoresis. Electrophoresis 27(2):461–467. doi:10.1002/elps.200500300
35. O'Shea TJ, Weber PL, Bammel BP, Lunte CE, Lunte SM, Smyth MR (1992) Monitoring excitatory amino acid release in vivo by microdialysis with capillary electrophoresis-electrochemistry. J Chromatogr 608(1–2):189–195
36. Javerfalk-Hoyes EM, Bondesson U, Westerlund D, Andren PE (1999) Simultaneous analysis of endogenous neurotransmitters and neuropeptides in brain tissue using capillary electrophoresis—microelectrospray-tandem mass spectrometry. Electrophoresis 20(7):1527–1532. doi:10.1002/(SICI)1522-2683(19990601)20:7<1527::AID-ELPS1527>3.0.CO;2-9
37. Presti MF, Watson CJ, Kennedy RT, Yang M, Lewis MH (2004) Behavior-related alterations of striatal neurochemistry in a mouse model of stereotyped movement disorder. Pharmacol Biochem Behav 77(3):501–507. doi:10.1016/j.pbb.2003.12.004
38. Guzman-Ramos K, Osorio-Gomez D, Moreno-Castilla P, Bermudez-Rattoni F (2010) Off-line concomitant release of dopamine and glutamate involvement in taste memory consolidation. J Neurochem 114(1):226–236. doi:10.1111/j.1471-4159.2010.06758.x
39. Tuma P, Soukupova M, Samcova E, Stulik K (2009) A determination of submicromolar concentrations of glycine in periaqueductal gray matter microdialyzates using capillary zone electrophoresis with contactless conductivity detection. Electrophoresis 30(19):3436–3441. doi:10.1002/elps.200900187
40. Robert F, Bert L, Parrot S, Denoroy L, Stoppini L, Renaud B (1998) Coupling on-line brain microdialysis, precolumn derivatization and capillary electrophoresis for routine minute sampling of O-phosphoethanolamine and excitatory amino acids. J Chromatogr A 817(1–2):195–203
41. O'Brien KB, Esguerra M, Miller RF, Bowser MT (2004) Monitoring neurotransmitter release from isolated retinas using online microdialysis-capillary electrophoresis. Anal Chem 76(17):5069–5074. doi:10.1021/ac049822v
42. Soukupova M, Binaschi A, Falcicchia C, Palma E, Roncon P, Zucchini S, Simonato M (2015) Increased extracellular levels of glutamate in the hippocampus of chronically epileptic rats. Neuroscience 301:246–253. doi:10.1016/j.neuroscience.2015.06.013
43. Calcagno E, Carli M, Invernizzi RW (2006) The 5-HT(1A) receptor agonist 8-OH-DPAT prevents prefrontocortical glutamate and serotonin release in response to blockade of cortical

NMDA receptors. J Neurochem 96(3):853–860. doi:10.1111/j.1471-4159.2005.03600.x

44. Cavus I, Kasoff WS, Cassaday MP, Jacob R, Gueorguieva R, Sherwin RS, Krystal JH, Spencer DD, Abi-Saab WM (2005) Extracellular metabolites in the cortex and hippocampus of epileptic patients. Ann Neurol 57(2):226–235. doi:10.1002/ana.20380
45. Cavus I, Widi GA, Duckrow RB, Zaveri H, Kennard JT, Krystal J, Spencer DD (2016) 50 Hz hippocampal stimulation in refractory epilepsy: higher level of basal glutamate predicts greater release of glutamate. Epilepsia 57(2):288–297. doi:10.1111/epi.13269
46. Nagai T, Takata N, Shinohara Y, Hirase H (2015) Adaptive changes of extracellular amino acid concentrations in mouse dorsal striatum by 4-AP-induced cortical seizures. Neuroscience 295:229–236. doi:10.1016/j.neuroscience.2015.03.043
47. Kehr J (1998) Determination of glutamate and aspartate in microdialysis samples by reversed-phase column liquid chromatography with fluorescence and electrochemical detection. J Chromatogr B Biomed Sci Appl 708(1–2):27–38
48. Paredes D, Granholm AC, Bickford PC (2007) Effects of NGF and BDNF on baseline glutamate and dopamine release in the hippocampal formation of the adult rat. Brain Res 1141:56–64. doi:10.1016/j.brainres.2007.01.018
49. Shah AJ, de Biasi V, Taylor SG, Roberts C, Hemmati P, Munton R, West A, Routledge C, Camilleri P (1999) Development of a protocol for the automated analysis of amino acids in brain tissue samples and microdialysates. J Chromatogr B Biomed Sci Appl 735(2):133–140
50. Roettger VR, Goldfinger MD (1991) HPLC-EC determination of free primary amino acid concentrations in cat cisternal cerebrospinal fluid. J Neurosci Methods 39(3):263–270
51. Peinado JM, McManus KT, Myers RD (1986) Rapid method for micro-analysis of endogenous amino acid neurotransmitters in brain perfusates in the rat by isocratic HPLC-EC. J Neurosci Methods 18(3):269–276
52. Qu Y, Li Y, Vandenbussche E, Vandesande F, Arckens L (2001) In vivo microdialysis in the visual cortex of awake cat: II. Sample analysis by microbore HPLC-electrochemical detection and capillary electrophoresis-laser-induced fluorescence detection. Brain Res Brain Res Protoc 7(1):45–51
53. Rowley HL, Martin KF, Marsden CA (1995) Determination of in vivo amino acid neurotransmitters by high-performance liquid chromatography with o-phthalaldehyde-sulphite derivatisation. J Neurosci Methods 57(1):93–99
54. Donzanti BA, Yamamoto BK (1988) An improved and rapid HPLC-EC method for the isocratic separation of amino acid neurotransmitters from brain tissue and microdialysis perfusates. Life Sci 43(11):913–922
55. Caringi D, Maher TJ, Chaiyakul P, Asmundsson G, Ishide T, Ally A (1998) Extracellular glutamate increases in rostral ventrolateral medulla during static muscle contraction. Pflugers Arch 435(4):465–471. doi:10.1007/s004240050540
56. Lominac KD, Quadir SG, Barrett HM, McKenna CL, Schwartz LM, Ruiz PN, Wroten MG, Campbell RR, Miller BW, Holloway JJ, Travis KO, Rajasekar G, Maliniak D, Thompson AB, Urman LE, Kippin TE, Phillips TJ, Szumlinski KK (2016) Prefrontal glutamate correlates of methamphetamine sensitization and preference. Eur J Neurosci 43(5):689–702. doi:10.1111/ejn.13159
57. Chen J, Nam HW, Lee MR, Hinton DJ, Choi S, Kim T, Kawamura T, Janak PH, Choi DS (2010) Altered glutamatergic neurotransmission in the striatum regulates ethanol sensitivity and intake in mice lacking ENT1. Behav Brain Res 208(2):636–642. doi:10.1016/j.bbr.2010.01.011
58. Santos-Fandila A, Zafra-Gomez A, Barranco A, Navalon A, Rueda R, Ramirez M (2013) Quantitative determination of neurotransmitters, metabolites and derivates in microdialysates by UHPLC-tandem mass spectrometry. Talanta 114:79–89. doi:10.1016/j.talanta.2013.03.082
59. Zhang X, Rauch A, Lee H, Xiao H, Rainer G, Logothetis NK (2007) Capillary hydrophilic interaction chromatography/mass spectrometry for simultaneous determination of multiple neurotransmitters in primate cerebral cortex. Rapid Commun Mass Spectrom 21(22):3621–3628. doi:10.1002/rcm.3251
60. Buck K, Voehringer P, Ferger B (2009) Rapid analysis of GABA and glutamate in microdialysis samples using high performance liquid chromatography and tandem mass spectrometry. J Neurosci Methods 182(1):78–84. doi:10.1016/j.jneumeth.2009.05.018
61. Song P, Mabrouk OS, Hershey ND, Kennedy RT (2012) In vivo neurochemical monitoring using benzoyl chloride derivatization and liquid chromatography-mass spectrometry. Anal Chem 84(1):412–419. doi:10.1021/ac202794q

62. Bredewold R, Schiavo JK, van der Hart M, Verreij M, Veenema AH (2015) Dynamic changes in extracellular release of GABA and glutamate in the lateral septum during social play behavior in juvenile rats: implications for sex-specific regulation of social play behavior. Neuroscience 307:117–127. doi:10.1016/j.neuroscience.2015.08.052

Chapter 14

In Vivo Determination of Glutamate Uptake by Brain Microdialysis

Sandrine Parrot, Monique Touret, and Luc Denoroy

Abstract

Glutamate uptake has a predominant role in the regulation of glutamate homeostasis mainly for the prevention of hyperexcitability and excitotoxicity which induce neuronal dysfunctions and neurological diseases. An evaluation of in vivo glutamate transport can enable to verify the functional impact of alterations previously found in in vitro and ex vivo preparations. The use of brain microdialysis to sample the extracellular medium has allowed to quantify the dynamic uptake of glutamate through the active transporters present on cell membranes which maintain glutamate homeostasis. We describe two quantitative methods based on the application of labeled or unlabeled glutamate through a microdialysis probe to measure the in vivo uptake of glutamate in a discrete brain area of freely moving rats, accompanied with step by step details. Advantages and drawbacks of the methods are also discussed on both methodological and physiological bases, by relying on previous studies carried out in experimental pathological or pharmacological models.

Key words Glutamate, Uptake, Brain, Microdialysis, In vivo, Rodent, Reverse dialysis

1 Introduction and Background

1.1 Background on Glutamate

Glutamate (Glu) is an important molecule in life science since it has significant roles in biochemistry and physiology. It is present in its both enantiomeric forms in nature, but L-Glu is predominant in vertebrates, even if D-Glu can be found in various organs of birds and rats [1, 2], mainly in young animals [3]. First, L-Glu is a non-essential proteinaceous amino acid; its synthesis pathways have been well described from mainly α-ketoglutarate and glutamine (Gln) [4], and from histidine, proline, and arginine in a minor extent [5]. Within the general metabolism, L-Glu is involved in glucose metabolism due to its link to Krebs cycle through α-ketoglutarate, and is regarded as an intermediate of synthesis for many amino acids such as γ-aminobutyric acid (GABA), L-aspartate (L-Asp), and alanine, through decarboxylation or transamination [4, 5]. L-Glutamate has also an important impact in brain cell

Sandrine Parrot and Luc Denoroy (eds.), *Biochemical Approaches for Glutamatergic Neurotransmission*, Neuromethods, vol. 130, DOI 10.1007/978-1-4939-7228-9_14,

maturation [6, 7]. Besides its role in general metabolism and cell maturation, L-Glu is a major excitatory neurotransmitter [8] and has been defined as a neurotransmitter according to the seven classical criterion in the 1980s [9]: (1) precursors (α-ketoglutarate and Gln) and synthesis enzymes (phosphate activated glutaminase (PAG), glutamate dehydrogenase (GluDH), and aspartate aminotransferase (AAT)) are localized in neurons; (2) glutamatergic neurons are distributed in specific regional areas; (3) Glu is concentrated in the terminal vesicles by an active transport, i.e., through vesicular Glu transporters (vGluT1-3); (4) neuronal stimulation induces Glu release in physiological quantities; (5) specific receptors, namely ionotropic (*N*-methyl-D-aspartate (NMDA) or α-amino-3-hydroxy-5-methyl-4-isoxazole propionate/kainate (AMPA/KA) receptors) and metabotropic (mGluR1-8) receptors are present on the targeted neurons [10–12], but also on astroglial cells [13, 14]; (6) interaction between Glu and its receptors modifies the ionic permeability of membrane; (7) specific inactivation mechanisms of extra stimulation, relying on either receptor internalization or removal by transporters, allow to disable or prevent long-lasting Glu-receptor interactions in a physiological timescale. L-Glutamate has been shown to be also released by astrocytic cells [15] and can be now defined too as a gliotransmitter [16]. So, astrocytes and neurons can both receive and send glutamatergic signals each other, and in turn respond. Glial release mechanisms proposed for the physiological conditions seem to rely mainly on Ca^{2+}-regulated vesicular exocytosis, like the neuronal release [17]. Now, a regulation concept called "tripartite synapse" based on the exchanges between presynaptic, postsynaptic, and glial cells has emerged [18]. Beside the vesicular release, a non-vesicular release can occur via membrane ion channels, such as connexin/pannexin hemichannels, purinergic P2X7 channels and volume-regulated anion channels, or through pumps, such as the Glu-cysteine exchanger or the reversal of Glu transporters [19].

Between those non-vesicular mechanisms of glial release, the reversal of Glu uptake transport was the first studied and both existence and properties of the plasma membrane transporters have been carefully described from the 1990s. In normal conditions, those Glu transporters act to specifically remove the excess of released Glu from the extracellular environment. They belong to three distinct families classified according to the ionic gradient generating the energy of transport: sodium/potassium-dependent, chloride-dependent, or calcium-dependent transporters. The most efficient ones are the sodium/potassium-dependent carriers and can be divided into five subtypes of excitatory amino acid transporters (EAAT-1 to EAAT-5); EAAT-1 is also known as GLAST (Glu/Asp transporter), EAAT-2 as GLT-1 (glial transporter-1), and EAAT-3 as EAAC-1 (excitatory amino acid carrier-1). They exhibit a high affinity towards L-Glu (Km = 2–20 μmol/L, as compared to the two other families of Glu transporters with low affinity, Km = 100 μmol/L), and are able to carry from ~15 molecules (EAAT-2; EAAT-1) to more than 90 molecules (EAAT-3) per second while EAAT-5 and EAAT-4 are slower transporters, with 1–3 cycles of transport per seconds, respectively [20]. The removal of Glu from the extracellular medium, mainly ensured by transporters on astrocytes, is important to guarantee homeostasis around both neurons and astrocytes in order to prevent hyperexcitability, but also to prevent excitotoxicity deleterious to neurons [21]. However, during a pathological situation such as ischemia, the transporters can work in reverse due to an increased level of extracellular K^+ levels, which leads to a non-exocytosis release of Glu in the extracellular space [17]. The importance of each transporter and the way to study each of them is still under debate [22]. However, it is now admitted that EAAT-1 (GLAST) is exclusively localized on the glial plasma membrane near synapses, whereas EAAT-3 (EAAC-1) and EAAT-4 are localized on the neuronal membrane, in synapses or extrasynaptically, EAAT-4 being expressed only on Purkinje cells. EAAT-5 is only expressed in retina on neurons and glial cells. Finally, EAAT-2 (GLT-1) is mainly expressed in astrocytes and seems to represent more than 95% of the total Glu uptake activity [22].

For a better understanding of the role of Glu transport, pharmacological tools have been developed to block Glu uptake in many experimental paradigms [23]. More precisely, D,L-threo-β-benzyloxyaspartate (TBOA) is commonly used as an ubiquitous Glu uptake blocker even if its K_i for each Glu transporter can widely vary (from 0.12 to 9 μmol/L) [23]. L-Trans-pyrrolidine-2,4-dicarboxylate (PDC) is another drug able to block Glu uptake but under a different mechanism as it can be transported by the Glu transporters as a Glu competitor [24], whereas TBOA cannot [25]. Dihydrokainate is also used as it is a Glu competitor for GLT-1 [23]. The use of transportable inhibitors of EAATs can lead to an additional increase of extracellular Glu levels due to a mechanism of hetero-exchange induced by the reverse transport of Glu from cells [26]. A last important feature of the high affinity L-Glu transporters is that they can also carry L-Asp, another excitatory amino acid able to act on NMDA-type receptors, and its enantiomer D-Asp, with higher affinities than for L-Glu [27].

In conclusion, Fig. 1 summarizes typical glutamatergic sources of release and the location of transporters involved in neurotransmission.

1.2 Methods to Study Glutamate Transport

As transporters are surety of Glu homeostasis, the appreciation of transport is crucial when studying Glu neurotransmission and several approaches are available to evaluate alterations of transport by measuring levels of mRNA, proteins, or uptake activity.

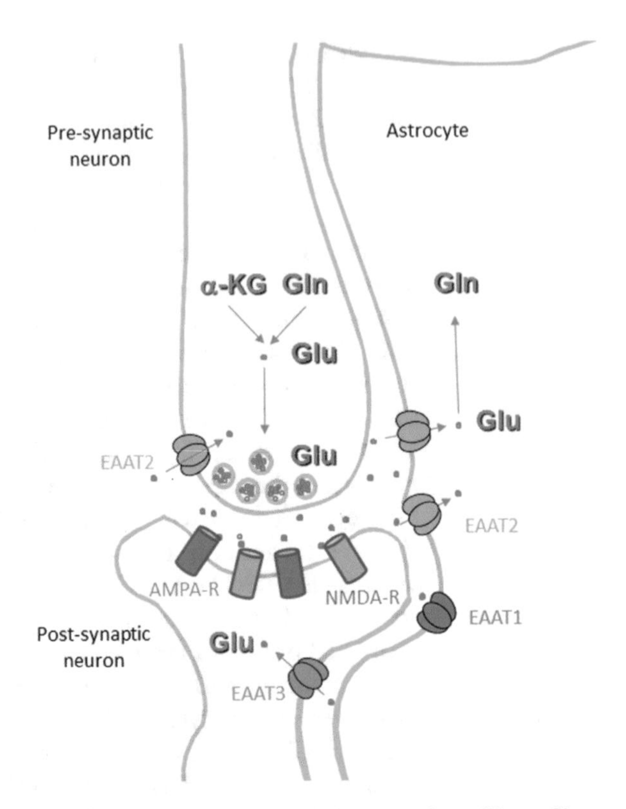

Fig. 1 Typical glutamatergic synapse. Under normal conditions, Glu can be synthesized from glutamine or from α-ketoglutarate. After being released into the synaptic cleft, Glu can act on the ionotropic AMPA/NMDA receptors and on metabotropic receptors (for sake of clarity, only synaptic AMPA/NMDA receptors are shown here). The excess of Glu in the synaptic cleft and extracellular space is removed by Glu transporters on the pre- and postsynaptic membranes and glial cells. EAAC-1 (EAAT-3) is located on the postsynaptic membrane, while GLAST (EAAT-1) and GLT-1 (EAAT-2) are located on glial cell membranes. Adapted from [20]

1.2.1 mRNA Quantification

After tissue extraction, levels of Glu transporters mRNAs can be quantitatively measured by the techniques of northern blotting, reverse transcription-polymerase chain reaction (RT-PCR) coupled to Southern blotting or quantitative PCR, the latter technique being more quantitative, easier to carry out, and the most used at present [28–31].

Northern blot gives information on size and sequence of the mRNA fragments of interest, denatured with formamide/formaldehyde, mRNA molecules being separated on an agarose gel. After a RNA transfer to positively charged nylon membranes, RNA is fixed to the membrane using UV crosslinking. Then, hybridization of the targeted sequence to a labeled, complementary RNA probe is performed. Detection is achieved by autoradiography or colorimetry depending on the type of probe labeling (isotopes; haptens

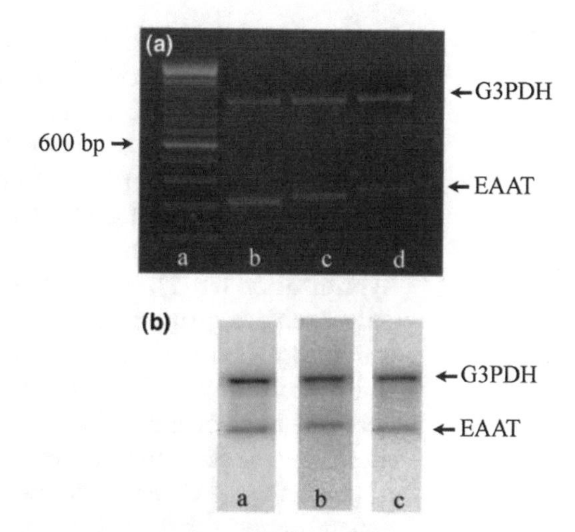

Fig. 2 RT-PCR detection of mRNAs, encoding G3PDH, GLT-1, GLAST, and EAAC-1 from frontoparietal cortex samples. (**a**) An ethidium bromide-stained gel showing: *lane a*, 100 bp DNA ladder (Life Technologies); *lanes b*, *c*, and *d*, RT-PCR products of the co-amplification of G3PDH (housekeeping gene, 983 bp) with the excitatory amino-acid transporter (EAAT) GLT-1 (*lane b*) (296 bp), GLAST (*lane c*) (331 bp), and EAAC-1 (*lane d*) (385 bp). (**b**) Southern blots of RT-PCR products of the co-amplification of G3PDH with GLT-1 (*lane a*), GLAST (*lane b*), and EAAC-1 (*lane c*) after hybridization with specific radiolabeled internal probes. From [34], with permission

as digoxigenin or biotin). For instance, the first works studying the expression of GLAST mRNA in brain used northern blotting [32, 33], but it requires large quantities of RNA preventing any accurate quantification with trace levels.

In RT-PCR technique, reverse transcription first generates a single-stranded DNA template (cDNA) from the mRNA through the use of reverse transcriptase. The targeted cDNA template can then be amplified as the same time as a housekeeping gene—G3PDH for instance—for the further quantification by Southern blotting. After the electrophoretic migration, amplified cDNAs are electro-transferred onto a positively charged nylon membrane and hybridized with specific oligonucleotide probes end-labeled by γ-^{32}P-ATP for detection and quantification targeting the genes of interest.

The use of RT-PCR allows a more powerful, sensitive, and quantitative detection of RNA levels than northern blot alone. For instance, mRNA of Glu transporters has been evaluated using RT-PCR and Southern blot in epileptic rats (Fig. 2), suspected to have an alteration in glutamatergic transmission [34].

As alternative, PCR can be used now as a quantitative method (qPCR) using the fluorogenic properties of dyes in presence of DNA, since the target cDNA can be amplified, quantified, and compared to a housekeeping cDNA by measuring the number of its amplicons generated and tagged in real time. The technique allowed to study the abundance of splicing variants suspected to be associated to multiple psychiatric and neurological disorders, as for the EAAT-2 (GLT-1) transporter [35].

One limit of both northern blot and PCR approaches is that they are performed on tissue or cell extracts. Alternatively, in situ hybridization (ISH), carried out on histological slides, permits not only a relative quantification of RNA levels, but also a precise anatomical localization due to the preservation of local cell organization. Briefly, this technique is also based on the complementarity between nucleobases, RNA being denatured with heat, allowing thereafter the fixation of a targeting labeled RNA probe. Detection is performed by autoradiography, fluorescence, or colorimetry depending on the type of probe labeling (isotopes, fluorochromes, haptens, or enzymes). The probed sample is then viewable by microscopy to identify where the mRNA is. For instance, this approach has been historically used to decipher the respective localization of the glial transporters GLAST and GLT-1 in rat brain using two different ^{35}S-labeled oligonucleotide probes [36]. This method has been applied on pathological brains e.g., on brain coronal slides of transgenic mice expressing an N-terminal fragment of mutant huntingtin (R6/2) in order to compare the expression of GLT-1, GLAST and/or EAAC-1 mRNA to wild-type littermate controls [37, 38] or on postmortem brains of patients suffering from Huntington's disease for the visualization of glial GLT-1 mRNA compared to the localization of neuronal NMDAR1/2B mRNA [39]. Moreover, this ISH approach has permitted to discriminate the abundance of the different spliced sequences of Glu transporters in a transgenic mouse model of amyotrophic lateral sclerosis for EAAT-2 transporter [40].

However, due to the delays between translation and transcription steps, the study of mRNA levels does not reflect the genuine functional alteration if it is not associated with the quantification of protein levels. As a consequence, mRNA quantification is often coupled to protein expression quantification [34, 37].

1.2.2 Protein Quantification

The most commonly used method for protein quantification is western blot. Classically, the proteins extracted by cell lysis are separated on a polyacrylamide gel (SDS-PAGE), transferred to a membrane and then probed with a specific antibody conjugated to a fluorophore or a peroxidase for detection. Quantification is rather accurate and also allows to bring information about modifications undergone by the targeted protein, as proteolysis, ubiquitination, and phosphorylation, that induce changes in its size. For instance, protein amounts using immunofluorescence detection were studied in a rat model of neuromyelitis optica [41] and in the physiopathological models of epileptic GAERS rats [34, 42, 43] as

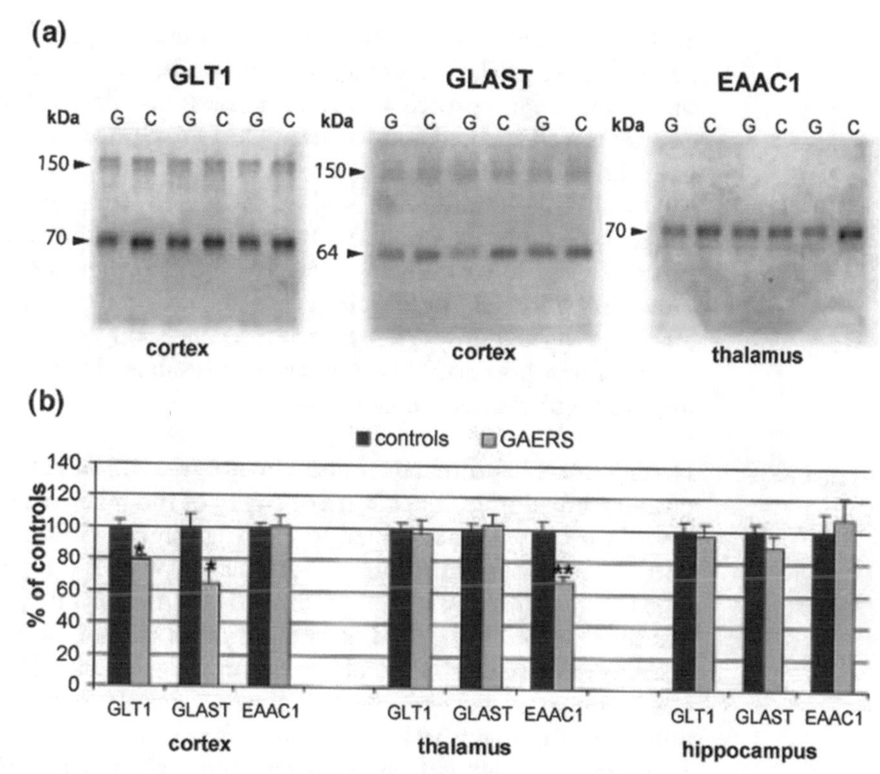

Fig. 3 Expression of glutamate transporter protein in 30-day-old GAERS. (**a**) Immunoblots showing the lower expressions of GLT-1, GLAST, and EAAC-1 in GAERS (G) compared to control rats (C) in cortex and thalamus. Sample homogenates were subjected to SDS-PAGE and immunoblotted with anti-GLT-1, anti-GLAST, or anti-EAAC-1 antibodies. A fluorescent secondary antibody allowed the detection of the immunoreactivity by a Fluorimager. Note the predominance of the monomeric bands in our experimental conditions. (**b**) Western blot relative quantification of the cortex, thalamus, and hippocampus of young control rats (*dark grey bars*; $n = 5$) and GAERS (*pale grey bars*; $n = 5$) for GLT-1, GLAST, and EAAC-1. The results are represented as a mean percentage of control rats ± SEM. GAERS values were compared to control rats values with a two-tailed Student's *t*-test (*$p < 0.05$, **$p < 0.01$). From [34], with permission

illustrated in Fig. 3. In the latter rat model, the approach was able to show differential decreases in GLT-1, GLAST, and EAAC-1 amounts down to 60% of controls in thalamus and cortex, two brain areas involved in both genesis and synchronization of epileptic spike-and-wave discharges (SWDs). Immunohistochemistry detection has been used to quantify glial EAAT-1 and EAAT-2 in postmortem human brains of patients suffering from mucopolysaccharidoses II and IIIB, also known as Sanfilippo's syndrome [44]. Alternatively, the enzyme-linked immunosorbent assay (ELISA) technique using antibodies immobilized on a microtiter plate can be also used to capture proteins of interest. Antibodies conjugated to horseradish peroxidase or alkaline phosphatase permit to quantify the trapped protein by the fluorometric/colorimetric detection. This type of approach seems well appropriate to aCSF samples as shown for glutamine synthetase [45] or glutamate decarboxylase [46], two proteins of Glu metabolism. For the study of the membrane proteins such as transporters, flow cytometric analysis is well adapted and has been used for GLAST and GLT-1 [41]. Briefly, cells were incubated with unconjugated anti-GLAST and anti-GLT-1 antibodies and stained with a fluorochrome-conjugated immunoglobulin G; the fluorescence intensity is measured using a flow cytometry analyzer (also known as FACScan). However, with all these approaches quantifying proteins, the efficiency of the transporters detected is still lacking and measurements of transport activity are necessary to evaluate both functioning and regulation of the Glu transporters.

1.2.3 Transport Activity

Uptake Measured Ex Vivo/In Vitro

Neither mRNA/protein quantification nor transporter localization gives information on the effective activity of the transport of Glu. Other studies were performed to evaluate directly the Glu transport on cells or synaptosomes. For that purpose, D-[^{3}H]-Asp or L-[^{3}H]-Glu uptake assays are performed. Historically, D-[^{3}H]-Asp was the first tool to be used on synaptosomes because of its specificity towards high affinity Glu transporters, whereas L-[^{3}H]-Glu can be regarded now as a better mime of endogenous Glu. The principle of the approach is as follows: synaptosomes or cell cultures are in contact with an exogenous extracellular substrate (Asp or Glu) labeled with isotope (e.g., D-[^{3}H]-Asp or L-[^{3}H]-Glu) in order to evaluate the uptake by quantifying the isotopic radioactivity inside the lyzed synaptosomes or cells [47]. This technique was used to study Glu transport in diseases, such as in Huntington's disease [37] and in amyotrophic lateral sclerosis (ALS) [48], or in aging [49]. However, criticisms can be pointed out. First, in case of synaptosomes, it has been originally assumed that only neuronal transport components were detected, but, in fact, the contamination by astrocytes is not negligible [22]. Recently, an alternative approach seems to overcome that type of inconvenience; Glu dynamics were quantified in situ using high-speed imaging of an intensity-based Glu-sensing fluorescent reporter (iGluSnFR) coupled to electrophysiological recordings [50]. The study, done in an animal model of Huntington's disease, compares the data on Glu uptake assays obtained on synaptosomes using [^{3}H] L-Glu to the data obtained on slices using iGluSnFR. The impairment of Glu uptake observed with synaptosomes was not at all evidenced using iGluSnFR on slices; in contrast, the authors observed a significant acceleration of Glu clearance in the R6/2 model of Huntington's disease, casting doubt on the physiological relevance of uptake assays that utilize exogenous Glu or Asp, but also on the potential use of Glu uptake enhancement as a therapeutic strategy for the treatment of HD [50]. In conclusion, the approaches using isotopic D-Asp and L-Glu done on ex vivo or in vitro preparations permit to obtain interesting data but do not take into account the integrity of the targeted brain structure.

Uptake Measured In Vivo

Two techniques allowing evaluation of Glu uptake in in vivo preparations are available. The main advantage is that the integrity of the brain is preserved and the experiments can be done on freely moving animals using techniques based on Glu monitoring. Voltammetry and microdialysis can be used to estimate Glu uptake. Voltammetry will be briefly explained in this paragraph while the rest of this chapter deals with the microdialysis approach.

Voltammetry is a technique based on an implantable electrode whose end is coated with a glutamate oxidase (GluOx) able to convert Glu to α-ketoglutarate as follows:

$$\text{L-glutamate} + H_2O + \text{GluOx} / \text{FAD} \rightarrow \alpha\text{-ketoglutarate} + NH_3 + \text{GluOx} / \text{FADH}_2.$$

$$\text{GluOx} / \text{FADH}_2 + O_2 \rightarrow \text{GluOx} / \text{FAD} + H_2O_2.$$

This two-step reaction generates the production of H_2O_2, which is oxidized at 0.7 V versus an Ag/AgCl reference electrode. The current generated on the surface of the working electrode corresponds to the concentration of Glu, determined by calibration using a standard solution. Calibration tests also allow to verify the selectivity for Glu and the limits of detection, classically ~2.5 μmol/L. In practice, those types of electrodes (both working and reference electrodes) can be implanted proximally in a target structure in rats or in some large brain areas of mice, thanks to a thin diameter (order of magnitude, 100 μm). Other technical and methodological information on a type of Glu sensors can be found in Chap. 11 in this book. To evaluate Glu uptake, an incremental extracellular Glu level is generated artificially, physiologically, optogenetically, or pharmacologically, the measured parameter is the time related to reuptake time of supplementary Glu, more precisely, and it can be estimated between the time of maximal Glu levels and the time corresponding to return to baseline. For illustration, the use of voltammetry has been applied to the study of Glu uptake in the dorsal region of spinal cord of anesthetized rats [51, 52]. However, this approach can be used also in vitro [53].

One caution to use this technique is the need to calibrate each sensor one by one to prevent experimental bias due to the fabrication of the electrode. Our alternative method to determine Glu uptake which is based on microdialysis sampling is presented below in this chapter.

1.3 Microdialysis Technique

1.3.1 Principle of the Technique

Microdialysis is a sampling technique consisting in infusing an artificial cerebrospinal fluid (aCSF) through a probe at a constant low flow rate. The probe whose end is made of a semipermeable membrane permits the sampling of low molecular weight endogenous compounds through their diffusion according to a concentration gradient (Fig. 4). Ten to twenty percent of the extracellular molecules

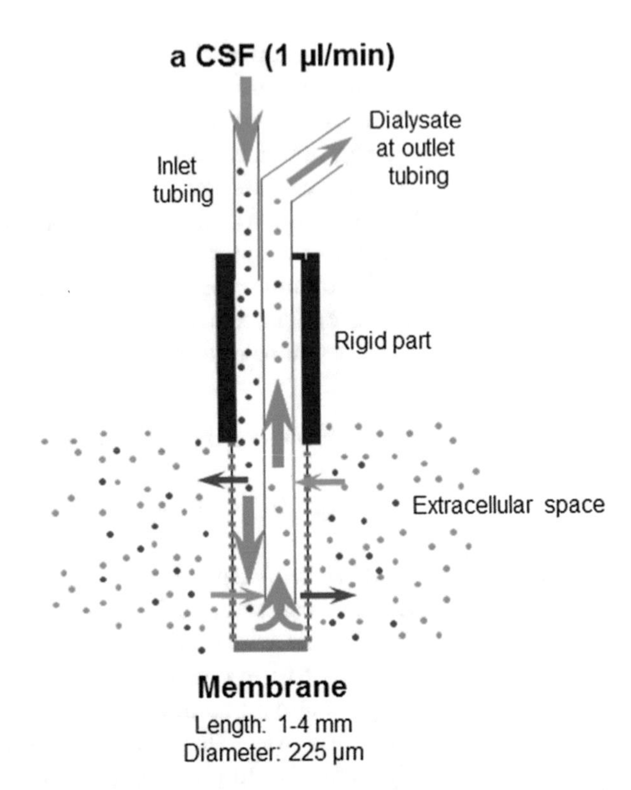

Fig. 4 Schematic representation of a microdialysis probe. The probe is continuously perfused by an isotonic artificial cerebrospinal fluid (aCSF, *green arrows*). Endogenous compounds (*orange circles*) diffuse through the dialysis membrane (*orange arrows*) and are sampled in the microdialysate collected at the outlet of the probe. Drugs (*blue circles*) can be added to the perfusion medium and can diffuse out of the probe into the extracellular medium (*blue arrows*) providing a means of local administration. Adapted from [97]

are collected at the outlet of the probe. By this means, monitoring of the molecules of interest can be done at a frequency between 1 to 30 min according to the experimental paradigm chosen by the neuroscientist. As a consequence, the higher is the sampling rate, the more timely precise is the monitoring of extracellular concentrations. The microdialysis probe can be also used as way of administration of drugs. For that purpose, the pharmacological drug is diluted in the aCSF and delivered by the infusate in the brain tissue. Exogenous compounds can diffuse by reverse dialysis according to the concentration gradient (Fig. 4).

Classically, the content of the samples collected and called microdialysates or dialysates is then analyzed using a separative micro-method such as high performance liquid chromatography (HPLC) or capillary electrophoresis (CE) in the case of Glu. The coupling between microdialysis and the analytical technique permits the monitoring of average concentrations over a given time interval, i.e., corresponding to the sampling rate. The higher is the sampling rate; the more timely precise is the monitoring of concentrations. However, one can note here that only a part of the neurotransmitters will diffuse through the membrane. Indeed, the diffusion would be full only at equilibrium between the medium inside the probe and the extracellular space, which is never reached since the delivering of "fresh" aCSF is continuous. The recovery depends on the flow rate. If the flow rate is low, the relative recovery is high, but the extraction is low. In contrast, if the flow rate is high, the equilibrium is difficult to reach but the extraction is high, so that the relative recovery is weak. This condition is not favorable because it leads to a depletion of neurotransmitters around neurons. The recovery also depends on body fluid temperature, the geometry of the membrane (molecular weight cut-off (MWCO), size), and the chemical properties of the membrane. In vivo, it can be also influenced by the mechanisms of uptake or inactivation of the neurotransmitters as compared to in vitro studies. Generally, recovery values are comprised between 5% and 30%. Microdialysis is then said to be semiquantitative and only variations of concentrations can be measured. For instance, a doubling in extracellular concentrations will be detected thanks to a doubling in microdialysate concentrations. So that, data are normally expressed as percent of baseline, since a recovery determined in vitro cannot perfectly reflect a concentration found in vivo.

1.3.2 History and Applications

First used in neuroscience in 1974 by the Swedish neuroscientist, Urban Ungerstedt [54], the microdialysis technique has been popular and has been extended from brain to all the organs in humans and animals, such as heart, kidney, adipose tissue, muscle, and uterus. Brain still remains the most targeted organ and 93% of brain microdialysis studies are done on laboratory animals [55]. Initially carried out on anesthetized animals, mainly rats, gerbils, or cats, the technique was adapted to freely moving animals during the 1980s, in order to study chemical variations in a context of behavior such as learning and vigilance, and pharmacological paradigms. Initially coupled to HPLC, microdialysis has been started to be also combined to CE from 1996 because, at that time, miniaturized HPLC was still in development and the high volumes (20–30 μL) required by classical HPLC hampered the application of microdialysis to high sampling rate experiments such as 1 min [56], needed when one wants to monitor rapid neurochemical changes. The miniaturization of the microdialysis probes also allows to study small structures from laboratory animals such as mice or rats while limiting brain lesions [57, 58].

Applications in microdialysis are nevertheless numerous, including monitoring, administration of drugs, pharmacokinetics and pharmacodynamics studies (PK/PD). However, several protocols have been designed to overcome the inconvenience of semiquantitative technique, as the approaches of zero net flux or no net flux for the determination of true extracellular concentrations theorized previously [59–61]. Briefly, the methods are based on the mathematical models of dialysis and play on the variation of flow rate or the difference of gradient concentration to extrapolate or interpolate the true concentration.

1.3.3 Outlines of this Chapter

Our chapter will describe the no net flux technique step by step since this technique, commonly used for monoamines [62], can be used for Glu [43], and permits to evaluate not only the true extracellular concentration, but also the uptake capacity using the extraction index obtained by increasing Glu concentration infusions. Besides, no net flux has been associated to a second approach based on reverse dialysis and isotope infusion in order to determine Glu uptake. The two methods will be fully described in this chapter and the results which can be obtained thanks to them will be carefully discussed in regard to both methodological and neurochemical contributions.

2 Materials

2.1 Animals

Male rats (Wistar or Oncins France Strain A, weighting 250–300 g) were used in our studies to evaluate Glu uptake. The animals were kept and used in experiments according to the current European Communities Council Directives and the protocol was approved by our local university committee.

2.2 Surgery

Materials needed for the surgery procedures are as follows: a device for volatile anesthesia (isoflurane, Aerrane, Baxter SAS, France) including an oxygen extractor or a set for fixed injection (ketamine/xylazine), a stereotaxic frame for rats (its XYZ movements should be verified from time to time to warrant both horizontality and verticality in the frame; David Kopf, Epinay-sur-Seine, France) equipped with a rat mask when using volatile anesthesia, a heating pad (Harvard Instruments, France), a small shaver, cleaning solutions for skin (Vetedine-like), cleaning solutions for skull (alcohol, H_2O_2 10 volumes), swabs, scalpel blades, sterile dispensers of saline solution, a small drill, a set of forceps and surgical tools adapted to rats, surgical threads, anchoring screws (timepiece-like), and guide cannulas (CMA12, CMA, Sweden) for a further microdialysis probe implantation.

2.3 Microdialysis Probes

Home-designed concentric microdialysis probes were constructed from regenerated cellulose dialysis tubing (MWCO 13,000 Da, 225 mm o.d., 2-mm active dialysis length) and fused-silica capillary tubing, the body of the probe being made of a 3-cm 26 G stainless steel tube, which was glued on a flat probe holder (Harvard, USA). The step by step details were previously described [63].

2.4 Microdialysis Experiments

When the implanted rats with full surgical recovery are obtained, the materials needed for the microdialysis are as follows: a 500-μL Hamilton syringe and polyethylene tubing to adapt on the probe, a pump for the infusion of the aCSF into the probe, an experimental chamber (a cylinder shape for instance) with a bottom of litter and a clean water bottle, a liquid swivel and its balancer (mouse model from Instech Solomon, USA), sterile PCR tubes for collection samples, and a −20 °C freezer to store the samples until analysis. Before implantation, the probes were perfused at a rate of 1 μL/min with a 0.22-μm filtered artificial cerebrospinal fluid (aCSF) (149 mmol/L NaCl, 2.80 mmol/L KCl, 1.0 mmol/L $MgCl_2$, 1.2 mmol/L $CaCl_2$, 2.78 mmol/L phosphate buffer, pH 7.4, made with sterile ultrapure water). This flow rate can be kept or slightly changed according to the type of experiments.

2.5 Glutamate Content Analysis

1. The true extracellular concentration in each rat was determined using the no net flux method, by infusing increasing concentrations of Glu through the probe and collecting the brain microdialysis sample obtained at the outlet of the probe. Glu levels in the dialysates were determined using capillary electrophoresis with laser-induced fluorescence detection (CE-LIFD) as follows: on the day of analysis, 4 μL of sample and 4 μL of standard solutions were derivatized at room temperature by adding 1.6 μL of a mixture (1:2:1 v/v/v) of (1) internal standard (100 μmol/L cysteic acid in 0.117 mol/L perchloric acid), (2) a borate/NaCN solution (100:20 v/v mixture of 500 mmol/L borate buffer, pH 8.7, and 87 mmol/L NaCN in water), and (3) a 2.925 mmol/L solution of naphthalene-2,3-dicarboxaldehyde in acetonitrile/water (50:50 v/v). The samples were then analyzed using an automatic capillary zone electrophoresis P/ACE™ MDQ system (Beckman, USA) equipped with a ZETALIF laser-induced fluorescence detector (Picometrics, France). Excitation was performed using a diode laser (Melles Griot, USA) at a wavelength of 410 nm. Separations were carried out on a 63 cm × 50 μm i.d. fused-silica capillary (Composite Metal Services, Worcester, England) with an effective length of 52 cm. Each day, before the analyses, the capillary was sequentially flushed with 0.25 mol/L NaOH (15 min), ultrapure water (15 min), and running buffer (75 mmol/L sodium borate, pH 9.20 ± 0.02, containing 10 mmol/L HP-β-CD and 70 mmol/L SDS) (5 min). The separation conditions were an applied voltage of 25 kV, hydrodynamic sample injection (10 s at 0.6 psi), and a temperature, between 36 and 38 °C. The capillary was sequentially flushed for 30 s each with 0.25 mol/L NaOH, ultrapure water, and running buffer between analyses. Electropherograms were acquired at 15 Hz using P/ACE™ MDQ software [64]. All the solutions are

made with highest analytical purity powders and sterile ultrapure water. Autoclave sterilization of water drastically reduces Glu contamination and other amino acids naturally present in ultrapure water. All glass materials should be cleaned with ultrapure water and the other plastic dispensers used should be bought as sterile materials. To avoid contamination, use weight spatulas only devoted to the CE-LIFD approach. All the solutions are kept at +4 °C, except derivatization buffer which must be kept at room temperature to prevent crystal precipitation due to a weak solubility of borate at such high concentration (500 mmol/L). Both cysteic acid and borate/NaCN solutions should be prepared every day of analysis. Derivatization borate buffer, NaCN, and NDA solutions can be kept several weeks provided a careful free contamination use. NDA solutions should be kept in a dark tube to prevent light oxidation. Non-limpid (NaCN) and/or pink colored solutions (NDA) should be discarded.

2. For radioactivity experiments, aCSF is needed to dilute the perfused compounds, i.e., a 1 mmol/L Glu solution [750 μmol/L of unlabeled Glu and 250 μmol/L ^{3}H-Glu (15 Ci/mmol)], containing 120 μmol/L ^{14}C-mannitol (MAN) (60 mCi/mmol) (PerkinElmer Life and Analytical Sciences, USA) and further containing or not 20 mmol/L threo-β-hydroxyaspartate (THA), a Glu uptake inhibitor, as well as a syringe and connections from syringe to the dialysis probe devoted to isotopic infusion, a dual label scintillation counting, detergent specific for radioactivity cleaning (TFD4, 5% diluted in warm water) to flush tubing and the syringe contaminated with radioactive traces.

2.6 Waste

The use of materials exhibiting biological, chemical, and radioactivity risks or coexisting risks in liquid or solid forms implies adequate and specific waste bins and treatments.

3 Methods

3.1 Animals

The animal model, in contrast to cell cultures, slices, or synaptosomes, takes into account all the complexity and diversity of the physiological regulations modulating Glu uptake. Even if many mammals commonly used in research laboratories could be used for the in vivo uptake studies (rats, mice, gerbils, guinea pigs, cats...), brain microdialysis for evaluating Glu uptake has been applied only to awake rats probably because they are rather small, calm, and easy to manipulate, which can be safer for the experimenters when radioactive isotopes are employed. The first group describing the method used male Wistar rats weighting 300–350 g using labeled D-Asp as a substrate to Glu uptake [65]. Other groups performed their isotope uptake experiments on male Sprague–Dawley rats [66, 67] using radioactive Glu.

1. In our laboratory, male Wistar (250–300 g) rats from the Genetic Absence Epilepsy Rats from Strasbourg (GAERS) strain displaying recurrent generalized absence seizures concomitant with behavioral arrests and their non-epileptic control (NEC) Wistar rats were used and uptake determination was performed with labeled Glu [43]. In a recent study, we used also male OFA (Oncins France Strain A, weighting 280 g) rats in a model of Devic's disease, an autoimmune astrocytopathy also known as neuromyelitis optica (NMO) [41].
2. Once delivered by the suppliers, the rats were acclimated to the housing environment and kindly handled for 7 days prior to the start of the surgery, with food and water ad libitum. Housing temperature was kept constant (21 °C ± 1 °C) and light/dark cycle was 12 h/12 h, light beginning at 7:00 am.

3.2 Surgery

1. After an induction of the isoflurane anesthesia using the box of the device (3–4%, 0.8–1.5 L/min for the oxygen flow, see precautions in Note 1), the rats are rapidly settled on the stereotaxic frame to avoid any waking between the induction and the prolongation of the anesthesia. The body temperature is maintained constant using a heating pad. A rat-adapted mask is fully integrated in the stereotaxic frame and permits to maintain the animals continuously anesthetized (<2% isoflurane) until the end of the surgery. In case of anesthesia with ketamine/xylazine (i.p. injection, 0.2 mL/100 g of a mixture of ketamine (100 mg/kg) and xylazine (10 mg/kg)) [68], the rats are mounted conventionally in the stereotaxic frame once the deep anesthesia is verified through the absence of hind paw withdrawal after pinching. The maintaining of anesthesia depth is ensured by additional doses of only ketamine (100 mg/kg) if the same test on hind paws reveals any withdrawal.
2. The scalp is shaved, cleaned with Vetedine®, then sagittally incised with a sterile #23 scalpel blade; the skin is pushed aside with swabs soaked in saline solution to prevent dehydration and is maintained removed by two forceps on each side. The skull is cleaned with dry swabs in order to remove the connective tissue, then with swabs soaked in H_2O_2 (10 volumes) to prevent bleeding.
3. Once the position of the skull is checked to be flat, when any bleeding has stopped, thin holes are drilled into the skull. The first hole is to anchor a screw in order to solid the final assembly; the second hole will allow the implantation of the guide cannula. The choice of the drill is important for both holes.

The best diameter must be a bit smaller than the screw in order to screw into the skull with an excellent maintaining. The tip is to drill the two holes as close as possible, but by taking into account the sizes of the cannula and screw, respectively. In this case, the anchoring will be stronger if the screw is placed close to the guide cannula. Besides, the diameter of the hole for the guide cannula must be a bit larger than the diameter of the guide cannula in order to avoid deviation from the verticality due to the presence of any remaining pieces of bone when implanting the guide.

4. Another important precaution is to verify the absence of remaining meninges before implanting the guide. A resection with the tip of a curved needle is enough to tear the meninges with minimal harm for the rat. This step is peculiarly crucial because the meninges insufficiently removed can be pushed into the guide tube while implanting it and they can be very painful after the probe implantation on the day of the microdialysis, even 7 or 15 days after surgical recovery.
5. Eventually, in case of bleeding induced by drilling, only swabs soaked with saline solution have to be used to clean the skull surface in order to not damage the brain. Homeostatic swabs can be used to prevent local bleeding. Once the bleeding is stopped, the surface skull must be dehydrated by alcohol, but one has to take care to not introduce alcohol in the holes. When this preparation is done, the guide can be lowered in the brain according to the coordinates chosen on the atlas of rat brain [69].
6. Once the guide is well placed in the stereotaxic frame, a thin layer of dental cement (Super-Bond C&B) is spread on the skull and the assembly is grouted by successive layers of dental cement (Paladur® like), sufficiently secure but not too heavy to avoid a weight superior to 2 g on the head of the rat. This is particularly important because the animal has to move normally after surgery until the microdialysis experiment.
7. In a recent study [41], we also placed a second guide cannula from an Alzet pump kit for brain infusion, which was lowered into the right lateral ventricle using stereotaxic coordinates (anteroposterior (AP) −0.7 mm; mediolateral (ML) −1.7 mm from bregma; depth, 5 mm from the skull surface) and strongly fixed to the skull with superglue just before guide cannula implantation and cementation. One of the aim of the study was to induce an experimental model of neuromyelitis optica (NMO) using a 7-day chronic infusion of aquaporin-4 (AQP4)-IgG, an antibody against the human aquaporin 4, and to study in vivo the alteration of Glu uptake, previously observed using an in vitro model [70]. That study showed the compatibility of the technique with a setup of intraventricular infusion.

8. After surgery, the animals recovered rapidly from anesthesia and normal postsurgical observations should be: normal aspect on scarring without any exuding biofluid, absence of pain, animal with no stereotypy or any withdrawal behavior in the cage, absence of eye porphyrins, clean hairs, absence of shivering or bristling hairs, weight regaining after 2–3 days, normal appetite and thirst, etc. In case of small blood leak, soak up with sterile dry cotton swabs, and avoid physiological serum and H_2O_2 because it often leads to itching; in case of dry blood, do not touch the scar; in the other cases, search veterinary assistance. Each animal is kept in a transparent individual cage and is maintained at 21 °C, on a 12-h light/dark cycle with food and water ad libitum for up to 21 days, and lairage duration depends on the conducted experiments. A weight loss is observed during the couple of days following the surgery and total weight regain is observed after 8 or 10 days, so that room temperature and animal weight should be looked after or checked out at least twice per day, or once if everything looks alright. The use of individual cages is necessary in order to prevent damage to the cannula assembly due to another congener. However, the sound and ultrasonic communication between the rats is not interrupted and, despite a natural bad vision observed for instance in albino rats, the visual communication is still present and the stress due to isolation is then greatly attenuated.
9. A last comment is that several brain structures can be targeted according to literature data: striatum (AP 0.3 mm and L +3 mm from bregma, V −7.2 mm below dura) [65], frontoparietal cortex (AP +1.7 mm and L +5.0 mm from bregma, V −4.1 mm from the top of the skull, with a +40° angle to the mid-sagittal plane) [66], ventrobasal thalamus (AP −3.5 mm and L +2 mm from bregma, V −6 mm below dura) and cortex (AP −3.3 mm and L +2.3 mm from bregma, V −2.5 mm below dura, with a +30° angle) [43] as illustrated in Fig. 5, nucleus accumbens (AP +1.6 mm and L ±1.1 mm from bregma, V −8 mm below the top of the skull) [67], dorsal hippocampus (AP −4.3 mm, L −2.5 mm, V −4.2 mm from bregma) [41], showing the potential versatility of the approach throughout the rat brain.

3.3 Probe Implantation and Microdialysis Sampling

1. A probe was implanted into the guide cannula of the freely moving animal, which must be habituated several times to both handling and experiment cage cylinder if the home cage is not used as experimental chamber.
2. In order to let the animals moving normally, the inlet and outlet of the probe were connected to a liquid swivel carried by its balancer (mouse model, Instech Solomon USA) attached on

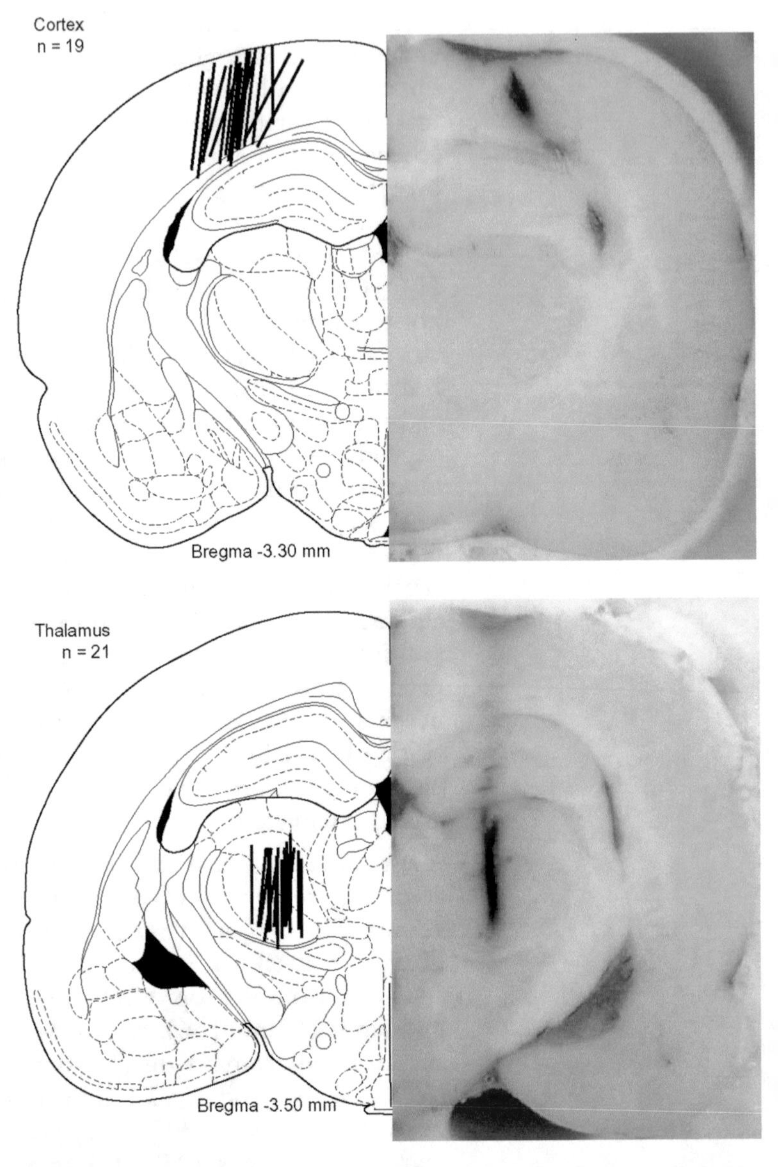

Fig. 5 Microdialysis probe placements in the cortex (*top*) or thalamus (*bottom*) in the experimental groups of adult GAERS rats used in [43]. On the left, each *black bar* represents the position and length of the microdialysis probe membrane implanted in an individual and reported on the hemisphere coronal sections, relative to bregma [69] (with permission to reproduce). On the right, one histological example of track left by a probe after reverse dialysis of methylene blue (brain collected and frozen after killing)

the top of the cylinder wall during a 3-h stabilization period, i.e., the elapsed time between the brain tissue modification induced by the probe placement and a normalization of basal levels (Fig. 6a).

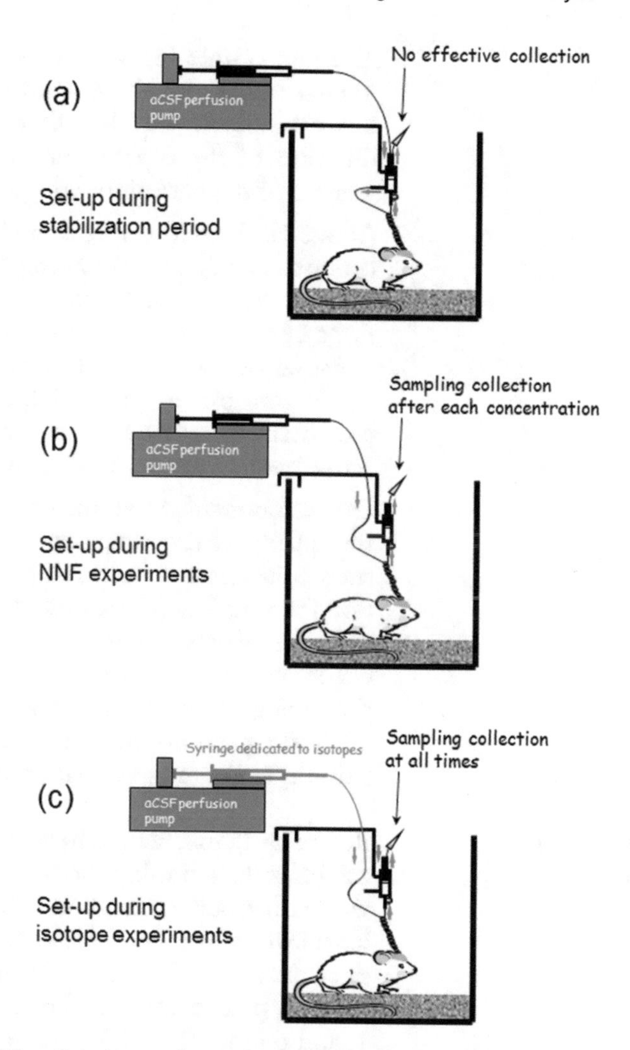

Fig. 6 Microdialysis experimental setup in awake rats. After probe implantation, the animal is placed in the experimental chamber which can be its home cage to limit stress due to novel environment, as here. aCSF is perfused using a pump and very thin tubing. (**a**) The rat is left at least 3 h before the beginning of the sample collection. A two-way liquid swivel is used to prevent kinking while the animal moves freely. No collection is needed. (**b**) After the 3-h stabilization period, successive concentrations of Glu are brought through the probe, and effective sampling is done 10 min after the beginning of each infusion. To avoid the dead volume at the inlet of the probe, only the outlet of the liquid swivel is used. (**c**) For the isotope experiments, a syringe and inlet tubing are devoted to limit radioactive contamination and subsequent cleaning. As for no net flux experiments, only the outlet of the liquid swivel is used. Adapted from [63]

The choice of the mouse model for the liquid swivel and its balancer is justified by the low dead volume inside the swivel which minimizes both flow perturbations for microdialysis sampling and internal layers contamination by the isotopes infusion.

3. The rate of infusion was adjusted to 1 or 2 μL/min for the in vivo evaluation of Glu uptake and 2 μL/min for the no net flux experiment [41, 43]. One has to keep in mind that the duration of the experiment should be compatible with the volume of the perfusion syringe, here a 500-μL syringe.
4. About 10–15 min before sampling collection, the inlet of the probe can be directly connected to the pump syringe (Fig. 6b). This can be necessary because the inlet of the liquid swivel has not a constant volume when rotating (a phenomenon getting worst and worst when the swivel is getting older and older), which can greatly affect high sampling rate and, thereafter, membrane recovery. Besides, it allows to have a minimal radioactive contamination of the swivel (Fig. 6c).

   One can note here that the dead volume has a great impact on the quality of the sampling collection, especially because the success in this approach relies on both high temporal resolution for sampling (every min or every other minute) and short duration of isotope application (10 min). An inappropriate dead volume would lead to dilution of the compounds within the tubing of the microdialysis probe [71] or change in recovery leading to both nonquantitative and nonqualitative results (Monique Touret, personal observations).

3.4 No Net Flux Studies

1. Before using the method based on isotopic infusion, Glu uptake can be evaluated using the no net flux quantitative microdialysis, a rather popular method initially designed to determine the basal extracellular concentrations of Glu in discrete brain areas, since the no net flux method relies on exogenous Glu infusions [60]. In practice, four different concentrations of Glu (0, 0.5, 1, and 5 μmol/L) (C_{in}), chosen to bracket the expected extracellular concentration, were passed through the microdialysis membrane (each for 16 min, separated by 10-min perfusion with aCSF for washing); no sample was collected during the first 10 min of Glu perfusion, allowing a period of equilibration, then three 2-min dialysis samples were collected and stored at −40 °C until analyzed for amino acid content.
2. The concentrations of neurotransmitter in the dialysate samples (C_{out}) obtained during perfusion with the various concentrations of neurotransmitter (C_{in}) were used to construct a linear regression plot of the net change in neurotransmitter ($\Delta C = C_{in} - C_{out}$) against C_{in} for each animal (Fig. 7). This linear regression can provide two pieces of information, namely the true extracellular level and the extraction fraction of the probe, as theorized previously [72]. The true extracellular concentration is equal to C_{in} when $\Delta C = 0$ while the extraction fraction (Ed) representing the probe extraction of Glu is determined from the slope of the linear regression. This extraction

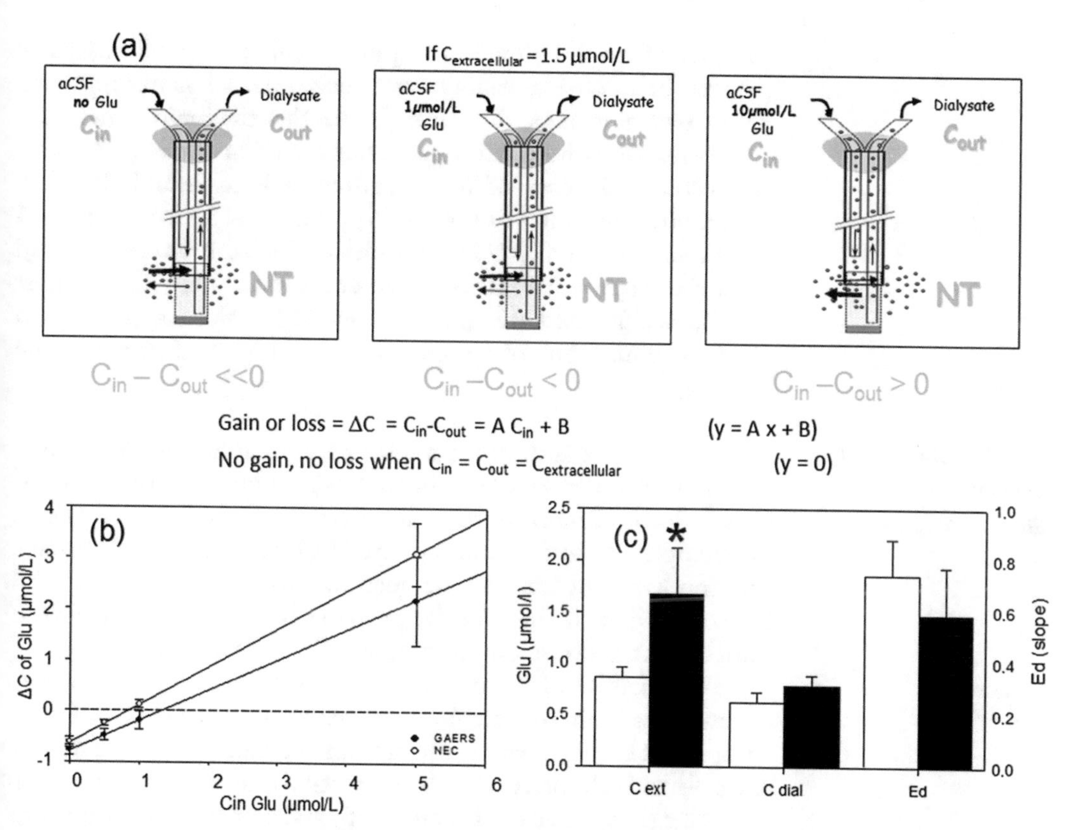

Fig. 7 No net flux experiment. (**a**) Principle of the no net flux method, based on microdialysis dynamics. Briefly, known concentrations of Glu are infused through the probe (C_{in}). Depending on the concentration present in the extracellular space ($C_{extracellular}$), the direction of Glu diffusion will modify the Glu concentration at the outlet of the probe (C_{out}). $C_{extracellular}$ is supposed to be reached when $C_{in} = C_{out}$, and a linear plot of C_{in}–C_{out} in function of C_{in} will allow to determine $C_{extracellular}$ with 3–4 different C_{in} points. From Dr. M.-F. Suaud-Chagny, Lyon Neuroscience Research Center, Lyon, France, with permission. (**b**) Typical linear curves obtained in the cortex of a GAERS (*black circles*) versus a non-epileptic rat (NEC, *white circles*) showing the gain or loss of Glu (ΔC) according to the concentration infused through the probe (C_{in}). (**c**) True extracellular concentrations (C_{ext}) of Glu and extraction fraction (Ed) determined respectively at $\Delta C = 0$ and from the slope of the linear curves obtained in cortex of GAERS (*black bars*, $n = 5$) as compared to NEC rats (white bars, $n = 5$). C_{ext} can be compared to the dialysate concentrations (C_{dial}) corresponding to a Glu-free perfusion in classical microdialysis. Adapted from [43]

fraction can be regarded as an index of Glu uptake. Figure 7 shows typical no net flux curves obtained from GAERS cortex as compared to their control rats (NEC). Data on slope and $\Delta C = 0$ reveal no difference between GAERS and NEC in thalamus (right), whereas a significant variation in standard deviation for Glu levels is seen in cortex of GAERS as compared to NEC, although there is no difference in the extracellular levels. Differences in extracellular concentrations or extraction fractions between GAERS and control rats were tested using the F-test followed by either Welch's test or Student's t-test, as appropriate, but did not reveal other differences.

3. Even if no net flux is a popular technique in neuroscience studies, some disadvantages have been pointed on this approach by several groups. First, it supposes that the brain structure is normal or without neuronal loss as underlined previously [61]. Second, the cost of the experiments is high and the time required included sampling [67], and analysis of at least 24 samples per rat [43]. Third, a pitfall which can be encountered is the absence of linearity for the curve ($\Delta C = C_{in} - C_{out}$) against C_{in}, as illustrated in Fig. 8 and discussed in Note 3, observed in two animals out of a recent series of 12 rats dialyzed in the hippocampus.

3.5 Measurements of in Vivo Glutamate Uptake

As an alternative and/or a complementary method to no net flux approach, Glu uptake was estimated using a method based on isotope infusion. Briefly, trace radioactive Glu diluted in cold Glu at physiological concentration is infused by reverse dialysis. Cells capture this excess of Glu and the quantification of labeled Glu in the infused fluid at the inlet of the probe vs the microdialysate at the outlet of the probe allows to estimate Glu uptake ability.

The quantity of radioactivity needed to carry out the in vivo experiments is an important parameter because the ability of cells to uptake Glu depends on the amount of both Glu contained in the cells and Glu present at the extracellular level, on the number of uptake sites, and on the activity of uptake sites. We have already mentioned the variety of transporters and their localization in brain [20, 22]. One must also take into account the animal used and the brain area targeted as variations in tissue contents were observed between species [73, 74] or brain structures [75]. Besides, many factors are known to modulate Glu contents such as anesthesia [76], drug intake [73], development [77], or epileptogenesis [78]. The in vivo extracellular concentrations of Glu greatly vary as well and in the rat, the extracellular concentrations are in the range of 1–20 μmol/L depending on the brain structures [79], and also on experimental conditions such as handling [79], anesthesia [80], or even the season when the experiments are carried out [81]. Previously, a concentration of 0.85 μmol/L ^{3}H-Glu has been chosen for an extracellular concentration of endogenous Glu equal to 1.96 μmol/L in frontoparietal cortex, after having verified the time necessary for Glu extraction, i.e., 10–15 min to reach steady state [66]. The choice of the Glu concentrations (labeled and unlabeled) in the experiments using radioactivity trace for uptake estimation is crucial to succeed in this approach. Indeed, Glu uptake per se is not the only one parameter that determines how much labeled Glu is extracted from the probe as experienced by Alexander and colleagues [66]. Indeed, Glu extraction by cells depends on: (a) the diffusion through membrane governed mainly by geometry, molecular cutoff, and flow rate and (b) the tissue diffusion and cellular membrane permeability. These authors minimized the

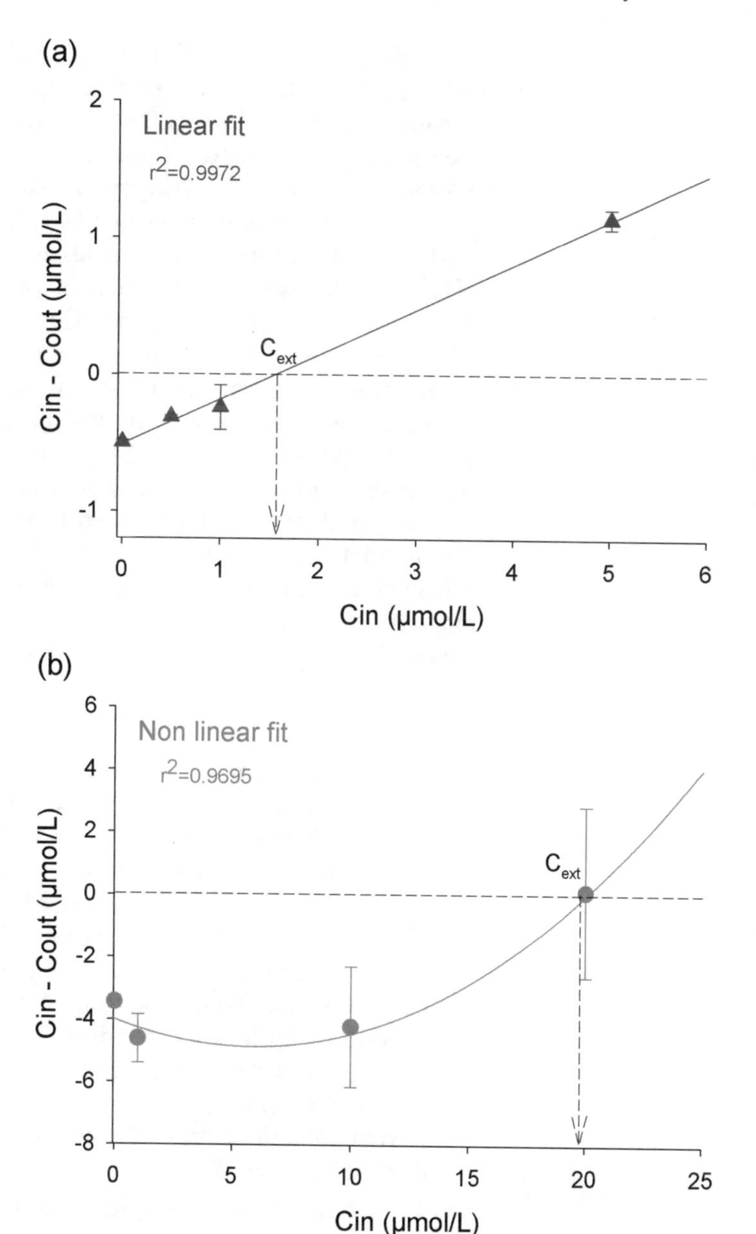

Fig. 8 Linear vs nonlinear fit in no net flux experiments. (**a**) Typical example of the data normally obtained in 95% of the experiments carried out on Glu concentration. (**b**) Typical example of the data obtained in less than 5% of our previous experiments. Extracellular concentration is extrapolated from the equation at "$C_{in}–C_{out}$ = 0" (*dashed arrow* versus *dashed line*)

resistance of probe and tissue in order to favor the uptake by using high concentration of Glu perfused in the probe with labeled Glu (0.85 µmol/L) and unlabeled Glu (>100 µmol/L) [66]. Besides, they showed that less than 2% and 6% of tritium of the initial labeled Glu molecules was detected after 10 or 45 min of perfusion, respectively, recommending that a short time of perfusion (<10 min) should be used to avoid any significant impact of Glu metabolism kinetics on the uptake experiments [66]. As already mentioned in this chapter, D-Asp has been widely used in vitro [37, 47–49] and once in vivo [65] to evaluate the activity of Glu transporters since it is not metabolized by cells. However, as the modulation of transport in the brain is different for Asp and Glu, the use of labeled Glu instead of labeled D-Asp can be better because (1) it is the same species as endogenous Glu and (2) Glu modulation by other compounds can be investigated [66]. Lastly, to take into account the passive diffusion through membranes and/or barriers, Glu uptake is compared to mannitol uptake as mannitol is an exogenous compound not taken by any active cell transporters [82] (Please see the limits for the use of labeled isotopes in Note 4).

Our method is partly adapted from a method developed for frontoparietal cortex [66] and we confronted the following protocols in three other brain areas, i.e., frontal cortex, thalamus, and hippocampus [41, 43]. We chose to increase the diffusion of exogenous Glu into the tissue by increasing the Glu concentration infused in the probe. By this means, the method can be convenient to brain areas containing high concentrations of Glu.

1. In our protocol, uptake experiments were performed by switching the microdialysis probe perfusion fluid for 10 min to aCSF containing 1 mmol/L Glu solution [750 μmol/L of unlabeled Glu and 250 μmol/L ^{3}H-Glu (15 Ci/mmol)] containing 120 μmol/L ^{14}C-mannitol (MAN) (60 mCi/mmol) (PerkinElmer Life and Analytical Sciences, USA), samples of microdialysis probe effluent being collected every 2 min for dual label scintillation counting. After perfusion with aCSF for 12 min to flush out the radioactive isotopes, the experiment was repeated in the presence of 20 mmol/L THA, a Glu uptake inhibitor on the first rats of a series to verify our experimental conditions for each structure or rat strain used [43], as illustrated in Fig. 9a.
2. As explained in Sect. 3.3, we used the outlet of a liquid swivel for mouse experiments to have minimal contamination by the isotopes used on internal surfaces of the swivel, but the risk of radioactive splashing due to the movements of the animals is possible (Fig. 6c). Indeed, the presence of the swivel balancer permits the rat moving freely in the cage with limited pulling on the tubings of the probe, preventing rupture of the flow leak and thereafter radioactive dissemination. The balancer also attenuates the effects of the lateral movements of the animal; however, the balancer is not efficient against rat rearing which can be the cause of potential radioactive splash in case of too much moving. To limit the risk of undesired movements the experimenter needs to habituate the rat to the cage, if it is

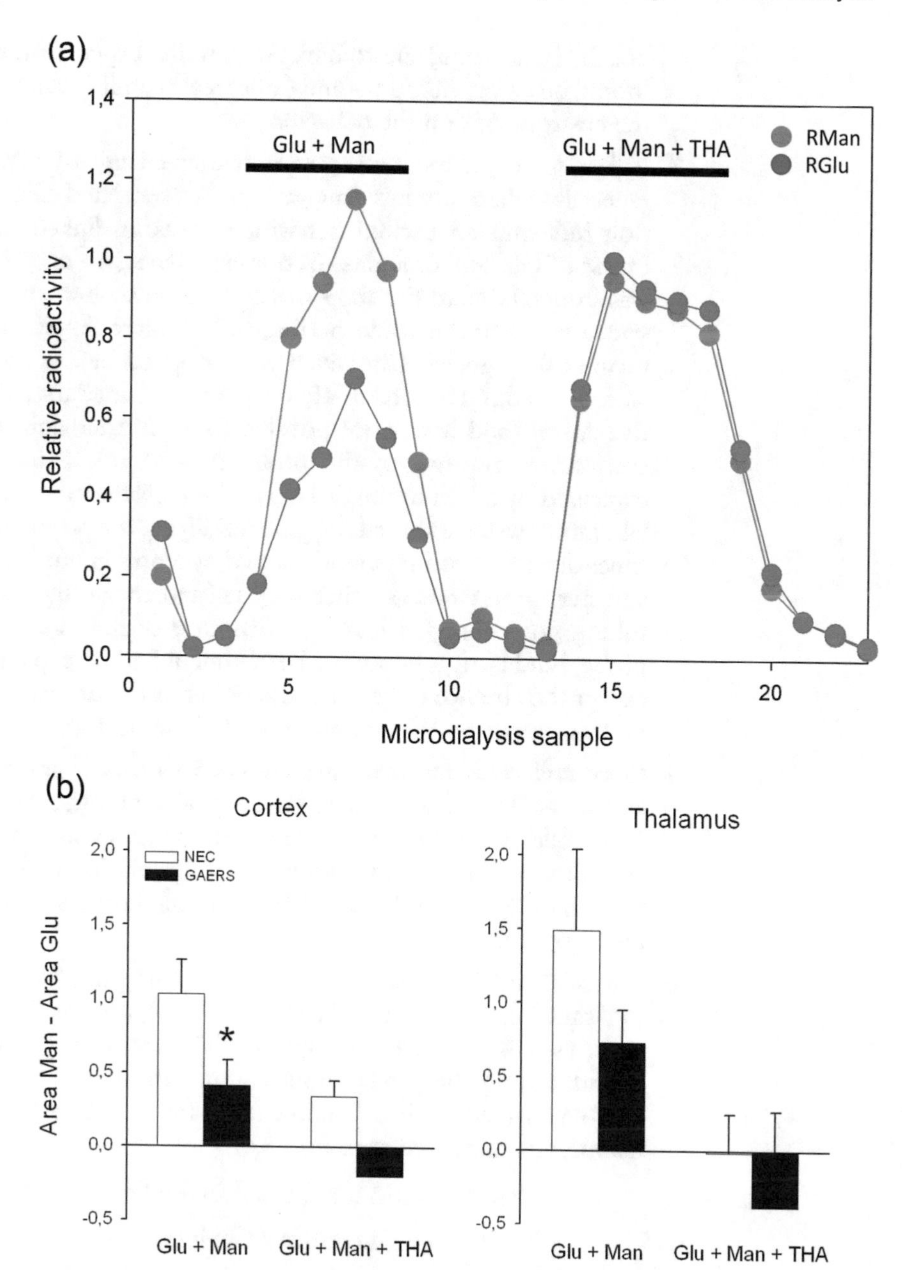

Fig. 9 Glutamate uptake is evaluated using radioactivity experiment. (**a**) An example of raw data exhibited the relative radioactivity recovery for mannitol (RMan, *red circles*) and Glu (RGlu, *blue circles*) throughout the experiment in a GAERS rat, in absence (*left*) or presence (*right*) of THA. (**b**) Differences in areas under the curve between L-[^{3}H]-Glu and [^{14}C]-mannitol recoveries in NEC rats (*white bars*) and GAERS rats (*black bars*) during the concomitant application of the two isotopes (Glu + Man) in absence or presence of a Glu uptake blocker, THA. Note that the uptake is altered in the cortex of GAERS as compared to the NEC rat. Other details are given in Sect. 3.5

not the home cage, and to him. Besides, the experiment room should be quiet, vibration- and odor-free to limit the impact of the environment on the behavior.

3. At last, the experimenter should note each change of behavior, especially when administering a high-concentrated Glu solution inducing an excited behavior or waking linked to the excess of Glu for some rats. In our experience, no experiment was stopped due to the abnormal animal movements or excessive movements due to the perfusion. The microdialysis experiments using labeled Glu which were stopped before the end were seven out of a total of 40: once was explained for a defective swivel (and it was not possible to go further due to the consecutive rupture of the probe membrane), twice were explained by a leak in the dialysis probe (solvent vapors in the laboratory to have altered the glue solidifying the probe), three times due to ultrasonic sounds caused by works in our building and perturbing rats (so that they twisted manically and the tubing got entangled leading to rupture of flow rate in the probe), and lastly, one stopped experiment had no explanation except the absence of liquid at the outlet of the probe, despite no leak was detectable at any stage of the perfusion.
4. Once collected, the dialysates enriched with isotopes can be kept at +4 °C until analysis. On the day of counting, caps of the PCR collection tubes are removed and the tubes are placed in counting vials filled with a liquid counting solution for quantification of ^{3}H-Glu and ^{14}C-MAN (counting during 4 min maximum per tube).
5. The relative Glu uptake was determined using the "recovery" vs. "time" curves for ^{3}H-Glu and ^{14}C-MAN, used as the reference, as it is not taken up by cells. The recovery (R) is the counts exiting the probe/input counts ratio (Eqs. 1 and 2). The cellular extraction fraction (EL) for Glu corrected for mannitol is calculated as in Eq. (3).

$$\text{RMAN} = \text{MAN outlet} / \text{MAN inlet} \tag{1}$$

$$\text{RGlu} = \text{Glu outlet} / \text{Glu inlet} \tag{2}$$

$$\text{EL Glu} = \text{RMAN} - \text{RGlu} / \text{RMAN} \tag{3}$$

The relative Glu uptake in the presence or absence of THA was calculated for each 2 min of the 10-min infusion period as shown in Fig. 9. Differences in uptake between the mean ± SEM for each group (*N* from 6–12 individuals) are tested using the unpaired Student's *t*-test and a level of significance of $p < 0.05$. When comparing Glu uptake in cortex and thalamus in GAERS rats and their NEC controls, we showed a specific decrease in Glu uptake in the cortex of GAERS (Fig. 9b), which was linked to an increase of expression for vGluT2 and synaptophysin, two markers of synaptic changes [43]. Our data showed an impairment of the glutamatergic terminal network, including both release and uptake. Unfortunately, neither the true extracellular Glu concentration nor Glu extraction was not clear-cut affected using the no net flux approach as we previously mentioned in Sect. 3.4 of this chapter, showing that the method using labeled Glu is much more sensitive and more powerful than the no net flux method experiments regarding uptake alteration. Indeed, a small number of animals (4–5) is sufficient to reveal significant Glu uptake changes.

6. Lastly, the no net flux method requires both a strict linearity which is observed with animals used in basal or normal conditions, and no traumatic brain tissue, which is not always the case in neuropathic animal models (see Note 1). As a consequence, the method based on radioactivity diffusion appears the most versatile in both basal and evoked Glu conditions.

3.6 Implantation Verification

At the end of the experiment, methylene blue (0.1%) can be injected via the probe to tag the probe tract in the brain. The rats are killed with a lethal dose of pentobarbital just after the removal of the microdialysis probe to limit the diffusion of the dye. The verification of the placement of the cannula does not need the fixation of the brain after an intracardiac paraformaldehyde perfusion. Classically, we verify the probe tracts on the brains frozen after their removal [43, 83, 84]. The issue is that the brains contain radioactivity and can contaminate the parts of the cryostat used to cut them, such as the knife, the bin collecting discarded cuts, the anti-roll slide and may be other parts in the interior of the cryostat. So that, we do not use a cryostat but a scalpel to section the brains just being removed, i.e., not frozen, in order to verify the right placement of the track left by the microdialysis probe (Fig. 10). This cutting can be easily done by a scientist highly skilled in rat brain anatomy, which greatly limits isotope contamination present in the dialyzed brain to the scalpel blade and the brain support (here, aluminum paper). By this means, a precise localization can be done before freezing the brain for archiving it or destroying it via appropriate radioactive waste.

3.7 Cleaning and Waste

The microdialysis material contaminated with radioactivity should be cleaned with TDF4, a specific detergent solution (5% in warm water), before being flushed by deionized water to remove the detergent. The other microdialysis material is flushed with deionized water. Rinsing solutions should be gathered in the waste bins for radioactivity elements or classical chemical bins, as appropriate. All the materials, except the microdialysis probes that are used once to guarantee reproducible data, can be reused.

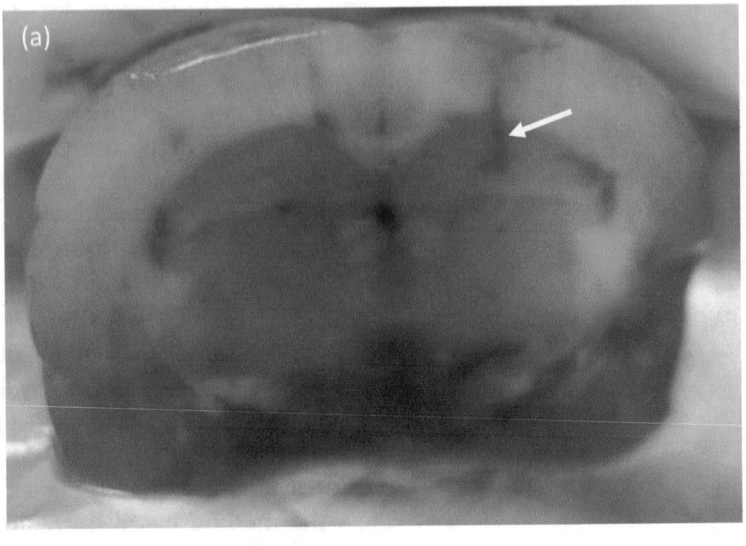

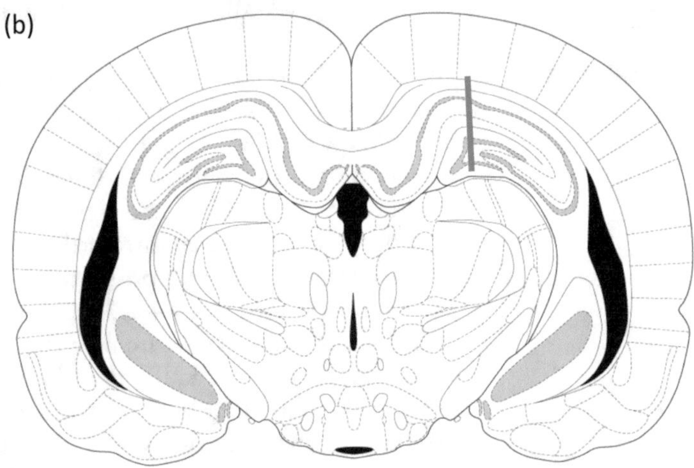

Fig. 10 Verification of probe placement on a brain collected just after the experiment of radioactivity. Note the real track left in the right hippocampus (**a**) by the microdialysis probe (indicated with the *white arrow*) as compared to the coronal section from the atlas by Paxinos and Watson [98] (with permission to reproduce) (**b**). The *blue line* indicates the placement reported on the atlas

4 Discussion and Conclusions

In vivo approaches are of crucial importance when confronting data obtained in vitro or ex vivo in individuals because of the complexity and diversity of physiological regulations observed in intact organisms. Brain microdialysis, which is a classical technique to monitor Glu levels, is also used for the in vivo study of Glu uptake using no net flux method or isotope infusion. The no net flux method can assess the uptake of Glu in addition to give the absolute extracellular concentration since the slope of the curve ΔC of Glu vs Glu C_{in} represents the probe extraction of Glu, thus providing information on the Glu uptake, but with a lower sensitivity as compared to the isotopic approach (see further in this discussion). The first advantage of the no net flux approach is that comparison of real concentrations between strains as shown on GAERS rats or after chronic treatments is possible. In fact, the method is very popular in both rats and mice for dopamine [62, 85, 86] or serotonin [87–90] or Glu [91, 92] or several neurotransmitters challenged successively on the same rat [43, 93]. As it can be coupled to capillary electrophoresis as in our no net flux study, collection time is 2 min, instead of 10–20 min when HPLC is used. The main convenience is that the total duration of the experiment is inferior to 2 h, instead of 6–10 h with longer sampling time. As a consequence, several compounds could be targeted on the same animal as previously done on GAERS rats with Glu, GABA, and Asp evaluated on the same animal [43]. This is of great importance to minimize some experimental bias as circadian rhythms, but to correlate data between other neurotransmitters interacting with Glu. A major inconvenience of the no net flux approach is the number of rats needed to reach significance. At least 8–12 animals are required to detect difference of concentrations between two groups. In contrast, the main advantage of the isotopic approach is that only 5–6 animals (or even just four rats) are sufficient to reveal great variations in in vivo uptake capacity and it exhibits better power as compared to no net flux approach for detecting Glu uptake alteration. However, the major limit of this approach is the cost of radioactive products containing ^{14}C and ^{3}H, including purchasing, storage, and waste elimination. In addition, the handling of radioactive compounds requires authorizations for rooms, materials and staff, identified rooms, and skilled experimenters. To exploit the strengths of the two approaches, an approach combining both no net flux and isotope infusion has been proposed recently [67] and it is based on the use of labeled Glu in no net flux experiments to overcome the inconveniences in classical no net flux, such as time-consumption due to the successive perfusions of at least four concentrations of Glu and the possible change in probe recovery throughout the long-time experiment [67]. Besides those methodological aspects, the method takes into account the changes of recovery due the administration of drugs which can modulate the dynamics of Glu, including transport [66]. One limit for a routine use is that detection by mass spectrometry is required for quantifying the labeled Glu, which serves as quantitative marker [67]. Those approaches based on isotope infusion are still poorly used but they could be useful for studying not only Glu, but also all the compounds taken by cells by active exchange transporters, as done for dopamine [67], and potentially almost all the neurotransmitters regulated by active transport, such as dopamine, norepinephrine, serotonin, GABA, and glycine. More precisely, the isotope uptake method has been applied on rat dopaminergic system using [^{3}H]-MPP+ as a neurotoxic competitor of dopamine transporters and [^{14}C]-mannitol as reference substance [94]. A similar approach has been used in vitro on serotonin in rat pineal glands [95], suggesting the versatility of the in vivo method for studying the uptake impairment of the other major neurotransmitters. Lastly, in order to get better information in kinetics, a high sampling rate (<1–2 min) is to be encouraged since low sampling rate can mask fine-tuned variations [56]. Moreover, this chapter also shows that the study can be rather easily carried out in freely moving rats provided a well-adapted material, which is a requisite when considering a fully physiological preparation, either in physiopathological conditions or pharmacological studies. As a conclusion, these isotope-based methods are very promising to study active exchange and are a powerful alternative to classical no net flux method due to its greater sensitivity and short microdialysis experiment time. The application on Glu is of great interest as Glu transporters are involved in many pathologies, such as ischemia, traumatic brain injury, autism, pain, neurodegenerative diseases, epilepsies, amyotrophic sclerosis, and depression, and the potential on the experimental models available is considerable.

5 Notes

1. Care for isoflurane anesthesia: temperature of the room and doses

   The surgery should be done in a controlled temperature room, upper than 17 °C and lower than 26 °C. Indeed, if the temperature is too low, the isoflurane cannot be vaporized and liquid isoflurane reaches the rat airways, causing immediate death. If the temperature is too high, the level of vaporization is high and the amount of isoflurane is not controlled leading to overdose and rapid death.

   The doses should be adjusted according to preliminary tests in the strain used. Take care to adjust properly the flow rate to evacuate isoflurane to avoid dissemination of the anesthetic in the room air, for the sake of the experimenters. Moreover, weight the filter absorbing the evacuated isoflurane after each use and change it if the weight of isoflurane reached the value authorized by the supplier.

2. Absence of linearity in some rare no net flux experiments

   A pitfall which can be encountered is the absence of linearity for the curve ($\Delta C = C_{in} - C_{out}$) against C_{in}, as illustrated in Fig. 8. This phenomenon was observed in our laboratory on two animals out of a recent series of 12 rats. In this series, the probe was implanted in the dorsal hippocampus of rats equipped with an Alzet pump delivering AQP4-antibodies which are able to induce an experimental NMO disease [41]. One explanation could have been the neuronal damage which can be observed in some areas of the central nervous system of NMO rats such as spinal cord or optic nerve; however, this is unlikely because one of the two rats was a control rat. As we used the method in two other brain areas, cortex and thalamus, without any matter on linearity, we can assume that the absence of linearity can be linked to the structure targeted itself. Indeed, the hippocampus has a complex and tortuous shape, with folded layers (Fig. 10). In that context, our microdialysis probe membrane cannot target a homogenous area due to its large dimensions (2-mm length and ~200 μm as external diameter). Moreover, the complexity of the structure could be linked not only to its shape and tortuosity but could be also due to the diversity of receptors types and synapses. As a consequence, it seems that the method can be applied throughout the brain not only using a rigorous stereotaxy but also using a rodent population with anatomical homogeneity.

In conclusion, this approach of no net flux should be taken with precaution for the estimation of Glu uptake and the approach testing more directly Glu uptake as the radioactivity approach should be considered as a gold standard method.

3. Stability of glutamate over time

The stability of unlabeled and labeled Glu is the same over time; the solutions can be kept in the refrigerator in acidic conditions during 2 years without significant alterations. Control experiments can be done easily for unlabeled Glu by quantification with an analytical method such as CE-LIFD, as briefly mentioned in this chapter. To check the concentration of labeled Glu, the use of CE-LIFD is not allowed in our laboratory, so the labeled Glu was validated using a very small set of animal controls, by infusing the labeled Glu twice, once in presence of mannitol as diffusion control, the second time in presence of mannitol and THA, a Glu uptake inhibitor. Data obtained are shown in Fig. 9a. An alternative animal-free means of controlling the quality of the labeled Glu solution would be to analyze its purity using thin layer chromatography (or TLC) which is commonly used in synthesis chemistry. Only one spot should be revealed (by ninhydrin for instance) and should contain radioactive atoms. In case of several spots exhibiting radioactivity, it would prove that the solution is partly degraded.

4. Limits on the use of radioactive isotopes

Preliminary experiments carried out on young rats (aged 30 days) showed that the recovery for labeled Glu was higher than the recovery of labeled mannitol. Two explanations are possible: first, Glu permeability is lower in young rats; second, mannitol permeability is higher in the young rats. Both assumptions are in agreement with a study by Al-Sarraf et al. [96] showing that the accumulation of [^{3}H]-mannitol and [^{14}C]-Glu (instead of [^{3}H]-Glu and [^{14}C]-mannitol, as in our study) by the choroid plexuses of the lateral ventricles is different in pups and adult rats. Indeed, the accumulation of [^{14}C]-Glu was found to be 2–3 times greater in the adult than in the neonatal rat while [^{3}H]-mannitol measured into choroidal tissue in the 1-week-old rat was significantly greater than those in the adult [96]. Thus, the use of mannitol as an extracellular marker seems to be limited to adult animals.

Acknowledgments

The authors thank the Institut National de la Santé et de la Recherche Médicale (INSERM), the Centre National de la Recherche Scientifique (CNRS), and the Université Claude Bernard Lyon 1.

References

1. Kera Y, Aoyama H, Matsumura H, Hasegawa A, Nagasaki H, Yamada R (1995) Presence of free D-glutamate and D-aspartate in rat tissues. Biochim Biophys Acta 1243(2):283–286
2. Kera Y, Aoyama H, Watanabe N, Yamada RH (1996) Distribution of D-aspartate oxidase and free D-glutamate and D-aspartate in chicken and pigeon tissues. Comp Biochem Physiol B Biochem Mol Biol 115(1):121–126
3. D'Aniello A, D'Onofrio G, Pischetola M, D'Aniello G, Vetere A, Petrucelli L, Fisher GH (1993) Biological role of D-amino acid oxidase and D-aspartate oxidase. Effects of D-amino acids. J Biol Chem 268(36):26941–26949
4. Palmada M, Centelles JJ (1998) Excitatory amino acid neurotransmission. Pathways for metabolism, storage and reuptake of glutamate in brain. Front Biosci 3:d701–d718
5. Kvamme E (1998) Synthesis of glutamate and its regulation. Prog Brain Res 116:73–85
6. Dulac O, Milh M, Holmes GL (2013) Brain maturation and epilepsy. Handb Clin Neurol 111:441–446. doi:10.1016/B978-0-444-52891-9.00047-6
7. Feldman DE, Knudsen EI (1998) Experience-dependent plasticity and the maturation of glutamatergic synapses. Neuron 20(6):1067–1071
8. Fonnum F (1984) Glutamate: a neurotransmitter in mammalian brain. J Neurochem 42(1):1–11
9. Orrego F (1979) Criteria for the identification of central neurotransmitters, and their application to studies with some nerve tissue preparations in vitro. Neuroscience 4(8):1037–1057
10. Wheal H, Thomson A (1991) Excitatory amino acids and synaptic transmission. Academic Press, San Diego
11. Lees GJ (2000) Pharmacology of AMPA/kainate receptor ligands and their therapeutic potential in neurological and psychiatric disorders. Drugs 59(1):33–78
12. Ottersen OP, Landsend AS (1997) Organization of glutamate receptors at the synapse. Eur J Neurosci 9(11):2219–2224
13. Verkhratsky A, Kirchhoff F (2007) NMDA receptors in glia. Neuroscientist 13(1):28–37. doi:10.1177/1073858406294270

14. Verkhratsky A, Burnstock G (2014) Purinergic and glutamatergic receptors on astroglia. Adv Neurobiol 11:55–79. doi:10.1007/978-3-319-08894-5_4
15. Parpura V, Basarsky TA, Liu F, Jeftinija K, Jeftinija S, Haydon PG (1994) Glutamate-mediated astrocyte-neuron signalling. Nature 369(6483):744–747. doi:10.1038/369744a0
16. Parpura V, Grubisic V, Verkhratsky A (2011) Ca(2+) sources for the exocytotic release of glutamate from astrocytes. Biochim Biophys Acta 1813(5):984–991. doi:10.1016/j.bbamcr.2010.11.006
17. Malarkey EB, Parpura V (2008) Mechanisms of glutamate release from astrocytes. Neurochem Int 52(1–2):142–154. doi:10.1016/j.neuint.2007.06.005
18. Araque A, Parpura V, Sanzgiri RP, Haydon PG (1999) Tripartite synapses: glia, the unacknowledged partner. Trends Neurosci 22(5):208–215
19. Sahlender DA, Savtchouk I, Volterra A (2014) What do we know about gliotransmitter release from astrocytes? Philos Trans R Soc Lond Ser B Biol Sci 369(1654):20130592. doi:10.1098/rstb.2013.0592
20. Vandenberg RJ, Ryan RM (2013) Mechanisms of glutamate transport. Physiol Rev 93(4):1621–1657. doi:10.1152/physrev.00007.2013
21. Rothstein JD, Dykes-Hoberg M, Pardo CA, Bristol LA, Jin L, Kuncl RW, Kanai Y, Hediger MA, Wang Y, Schielke JP, Welty DF (1996) Knockout of glutamate transporters reveals a major role for astroglial transport in excitotoxicity and clearance of glutamate. Neuron 16(3):675–686
22. Danbolt NC, Furness DN, Zhou Y (2016) Neuronal vs glial glutamate uptake: resolving the conundrum. Neurochem Int 98:29–45. doi:10.1016/j.neuint.2016.05.009
23. Alexander SP, Mathie A, Peters JA (2008) Guide to receptors and channels (GRAC), 3rd edition. Br J Pharmacol 153(Suppl 2):S1–209. doi:10.1038/sj.bjp.0707746
24. Waagepetersen HS, Shimamoto K, Schousboe A (2001) Comparison of effects of DL-threo-beta-benzyloxyaspartate (DL-TBOA) and L-trans-pyrrolidine-2,4-dicarboxylate (t-2,4-PDC) on uptake and release of [3h]D-aspartate in astrocytes and glutamatergic neurons. Neurochem Res 26(6):661–666
25. Jabaudon D, Shimamoto K, Yasuda-Kamatani Y, Scanziani M, Gahwiler BH, Gerber U (1999) Inhibition of uptake unmasks rapid extracellular turnover of glutamate of nonvesicular origin. Proc Natl Acad Sci U S A 96(15):8733–8738
26. Volterra A, Bezzi P, Rizzini BL, Trotti D, Ullensvang K, Danbolt NC, Racagni G (1996) The competitive transport inhibitor L-trans-pyrrolidine-2, 4-dicarboxylate triggers excitotoxicity in rat cortical neuron-astrocyte co-cultures via glutamate release rather than uptake inhibition. Eur J Neurosci 8(9):2019–2028
27. Nicholls D, Attwell D (1990) The release and uptake of excitatory amino acids. Trends Pharmacol Sci 11(11):462–468
28. Ketheeswaranathan P, Turner NA, Spary EJ, Batten TF, McColl BW, Saha S (2011) Changes in glutamate transporter expression in mouse forebrain areas following focal ischemia. Brain Res 1418:93–103. doi:10.1016/j.brainres.2011.08.029
29. Chen KH, Reese EA, Kim HW, Rapoport SI, Rao JS (2011) Disturbed neurotransmitter transporter expression in Alzheimer's disease brain. J Alzheimers Dis 26(4):755–766. doi:10.3233/JAD-2011-110002
30. O'Donovan SM, Hasselfeld K, Bauer D, Simmons M, Roussos P, Haroutunian V, Meador-Woodruff JH, McCullumsmith RE (2015) Glutamate transporter splice variant expression in an enriched pyramidal cell population in schizophrenia. Transl Psychiatry 5:e579. doi:10.1038/tp.2015.74
31. Llorente IL, Landucci E, Pellegrini-Giampietro DE, Fernandez-Lopez A (2015) Glutamate receptor and transporter modifications in rat organotypic hippocampal slice cultures exposed to oxygen-glucose deprivation: the contribution of cyclooxygenase-2. Neuroscience 292:118–128. doi:10.1016/j.neuroscience.2015.02.040
32. Storck T, Schulte S, Hofmann K, Stoffel W (1992) Structure, expression, and functional analysis of a Na(+)-dependent glutamate/aspartate transporter from rat brain. Proc Natl Acad Sci U S A 89(22):10955–10959
33. Velaz-Faircloth M, McGraw TS, Alandro MS, Fremeau RT Jr, Kilberg MS, Anderson KJ (1996) Characterization and distribution of the neuronal glutamate transporter EAAC1 in rat brain. Am J Phys 270(1 Pt 1):C67–C75
34. Dutuit M, Touret M, Szymocha R, Nehlig A, Belin MF, Didier-Bazes M (2002) Decreased expression of glutamate transporters in genetic absence epilepsy rats before seizure occurrence. J Neurochem 80(6):1029–1038
35. Lauriat TL, Richler E, McInnes LA (2007) A quantitative regional expression profile of EAAT2 known and novel splice variants reopens the question of aberrant EAAT2 splicing in disease. Neurochem Int 50(1):271–280. doi:10.1016/j.neuint.2006.08.014

36. Torp R, Danbolt NC, Babaie E, Bjoras M, Seeberg E, Storm-Mathisen J, Ottersen OP (1994) Differential expression of two glial glutamate transporters in the rat brain: an in situ hybridization study. Eur J Neurosci 6(6):936–942
37. Behrens PF, Franz P, Woodman B, Lindenberg KS, Landwehrmeyer GB (2002) Impaired glutamate transport and glutamate-glutamine cycling: downstream effects of the Huntington mutation. Brain 125(Pt 8):1908–1922
38. Lievens JC, Woodman B, Mahal A, Spasic-Boscovic O, Samuel D, Kerkerian-Le Goff L, Bates GP (2001) Impaired glutamate uptake in the R6 Huntington's disease transgenic mice. Neurobiol Dis 8(5):807–821. doi:10.1006/nbdi.2001.0430
39. Arzberger T, Krampfl K, Leimgruber S, Weindl A (1997) Changes of NMDA receptor subunit (NR1, NR2B) and glutamate transporter (GLT1) mRNA expression in Huntington's disease—an in situ hybridization study. J Neuropathol Exp Neurol 56(4):440–454
40. Munch C, Ebstein M, Seefried U, Zhu B, Stamm S, Landwehrmeyer GB, Ludolph AC, Schwalenstocker B, Meyer T (2002) Alternative splicing of the 5′-sequences of the mouse EAAT2 glutamate transporter and expression in a transgenic model for amyotrophic lateral sclerosis. J Neurochem 82(3):594–603
41. Marignier R, Ruiz A, Cavagna S, Nicole A, Watrin C, Touret M, Parrot S, Malleret G, Peyron C, Benetollo C, Auvergnon N, Vukusic S, Giraudon P (2016) Neuromyelitis optica study model based on chronic infusion of autoantibodies in rat cerebrospinal fluid. J Neuroinflammation 13(1):111. doi:10.1186/s12974-016-0577-8
42. Dutuit M, Didier-Bazes M, Vergnes M, Mutin M, Conjard A, Akaoka H, Belin MF, Touret M (2000) Specific alteration in the expression of glial fibrillary acidic protein, glutamate dehydrogenase, and glutamine synthetase in rats with genetic absence epilepsy. Glia 32(1):15–24
43. Touret M, Parrot S, Denoroy L, Belin MF, Didier-Bazes M (2007) Glutamatergic alterations in the cortex of genetic absence epilepsy rats. BMC Neurosci 8:69. doi:10.1186/1471-2202-8-69
44. Hamano K, Hayashi M, Shioda K, Fukatsu R, Mizutani S (2008) Mechanisms of neurodegeneration in mucopolysaccharidoses II and IIIB: analysis of human brain tissue. Acta Neuropathol 115(5):547–559. doi:10.1007/s00401-007-0325-3
45. Herbert MK, Kuiperij HB, Verbeek MM (2012) Optimisation of the quantification of glutamine synthetase and myelin basic protein in cerebrospinal fluid by a combined acidification and neutralisation protocol. J Immunol Methods 381(1-2):1–8. doi:10.1016/j.jim.2012.04.001
46. Schlosser M, Hahmann J, Ziegler B, Augstein P, Ziegler M (1997) Sensitive monoclonal antibody-based sandwich ELISA for determination of the diabetes-associated autoantigen glutamic acid decarboxylase GAD65. J Immunoass 18(4):289–307
47. Fang J, Han D, Hong J, Tan Q, Tian Y (2012) The chemokine, macrophage inflammatory protein-2gamma, reduces the expression of glutamate transporter-1 on astrocytes and increases neuronal sensitivity to glutamate excitotoxicity. J Neuroinflammation 9:267. doi:10.1186/1742-2094-9-267
48. Goursaud S, Maloteaux JM, Hermans E (2009) Distinct expression and regulation of the glutamate transporter isoforms GLT-1a and GLT-1b in cultured astrocytes from a rat model of amyotrophic lateral sclerosis (hSOD1G93A). Neurochem Int 55(1-3):28–34. doi:10.1016/j.neuint.2009.02.003
49. Potier B, Billard JM, Riviere S, Sinet PM, Denis I, Champeil-Potokar G, Grintal B, Jouvenceau A, Kollen M, Dutar P (2010) Reduction in glutamate uptake is associated with extrasynaptic NMDA and metabotropic glutamate receptor activation at the hippocampal CA1 synapse of aged rats. Aging Cell 9(5):722–735. doi:10.1111/j.1474-9726.2010.00593.x
50. Parsons MP, Vanni MP, Woodard CL, Kang R, Murphy TH, Raymond LA (2016) Real-time imaging of glutamate clearance reveals normal striatal uptake in Huntington disease mouse models. Nat Commun 7:11251. doi:10.1038/ncomms11251
51. Liaw WJ, Stephens RL Jr, Binns BC, Chu Y, Sepkuty JP, Johns RA, Rothstein JD, Tao YX (2005) Spinal glutamate uptake is critical for maintaining normal sensory transmission in rat spinal cord. Pain 115(1-2):60–70. doi:10.1016/j.pain.2005.02.006
52. Binns BC, Huang Y, Goettl VM, Hackshaw KV, Stephens RL Jr (2005) Glutamate uptake is attenuated in spinal deep dorsal and ventral horn in the rat spinal nerve ligation model. Brain Res 1041(1):38–47. doi:10.1016/j.brainres.2005.01.088
53. Hacimuftuoglu A, Tatar A, Cetin D, Taspinar N, Saruhan F, Okkay U, Turkez H, Unal D, Stephens RL Jr, Suleyman H (2016) Astrocyte/neuron ratio and its importance on glutamate toxicity: an in vitro voltammetric study. Cytotechnology 68(4):1425–1433. doi:10.1007/s10616-015-9902-9

54. Ungerstedt U (1991) Microdialysis—principles and applications for studies in animals and man. J Intern Med 230(4):365–373
55. Kehr J (2010) Monitoring molecules in neuroscience: 50 years. In: Westerink B, Clinkers R, Smolders I, Sarre S, Michotte Y (eds) Monitoring molecules in neuroscience, Brussels, Belgium, 2010. Proceedings of 13th international conference on in vivo methods
56. Parrot S, Bert L, Mouly-Badina L, Sauvinet V, Colussi-Mas J, Lambas-Senas L, Robert F, Bouilloux JP, Suaud-Chagny MF, Denoroy L, Renaud B (2003) Microdialysis monitoring of catecholamines and excitatory amino acids in the rat and mouse brain: recent developments based on capillary electrophoresis with laser-induced fluorescence detection—a mini-review. Cell Mol Neurobiol 23(4-5):793–804
57. Lee WH, Ngernsutivorakul T, Mabrouk OS, Wong JM, Dugan CE, Pappas SS, Yoon HJ, Kennedy RT (2016) Microfabrication and in vivo performance of a microdialysis probe with embedded membrane. Anal Chem 88(2):1230–1237. doi:10.1021/acs.analchem.5b03541
58. Drew KL, Pehek EA, Rasley BT, Ma YL, Green TK (2004) Sampling glutamate and GABA with microdialysis: suggestions on how to get the dialysis membrane closer to the synapse. J Neurosci Methods 140(1-2):127–131. doi:10.1016/j.jneumeth.2004.04.039
59. Bungay PM, Morrison PF, Dedrick RL (1990) Steady-state theory for quantitative microdialysis of solutes and water in vivo and in vitro. Life Sci 46(2):105–119
60. Justice JB Jr (1993) Quantitative microdialysis of neurotransmitters. J Neurosci Methods 48(3):263–276
61. Chen KC (2006) Effects of tissue trauma on the characteristics of microdialysis zero-net-flux method sampling neurotransmitters. J Theor Biol 238(4):863–881. doi:10.1016/j.jtbi.2005.06.035
62. Brun P, Begou M, Andrieux A, Mouly-Badina L, Clerget M, Schweitzer A, Scarna H, Renaud B, Job D, Suaud-Chagny MF (2005) Dopaminergic transmission in STOP null mice. J Neurochem 94(1):63–73. doi:10.1111/j.1471-4159.2005.03166.x
63. Parrot S (2000) Intérêts et limites de la microdialyse intracérébrale couplée à l'électrophorèse capillaire pour l'étude des acides aminés excitateurs cérébraux. Thèse de doctorat Sciences. Biochimie, Université Claude Bernard Lyon 1
64. Sauvinet V, Parrot S, Benturquia N, Bravo-Moraton E, Renaud B, Denoroy L (2003) In vivo simultaneous monitoring of gamma-aminobutyric acid, glutamate, and L-aspartate using brain microdialysis and capillary electrophoresis with laser-induced fluorescence detection: analytical developments and in vitro/in vivo validations. Electrophoresis 24(18):3187–3196. doi:10.1002/elps.200305565
65. Bruhn T, Christensen T, Diemer NH (1995) Microdialysis as a tool for in vivo investigation of glutamate transport capacity in rat brain. J Neurosci Methods 59(2):169–174
66. Alexander GM, Grothusen JR, Gordon SW, Schwartzman RJ (1997) Intracerebral microdialysis study of glutamate reuptake in awake, behaving rats. Brain Res 766(1-2):1–10
67. Hershey ND, Kennedy RT (2013) In vivo calibration of microdialysis using infusion of stable-isotope labeled neurotransmitters. ACS Chem Neurosci 4(5):729–736. doi:10.1021/cn300199m
68. Ferry B, Parrot S, Marien M, Lazarus C, Cassel JC, McGaugh JL (2015) Noradrenergic influences in the basolateral amygdala on inhibitory avoidance memory are mediated by an action on alpha2-adrenoceptors. Psychoneuroendocrinology 51:68–79. doi:10.1016/j.psyneuen.2014.09.010
69. Paxinos G, Watson C (1998) The rat brain in stereotaxic coordinates, 4th edn. Academic Press, San Diego
70. Marignier R, Nicolle A, Watrin C, Touret M, Cavagna S, Varrin-Doyer M, Cavillon G, Rogemond V, Confavreux C, Honnorat J, Giraudon P (2010) Oligodendrocytes are damaged by neuromyelitis optica immunoglobulin G via astrocyte injury. Brain 133(9):2578–2591. doi:10.1093/brain/awq177
71. Parrot S, Sauvinet V, Riban V, Depaulis A, Renaud B, Denoroy L (2004) High temporal resolution for in vivo monitoring of neurotransmitters in awake epileptic rats using brain microdialysis and capillary electrophoresis with laser-induced fluorescence detection. J Neurosci Methods 140(1-2):29–38. doi:10.1016/j.jneumeth.2004.03.025
72. Parsons LH, Justice JB Jr (1994) Quantitative approaches to in vivo brain microdialysis. Crit Rev Neurobiol 8(3):189–220
73. Westerberg E, Chapman AG, Meldrum BS (1983) Effect of 2-amino-7-phosphonoheptanoic acid on regional brain amino acid levels in fed and fasted rodents. J Neurochem 41(6):1755–1760
74. Wolfersberger MG, Tabachnick J, Finkelstein BS, Levin M (1973) L-pyrrolidone carboxylic acid content in mammalian epidermis and other tissues. J Invest Dermatol 60(5):278–281

75. El-Khoury R, Panayotis N, Matagne V, Ghata A, Villard L, Roux JC (2014) GABA and glutamate pathways are spatially and developmentally affected in the brain of Mecp2-deficient mice. PLoS One 9(3):e92169. doi:10.1371/journal.pone.0092169
76. Dobkin J (1972) Effects of excitation and anaesthesia on the glutamine content of the rat brain with a reference to the administration of glutamine. J Neurochem 19(4):1195–1202
77. Agrawal HC, Davis JM, Himwich WA (1968) Developmental changes in mouse brain: weight, water content and free amino acids. J Neurochem 15(9):917–923
78. Emson PC, Joseph MH (1975) Neurochemical and morphological changes during the development of cobalt-induced epilepsy in the rat. Brain Res 93(1):91–110
79. Timmerman W, Cisci G, Nap A, de Vries JB, Westerink BH (1999) Effects of handling on extracellular levels of glutamate and other amino acids in various areas of the brain measured by microdialysis. Brain Res 833(2):150–160
80. Moghaddam B, Adams B, Verma A, Daly D (1997) Activation of glutamatergic neurotransmission by ketamine: a novel step in the pathway from NMDA receptor blockade to dopaminergic and cognitive disruptions associated with the prefrontal cortex. J Neurosci 17(8):2921–2927
81. Parrot S, Bert L, Renaud B, Denoroy L (2001) Large inter-experiment variations in microdialysate aspartate and glutamate in rat striatum may reflect a circannual rhythm. Synapse 39(3):267–269. doi:10.1002/1098-2396(20010301)39:3<267::AID-SYN1008>3.0.CO;2-W
82. Wise BL, Perkins RK, Stevenson E, Scott KG (1964) Penetration of C14-labelled mannitol from serum into cerebrospinal fluid and brain. Exp Neurol 10:264–270
83. Bert L, Favale D, Jego G, Greve P, Guilloux JP, Guiard BP, Gardier AM, Suaud-Chagny MF, Lestage P (2004) Rapid and precise method to locate microdialysis probe implantation in the rodent brain. J Neurosci Methods 140(1-2):53–57. doi:10.1016/j.jneumeth.2004.04.042
84. Hegoburu C, Sevelinges Y, Thevenet M, Gervais R, Parrot S, Mouly AM (2009) Differential dynamics of amino acid release in the amygdala and olfactory cortex during odor fear acquisition as revealed with simultaneous high temporal resolution microdialysis. Learn Mem 16(11):687–697. doi:10.1101/lm.1584209
85. Engleman EA, Ingraham CM, McBride WJ, Lumeng L, Murphy JM (2006) Extracellular dopamine levels are lower in the medial prefrontal cortex of alcohol-preferring rats compared to Wistar rats. Alcohol 38(1):5–12. doi:10.1016/j.alcohol.2006.03.001
86. Unger EL, Bianco LE, Jones BC, Allen RP, Earley CJ (2014) Low brain iron effects and reversibility on striatal dopamine dynamics. Exp Neurol 261:462–468. doi:10.1016/j.expneurol.2014.06.023
87. Guiard BP, David DJ, Deltheil T, Chenu F, Le Maitre E, Renoir T, Leroux-Nicollet I, Sokoloff P, Lanfumey L, Hamon M, Andrews AM, Hen R, Gardier AM (2008) Brain-derived neurotrophic factor-deficient mice exhibit a hippocampal hyperserotonergic phenotype. Int J Neuropsychopharmacol 11(1):79–92. doi:10.1017/S1461145707007857
88. Mathews TA, Fedele DE, Coppelli FM, Avila AM, Murphy DL, Andrews AM (2004) Gene dose-dependent alterations in extraneuronal serotonin but not dopamine in mice with reduced serotonin transporter expression. J Neurosci Methods 140(1-2):169–181. doi:10.1016/j.jneumeth.2004.05.017
89. Shippenberg TS, Hen R, He M (2000) Region-specific enhancement of basal extracellular and cocaine-evoked dopamine levels following constitutive deletion of the Serotonin(1B) receptor. J Neurochem 75(1):258–265
90. Luellen BA, Bianco LE, Schneider LM, Andrews AM (2007) Reduced brain-derived neurotrophic factor is associated with a loss of serotonergic innervation in the hippocampus of aging mice. Genes Brain Behav 6(5):482–490. doi:10.1111/j.1601-183X.2006.00279.x
91. Pati D, Kelly K, Stennett B, Frazier CJ, Knackstedt LA (2016) Alcohol consumption increases basal extracellular glutamate in the nucleus accumbens core of Sprague-Dawley rats without increasing spontaneous glutamate release. Eur J Neurosci 44(2):1896–1905. doi:10.1111/ejn.13284
92. Miller BR, Dorner JL, Shou M, Sari Y, Barton SJ, Sengelaub DR, Kennedy RT, Rebec GV (2008) Up-regulation of GLT1 expression increases glutamate uptake and attenuates the Huntington's disease phenotype in the R6/2 mouse. Neuroscience 153(1):329–337. doi:10.1016/j.neuroscience.2008.02.004
93. Katner SN, Weiss F (2001) Neurochemical characteristics associated with ethanol preference in selected alcohol-preferring and -non-preferring rats: a quantitative microdialysis study. Alcohol Clin Exp Res 25(2):198–205
94. Zimmer L, Kodas E, Guilloteau D, Garreau L, Besnard J, Chalon S (2000) Microdialysis as a tool for in vivo study of dopamine transporter

function in rat brains. J Neurosci Methods 103(2):137–144

95. Aloyo VJ, Walker RF (1987) Noradrenergic stimulation of serotonin release from rat pineal glands in vitro. J Endocrinol 114(1):3–9
96. al-Sarraf H, Preston JE, Segal MB (1997) Acidic amino acid accumulation by rat choroid plexus during development. Brain Res Dev Brain Res 102(1):47–52
97. Denoroy L, Bert L, Parrot S, Robert F, Renaud B (1998) Assessment of pharmacodynamic and pharmacokinetic characteristics of drugs using microdialysis sampling and capillary electrophoresis. Electrophoresis 19(16-17):2841–2847. doi:10.1002/elps.1150191609
98. Paxinos G, Watson C (2005) The rat brain in stereotaxic coordinates, 5th edn. Academic Press, San Diego

Chapter 15

Enzymes of Glutamate System

Irina S. Boksha, Olga K. Savushkina, Elena B. Tereshkina, Tatyana A. Prokhorova, and Elizabeta B. Mukaetova-Ladinska

Abstract

In this chapter, we discuss the role and methodological approaches for extraction and quantification of four major enzymes involved in human brain glutamate metabolism, glutamine synthetase (GS), glutamic acid decarboxylase (GAD), glutamate dehydrogenase (GDH), and phosphate-activated glutaminase (PAG). We describe their quantification in extracts of postmortem brain tissue from healthy subjects (controls) and patients with mental and neurodegenerative disorders. We discuss the presence of multiple isoforms for each of these enzymes. We detail the use of Western blotting with chemiluminescent enhancement of signal (ECL) as the most informative method for studying glutamate-metabolizing enzymes. Several examples are given for determining relative quantities of GS, GDH, GAD, and PAG in extracts from brain samples coming from various brain areas and obtained from control cases and subjects with schizophrenia and Alzheimer's disease.

Key words Glutamine synthetase, Glutamic acid decarboxylase, Glutamate dehydrogenase, Phosphate-activated glutaminase, Human brain, Schizophrenia, Alzheimer's disease, Enzyme extraction

1 Introduction: Metabolism of Glutamate in the Brain

Glutamate (Glu) is a multifunctional amino acid. It plays a key role linking carbohydrate and amino acid metabolism via tricarboxylic acid cycle, as well as nitrogen trafficking and ammonia homeostasis in brain. Moreover, the Glu conversion pathways are linked to brain energy metabolism [1–3].

Glutamate serves as the main excitatory neurotransmitter and precursor of γ-aminobutyric acid (GABA, another inhibiting neurotransmitter) in the brain tissue and has a complex and highly compartmentalized metabolism [4–6]. The dynamic concentration of Glu depends on enzymatic activities of all enzymes involved in its metabolism [5, 7] (Figs. 1 and 2).

Sandrine Parrot and Luc Denoroy (eds.), *Biochemical Approaches for Glutamatergic Neurotransmission*, Neuromethods, vol. 130, DOI 10.1007/978-1-4939-7228-9_15,

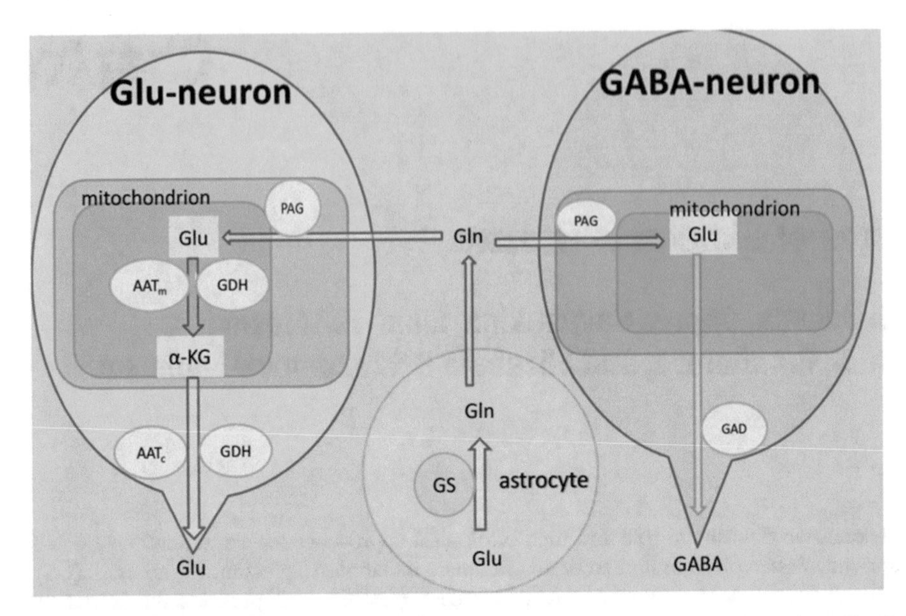

Fig. 1 Compartmentalization of glutamate and GABA pathways. Abbreviations: *Gln* glutamine, *Glu* glutamate, *AATc and AATm* cytoplasmic and mitochondrial aspartate aminotransferases, *α-KG* α-ketoglutarate, *PAG* phosphate-activated glutaminase, *GAD* glutamic acid decarboxylase, *GS* glutamine synthetase

In this review, we will concentrate on four major enzymes involved in brain Glu metabolism, namely glutamine synthetase (GS), glutamate dehydrogenase (GDH), glutamic acid decarboxylase (GAD), and phosphate-activated glutaminase (PAG).

1.1 Glutamine Synthetase

Human brain glutamine synthetase (GS, EC 6.3.1.2) is a glial (Figs. 1 and 2) octameric enzyme, which is a subject to complex regulatory mechanisms. GS catalyzes the conversion of Glu and ammonium to glutamine (Gln) driven by the hydrolysis of ATP:

$$\text{L}-\text{Glutamate}+\text{ATP}+\text{NH}_4^{\ +} \rightarrow L-\text{Glutamine}+\text{ADP}+\text{phosphate}_\text{i}.$$

In vitro GS activity can be determined by measuring γ-glutamylhydroxamate formation either in "synthetase" reaction (ATP + L-glutamate + NH_2OH) or, similarly, in "transferase" reaction (ADP + L-glutamine + NH_2OH) [8], see Fig. 3.

In animal and human brain, GS plays an important role in regulation of Glu concentration, in termination of Glu neurotransmitting signal and in ammonium assimilation and detoxification [9–11].

Due to the GS important metabolic role, the functional impairment of GS (as judged both from GS enzymatic activity and immunoreactivity) has been investigated in brain pathologies, including Alzheimer's disease (AD) [12–14], whereas its fine regulation has been well documented in mammal models of brain injury and mental disorders [15, 16].

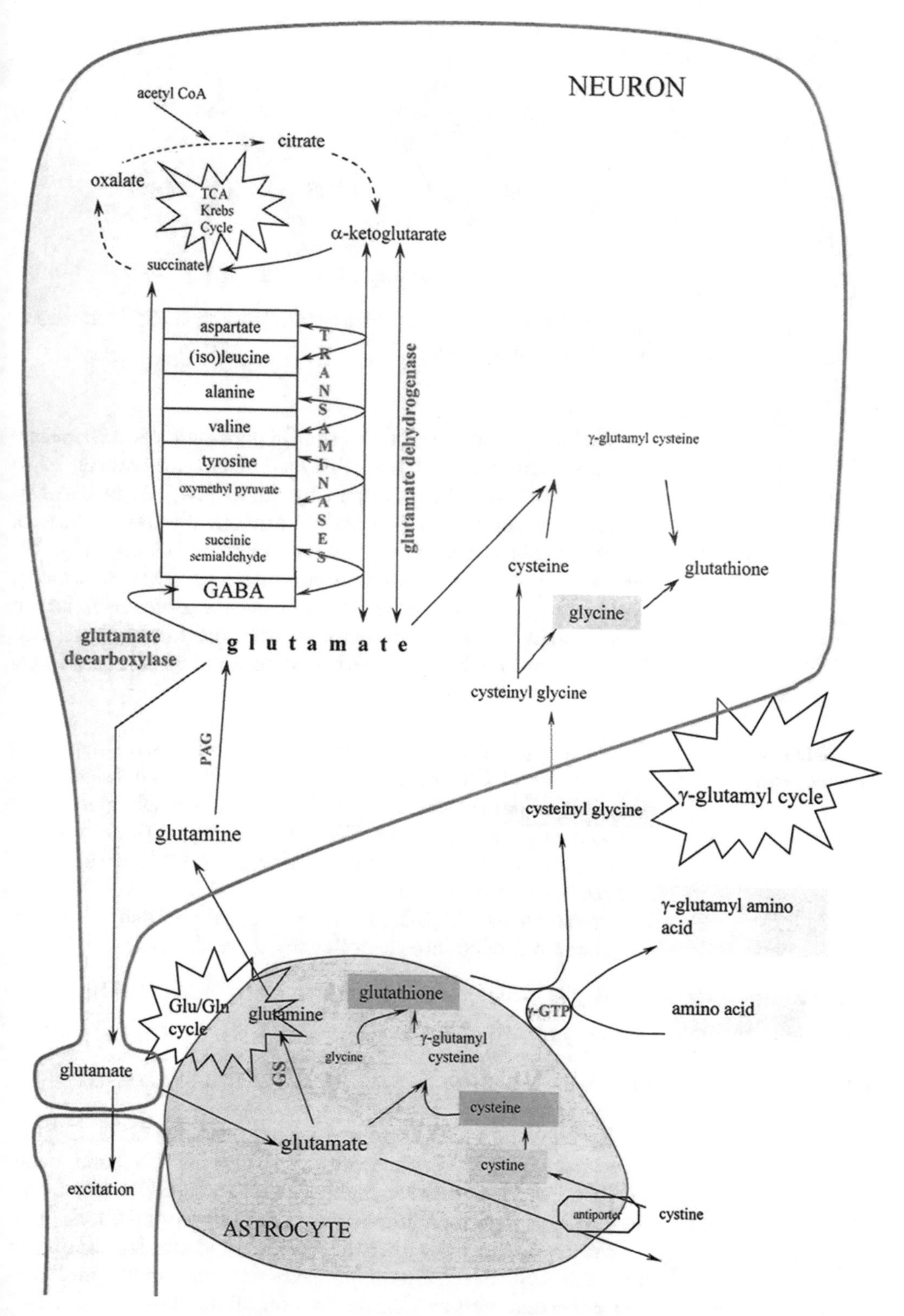

Fig. 2 Scheme of glutamate pathways. Abbreviations: *Gln* glutamine, *Glu* glutamate, *AATc and AATm* cytoplasmic and mitochondrial aspartate aminotransferases; *α-KG* α-ketoglutarate, *PAG* phosphate-activated glutaminase, *GS* glutamine synthetase, *γ-GTP* γ-glutamyltranspeptidase

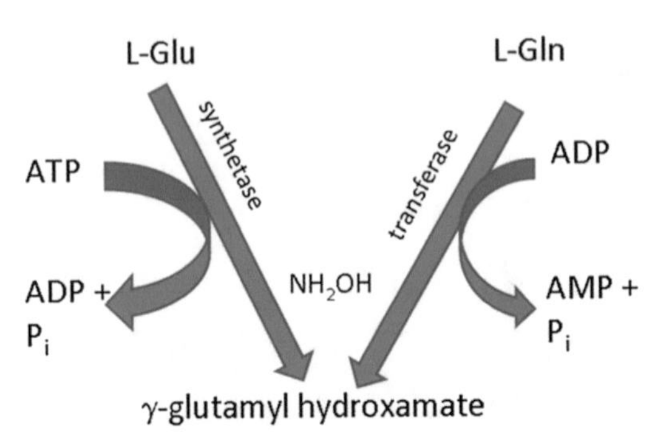

Fig. 3 Scheme of glutamine synthetase enzymatic activity in vitro. Abbreviations: *Gln* glutamine, *Glu* glutamate

We have previously reported the GS purification and characterization in human brain tissue [8, 17]. The enzyme was similar to those purified from human brain by others [10, 11] in terms of subunit molecular mass (~44 kDa), Mn^{2+}/Mg^{2+}-dependences of activity, specific activity, and subunit oligomerization [17]. We have also reported the isolation and characterization of a novel GS-like protein (GSLP) from human brain: the protein is similar to human brain GS in its enzymatic activity and immunoreactivity. The GSLP protein has oligomeric structure and consists of 54 kDa subunits [18].

1.2 Glutamate Dehydrogenase

Glutamate dehydrogenase (GDH, EC 1.4.1.2–1.4.1.4) is one of key enzymes of Glu metabolism (Figs. 1 and 2). GDH catalyzes reversible oxidative deamination of L-Glu to α-oxoglutarate using NAD/NADP as coenzymes. The native mammal GDH enzyme has oligomeric (hexameric) structure and, like GS, is a subject of fine regulation with cofactors.

Based on which cofactor is used, glutamate dehydrogenase enzymes are divided into the following three classes:

EC1.4.1.2 $\text{L}-\text{glutamate} + H_2O + \text{NAD}+ \leftrightarrow 2-\text{oxoglutarate} + NH_4^+ + \text{NADH} + H^+$

EC1.4.1.3 $\text{L}-\text{glutamate} + H_2O + \text{NAD(P)}+ \leftrightarrow 2-\text{oxoglutarate} + NH_4^+ + \text{NAD(P)H} + H^+$

EC1.4.1.4 $\text{L}-\text{glutamate} + H_2O + \text{NADP}+ \leftrightarrow 2-\text{oxoglutarate} + NH_4^+ + \text{NADPH} + H^+$

In humans, GDH isoforms are encoded by GLUD1 gene expressed widely (housekeeping), and also by a second gene, GLUD2, which encodes a highly homologous GDH isoenzyme (hGDH2) expressed predominantly in retina, brain, and testis [19].

We have reported earlier the isolation and characterization of three novel GDH isoforms in human brain: two readily soluble and one associated with particulate fraction GDH [20]. These GDH isoforms may represent either products of different genes, or, alternatively, they may be translated from different GDH mRNA species generated by alternative splicing, or due to post-translation modifications.

Study of brain GDH is of a particular interest because the enzyme has been found to be deficient in patients with some neurodegenerative disorders, including AD and Huntington's disease [21–23], as well as schizophrenia [24, 25].

1.3 Phosphate-Activated Glutaminase

Phosphate-activated glutaminase (PAG, EC3.5.1.2) hydrolyzes glutamine to glutamate:

$$\text{L}-\text{Glutamine}+\text{H}_2\text{O}\rightarrow\text{L}-\text{Glutamate}+\text{NH}_4^{\ +}.$$

The rat brain enzyme, purified and extensively characterized by Kvamme and coworkers, is highly enriched in neurons [3, 26]. PAG is essential for synthesis of neurotransmitter Glu in glutamatergic neurons [3]. PAG is involved significantly in synthesis of GABA neurotransmitter as well (Figs. 1 and 2). This has been now confirmed in nuclear magnetic resonance (NMR) studies that have shown that Gln synthesized in astrocytes and deamidated by neuronal PAG is preferentially used for GABA synthesis [27].

One of the most important roles of glutaminase is found in the axonal terminals of neurons in the central nervous system. After being released into the synapse for neurotransmission, Glu is rapidly taken up by nearby astrocytes, which convert it to Gln (Figs. 1 and 2). This Gln is then supplied to the presynaptic terminals of the neurons, where PAG converts it back to Glu for loading into synaptic vesicles. Both "kidney-type" (GLS1) and "liver-type" (GLS2) glutaminases are expressed in brain. GLS2 has been reported to exist only in cellular nuclei in CNS neurons [28].

1.4 Glutamic Acid Decarboxylase

Glutamic acid decarboxylase or glutamate decarboxylase (GAD, EC 4.1.1.15) is the central enzyme producing GABA via decarboxylation of Glu in the brain (Figs. 1 and 2) [29].

GAD is an enzyme that catalyzes the decarboxylation of Glu to GABA with releasing CO_2. The reaction can be presented as follows:

$$HOOC-CH_2-CH_2-CH(NH_2)-COOH\rightarrow CO_2+HOOC-CH_2-CH_2-CH_2-NH_2$$

GAD uses PLP (pyridoxal 5′-phosphate, the active form of vitamin B6) as a cofactor. In mammals, GAD exists in two isoforms encoded by two different genes GAD1 and GAD2. These isoforms are GAD67 and GAD65 with molecular masses of 67 and 65 kDa, respectively. GAD1 and GAD2 are expressed in the brain where GABA is used as a neurotransmitter, GAD2 is also expressed in the pancreas.

The two isoforms, GAD65 and GAD67, have different localization and functions [30]. Being encoded by GAD2 gene whose expression is induced under acute demand for GABA, GAD65 is mainly responsible for neurotransmission, and this isoform is associated with synaptic membranes in nerve terminals. GAD67 is encoded by a constitutively expressed GAD1 gene and is more evenly distributed in cells than GAD65. GAD67 catalyzes basic GABA formation for housekeeping neuronal functions that are not directly associated with neurotransmission.

The study of this Glu metabolizing enzyme is of special interest in AD [31]. GAD links two neurotransmitter systems (Glu and GABA), both implicated in AD pathogenesis. Although alterations in Glu metabolism may play a role in the cognitive impairment seen in AD, these changes are likely to occur together with other pathological events, namely amyloid beta peptide (Aβ) and apolipoprotein E (ApoE), and are, thus, closely linked to them [32–34].

Impairment of Glu neurotransmission system may also play an important role in the pathology of schizophrenia and other psychoses. Several lines of evidence indicate that changes in levels of different GAD isoforms in the brain cortex are associated with schizophrenia and other mental diseases, including affective disorders [35, 36]. Schizophrenic patients have been shown to have lower amounts of GAD67 in the dorsolateral prefrontal cortex (DLPFC) compared to healthy controls [36], but mechanism underlying the decreased levels of GAD67 in schizophrenic patients remains unclear. Some authors have proposed that expression of an immediate early gene, Zif268, which normally binds to the promoter region of GAD67 and increases transcription of GAD67, is lower in schizophrenic patients, thus contributing to decreased levels of GAD67 [37]. Since DLPFC is involved in working memory, and GAD67 and Zif268 mRNA levels are lower in the DLPFC of schizophrenic patients, this molecular alteration may account, at least in part, for the working memory impairments associated with schizophrenia.

In the present study, we describe methodological approaches for comparative study of Glu metabolizing enzymes, such as GS, GDH, PAG, and GAD, in postmortem human brain tissue from control subjects and from patients with AD and schizophrenia.

The study of brain glutamate-metabolizing enzymes in health and disease is complicated by both the existence of more than one isoform of each enzyme as well as the oligomeric structure of the native enzymes, which is necessary for displaying their enzymatic activity.

1.5 Some Molecular Characteristics of GS, GSLP, and GDH Isoforms

Molecular characteristics of enzymes isolated and purified from human brain demonstrate a variety of their isoforms, and suggest possible differences in distribution of the isoforms in different cellular compartments.

Further characterization of GS, GSLP, GDH I, II and GDH III isoforms subunits was obtained using two-dimensional gel electrophoresis. GS is represented by a train of several (at least five) spots corresponding to MW of 44 ± 1 kDa with p*I* in a range 6.4–6.7. GSLP is represented by a train of several spots (at least three) corresponding to MW of 54 ± 1 kDa with p*I* in a range 5.9–6.2. GDH I is represented by a train of several spots (at least five) corresponding to MW of 58 ± 1 kDa with p*I* in a range 7.5–6.6. GDH II is represented by a train of at least six spots corresponding to MW of 56 ± 1 kDa with p*I* ranging from 7.3 to 6.6. GDH III shows also microheterogeneity: at least five spots corresponding to MW of 56 ± 1 kDa ranged in p*I* range from 7.5 to 6.6.

Apparent molecular mass of native isolated and purified GS as determined by gel-filtration calibrated column is ca. 140 ± 20 kDa; further characterization of purified GS by analytical ultracentrifugation (sedimentation equilibrium analysis) gave overall average molecular mass of 351 ± 8 kDa which distinctly corresponds to octameric quaternary structure of the enzyme in a solution. As known from literature, native mammal GDH has also oligomeric (hexameric) structure [38].

2 Materials

2.1 Human Brain Samples

In the current study, we used the brain tissue collection stored at the Federal State Budgetary Scientific Institution «Mental Health Research Centre» (MHRC, Moscow). The collection consists of postmortem human brain tissue obtained from several Moscow Psychiatric Hospitals and the Moscow Highest Medical School. Consent for autopsy was obtained from family members, and all research projects were approved by the Ethics Committee of MHRC.

Chronically hospitalized patients with schizophrenia or AD were diagnosed with the ICD-10 criteria prior their death. The definite diagnosis of AD was confirmed neuropathologically [39] in the MHRC Laboratory of Clinical Neuromorphology. The samples of brain tissue (Brodmann areas, BA 1–3, 10, 23, 24) from the left hemispheres, caudate nucleus, and cerebellar cortex from patients with schizophrenia and AD, as well as control subjects (devoid of history of mental or neurological disorders and with no drug history for treatment for psychosis or dementia) were dissected and processed at 4 °C, placed into 5 mL vials (Corning Cool Cell 5 mL Freezing container LX), frozen in liquid nitrogen and stored at −80 °C until use avoiding defrosting/freezing cycles. Small pieces (50–200 mg) of brain tissue were chopped before the experiment from each cylinder of frozen tissue avoiding thawing.

The two analyzed groups (patients and controls) were age and sex matched, and had similar short postmortem interval (PMI; under 8 h, see Note 1) and brain storage duration. The causes of death were also similar in both analyzed groups, and they included cardiovascular failure, acute cardiac ischemia, pulmonary embolism,

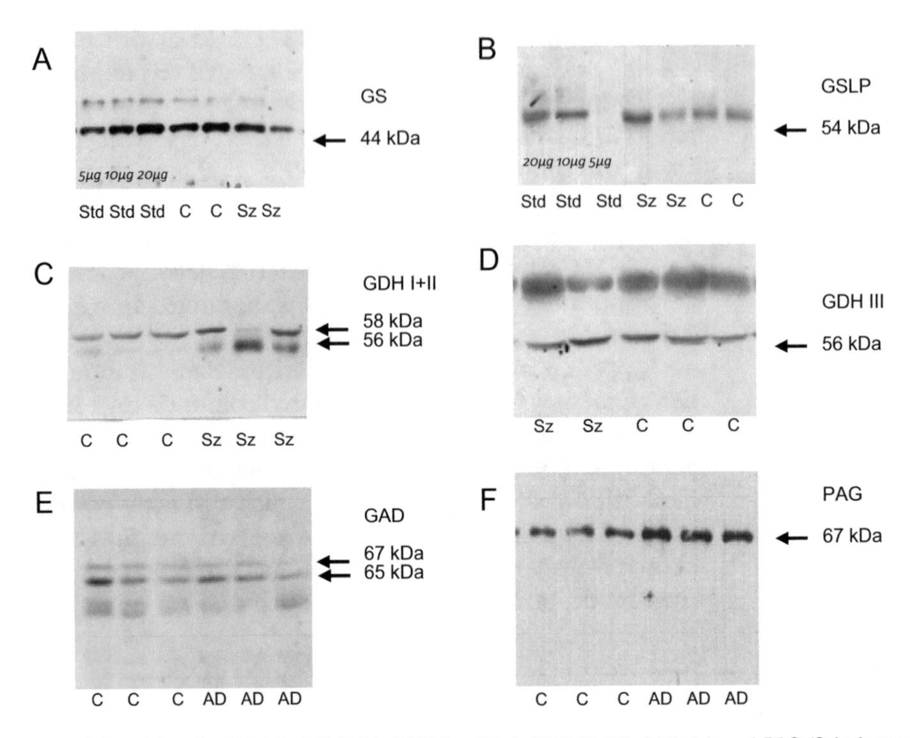

Fig. 4 Patterns of staining for GS (**a**), GSLP (**b**), GDH I + II (**c**), GDH III (**d**), GAD (**e**) and PAG (**f**) in human brain tissue extracts. *Arrows* and *numbers* indicate respective molecular masses for proteins/enzyme subunits. Total protein amount is 10 μg per lane, except for the inner standards (5, 10, and 20 μg of total protein, respectively). Abbreviations: *Std* standard for internal standards, *C* control, *Sz* schizophrenia, *AD* Alzheimer's disease. Please note that samples (C, Sz, and AD) are all from different subjects

bronchitis, and/or pneumonia. Cases with comorbid alcoholism and/or drug abuse were excluded from analyses.

2.2 Immunoprobes

All immunoprobes used in the current study were aliquoted and stored at −20 °C, avoiding defrosting/freezing cycles. Each aliquot was thawed only once before the experiment. Titer (working dilution) may decrease under prolonged (more than 5 years storage under −20 °C), and the titer must be determined again.

The specificity of each immunoprobe was evaluated by Western blotting with chemiluminescent detection (ECL-immunoblotting).

GS immunoreactive protein was detected using a monoclonal antibody to GS (No. 4673-5007, Clone No. 49G3, Biogenesis, UK) at 1:400,000 working dilution. An example for GS staining (44 kDa subunit) is given in Fig. 4a.

GSLP and GDH isoforms were determined using highly specific polyclonal rabbit antisera raised in the Laboratory of Neurochemistry MHRC.

Antisera against purified GSLP were obtained by immunization of rabbits. Three rabbits (males, Chinchilla) were immunized by subcutaneous injections: the first immunization −100 μg of

GSLP with complete Freund's adjuvant, the second—1 month after the first—100 μg of the protein with incomplete Freund's adjuvant and the third—100 μg of the protein after 10 days.

The specificity of each antiserum was evaluated by ECL-immunoblotting using the purified GSLP protein sample, titers varied insignificantly therewith.

The amount of GSLP immunoreactive protein in brain tissue extracts was evaluated with the obtained rabbit antisera used at 1:15,000 working dilution. The immunoprobe immunolabelled the major protein band corresponding to GSLP (54 ± 1 kDa). An example of staining for GSLP (54 kDa subunit) is given in Fig. 4b.

The anti-GDH rabbit antiserum was raised following a similar protocol as above: the first immunization of three rabbits, 100 μg of GDH was injected subcutaneously with complete Freund's adjuvant, the second—2 weeks after the first—100 μg of GDH with incomplete Freund's adjuvant, and the third—100 μg of GDH after 2 weeks. Polyclonal rabbit antisera against GDH I + II were raised to GDHI + GDHII (mixture of two proteins), whereas the rabbit GDH III antisera were raised to purified single GDH III protein.

The specificity of each antiserum was evaluated by ECL-immunoblotting using the purified GDH I + II and GDH III protein samples. The rabbit antiserum against GDH I + II (used at 1:15,000 working dilution) immunostained two protein bands corresponding to GDH I (58 ± 1 kDa) and GDH II (56 ± 1 kDa) (Fig. 4c); the rabbit antiserum against GDH III (used at 1:10,000 working dilution) immunolabelled a single protein band corresponding to GDH III (56 ± 1 kDa) (Fig. 4d) and an additional, yet unidentified, 66 kDa band, which was not accounted. The 56 kDa and this additional 66 kDa protein band were both recognized by polyclonal antiserum purchased from Biogenesis (Cat. No. 4670-5488) (data not shown).

As for 66 kDa protein recognizable by polyclonal antisera to GDH III and commercial antiserum, the 66 kDa protein was separated from protein fractions possessing GDH activity after the first step of GDH purification [20] and was eluted in fractions with no GDH enzymatic activity. Also, the monitoring of GDH enzymatic activity during GDH chromatography showed that 66 kDa protein immunoreactivity was absent in fractions with GDH enzymatic activity during chromatography steps. Hence, 66 kDa protein is not likely to be a GDH isoform. However, GDH and 66 kDa protein carry some epitope(s) with similar structure.

The isoforms of GDH described here can be scarcely matched with the GDH isoforms found by others. The proteins corresponding to these GDH isoforms were not sequenced, and we cannot rule out the possibility that these proteins may be not the same as described in literature (the protein products of GLUD1 and GLUD2 genes) [40].

Amounts of GAD 65/67 isoforms were determined using commercially available rabbit polyclonal antibody Ab1511 (Chemicon) staining both GAD65 (65 kDa) and GAD67 (67 kDa) isoforms (at 1:10,000 working dilution); an example of GAD staining is given in Fig. 4e.

PAG was determined with a polyclonal rabbit antibody raised against the C-terminal peptide fragment of PAG (rat kidney PAG isoform, kindly provided by Prof. O.P. Ottersen and Dr. I.A. Turgner, Norway) at 1:20,000 working dilution; an example of PAG staining is given in Fig. 4f.

Secondary antibody conjugates with horseradish peroxidase were used at following dilutions: 1:10,000 Anti-Rabbit IgG–F(ab′)$_2$-fragment donkey antibody NA 9340 (Amersham Biosciences) for GSLP, GDH I + II, GDH III, PAG, and GAD immunolabelling, and Anti-Mouse–F(ab′)$_2$-fragment sheep antibody NA 9310 (Amersham), used at 1:3000 dilution for GS visualization.

2.3 Chemicals

ADP, ATP, NADH, NADPH, L-Glu, L-Gln, 2-oxoglutarate (α-ketoglutarate), γ-glutamylhydroxamate, NH_2OH, salts, and chemicals for electrophoresis and Western blotting, and metal ion salts were all purchased from Sigma-Aldrich, USA; Bio-Rad DC Protein Assay kit, USA, was used for determination of protein concentration.

2.4 Equipment and Supplies

Protean II xi 2-D Cell (Bio-Rad, USA);

Semidry transfer Unit Hoefer TE 70 Amersham Biosciences (GE Healthcare Life Sciences) with Electrophoresis Power Supply EPS 601 and supplies;

Hypercassettes, Nitrocellulose membranes Hybond-ECL, Hybond Blotting Paper, films Hyperfilm ECL for exposition of nitrocellulose membranes were from Amersham Biosciences (GE Healthcare Life Sciences);

Kodak Image Station 2000R (USA);

Beckman DU800 spectrophotometer (USA);

Hidex Chameleon microplate reader, Pribori-Oy (Finland).

3 Methods

3.1 Brain Protein Extraction for Biochemical Analyses

Different extraction methods were used to accommodate for variations in enzymatic activities, detection levels of immunoreactive proteins by ECL Western blotting, and amount of accessible brain sample material.

Extraction "Method 1" (Sects. 3.1.1 and 3.1.2) was used when determining the enzymatic activity and level of immunoreactive proteins, whereas extraction "Method 2" (Sect. 3.1.3) was implemented when determining level of immunoreactive proteins only.

3.1.1 Extraction Method 1A

The preparation of brain samples for the comparative study of enzymatic activities and quantities of immunoreactive enzymes (proteins) included homogenization of 200 mg of tissue in 1 mL 50 mmol/L Tris–HCl buffer, pH 7.5, with 1.4 mmol/L 2-mercaptoethanol using a glass-Teflon homogenizer at 4 °C on ice, with subsequent centrifugation at 1000 × *g* for 15 min, followed by centrifugation at 60,000 × *g* for 1 h. All centrifugation steps were carried out at 2–4 °C. The final supernatant "cytoplasm" was immediately used for measurement of GS + GSLP "transferase" and GDH enzymatic activity and quantitative determination of immunoreactive proteins by ECL-Western immunoblotting. Protein concentration was determined (see Examples 4.1). Samples for ECL-immunoblotting were prepared. For this, the "reducing sample buffer" for sodium dodecylsulphate (SDS) electrophoresis in polyacrylamide gel was added to each sample (1:10, v/v, the composition of the "reducing buffer" can be found in Sect. 3.3.1), boiled for 5 min and used for protein electrophoresis in polyacrylamide gel with subsequent analyses of GS, GSLP, and GDH I + II by ECL-immunoblotting. Using the total protein concentration data, the required volume of the sample in microliters (μL) was calculated for each sample containing 10 μg of total protein amount. This volume of sample was loaded to polyacrylamide gel lane (i.e., all samples were matched by protein amount loaded on the polyacrylamide gel, namely 10 μg).

Pellets obtained after the second centrifugation step—"membranes," or particulate fractions—were suspended in 1 mL of cold buffer (50 mmol/L Tris–HCl, pH 6.8), followed by centrifugation at 60,000 × *g* for 1 h. The obtained precipitate was suspended in 100 μL of the same buffer. Protein concentration was determined in the samples. The "reducing sample buffer" for SDS electrophoresis was added to each sample, boiled for 5 min. The volume of each sample applied on a lane for protein electrophoresis in polyacrylamide gel with subsequent analyses of GDH III isoform and GSLP by ECL-immunoblotting was calculated like in the case of "cytoplasm," the samples were matched by 10 μg of total protein.

GDH I and GDH II can be detected only in the "cytoplasm" fraction, whereas the GDH III is only located in the "membranes." GS and GSLP can be found mainly in "cytoplasm" fraction prepared from frozen brain tissue samples, used for the present study.

3.1.2 Extraction Method 1B

This extraction protocol was as follows: 100 mg of brain tissue was homogenized using glass-Teflon homogenizer in 1 mL of homogenization medium (0.35 mol/L sucrose in 62 mmol/L Tris–HCl buffer, pH 6.8, containing the Complete, EDTA-free Protease

Inhibitor Cocktail, according to "Roche Diagnostics GmbH," Germany, protocol) followed by a centrifugation at 1000 × *g* for 15 min to discharge nuclear fragments and cell debris and to obtain "supernatant 1," and another one at 60,000 × *g* for 1 h. Resulting supernatants ("supernatant 1" and "cytoplasm") and pellets ("membranes") were treated separately as follows: protein concentration was determined in aliquots of the "supernatant 1" and "cytoplasm"(see Examples 4.2). The "reducing sample buffer" (for the composition of the "reducing buffer" see Sect. 3.3.1) for SDS electrophoresis in polyacrylamide gel was added to each sample (1:10, v/v), boiled for 5 min, the sample volumes applied to polyacrylamide gel were matched for total protein amount (see Sect. 3.1.1) and used for protein electrophoresis with subsequent immunochemical analyses of GS, GSLP, and GDH I + II isoenzymes by ECL-immunoblotting, as described below in Sect. 3.3.2.

"Membranes" were suspended in 1 mL of cold homogenization medium (0.35 mol/L sucrose in 62 mmol/L Tris–HCl buffer, pH 6.8), followed by centrifugation at 60,000 × *g* for 1 h, and the obtained pellet was suspended in 100 μL of the same medium. Protein concentration was determined in the resulting samples (see Examples 4.2), and then the "reducing sample buffer" for SDS electrophoresis was added to each sample (1:10, v/v), boiled for 5 min, the sample volumes applied to polyacrylamide gel were matched for total protein amount (see Sect. 3.1.1) and used for protein electrophoresis in polyacrylamide gel with subsequent analyses of GDH III isoform, GSLP, and PAG by ECL-immunoblotting.

All procedures described in Method 1A and Method 1B were carried out at 2 °C.

These methods were used in our comparative studies of postmortem brain tissues from patients with AD [41, 42] and schizophrenia [24, 43].

3.1.3 Extraction Method 2

To optimize extraction of GS, GSLP, GDH isoenzymes and GAD, we developed the following protocol. Brain sample (50 mg) was homogenized (by glass-glass pestle Potter homogenizer) in 1 mL of 50 mmol/L Tris–HCl buffer, pH 7.0, with 1% (w/v) SDS in a boiling water bath for 5 min, followed by centrifugation at 11,000 × *g* (Eppendorf centrifuge) for 30 min at room temperature (RT). The obtained supernatant was used for further analysis. Protein concentration was measured using Bio-Rad DC assay (micromethod). As detailed in the Sect. 3.1.1, using the total protein concentration data, the required volume of the sample (μL) was calculated for each sample containing 10 μg of total protein amount. This volume of sample was loaded to polyacrylamide gel lane (i.e., all samples were matched by protein amount loaded on the polyacrylamide gel, namely 10 μg). 2-Mercaptoethanol (to its final concentration of 5%) was added to all samples before their application on polyacrylamide gel (see Examples 4.3).

This method was used in our comparative studies of enzymes of Glu system in postmortem brain of patients with schizophrenia [25].

3.2 Determination of Enzymatic Activities

3.2.1 GS and GSLP Enzymatic Activity

GS and GSLP enzymatic activity was determined by measuring γ-glutamylhydroxamate formation using NH_2OH as the substrate according to Iqbal and Ottaway [44], with minor modifications (i.e., all reaction mixture volumes were reduced tenfold to 100 μL to use a microplate reader for photometric assay).

GS activity can be assayed in the "synthetase" (ATP + L-glutamate + NH_2OH) or "transferase" (ADP + L-glutamine + NH_2OH) reactions by measuring γ-glutamylhydroxamate formation (see Sect. 1.1). GSLP catalyzes "transferase" reaction only (see Fig. 3).

The reaction mixtures for "synthetase" activity assay (100 μL) were as follows: imidazole-HCl, 50 mmol/L, pH 6.8; NH_2OH, 100 mmol/L; L-glu, 50 mmol/L; $MgCl_2$, 20 mmol/L; 2-mercaptoethanol, 25 mmol/L; ATP, 10 mmol/L; enzyme solution in Tris–HCl, pH 6.8.

The reaction mixtures (100 μL) for "transferase" activity assay consisted of: imidazole-HCl, 50 mmol/L, pH 6.8; NH_2OH, 50 mmol/L; L-Gln, 100 mmol/L; $MnCl_2$, 0.5 mmol/L; KH_2AsO_4, 25 mmol/L; ADP, 0.2 mmol/L; enzyme solution in Tris–HCl, pH 6.8.

After incubation at 37 °C for 30 min, an equal volume (100 μL) of stop solution (0.37 mol/L $FeCl_3$, 0.3 mol/L trichloroacetic acid, 0.6 mol/L HCl) was added to the reaction mixture in both cases.

The samples were clarified by centrifugation at 5000 × *g* (Eppendorf centrifuge) for 2 min at RT, when a pellet was formed.

The absorbance of the colored complex formed by γ-glutamylhydroxamate with iron was measured at 505 nm; microplate photometer (reader) was used for the photometric assay.

A solution of 1.0 μmol/L γ-glutamylhydroxamate was used as a standard.

The specific activity of GS and GSLP is given in μmol of γ-glutamylhydroxamate formed per 1 mg of protein in 1 min [8, 44].

Note, the "synthetase" activity is highly specific for GS (but not GSLP), but it can be measured only in isolated and purified GS. "Synthetase" activity determined in crude extracts gives underestimations of the enzyme activity and false results because of numerous enzymatic side reactions consuming ATP from the reaction mixture.

The "transferase" assay being applied in crude brain tissue extract gives a sum of GS and GSLP activities. For an example of obtained result, see Table 1. Separate determination of GS and GSLP activities is not possible due to high similarity in their affinities to the substrates [18].

Table 1
GS and GSLP enzymatic activity (U/mg) in "transferase" reaction in "cytoplasm" fraction in control and schizophrenia brain samples

Brodmann area	n	Control (U/mg) median (min–max)	n	Schizophrenia (U/mg) median (min–max)	Between-group difference statistical significance
23	14	15.4 (7.2–23.6)	13	22.0 (7.2–36.8)	No
24	15	13.7 (7.8–19.6)	15	15.6 (8.2–23.0)	No
10	15	9.8 (6.2–13.4)	9	8.2 (5.9–10.5)	No

For the enzymatic assay, see Sect. 3.2.1 and Fig. 3

Table 2
GDH enzymatic activity (U/mg) in "cytoplasm" fraction in control and schizophrenia brain samples

Brodmann area	n	Control (U/mg) median (min–max)	n	Schizophrenia (U/mg) median (min–max)	Between-group difference statistical significance
23	14	98.2 (45.7–150.7)	13	161.1 (83.2–239.0)	$p < 0.05$
24	15	103.3 (67.1–139.5)	15	139.5 (79.7–199.3)	No
10	15	65.5 (44.1–86.9)	11	80.2 (40.1–120.3)	$p < 0.01$

For the enzymatic assay, see Sect. 3.2.2

3.2.2 GDH Enzymatic Activity

GDH in vitro assay in human brain extracts is based on the reaction corresponding to the following chemical equation:

$$2-\text{Oxoglutarate}+\text{NH}_4^+ +\text{NADH}+\text{H}^+ \rightarrow \text{L}-\text{glutamate}+\text{H}_2\text{O}+\text{NAD}^+.$$

according to Fahien and Cohen [45]. GDH enzymatic activity was determined spectrophotometrically (using 1-cm quartz cells, at 25 °C) (for an example of obtained results see Table 2). The reaction was evaluated by monitoring the decrease of absorbency at 340 nm in reaction mixture containing 10 mmol/L Tris–HCl, pH 8.0, 2.6 mmol/L EDTA, 100 mmol/L ammonium chloride, 0.1 mmol/L NADH, and 1 mmol/L ADP. The reaction was started with the addition of 2-oxoglutarate to 10 mmol/L final concentration. One unit of enzyme was defined as the amount of GDH enzyme required to oxidize 1 mmol of NADH per 1 min.

In the current study, GAD and PAG enzymatic activities were not determined.

3.3 Protein Electrophoresis and ECL-Western Immunoblotting

3.3.1 Protein Electrophoresis

Electrophoresis in polyacrylamide gel under denaturing conditions (SDS-PAGE) was performed in 16 × 16-cm slab gels of 0.75 mm thickness containing 10% polyacrylamide in accordance with Laemmli [46], using 25 sample well comb.

Reducing sample buffer (1:10 buffer to sample ratio, v/v) was used and consisted of: 10% SDS, 20% 2-mercaptoethanol, 0.05% bromophenol Blue, 62.5 mmol/LTris–HCl, pH 6.8.

In all experiments, the explored protein samples (usually up to eight from each group—controls and disease, with each sample containing 10 μg protein) were randomly applied to gel lanes. Internal standard—see Sect. 3.3.3—(three different total protein amounts—5, 10, and 20 μg per lane), and molecular weight protein standards were run simultaneously on the same polyacrylamide gel.

3.3.2 ECL-Western Immunoblotting

After SDS-PAGE, proteins were transferred to Hybond ECL membrane and used immediately for analysis. Hybond ECL membrane Amersham Biosciences (GE Healthcare Life Sciences) was equilibrated in transfer buffer (0.02 mol/L Tris, 0.15 mol/L glycine, 20% CH_3OH, pH 8.3) at least 5–10 min before blotting.

Current and transfer time are important parameters: insufficient current or voltage, as well as insufficient transfer time can lead to incomplete protein transfer. Semidry transfer was run at a constant current of 0.8 mA/cm^2, for 1 h, at 4 °C.

Ponceau S 0.5% (w/v) in 1% (v/v) acetic acid was used for locating the transferred protein bands on nitrocellulose after the blotting (see Note 2). Nitrocellulose membrane was stained for 1–5 min, and washed with 1% acetic acid until the protein bands appeared and a photo image was taken. Molecular mass protein markers were marked with a pencil. Then, nitrocellulose was quickly washed with 0.2 mol/L NaOH until the protein bands disappeared, and washed in distilled water.

Membrane was then placed with protein side up on a Teflon pad (plate) for 1 h at 25 °C, at RT, on an orbital shaker for blocking nonspecific binding sites. For this, the membrane was covered with 5% fetal serum diluted in washing buffer (0.01 mol/L Tris–HCl, 0.2% Tween 20 (v/v), 0.15 mol/L NaCl, pH 7.5) in volume of 0.1 mL/cm^2.

Following this, the membrane was briefly rinsed using two changes of washing buffer, washed in more than 4 mL/cm^2 volume of washing buffer for 15 min at RT on an orbital shaker, and the membrane was finally washed in fresh washing buffer twice with the same volume for 5 min at RT on an orbital shaker.

The primary antibodies used for visualization of the protein products were diluted in the washing buffer containing 0.5% fetal serum (for primary antibodies see Sect. 2.2).

The membrane was incubated in diluted primary antibody for 2 h at RT on an orbital shaker.

After this step, the membrane was briefly rinsed using two changes of washing buffer and then washed in volume more than 4 mL/cm^2 of washing buffer for 15 min at RT on an orbital shaker, followed by a final wash in fresh washing buffer twice with the same volume for 5 min at RT on an orbital shaker.

The horseradish peroxidase (HRP) labeled secondary antibody (HPR conjugate) was diluted in the washing buffer (for secondary antibodies, see Sect. 2.2). The membrane was incubated in secondary antibody for 1 h at RT on an orbital shaker.

The membrane was washed in volume more than 4 mL/cm^2 of washing buffer for 15 min at RT an orbital shaker, then the membrane was washed in fresh washing buffer twice, the same volume, for 5 min at RT on an orbital shaker.

Equal volumes of the detection solution 1 and detection solution 2 (ECL reagents, Amersham Biosciences) were mixed to the final volume of 0.125 mL/cm^2 membrane, and the mixture was equilibrated to RT.

The membrane was placed on a Teflon pad (plate) protein side up and the excess of washing buffer was wiped from sides of the membrane (and fluid was removed with filter paper on the sides). The membrane was incubated in the mixture of solutions 1 and 2 for 1 min at RT. Excess of detection reagent was dried off with filter paper and the membrane was placed the blots protein side down on to a piece of wrap. The blots were wrapped and air bubbles were gentle smoothed out. The wrapped blots were placed protein side up in an X-ray film cassette Hypercassettes, a sheet of autoradiography film Hyperfilm ECL (Amersham Biosciences) was placed on the top of membrane and exposed for 30 min (exposition time can vary and should be determine experimentally). The film was processed as required for X-ray films.

Film fragments demonstrating typical examples of staining are given in Fig. 4.

Quantitative analysis of films after ECL-immunoblotting was performed using “Kodak” Image Station 2000R. The value of Net Intensity obtained for individual protein spot after the film scanning (the Net Intensity is the sum of background-subtracted pixel values in the band rectangle) was divided by the value of the internal standard’s Net Intensity (respective amount of total protein, 10 μg) for each sample and expressed in percentages (“Relative Units”) see Note 3.

3.3.3 Internal Standard

A calibration curve (three various amounts of total protein, 5, 10, and 20 μg) for internal standard was plotted in every experiment (every polyacrylamide gel and blot). Figure 5 shows an example of calibration curve obtained for GS protein using the internal standard for brain extract “cytoplasm” from BA 1–3 prepared by Method 1A (Fig. 5a) or Method 1B (Fig. 5b).

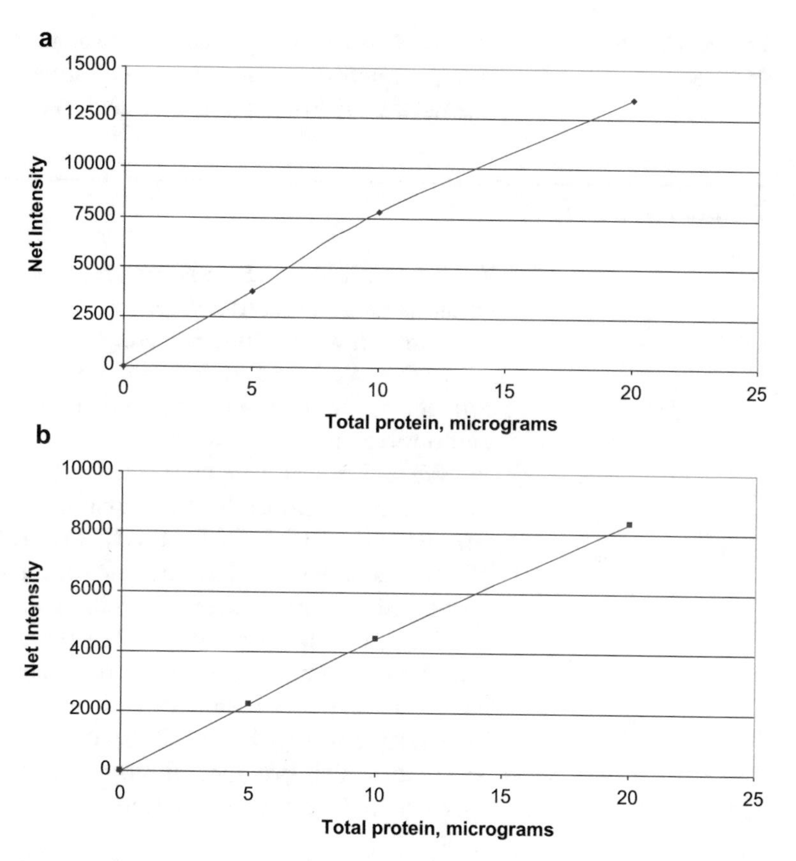

Fig. 5 Calibration curves for GS protein with the internal standard. Total protein amount is 5, 10, and 20 μg per lane, respectively; "cytoplasm" fraction from BA 1–3 prepared by Method 1A (**a**) or Method 1B (**b**)

A sample of a control brain (BA 1–3 from left hemisphere of mentally healthy man, 70 years old) was available in sufficient amount. The protein extract obtained from this brain sample with the same extraction Method (1A, 1B, or 2), which is employed for studied samples (Sects. 4.1–4.3) has been used as an internal standard for comparative analyses. Other protein standards scarcely can be used with this aim—see Note 4. The total protein amounts of the internal standards were usually 5, 10, and 20 μg loaded per lane, and this standard was applied on each polyacrylamide gel—see Fig. 4a, b—Std.

The relative amounts of GS and GSLP, GDH isoforms, PAG, and GAD isoforms were evaluated and expressed in relative optical density units (RU) relative to an internal standard. The data are presented in the diagrams in Figs. 7, 8, 10, 11, 12, 14, 15, and 17.

Relative amounts of all proteins were measured in triplicate. For reproducibility and robustness of the data, see Note 3.

3.4 Statistical Analysis

Statistical analysis was performed with Statistica, Version 6.0 software (nonparametric module, Mann-Whitney *U*-test, because of non-normal data distribution for all analyzed parameters).

4 Examples

We used different methods for preparation of human brain tissue extracts, as highlighted above.

Our aim was to compare enzymes' levels in "controls vs. mental pathology," and determine enzymatic activities in protein extracts of brain tissues, as well as amounts of corresponding immunoreactive proteins.

We and others have confirmed the presence of multiple isoenzyme forms responsible for enzymatic activity for every studied enzyme (GS, GDH, GAD, PAG), whereas the sole measurement of enzymatic activity does not provide per se information about specific alterations of isoenzymes in mental diseases [20, 42, 43].

Thus, the methods of sample preparation (protein extraction from brain tissue) were adapted to the aims of this study, i.e., leading to the extraction of various isoforms by taking into account their association with subcellular compartments and their hydrophobicity which provides different extent of association with the "particulate" membrane fraction.

4.1 Extraction Method 1A

We used the extraction Method 1A (see Sect. 3.1.1) to explore enzymatic activities of GS and GSLP, as well as GDH in "cytoplasm" fractions prepared from brain samples coming from BA 10, 23, 24 of control subjects and patients with schizophrenia.

First, protein concentration was determined in the prepared extracts (for instance, in "cytoplasm" extracts the following data was obtained: mean of 1.8 mg/mL, min = 1.3 mg/mL, max = 2.8 mg/mL, SD = 0.4, $n = 33$). Total protein concentration in extracts prepared from control samples and samples from brain with pathology did not differ significantly. Total protein concentration in extracts prepared from different brain cortex areas studied in the present work did not differ significantly.

4.1.1 Specific Enzymatic Activities

The specific L-glutamine hydroxylamine glutamyl "transferase" activity (U/mg) (which is attributed to GS and GSLP, see Sect. 3.2.1 and Fig. 3) was determined in these "cytoplasm" fractions, and comparison of this activity in control versus schizophrenia brain samples was drawn (Table 1). There was no between-group difference in the activity ($p > 0.05$; Mann-Whitney *U*-test, nonparametric statistics).

We also compared the specific GDH activity (U/mg) (see Sect. 3.2.2 for GDH enzymatic activity assay) between control and schizophrenia brain samples ("cytoplasm" fractions). Subjects with schizophrenia had significantly increased GDH enzymatic activity in brain extracts from BA 23 ($p < 0.05$) and BA 10 ($p < 0.01$), with a similar tendency in BA 24 (Mann-Whitney *U*-test, nonparametric statistics) (Table 2).

Thus, the bulk GS + GSLP "transferase" enzymatic activity attributed to both, GS and GSLP (similarly, GDH bulk activity), are the sum of enzymatic activities of corresponding isoenzymes. Similar levels of GS + GSLP ("transferase") enzymatic activity in both analyzed groups (controls and schizophrenia) may be due to the opposite alterations in amounts of enzymes—GS and GSLP—displaying this activity in protein extracts "cytoplasm" prepared according to Method 1A.

The bulk GDH activity is significantly elevated in schizophrenia in BA 10 and 23, and this is due to the significant elevation of all GDH isoenzymes' levels, as demonstrated further analysis by ECL-Western immunoblotting.

4.1.2 Immunoreactive Protein Quantities

We used the extraction Method 1A (Sect. 3.1.1) to explore relative amounts of GS, GSLP, and GDH I + II isoforms by ECL-immunoblotting in "cytoplasm" fraction and GDH III in "membranes" of various BA from controls and patients with schizophrenia (BA 10, BA 23, and BA 24).

Figure 6 shows examples of staining for GS, GSLP, and GDH I + II in "cytoplasm" (Fig. 6a–c), and GDH III in "membranes" (Fig. 6d) prepared by the Method 1A from BA 10, and BA 23.

Results on determination of relative quantities of immunoreactive GS protein in "cytoplasm" fraction of BA 10, BA 23, and BA 24 in control and schizophrenia subjects are presented in Fig. 7a (see Note 3). Data in this figure and other diagrams represent relative optical density units (RU, i.e., results of film scanning and image processing related to an internal standard, see Sect. 3.3.3).

Results on determination of relative quantities of immunoreactive GSLP protein in "cytoplasm" fraction of BA 10, BA 23, and BA 24 in control and schizophrenia subjects are presented in Fig. 7b.

Results on determination of relative quantities of immunoreactive GSLP in "cytoplasm" fraction by extraction Method 1A (Sect. 3.1.1) may vary dependently on whether the tissue was frozen or not—see also Note 5.

Results on determination of relative quantities of immunoreactive GDH I and GDH II proteins in "cytoplasm" fraction by extraction Method 1A (Sect. 3.1.1) of BA 10, BA 23, and BA 24 in control and schizophrenia subjects are presented in Fig. 8a and b.

4.1.3 Discussion of Data Obtained with Method 1A

In summary, when using extraction Method 1A, we found statistically significant changes ($p < 0.01$ by Mann-Whitney *U*-test) in relative amounts of target proteins (GS, GSLP, GDH I + II, GDH

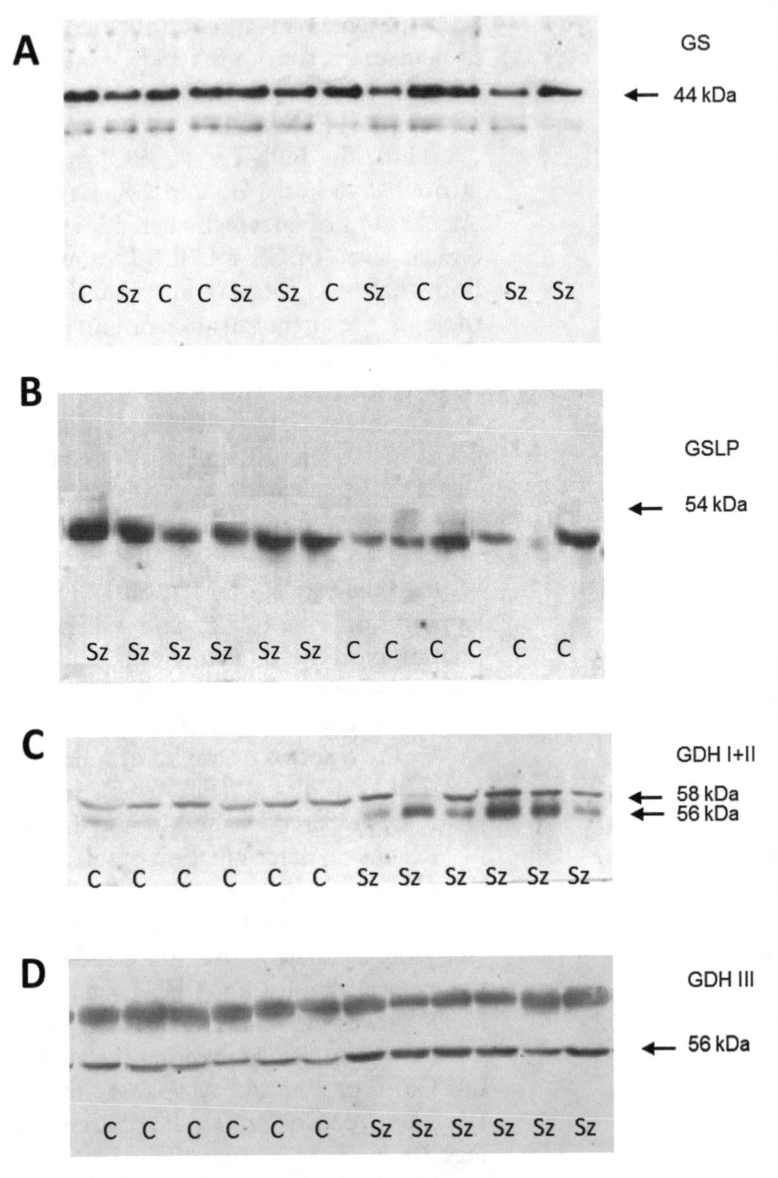

Fig. 6 Patterns of staining for GS (**a**), GSLP (**b**), GDH I + II (**c**) in "cytoplasm" fraction and GDH III (**d**) in "membranes" prepared by Method 1A from BA 10 (**a**, **d**) and BA 23 (**b**, **c**). Total protein amount is 10 μg per lane. Abbreviations: *C* controls, *Sz* schizophrenia. The samples (C and Sz) are from different subjects

III, evaluated by immunoblotting) in BA 10, 23, and 24 in controls and patients with schizophrenia (statistics is presented for 15 subjects in each group). Relative amounts of GSLP, GDH I + II, and GDH III were increased in all studied brain areas in schizophrenia patients. As for GS, its amount was decreased in the "cytoplasm" fraction in BA 10 in patients with schizophrenia, increased in BA 24, whereas in BA 23, the GS quantity did not differ between control and schizophrenia subjects.

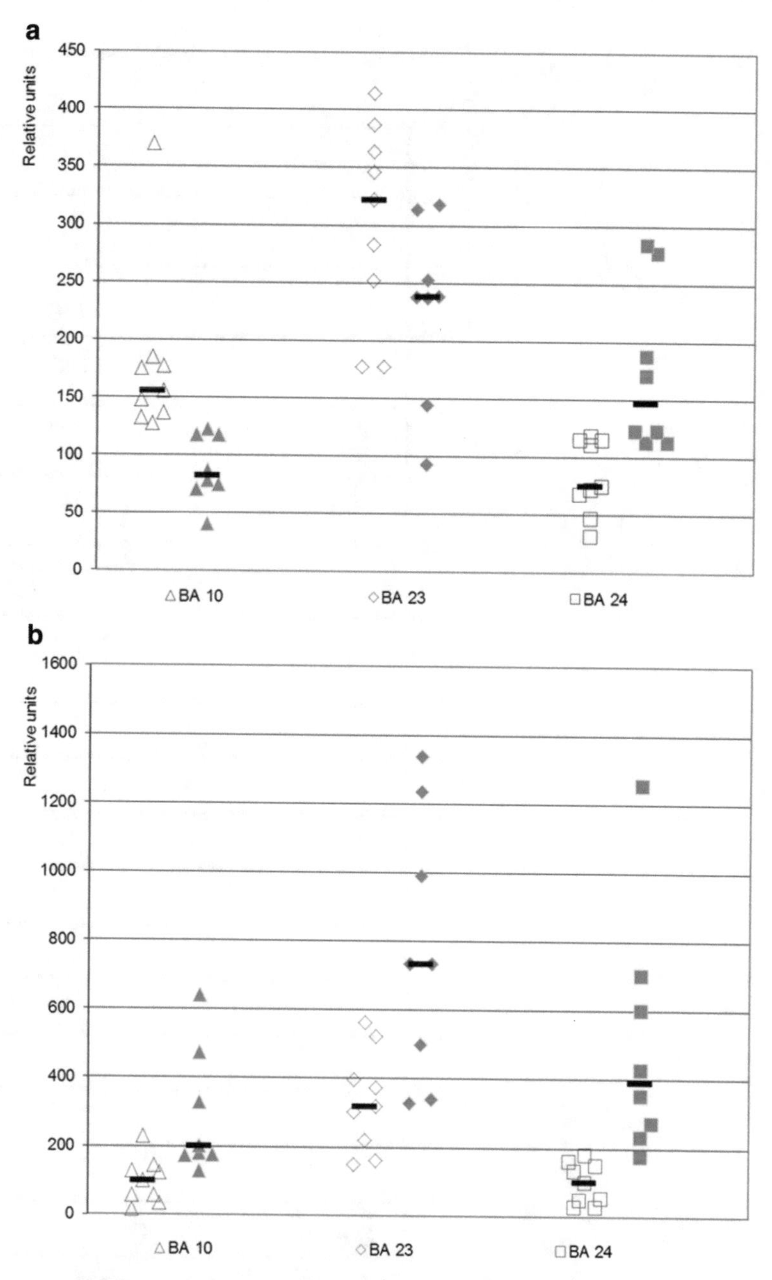

Fig. 7 Relative quantities of GS (**a**) and GSLP (**b**) in "cytoplasm" fraction prepared by the Methods 1A from BA 10, BA 23, and BA 24 samples from controls and patients with schizophrenia. *Open symbols*—controls, *solid symbols*—patients with schizophrenia. *Horizontal black lines*—medians for groups. $^{*}p < 0.01$

In conclusion, the measurement of the bulk enzymatic activities in brain tissue protein extracts "cytoplasm" prepared according to Method 1A gives little information about the quantitative alterations of isoforms for these enzymes in the brain tissue of subjects with schizophrenia. Hence, evaluation of relative amounts of all

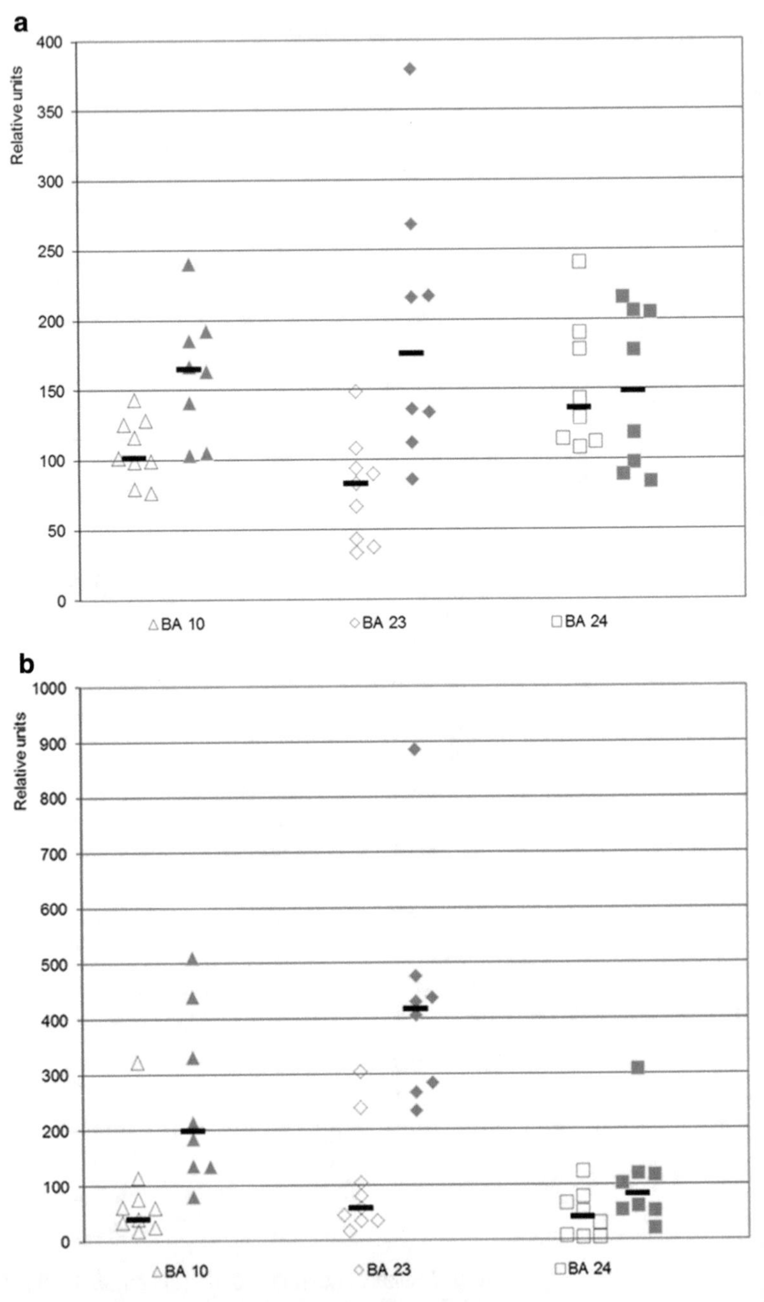

Fig. 8 Relative quantities of GDH I (58 kDa) and GDH II (56 kDa) in "cytoplasm" fraction prepared by the Methods 1A from BA 10, BA 23, and BA 24 samples from controls and patients with schizophrenia. *Open symbols*—controls, *solid symbols*—patients with schizophrenia. *Horizontal black lines*—medians for groups $^*p < 0.01$

isoforms present in the human brain was undertaken separately by establishing levels of corresponding immunoreactive proteins.

For general remark regarding extraction Method 1A, see Note 6.

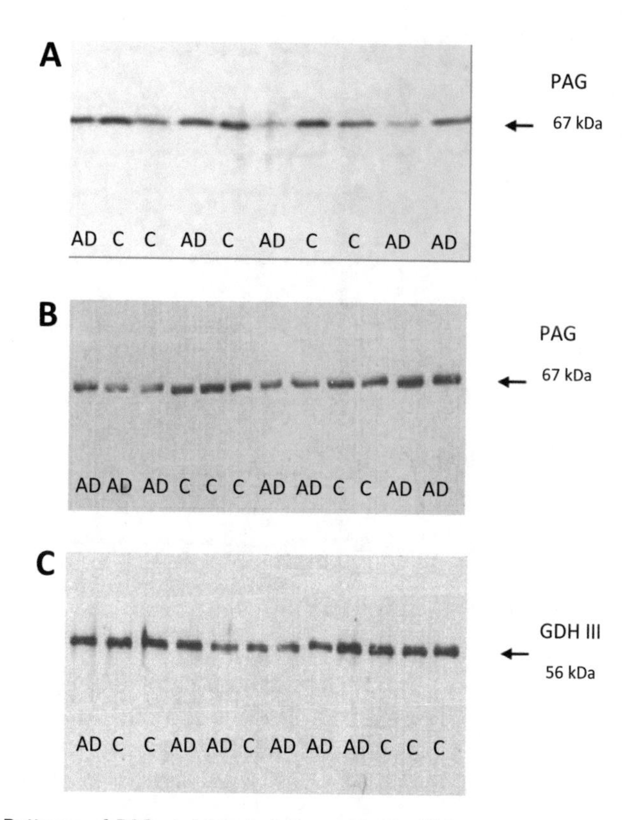

Fig. 9 Patterns of PAG staining in "supernatants 1" from cerebellar cortex (**a**), PAG staining in "membranes" from BA 10 (**b**); GDH III staining in "membranes" from BA 10 (**c**)—all extracts are prepared by Method 1B. Total protein amount is 10 μg per lane. Abbreviations: *C* controls, *AD* Alzheimer's disease cases. Samples (C and AD) are all from different subjects

4.2 Extraction Method 1B

We used the extraction method 1B (see Sect. 3.1.2) to determine relative amounts of target enzymes (immunoreactive proteins) in extracts from cerebellar cortex or BA 10 from patients with AD and controls by ECL-immunoblotting using immunoprobes.

First, protein concentration was determined (for instance, in "supernatant 1" extracts prepared from cerebellar cortex the following data was obtained: mean of 1.9 mg/mL, min = 1.4 mg/mL, max = 2.4 mg/mL, SD = 0.4, $n = 17$).

An example of immunoreactive GAD65 and GAD67 isoform staining in "supernatants 1" from cerebellar cortex from brain samples of controls and patients with AD is given in Fig. 4e.

An example of immunoreactive PAG staining in "supernatants 1" from cerebellar cortex from brain samples of controls and patients with AD is given in Fig. 9a.

Since "supernatant 1" fractions contained membrane-associated proteins, apparently, PAG and GAD isoenzymes associated with membrane cellular structures can be detected in "supernatants 1" thus enabling to quantify them.

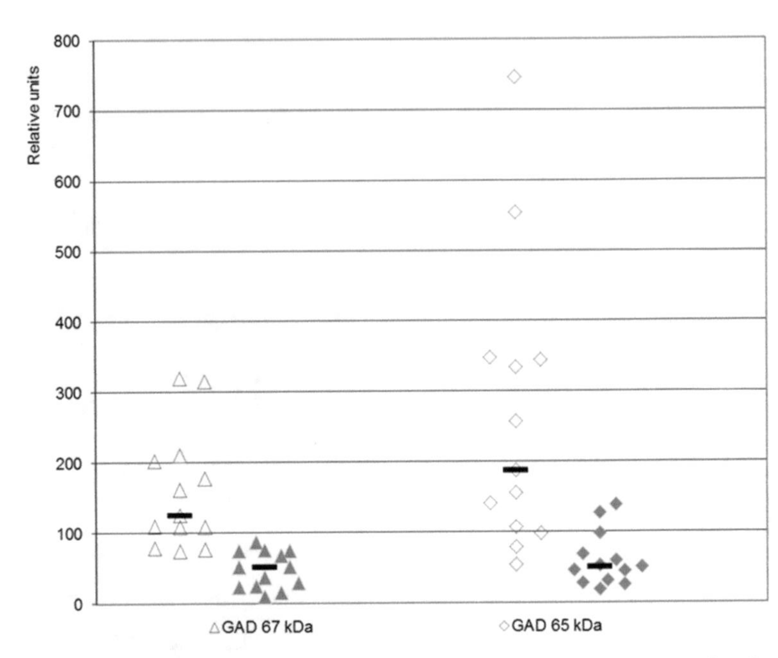

Fig. 10 Relative quantities of GAD 65 and GAD 67 in "supernatant 1" fraction prepared by the Method 1B from cerebellar cortex samples from controls and patients with AD. *Open symbols*—controls, *solid symbols*—patients with AD. *Horizontal black lines*—medians for groups. $^{*}p < 0.01$

Results on determination of relative quantities of immunoreactive GAD65 and GAD67 in "supernatant 1" from cerebellar cortex from controls and patients with AD (n = 13 in each group) are given in Fig. 10.

Results on determination of relative quantities of immunoreactive PAG in "supernatants 1" and "membranes" prepared by the Method 1B from cerebellar cortex of controls and patients with AD (n = 13 in each group) are presented in Fig. 11.

In addition, samples of BA 10 from controls and patients with AD were used for preparation of "membranes" and "cytoplasms" fractions by the Method 1B (see Sect. 3.1.2), and immunoreactive enzymes were detected: GSLP, GS, GDH I, and GDH II in "cytoplasms," and PAG, GDH III, GSLP—in "membranes."

Patterns of PAG and GDH III staining are given for "membranes" from BA 10 (of controls and patients with AD, n = 9 and 11, respectively) prepared in accordance with Method 1B (see Sect. 3.1.2): see Fig. 9b and c.

Protein concentration was determined (for instance, in "membranes" extracts prepared from BA 10 of controls and patients with AD), and the following data was obtained: mean of 3.2 mg/mL, min = 1.5 mg/mL, max = 5.4 mg/mL, SD = 1.1, n = 21.

Results on determination of relative quantities of immunoreactive GDH I and GDH II in "cytoplasm" fraction and GDH III in "membranes" are presented in Fig. 12.

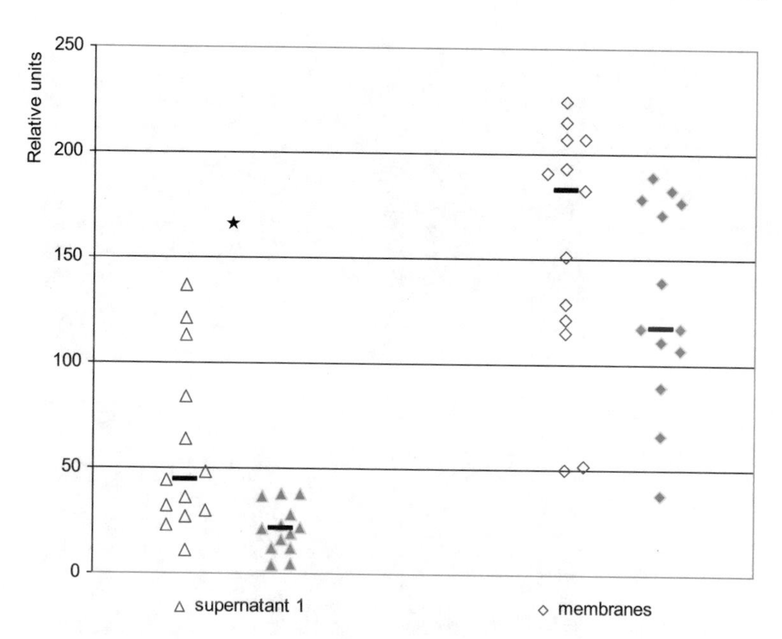

Fig. 11 Relative quantities of PAG in “supernatant 1” fraction and “membranes” fraction prepared by the Method 1B from cerebellar cortex samples from controls and patients with AD. *Open symbols*—controls, *solid symbols*—patients with AD. *Horizontal black lines*—medians for groups. $^{*}p < 0.01$

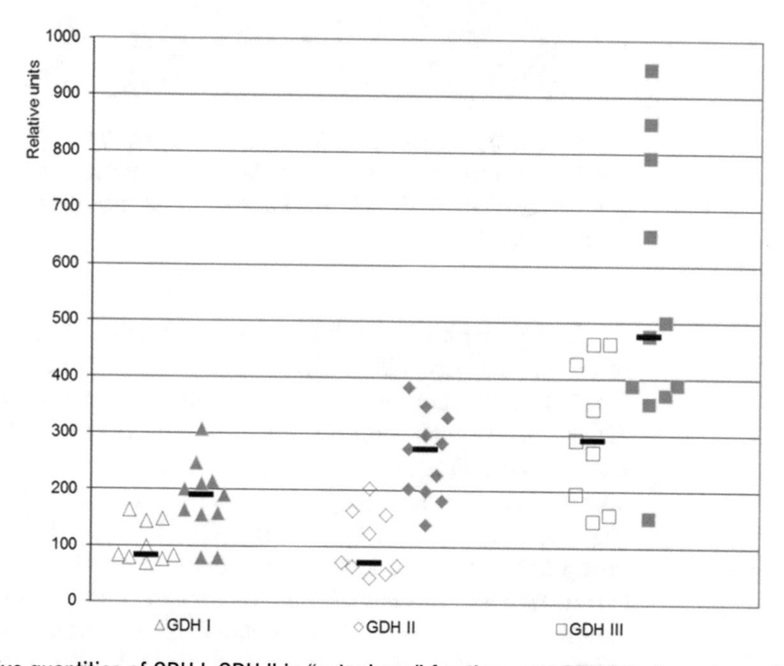

Fig. 12 Relative quantities of GDH I, GDH II in “cytoplasm” fractions and GDH III in “membranes” prepared by the Method 1B from BA 10 samples from controls and patients with AD. *Open symbols*—controls, *solid symbols*—patients with AD. *Horizontal black lines*—medians for groups. GDH I, GDH II—$^{*}p < 0.01$, and GDH III—$^{*}p < 0.025$

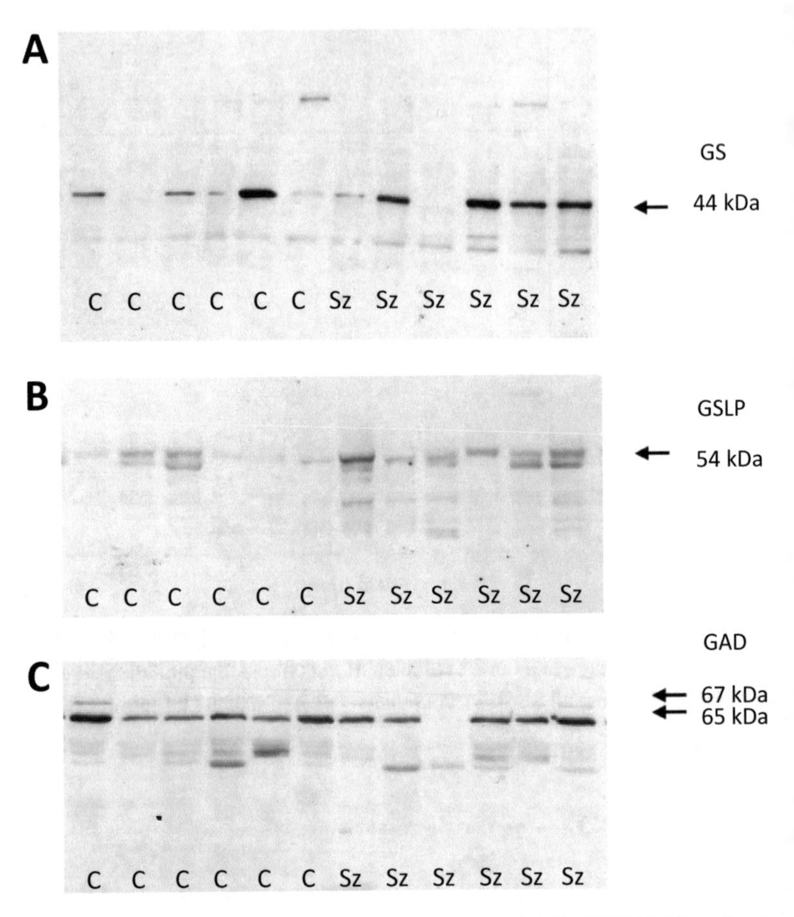

Fig. 13 Patterns of staining for GS (**a**), GSLP (**b**) and GAD 65/67 (**c**) in extracts from caudate nucleus prepared by Method 2. Total protein amount is 10 μg per lane. Abbreviations: *C* controls, *Sz* schizophrenia. Samples (C and Sz) are all from different subjects

In summary, when using extraction Method 1B, the quantities of target proteins (GS, GSLP, GDH I + II, GDH III, PAG, and GAD) in extracts prepared by Method 1B from BA 10 and cerebellar cortex were significantly different between the analyzed control and AD groups. Namely, in BA 10 the AD group had significant increase in "cytoplasm" fraction: GS ($p < 0.003$), GSLP ($p < 0.0005$), GDH I ($p < 0.01$), GDH II ($p < 0.001$); the AD group had significant increase in "membranes" fraction: GDH III ($p < 0.025$) and PAG ($p < 0.04$). As for cerebellar cortex, in "supernatant 1" the levels of PAG, GAD65, and GAD67 were significantly decreased ($p < 0.004$, $p < 0.0002$, and $p < 0.00004$, respectively). See Note 7.

4.3 Extraction Method 2

The extraction Method 2 (see Sect. 3.1.3) was used to determine relative amounts of target enzymes (proteins) GS, GSLP, GDH,

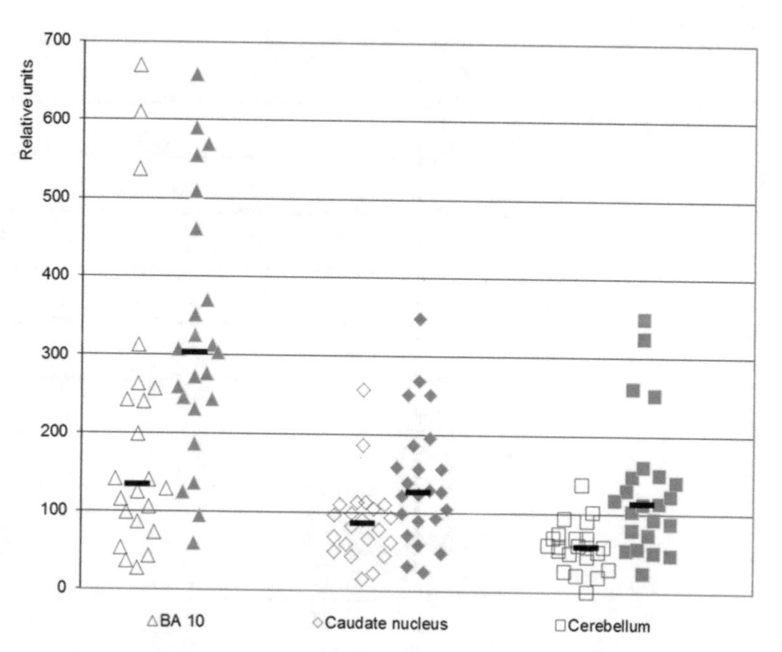

Fig. 14 Relative quantities of GS in extracts from BA 10, caudate nucleus, and cerebellum prepared by Method 2. *Open symbols*—controls, *solid symbols*—patients with schizophrenia. *Horizontal black lines*—medians for groups. $^{*}p < 0.01$

and GAD 65/67 in extracts from BA 10, caudate nucleus, and cerebellar cortex of controls and patients with schizophrenia (n = 22 and n = 23, respectively) by ECL-immunoblotting using respective immunoprobes.

First, protein concentration was determined (for instance, in extracts prepared from caudate nucleus) the following data was obtained: mean of 4.5 mg/mL, min = 3.2 mg/mL, max = 6.6 mg/mL, SD = 0.6, n = 45.

An example of GS staining in extracts from caudate nucleus is given in Fig. 13a. An example of GSLP staining in extracts from caudate nucleus is given in Fig. 13b.

Relative quantities of immunoreactive GS were determined in extracts of three different brain areas—BA 10, caudate nucleus and cerebellum—from controls and patients with schizophrenia, and comparative results on GS relative quantities in extracts from these brain areas prepared by Method 2, are presented in the Fig. 14 .

For comparison of relative quantities of immunoreactive GSLP in extracts prepared by the extraction Method 2 with the results obtained by extraction Method 1A and 1B, see Note 5.

An attempt was made to quantify GAD 65/67 simultaneously by ECL-immunoblotting in extracts prepared by the Method 2 from caudate nucleus in control cases and patients with schizophrenia. An example of GAD 65/67 staining is given in Fig. 13c.

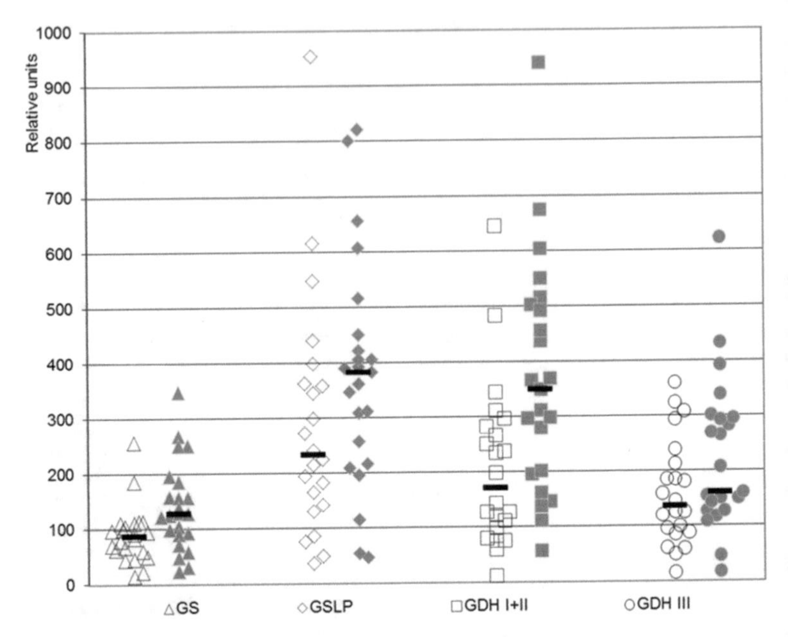

Fig. 15 Relative quantities of GS, GSLP, GDH I + II, and GDH III in extracts prepared by the Method 2 from caudate nucleus samples from controls and patients with schizophrenia. *Open symbols*—controls, *solid symbols*—patients with schizophrenia. *Horizontal black lines*—medians for groups. *$p < 0.025$

Only one sample (lane 1, a control case) contains detectable amount of GAD 67; for possible explanation, see Note 8. So, simultaneous quantification of GAD65 and GAD67 in not possible in this case.

When using extraction Method 2, patients with schizophrenia had statistically significantly elevated relative quantities of target proteins (GS, GSLP, GDH I + II) in BA 10, caudate nucleus and cerebellar cortex ($p < 0.025$, Mann-Whitney *U*-test), whereas the GDH III measures were similar between controls and subjects with schizophrenia.

Results on relative quantities of the studied immunoreactive proteins in extracts prepared by the Method 2 from caudate nucleus samples from controls and patients with schizophrenia are presented in Fig. 15.

In summary, the extraction Method 2 enables to use minimal volumes of brain samples (50 mg) and achieves good reproducibility due to using strong detergent (SDS) and boiling extracts during preparation.

5 Notes

Possible artifacts and difficulties in these studies can arise from multistage experiments and these include problems associated with biological material used in biochemical analyses (autopsied brain

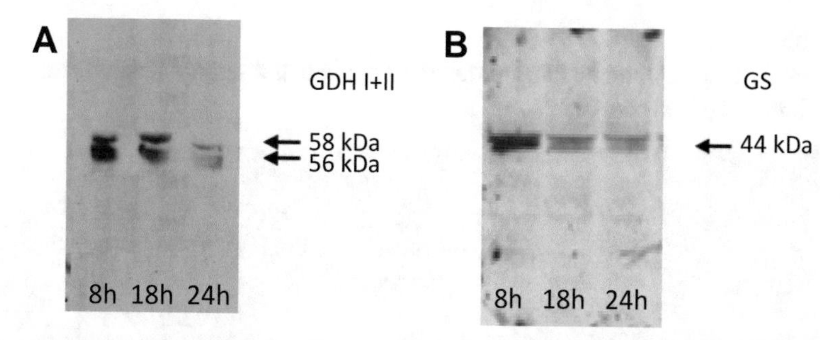

Fig. 16 ECL-immunoblotting with specific antibodies for GDH I + II (**a**) and GS (**b**), in extracts (supernatants 1) prepared from the same brain area (BA 1–3) by Method 1A after incubation of the brain for 8, 18, and 24 h. Total protein amount is 10 mkg per lane

samples), various methods for protein extraction, and enzymes assay (i.e., presence of different isoforms).

1. To avoid artifacts associated with tissue storage duration, samples should be matched, whenever possible, for this parameter. Our experience suggests that 5 mL vials with tightly closed caps or 1.8 mL cryogenic tubes Nunc CryoTubes Vials, internal thread with screw caps, provide sufficient tissue quality after (up to) 10 years storage at −80 °C.

   Postmortem interval (PMI) influences the detection of immunoreactive isoforms of target enzymes in brain sample extracts. We confirm this in the following experiment: autopsied human brain from a control case was incubated for 24 h at 4 °C, and 200 mg BA 1–3 samples were taken from the left hemisphere at 8, 18, 24 h of incubation.

   Substantial decrease in protein measures was observed at 18 h PMI for GSLP and GDH I + II. Example of GSLP (A) and GDH I + II (B) staining by ECL-immunoblotting in "cytoplasm" extracts from BA 1–3 prepared in accordance with the Method 1A is given in Fig. 16a and b.

2. The staining with Ponceau S is a necessary step for control of protein transfer. It enables to save time by avoiding further manipulations when the transfer is of low quality. However, the staining with Ponceau S cannot visualize the target enzymes (minor proteins) due to low sensitivity, hence, the quality of transfer can be estimated only for major proteins.

3. The ECL-immunoblotting method is semiquantitative; it can be used for comparative evaluations of relative amounts of proteins, and the data are expressed in relative units (RU). The working range of the method in many cases is rather narrow, and linear range is yet narrower—see, for instance, Fig. 4b, where standard (Std) sample containing 5 μg of total protein is

Table 3
GSLP quantification by ECL-immunoblotting in supernatants ("cytoplasm" fraction) prepared by Method 1B from BA 10

Sample	Experiment 1		Experiment 2		Experiment 3		Net intensity		Relative units (average value)
	Net intensity	Relative units	Net intensity	Relative units	Net intensity	Relative units	Average value	SD	
1	2	3	4	5	6	7	8	9	10
Standard 5 mkg	2165		1794		1981		1980	186	
Standard 10 mkg	6039		5015		5543		5532	512	
Standard 20 mkg	8340		7572		8496		8136	495	
Sample (C)	14,668	243	12,832	256	12,238	221	13,246	1267	240
Sample (C)	13,773	228	9272	185	11,753	212	11,599	2254	208
Sample (C)	12,246	203	8617	172	9493	171	10,119	1894	182
Sample (Sz)	6789	112	6754	135	7449	134	6997	391	127
Sample (Sz)	6656	110	7325	146	7200	130	7060	356	129
Sample (Sz)	7212	119	9298	185	8007	144	8172	1053	150

Total protein amounts for samples are 10 μg per lane
Abbreviations: *C* controls, *Sz* schizophrenia. The samples (C and Sz) are from different subjects

stained extremely faint, or Fig. 13a, lanes 2, 6, and 9 with very weak staining. In these cases, more total protein load is required (for instance, 20–30 μg per lane).

An example of Relative Units calculating for GSLP in supernatants ("cytoplasm" fraction) prepared by Method 1B from BA 1–3 is represented in the Table 3. The value of Net Intensity obtained for individual protein spot after the film scanning (see columns 2, 4, 6) was divided by the value of the internal standard's Net Intensity (respective amount of 10 μg in the same column) for each sample and expressed in percentages—Relative Units (RU)—see columns 3, 5, 7.

The results of repeatability of internal standard and samples between different gel slabs (blots) performed on different days are presented in columns 8, 9 in Table 3. Average RU for three experiments are presented in column 10.

In our work, the most reproducible and robust results were obtained in calculations with 5, 10, and 20 μg of total protein for internal standard, and 10 μg of total protein for samples. Loading of larger amounts of total protein on a gel lane leads

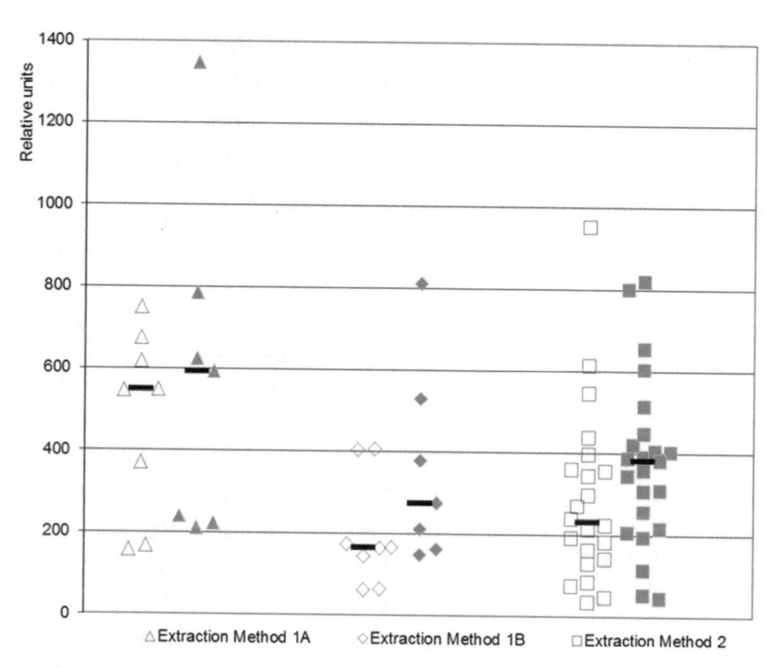

Fig. 17 Relative quantities of GSLP in "cytoplasm" fraction prepared by the Methods 1A, 1B, and extracts prepared by the Method 2 from caudate nucleus samples from controls and patients with schizophrenia. *Open symbols*—controls, *solid symbols*—patients with schizophrenia. *Horizontal black lines*—medians for groups

to more enhancement of weaker protein spots than bright ones. On the other hand, decreasing amounts of total protein in samples loaded on each lane leads to bleaching (up to disappearance) of weaker spot intensities, resulting in poorer reproducibility and decreasing robustness and sensitivity of the method revealing between-group differences.

To check reproducibility, in addition to internal standard (usually, 5, 10, and 20 μg of total protein, see Sect. 3.3.3), two samples (randomly selected from previous experiments) were repeatedly tested every time in independent experiments. When data for internal standard and for these samples were not reproduced (SD above 20%), the experiment was not validated and was excluded from further analyses.

4. The usage of purified proteins as internal standards cannot be applied since it gives no information about estimation of level of total proteins (of brain tissue extracts) required for obtaining calibration curves and determining the range of validity of the semiquantitative method we used (ECL-immunoblotting).

5. Figure 17 shows comparative results on relative quantities of immunoreactive GSLP in "cytoplasm" fraction (Method 1A, see Sect. 3.1.1), "cytoplasm" fraction (Method 1B, see Sect. 3.1.2),

and extracts (Method 2, see Sect. 3.1.3) from the same BA 10 samples obtained from controls and patients with schizophrenia.

Warnings and considerations arising from these comparative results relate to distribution of isoforms among subcellular compartments. In fact, when studying distribution of target enzymes' isoforms between "cytoplasm" and "membrane" fractions using extraction Methods 1A and 1B, one should take into account that the obtained data does not reflect the true distribution of the enzymes among subcellular compartments. Usage of material such as frozen brain samples in biochemical studies is associated with distinct limitations because freezing of brain tissue leads to disruption of intracellular membrane structures and solubilization of mitochondrial proteins. This aspect is particularly addressed in our works [17, 18].

6. Method 1A requires relatively larger amounts of brain samples because it gives poor reproducible data when using sample quantity of less than 200 mg.
7. The extraction Method 1B provides reproducible results when determining relative amounts of isoforms of the target enzymes when using 100 mg amounts of biological material, and when small groups of control and diseased cases are available.
8. Absence of GAD 67 protein band and additional minor bands seen under the main band (GAD 65) probably, due to proteolytic degradation of GAD 67, do not allow semiquantitative analysis of GAD isoforms in this brain area. Sensitivity of GAD 67 to proteolysis was noted in literature [47].
9. Difficulties and issues associated with ECL Western blotting, as the main tool for assessment of relative amounts of enzyme isoforms, can be overcome via consulting appropriate manuals and protocols. However, the following needs careful consideration: various antibodies against GS, GDH, and GAD are commercially available. Their specificity and sensitivity may significantly vary, and their usage requires adaptation of the ratio between the total protein loaded on a lane of polyacrylamide gel during electrophoresis and antibody (antiserum) dilution for ECL-Western immunoblotting.

In this chapter, a standard ECL-immunoblotting system is discussed consisting of two components, namely primary polyclonal (or monoclonal) specific antibody and secondary antibody conjugated with horseradish peroxidase. Application of this system is a complex multistage experimental process, requiring titer determination for the primary and secondary antibody. Multistage process may bring artifacts decreasing reproducibility. One possible solution is application of primary monoclonal (or affinity purified polyclonal) antibodies, conjugated with peroxidase. This approach enables to decrease the duration of the analysis due to minimization of stages, as well as increase the signal/noise ratio (due to the minimization of background staining).

6 Discussion and Conclusion

Our studies generally aimed to reveal between-group differences (control group vs. mental pathology, namely AD or schizophrenia) in amounts of target enzymes involved in Glu metabolism. Taking into account significant ultrastructural alterations of glial membrane structures indicating substantial impairment of the brain lipid and energy metabolism in the studied mental pathologies (schizophrenia, [48] and AD [49]), we used various extraction methods, assuming possible different extent of solubility/association of the target proteins with membranes and other cellular structures.

For instance, the extraction Methods 1A and 1B differ in the presence of sucrose added in the second case. Sucrose was introduced for stabilization of possible association between membranes and proteins and to prevent disruption of intracellular membrane structures (particularly mitochondria) by osmotic shock.

Because of limited amounts of brain sample tissues from patients with AD, brain samples from patients with schizophrenia and respective controls were used for direct comparison of data obtained by various extraction Methods.

The relative quantities of immunoreactive GS in "cytoplasm" fraction obtained by the extraction Method 1A (Sect. 3.1.1) from BA 10 samples from controls and patients with schizophrenia ($n = 8$ and $n = 7$, respectively) were compared with the results obtained on the same brain samples treated by the extraction Method 1B (Sect. 3.1.2).

Figure 18 demonstrates results of comparison drawn between the data obtained with different extraction methods (Method 1A and 1B) applied for evaluation of immunoreactive GS contents (in Net Intensity).

Although individual variations between the samples are obvious, regularities of differences between control groups and patients with mental pathology are common therewith. To achieve reproducibility in every experiment (for each PAGE run and blot), relative quantities of the target proteins (enzymes) were determined in relation to internal standard (wherein samples and the standard were prepared by identical method).

Figure 17 demonstrates results of comparison drawn between the data obtained with extraction methods 1A, 1B, and Method 2, applied for evaluation of GSLP quantities. One can see common regularities for Methods 1A and 1B ("cytoplasm" fraction) although Method 2 ("total lysates") gives another tendency, and this fact suggests different association of this enzyme with subcellular structures in control and in patient group.

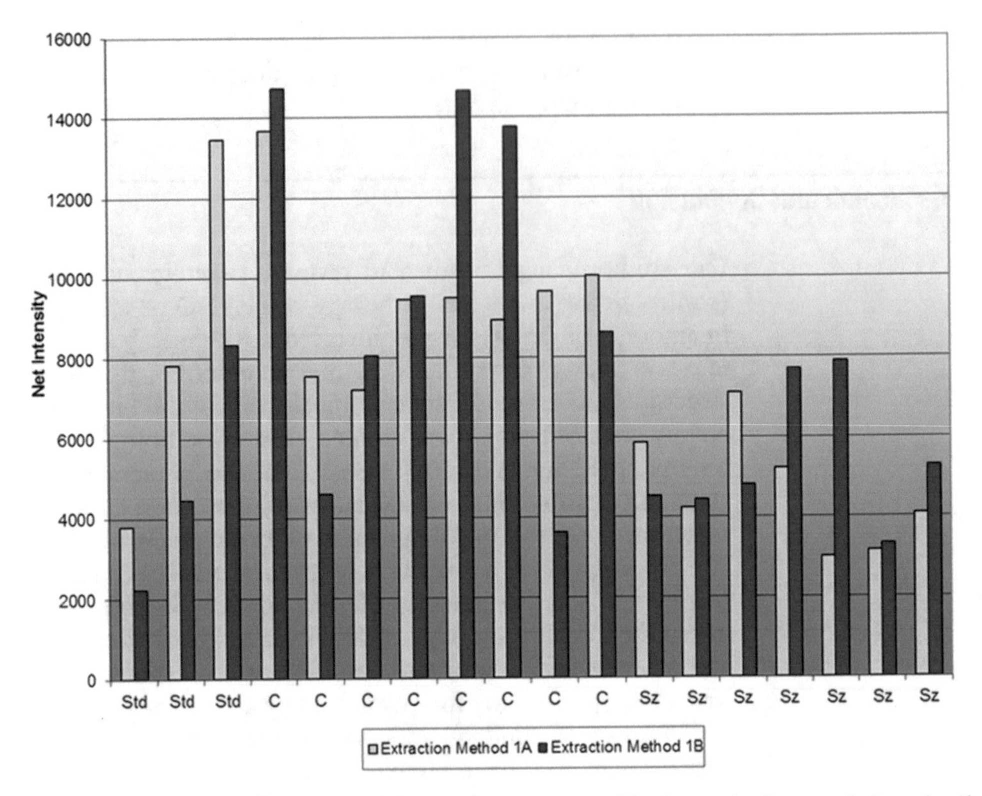

Fig. 18 Net Intensities for GS in "cytoplasm" fraction prepared from BA 10 samples from controls and patients with schizophrenia by the Methods 1A (*light columns*), or 1B (*dark columns*). Total protein amount is 10 μg per lane, except for the internal standards (5, 10, and 20 μg of total protein, respectively). Abbreviations: *C* controls, *Sz* schizophrenia. Samples (C and Sz) are all from different subjects

So, when choosing extraction method in experiments with brain tissue from persons with brain pathology, one should take into consideration not only possible individual features (variety) of samples, but also etiology of the pathological process, i.e., probable alteration in extent of association between proteins and membranes (and other cellular structures). Hence, when comparing own results with data obtained by others, one should consider the extraction methods used in each given study.

Different approaches are currently used in quantitative proteomics when studying alteration/regulation of protein amounts associated with a pathology: total protein extracts can be normalized by a major protein (which is supposedly unchanged in pathology) or, alternatively, all samples can be normalized by total protein amount. We used the last technique, because a protein, the level of which was chosen as a "standard," could be altered due to significant glial and neuronal alterations in studied mental pathologies (as we and others have previously described for glial fibrillary acidic protein in subjects with schizophrenia) [24, 50]. Total protein concentrations measured in brain extracts prepared by the same methods did not differ significantly from one brain area to another and did not differ in control and patient groups. However, we should admit that total protein content (concentration) is not universal "gold standard" and also can be a variable value (at least in animal models).

In proteomics, generally, the methods employed for sample preparations (protein extraction) enable to achieve "as more as possible" extraction and solubilization of cellular proteins (via using high concentrations of urea and detergents). This approach does not address the redistribution of isoenzymes among subcellular compartments associated with brain pathology.

This means that different isoenzymes are differently solubilized (particularly, depending on pathology) under varying conditions (we assume here relatively "soft" extraction conditions—Extraction Methods 1A and 1B, avoiding high ionic strength, detergents, strong chaotropic agents, etc.). In this case, high reliability can be achieved only under multiple reproductions of the experiments and use of larger quantity of brain samples (at least 200 mg of brain tissue).

In contrast, preparation of "total extracts" (Method 2) using strong ionic detergents (i.e., SDS, protein denaturing agents, boiling) enables to achieve good reproducibility, especially when using small quantity of brain material (volume of brain sample about 50 mg). However, this approach fails to address the fine redistribution patterns of isoenzymes and, thus, provides little information on the extent of association of different isoenzymes with both subcellular components (i.e., membrane structures) and fine (ultra)cellular alterations accompanying brain pathology in mental health disorders. In other words, different isoenzymes can become more or less "soluble," depending on pathological conditions within the brain.

In conclusion, we note that not only quantities of enzymes per se, but correlative links between amounts of various enzymes involved in Glu metabolism differ in patients with mental pathology and in controls [25]. For calculation of these correlations, the quantities of these enzymes can be measured by the methods described in the present work.

The data described in this chapter provide a strong support for Glu metabolism impairment playing an important role in pathological events associated with the pathogenesis of AD and schizophrenia.

The role of Glu in AD has been extensively investigated, and Glu excitotoxicity (due to uncontrolled elevation of its local concentrations) has been recognized as the main mechanism leading to neurodegeneration [34]. Our work demonstrates further changing levels, recoupling and relocation of all main enzymes converting Glu in the brain tissue of people with AD. Thus, not only glutamatergic system is impaired in AD (i.e., Glu uptake and reception are imbalanced due to deficits in corresponding transporters and receptors), but also Glu recycling by Glu metabolizing enzymes is substantially altered in AD.

These findings have a practical translational implication with a view of potential development of novel AD treatment(s). The decrease in quantity of GAD isoforms we describe in the current study complements previous work on insufficiency of GABA-ergic inhibiting system in AD and the consequent imbalance between excitation/inhibition processes [42]. We propose that GAD activity may be elevated and thus normalized. For instance, this can be achieved by means of gene therapy: GAD "insufficiency" could be compensated by means of vector construction (DNA sequence encoding GAD and thus enlarging GAD expression) and incretion in the brains of patients with AD [51].

In schizophrenia, the currently overwhelming work on only one brain neurotransmitter (dopamine) may easily oversimplify our understanding of the disease molecular pathogenesis and thus hinders the search for effective treatment. Like the situation with the research of Alzheimer's disease, the main focus regarding glutamate involvement in schizophrenia remains glutamate receptors and transporters, whereas glutamate recycling by enzymes has been largely neglected. As a result, the novel generation of antipsychotic drugs, currently in development, are directed towards receptors and transporters, but none of them addresses the regulation of Glu recycling enzymes. Further work is now needed to both elucidate the Glu metabolizing enzymes, their role in the pathogenesis of both AD in schizophrenia and their applications in the clinical treatment of these disabling diseases.

Acknowledgments

We would like to thank Prof. O.P. Ottersen and Dr. I.A. Turgner, from the Centre for Molecular Biology and Neuroscience, and Department of Anatomy, University of Oslo and Neurochemical Laboratory, University of Oslo, Norway, for kindly providing antibodies to C-terminal peptide fragment of PAG.

References

1. Tapiero H, Mathe G, Couvreur P et al (2002) Glutamine and glutamate. Biomed Pharmacother 56:446–457
2. Tsakopoulos M, Magistretti PJ (1996) Metabolic coupling between glia and neurons. J Neurosci 16:877–885
3. Laake JH, Takumi Y, Eidet J et al (1999) Postembeddingimmunogold labelling reveals subcellular localization and pathway-specific enrichment of phosphate activated glutaminase in rat cerebellum. Neuroscience 88:1137–1151
4. Cooper AJ, Jeitner TM (2016) Central role of glutamate metabolism in the maintenance of nitrogen homeostasis in normal and hyperammonemic brain. Biomolecules 6(2):16–49. pii: E16. doi:10.3390/biom6020016
5. Bernstein HG, Tausch A, Wagner R et al (2013) Disruption of glutamate-glutamine-GABA cycle significantly impacts on suicidal behaviour: survey of the literature and own findings on glutamine synthetase. CNS Neurol Disord Drug Targets 12:900–913

6. Schousboe AA, Tribute to Mary C. McKenna (2015) Glutamate as energy substrate and neurotransmitter-functional interaction between neurons and astrocytes. Neurochem Res. 31 Dec 2015, published online, 1–6
7. Boksha IS (2004) Coupling between neuronal and glial cells via glutamate metabolism in brain of healthy persons and patients with mental disorders. Biochem Mosc 69:705–719
8. Derouiche A, Frotscher M (1991) Astroglial processes around identified glutamatergic synapses contain glutamine synthetase, evidence for transmitter degradation. Brain Res 552(2):346–350
9. Yamamoto H, Konno H, Yamamoto T et al (1987) Glutamine synthetase of the human brain, purification and characterization. J Neurochem 49(2):603–609
10. Tumani H, Shen GQ, Peter JB (1995) Purification and immunocharacterization of human brain glutamine synthetase. J Immunol Methods 188:155–163
11. Gunnersen D, Haley B (1992) Detection of glutamine synthetase in the cerebrospinal fluid of Alzheimer diseased patients, a potential diagnostic biochemical marker. Proc Natl Acad Sci U S A 89:11949–11953
12. Carney JM, Smith CD, Carney AM et al (1994) Aging- and oxygen-induced modifications in brain biochemistry and behavior. Ann N Y Acad Sci 738:44–53
13. Le Prince G, Delaere P, Fages C et al (1995) Glutamine synthetase expression is reduced in senile dementia of the Alzheimer type. Neurochem Res 20:859–862
14. Rao C, Shi H, Zhou C, et al (2016) Hypothalamic proteomic analysis reveals dysregulation of glutamate balance and energy metabolism in a mouse model of chronic mild stress-induced depression. Neurochem Res. May 26. Published online, 1–14
15. Khairova VR, Safarov MI (2015) Dynamics of glutamine synthase activity in rat brain in prenatal hypoxia model. FiziolZh 61(5):65–70. [Article in Russian]
16. Boksha IS, Tereshkina EB, Burbaeva GS (1995) Purification and some properties of glutamine synthetase from human brain. Biochem Mosc 60:1299–1303
17. Boksha IS, Schönfeld H-J, Langen H et al (2002) Glutamine synthetase isolated from human brain, octameric structure and homology of partial primary structure with human liver glutamine synthetase. Biochem Mosc 67:1012–1020
18. Boksha IS, Tereshkina EB, Burbaeva GS (2000) Glutamine synthetase and glutamine synthetase-like protein from human brain, purification and comparative characterization. J Neurochem 75:2574–2582
19. Zaganas I, Kanavouras K, Mastorodemos V et al (2009) The human GLUD2 glutamate dehydrogenase, localization and functional aspects. Neurochem Int 55(1–3):52–63
20. Burbaeva GS, Turishcheva MS, Vorobyeva EB et al (2002) Diversity of glutamate dehydrogenase in human brain. Prog Neuropsychopharmacol Biol Psychiatry 26:427–435
21. Plaitakis A, Berl S, Yahr MD (1984) Neurological disorders associated with deficiency of glutamate dehydrogenase. Ann Neurol 15:144–153
22. Plaitakis A, Flessas P, Natsiou AB et al (1993) Glutamate dehydrogenase deficiency in cerebellar degenerations, clinical, biochemical and molecular genetic aspects. Can J Neurol Sci 3(Suppl):S109–S116
23. Hussain MH, Zannis VI, Plaitakis A (1989) Characterization of glutamate dehydrogenase isoproteins purified from the cerebellum of normal subjects and patients with degenerative neurological disorders, and from human neoplastic cell lines. J Biol Chem 264:20730–20735
24. Burbaeva G, Boksha I, Turishcheva M et al (2003) Glial glutamate metabolizing enzymes in cingulate cortex in schizophrenia. Glia S2:65
25. Burbaeva GS, Boksha IS, Tereshkina EB et al (2007) Systemic neurochemical alterations in schizophrenic brain, glutamate metabolism in focus. Neurochem Res 32(9):1434–1444
26. Kvamme E, Torgner IA, Roberg B (2001) Kinetics and localization of brain phosphate activated glutaminase. J Neurosci Res 66:951–958
27. Schousboe A, Bak LK, Waagepetersen HS (2013) Astrocytic control of biosynthesis and turnover of the neurotransmitters glutamate and GABA. Front Endocrinol 4:102–113
28. Olalla L, Gutierrez A, Campos JA et al (2002) Nuclear localization of L-typeglutaminase in mammalian brain. J Biol Chem 277:38939–38944
29. Fenalti G, Law RHP, Buckle AM et al (2007) GABA production by glutamic acid decarboxylase is regulated by a dynamic catalytic loop. Nat Struct Mol Biol 14:280–286
30. Wei J, Wu J-Y (2008) Post-translational regulation of L-glutamic acid decarboxylase in the brain. Neurochem Res 33:1459–1465
31. Schwab C, Yu S, Wong W et al (2013) GAD65, GAD67, and GABAT immunostaining in human brain and apparent GAD65

loss in Alzheimer's disease. J Alzheimers Dis 33(4):1073–1088

32. Hoey SE, Buonocore F, Cox CJ et al (2013) AMPA receptor activation promotes non-amyloidogenicamyloid precursor protein processing and suppresses neuronal amyloid-β production. PLoS One 8(10):e78155
33. Wang ZC, Zhao J, Li S (2013) Dysregulation of synaptic and extrasynaptic *N*-methyl-D-aspartate receptors induced by amyloid-β. Neurosci Bull 29(6):752–760
34. Danysz W, Parsons CG (2012) Alzheimer's disease, β-amyloid, glutamate, NMDA receptors and memantine – searching for the connections. Br J Pharmacol 167(2):324–352
35. Davis KN, Tao R, Li C et al (2016) GAD2 alternative transcripts in the human prefrontal cortex, and in schizophrenia and affective disorders. PLoS One 11(2):e0148558. doi:10.1371/journal.pone.0148558. eCollection 2016
36. Guidotti A, Auta J, Davis JM et al (2000) Decrease in reelin and glutamic acid decarboxylase 67 (GAD67) expression in schizophrenia and bipolar disorder, a postmortem brain study. Arch Gen Psychiatry 57:1061–1069
37. Kimoto S, Bazmi HH, Lewis DA (2014) Lower expression of glutamic acid decarboxylase 67 in the prefrontal cortex in schizophrenia: contribution of altered regulation by Zif268. Am J Psychiatry 171:969–978
38. Bailey J, Powell L, Sinanan L et al (2011) A novel mechanism of V type zinc inhibition of glutamate dehydrogenase results from disruption of subunit interactions necessary for efficient catalysis. FEBS J 278(17):3140–3151
39. Kiktenko AI, Uranova NA, Denisov DV (1997) Quantitative characteristics of changes in synaptic contacts in the hippocampus in Alzheimer's disease. Neurosci Behav Physiol 27(6):681–682
40. Plaitakis A, Metaxari M, Shashidharan P (2000) Nerve tissue-specific (GLUD2) and housekeeping (GLUD1) human glutamate dehydrogenases are regulated by distinct allosteric mechanisms. J Neurochem 75:1862–1869
41. Burbaeva GS, Boksha IS, Tereshkina EB et al (2005) Glutamate metabolizing enzymes in prefrontal cortex of Alzheimer's disease patients. Neurochem Res 30(11):1443–1451
42. Burbaevan GS, Boksha IS, Tereshkina EB et al (2014) Glutamate and GABA-metabolizing enzymes in post-mortem cerebellum in Alzheimer's disease, phosphate-activated glutaminase and glutamic acid decarboxylase. Cerebellum 13(5):607–615
43. Burbaeva GS, Boksha IS, Turishcheva MS et al (2003) Glutamine synthetase and glutamate dehydrogenase in the prefrontal cortex of patients with schizophrenia. Prog Neuro-Psychopharmacol Biol Psychiatry 27:675–680
44. Iqbal K, Ottaway JH (1970) Glutamine synthetase in muscle and kidney. Biochemist 119:145–156
45. Fahien LA, Wiggert BO, Cohen PP (1965) Crystallization and kinetic properties of glutamate dehydrogenase from frog liver. J Biol Chem 240:1083–1090
46. Laemmli U (1970) Cleavage of structural proteins during the assembly of the head of bacteriophage T4. Nature 227:680–685
47. Martin SB, Waniewski RA, Battaglioli G et al (2003) Post-mortem degradation of brain glutamate decarboxylase. Neurochem Int 42:549–554
48. Vikhreva OV, Rakhmanova VI, Orlovskaya DD, Uranova NA (2016) Ultrastructural alterations of oligodendrocytes in prefrontal white matter in schizophrenia: a post-mortem morphometric study. Schizophr Res 177:28–36
49. Islam MT (2016) Oxidative stress and mitochondrial dysfunction-linked neurodegenerative disorders. Neurol Res 3:1–10
50. Steffek AE, McCullumsmith RE, Haroutunian V et al (2008) Cortical expression of glial fibrillary acidic protein and glutamine synthetase is decreased in schizophrenia. Schizophr Res 103(1–3):71–82
51. Bland R, Fitzsimons H (2009). Novel glutamic acid decarboxylase (GAD) chimera and methods of use. US Patent US8071563 B2

Chapter 16

The Glutamatergic System as Potential Clinical Biomarkers for Blood and Cerebrospinal Fluid Monitoring

Kenji Hashimoto

Abstract

Glutamate is the most abundant excitatory amino acid in the brain. In addition to the protein structure, it plays important roles in metabolism, nutrition, and signaling via glutamate receptors. Glutamate is synthesized from glutamine by glutaminase, while it is metabolized to the inhibitory amino acid γ-aminobutyric acid (GABA) by glutamic decarboxylase. Thus, the glutamine-glutamate-GABA cycle plays an important role in both excitatory and inhibitory neurotransmissions via glutamate and GABA receptors. Accumulating evidence suggests that abnormalities in glutamatergic neurotransmission via ionotropic and metabotropic glutamate receptors may play crucial roles in the pathophysiology of a variety of psychiatric, neurologic as well as other peripheral disorders. In this chapter, the author discusses the glutamatergic system as potential clinical biomarkers for human blood and cerebrospinal fluid monitoring.

Key words Amino acids, Blood, Cerebrospinal fluid, Depression, GABA, Glutamate, Glutamine, Glycine, NMDA receptor, Schizophrenia, Serine

1 Introduction: Biosynthesis and Glutamatergic System

Glutamate (Glu), one of the 20 amino acids (L-forms) used to construct proteins, is found in high concentrations in every part of the body. In humans, dietary proteins are broken down by digestion into amino acids, which serve as metabolic fuel for other functional roles in the body. Transamination plays a key role in the amino acid degradation of proteins. The amino group of Glu is metabolized to α-ketoglutaric acid by glutamate dehydrogenase or transaminase. Glutamate can be synthesized from α-ketoglutaric acid, which is produced as part of the citric acid cycle (Fig. 1) [1].

In the mammalian central nervous system (CNS), Glu is one of the major excitatory neurotransmitters, whereas γ-aminobutyric acid (GABA) represents the predominant inhibitory neurotransmitter. Glutamate is synthesized in CNS from glutamine (Gln) by glutaminase as part of the Gln-Glu cycle. Furthermore, Glu serves as the precursor for GABA, via glutamate decarboxylase (Fig. 1) [2–6].

Sandrine Parrot and Luc Denoroy (eds.), *Biochemical Approaches for Glutamatergic Neurotransmission*, Neuromethods, vol. 130, DOI 10.1007/978-1-4939-7228-9_16,

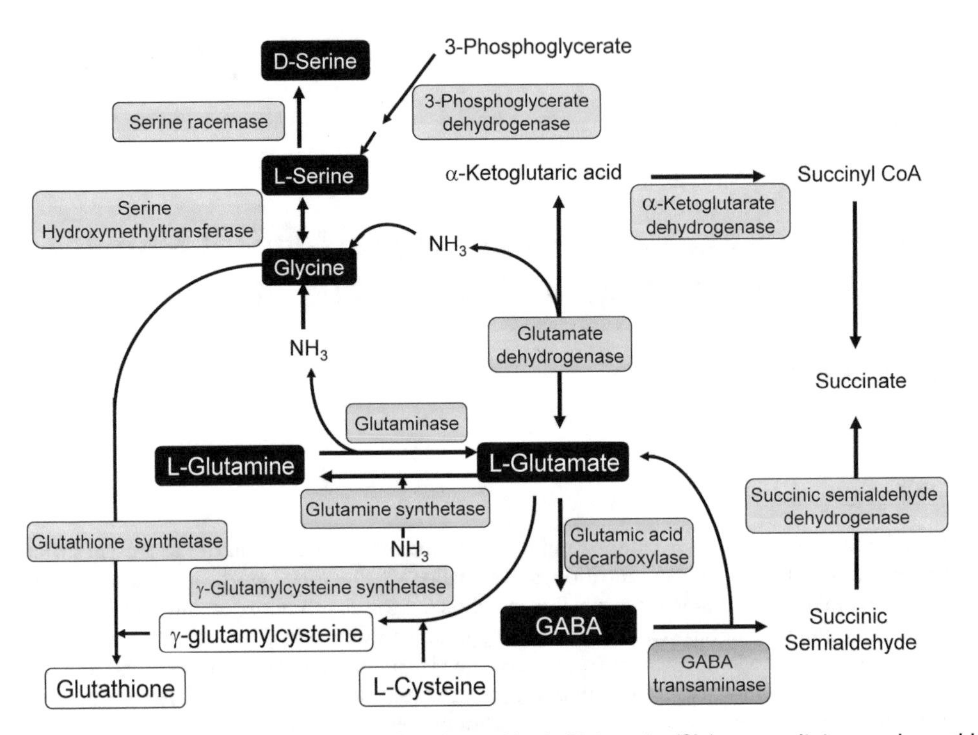

Fig. 1 Synthetic and metabolic pathways of amino acids. L-Glutamate (Glu), an excitatory amino acid, is synthesized from L-glutamine (Gln) by glutaminase, and metabolized to L-Gln by glutamine synthetase. In addition, L-Glu is metabolized to γ-aminobutyric acid (GABA), an inhibitory amino acid, by glutamic acid decarboxylase. GABA is metabolized to succinic semialdehyde by GABA transaminase. D-Serine is synthesized from L-Ser by serine racemase. L-Serine is converted to glycine (Gly) by serine hydroxymethyltransferase. Glutathione is synthesized via γ-glutamylcysteine from L-Glu, L-cysteine, and Gly

Thus, the Gln-Glu-GABA cycle in glia–neuron communication plays an important role in both excitatory and inhibitory neurotransmission [7–9].

Mammalian Glu receptors are classified based on their pharmacology. Ionotropic Glu receptors include the *N*-methyl-D-aspartate receptors (NMDARs), α-amino-3-hydroxyl-5-methyl-4-isoxazole-propionate (AMPA) receptors, and kainate (KA) receptors. In addition, metabotropic Glu receptors also exist. Therefore, glutamatergic neurotransmission via these ionotropic and metabotropic Glu receptors plays a key role in the pathophysiology of a variety of psychiatric and neurological disorders [5, 7, 8, 10–12].

Significant correlation between the cerebrospinal fluid (CSF) and serum levels of Glu or Gln have been reported [13]. Sodium-dependent neutral amino acid transporters, located in the abluminal membranes of the blood–brain barrier, are capable of actively removing neutral amino acids from the brain [14]. These findings suggest that the concentration of neutral amino acids, including Glu and Gln, in the extracellular fluid of brain are maintained at approximately 10% of those of the blood [13–16].

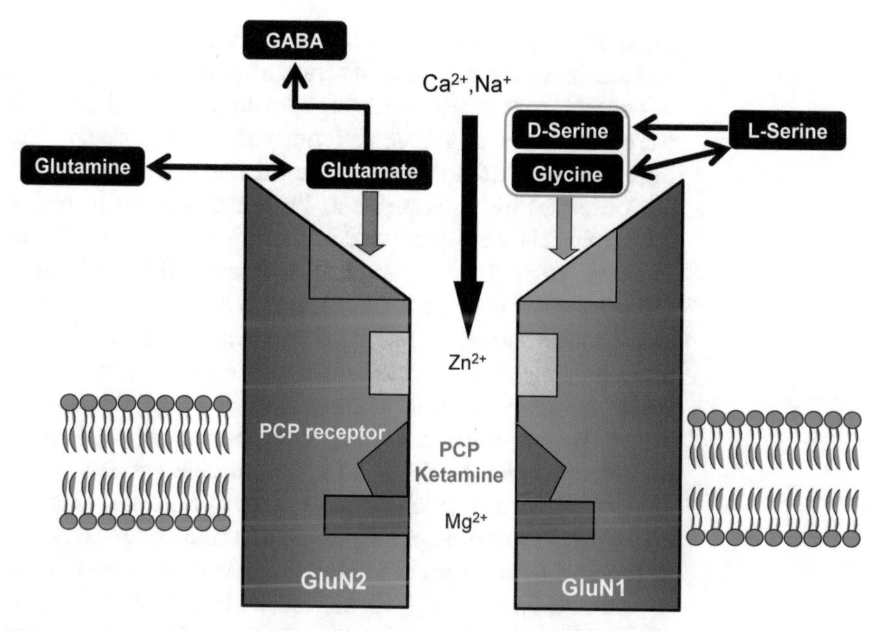

Fig. 2 NMDA receptors in the brain. Phencyclidine (PCP) and ketamine are ion-channel blockers of the NMDA receptor. Glycine and D-Ser are endogenous co-agonists of the Gly-modulatory site on the GluN1 subunit. Glutamate is an endogenous agonist at Glu sites on the GluN2 subunit. Glutamate is synthesized from Gln, and it is metabolized to GABA. Both D-Ser and Gly are synthesized from L-Ser. Thus, Glu-Gln-GABA cycle plays a role in the neurotransmission via the NMDAR

2 Methods to Study Glutamate and Its Related Amino Acids in Biological Samples

Glutamate binds to the Glu site of the GluN2 subunits in NMDAR complexes, while D-serine (D-Ser) and glycine (Gly) bind to the Gly-modulatory site of the GluN1 subunit in NMDAR complexes; these binding patterns may be involved in psychiatric diseases [5, 7, 17]. D-Ser and Gly are synthesized from L-Ser by serine racemase and serine hydroxymethyltransferase, respectively (Figs. 1 and 2). In this chapter, we would like to discuss the glutamatergic system as clinical biomarkers in blood and CSF samples from humans because abnormalities in the Gln-Glu-GABA cycle may be involved in the pathophysiology of a variety of psychiatric and neurological disorders.

3 Materials

3.1 Blood Levels of Glutamate and Related Amino Acids in Schizophrenia

Macciardi et al. [18] reported that blood levels of Glu, Gly, and Ser in patients with schizophrenia were higher than those of control subjects. In contrast, plasma levels of Glu in patients with first-episode schizophrenia were lower than those of control subjects [19].

Furthermore, Tomiya et al. [20] reported that serum levels of Glu and Ser in male patients with schizophrenia were higher than those of control male subjects although serum levels of these amino acids were not altered in female patients with schizophrenia. These findings suggest gender differences in the alterations in serum Glu and Ser observed in schizophrenia. Plasma Glu levels in patients with schizophrenia were increased with antipsychotic medication [21]. Patients treated with atypical antipsychotics had significantly increased plasma Glu concentrations in the "remission stage" compared to the "acute stage." Patients treated with conventional antipsychotics also had increased Glu concentrations in the "remission stage" compared to the "acute stage"; however, this difference was not statistically significant [21]. Fukushima et al. [22] also reported higher serum levels of Glu in schizophrenia patients compared to controls. Recently, Nishimura et al. [23] reported that the differences in mismatch negativity latency in monozygotic twins may be influenced by changes in plasma Glu levels and Glu-Gln cycling.

In addition to Glu, D-Ser, an endogenous co-agonist at the NMDAR, is believed to play a role in the pathophysiology of schizophrenia. Previously, we reported that serum levels of D-Ser in patients with schizophrenia were lower than those of control subjects although serum levels of L-Ser in patients with schizophrenia were higher than those of control subjects [24, 25]. These findings support a hypofunction of NMDAR in schizophrenia. Subsequently, our findings have been replicated by other groups [22, 26], and a recent meta-analysis showed lower levels of D-Ser in schizophrenia [27]. In addition, another study demonstrated that D-Ser/L-Ser and Gly/L-Ser ratios could be used as biomarkers of the therapeutic efficacy of the atypical antipsychotic drug clozapine or the clinical state in treatment-resistant schizophrenia [28].

Sumiyoshi et al. [29] reported that plasma levels of DL-Ser were increased in drug-free patients with schizophrenia compared to normal subjects, which is consistent with our reports of DL-Ser [24, 25]. Glycine is also an endogenous co-agonist at the Gly-modulatory site of the NMDAR (Fig. 2). Plasma Gly levels and Gly/DL-Ser ratios were decreased in patients with schizophrenia relative to control subjects [29]. Taken together, it is likely that abnormalities in the NMDAR-related amino acids (e.g., Glu, Gln, D-Ser, L-Ser, and Gly) play a role in the pathophysiology of schizophrenia [5, 7] (Table 1).

3.2 Blood Levels of Glutamate and Related Amino Acids in Mood Disorders

Kim et al. [39] reported that serum levels of Glu in patients with major depressive disorder (MDD) are significantly higher than those of healthy controls. Furthermore, Altamura et al. [40] reported that plasma levels of Glu in patients with mood disorders are significantly higher than those of the control group. Increased levels of Glu in patients with MDD have been confirmed by other

Table 1
Alterations in glutamate and related amino acids in schizophrenia

First author [Ref]	Source	Increase	Decrease	No change
Macciardi [18]	Plasma	Glu, Gly, DL-Ser		
Palomino [19]	Plasma		Glu	
Calcia [26]	Plasma		D-Ser, D-Ser/DL-Ser ratio	DL-Ser, L-Ser
Sumiyoshi [29]	Plasma	DL-Ser		
Tomiya [20]	Serum	Glu, DL-Ser (male), Pro (female)		Glu, DL-Ser (female)
Fukushima [22]	Serum	Glu, L-Ser	D-Ser	
Hashimoto [24]	Serum	DL-Ser, L-Ser	D-Ser	
Yamada [25]	Serum	DL-Ser, L-Ser	D-Ser	
Kim [30]	CSF		Glu	
Kim [31]	CSF		Glu	
Perry [32]	CSF			Glu
Do [33]	CSF	Iso-Leucine	Taurine	Glu
Tsai [34]	CSF			Glu, Gln, Asp,Gly, NAA, NAAG
Hashimoto [35]	CSF	Gln/Glu ratio		Glu, Gln
Hashimoto [36]	CSF		D-Ser/DL-Ser ratio	D-Ser, L-Ser
Bendikov [37]	CSF		D-Ser, D-Ser/L-Ser ratio	Glu, Gln, L-Ser
Fuchs [38]	CSF			D-Ser, L-Ser, Gly

Abbreviations: *Glu* Glutamate, *Gln* Glutamine, *Gly* Glycine, *DL-Ser* DL-Serine, *L-Ser* L-Serine, *D-Ser* D-Serine, *NAA* *N*-acetyl-L-aspartate, *NAAG* *N*-acetylaspartylglutamate

groups [41, 42]. There is also a positive correlation between plasma Glu levels and the severity of depressive symptoms in MDD patients [42].

Glutamate is stored in platelet dense granules and large amounts (>400 μmol/L) are released during thrombus formation [43–45]. Although the NMDAR has been shown in platelets, its precise roles are unclear [46]. A recent study demonstrated that human platelets contain four subunits, namely, GluN1, GluN2A, GluN2D, and GluN3A [45]. Furthermore, the NMDAR in platelets plays a role in a process that involves activation-dependent receptor relocation toward the platelet surface [45]. Interestingly, the MDD group showed a significantly greater platelet intracellular calcium response to Glu stimulation than the control group [47], suggesting that platelets may be a possible peripheral marker of Glu function in MDD.

Patients with MDD have increased plasma levels of DL-Ser and decreased Gly/Ser ratios compared to control subjects [29]. Subsequently, we reported that serum levels of D-Ser and L-Ser in MDD patients were higher than those of healthy controls, and that the ratio of L-Ser to Gly in MDD patients was significantly higher than that of healthy controls [48]. These findings suggest that Ser enantiomers may be peripheral biomarkers for depression, and that an abnormality in the D-Ser-L-Ser-Gly cycle may play a role in the pathophysiology of depression [48].

Serum levels of Gln, Gly, and D-Ser were significantly higher, whereas L-Ser levels were lower, in patients with bipolar disorder (BD) as compared with controls [49]. A study using metabolomics analysis showed abnormalities in amino acid metabolisms in BD [50]. Previously, we reported that serum levels of D-Ser in patients with schizophrenia were lower than those of control subjects, and that serum levels of L-Ser in patients were higher than those of control subjects [24, 25]. Taken together, it is likely that Ser enantiomers may be diagnostic biomarkers for schizophrenia and BD [51] (Table 2).

3.3 CSF Levels of Glutamate and Related Amino Acids in Schizophrenia

Because the findings of decreased CSF levels of Glu in patients with schizophrenia [30, 31], the Glu hypothesis of schizophrenia has gained impact in neurobiological research [5, 7, 59–64]. However, the findings of subsequent studies have been inconsistent, with several reports of no alterations in the CSF levels of Glu [32–35]. Interestingly, we reported an increased ratio of Gln to Glu in the CSF of first-episode, drug-naïve patients with schizophrenia although the CSF levels of these amino acids were unaltered [35]. This study suggests that abnormalities in the Gln-Glu cycle play a role in the pathophysiology of schizophrenia [7, 35].

In addition, we reported reduced ratios of D-Ser to DL-Ser in the CSF from first-episode drug-naïve patients with schizophrenia, suggesting that abnormalities in the synthetic or metabolic pathways of D-Ser may play a role in the pathophysiology of schizophrenia [36]. A subsequent study replicated the findings of decreased CSF levels of D-Ser in schizophrenia [37] although there has also been a report of no changes in CSF D-Ser levels in schizophrenia [38]. Collectively, it is likely that decreased D-Ser in the brain may contribute to hypofunction of the NMDAR in schizophrenia because D-Ser is effective at relieving several symptoms of schizophrenia [65, 66] (Table 1).

3.4 CSF Levels of Glutamate and Related Amino Acids in Mood Disorders

Accumulating evidence suggests that abnormalities in glutamatergic neurotransmission via the NMDAR play a key role in the pathophysiology of mood disorders, including MDD and BD [2–5, 9, 67–70]. We reported increased levels of Glu in the prefrontal cortex in postmortem brain samples from MDD and BD patients, suggesting a role of the glutamatergic system in mood disorders [71].

Table 2
Alterations in glutamate and related amino acids in mood disorders

First author [Ref]	Source	Increase	Decrease	No change
Mauri [41]	Platelets	Asp, DL-Ser, Lysine (MDD)		
	Plasma	Glu, Taurine, Lysine (MDD)	Tryptophan (MDD)	
Altamura [40]	Plasma	Glu (MDD)		
Mitani [42]	Plasma	Glu, Gln, Gly, Taurine (MDD)		Asp, Ala, L-Ser, Arg, D-Ser (MDD)
Sumiyoshi [29]	Plasma	DL-Ser (MDD)	Gly/L-Ser ratio (MDD)	
Hashimoto [48]	Serum	D-Ser, L-Ser (MDD)	Gly/L-Ser ratio (MDD)	Glu, Gln, Gly (MDD)
Kim [39]	Serum	Glu (MDD)		
Pålsson [49]	Serum	Gln, Gly, D-Ser (BD)	L-Ser (BD)	
	CSF			Glu, Gln, Gly, D-Ser, L-Ser (MDD)
Levin [52]	CSF	Glu (MDD)		
Frye [53]	CSF		Glu (MDD)	
Garakani [54]	CSF			Glu, Gln (MDD)
Hashimoto [55]	CSF	Gln/Glu ratio (MDD)		Glu, Gln, Gly, D-Ser, L-Ser (MDD)
Gerner [56]	CSF		GABA (MDD)	
Vieira [57]	CSF		GABA (MDD)	
Mann [58]	CSF		GABA (MDD)	

Abbreviations: *Ala* Alanine, *Asp* Aspartate, *Arg* Arginine, *Glu* Glutamate, *Gln* Glutamine, *Gly* Glycine, *DL-Ser* DL-Serine, *L-Ser* L-Serine, *D-Ser* D-Serine, *MDD* Major depressive disorder, *BD* Bipolar disorder

A noninvasive in vivo proton magnetic resonance spectroscopy (^{1}H-MRS) study revealed increased levels of Glu in the occipital cortex of patients with MDD [72] although other ^{1}H-MRS studies found decreased levels of Glx (Glu and Gln) in the anterior cingulate cortex [73] and dorsomedial/dorsal anterolateral prefrontal cortex [74] in patients with MDD. In contrast, there were no differences in Glu levels in the occipital cortex (or hippocampus) between MDD patients and controls [75, 76]. A recent meta-analysis showed a significant reduction in Glx levels, but not Glu levels alone, in the brains of depressed patients compared with controls across all studies included [77]. Furthermore, a recent study demonstrated reduced levels of Glu in the medial prefrontal cortex in drug-naïve MDD patients compared with controls, and a correlation between Glu levels in this region and executive function [78].

Levin et al. [52] reported higher CSF levels of Gln in unmedicated patients with MDD. In contrast, a study showed low Glu levels in the CSF of refractory patients with affective disorder [53]. A recent study showed no significant differences between the control group and the MDD patient group in baseline CSF levels of Glu and Gln [54]. Thus, the overall results on CSF levels of Gln and Glu in MDD patients are inconsistent. Recently, we reported that the ratio of Gln to Glu was significantly higher at baseline in elderly patients with MDD than in controls although CSF levels of amino acids did not differ across MDD patient and control groups. The ratio was shown to decrease in individuals with MDD over the 3-year follow-up, and this decrease correlated with a decrease in severity of depression [55].

Depressed patients, particularly those over 40 years of age, had lower CSF levels of GABA than the control group [56]. Furthermore, CSF levels of GABA were lower in MDD patients compared with BD patients and control subjects [57, 58]. In contrast, no differences between the patient and control group in CSF levels of amino acids (Glu, Gln, Gly, D-Ser, L-Ser) were observed [49]. These findings suggest that decreased CSF levels of GABA may play a role in the pathophysiology of depression (Table 2).

Taken together, abnormalities in the Gln-Glu-GABA cycle in glia–neuron communication may play a role in the pathophysiology of depression.

3.5 CSF Levels of Glutamate and Related Amino Acids in Other Disorders

Kim et al. [30] reported decreased CSF levels of Glu in Huntington's disease. Furthermore, Jeon et al. [79] reported higher CSF levels of Gln and lower CSF levels of Glu in patients with unilateral moyamoya disease, suggesting that elevated CSF levels of Gln may be associated with pathogenesis of moyamoya disease. Kashiwagi et al. [80] reported that CSF levels of Glu, Gly, and taurine in the encephalopathy group were significantly higher than those in the cluster group. Patients with calcified cysts who did not present seizures revealed significantly elevated CSF levels of Gln and GABA [81]. A recent meta-analysis showed that CSF levels of Glu were higher in subjects with migraines [82], suggesting CSF Glu as a potential migraine biomarker.

CSF levels of D-Ser were higher in probable Alzheimer's disease (AD) patients than in noncognitively impaired subject groups, suggesting that increased CSF D-Ser levels are associated with AD [83]. However, a recent study showed that CSF levels of D-Ser in AD patients did not differ from other dementias such as Lewy bodies and frontotemporal dementia [84].

Hepatic encephalopathy encompasses a spectrum of neurological and neuropsychiatric abnormalities observed in patients with liver disease and/or portosystemic shunt, after exclusion of other

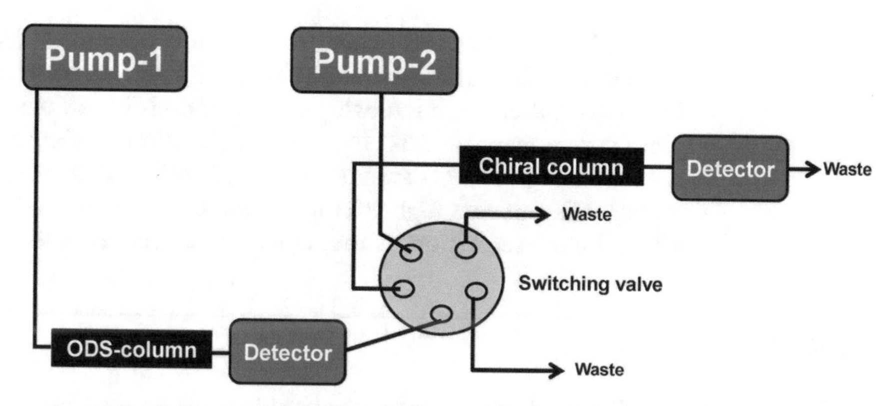

Fig. 3 Diagram of column-switching HPLC system. A column-switching HPLC system consisting of an octadecylsilica (ODS) column and a tandem series of two chiral columns, Sumichiral OA-2500 (*S*), was used for measurement of the fluorescence derivatives, NBD-D-Ser and NBD-L-Ser in biological samples from humans

known brain diseases. A metabolomics analysis of CSF sample showed that CSF levels of Glu and Gln were altered in patients with hepatic encephalopathy [85].

Furthermore, CSF metabolomics analysis demonstrated increased CSF levels of Glu, *N*-acetylaspartate, and *N*-acetylaspartylglutamic acid in Human Immunodeficiency Virus (HIV) patients on antiretroviral therapy compared with HIV-negative controls [86], suggesting a role of the glutamatergic system in HIV-associated cognitive impairments.

3.6 Methodologies for Measuring Glutamate and Its Related Amino Acids

Glutamate and its related amino acids (e.g., Gln, Gly, and Ser) in human samples are measured using high-performance liquid chromatography (HPLC) after a fluorescent derivation of NBD-F (4-fluoro-7-nitro-2,1,3-benzoxadiazole) [87], and LC with tandem mass spectrometry (LC/MS/MS) [88].

LC methods for D-Ser determinations are divided into two types: chiral derivatization with *o*-phthalaldehyde in the presence of chiral thiols such as cysteine analogs, followed by HPLC separation on a non-chiral stationary phase, and non-chiral derivatization with fluorescent reagents followed by HPLC separation with chiral columns [89, 90]. The latter method was improved by using a column-switching system [91] (Fig. 3) and has been used to detect D-Ser and L-Ser in human samples (serum and CSF) [24, 25, 28, 48, 49, 55]. Due to the close retention times of L-serine, L-Gln, and D-Ser, the LC method may not be suitable for the quantification of low levels of endogenous D-Ser in human blood. In contrast, a column-switching HPLC with fluorimetric detection and chiral LC/MS/MS have been used for determination of low D-Ser (or other D-amino acid) concentrations in biological samples [91]. Although electrophoretic and enzymatic methods to measure D-amino acids in mammalian tissues have been also reported [90, 92], these methods are not popular. Taken together, it is likely that column-switching HPLC with fluorimetric detection and chiral LC/MS/MS would be useful for the determination of D-Ser (or other D-amino acid) concentrations in biological samples from human subjects. In addition, the storage of samples in a deep freezer (−80 °C) are important for amino acids because the storage of samples at higher temperature (e.g., −40 or −20 °C) can affect the concentration of amino acids in human samples.

4 Conclusions

As discussed above, glutamatergic neurotransmission via the Gln-Glu-GABA cycle plays a key role in the pathophysiology of a variety of psychiatric and neurological disorders, as well as other disorders in the periphery. Therefore, it seems that measurement of these amino acids in blood and CSF samples from these patients is of great interest.

Anti-NMDAR encephalitis is a recently discovered synaptic autoimmune disorder in which auto-antibodies target NMDARs in the brain [93]. Accumulating evidence suggests that anti-NMDAR encephalitis may be involved in several psychiatric and neurological disorders [92–96]. Measurement of NMDAR-related amino acids (e.g., Glu, Gln, D-Ser, L-Ser, Gly, and GABA) in blood and CSF samples is necessary to further understand this disorder because anti-NMDARs were detected in both samples.

Beyond the brain, NMDARs are also expressed in the peripheral tissues, including kidney, heart, pancreas, lung, lymphocytes, skin, and osteoblasts [97]. In addition, mGluRs have a widespread distribution outside CNS, suggesting diverse roles of mGluRs in a variety of processes in both health and disease [98]. Therefore, the measurement of Glu and its related amino acids in blood and CSF samples from patients with a variety of peripheral disorders is also of interest.

In summary, Glu is the most abundant excitatory amino acid in the body, and the Gln-Glu-GABA cycle plays a key role in glutamatergic neurotransmission in CNS and peripheral tissues. Therefore, the measurements of Glu and related amino acids may be important for use as potential biomarkers of a variety of disorders.

Acknowledgements

This study was supported by a Grant-in-Aid for Scientific Research on Innovative Areas of the Ministry of Education, Culture, Sports, Science and Technology, Japan, and Strategic Research Program for Brain Sciences, from Japan Agency for Medical Research and Development, AMED.

References

1. Kvamme E (1998) Synthesis of glutamate and its regulation. Prog Brain Res 116:73–85
2. Hashimoto K (2009) Emerging role of glutamate in the pathophysiology of major depressive disorder. Brain Res Rev 61:105–123
3. Hashimoto K (2011) The role of glutamate on the action of antidepressants. Prog Neuro-Psychopharmacol Biol Psychiatry 35: 1558–1568
4. Tokita K, Yamaji T, Hashimoto K (2012) Roles of glutamate signaling in preclinical and/or mechanistic models of depression. Pharmacol Biochem Behav 100:688–704
5. Hashimoto K, Malchow B, Falkai P, Schmitt A (2013) Glutamate modulators as potential therapeutic drugs in schizophrenia and affective disorders. Eur Arch Psychiatry Clin Neurosci 263:367–377
6. Hertz L (2013) The glutamate-glutamine (GABA) cycle: importance of late postnatal development and potential reciprocal interactions between biosynthesis and degradation. Front Endocrinol (Lausanne) 4:59
7. Hashimoto K (2014) Targeting of NMDA receptors in new treatments for schizophrenia. Expert Opin Ther Targets 18:1049–1063
8. Hashimoto K (2014) Abnormalities of the glutamine-glutamate-GABA cycle in the schizophrenia brain. Schizophr Res 156:281–282
9. Ohgi Y, Futamura T, Hashimoto K (2015) Glutamate signaling in synaptogenesis and NMDA receptors as potential therapeutic targets for psychiatric disorders. Curr Mol Med 15:206–221
10. Machado-Vieira R, Henter ID, Zarate CA Jr (2015) New targets for rapid antidepressant action. Prog Neurobiol 2015 Dec 23. pii:S0301-0082(15)30038-1. doi:10.1016/j.pneurobio.2015.12.001
11. Ribeiro FM, Vieira LB, Pires RG, Olmo RP, Ferguson SS (2016) Metabotropic glutamate receptors and neurodegenerative diseases. Pharmacol Res 115:179–191
12. Dore K, Aow J, Malinow R (2016) The emergence of NMDA receptor metabotropic function: insights from imaging. Front Synaptic Neurosci 8:20
13. Alfredsson G, Wiesel FA, Tylec A (1988) Relationships between glutamate and monoamine metabolites in cerebrospinal fluid and serum in healthy volunteers. Biol Psychiatry 23:689–697
14. O'Kane RL, Viña JR, Simpson I, Hawkins RA (2004) Na^{+}-dependent neutral amino acid transporters A, ASC, and N of the blood-brain barrier: mechanisms for neutral amino acid removal. Am J Physiol Endocrinol Metab 287:E622–E629
15. Hawkins RA, O'Kane RL, Simpson IA, Viña JR (2006) Structure of the blood-brain barrier and its role in the transport of amino acids. J Nutr 136(1 Suppl):218S–226S
16. Hawkins RA, Viña JR (2016) How glutamate is managed by the blood-brain barrier. Biology (Basel) 5:E37
17. Fujita Y, Ishima T, Hashimoto K (2016) Supplementation with D-serine prevents the onset of cognitive deficits in adult offspring after maternal immune activation. Sci Rep 6:37261
18. Macciardi F, Lucca A, Catalano M, Marino C, Zanardi R, Smeraldi E (1990) Amino acids patterns in schizophrenia: some new findings. Psychiatry Res 32:63–70
19. Palomino A, González-Pinto A, Aldama A, González-Gómez C, Mosquera F, González-García G, Matute C (2007) Decreased levels of plasma glutamate in patients with first-episode schizophrenia and bipolar disorder. Schizophr Res 95:174–178
20. Tomiya M, Fukushima T, Watanabe H, Fukami G, Fujisaki M, Iyo M, Hashimoto K, Mitsuhashi S, Toyo'oka T (2007) Alterations in serum amino acid concentrations in male and female schizophrenic patients. Clin Chim Acta 380:186–190
21. Maeshima H, Ohnuma T, Sakai Y, Shibata N, Baba H, Ihara H, Higashi M, Ohkubo T, Nozawa E, Abe S, Ichikawa A, Nakano Y, Utsumi Y, Suzuki T, Arai H (2007) Increased plasma glutamate by antipsychotic medication and its relationship to glutaminase 1 and 2 genotypes in schizophrenia – Juntendo University Schizophrenia Projects (JUSP). Prog Neuro-Psychopharmacol Biol Psychiatry 231:1410–1418
22. Fukushima T, Iizuka H, Yokota A, Suzuki T, Ohno C, Kono Y, Nishikiori M, Seki A, Ichiba H, Watanabe Y, Hongo S, Utsunomiya M, Nakatani M, Sadamoto K, Yoshio T (2014) Quantitative analyses of schizophrenia-associated metabolites in serum: serum D-lactate levels are negatively correlated with gamma-glutamylcysteine in medicated schizophrenia patients. PLoS One 9:e101652
23. Nishimura Y, Kawakubo Y, Suga M, Hashimoto K, Takei Y, Takei K, Inoue H, Yumoto M, Takizawa R, Kasai K (2016) Familial influences on mismatch negativity and its association with plasma glutamate level: a magnetoencephalographic study in twins. Mol Neuropsychiatry 2:161–172

24. Hashimoto K, Fukushima T, Shimizu E, Komatsu N, Watanabe H, Shinoda N, Nakazato M, Kumakiri C, Okada S, Hasegawa H, Imai K, Iyo M (2003) Decreased serum levels of D-serine in patients with schizophrenia: evidence in support of the N-methyl-D-aspartate receptor hypofunction hypothesis of schizophrenia. Arch Gen Psychiatry 60: 572–576
25. Yamada K, Ohnishi T, Hashimoto K, Ohba H, Iwayama-Shigeno Y, Toyoshima M, Okuno A, Takao H, Toyota T, Minabe Y, Nakamura K, Shimizu E, Itokawa M, Mori N, Iyo M, Yoshikawa T (2005) Identification of multiple serine racemase (SRR) mRNA isoforms and genetic analyses of SRR and DAO in schizophrenia and D-serine levels. Biol Psychiatry 57:1493–1503
26. Calcia MA, Madeira C, Alheira FV, Silva TC, Tannos FM, Vargas-Lopes C, Goldenstein N, Brasil MA, Ferreira ST, Panizzutti R (2012) Plasma levels of D-serine in Brazilian individuals with schizophrenia. Schizophr Res 142: 83–87
27. Cho SE, Na KS, Cho SJ, Kang SG (2016) Low D-serine levels in schizophrenia: a systemic review and meta-analysis. Neurosci Lett 634: 42–51
28. Yamamori H, Hashimoto R, Fujita Y, Numata S, Yasuda Y, Fujimoto M, Ohi K, Umeda-Yano S, Ito A, Ohmori T, Hashimoto KM (2014) Changes in plasma D-serine, L-serine, and glycine levels in treatment-resistant schizophrenia before and after clozapine treatment. Neurosci Lett 582:93–98
29. Sumiyoshi T, Anil AE, Jin D, Jayathilake K, Lee M, Meltzer HY (2004) Plasma glycine and serine levels in schizophrenia compared to normal controls and major depression: relation to negative symptoms. Int J Neuropsychopharmacol 7:1–8
30. Kim JS, Kornhuber HH, Holzmüller B, Schmid-Burgk W, Mergner T, Krzepinski G (1980) Reduction of cerebrospinal fluid glutamic acid in Huntigton's chorea and in schizophrenic patients. Arch Psychiatry Nervenkr 228:7–10
31. Kim JS, Kornhuber HH, Schmid-Burgk W, Holzmuller B (1980) Low cerebrospinal fluid glutamate in schizophrenic patients and a new hypothesis on schizophrenia. Neurosci Lett 20:379–382
32. Perry TL (1982) Normal cerebrospinal fluid and brain glutamate levels in schizophrenia do not support the hypothesis of glutamatergic neuronal dysfunction. Neurosci Lett 28: 81–85
33. Do KQ, Lauer CJ, Schreiber W, Zollinger M, Gutteck-Amsler U, Cuenod M, Holsboer F (1995) γ-Glutamylglutamine and taurine concentrations are decreased in the cerebrospinal fluid of drug-naive patients with schizophrenic disorders. J Neurochem 65:2652–2662
34. Tsai G, van Kammen DP, Chen S, Kelley ME, Grier A, Coyle JT (1998) Glutamatergic neurotransmission involves structural and clinical deficits of schizophrenia. Biol Psychiatry 44:667–674
35. Hashimoto K, Engberg G, Shimizu E, Nordin C, Lindström LH, Iyo M (2005) Elevated glutamine/glutamate ratio in cerebrospinal fluid of first episode and drug naive schizophrenic patients. BMC Psychiatry 5:6
36. Hashimoto K, Engberg G, Shimizu E, Nordin C, Lindström LH, Iyo M (2005) Reduced D-serine to total serine ratio in the cerebrospinal fluid of drug naive schizophrenic patients. Prog Neuro-Psychopharmacol Biol Psychiatry 29:767–769
37. Bendikov I, Nadri C, Amar S, Panizzutti R, De Miranda J, Wolosker H, Agam G (2007) A CSF and postmortem brain study of D-serine metabolic parameters in schizophrenia. Schizophr Res 90:41–51
38. Fuchs SA, De Barse MM, Scheepers FE, Cahn W, Dorland L, de Sain-van d, Velden MG, Klomp LW, Berger R, Kahn RS, de Koning TJ (2008) Cerebrospinal fluid D-serine and glycine concentrations are unaltered and unaffected by olanzapine therapy in male schizophrenic patients. Eur Neuropsychopharmacol 18:333–338
39. Kim JS, Schmid-Burgk W, Claus D, Kornhuber HH (1982) Increased serum glutamate in depressed patients. Arch Psychiatr Nervenkr 232:299–304
40. Altamura CA, Mauri MC, Ferrara A, Moro AR, D'Andrea G, Zamberian F (1993) Plasma and platelet excitatory amino acids in psychiatric disorders. Am J Psychiatry 150: 1731–1713
41. Mauri MC, Ferrara A, Boscati L, Bravin S, Zamberlan F, Alecci M, Invernizzi G (1998) Plasma and platelet amino acid concentrations in patients affected by major depression and under fluvoxamine treatment. Neuropsychobiology 37:124–129
42. Mitani H, Shirayama Y, Yamada T, Maeda K, Ashby CR Jr, Kawahara R (2006) Correlation between plasma levels of glutamate, alanine and serine with severity of depression. Prog Neuro-Psychopharmacol Biol Psychiatry 30: 1155–1158

43. Begni B, Tremolizzo L, D'Orlando C, Bono MS, Garofolo R, Longoni M, Ferrarese C (2005) Substrate-induced modulation of glutamate uptake in human platelets. Br J Pharmacol 145:792–799
44. Hoogland G, Bos IW, Kupper F, van Willigen G, Spierenburg HA, van Nieuwenhuizen O, de Graan PN (2005) Thrombin-stimulated glutamate uptake in human platelets is predominantly mediated by the glial glutamate transporter EAAT2. Neurochem Int 47: 499–506
45. Kalev-Zylinska ML, Green TN, Morel-Kopp MC, Sun PP, Park YE, Lasham A, During MJ, Ward CM (2014) *N*-Methyl-D-aspartate receptors amplify activation and aggregation of human platelets. Thromb Res 133:837–847
46. Hitchcock IS, Skerry TM, Howard MR, Genever PG (2003) NMDA receptor-mediated regulation of human megakaryocytopoiesis. Blood 102:1254–1259
47. Berk M, Helene P, Ferreira D (2001) Platelet glutamate receptor supersensitivity in major depressive disorder. Clin Neuropharmacol 24: 129–132
48. Hashimoto K, Yoshida T, Ishikawa M, Fujita Y, Niitsu T, Nakazato M, Watanabe H, Sasaki T, Shiina A, Hashimoto T, Kanahara N, Hasegawa T, Enohara M, Kimura A, Iyo M (2016) Increased serum levels of serine enantiomers in patients with schizophrenia. Acta Neuropsychiatr 28:173–178
49. Pålsson E, Jakobsson J, Södersten K, Fujita Y, Sellgren C, Ekman CJ, Ågren H, Hashimoto KM (2015) Markers of glutamate signaling in cerebrospinal fluid and serum from patients with bipolar disorder and healthy controls. Eur Neuropsychopharmacol 25:133–140
50. Yoshimi N, Futamura T, Kakumoto K, Salehi AM, Sellgren CM, Holmén-Larsson J, Jakobsson J, Pålsson E, Landén M, Hashimoto K (2016) Blood metabolomics analysis identifies abnormalities in the citric acid cycle, urea cycle, and amino acid metabolism in bipolar disorder. BBA Clin 5:151–158
51. Hashimoto K (2016) Serine enantiomers as diagnostic biomarkers for schizophrenia and bipolar disorder. Eur Arch Psychiatry Clin Neurosci 266:83–85
52. Levine J, Panchalingam K, Rapoport A, Gershon S, McClure RJ, Pettegrew JW (2000) Increased cerebrospinal fluid glutamine levels in depressed patients. Biol Psychiatry 47: 586–593
53. Frye MA, Tsai GE, Huggins T, Coyle JT, Post RM (2007) Low cerebrospinal fluid glutamate and glycine in refractory affective disorder. Biol Psychiatry 61:162–166
54. Garakani A, Martinez JM, Yehuda R, Gorman JM (2013) Cerebrospinal fluid levels of glutamate and corticotropin releasing hormone in major depression before and after treatment. J Affect Disord 146:262–265
55. Hashimoto K, Bruno D, Nierenberg J, Marmar CR, Zetterberg H, Blennow K, Pomara N (2016) Abnormality in glutamine-glutamate cycle in the cerebrospinal fluid of cognitively intact elderly individuals with major depressive disorder: a 3-year follow-up study. Transl Psychiatry 6:e744
56. Gerner RH, Fairbanks L, Anderson GM, Young JG, Scheinin M, Linnoila M, Hare TA, Shaywitz BA, Cohen DJ (1984) CSF neurochemistry in depressed, manic, and schizophrenic patients compared with that of normal controls. Am J Psychiatry 141:1533–1540
57. Vieira DS, Naffah-Mazacoratti MG, Zukerman E, Senne Soares CA, Alonso EO, Faulhaber MH, Cavalheiro EA, Peres MF (2006) Cerebrospinal fluid GABA levels in chronic migraine with and without depression. Brain Res 1090:197–201
58. Mann JJ, Oquendo MA, Watson KT, Boldrini M, Malone KM, Ellis SP, Sullivan G, Cooper TB, Xie S, Currier D (2014) Anxiety in major depression and cerebrospinal fluid free gamma-aminobutyric acid. Depress Anxiety 31: 814–821
59. Javitt DC, Zukin SR (1991) Recent advances in the phencyclidine model of schizophrenia. Am J Psychiatry 148:1301–1308
60. Olney JW, Farber NB (1995) Glutamate receptor dysfunction and schizophrenia. Arch Gen Psychiatry 52:998–1007
61. Coyle JT (1996) The glutamatergic dysfunction hypothesis for schizophrenia. Harv Rev Psychiatry 3:241–253
62. Krystal JH, D'Souza DC, Petrakis IL, Belger A, Berman RM, Charney DS, Abi-Saab W, Madonick S (1999) NMDA agonists and antagonists as probes of glutamatergic dysfunction and pharmacotherapies in neuropsychiatric disorders. Harv Rev Psychiatry 7: 125–143
63. Hashimoto K, Okamura N, Shimizu E, Iyo M (2004) Glutamate hypothesis of schizophrenia and approach for possible therapeutic drugs. Curr Med Chem CNS Agents 4:147–154
64. Hashimoto K, Shimizu E, Iyo M (2005) Dysfunction of glia-neuron communication in pathophysiology of schizophrenia. Curr Psychiatr Rev 1:151–163

65. Tsai G, Lin PY (2010) Strategies to enhance N-methyl-D-aspartate receptor-mediated neurotransmission in schizophrenia, a critical review and meta-analysis. Curr Pharm Des 16:522–537
66. Singh SP, Singh V (2011) Meta-analysis of the efficacy of adjunctive NMDA receptor modulators in chronic schizophrenia. CNS Drugs 25:859–885
67. Sanacora G, Zarate CA, Krystal JH, Manji HK (2008) Targeting the glutamatergic system to develop novel, improved therapeutics for mood disorders. Nat Rev Drug Discov 7: 426–437
68. Krystal JH, Sanacorra G, Duman RS (2013) Rapid-acting glutamatergic antidepressants: the path to ketamine and beyond. Biol Psychiatry 73:1133–1141
69. Dang YH, Ma XC, Zhang JC, Ren Q, Wu J, Gao CG, Hashimoto K (2014) Targeting of NMDA receptors in the treatment of major depression. Curr Pharm Des 20:5151–5159
70. Monteggia LM, Zarate C Jr (2015) Antidepressant actions of ketamine: from molecular mechanisms to clinical practice. Curr Opin Neurobiol 30:13–143
71. Hashimoto K, Sawa A, Iyo M (2007) Increased levels of glutamate in brains from patients with mood disorders. Biol Psychiatry 62: 1310–1316
72. Sanacora G, Gueorguieva R, Epperson CN, Wu YT, Appel M, Rothman DL, Krystal JH, Mason GF (2004) Subtype-specific alterations of gamma-aminobutyric acid and glutamate in patients with major depression. Arch Gen Psychiatry 61:705–713
73. Auer DP, Pütz B, Kraft E, Lipinski B, Schill J, Holsboer F (2000) Reduced glutamate in the anterior cingulate cortex in depression: an in vivo proton magnetic resonance spectroscopy study. Biol Psychiatry 47:305–313
74. Hasler G, van der Veen JW, Tumonis T, Meyers N, Shen J, Drevets WC (2007) Reduced prefrontal glutamate/glutamine and gamma-aminobutyric acid levels in major depression determined using proton magnetic resonance spectroscopy. Arch Gen Psychiatry 64:193–200
75. Godlewska BR, Near J, Cowen PJ (2015) Neurochemistry of major depression: a study using magnetic resonance spectroscopy. Psychopharmacology 232:501–507
76. Hermens DF, Naismith SL, Chitty KM, Lee RS, Tickell A, Duffy SL, Paquola C, White D, Hickie IB, Lagopoulos J (2015) Cluster analysis reveals abnormal hippocampal neurometabolic profiles in young people with mood disorders. Eur Neuropsychopharmacol 25: 836–845
77. Arnone D, Mumuni AN, Jauhar S, Condon B, Cavanagh J (2015) Indirect evidence of selective glial involvement in glutamate-based mechanisms of mood regulation in depression: meta-analysis of absolute prefrontal neurometabolic concentrations. Eur Neuropsychopharmacol 25:1109–1117
78. Shirayama Y, Takahashi M, Osone F, Hara A, Okubo T (2017) Myo-inositol, glutamate and glutamine in the prefrontal cortex, hippocampus and amygdala in major depression. Biol Psychiatry Cog Neurosci Neuroimaging 2: 196–204
79. Jeon JP, Yun T, Jin X, Cho WS, Son YJ, Bang JS, Kang HS, Oh CW, Kim JE, Park S (2015) ^{1}H-NMR-based metabolomic analysis of cerebrospinal fluid from adult bilateral moyamoya disease: comparison with unilateral moyamoya disease and atherosclerotic stenosis. Medicine (Baltimore) 94:e629
80. Kashiwagi Y, Kawashima H, Suzuki S, Nishimata S, Takekuma K, Hoshika A (2015) Marked elevation of excitatory amino acids in cerebrospinal fluid obtained from patients with rotavirus-associated encephalopathy. J Clin Lab Anal 29:328–333
81. Camargo JA, Bertolucci PH (2015) Quantification of amino acid neurotransmitters in cerebrospinal fluid of patients with neurocysticercosis. Open Neurol J 9:15–20
82. van Dongen RM, Zielman R, Noga M, Dekkers OM, Hankemeier T, van den Maagdenberg AM, Terwindt GM, Ferrari MD (2016) Migraine biomarkers in cerebrospinal fluid: a systematic review and meta-analysis. Cephalagia 37:49–63
83. Madeira C, Lourenco MV, Vargas-Lopes C, Suemoto CK, Brandão CO, Reis T, Leite RE, Laks J, Jacob-Filho W, Pasqualucci CA, Grinberg LT, Ferreira ST, Panizzutti R (2015) D-serine levels in Alzheimer's disease: implications for novel biomarker development. Transl Psychiatry 5:e561
84. Biemans EA, Verhoeven-Duif NM, Gerrits J, Claassen JA, Kuiperij HB, Verbeek MM (2016) CSF D-serine concentrations are similar in Alzheimer's disease, other dementias, and elderly controls. Neurobiol Aging 42:213–216
85. Weiss N, Barbier Saint Hilaire P, Colsch B, Isnard F, Attala S, Schaefer A, Amador MD, Rudler M, Lamari F, Sedel F, Thabut D, Junot C (2016) Cerebrospinal fluid metabolomics highlights dysregulation of energy metabolism in overt hepatic encephalopathy. J Hepatol 65:1120–1130
86. Cassol E, Misra V, Dutta A, Morgello S, Gabuzda D (2014) Cerebrospinal fluid metabolomics reveals altered waste clearance and accelerated aging in HIV patients with

neurocognitive impairment. AIDS 28: 1579–1591

87. Aoyama C, Santa T, Tsunoda M, Fukushima T, Kitada C, Imai K (2004) A fully automated amino acid analyzer using NBD-F as a fluorescent derivatization reagent. Biomed Chromatogr 18:630–636
88. Eckstein JA, Ammerman GM, Reveles JM, Ackermann BL (2008) Analysis of glutamine, glutamate, pyroglutamate, and GABA in cerebrospinal fluid using ion pairing HPLC with positive electrospray LC/MS/MS. J Neurosci Methods 171:190–196
89. Hamase K, Morikawa A, Zaitsu K (2002) D-Amino acids in mammals and their diagnostic value. J Chromatogr B Analyt Technol Biomed Life Sci 781:73–91
90. Hamase K, Konno R, Morikawa A, Zaitsu K (2006) Sensitive determination of D-amino acids in mammals and the effect of D-amino acid oxidase activity on their amounts. Biol Pharm Bull 28:1578–1584
91. Fukushima T, Kawai J, Imai K, Toyo'oka T (2004) Simultaneous determination of D- and L-serine in rat brain microdialysis sample using a column-switching HPLC with fluorimetric detection. Biomed Chromatogr 18:813–819
92. Koval D, Jiraskova J, Strisovsky, Konvalinka J, Kasicka V (2006) Capillary electrophoresis method for determination of D-serine and its application for monitoring of serine racemase activity. Electrophoresis 27:2558–2566
93. Kayser MS, Dalmau J (2016) Anti-NMDA receptor encephalitis, autoimmunity, and psychosis. Schizophr Res 176:36–40
94. Barry H, Byrne S, Barrett E, Murphy KC, Cotter DR (2015) Anti-N-methyl-D-aspartate receptor encephalitis: review of clinical presentation, diagnosis and treatment. BJPsych Bull 39:19–23
95. Kiani R, Lawden M, Eames P, Critchley P, Bhaumik S, Odedra S, Gumber R (2015) Anti-NMDA-receptor encephalitis presenting with catatonia and neuroleptic malignant syndrome in patients with intellectual disability and autism. BJPsych Bull 39:32–35
96. Dalmau J (2016) NMDA receptor encephalitis and other antibody-mediated disorders of the synapse: the 2016 Cotzias Lecture. Neurology 87:2471–2482
97. Hogan-Cann AD, Anderson CM (2016) Physiological roles of non-neuronal NMDA receptors. Trends Pharmacol Sci 37:750–767
98. Julio-Pieper M, Flor PJ, Dinan TG, Cryan JF (2011) Exciting times beyond the brain: metabotropic glutamate receptors in peripheral and non-neural tissues. Pharmacol Rev 63:35–58

Chapter 17

Clinical CNS Microdialysis of Glutamate with a Special Methodological Focus on Human Spinal Cord

Bernard Renaud, Luc Denoroy, Delphine Collin-Chavagnac, Patrick Mertens, and Sandrine Parrot

Abstract

Glutamate has a paramount role in central nervous system since it is the main excitatory neurotransmitter in both the brain and the spinal cord. In clinical studies, the monitoring of glutamate has revealed tight links between the variations of its interstitial concentration and those of other metabolic biomarkers and partial pressure in oxygen within tissue. The use of microdialysis has allowed to monitor glutamate over several days in patients with brain injury, epilepsy, or Parkinson's disease. We describe here a neurochemical method carried out in the spinal cord of patients suffering from chronic pain compared to patients suffering from spasticity, with a stress on peculiarities on such a monitoring. We provide also a detailed discussion on the significant usefulness of microdialysis and its technical and methodological limits in the specific field of clinical research.

Key words Glutamate, Central nervous system, Spinal cord, Brain, Microdialysis, Human, Patient

1 Introduction

1.1 Background on Glutamate in Brain Clinical Microdialysis

The brain microdialysis technique allows a continuous estimation of the extracellular concentration of glutamate (Glu) and has proven to be an efficient approach for the in vivo monitoring of Glu in laboratory animals [1]. The previous chapters of this book, devoted to microdialysis in preclinical studies (Chaps. 12–14) illustrate the high biochemical selectivity, associated with a satisfactory anatomical accuracy, offered by that approach in the study of the neurobiology and neuropharmacology of glutamatergic neurons.

Moreover, cerebral microdialysis can also be used in humans in specific circumstances needing a neurosurgical procedure for therapeutic, diagnostic, or prognostic purposes.

Indeed, for ethical reasons, a microdialysis probe can be implanted in the human central nervous system (CNS), only in areas which need to be resected or lesioned, or in which a probe for

Sandrine Parrot and Luc Denoroy (eds.), *Biochemical Approaches for Glutamatergic Neurotransmission*, Neuromethods, vol. 130, DOI 10.1007/978-1-4939-7228-9_17, © Springer Science+Business Media LLC 2018

intracranial pressure monitoring or an electrode for intracranial electroencephalography, or for deep brain stimulation will be placed. In these specific cases, microdialysis allows a direct and continuous monitoring of CNS Glu levels in the patients, who can be anesthetized or awake, depending on the clinical context. Interestingly, such a monitoring can be performed for a long period of time (up to several (i.e., 8–10) days), in contrast with animal studies in which brain microdialysis commonly lasts 1 or 2 days.

This chapter describes an original method for the microdialysis monitoring of spinal cord Glu in humans, performed during functional neurosurgery aiming to treat a chronic pharmacoresistant pain. This chapter also contains a short review focused on the use and on the advantages and limits of brain Glu microdialysis in humans.

1.2 Glutamate, Pain, and Spinal Cord

Nociceptive afferent fibers Aδ and C coming from periphery project to the spinal cord at the level of the dorsal horn (DH) (Fig. 1) where they release Glu and other transmitters, including γ-aminobutyric acid (GABA) and neuropeptides, within several layers called Rexed's laminae I–V of the dorsal horn of the spinal cord (identified in the early 1950s by Bror Rexed). The action of Glu on ionotropic and metabotropic receptors depends on the intensity of the nociceptive stimulus. *N*-Methyl-D-aspartate (NMDA) receptors are mainly activated when the stimulus is strong and lengthy, i.e., in case of prolonged activation of C fibers, but not in physiological conditions. The subsequent cellular activation will lead to long-lasting modifications of functional neuronal properties, through the activation of calcium-dependent kinases and NO synthesis, causing hyperexcitability. Notably, phosphorylation of membrane proteins, including NMDA receptors, induces an amplification of this activation processes. NMDA receptors are regarded as paramount in central hyperalgesia and spinal chronic pain [2]. Metabotropic receptors are also involved since some of them can amplify the effects of NMDA receptors [3]. Other messengers involved in pain mechanisms at the spinal level, including prostaglandins synthetized by cyclooxygenase 2, can also increase Glu release. Therefore, Glu in the spinal DH is strongly linked to nociception.

Analgesics can relieve chronic pain most of the time as they can cure acute pain. Although some compounds acting on Glu metabotropic receptors or on Glu transporters (see Chaps. 7 and 8) have proven to be analgesic in animals, their use in humans has been, up to now, limited to some clinical trials [4–7]. Thus, in some cases, when classical pharmacological analgesia is inefficient or becomes less and less efficient, alternative therapies are needed. In humans in this case, the DH of the spinal cord can be the target of such non-pharmacological therapies aiming at modulating the neurochemical signals from nociceptive afferences.

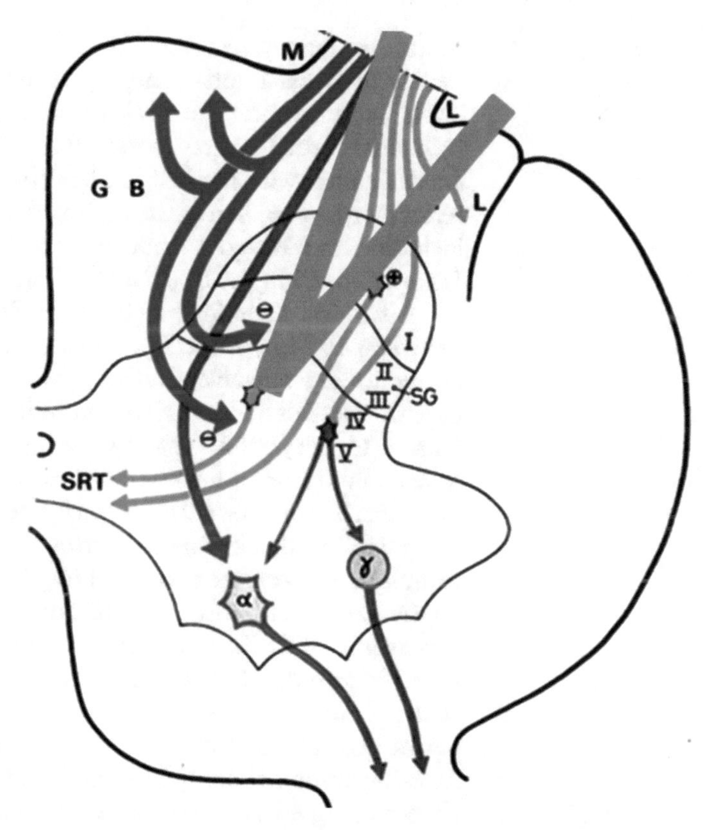

Fig. 1 Schematic representation of the afferent fibers in the dorsal root entry zone (DREZ) and of the Drezotomy lesion in the dorsal horn (DH) on a drawing of half spinal cord. The afferent fibers are penetrating the spinal cord from the dorsal root. They are mainly non-nociceptive (Aα and Aβ fibers in *blue*), myotatic (IA, IB fibers in *brown*), and nociceptive fibers (Aδ and C fibers in *orange*). The Drezotomy (*red arrow head*) aims at selectively interrupt the fibers in the ventrolateral part of the DREZ including the nociceptive and myotatic fibers and spare the non-nociceptive fibers going to the dorsal column and giving an inhibitory collateral branch to the nociceptive inputs. The DREZ lesion is performed 4-mm deep in the head of the dorsal horn (corresponding to Rexed's laminae I–V) to destroy, also if necessary, the neurons of the DH that are abnormally hyperactive in case of deafferentation pain. The microdialysis probe is inserted in the same location, in the Rexed's laminae I–V, before DREZ lesioning

Since the dorsal root entry zone of DH is both the converging area of nociceptive nerve impulses from periphery, the relay towards ascending pathways to pain-integrating networks and an important target for supraspinal top-down modulation system, it can modulate nociceptive transmission as the site of pain control. For this reason, the DH can be targeted in the treatment of pharmacoresistant chronic pain.

1.3 DREZ as a Study Model and DREZotomy

The DH has a well-known organization with several layers receiving the periphery fibers (Fig. 1). Periphery fibers (myelinated Aβ, projecting mainly to layers III and V; myelinated Aδ, projecting mainly to layers I and V; nonmyelinated C, projecting mainly to layers I and II) enter the CNS through the posterior roots to reach the spinal cord. The gray matter of the cord is divided into ten layers or laminae with the first five laminae corresponding to the posterior cord. The Dorsal Root Entry Zone (DREZ) is the area including the Rexed's laminae from I to V and the tractus of Lissauer covering the head of DH (Fig. 1).

A surgical technique called DREZotomy, designed in Lyon, France in 1974 by Marc Sindou [8], is a useful way to relieve pain by sectioning selectively nociceptive afferences in the superficial part of the DREZ and also myotatic afferences responsible for muscle tonicity while sparing the non-nociceptive fibers which have an inhibitory influence on nociceptive input [9–11]. If necessary, microcoagulation can be performed more deeply in the head of the DH itself, allowing to destroy the nociceptive neurons. This technique is therefore successful for the treatment of well-defined pharmacoresistant chronic pain, but also in some cases of spastic hypertonia. DREZotomy can be performed in patients with neuropathic pain due to lesion of peripheral nerves or avulsion/lesion of the plexus (brachial or lumbosacral). This technique can also be performed for patients suffering from severe diffuse spasticity in a superior or inferior limb. In these latter cases, DREZotomy allows to reduce significantly the excessive muscular tone.

For such a surgery, the patient is lying in ventral decubitus position, under general anesthesia and the neurosurgeon approaches the surgery of the spinal cord by a laminectomy and a meningeal opening at the targeted level (cervical or lumbar or sacral). DREZotomy consists in lesioning the latero-ventral part of the DREZ by microcoagulation (Fig. 1, red double head). This surgical procedure, performed on both anatomically and functionally individualized nociceptive pathways, provides a direct access to biochemical studies before performing lesion in the DH. Thus, it gave us the opportunity to monitor spinal neurotransmitters including Glu, for the first time in human spinal cords, by microdialysis sampling, while respecting the ethical rules, since the microdialysis probe is implanted in an area which will be fully destroyed secondarily by DREZotomy and so don't create any additional tissular lesion than the surgery itself. Thanks to this approach, our group found a significant difference in the balance of excitatory and inhibitory transmissions between patients suffering from chronic pain and those suffering from spasticity, without chronic pain. This was one of the first demonstrations in humans of a concept previously hypothesized from animal studies [9].

1.4 Outlines of the Chapter

This chapter will not fully describe DREZotomy or methods to quantify Glu, but rather the microdialysis steps allowing to sample Glu in spinal cord just before DREZotomy. A scientific discussion on the studies carried out in humans will show the data obtained by using the microdialysis of CNS Glu in various pathologies including CNS traumatic injury, epilepsy, ischemia, and Parkinson's disease, since literature on clinical Glu microdialysis is far more documented for brain than for spinal cord. An emphasis on methodological aspects will also be provided to show the advantages and the pitfalls to monitor Glu in CNS of human patients.

2 Materials

2.1 Patients

1. We included in this study ten patients who had to undergo a microsurgical DREZotomy (DREZO) for different therapeutic goal: five for chronic pain and five for harmful spasticity. The aim of DREZO is to selectively interrupt the nociceptive (for pain) and myotatic (for spasticity) afferences in the lateral part of the dorsal root entry zone (DREZ) and to destroy the likely hyperactive neurons located in the superficial layers of the DH of spinal cord by microcoagulation [8, 9] (Fig. 1). The protocol of intraoperative microdialysis sampling performed during DREZO [12] was approved by the Ethics Committee of the University Hospital and the Consultative Committee for Human Protection in Biomedical Research in Lyon.
2. Prior to surgery, informed consent was signed by the patient or his (her) next of kin, as required when studying human beings as subjects of scientific studies (Note 1).

2.2 Surgery and Microdialysis Experiments

1. Sterile materials needed for the surgery procedures for DREZO per se are not detailed in this chapter, because it is beyond the scope of the chapter. A surgical microscope is required for probe implantation as for surgical intervention.
2. Sterile materials needed for the microdialysis experiments are as follows: a precise delivering pump for the infusion of the artificial spinal fluid (aCSF) (Harvard, models#22/PHD), two 2.5 mL syringes, floating sterile microdialysis home-made probes (see next subparagraph) adapted to the DH, commercial sterile Ringer solutions, samples tubes for collection adapted to the analytical system used, a long tubing between one syringe and the probe inlet.
3. A specific concentric microdialysis probe was designed for the DH studies (Figs. 2 and 3a). In order to be adapted for the in vivo implantation inside the human spinal cord, this probe was made with flexible, noncompliant, light, and biocompatible materials: Tygon (Bioblock Scientific, Illkirch, France) for the inlet tubing and fused silica coated with polyimide (PolymicroTechnologies, Phoenix, AZ, USA) for the outlet tubing. The dialysis membrane was made of Cuprophan

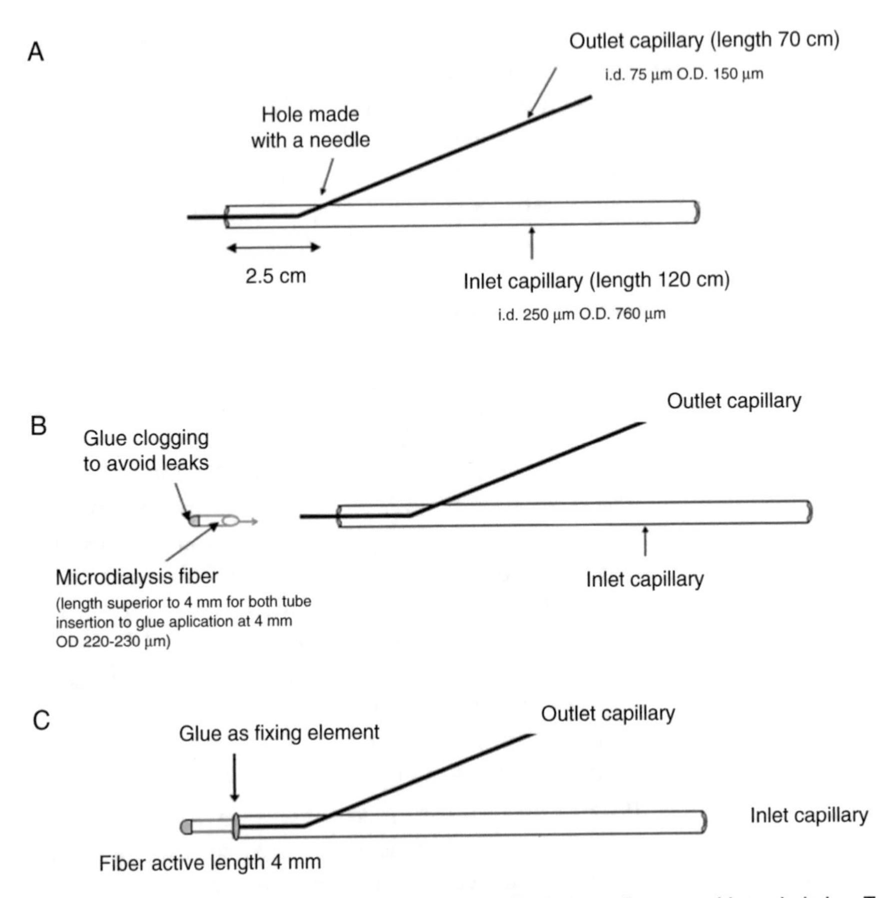

Fig. 2 Schema showing the process of probe making. (**a**) The first step relies on making a hole in a Tygon tubing (inlet of the probe) and inserting a capillary (outlet of the probe). The hole is clogged with a bi-component glue to ensure sealing. (**b**) A membrane fiber, previously cut as the desirable length and clogged with the glue to avoid further leaks, is inserted in the Tygon tubing. (**c**) The fiber is glued with a very small quantity of glue to ensure both sealing and warrant the totality of the dialyzing surface

(Hospal Industrie, Meyzieu, France; O.D. 216 μm in dry conditions) with a molecular cut-off of 6000 Da. A medical glue (Eccobond, Grace Electronic Materials) was used to consolidate the three components of the probe and to prevent leaks of the perfusing fluid. The weight of the probe itself was 20 mg. Thus, this light microdialysis probe was designed to be a floating probe able to pursue the natural movement of the human spinal cord. The probe and the tubes were considered as single-use devices and were sterilized by ethylene-oxide gas [9] or hydrogen peroxide (0.66 bar, 45 °C, 45 min) [13].

2.3 Glutamate Content Analysis

Glutamate analysis can be performed by high performance liquid chromatography with laser-induced fluorescence detection (HPLC-LIFD) or capillary electrophoresis (CE)-LIFD after sample derivatization with the following reagents, sodium cyanide as

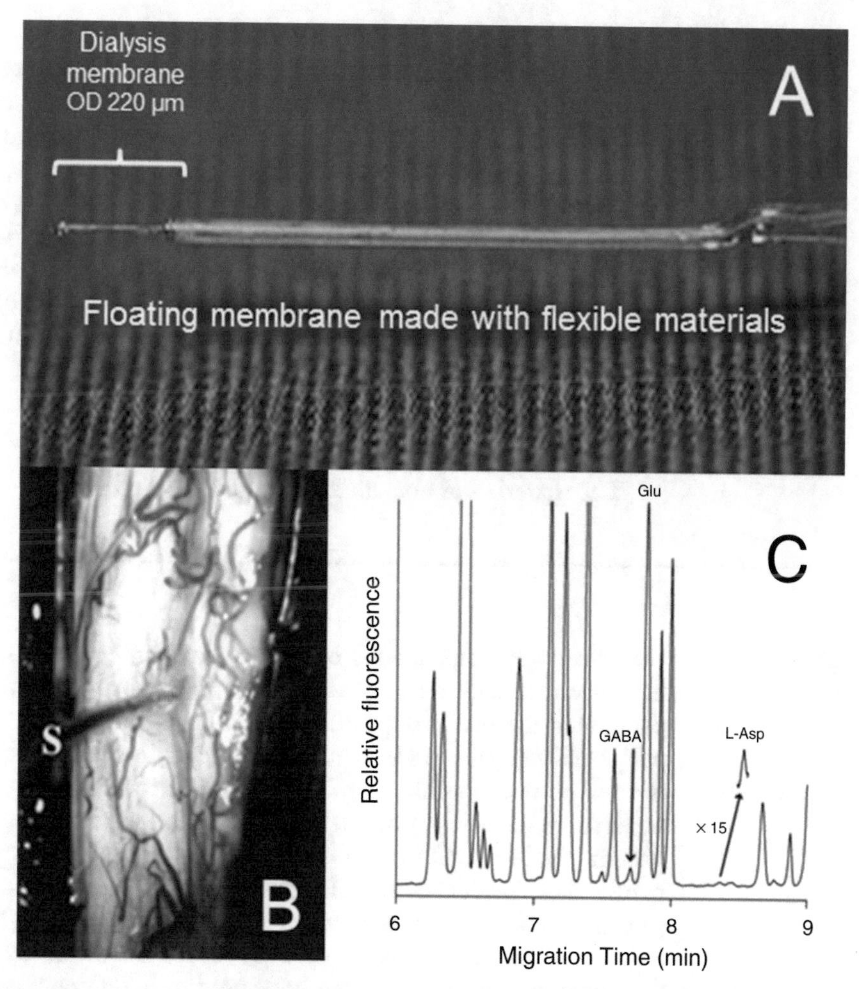

Fig. 3 Microdialysis experiments carried out in patients' DH. (**a**) The microdialysis probe has been especially designed for size and safety to target the DH of the spinal cord. (**b**) Example of an intraoperative photograph showing the microdialysis probes (S) implanted in the ventrolateral part of the DREZ in a patient suffering from chronic neuropathic pain due to left brachial plexus avulsion (adapted from [12]). (**c**) Typical analysis of a microdialysate collected in the spinal cord of a patient about to undergo a DREZO surgery, using off-line capillary electrophoresis with laser-induced fluorescence detection (volume injected, 15 nL)

nucleophile agent, borate buffer, naphthalene-2,3-dicarboxaldehyde as fluorogenic agent, and cysteic acid as internal standard.

1. HPLC-LIFD

   One microliter of internal standard was added to 10 μL of microdialysate. The mixture was derivatized with 3 μL of a solution containing 2 volumes of NaCN-Borate (borate buffer 0.5 mol/L pH 8.7 + NaCN 87 mmol/L, v:v, 5:1) and 1 volume of NDA 2.925 mmol/L). Samples were frozen at −80 °C until analysis.

   The HPLC system consisted in two HPLC pumps (model 420), an automatic autosampler (model 460), and a spectrofluorimeter (model SFM-25, Kontron Instrument System, Zürich, Switzerland). Excitation and emission wavelengths were 419 and 483 nm, respectively. Separations were carried out at 50 °C (temperature controller, Waters Corporation, Milford, USA) on a C18 ODS-3 Inertsil® column (GL Sciences Inc., Tokyo, Japan). Mobile phase run according to the following components: (A) phosphate buffer 50 mmol/L pH 6.8–THF/97:3 (v: v), and (B) phosphate buffer 50 mmol/L pH 6.8–Methanol–Acetonitrile/35:10:55 (v:v:v), under the following gradient: at 0 min 88% A–22% B, from 0 to 75 min, linear gradient to 53% A–- 47% B, followed by 100% B from 76 min to the end of each run (90 min).

2. CE-LIFD

   The materials are detailed in Chap. 13.

3 Methods

3.1 Groups of Patients

Two homogeneous groups of five patients undergoing DREZO have been studied in this study on the DH of the spinal cord. Five patients in the first group suffered from neuropathic pain due to a peripheral nervous lesion while five patients in the second group, devoid of neuropathic pain, were affected by disabling muscle hypertonia (spasticity) due to cerebral lesion sparing the sensory system, the DREZO surgery being the only way to relieve permanently their persistent muscle spasms. In this study, no control (healthy) patients could be used for obvious ethical reasons (Note 1). As a consequence, as this study was focused on the study of pain mechanisms, the spastic group was used as a reference group as patients from this group did not exhibit any sensory disturbances and so, no chronic peripheral neuropathic pain.

A standardized surgical procedure was performed under general anesthesia through a posterior approach of the spinal cord, via a laminectomy at the level of the cord segments involved in hyperspasticity or pain. The surgical targets were the ventrolateral part of the DREZ and the DH. Neither intraoperative nor postoperative complications were reported in any of the patients with a follow-up of 18 months on average. The expected clinical improvements, decrease of pain or spasticity, were obtained postoperatively as expected for each patient.

Even if the practitioner selects the patients with similar symptoms and that the surgery is standardized and well-mastered by the Lyon neurosurgical team, variability on microdialysis data is possible as discussed in Sect. 3.3.

3.2 Surgery Constraints in Clinical Microdialysis Studies

1. The microdialysis studies must first warrant continuity in sterility in the surgical room. For that purpose, an additional operating field, close to the patient covered by his (her) usual

surgical drape, allowed the neurochemist to connect the microdialysis probe to a pump delivering a sterile aCSF infusion, and to collect microdialysis samples for Glu monitoring, (a) far enough from the patient to avoid to interfere with the surgical procedure performed by the medical staff, but (b) not too far to avoid to generate high dead volumes in the microdialysis probe compromising the quality of monitoring (see subparagraph 8 of this part).

2. The microdialysis probe was continuously perfused with a sterile Ringer's solution (Aguettant, Lyon, France) with the following composition (mmol/L): 147.16 Na^+, 4.02 K^+, 2.23 Ca^{2+}, 155.54 Cl^-. During the implantation and before sampling, the flow rate was set to 10 μL/min [3, 4]. The sampling procedure was performed over a period of 60 min on average, after a 45-min stabilization period. During sampling the flow rate was 2 μL/min. The fractions were collected in sealed glass vials every 5 min (sample volume, 10 μL) [12] or in PCR tubes every minute (sample volume: 2 μL) [13] and immediately stored on ice for later analysis.

3. In contrast with experimental studies in animals, it was not possible to wait 2–3 h between probe implantation and sample collection. This 2- or 3-h period normally allows to flush excess of compounds released in the extracellular space following the tissue damage induced by the insertion of the probe. Indeed, the probe insertion is likely to break the blood–brain barrier (BBB) and BBB is expected to be restored around 2-h after the probe implantation, which could justify this delay before sample collection. However, such a delay cannot comply with surgical procedures. Indeed, the duration of anesthesia should be as short as possible to include all the surgical steps: pre-preparation of the patient (e.g., skin disinfection), incision, and preparation of the body part (here, clearance of the DH of the spinal cord) for the surgical handling (here, DREZO). For that reason, the scientific goal cannot prevail over the therapeutic purpose of the surgery. Consequently, the postimplantation stabilization period was set at 45 min as presented in Fig. 4a showing the typical decrease in Glu levels in a patient's DH monitored every 5 min with an almost stabilized level after 45 min of implantation. This decline in spinal Glu concentrations until baseline was judged rapid in contrast with another result found in rat spinal cord [14], but similar to our data obtained in rat brain striatum (Fig. 4b). The monitoring of rat brain glycerol (Fig. 4c) and rat brain phosphoethanolamine (PEA, Fig. 4d), which are regarded as indexes of membrane integrity, suggested that the release due to the probe was maximal during the 20 first minutes following its implantation. So that, in human spinal cord, even if the BBB was not totally

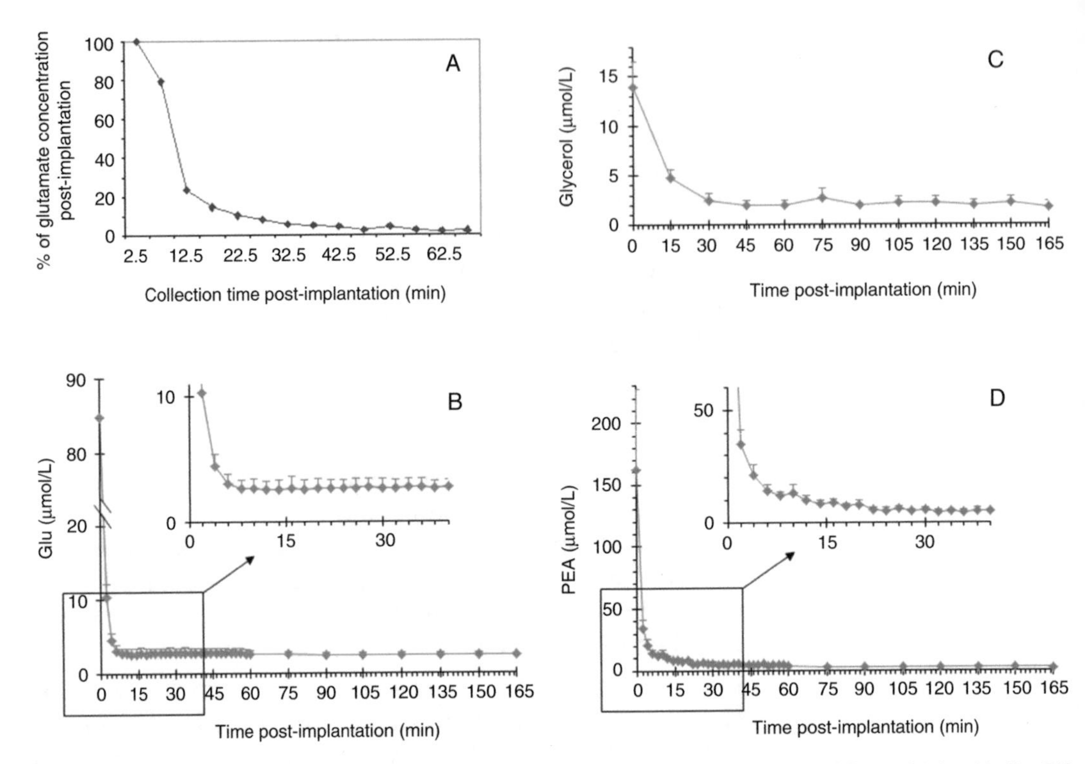

Fig. 4 Monitoring of glutamate and other compounds following probe implantation. (**a**) Data obtained in the DH in a patient for whom the time to prepare accessibility to DREZ was lower than in average and microdialysis monitoring was possible up to more than 1-h postimplantation. Concentrations (*blue symbols*) were expressed as percentage of the initial concentration at $t = 0$ min, i.e., 147 μmol/L. (**b**) Brain Glu 2-min monitoring in the striatum of rats just after probe implantation, following by a 15-min monitoring from 1 to 3-h postimplantation (*red symbols*). Concentrations were expressed in μmol/L ($n = 5$). (**c**) Brain phosphoethanolamine (PEA) monitoring simultaneously to that of Glu in B ($n = 5$, *dark green symbols*). (**d**) Brain glycerol 1 and 15-min monitoring carried out in the striatum of rats ($n = 4$, *green symbols*)

restored 45 min after implantation, we can hypothesize that BBB was largely restored when the collection was finished, less than 1 h after probe insertion into the spinal cord. We can note that a slower implantation made over 1 h would not damage less the tissue, but would lead to far less releases of compounds from tissue at the end of the insertion and to a 20-fold more rapid normalization of the Glu, PEA and glycerol levels as previously shown in the rat [15]. However, this type of slow insertion is not easy to perform without an automated descending device and very difficult in human DH, due to its poor accessibility. Even if DREZO provides an easier accessibility to the DH, probes had to be implanted with a 35° angle in a few seconds inside the gray matter by the surgeon's hand, with appropriate magnification under a surgical microscope.

4. As commercial probes were not adapted to spinal cord at the time the protocol was started, because they were designed for

intracranial dialysis (10 mm of active dialysis membrane), our group customized its own probes to target Rexed's laminae I–V of the DH. Two experimental constraints were taken into account: the 4-mm membrane and flexibility as discussed in the following paragraph. Moreover, as for commercial probes, our probes should be easy to make, safe, ergonomic, robust, biocompatible, addressed to human use and, finally, resistant to sterilization procedures and tissue insertion.

5. An important probe feature is its flexibility allowing it to "float" in the DH, implying that the probe was not maintained in a frame and that the membrane could easily follow the normal spontaneous movements of the spinal cord and cerebrospinal fluid (related mainly to respiratory and cardiac rhythms) (Fig. 3b). Besides, the flexibility allowed the introduction of a probe into the tissue that will be therapeutically micro-lesioned during the surgical intervention. As the size of the microdialysis probe was smaller than the microsurgical lesion, a biochemical investigation could be performed with no additional mechanical damage to the human spinal cord tissue. Indeed, after dialysis collection, the DREZO lesion was performed by microcoagulations 4-mm deep in the DH. It represented a lesion volume 10- to 15-fold larger than the lesion volume induced by the transient probe insertion.
6. The probe and the tubes were considered as single-use devices and were sterilized by ethylene-oxide gas in the human spinal cord studies carried out by our group. However, the glue that we used for the first home-made probes became unavailable, and we had to find another biocompatible glue to assemble the parts of our probes. We chose to work with a bi-component glue (Eccobond, Grace Electronic Materials), but we had to change the sterilization procedure for a sterilization using hydrogen peroxide, to avoid the dried glue to lose its ability to stick, which was the cause of perfusion leaks occurring at the junction between the dialysis membrane and the Tygon tubing [2].
7. In order to preserve the sterility of the microdialysis probe before surgery, the determination of the probe recovery was not performed before implantation, but after the microdialysate collection, in contrast with the animal studies. Thus, the in vitro relative recovery of each probe was assessed immediately after surgery using exactly the same experimental conditions as those used in the operating room except for the probe which was immersed at 37 °C in vials containing standard solutions. The probe, infused by the aCSF at the same rate as during the collection of human microdialysates, was plunged into a synthetic solution with a known Glu concentration (1 μmol/L). Three successive 5-min samples were collected after waiting 3 min and immediately stored at −80 °C until

analysis. The recovery was calculated according to the formula: $R = C_{out}/C_{std}$ with C_{out}, the concentration at the outlet of the probe, and C_{std}, the concentration of the synthetic solution. The in vitro recoveries were very similar for all the probes (~17.5%, [9]). This information, as for preclinical studies, was not used to correct the concentration values of the human microdialysates, since in vitro recovery is different from the in vivo recovery because an aqueous solution cannot mimic a tissue matrix [16]. However, the values found for the recoveries were useful to assess the repeatability and the quality of our probe microfabrication process.

8. Another last important feature to take into account was the dead volume of the probe because it takes time to bring a volume section from the dialysis membrane, where the neurochemical exchanges occur, to the probe outlet where the samples are collected (flow rate, 2 μL/min). Thus, it was important to reduce the dead volume as much as possible and to determine it precisely in order to correlate the neurochemical variations to surgical events, for instance if cutaneous sensitivity in ghost arms is investigated for patients with root avulsion. In this respect, the determination of dead volume could also be performed at the same time as the determination of probe recovery, i.e., after surgery (see Fig. 5 for the details). In a work aiming at monitoring every minute the concentration of Glu, we were able to estimate the corresponding dead volume time, and we found a value of 2 min 50 s, i.e., less than 3 min, using a 1-min sampling rate (Fig. 5). In other words, the variations occurring near the probe at time 0 min, were detectable 2 min 50 s later. This has been confirmed on a patient in one pilot study performed with a 1-min sampling rate: an electrical stimulation of the dorsal rootlets of one patient with chronic neuropathic pain was performed (2.5 Hz, 0.06 V, 1 mA, 30 s, bipolar electrode) and induced no increase of microdialysate Glu, but a rapid, transient increase of GABA level which was observed in the third fraction collected after the stimulation [13]. This pilot study points out that our approach is suitable for accurately monitoring a short change in neurotransmitter release linked to a fast-acting neurobiological event.

9. Data were expressed generally as microdialysate concentration without correction by the measured "in vitro" recovery, as discussed just above, in subparagraph 7. However, we have to keep in mind that data representation can take into account the dead volume. In this respect, the time scale in the figures showing microdialysis monitoring can correspond to the actual time of collection of the fractions, like in Fig. 4b–d, or either to the corrected time of collection by taking into account dead volumes

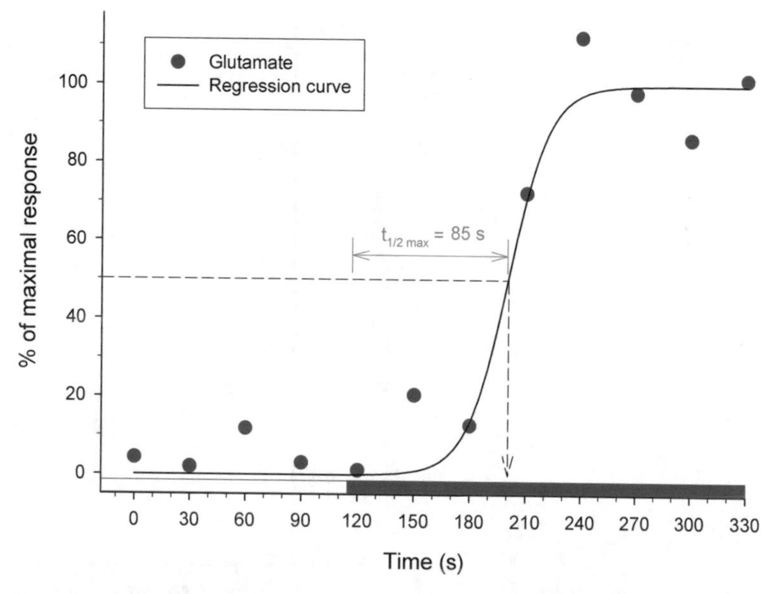

Fig. 5 Determination of the dead time of a spinal microdialysis probe. A sigmoidal regression curve fitted to the monitoring of Glu concentration (*blue circles*) before (*open bar*) and after (*blue bar*) the probe was plunged in a standard solution containing 10 μmol/L Glu. Data were expressed as percent of the maximal response, based on the average of the four last values (plateau). $t_{1/2\ max}$ corresponded to the half-time between the beginning of the dialysis in Glu solution (precisely at 115 s) and the reach of the maximal response. Samples were collected every 30 s for that determination (adapted from [13])

if estimated theoretically or determined experimentally [13, 17, 18]. Finally, we have to mention that some authors can represent the mean time (i.e., 2.5 min) between two sampling times (i.e., collection at 0 and 5 min) to position the points in graphics, like in Fig. 4a, or with histograms [13].

3.3 Surgery and Variability of Data

Several causes of variability can be mentioned due to the surgical constraints.

1. According to the patients' damage in their CNS or DREZ circuitry, various metameric levels can be the targets for the surgeon (e.g., L2–S3 for cerebral trauma in spastic patients, or C5–T1 for a brachial plexus avulsion in a patient suffering from a chronic neuropathic pain). One damaged metamer was chosen for Glu monitoring while DREZO would destroy all the targeted and defective metameric levels. Actually, C-8, T-1, L-2, and S-1 were the levels the most often dialyzed. Despite the fact that the surgeon chose different metamers, depending on the patients, the probe was always implanted into the gray matter under the control of the surgical microscope.

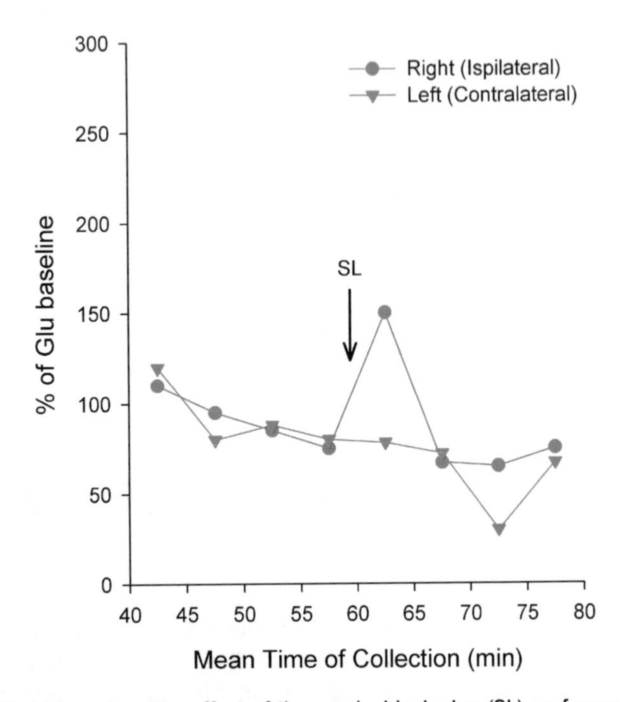

Fig. 6 Case Report on the effect of the surgical lesioning (SL) performed on the right side in patient in whom bilateral probes were implanted in the DH at L-2 level. Note the transient increase in Glu concentration in the ipsilateral, but not in the contralateral side to the SL. Concentrations were expressed as % of baseline (i.e., 1.989 μmol/L) taken between 45th and 50th min postimplantation (adapted from [12])

2. Data variability can be explained by the presence of Glu both in primary afferent terminals but also in intrinsic neurons, and/or in terminals of descending pathways and/or in glial cells. It can also be explained by the probe placement. Indeed, a test was carried out in one patient for whom a bilateral sampling was possible over 1 h 10, instead of 50–60 min, using two probes implanted at L2-level DH. A surgical lesioning performed on the right side induced a transient increase in Glu concentration in the ipsilateral, but not in the side contralateral to the surgical lesion, showing a great anatomical specificity in a neurochemical event when a response was detected [12] (Fig. 6).

3.4 Data Analysis

1. The Glu analysis was performed off-line, so that a good preservation was required. Several days can timely separate collection and analysis.
2. The use of HPLC vs. CE was conditioned by the time resolution for dialysis sampling. For the finest monitoring (<5 min), CE-LIFD was required. A typical CE electropherogram is shown in Fig. 3c.

3. The detection limits of Glu, calculated for 2 μL of derivatized dialysate, ranged around 13 nmol/L with our HPLC system and 3.7 nmol/L with our CE apparatus.
4. Other considerations on Glu analysis can be seen in Chap. 13.

4 Discussion

The use of microdialysis for sampling and monitoring the extracellular Glu concentration in the human central nervous system (CNS) has been found of great interest in clinical investigations dealing with several neurological disorders, being a valuable approach for elucidating the pathophysiological mechanisms, as well as being a pertinent tool for diagnostic or prognostic purposes. The main advances brought by microdialysis monitoring of human CNS Glu in the understanding of neurological diseases or disorders are reviewed below.

4.1 Traumatic Spinal Cord and Brain Injury

The only microdialysis monitoring of Glu in patients with traumatic spinal cord injury (TSCI) has been recently performed for up to 9 days after the initial injury [19, 20]. The microdialysis probe was placed on the surface of the injured spinal cord together with a probe to measure intraspinal pressure and to give indirectly the spinal cord perfusion pressure. The extrapolation to zero-flow method was used to estimate the absolute extracellular Glu concentration which was found to be around 30 μmol/L, an elevated value which was even found increased when the spinal cord perfusion pressure was decreasing to values equal or inferior to 50 mmHg. Thus, high extracellular Glu levels are likely to be concomitant with a local ischemia. Accordingly, Glu level was found higher up to 3 days after surgery in patients with a large neurological deficit.

In contrast with TSCI, more data are available on microdialysis Glu monitoring in traumatic brain injury (TBI). In one early pilot study performed on a single patient with severe head trauma in 1994, very high brain extracellular concentrations of Glu (300 μmol/L, when using a flow rate of 2 μL/min for perfusing the probe) were recorded in the microdialysis samples collected over 3 h [21]. In a further study [22], the cerebral microdialysate Glu levels were evaluated for 9 days in 17 TBI patients. In every patient, transient elevations of Glu were observed every day, reaching 25 μmol/L and with a mean duration of 4.4 h. These increases were observed in conjunction with seizure activity and/or when cerebral perfusion pressure (CPP) was falling below 70 mmHg. This suggests that, with a reduced CPP, neurons may become ischemic, releasing Glu into the extracellular fluid. Thus, a reduction in CPP to less than 70 mmHg is likely to be associated with excitotoxic events which may, in turn, further damage the brain.

In another study performed on seven TBI patients, increased Glu brain microdialysate levels, characteristic of a cerebral ischemia, could be seen when the probe was implanted in the cortex just surrounding the contusion. If the probe was positioned in a cortex area far from the contusion, an increase in microdialysate Glu was usually observed after an increase in intracerebral pressure (ICP), the issue of which was fatal [23]. However, microdialysate Glu levels themselves may be of poor predictive value for ICP rise above the normal level. Indeed, in a study performed on a larger number of TBI patients (i.e., 25), a high Glu concentration was found not significantly associated with a greater risk of subsequent ICP rise, although it was the case for an abnormal lactate/pyruvate ratio [24].

The link between microdialysate Glu level and the hypoxic status of the brain was further supported by a study performed on 57 TBI patients in which microdialysate Glu and brain tissue pO_2 ($PbtO_2$) were simultaneously monitored. Microdialysate Glu was found significantly elevated only when $PbtO_2$ decreased to very low levels (<10 mmHg and further) [25]. Similar results were obtained in another study performed on 41 TBI patients and reporting increased Glu during impending hypoxia ($PbtO_2$ < 10–15 mmHg) and a further increase in Glu during manifest hypoxia ($PbtO_2$ < 10 mmHg) [26]. Moreover, impending hypoxia was the most frequently associated with intracranial hypertension and Glu, measured in non-lesioned cerebral white matter, was identified as a sensitive and early marker of impending cerebral hypoxia, although the energy metabolites remained stable. In contrast, manifest cerebral hypoxia was characterized by elevated extracellular Glu and lactate concentrations and was most often associated to hyperventilation. Such a link between hyperventilation and higher microdialysate Glu has also been shown in a study by Marion et al. [27]: when the microdialysis probe was implanted in the brain tissue adjacent to the cerebral contusion, increased microdialysate Glu level was observed during brief periods of hyperventilation. Among 20 TBI patients, these hyperventilation-induced changes are much more common during the first 24–36 h after injury than later (3–4 days). This result may question the use of hyperventilation therapy, traditionally used in the treatment of intracranial hypertension during the first 24 h after injury because of the risk of hyperventilation-induced ischemia. In accordance with these studies, a more recent work performed on 36 patients with severe TBI underlines the negative effect of high FiO_2 on brain microdialysate Glu [28]. Indeed, an increase of FiO_2 through normobaric oxygen therapy could be used, in a first attempt, to treat compromised $PbtO_2$ in TBI patients, but high levels of oxygen could be associated with worse prognosis in severe TBI patients. This may involve Glu since a significant and progressive increase in brain microdialysate Glu was associated with an increased level of FiO_2. Hyperoxia (PaO_2 > 150 mmHg) was also associated with higher brain microdialysate Glu.

Moreover, microdialysis measurements performed on 20 TBI patients have revealed a link between brain Glu and brain glucose, the brain Glu concentrations being the lowest for brain glucose in the 3–5 mmol/L range and increased when brain glucose was below 3 mmol/L or higher than 6 mmol/L [29].

Furthermore, increasing arterial blood glucose levels was associated with a significant increase in brain Glu, and insulin-induced hypoglycemia was associated with a significantly elevated brain Glu, in agreement with a previous study [30]. The elevated extracellular Glu at a low brain glucose concentration may be due to a depolarization-induced release of Glu caused by a lack of glucose (ischemia-like effect), while the increased Glu at high brain glucose may result from an enhanced synthesis of Glu from metabolized glucose (metabolic effect). These results support the consensus guidelines which recommend to avoid both hypoglycemia and hyperglycemia to prevent a worsening of the underlying brain damage in TBI patients and may partly explain the poor outcome which is likely associated with hyperglycemia in such patients.

Value of brain microdialysate Glu in TBI patients may depend on the age. For example, among 69 patients, a significantly higher microdialysate concentration of Glu was found in patients older than 65, as compared with younger patients [31]. This enhanced brain Glu was associated with an increased concentration of glycerol and could correspond to more extensive damaging processes in the elderly. It is also consistent with the expected poor prognosis associated with increasing age in TBI patients.

The predictive, i.e., prognostic, value of brain microdialysate Glu in TBI patients has been discussed in the reference studies of Chamoun et al. [32] and Timofeev et al. [33] performed on 165 and 223 patients, respectively. Both studies show that brain Glu level is correlated with the mortality rate and the 6-month functional outcome. If a tendency of microdialysate Glu levels to normalize over the monitoring period is likely to be indicative of a recovery, along the same line, a sustained high Glu was more frequent in patients who died. Nevertheless, both studies underline that if microdialysate Glu reflects the severity of brain injury, it is obvious that, being a biomarker, it does not assess, by definition, the clinical status as does the widely recognized Glasgow Coma Scale (GCS) or ICP monitoring, and must, therefore, be used in conjunction with these other independent prognostic factors. However, from a more general point of view, one can suggest that Glu concentration in cerebral microdialysates may be considered for appropriate clinical investigations in order to determine its possible interest as one of the neurochemical biomarkers contributing to the biochemical definition of the cerebral death in the future.

The link between high microdialysate Glu and a less favorable outcome was also supported by studies from Tolias et al. [34] and Richards et al. [35] performed on a small number of children.

Moreover, taking into account the relationship between glutamine (Gln) microdialysate and GCS, these studies have further proposed that the molar ratio of Gln to Glu, may be a better prognostic indicator in the assessment of TBI patients than Glu alone, a higher ratio being associated with a more favorable outcome.

Glutamate monitoring by brain microdialysis has been employed in a few pilot studies aiming at assessing the neuroprotective effect of pharmacological agents in TBI patients. In a study performed on 20 patients, Alves et al. [36] have shown that the systemic administration of antiepileptic drug topiramate, a Glu release inhibitor, induced a lowering of brain extracellular Glu. Another study performed on 50 patients was focused on the effect of Cyclosporin A. This compound, which may be neuroprotective through its ability to preserve mitochondrial bioenergetic status, is able to induce a moderate lowering of brain extracellular Glu for the 2 days following a 24 h intravenous (i.v.) infusion [37]. Finally, in a third study, i.v. administration of prostacyclin was started in a patient when his brain microdialysis monitoring indicated a progressive secondary ischemia (increase in lactate/pyruvate ratio, decrease in glucose and increase in Glu) and led to a normalization of the microdialysate concentration of Glu [38].

Gln supplementation has been shown to improve the outcome of intensive care unit patients. However, there was reluctance to use Gln-containing i.v. infusion in TBI or subarachnoid hemorrhage (SAH, see below) patients, since it cannot be ruled out that such an infusion may increase brain extracellular Glu, possibly resulting in detrimental effects upon clinical outcome. However, two microdialysis-based studies have shown that this is not the case. Indeed, intravenous Gln, in clinically relevant doses, did not affect microdialysate Glu level in nine TBI and six SAH patients, whatever the basal pre-supplementation Glu was normal or elevated [39]. In the other study performed on three TBI and seven SAH patients, the i.v. infusion of a Gln-containing amino acids mixture led to a modest although significant increase in microdialysate Gln, while the Glu level was unaffected [40].

4.2 Aneurysmal SAH and Other Cerebral Hemorrhages

An unfavorable outcome may occur after SAH or aneurysmal SAH. This may be related to the volume of the initial bleeding, to surgical complications or to the occurrence of a cerebral vasospasm, and to a subsequent delayed cerebral ischemia. Vasospasm may lead to serious focal neurological deficits or cognitive impairments in SAH patients and arterial narrowing can be seen in up to 75% of the cerebral angiograms of these patients. The maximal vasoconstriction is reported to occur around 7 days after SAH and arteries size is believed to be back to near normal in about 14 days. However, only 30% of the patients experiencing a vasospasm will exhibit delayed ischemic neurological deficits.

An abnormal release of Glu may be of particular importance in the cerebral ischemia associated with SAH. Indeed, in a study performed in seven patients, Saveland et al. [41] reported that microdialysate Glu can rise to high levels after surgery and that the increased levels of Glu correlated well with the clinical course and the neurological symptoms of the patients. These findings were confirmed in a further study from the same group performed on another group of ten patients, showing that Glu (as lactate) may be the most sensitive and earliest indicator of impending ischemia [42]. In a cohort of 40 SAH patients monitored in another study, 12 patients encountered at least one episode (1–6 per patient) of microdialysate Glu levels higher than 10 μmol/L, lasting for more than 1 h. An unfavorable outcome, with infarction in the region monitored by the probes, was associated with such episodes of high Glu. Only half of the episodes of increased Glu were preceded by a $PbtO_2$ lower than 8.2 mmHg, whereas in the majority of these episodes, CPP was normal. These data further strengthen the sensitivity of microdialysate Glu for detecting delayed cerebral ischemia in SAH patients [43].

Another study, performed on 97 SAH patients, distinguished the patients with delayed ischemic neurological deficit (DIND) or with acute ischemic neurological deficit (AIND) from the asymptomatic ones. The temporal courses of the microdialysate concentration of Glu were different between the three groups [44]. If microdialysate Glu normalized and remained low from the third day of monitoring in the asymptomatic patients, it was eight- to tenfold higher in DIND patients and did not normalize during the first 7 days following the SAH. Furthermore, AIND patients had highest Glu concentrations (>70 μmol/L) that decreased only slightly within these 7-day period. Interestingly, the increase in Glu occurred before the development of symptoms (7.8 h in average) in 83% of DIND patients showing again that microdialysate Glu is an early marker of impending ischemia in SAH patients. Further studies from the same group, one on 44 patients and the other one on 17 patients, were more focused on AIND and confirmed the higher microdialysate Glu in these SAH patients [45, 46]. Indeed, a significant difference ($p < 0.001$) was found in baseline Glu between patients with AIND (86.4 μmol/L) and those without (2.8 μmol/L). These studies strengthened the interest of microdialysis monitoring of brain Glu, by reporting that a further deterioration of Glu (together with lactate and lactate/pyruvate ratio) was a predictor of a subsequent neurological worsening (likely resulting from brain swelling or cerebral vasospasm) and highly indicative of the development of cerebral infarction and permanent neurological deficits. Our group reported similar findings on SAH patients, i.e., increased microdialysate Glu level and lactate/pyruvate ratio associated to a large ischemic area as revealed by cerebral computerized tomography [47].

Moreover, microdialysate Glu correlated best with regional cerebral blood flow (rCBF, measured by ^{15}O-H_2O positron emission tomography (PET)) in 13 patients as shown in two studies [48, 49]. However, if Glu was rather sensitive in detecting critically reduced rCBF, appearing as an early marker of impending ischemia, it could also be unspecific, being increased above the threshold of ischemia, in response to other mechanisms such as transmitter release and unspecific leakage from injured cells. In this respect, the ratio Gln/Glu was suggested to be a more specific biomarker than Glu in the studies from Samuelsson et al. [50, 51] each performed on 33 SAH patients. They showed that the Gln/Glu ratio was negatively correlated with intracranial pressure (ICP), this ratio being lower in periods of elevated ICP (>10 mmHg, potentially reducing CBF) than when the ICP was below 10 mmHg.

A higher microdialysate Glu level has also been reported in spontaneous non-aneurysmal intracerebral hemorrhage (ICH). In two studies, microdialysis monitoring was performed after the surgical evacuation of ICH in 22 or 12 patients, respectively [52, 53], the probe being placed in an area close to the evacuated ICH. Glu was found increased during 1–12 h after surgery before normalizing within 24–48 h, suggesting that a penumbra zone surrounds large primary ICH, similarly to the penumbra surrounding focal traumatic brain lesions. However, as an unaltered lactate/pyruvate ratio was associated to this enhancement in Glu, the neurological deterioration which may occur secondary to ICH may be more likely due to excitotoxicity than to ischemia. On the other hand, a higher microdialysate Glu in the peri-hematomal area of patients with primary basal ganglia hemorrhage could be reduced by intensive insulin therapy aimed at maintaining a low blood glucose level [54]. As such a lowering in Glu was associated with a higher $PbtO_2$ and with a lower lactate/pyruvate ratio, the occurrence of a local hypoxia in the penumbra region cannot be ruled out.

4.3 Middle Cerebral Artery Infarction and Stroke

A pilot study [55] was performed on a patient suffering from occlusive stroke in the territories of the middle and anterior cerebral arteries and who underwent a partial resection of the infarcted tissue. A microdialysis probe was placed in the remaining infarcted tissue and a massive increase in Glu concentration was detected (up to 250 μmol/L at the beginning of the monitoring) which progressively declined within 48 h. A further study performed on ten patients suffering from a large middle cerebral artery (MCA) infarction described the dynamics of microdialysate Glu in accordance with clinical course and the development of a local edema. Patients exhibiting enhanced Glu associated with increased ICP underwent massive brain edema (as assessed by CT scans) whereas those without edema exhibited a pattern of low and stable concentrations of Glu and stable ICP values [56]. Interestingly, the probes were placed in a brain area which was, according to CT findings, not involved in the ischemic process. Increases of Glu in areas outside the primary ischemic territory are likely to indicate secondary changes related to the formation of a malignant brain edema. Data were refined in a further study from the same group in which Glu and $PbtO_2$ were monitored in 34 patients suffering from MCA infarction while the volume of cerebral damage and CBF were assessed by PET scan. As compared to patients with a benign course, patients with a malignant course (i.e., large edema formation with midline shift), exhibited a decrease in cerebral perfusion pressure and $PbtO_2$, associated with higher microdialysate Glu levels, indicating a secondary ischemia. Significant correlations between outcome and peak values for Glu were found [57].

The link between elevated microdialysate Glu and delayed infarct progression in patients suffering from MCA infarction was confirmed in a more recent study. Among 18 patients in whom surgery was performed to prevent herniation, nine of them exhibited an infarct progression associated with elevated Glu levels, using a probe placed at a 15-mm distance from the initial infarct, suggesting that altered Glu levels can be detected in a rather distant area surrounding the infarct [58].

Berger et al. [59] have studied the effect of hypothermia (33–34 °C) on microdialysate Glu in patients suffering from MCA infarction. If hypothermia can significantly decrease the enhanced Glu levels from peri-infarct tissue, it does not affect the dramatically increased Glu (453 μmol/L) in the area of irreversibly damaged tissue. Thus, hypothermia decreased Glu in the "tissue at risk" area of the infarct but not within the core of the infarct. Moreover, a positive correlation was found between Glu and glycerol in the infarcted tissue, whereas this could not be found in the nonischemic tissue. This suggests that Glu could be released through autolysis of neuronal cells because glycerol is one of the end products of membrane phospholipids degeneration and is, therefore, a marker of the disruption of cellular membranes leading to neuronal autolysis. This is further supported by the associated decreases in microdialysate Glu and glycerol in peri-infarct tissue when performing hypothermia or hemicraniectomy (to prevent brain herniation due to edema formation) [60]. On the other hand, Glu level does not seem to depend on ICP: no difference in ICP has been found between peri-infarct and non-infarcted tissues whereas differences in Glu were marked [59]; furthermore, treatment of brain edema by an osmotic agent in patients suffering from MCA infarction induced a rapid decrease in ICP, but does not affect microdialysate Glu during the monitoring period [61].

4.4 Ischemia During Neurosurgery

Further information on the effect of brain ischemia on the extracellular level of Glu has been brought by microdialysis studies performed when a transient focal ischemia was induced by a surgical procedure or in diseases in which a chronic ischemia is thought to occur.

In that respect, temporal lobe resection for intractable epilepsy surgery leads to an acute ischemic situation in the resected brain area and could be considered as a model of focal ischemia. In this case, if a microdialysis probe was inserted intraoperatively into the area to be excised, the Glu level started to increase at the beginning of the devascularization and resection of the temporal lobe and further peaked with the total isolation of the temporal pole, which corresponds to a complete ischemia [62, 63].

Moreover, brain retraction may be necessary for adequate exposure during intracranial surgery, producing a certain pressure on the brain surface which may lead to ischemia due to the reduction of regional perfusion pressure. In this respect, it is of interest to note that Xu et al. [64] performed brain microdialysis in six patients during brain retraction for subfrontal operation for a pituitary adenoma and found a marked increase in Glu level during retraction, likely due to an incomplete cerebral ischemia.

Mendelowitsch et al. [65] monitored the extracellular concentration of Glu by microdialysis in ten patients undergoing high flow extracranial-intracranial bypasses for cranial base tumors or aneurysms operations, which are among the most risky procedures for ischemic complications in neurosurgery. Indeed, three patients had significant intraoperative Glu increases suggesting ischemia and two of them awoke with a neurological deficit (hemiparesis). This study underlines again that Glu, as monitored by microdialysis, is an early sensitive marker of neuronal damage which may occur during neurosurgery. This is illustrated in another article from the same group [66] reporting the use of microdialysis to measure metabolic indices of ischemia during profound hypothermia and circulatory arrest in three patients undergoing surgery for basilar aneurysms. Glu appeared as a good predictor of postoperative outcome since a neurological deficit (hemiparesis) occurred only in the sole patient who exhibited high Glu microdialysate concentrations during surgery.

In the idiopathic adult hydrocephalus syndrome, a chronic ischemia is thought to be caused by an arteriosclerotic process associated with a CSF hydrodynamic disturbance. After CSF drainage performed to influence this chronic ischemia, the microdialysate Glu tended to be decreased [67].

4.5 Epilepsy

In two pilot studies, microdialysis was performed in the epileptogenic cortical region of patients suffering from spontaneous partial seizures (during ~8-min periods) and selected for epilepsy surgery. A marked elevation of Glu extracellular cortical concentration associated to higher levels of aspartate, glycine, and serine, occurred when the seizures started, and then, these concentrations decreased with ongoing epileptic activity, returning to pre-seizure levels in the 4 min following the termination of the electrocorticography (ECoG) seizure pattern. These altered levels of amino acids were detected in relation to both spontaneous and electrically induced seizures [68, 69].

In another study, a microdialysis probe was inserted into the hippocampus of 12 epileptic patients and confirmed the association of enhanced extracellular Glu with the seizure onset [70]. Among four patients having both a recording of their hippocampal EcoG and a microdialysis sampling during temporal lobe epilepsy surgery, two of them exhibited a spontaneously vigorous hippocampal epileptiform activity accompanying an elevated basal microdialysate concentration of Glu, while the other two ones exhibited a minimal epileptiform activity together with low levels of microdialysate Glu [71]. Such a result was confirmed in further two studies from the same group, each performed on seven patients [72, 73]. Thus, it appears that, in patients whose sclerotic anterior hippocampi are considered as epileptogenic, the onset of seizures is associated with high microdialysate levels of Glu. Moreover, this enhanced extracellular Glu seems to be linked to the hyperexcitable neuronal status rather than to an alteration in local cellular and/or histopathological architecture.

These studies which report the concentration of Glu in microdialysate have been performed acutely after probe implantation in anesthetized patients who were undergoing surgery for epilepsy. They strongly suggest that the extracellular concentration of Glu is not altered during the interictal periods in the epileptogenic brain areas. They have been challenged by works from Cavus's group performed on awake patients and based on the measurement of the actual extracellular concentration of Glu and Gln, using the zero-flow microdialysis method, an approach which eliminates the effects of possible probe-to-probe variations on dialysate levels. In the first work, Glu and Gln levels were determined, a few days after probe implantation, in the epileptogenic and in the non-epileptogenic cortex and hippocampus during the interictal periods of 38 awake patients undergoing intracranial electroencephalogram evaluation for seizure focus localization before a possible resective treatment [74]. A high Glu level associated to a low Gln/Glu ratio has been found in the epileptogenic hippocampus, as compared to the non-epileptogenic one, whereas the Glu level was only marginally increased in the epileptogenic cortex. Moreover, Glu correlates positively with Gln only in the epileptogenic hippocampus, a result which was interpreted as reflecting a disruption in the Glu-Gln cycling. Moreover, as an increase in extracellular lactate level was also present in the epileptogenic hippocampus, the authors have proposed that an interictal energetic deficiency could lead to impaired Glu uptake and Glu-Gln cycling in the epileptogenic hippocampus, resulting in a persistent increase in extracellular Glu. In a further study performed on 17 patients, the measurement of hippocampal volume by quantitative MRI volumetrics was associated to the determination of extracellular interictal Glu levels [75]. This study shows that the increased Glu in the epileptogenic hippocampus was significantly related to a smaller hippocampal volume, even when corrected for disease duration and seizure frequency. Glutamate in the atrophic hippocampus was higher with the threshold for hippocampal atrophy estimated as 5 μmol/L. In a third study, the effect of the electrical stimulation of the hippocampus on the extracellular Glu was determined in ten epileptic patients [76]. The stimulation of hippocampus causes increases in Glu and the extent of the stimulation-induced increase in Glu was related to the elevation in interictal basal Glu level. Only a subset of patients experienced brief stimulated seizures. Such a dissociation between stimulus-induced Glu efflux and stimulus-induced seizures suggests that divergent mechanisms may exist for seizure induction and increased Glu in patients with epilepsy.

4.6 Parkinson's Disease

Microdialysis studies were performed in patients suffering from Parkinson's disease concomitantly to the surgical implantation of electrodes for deep brain stimulation (DBS) aiming at relieving their symptoms. If the changes in microdialysate Glu from the internal globus pallidus (GPi) are negligible during an effective DBS of its afferences located in the subthalamic nucleus (STN), such stimulation induces an increase in the microdialysate concentration of cyclic guanosine monophosphate (cGMP), which is a marker of the activity of the Glu receptor/nitric oxide synthase/soluble guanylyl cyclase pathway. Thus, this increase in cGMP may indirectly reflect an increase in Glu release from STN fibers terminating in the GPi [77].

Kilpatrick et al. [78] have shown that the basal level of Glu could be monitored in the STN and such a measurement has been used in a study aiming at investigating the neurochemical mechanisms underlying the dysfunction in basal ganglia network in patients with Parkinson's disease. In that respect, microdialysate Glu was monitored during cognitive tasks on which the performance depends on basal ganglia. In five patients, a significant decrease in microdialysate Glu of the STN was found during the performance of an implicit memory task (weather prediction task) while no difference was found when were performed declarative memory tasks [79]. In a further study performed on two patients, the microdialysis probe was implanted into the GPi. Performance of the implicit memory task was associated with an increased level of Glu compared to baseline, whereas decreased levels of Glu was found when performing a declarative memory task (verbal learning task) [80]. These first data provide a corroborative evidence for the direct involvement of basal ganglia in cognitive functions and are in accordance with the hypothesized dysfunction of this network in Parkinson's disease.

4.7 Brain Tumors

The microdialysis approach has also been applied to the study of brain tumors and has given in vivo data about the involvement of amino acids in the proliferative process. For that purpose, the microdialysis probe was implanted in the tumoral and/or the peritumoral tissue during surgery for tumor resection. Indeed, microdialysate sampling in peritumoral in comparison to tumoral or normal parenchyma represents an opportunity to detect early neurochemical changes possibly predictive of the tumor growth and/or severity, in regions which could not be distinguished from normal areas by classical neuroradiological techniques.

In a study performed on 21 patients, the microdialysate Glu concentration was found to be decreased in the grade III glioma and in the peritumoral tissue, in contrast to other amino acids which were increased [81]. In addition, in the case of grade IV tumors (glioblastoma), a significant decrease in Glu was only observed in one epileptic subject. However, as patients with grade II tumors (i.e., anaplastic astrocytoma or oligodendroglioma) are also epileptic, epilepsy may be a confounding factor, to take into account when tumoral microdialysate Glu is considered.

In another study, microdialysis probes were inserted into the glioblastoma and in the adjacent peritumoral region from 11 patients. The microdialysates were collected before and during the first 5 days of radiotherapy treatment and further analyzed by gas chromatography coupled to mass spectrometry [82]. First, this metabolomic approach allowed to detect 151 compounds, the levels of 51 of them being different between the tumoral and the peritumoral regions. In that respect, extracellular Glu was found more abundant in tumor compared to peritumoral tissue. Second, 33 metabolites were affected by the radiotherapy treatment. Among them both Glu and Gln were found increased. As Gln is used in several metabolic and proliferative processes, its changes may be a sign of reduced proliferation following radiotherapy. Besides, the increase in Glu may reflect a release from tumor or astrocytic cells damaged by radiation. Interestingly, many of the findings in this microdialysis study by Wibom et al. [82] are in accordance with the findings from ^{1}H-magnetic resonance spectroscopy performed on a tissue sample. This further validates the microdialysis approach.

4.8 Methodological Consideration on Microdialysis in Humans

4.8.1 Effect of Change in Probe Efficiency

Brain microdialysis allows a quantitative sampling (and monitoring) of small molecules in the cerebral extracellular space provided that the efficiency of the probes remains the same over the entire duration of the monitoring period. If this condition is fulfilled, one can assume that the changes of the concentration of a given molecule in microdialysate are likely to be due to actual changes in the extracellular concentration. Vespa et al. [22] have performed an end-of-study testing for the probe efficiency using an inert compound such as urea and have reported that little change occurs over as many as 7 days of continuous microdialysis. Moreover, when studying a population of patients, one must be sure that the efficiency is similar along the use of successive individual probes. Thus, the investigator can be confident that the possible differences between the probes and in the same probe across the duration of its use do not contribute significantly to the variability of the results.

There is however an experimental condition which may induce changes in probe recovery during the monitoring period. Indeed, Mendelowistsh et al. [66] reported that the in vitro recovery for Glu was 11% and that it was reduced to 8% under profound hypothermic conditions (15.3 °C). This must be taken into account when interpreting microdialysis measurements performed in patients receiving therapeutic profound hypothermia.

4.8.2 Effect of Membrane Pore Size and of Perfusion Medium

Two microdialysis probes are commercially available from CMA for human brain microdialysis: the standard one with a 20 kDa cut-off membrane (CMA 70) and a probe with a high cut-off membrane (100 kDa) allowing the sampling of peptides and small proteins together with small molecules (CMA 71).

With a standard membrane, only very small net filtration of standard perfusion fluid occurs in the brain. However, with the larger pores of the 100 kDa membrane, there is an excessive loss of standard perfusion fluid to the surrounding tissue. In order to counterbalance the outward hydrostatic force over the membrane segment, the oncotic pressure of the perfusate could be increased by adding a polymer to the standard perfusion medium.

In one study, the effects of two modified perfusion medium on the 100 kDa probe efficiency were compared on TBI or SAH patients. For that purpose, two probes were implanted, always positioned in close proximity each other to allow a comparison of their properties in areas with comparable biochemical composition, one probe being perfused with a standard solution containing 3.5% albumin, the other containing Ringer-dextran 60. The in vivo levels of Glu in the microdialysates from the two probes were found similar. In contrast, the in vitro recovery of Glu was better using Ringer-dextran 60 compared to albumin [83]. The same approach was used for comparing 20 kDa (perfused with standard fluid) and 100 kDa (perfused with Ringer-dextran 60) membrane probes, both implanted in TBI or SAH patients. The in vivo level of Glu was found slightly higher with 100 kDa membrane probe, but the pattern of Glu concentration changes over time was very consistent between the two types of probes [84]. As the Ringer-dextran is no longer commercially available, the same group is now using 0.9% sodium chloride containing poly-(0-2-hydroxyethyl) starch as a perfusion medium for 100 kDa membrane probe and obtain the same pattern of Glu recovery as with Ringer-dextran [31].

However, the use of a perfusion fluid containing a polymer has been challenged by Hutchinson et al. [85]. Indeed, equivalent in vivo Glu levels were obtained when implanting paired 20 and 100 kDa membrane probes in TBI, both probes being perfused with standard solution without dextran. Moreover, the microdialysate volume was constant at the outlet of the 100 kDa membrane probe showing that no leakage of the perfusion fluid was occurring despite the omission of dextran and that paper did not favor the addition of dextran when using a large pore membrane.

4.8.3 Positioning of the Probe

Since microdialysis gives information on extracellular biochemistry of the tissue surrounding the microdialysis membrane (i.e., within the order of a few millimeters), the positioning of the probe is crucial [23]. First, care should be taken to avoid insertion directly into a clot [45]. Secondly, the site of probe implantation in a traumatized brain will have considerable effect on the neurochemical data obtained and on their interpretation. A choice between the focal monitoring of the tissue adjacent to the damaged area or the monitoring of global effects from more remote sites needs to be made [35]. Even placed in two pericontusional areas, the Glu levels given by the two probes can be different [38]. In SAH patients, simultaneous sampling from two vascular territories (MCA and anterior cerebral artery) also showed that a rise in Glu in one territory was not necessarily accompanied with a rise in the other [41]. In that respect, the neurochemical responses to ischemia in gray and in white matter may be different, histopathological findings suggesting a greater susceptibility for infarction of the gray than of the white matter for a given degree of rCBF reduction [48].

Regarding MCA infarction, the placement of the microdialysis probe in the peri-infarct tissue is crucial since it is in the penumbra that significant changes in the energy metabolism and in excitotoxic amino acids can be observed postischemically [60]. However, even when inserting the microdialysis probe into the vascular territory with a high risk of occurrence of secondary pathophysiological events, i.e., regional ischemia, the affected brain region might be missed in some cases, i.e., when ischemia is focal and not detected by the microdialysis probe [49].

4.8.4 Effect of Changes in the Extracellular Volume

The degree to which microdialysate concentration of Glu reflects the Glu release may depend on changes in the extracellular space volume. For instance, a severe TBI often causes cerebral edema and an increase in ICP. The resulting changes in tissue water over time will influence the concentration of soluble molecules such as Glu. Moreover, there may be changes occurring in the extracellular space, such as changes in the size of the interstitial compartment and in the extracellular water content, which could affect the interstitial diffusion characteristics of Glu and could alter the dynamics of the diffusion through the membrane probe [22, 27].

On the other hand, a significant (20–50%) decrease in the extracellular space may occur during seizure activity, which may transiently increase the concentration of the extracellular Glu. Besides, the diffusion of Glu in the brain may change by a decrease in the volume of the extracellular fraction lowering mass transport into the microdialysate. Such considerations must be taken into account when interpreting the changes in microdialysate Glu concentrations.

4.8.5 Limitation Due to the Collection Time

As in any preclinical study, the interpretation of clinical microdialysis data depends on the sample collection time. While taking into account the performances and the sample volume needed by the analytical method coupled to microdialysis, the sample collection time must be suitable to accurately monitor a given pathophysiological mechanism. In a pathological condition for which we can expect a slow evolution, in TBI for instance, one sample collected every hour is judged satisfactory although it may not allow discriminating short-lasting events [22]. In contrast, microdialysates must be sampled with a higher temporal resolution (i.e., every 5 min) when monitoring neurochemical changes associated to cognitive tasks in parkinsonian patients [79, 80].

4.8.6 Effect of Anesthesia and Activity

When interpreting brain microdialysate Glu data obtained in anesthetized patients, the effect of anesthesia must be well documented and should be considered. Furthermore, variations in Glu concentrations between different studies might be explained by the use of different sedatives drugs [59]. In their study aiming at obtaining baseline cortical microdialysate values in patients undergoing tumor surgery of the cerebral posterior fossa, Reinstrup et al. [86] reported that basal Glu values tend to increase during anesthesia as compared to the awake state (measured after surgery). They considered these data as unexpected because isoflurane, which was used for continued anesthesia, is known to attenuate Glu release and to increase its uptake.

When possible, performing microdialysis study on awake patients avoids such influences. When performing their interictal microdialysis measurements on epileptic patients, Cavus et al. [74] conducted their study for 2–5 days after probe implantation in order to be free of the anesthetics and behavioral stimuli induced by the acute probe implantation.

4.8.7 Comparison of Microdialysis Values Between Studies

The microdialysis approach generally gives Glu concentration in microdialysate. Although the microdialysate Glu comes from the extracellular Glu, its concentration can deviate from the actual extracellular level because it depends on several experimental factors such as the probe recovery, the flow rate of the perfusion fluid in the probe, and the temperature. As a consequence, it is not possible to compare absolute microdialysate levels among different studies unless exactly the same equipment and procedures are used. Otherwise, as the levels reported by different authors cannot be directly compared, only values relative to a baseline can be compared [41, 42, 66].

4.8.8 How Cerebral Microdialysate Reflects Brain Extracellular Concentration

The degree to which microdialysate Glu concentration accurately reflects its actual brain extracellular concentration depends on predictable variables (probe characteristics and perfusion fluid flow rate) as well as on unpredictable or dynamic variables due to the probe environment (temperature of the tissue, interstitial diffusion characteristics, inflammation, etc.). As a consequence, the estimation of the "true" (i.e., actual) extracellular concentration based on knowledge of in vitro probe recovery and microdialysate concentration alone is misleading for the human studies [27, 41]. The same observation can also be made for the animal studies.

The general consensus is that the absolute brain extracellular concentration of an analyte can only be determined through the "no-net flux" or the "zero-flow" quantitative microdialysis method. However, the no-net flux method is not suitable for human research since it requires perfusion of exogenous neurochemicals (see Chap. 14). Moreover, the no-net flux method can give valid data only in non-lesioned tissue [87]. Thus, only the zero-flow method can be used on human subjects, eliminating the effects of the possible probe-to-probe variation on microdialysate levels. It has been successfully used by Cavus et al. [74, 75] for the determination of the absolute concentration of extracellular Glu and Gln levels in the hippocampus and in the cortex of epileptic patients and by Phang et al. [19] and Chen et al. [20] for determining absolute concentrations of extracellular Glu in the spinal cord of TSCI patients.

4.8.9 Interpretation of Microdialysis Data: Focal Aspect

One must keep in mind that microdialysis is a focal technique, the volume of brain tissue monitored by the microdialysis probe being within the order of a few millimeters from the membrane. The small volume of brain sampled means that the probe can be insensitive to metabolic changes in remote areas. On the other hand, the microdialysis data can give information regarding the local "tissue outcome" since the local changes which are monitored may or may not predict the patient outcome [88]. In some conditions, such as TBI or SAH, biochemical changes in some areas (penumbra zones, regions subjected to vasospasm, etc.) may precede clinical deterioration. This underlines the importance of choosing the right zone for implanting the probe, the clinical interest of microdialysis being related to an early detection and, thus, treatment of the secondary adverse events before they affect global variables [33, 43, 49, 52].

4.8.10 Interpretation of Microdialysis Data: Interindividuals Variations and Lack of Controls

Unlike the preclinical microdialysis investigations, any clinical microdialysis study may encounter a large variability between subjects. Consequently, caution must be exercised when drawing firm conclusions from the data obtained. Indeed, the human subject may be diverse in terms of age, cause, and severity of the pathology, the presence or absence of comorbidities, the time and the location of the probe implantation, etc. Consequently, large variations in microdialysis results could occur not only between different subjects, but also within the data from the same individual subject over time, different patterns of variation being able to reflect different pathophysiological changes and pharmacological treatments [24, 35, 56].

Besides, in contrast to blood, urine or CSF samples, microdialysate samples cannot, for ethical reasons, be collected from a normal population [79]. Therefore, the net contribution of Glu to the pathology cannot be assessed completely. Conclusions can only be drawn from variations in microdialysates over time and from their correlations with other data (such as imaging) and clinical status, as well as from extrapolations from laboratory animal-based studies.

4.8.11 Interpretation of Microdialysis Data: Mechanisms of Glutamate Variations

When discussing increased levels of microdialysate Glu, one has to avoid interpreting them too early as an enhanced Glu release. One must keep in mind that beside transmitter release, increased Glu may come from an unspecific leakage from injured cells, lower cellular uptake or diffusion from the blood through an injured blood–brain barrier. Moreover, Glu concentration in the synaptic cleft which directly indicates neurotransmitter release may not be reflected by the extracellular level [44, 50]. Therefore, additional measurement of compounds linked to Glu neurotransmission such as Gln or cGMP [77] or of energy metabolites may help to interpret the data.

4.8.12 Needs for a Better Human Microdialysis Approach

Despite its difficulties and limitations, human brain microdialysis plays and will continue to play an important role in clinical neuroscience research.

Regarding the pharmacology of neuroprotection, for example, if the use of animal models of TBI or of ischemia has led to the demonstration of the efficacy of novel pharmacological agents, such as Glu receptors antagonists, this has not yet been translated into success in clinical studies. The neuroprotection trials failure in human TBI and ischemia research may be due to: (a) a limited validity of animal models, requiring further neurochemical investigations in man and (b) to a non-optimized protocol for drug administration due to a lack of knowledge about the pharmacokinetics of potentially neuroprotective drugs in the “living” human brain. Indeed, in TBI and ischemia, the injured brain is characterized by an increased diffusion distance, due to edema, and by a reduced CBF that modulates drug transport across the blood–brain barrier, which may both contribute to a decreased drug delivery. In the past, several drugs may have been identified as inefficient, and even withdrawn from expensive clinical trials, without knowing the actual penetration of these drugs into the injured human brain [36].

Neuroimaging, as PET, may give pertinent data on the drug penetration into the brain with two major advantages over microdialysis: (a) the measurement of the drug in all brain regions and not only in a focal area as microdialysis does and (b) the possibility to perform this investigation on a normal control population. Nevertheless, the use of PET is restricted by the availability of labeled drugs. Moreover, the assessment of neurotransmission by the PET approach is only performed at the receptor level. Furthermore, neuroimaging is not a monitoring approach. In contrast, microdialysis allows to monitor directly and quantitatively the chemical constituents of the extracellular space of the brain [79].

In that respect, microdialysis, although being invasive, represents an interesting tool for the per/postoperative human brain monitoring. Somatosensory evoked potentials (SSEP) and electroencephalography (EEG) are noninvasive techniques which are currently used to detect neuronal damage during surgery. However, SSEP and EEG may remain normal when a cerebral ischemia occurs. Thus, brain clinical microdialysis may be a safe and pertinent approach for detecting metabolic alterations indicative of impending secondary ischemia, especially in patients suffering from TBI, SAH, or stroke. It offers a direct, continuous, and long-term (up to 8–14 days) measurement of ischemia and brain injury biomarkers (Glu, Gln, lactate, lactate/pyruvate ratio), which may give an alert signal before the onset of clinical symptoms [44, 65]. Moreover, as described above, the microdialysis monitoring is not restricted to TBI, SAH, and stroke and could be a pertinent research tools in other pathophysiological states, such as epilepsy, Parkinson's disease, brain tumors, and spinal cord injuries.

5 Notes

1. Information consent and ethical considerations.

   This chapter was mostly focused on patients with spinal cord alterations; however, in the case of a brain microdialysis carried out on patients suffering from coma, the information consent should be signed by the next of kin. The Ethical Committee won't approve protocols with probe implantations in CNS areas other than those lesioned and having lost their functionality (traumatic brain) or needing to be lesioned (DREZ for instance) for a therapeutic reason.

2. Duration sampling.

   In some experiments targeting brain in unconscious patients, microdialysis monitoring can be performed several days (see Sect. 4).

References

1. Baker DA, Kalivas PW (2007) Insights into glutamate physiology: contribution of studies utilizing in vivo microdialysis. In: Westerink B, Cremers T (eds) Handbook of microdialysis. Elsevier, Amsterdam, pp 33–46
2. Chavagnac D (2002) Suivi par microdialyse intraspinale per-opératoire de la libération des acides aminés neurotransmetteurs chez les patients douloureux chroniques Thèse d'exercice. Pharmacie (Pharm. D. thesis), Université Claude Bernard Lyon 1, France
3. Lorrain DS, Correa L, Anderson J, Varney M (2002) Activation of spinal group I metabotropic glutamate receptors in rats evokes local glutamate release and spontaneous nociceptive behaviors: effects of 2-methyl-6-(phenylethynyl)-pyridine pretreatment. Neurosci Lett 327(3):198–202
4. Schley M, Topfner S, Wiech K, Schaller HE, Konrad CJ, Schmelz M, Birbaumer N (2007) Continuous brachial plexus blockade in combination with the NMDA receptor antagonist memantine prevents phantom pain in acute traumatic upper limb amputees. Eur J Pain 11(3):299–308. doi:10.1016/j.ejpain.2006.03.003
5. Zhou Q, Price DD, Callam CS, Woodruff MA, Verne GN (2011) Effects of the *N*-methyl-D-aspartate receptor on temporal summation of second pain (wind-up) in irritable bowel syndrome. J Pain 12(2):297–303. doi:10.1016/j.jpain.2010.09.002
6. Petersen KL, Iyengar S, Chappell AS, Lobo ED, Reda H, Prucka WR, Verfaille SJ (2014) Safety, tolerability, pharmacokinetics, and effects on human experimental pain of the selective ionotropic glutamate receptor 5 (iGluR5) antagonist LY545694 in healthy volunteers. Pain 155(5):929–936. doi:10.1016/j.pain.2014.01.019
7. Truini A, Piroso S, Pasquale E, Notartomaso S, Di Stefano G, Lattanzi R, Battaglia G, Nicoletti F, Cruccu G (2015) N-acetyl-cysteine, a drug that enhances the endogenous activation of group-II metabotropic glutamate receptors, inhibits nociceptive transmission in humans. Mol Pain 11:14. doi:10.1186/s12990-015-0009-2
8. Sindou M, Quoex C, Baleydier C (1974) Fiber organization at the posterior spinal cord-rootlet junction in man. J Comp Neurol 153(1):15–26. doi:10.1002/cne.901530103
9. Mertens P, Ghaemmaghami C, Bert L, Perret-Liaudet A, Sindou M, Renaud B (2000) Amino acids in spinal dorsal horn of patients during surgery for neuropathic pain or spasticity. Neuroreport 11(8):1795–1798
10. Sindou M, Turano G, Pantieri R, Mertens P, Mauguiere F (1994) Intraoperative monitoring of spinal cord SEPs during microsurgical DREZotomy (MDT) for pain, spasticity and hyperactive bladder. Stereotact Funct Neurosurg 62(1–4):164–170
11. Sindou M, Georgoulis G, Mertens P (2014) Neurosurgery for spasticity. A practical guide for treating children and adults. Springer, New York
12. Mertens P, Ghaemmaghami C, Bert L, Perret-Liaudet A, Guenot M, Naous H, Laganier L, Later R, Sindou M, Renaud B (2001) Microdialysis study of amino acid neurotransmitters in the spinal dorsal horn of patients undergoing microsurgical dorsal root entry zone lesioning. Technical note. J Neurosurg 94(1 Suppl):165–173
13. Parrot S, Sauvinet V, Xavier JM, Chavagnac D, Mouly-Badina L, Garcia-Larrea L, Mertens P, Renaud B (2004) Capillary electrophoresis combined with microdialysis in the human spinal cord: a new tool for monitoring rapid peroperative changes in amino acid neurotransmitters within the dorsal horn. Electrophoresis 25(10–11):1511–1517. doi:10.1002/elps.200305852
14. Skilling SR, Smullin DH, Beitz AJ, Larson AA (1988) Extracellular amino acid concentrations in the dorsal spinal cord of freely moving rats following veratridine and nociceptive stimulation. J Neurochem 51(1):127–132
15. Parrot S (2000) Intérêts et limites de la microdialyse intracérébrale couplée à l'électrophorèse capillaire pour l'étude des acides aminés excitateurs cérébraux. Thèse de doctorat Sciences. Biochimie (Ph.D. thesis), Université Claude Bernard Lyon 1
16. Lindefors N, Amberg G, Ungerstedt U (1989) Intracerebral microdialysis: I. Experimental

studies of diffusion kinetics. J Pharmacol Methods 22(3):141–156
17. Bert L, Parrot S, Robert F, Desvignes C, Denoroy L, Suaud-Chagny MF, Renaud B (2002) In vivo temporal sequence of rat striatal glutamate, aspartate and dopamine efflux during apomorphine, nomifensine, NMDA and PDC in situ administration. Neuropharmacology 43(5):825–835
18. Parrot S, Sauvinet V, Riban V, Depaulis A, Renaud B, Denoroy L (2004) High temporal resolution for in vivo monitoring of neurotransmitters in awake epileptic rats using brain microdialysis and capillary electrophoresis with laser-induced fluorescence detection. J Neurosci Methods 140(1–2):29–38. doi:10.1016/j.jneumeth.2004.03.025
19. Phang I, Zoumprouli A, Papadopoulos MC, Saadoun S (2016) Microdialysis to optimize cord perfusion and drug delivery in spinal cord injury. Ann Neurol 80(4):522–531. doi:10.1002/ana.24750
20. Chen S, Phang I, Zoumprouli A, Papadopoulos MC, Saadoun S (2016) Metabolic profile of injured human spinal cord determined using surface microdialysis. J Neurochem 139(5):700–705. doi:10.1111/jnc.13854
21. Kanthan R, Shuaib A (1995) Clinical evaluation of extracellular amino acids in severe head trauma by intracerebral in vivo microdialysis. J Neurol Neurosurg Psychiatry 59(3):326–327
22. Vespa P, Prins M, Ronne-Engstrom E, Caron M, Shalmon E, Hovda DA, Martin NA, Becker DP (1998) Increase in extracellular glutamate caused by reduced cerebral perfusion pressure and seizures after human traumatic brain injury: a microdialysis study. J Neurosurg 89(6):971–982. doi:10.3171/jns.1998.89.6.0971
23. Stahl N, Mellergard P, Hallstrom A, Ungerstedt U, Nordstrom CH (2001) Intracerebral microdialysis and bedside biochemical analysis in patients with fatal traumatic brain lesions. Acta Anaesthesiol Scand 45(8):977–985
24. Belli A, Sen J, Petzold A, Russo S, Kitchen N, Smith M (2008) Metabolic failure precedes intracranial pressure rises in traumatic brain injury: a microdialysis study. Acta Neurochir 150(5):461–469.; discussion 470. doi:10.1007/s00701-008-1580-3
25. Hlatky R, Valadka AB, Goodman JC, Contant CF, Robertson CS (2004) Patterns of energy substrates during ischemia measured in the brain by microdialysis. J Neurotrauma 21(7):894–906. doi:10.1089/0897715041526195
26. Sarrafzadeh AS, Kiening KL, Callsen TA, Unterberg AW (2003) Metabolic changes during impending and manifest cerebral hypoxia in traumatic brain injury. Br J Neurosurg 17(4):340–346
27. Marion DW, Puccio A, Wisniewski SR, Kochanek P, Dixon CE, Bullian L, Carlier P (2002) Effect of hyperventilation on extracellular concentrations of glutamate, lactate, pyruvate, and local cerebral blood flow in patients with severe traumatic brain injury. Crit Care Med 30(12):2619–2625. doi:10.1097/01.CCM.0000038877.40844.0F
28. Quintard H, Patet C, Suys T, Marques-Vidal P, Oddo M (2015) Normobaric hyperoxia is associated with increased cerebral excitotoxicity after severe traumatic brain injury. Neurocrit Care 22(2):243–250. doi:10.1007/s12028-014-0062-0
29. Meierhans R, Bechir M, Ludwig S, Sommerfeld J, Brandi G, Haberthur C, Stocker R, Stover JF (2010) Brain metabolism is significantly impaired at blood glucose below 6 mM and brain glucose below 1 mM in patients with severe traumatic brain injury. Crit Care 14(1):R13. doi:10.1186/cc8869
30. Vespa P, Boonyaputthikul R, McArthur DL, Miller C, Etchepare M, Bergsneider M, Glenn T, Martin N, Hovda D (2006) Intensive insulin therapy reduces microdialysis glucose values without altering glucose utilization or improving the lactate/pyruvate ratio after traumatic brain injury. Crit Care Med 34(3):850–856. doi:10.1097/01.CCM.0000201875.12245.6F
31. Mellergard P, Sjogren F, Hillman J (2012) The cerebral extracellular release of glycerol, glutamate, and FGF2 is increased in older patients following severe traumatic brain injury. J Neurotrauma 29(1):112–118. doi:10.1089/neu.2010.1732
32. Chamoun R, Suki D, Gopinath SP, Goodman JC, Robertson C (2010) Role of extracellular glutamate measured by cerebral microdialysis in severe traumatic brain injury. J Neurosurg 113(3):564–570. doi:10.3171/2009.12.JNS09689
33. Timofeev I, Carpenter KL, Nortje J, Al-Rawi PG, O'Connell MT, Czosnyka M, Smielewski P, Pickard JD, Menon DK, Kirkpatrick PJ, Gupta AK, Hutchinson PJ (2011) Cerebral extracellular chemistry and outcome following traumatic brain injury: a microdialysis study of 223 patients. Brain 134(Pt 2):484–494. doi:10.1093/brain/awq353
34. Tolias CM, Richards DA, Bowery NG, Sgouros S (2002) Extracellular glutamate in the brains of children with severe head injuries: a pilot microdialysis study. Childs Nerv Syst 18(8):368–374. doi:10.1007/s00381-002-0623-y

35. Richards DA, Tolias CM, Sgouros S, Bowery NG (2003) Extracellular glutamine to glutamate ratio may predict outcome in the injured brain: a clinical microdialysis study in children. Pharmacol Res 48(1):101–109
36. Alves OL, Doyle AJ, Clausen T, Gilman C, Bullock R (2003) Evaluation of topiramate neuroprotective effect in severe TBI using microdialysis. Ann N Y Acad Sci 993:25–34. discussion 48-53
37. Mazzeo AT, Alves OL, Gilman CB, Hayes RL, Tolias C, Niki Kunene K, Ross Bullock M (2008) Brain metabolic and hemodynamic effects of cyclosporin A after human severe traumatic brain injury: a microdialysis study. Acta Neurochir 150(10):1019–1031.; discussion 1031. doi:10.1007/s00701-008-0021-7
38. Reinstrup P, Nordstrom CH (2011) Prostacyclin infusion may prevent secondary damage in pericontusional brain tissue. Neurocrit Care 14(3):441–446. doi:10.1007/s12028-010-9486-3
39. Berg A, Bellander BM, Wanecek M, Gamrin L, Elving A, Rooyackers O, Ungerstedt U, Wernerman J (2006) Intravenous glutamine supplementation to head trauma patients leaves cerebral glutamate concentration unaffected. Intensive Care Med 32(11):1741–1746. doi:10.1007/s00134-006-0375-3
40. Ronne Engstrom E, Hillered L, Enblad P, Karlsson T (2005) Cerebral interstitial levels of glutamate and glutamine after intravenous administration of nutritional amino acids in neurointensive care patients. Neurosci Lett 384(1–2):7–10. doi:10.1016/j.neulet.2005.04.030
41. Saveland H, Nilsson OG, Boris-Moller F, Wieloch T, Brandt L (1996) Intracerebral microdialysis of glutamate and aspartate in two vascular territories after aneurysmal subarachnoid hemorrhage. Neurosurgery 38(1):12–19. discussion 19–20
42. Nilsson OG, Brandt L, Ungerstedt U, Saveland H (1999) Bedside detection of brain ischemia using intracerebral microdialysis: subarachnoid hemorrhage and delayed ischemic deterioration. Neurosurgery 45(5):1176–1184. discussion 1184–1185
43. Kett-White R, Hutchinson PJ, Al-Rawi PG, Gupta AK, Pickard JD, Kirkpatrick PJ (2002) Adverse cerebral events detected after subarachnoid hemorrhage using brain oxygen and microdialysis probes. Neurosurgery 50(6):1213–1221. discussion 1221–1222
44. Sarrafzadeh AS, Sakowitz OW, Kiening KL, Benndorf G, Lanksch WR, Unterberg AW (2002) Bedside microdialysis: a tool to monitor cerebral metabolism in subarachnoid hemorrhage patients? Crit Care Med 30(5):1062–1070
45. Sarrafzadeh A, Haux D, Sakowitz O, Benndorf G, Herzog H, Kuechler I, Unterberg A (2003) Acute focal neurological deficits in aneurysmal subarachnoid hemorrhage: relation of clinical course, CT findings, and metabolite abnormalities monitored with bedside microdialysis. Stroke 34(6):1382–1388. doi:10.1161/01.STR.0000074036.97859.02
46. Sakowitz OW, Santos E, Nagel A, Krajewski KL, Hertle DN, Vajkoczy P, Dreier JP, Unterberg AW, Sarrafzadeh AS (2013) Clusters of spreading depolarizations are associated with disturbed cerebral metabolism in patients with aneurysmal subarachnoid hemorrhage. Stroke 44(1):220–223. doi:10.1161/STROKEAHA.112.672352
47. Mathieu L, Duclos AS, Limpar P, Grousson S, Convert J, Dailler F, Perret-Liaudet A, Xavier JM, Burel E, Artru F, Renaud B (2003) In vivo biochemical monitoring of cerebral ischemia by intracerebral microdialysis: a case report. In: Kehr J, Fuxe K, Ungerstedt U, Svensson T (eds) Monitoring molecules in neuroscience, Stockholm, Sweden, pp 225–227
48. Sarrafzadeh AS, Haux D, Ludemann L, Amthauer H, Plotkin M, Kuchler I, Unterberg AW (2004) Cerebral ischemia in aneurysmal subarachnoid hemorrhage: a correlative microdialysis-PET study. Stroke 35(3):638–643. doi:10.1161/01.STR.0000116101.66624.F1
49. Sarrafzadeh A, Haux D, Plotkin M, Ludemann L, Amthauer H, Unterberg A (2005) Bedside microdialysis reflects dysfunction of cerebral energy metabolism in patients with aneurysmal subarachnoid hemorrhage as confirmed by 15 O-H2 O-PET and 18 F-FDG-PET. J Neuroradiol 32(5):348–351
50. Samuelsson C, Hillered L, Zetterling M, Enblad P, Hesselager G, Ryttlefors M, Kumlien E, Lewen A, Marklund N, Nilsson P, Salci K, Ronne-Engstrom E (2007) Cerebral glutamine and glutamate levels in relation to compromised energy metabolism: a microdialysis study in subarachnoid hemorrhage patients. J Cereb Blood Flow Metab 27(7):1309–1317. doi:10.1038/sj.jcbfm.9600433
51. Samuelsson C, Howells T, Kumlien E, Enblad P, Hillered L, Ronne-Engstrom E (2009) Relationship between intracranial hemodynamics and microdialysis markers of energy metabolism and glutamate-glutamine turnover in patients with subarachnoid hemorrhage. Clinical article. J Neurosurg 111(5):910–915. doi:10.3171/2008.8.JNS0889

52. Nilsson OG, Polito A, Saveland H, Ungerstedt U, Nordstrom CH (2006) Are primary supratentorial intracerebral hemorrhages surrounded by a biochemical penumbra? A microdialysis study. Neurosurgery 59(3):521–528.; discussion 521–528. doi:10.1227/01.NEU.0000227521.58701.E5
53. Miller CM, Vespa PM, McArthur DL, Hirt D, Etchepare M (2007) Frameless stereotactic aspiration and thrombolysis of deep intracerebral hemorrhage is associated with reduced levels of extracellular cerebral glutamate and unchanged lactate pyruvate ratios. Neurocrit Care 6(1):22–29. doi:10.1385/NCC:6:1:22
54. Ho CL, Ang CB, Lee KK, Ng IH (2008) Effects of glycaemic control on cerebral neurochemistry in primary intracerebral haemorrhage. J Clin Neurosci 15(4):428–433. doi:10.1016/j.jocn.2006.08.011
55. Bullock R, Zauner A, Woodward J, Young HF (1995) Massive persistent release of excitatory amino acids following human occlusive stroke. Stroke 26(11):2187–2189
56. Schneweis S, Grond M, Staub F, Brinker G, Neveling M, Dohmen C, Graf R, Heiss WD (2001) Predictive value of neurochemical monitoring in large middle cerebral artery infarction. Stroke 32(8):1863–1867
57. Dohmen C, Bosche B, Graf R, Staub F, Kracht L, Sobesky J, Neveling M, Brinker G, Heiss WD (2003) Prediction of malignant course in MCA infarction by PET and microdialysis. Stroke 34(9):2152–2158. doi:10.1161/01.STR.0000083624.74929.32
58. Woitzik J, Pinczolits A, Hecht N, Sandow N, Scheel M, Drenckhahn C, Dreier JP, Vajkoczy P (2014) Excitotoxicity and metabolic changes in association with infarct progression. Stroke 45(4):1183–1185. doi:10.1161/STROKEAHA.113.004475
59. Berger C, Schabitz WR, Georgiadis D, Steiner T, Aschoff A, Schwab S (2002) Effects of hypothermia on excitatory amino acids and metabolism in stroke patients: a microdialysis study. Stroke 33(2):519–524
60. Berger C, Kiening K, Schwab S (2008) Neurochemical monitoring of therapeutic effects in large human MCA infarction. Neurocrit Care 9(3):352–356. doi:10.1007/s12028-008-9093-8
61. Berger C, Sakowitz OW, Kiening KL, Schwab S (2005) Neurochemical monitoring of glycerol therapy in patients with ischemic brain edema. Stroke 36(2):e4–e6. doi:10.1161/01.STR.0000151328.70519.e9
62. Kanthan R, Shuaib A, Griebel R, Miyashita H (1995) Intracerebral human microdialysis. In vivo study of an acute focal ischemic model of the human brain. Stroke 26(5):870–873
63. Kanthan R, Shuaib A, Griebel R, Miyashita H, Kalra J (1996) Glucose-induced decrease in glutamate levels in ischemic human brain by in-vivo microdialysis. Neurosci Lett 209(3):207–209
64. Xu W, Mellergard P, Ungerstedt U, Nordstrom CH (2002) Local changes in cerebral energy metabolism due to brain retraction during routine neurosurgical procedures. Acta Neurochir 144(7):679–683. doi:10.1007/s00701-002-0946-1
65. Mendelowitsch A, Sekhar LN, Wright DC, Nadel A, Miyashita H, Richardson R, Kent M, Shuaib A (1998) An increase in extracellular glutamate is a sensitive method of detecting ischaemic neuronal damage during cranial base and cerebrovascular surgery. An in vivo microdialysis study. Acta Neurochir 140(4):349–355. discussion 356
66. Mendelowitsch A, Mergner GW, Shuaib A, Sekhar LN (1998) Cortical brain microdialysis and temperature monitoring during hypothermic circulatory arrest in humans. J Neurol Neurosurg Psychiatry 64(5):611–618
67. Agren-Wilsson A, Roslin M, Eklund A, Koskinen LO, Bergenheim AT, Malm J (2003) Intracerebral microdialysis and CSF hydrodynamics in idiopathic adult hydrocephalus syndrome. J Neurol Neurosurg Psychiatry 74(2):217–221
68. Carlson H, Ronne-Engstrom E, Ungerstedt U, Hillered L (1992) Seizure related elevations of extracellular amino acids in human focal epilepsy. Neurosci Lett 140(1):30–32
69. Ronne-Engstrom E, Hillered L, Flink R, Spannare B, Ungerstedt U, Carlson H (1992) Intracerebral microdialysis of extracellular amino acids in the human epileptic focus. J Cereb Blood Flow Metab 12(5):873–876. doi:10.1038/jcbfm.1992.119
70. Wilson CL, Maidment NT, Shomer MH, Behnke EJ, Ackerson L, Fried I, Engel J Jr (1996) Comparison of seizure related amino acid release in human epileptic hippocampus versus a chronic, kainate rat model of hippocampal epilepsy. Epilepsy Res 26(1):245–254
71. Thomas PM, Phillips JP, Delanty N, O'Connor WT (2003) Elevated extracellular levels of glutamate, aspartate and gamma-aminobutyric acid within the intraoperative, spontaneously epileptiform human hippocampus. Epilepsy Res 54(1):73–79
72. Thomas PM, Phillips JP, O'Connor WT (2005) Microdialysis of the lateral and medial temporal lobe during temporal lobe epilepsy

surgery. Surg Neurol 63(1):70–79.; discussion 79. doi:10.1016/j.surneu.2004.02.031

73. Thomas PM, Phillips JP, O'Connor WT (2004) Hippocampal microdialysis during spontaneous intraoperative epileptiform activity. Acta Neurochir 146(2):143–151. doi:10.1007/s00701-003-0189-9
74. Cavus I, Kasoff WS, Cassaday MP, Jacob R, Gueorguieva R, Sherwin RS, Krystal JH, Spencer DD, Abi-Saab WM (2005) Extracellular metabolites in the cortex and hippocampus of epileptic patients. Ann Neurol 57(2):226–235. doi:10.1002/ana.20380
75. Cavus I, Pan JW, Hetherington HP, Abi-Saab W, Zaveri HP, Vives KP, Krystal JH, Spencer SS, Spencer DD (2008) Decreased hippocampal volume on MRI is associated with increased extracellular glutamate in epilepsy patients. Epilepsia 49(8):1358–1366. doi:10.1111/j.1528-1167.2008.01603.x
76. Cavus I, Widi GA, Duckrow RB, Zaveri H, Kennard JT, Krystal J, Spencer DD (2016) 50 Hz hippocampal stimulation in refractory epilepsy: higher level of basal glutamate predicts greater release of glutamate. Epilepsia 57(2):288–297. doi:10.1111/epi.13269
77. Stefani A, Fedele E, Galati S, Pepicelli O, Frasca S, Pierantozzi M, Peppe A, Brusa L, Orlacchio A, Hainsworth AH, Gattoni G, Stanzione P, Bernardi G, Raiteri M, Mazzone P (2005) Subthalamic stimulation activates internal pallidus: evidence from cGMP microdialysis in PD patients. Ann Neurol 57(3):448–452. doi:10.1002/ana.20402
78. Kilpatrick M, Church E, Danish S, Stiefel M, Jaggi J, Halpern C, Kerr M, Maloney E, Robinson M, Lucki I, Krizman-Grenda E, Baltuch G (2010) Intracerebral microdialysis during deep brain stimulation surgery. J Neurosci Methods 190(1):106–111. doi:10.1016/j.jneumeth.2010.04.013
79. Buchanan RJ, Darrow DP, Meier KT, Robinson J, Schiehser DM, Glahn DC, Nadasdy Z (2014) Changes in GABA and glutamate concentrations during memory tasks in patients with Parkinson's disease undergoing DBS surgery. Front Hum Neurosci 8:81. doi:10.3389/fnhum.2014.00081
80. Buchanan RJ, Gjini K, Darrow D, Varga G, Robinson JL, Nadasdy Z (2015) Glutamate and GABA concentration changes in the globus pallidus internus of Parkinson's patients during performance of implicit and declarative memory tasks: a report of two subjects. Neurosci Lett 589:73–78. doi:10.1016/j.neulet.2015.01.028
81. Bianchi L, De Micheli E, Bricolo A, Ballini C, Fattori M, Venturi C, Pedata F, Tipton KF, Della Corte L (2004) Extracellular levels of amino acids and choline in human high grade gliomas: an intraoperative microdialysis study. Neurochem Res 29(1):325–334
82. Wibom C, Surowiec I, Moren L, Bergstrom P, Johansson M, Antti H, Bergenheim AT (2010) Metabolomic patterns in glioblastoma and changes during radiotherapy: a clinical microdialysis study. J Proteome Res 9(6):2909–2919. doi:10.1021/pr901088r
83. Hillman J, Aneman O, Anderson C, Sjogren F, Saberg C, Mellergard P (2005) A microdialysis technique for routine measurement of macromolecules in the injured human brain. Neurosurgery 56(6):1264–1268. discussion 1268–1270
84. Hillman J, Milos P, Yu ZQ, Sjogren F, Anderson C, Mellergard P (2006) Intracerebral microdialysis in neurosurgical intensive care patients utilising catheters with different molecular cut-off (20 and 100 kDa). Acta Neurochir 148(3):319–324.; discussion 324. doi:10.1007/s00701-005-0670-8
85. Hutchinson PJ, O'Connell MT, Nortje J, Smith P, Al-Rawi PG, Gupta AK, Menon DK, Pickard JD (2005) Cerebral microdialysis methodology—evaluation of 20 kDa and 100 kDa catheters. Physiol Meas 26(4):423–428. doi:10.1088/0967-3334/26/4/008
86. Reinstrup P, Stahl N, Mellergard P, Uski T, Ungerstedt U, Nordstrom CH (2000) Intracerebral microdialysis in clinical practice: baseline values for chemical markers during wakefulness, anesthesia, and neurosurgery. Neurosurgery 47(3):701–709. discussion 709–710
87. Chen KC (2006) Effects of tissue trauma on the characteristics of microdialysis zero-net-flux method sampling neurotransmitters. J Theor Biol 238(4):863–881. doi:10.1016/j.jtbi.2005.06.035
88. Tholance Y, Barcelos G, Dailler F, Perret-Liaudet A, Renaud B (2015) Clinical neurochemistry of subarachnoid hemorrhage: toward predicting individual outcomes via biomarkers of brain energy metabolism. ACS Chem Neurosci 6(12):1902–1905. doi:10.1021/acschemneuro.5b00299

INDEX

Sandrine Parrot and Luc Denoroy (eds.), *Biochemical Approaches for Glutamatergic Neurotransmission*, Neuromethods, vol. 130, DOI 10.1007/978-1-4939-7228-9, © Springer Science+Business Media LLC 2018

J

K

L

M

Q

R

S

T

V

W

X

Z

Printed in the United States
By Bookmasters